.

ISBN 978-0-265-59032-4
PIBN 10871909

THE

JOURNAL

OF

ANATOMY AND PHYSIOLOGY

CONDUCTED BY

G. M. HUMPHRY, M.D. F.R.S.
PROFESSOR OF ANATOMY IN THE UNIVERSITY OF CAMBRIDGE,
HONORARY FELLOW OF DOWNING COLLEGE;

WM. TURNER, M.B.
PROFESSOR OF ANATOMY IN THE UNIVERSITY OF EDINBURGH;

MICHAEL FOSTER, M.D. F.R.S.
PRÆLECTOR IN PHYSIOLOGY IN TRINITY COLLEGE, CAMBRIDGE;

WM. RUTHERFORD, M.D.
PROFESSOR OF THE INSTITUTES OF MEDICINE IN THE UNIVERSITY OF EDINBURGH.

VOLUME X.

MACMILLAN AND CO.
Cambridge and London.
1876.

Cambridge:
PRINTED BY C. J. CLAY, M.A.
AT THE UNIVERSITY PRESS.

CONTENTS.

THIRD PART. APRIL, 1876.

FOURTH PART. JULY, 1876.

1.

Journal of Anatomy and Physiology.

ON THE PRIMARY VASCULAR DILATATION IN ACUTE INFLAMMATION. By FRANCIS DARWIN, M.B., *Cantab.*[1]

THE work of which I shall give an account was mainly conducted in the Laboratory of the Brown Institution, under the supervision of Dr Klein, to whose kind assistance I am much indebted. As I venture to bring forward a view which is opposed to that of Dr Cohnheim, I think it will be necessary to begin with a short account of his experimental work and the conclusions he draws from it.

He has, as is well known, studied the inflammatory process by observing the effects of injury on various organs which are sufficiently transparent to allow the changes which occur to be viewed by transmitted light. The experiment which now goes by his name consists in spreading out the tongue of a curarised frog on a plate of glass, which at the same time supports the animal lying on its belly ; the glass plate is then fastened to the stage of the microscope and the outspread tongue brought into view. The tongue is then pinched, or irritated in some way, and the effect on the vessels is directly observed. Dr Cohnheim has in this way studied the inflammatory process in the web of the frog's foot, and in the frog's mesentery. The cornea and the membrana nictitans of the frog and the ear of the rabbit were also observed by him with reflected light.

The following[2] is his account of an acute circumscribed inflammation induced by cauterizing the tongue of the frog (*R. esculenta*) with a small crystal of nitrate of silver. The first thing that happens is the rapid dilatation of the arteries in the neighbourhood, and this is quickly followed by the dilatation of the veins, so that the cauterized place is soon surrounded by a zone of hyperæmia. At the same

---

[1] The chief part of the present paper was presented as a Thesis for the degree of M.B.
[2] *Neue Untersuchungen über die Entzündung.* Dr Julius Cohnheim, Berlin, 1873, p. 12.

time the current in the vessels of this zone increases greatly in velocity. It may be seen that the quickening of the circulation is a purely local effect by comparing the hyperæmic region with other parts of the tongue where the blood-stream will be found to be of normal velocity. This state of things is not of long continuance. The arterial and subsequently the venous current begin to get slow at the cauterized place, and ultimately stop altogether. We have therefore at the spot where the cautery actually took effect a small region of complete stasis, surrounded by a region in which the blood-stream is flowing quickly through dilated vessels. After a time the dilated arteries in the more peripheral parts of this hyperæmic zone begin to contract and the current becomes slower, and ultimately both the velocity of the circulation and the calibre of the vessels return to the normal. The venous current necessarily becomes slower, depending as it does on the supply of blood from the arteries. The dilated veins do not begin to contract till after the arteries, but they also at last regain their normal condition. Between the region of complete stasis, and the peripheral zone where the vessels have recovered and are practically as they were at first, there is a zone of vessels which have not recovered—which have not contracted—but remain permanently dilated. And it is of importance to note, that in these permanently dilated vessels the velocity of the current instead of being above the normal as it was at first, has now fallen below it.

The process of extravasation now commences. This does not affect the arteries, only the capillaries and veins. It is chiefly the coloured corpuscles which leave the capillaries (diapedesis) while only colourless ones "emigrate" through the walls of the veins. In the latter the process is preceded by the phenomenon known as "adhesion." The colourless (amœboid) corpuscles begin one by one to stick to the walls till the whole inner surface of the veins is coated with them. The same process takes place in the arteries, but, as before remarked, it never passes into the stage of emigration. Besides the emigration and "diapedesis" of the formed elements, there is a considerable extravasation of the liquid part of the blood. So that the lymphatic sack at the base of the tongue becomes filled with fluid and with corpuscles. This short account is, I believe, a fair abstract of Dr Cohnheim's account of acute inflammation; and I have observed for my own satisfaction all the chief phenomena of the series.

Dr Cohnheim believes the dilatation which comes on in the arteries directly after the application of an irritant to be due to the paralysis of the muscles in the vascular walls, and that this loss of power is the direct effect of the injury on the tissues of the muscular coat. He does not believe this temporary paralysis to be an essential part of the true process of inflammation, but merely an incidental accompaniment of the injury inflicted. The subsequent changes which the injury produces, viz. the *permanent dilatation*, the slowing of the current and the adhesion and emigration of corpuscles, constitute, according to his views, the true process of inflammation. He insists especially that they are caused by, so to speak, chemical changes which the irritation brings on in the walls of the vessels.

He remarks that in the condition of primary dilatation we have a rapid current flowing through dilated arteries, and since in the state of permanent dilatation the current becomes slow although no alteration has taken place in the calibre of the vessels, the diminution or velocity of the arterial current can only be accounted for by assuming that some alteration has taken place in the walls of the vessel. He also remarks that the passage of formed elements into the surrounding tissues shows a loss of physiological integrity which can only be due to some quasi-chemical change in the vascular coats. Would it not, he asks, be a remarkable thing if an irritant such as nitrate of silver did *not* produce some alteration in such a structure as the wall of an artery?

Dr Cohnheim appears to believe (if I do not misinterpret him) that if his theory be not accepted the only conceivable one is, that the dilatation of the vessels is due to central reflex action. Against any explanation of the phenomena by reflex action, he brings several objections. The most important of these is embodied in the result of the following experiment. The central nervous system of a frog is destroyed, or else the tongue is divided in such a way that it is only attached to the body by the lingual arteries and veins ; he then repeats the experiment of irritating the tongue, and finds that the results are identical with those obtained with a normal tongue. His argument is, that the central nervous system being destroyed (or cut off from communication with the tongue), the only reflex action that could take place would be from peripheral vaso-motor ganglia situated in the tongue itself. And this, he concludes, is impossible, because no ganglia have been found in the tongue, and because no peripheral reflex action is known to occur anywhere.

It does not appear to me that all conceivable opposition is destroyed by the above experiment. I shall endeavour to show that the *primary* arterial dilatation which is accompanied by quickened circulation is due to some kind of nervous activity, and not, as Dr Cohnheim believes, to the direct effect of the irritant on the muscular coat of the arteries.

Paralysis may be due either to a lesion of the muscular tissue itself, as in progressive muscular atrophy[1], or it may be due to a loss of nerve-power, as in the paralysis of the lower extremities resulting from injuries to the spinal cord. Since Dr Cohnheim denies the existence of local vaso-motor ganglia, and leaves out of consideration the peripheral terminations of the vaso-motor nerves, he is forced to assume that the dilatation produced by local irritation is due *solely* to the direct effect of the irritant on the *muscular tissue* contained in the walls of the vessels.

When an irritant is applied to a curarised frog's tongue it

[1] Niemeyer's *Text-Book of Practical Medicine* (American Trans.), Vol. II. p. 519.

must act both on the intrinsic muscles of the organ and on the arteries contained in it. As is well known, curare paralyses the peripheral terminations of the ordinary motor nerves, but not those of the vaso-motor nerves; therefore any effect the irritant may produce on the intrinsic muscles of the tongue must be wholly due to its direct effect on the muscular tissue itself. This however is not necessarily the case with the arteries, since the irritant may primarily affect the vaso-motor nerves, and through their peripheral terminations may act on the muscular coat of the vessels. But according to Dr Cohnheim's views an irritant produces its results by its direct effect on the muscular coat of an artery; if this is so we have in a curarised frog's tongue an irritant acting *directly* on the striped muscle of the tongue, and *directly* on the unstriped muscle of the arteries, and we may expect that similar results will follow in the two cases. But this is not the case, for although irritation of the tongue produces dilatation of the arteries, it produces contraction (as I have often observed) in the intrinsic muscles of the tongue. Yet Dr Cohnheim uses this very example of the effects of irritation on the muscular tissue of the tongue to illustrate his theory of paralytic dilatation. He says (*Op. cit.* p. 26) " ......as the muscular twitchings (muskelzuckungen) which make their appearance on pinching or irritating a part of the tongue, after a time subside, so also the vessels and their muscles gradually recover from the effect of a sudden pinch."

If an irritant acts directly and solely on the muscles of the arterial walls, it ought to produce the same effect in whatever part of the body the artery acted on may be situated. At any rate in superficial organs like the tongue and the membrana nictitans, arteries considered merely as muscular tubes ought to behave in the same way under similar circumstances. Yet Dr Cohnheim says that in the membrana nictitans the primary dilatation is absent, while in the tongue it is well marked. It appears that in the membrana nictitans the arteries dilate after an interval of some hours, and this corresponds to the permanent dilatation in the vessels of the tongue, and marks the commencement, according to Dr Cohnheim, of the inflammatory process. Dr Cohnheim accordingly concludes, from his own experiments and those of others, that dilatation is the universal result of irritating an artery[1]. He quotes however Saviotti[2], who states that ammonia and some other irritants produce contraction of the arteries in the frog's foot; but these results are neglected, because on Saviotti's own admission they are somewhat variable, and because Dr Cohnheim believes that researches for which the more tender species of frog, *Rana temporaria*, is employed, are not trustworthy[3].

I worked with *Rana temporaria* before I was aware of Saviotti's results, and I can certainly confirm his assertion, that a solution of ammonia applied to the web of the frog's foot causes contraction of the arteries. I have observed the phenomenon so often that it is impossible to doubt that it is a genuine one. The same result has been produced by scratching the web with the broken end of a capillary pipette and by cauterising with nitrate of silver. The contraction is so intense that the arteries fade out of sight and complete venous stasis supervenes in consequence of the cessation of the arterial current. This condition of intense contraction has been observed to continue for as long as 13 or 14 minutes; as the artery recovers and the current returns, it dilates up to or sometimes beyond the normal calibre. A curious spasmodic condition is sometimes observed when the artery is recovering; the intense contraction coming on and disappearing again rapidly once or twice before the normal tone is fully re-established.

An interval of a few seconds usually intervenes between

---

[1] The following observers have recorded the fact that *contraction* may result from the irritation of an artery, Mr Lister (*Phil. Trans.* 1858, both papers), Sir James Paget (*Lectures on Surgical Pathology*), Dr Lionel Beale (*Monthly Microscopical Journal*, VIII. p. 58), Mr Wharton Jones (*Guy's Hospital Reports*, 1850, quoted in Carpenter's *Physiology*).

[2] Virchow's *Arch.* L. 592.

[3] Saviotti states (*loc. cit.* p. 610) that if the sciatic nerve be divided no contraction can afterwards be produced by local irritation of the web. This is quite at variance with Lister's, Cohnheim's and my own observations, and implies that reflex action from the central nervous system is essential to the production of the phenomena, and this cannot be maintained.

Again, Saviotti's statement that local irritation produces slowing of the current is considered by Riegel (*Mediz. Jahrbüch.* 1871, p. 99) to be at variance with his own observations. Riegel finds that stimulation of a sensory nerve always produces quickening of the current. With all due deference to the opinions of Stricker (see Riegel, note, p. 102) and Burdon Sanderson (Holmes' *System of Surgery*, v. 738), I believe that the slowing observed by Saviotti was merely the *mechanical* result of the diminution of the blood-supply consequent on the diminished arterial calibre, for he notices that, before the arteries contract, the current is *quickened* (*loc. cit.* p. 595), and he does not distinguish very clearly between the velocities of the arterial and venous currents.

the application of the irritant and the appearance of the contraction. On several occasions I have distinctly observed a transitory dilatation preceding the contraction, and occupying the short interval of time just mentioned. This dilatation is slight and is not a constant phenomenon, for I have several times made quite sure that no such dilatation preceded the contraction of the artery under observation. The variability of Saviotti's and my own results would be difficult to reconcile with Dr Cohnheim's theory of paralytic dilatation. But I hope to show that at any rate this slight variability in my own results can be reconciled with another theory.

In examining into the cause of the contraction produced by local irritation of the frog's foot, we may first exclude both *central reflex action* and the effect on the *heart* of irritating a sensory nerve, because the phenomena may be observed *after the division of the sciatic nerve and destruction of the spinal marrow.*

We must next consider whether the irritant acts directly on the muscular coat of the vessels or on the nerves which supply those muscles. As is well known, curare paralyses the peripheral terminations of the motor nerves, but has no such effect on the vaso-motor nerves; there is therefore no *à priori* reason for supposing that an irritant would act exclusively on the muscular coat and not also on the nervous mechanism of the arteries. The following reasons induce me to believe that the contraction is in great measure the result of nerve-stimulation. If the sciatic nerve is divided and the peripheral extremity irritated with an induced current, the arteries of the web are seen to contract. This contraction, like that resulting from local stimulation, is preceded by a short pause and is of the same intense character, causing disappearance of the arterial current and stasis in the veins. Now it is obviously unimportant at what point we divide the sciatic nerve, and of what length the peripheral piece may be to which the irritation is applied[1]; it therefore seems illogical to deny the possibility that the ammonia placed on the web acts primarily on the peripheral terminations of the vaso-motor nerves, the stimulus being subsequently con-

[1] Unimportant at least as far as the intrinsic nature of the result is concerned, although its *intensity* may be altered (Pflüger) quoted in Wundt's *Physiologie*.

veyed to the constrictor muscles in the arterial walls. Professor
Maurice Schiff[1] considers it highly probable that local irritants
do in this way take effect on the peripheral terminations of the
vaso-motor nerves. This one fact, that local irritation may
produce *contraction*, makes it impossible for me to accept Dr
Cohnheim's theory, that dilatation when produced by local irritants is the result of the direct paralysis of the arterial walls[2].
He tacitly assumes that the dilatation of a vessel is always due
to simple paralysis, and thus neglects the theory of *active dilatation of vessels*[3]. This theory is admirably stated and upheld by
Schiff (*Op. cit.*, Leçons 11 and 12). It appears that Claude
Bernard and Schiff agree in affirming the existence of a dilatation effected in some way by the activity of a nerve. Schiff
supposes that some hitherto unknown mechanism comes into
play, whereas Bernard assumes the existence in every case of
local vaso-motor ganglia, and believes that the dilatation is the
result of "inhibition." As an example of inhibition I cannot
do better than quote a passage from Dr Brunton's paper "On
inhibition peripheral and central[4]." "Thus the blood-vessels of
the penis are kept in a state of moderate contraction by the
stimulus which the vaso-motor nerves supply to their muscular
coats. This stimulus is derived at least in part from ganglia
lying close to the vessels, and to these ganglia proceed certain
nerves, the *nervi erigentes* which arise from the sacral plexus.
When the nerves are irritated the ganglia cease to stimulate
the vascular walls, and these consequently relax and yield to
the pressure of the blood which pours into and distends them."
Local vaso-motor ganglia have usually been considered essential
to active dilatation by inhibition. Thus Dr Brunton considers
the ganglia on the nervi erigentes essential to the inhibition of
the arteries which he describes (*loc. cit.*) in erection.

On the other hand, Dr Michael Foster and Mr Dew Smith
have shown[5] that the snail's heart, which seems to be destitute of both nerves and ganglion-cells, is capable of true inhi-

[1] *Leçons sur physiologie de la Digestion.* Paris, 1868, T. I. p. 248.
[2] Dr Sanderson says (Holmes, *Syst. Surgery*, v. 740) that the effect of irritation is certainly *not* to paralyse the arteries.
[3] John Hunter speaks of "active dilatation," and compares it to the dilatation of the os uteri, Paget's *Lectures on Surg. Path.* 1853.
[4] *West Riding Asylum Reports*, IV. 1874, p. 182.
[5] *Proc. R. Soc.* XXIII., No. 160.

bition. These observers also conclude that the snail's heart, like that of the vertebrate embryo, is capable of rhythmical contraction without the "automatic" ganglia. They suggest that in the fully developed vertebrate heart the ganglia are not essential to the production of the rhythm, but are chiefly of use in co-ordinating its complex movements, &c. Any part of the body stands to the rest of the body in the relation of an organism to its environments. Now those organisms are the most developed which are best able to adapt themselves by internal changes to alteration in their environment. And this higher development is connected with the acquirement of structures which intensify or magnify the effects of changes going on in the external world; that is to say, it is connected with the acquirement of *ganglionated* sense-organs. In the same way the heart will be more able to adapt itself to *its* environment when it possesses ganglia; and since the only way in which the heart can respond to changes in its environment is by altering the strength or rapidity of its beats, it seems logical that ganglia should be intimately connected with the production of the rhythm. It appears indeed that minute portions of the heart which contain ganglia differ from those which do not, in this very particular, for they alone are capable of rhythmic contraction[1]. We have then a simple contractile organ (snail's heart) apparently capable of contracting rhythmically without the help of ganglion cells, and we have a complex contractile organ (frog's heart) in which ganglia are developed, which are apparently essential to the production of rhythm. In the same way, is it not possible that ganglia may be necessary for the inhibition of a complex organ in spite of the possibility of inhibiting a *simple* organ where no such ganglia exist? This supposition is supported by the fact, that in the most perfect and complex cases of inhibition, for instance, in the vertebrate heart, the penis and the submaxillary gland ganglia are found to exist.

We must enquire what evidence there is for the existence of local vaso-motor mechanism in the frog's web, and in some other parts of the body[2] in which we are considering the phenomena of inhibition.

[1] Dr Sanderson in *Handbook for the Phys. Lab.* p. 204.
[2] Since Dr Cohnheim denies the existence of local vaso-motor ganglia, some

Some physiological evidence on this point is contained in a paper by Putzeys and Tarchanoff (*Centralblatt*, Aug. 29, 1874). They begin by referring to an experiment by Goltz which is as follows: A dilatation and a consequent rise of temperature is produced in the feet of various animals (frog among the number) by the section of the sciatic nerve. After a short time the vessels regain their tonus, and the temperature falls to the normal[1]. Goltz argues from this result that vaso-motor ganglia must exist in the feet; because if the maintenance of the arterial tonus were due to stimulus supplied entirely by the central nervous system, the vessels ought to remain permanently dilated when deprived of such stimulus by the section of the sciatic nerve. Dr Cohnheim has himself performed an experiment of this kind, without being aware, as it appears, of the conclusions which must be drawn from his results. He says (*Entzündung*, p. 24) that dividing the tongue so that it remains attached only by the lingual arteries and veins, or that completely destroying the central nervous system, produces only a slight and transitory dilatation of the vessels, which almost at once regain their tonus. In Mr Lister's paper on the vaso-motor nerves of the frog he records experiments of the same nature as those of Putzeys and Tarchanoff, but which are more carefully done, and from which he drew similar conclusions as to the presence of a local vaso-motor apparatus.

Judging from my own observations, I should say that dividing the sciatic nerve has very little effect on the tonus of the vessels[2]. I have seen dilatation ensue; but on other occasions it has been of so slight and doubtful a character that I have not been sure whether or not the artery had dilated. This points to the same conclusion as Goltz's experiment, *and shows moreover how slight in some cases is the share taken by the central nervous system in the maintenance of the tonus.* From the above

evidence in favour of an opposite opinion may be given. In the first place Dr Cohnheim (*Embolischer Processe*, p. 28), in describing the complexity of structure of the walls of an artery, mentions the fact that *ganglion-cells* are found in them. Dr Beale (*Monthly Microscop. Journal*, Aug. 1872, p. 57) says, "in the bladder of the frog I have been able to follow fine nerve fibres from the ganglia both to arteries and capillary vessels." Dr Beale has also described (*Philosophical Transactions*, 1868) a portion of the coat of a branch of the iliac artery of the frog; upon the surface external to the muscular fibres are seen some ganglion-cells in process of developement with their fibres which ramify upon the muscular coat." Lastly, I have described and figured (*Quarterly Journal of Micro. Sc.* XIV. 109) rich plexus of ganglia and nerves accompanying and supplying the vessels in the external coat of the bladders of dogs and rabbits.

[1] A similar experiment has been performed by Schiff, *Op. cit.* I. 258, on the ear of the rabbit. It is hardly necessary to remark that he has not drawn similar conclusions from it.

[2] Dr Sanderson points out (Holmes, *System of Surgery*, V. 786) that there is no single trunk whose division completely paralyses the web. Hence the experiments of Putzeys and Tarchanoff are not complete. Lister's results, however, suffice for my point.

evidence we may conclude that when the vessels of a frog's foot are paralysed, *i. e.* deprived of the stimulus of the central nervous system, they are brought back to a tonic condition by naturally existing internal stimuli, whose effect is probably intensified by a local vaso-motor apparatus. In order to investigate the nature of this apparatus I have made a careful search through a considerable number of webs, by staining them with chloride of gold and cutting horizontal sections. No distinct ganglia have been found, so that if any exist, they must be of such rarity as not to be worth considering. In a few instances I have found places where the nerves possess more or less distinct swellings due to the presence of masses of nuclei. These do not resemble the clear oval nuclei of ganglion-cells as seen, *e.g.* in Auerbach's plexus or the sympathetic plexus in the bladder, &c.; they are, however, very much like the small nuclei which form the swellings figured by Lovén in the nervi erigentes, and regarded by him as ganglionic apparatus[1]. It is possible that nuclei may to some extent perform the functions of ganglia. Beale maintains[2] that ganglion-cells are developed from nuclei, "which cannot at first be distinguished from ordinary nuclei in connection with nerve-fibres." This of course is not conclusive, since the peculiar power of the ganglion-cell might be developed only *pari passu* with the growth of the granular cell-substance around the nucleus. Dr Drysdale[3], however, in speaking of the increased effect produced on the muscle by stimulating the nerve at an increased distance from it says, "the explanation of Pflüger is, that the molecules of the nerve in succession disengage active energy, and each stimulates its successor; but the increase of action (avalanche like) shows that each molecule disengages more force than the one before. Beale's is evidently a much more natural explanation, for in it each little battery of protoplasm is set in action by the passing current, and contributes its quota to the current which is thus really swelled like an avalanche." If by "current" we mean nothing more than the passage of a molecular disturbance along a nerve, I believe that this explanation of Pflüger's observation will be assented to, and

---

[1] *Leipsig Berichte*, 1866.
[2] Nerve-cells of frog. *Phil. Trans.* 1868, II. 550.
[3] *Protoplasmic Theory of Life*, by John Drysdale, M.D. (Edin.) 1874, p. 121.

that it will not appear improbable that the function of ganglia is performed in the frog's web by nuclei.

Schiff made a careful examination of the rabbit's ear, but found no ganglia. He therefore rejected the inhibitory theory, and concluded that "active dilatation" is effected by some unknown mechanism. But since his active dilatation is identical with that which is usually called inhibitory, I shall mention one of his most striking experiments (*Leçons*, &c., I. 260). He paralyses the ear or the foot of the rabbit by dividing in one case the auricular branch of the fifth, in the other the sciatic nerve. The result is a dilatation of the vessels and a local rise of temperature in the paralysed part. He then produces fever either by injecting pus into the veins or by injuring the pleura. "The first effect of the fever is an increase in the general temperature of the body; this increase shows itself more rapidly in the normal ear, and its temperature is soon higher than that of the paralysed one, which only participates feebly in the febrile rise of temperature." He argues that this experiment proves the possibility of dilatation produced by the *activity* of a nerve, and not by paralysis, because the effect of the loss of nervous supply is not merely to make the vessels dilate, but to prevent them being able to do so. The conclusion becomes inevitable when he shows that by irritating the peripheral end of the auricular branch of the fifth nerve the vessels of the ear can be made to dilate[1].

For the sake of clearness I may be allowed to recapitulate very briefly.

The question under discussion is the cause of the primary dilatation of vessels produced by local irritants.

I. Several reasons were given why the phenomenon cannot be paralytic in nature; the most important being that *contraction* may in certain cases be produced by local irritation. Having shown that local irritants may be supposed to act on the peripheral terminations of the vaso-motor nerves, I passed on to show :—

II. The possibility of dilatation dependent on the activity of a nerve, and produced by the inhibition of the local vaso-motor apparatus. My proposition is, that the primary dilatation is the result of the stimulation of the peripheral terminations of the dilatatory, that is, of the inhibitory nerves. I shall at present confine myself to the result of irritating the frog's web.

In the first place, there can be no doubt that the vessels in this organ are capable of dilating on its being locally irritated.

[1] Lovén, *Leipzig Berichte*, 1866, confirms this result.

I have described the dilatation which I have observed to precede contraction, and Dr Cohnheim believes dilatation to be the universal result of irritation. If these phenomena are to be explained by the stimulation of inhibitory nerve-fibres, it must be shown that the sciatic nerve contains such fibres. Goltz concluded from his experiments that this was the case; but Putzeys and Tarchanoff on continuing the research did not find the assumption of inhibitory nerves necessary, but considered the dilatation to be the result of the exhaustion of the constrictor, or ordinary vaso-motor, nerves.

A similar question has been discussed by Schiff, who found that by dividing the auricular branch of the *cervical plexus* and irritating the central end, he could produce dilatation of the vessels of the ear. In this experiment a central inhibitory ganglion is supposed to be stimulated, and the vessels of the ear are made to dilate by reflex inhibition conveyed to them by the auricular branch of the *fifth nerve*, which remains intact. Donders however (quoted in *Leçons*, T. I. p. 244) imagined the dilatation to be the result of exhausting the ordinary vaso-motor or constrictor nerves. Schiff replies that he finds it impossible to produce dilatation by the most prolonged and violent irritation of the *peripheral* end of the auricular branch of the cervical plexus, *i. e.* of the constrictor nerve.

This is at variance with the results obtained by Putzeys and Tarchanoff, who state that long and protracted irritation of the peripheral end of a divided sciatic produced dilatation in the arteries of the web. On the other hand, it is in agreement with what I have myself observed. I began by irritating the peripheral end of the sciatic with a moderately weak current from a Du Bois inductive coil. The artery under observation contracted, and in a short time dilated in spite of the continuance of the stimulus. I am inclined to think that it was this dilatation which Putzeys and Tarchanoff ascribed to the exhaustion of the constrictor nerves. But it is evident that it is not due to this cause, for while this supposed state of dilatation from exhaustion continued, *the artery contracted on the web being irritated.* The vessels having recovered from the effects of local stimulation, I made the current stronger, and the artery again contracted, but after a time it seemed to become accustomed to the stimulation as in the first instance, and contracted on local irritation being applied. I strengthened the current several times with the same result, and the very powerful current which I ultimately made use of did not, as I believe, dilate the arteries beyond the normal.

It seems as if the peripheral nervous mechanism has the power of accommodating itself to an excess of stimulation received through the nerves from the central nervous system, and this is in accordance with the independent way in which it can maintain the tonus of the arteries when *deprived* of central nervous stimulation.

It seems then that Putzeys' and Tarchanoff's exhaustion hypothesis is not satisfactory, and that we must return to Goltz's view that dilatatory, *i.e.* inhibitory fibres exist in the sciatic.

In the rabbit's ear, as Schiff has shown, the dilator fibres are collected into one nerve, but in the sciatic, inhibitory and constrictor fibres are mingled together, so that when the central vaso-motor ganglia are irritated both kinds of fibres will be called into action. Dr Sanderson[1] states that contraction in the arteries of the web is produced by irritating the central end of the divided sciatic of the opposite side, that is, when the central vaso-motor ganglia are stimulated. I have performed this experiment with a like result. But on one occasion both Dr Klein and myself several times observed distinct dilatation, and I have confirmed this result by a subsequent experiment.

Having now shown the probability of the existence of inhibitory fibres in the sciatic I shall consider the effects of irritation on this nerve.

It appears to me that when a stimulus is applied to the vaso-motor mechanism of the sciatic nerve, either by irritating the central ganglia, the peripheral end of the divided nerve, or by local irritation of the web (and this is what especially relates to the question under discussion), a struggle commences between the dilatatory (inhibitory) and the constrictor nerves. In the case of local irritation of the web, this struggle usually ends in the victory of the constrictors, in the case of the tongue always in that of the dilators. In the web these are sometimes

[1] Holmes, *System of Surgery*, v. 787.

victorious for a short time, hence the temporary widening of the arteries which I have mentioned as occasionally preceding the contraction. But there are usually no signs of the struggle except the short pause which intervenes between the application of the stimulus and the consequent contraction, and the spasmodic contraction already mentioned. I believe that the primary dilatation in the tongue is in the same manner due to the stimulation of the peripheral terminations of the vaso-motor nerves. This view is rendered probable by Vulpian's results[1]; he finds that the vessels of the tongue (in what animal is not mentioned) are made to dilate where the *peripheral* extremities of the divided lingual and glossopharyngeal nerves are stimulated. The fact that stimulation of the nerves and local irritants both produce dilatation in the tongue, when contrasted with the fact that these agents cause *contraction* in the web, supports my conclusion, that irritants act through the peripheral terminations of the nerves.

The variable results of Saviotti's experiments (so described by Cohnheim, who neglects them for this reason) are exactly what, according to my view of the case, we should expect. As Schiff remarks (*Leçons*, T. I. p. 253), "Les exceptions du reste n'ont rien d'étonnant si l'on considère l'inconstance du trajet des nerfs vaso-moteurs en général, pour tout organe recevant des nerfs vasculaires de plusieurs sources distinctes." As an example, he mentions the fact, that even the section of the sympathetic in the neck does not produce always the same effect on the vessels of the ear of the rabbit.

Let us once more consider the inhibitory dilatation in the rabbit's ear—Schiff asserts (and he is confirmed by Lovén, *Op. cit.*) that irritating the *peripheral* extremity of one of the auricular nerves (A) causes contraction, while irritating the extremity of another nerve (B) causes dilatation. This cannot be accounted for by different intensities of stimulation, nor by saying that the "nature" of the nerves is different. All we can say is that in passing from nerve (A) to the artery the stimulus remains a stimulus in the ordinary sense of the word[2], but that in passing

---

[1] *Gazette Hebdom.* No. 52, 1874 (quoted in *Centralblatt*, May 22, 1875).

[2] Brücke (Vienna *Denkschriften*, IV. 1852) in speculating why it is that light which should act as a stimulus produces "paralysis" of the pigment-cells of the chameleon, whereas darkness makes them "contract," remarks that "we are in certain relations to the external world, changes in it produce changes in us;" he goes on to say that anything producing a change may be called a stimulus (reiz),

from nerve (B) to the artery it changes its character and only remains a stimulus in the sense that it produces *some* change in the artery. And the conclusion we seem driven to is, that the means by which the stimulus passes from A to the artery are different from those by which it passes from B to it. That is to say, there must in the latter case be some local mechanism interposed between the nerve and the artery, which has the power of altering the stimuli it receives and passing them on to the artery changed in some way. And we can best conceive such a mechanism taking the form of an arrangement of ganglion-cells or nuclei.

In fact we suppose that an artery contracts under a certain stimulus, but dilates when this stimulus is altered by some local mechanism—or in other words, we assume that its contractile tissue possesses two kinds of excitability. This appears to me to be in accordance with Dr Foster's and Mr Dew Smith's supposition[1], that the nature of the tissue of the snail's heart is such that inductive shocks of different intensities produce exactly opposite results. For this is equivalent to endowing the tissue with two kinds of excitability. By stimulating certain nerves we can produce either flexion or extension of a vertebrate animal's limb, but because by the application of appropriate stimuli we can also produce contraction and extension of a nerveless mass of protoplasm[2], we do not deny that distinct nerves for flexion and extension may exist in more highly developed organisms. This analogy is exaggerated and false, but it may serve to show how the supposition of the existence of distinct nerves for inhibition and contraction is consistent with the simple form of inhibition described by Dr Foster and Mr Dew Smith.

In the course of this paper I have ventured to state the way in which the difficulty of inhibition presents itself to me; and I have tried to show that the only "explanation" or generalization of the facts which I can form, seems to accord with the view that certain forms of inhibition (where the nature of the result

but that "if this word is to be useful in physiological language we must limit it to those agents which when acting in a motor nerve produce contraction (of a muscle), and when acting on a sensory nerve produce sensation........."

[1] *Op. cit.* p. 823.

[2] Stricker, *Handbuch*, &c., Lieferung, I.

is not determined by the strength of the stimulus applied)
require a local vaso-motor apparatus. Also that the supposition
of the existence of distinct inhibitory and constrictor nerves
is not at variance with the simple form of inhibition of Dr
Foster and Mr Dew Smith. I may be permitted to restate the
difficulty very briefly. A mere contractile mass of the nature
of protoplasm can be inhibited, or the reverse, by varying the
intensity of the stimulus. We "explain" this by saying that the
tissue has two opposite excitabilities. But in the case of the
rabbit's ear, (Schiff, *Leçons* 11 and 12,) the same artery can be
made to dilate or to contract by applying the same stimulus to
different nerves. Here we cannot explain anything by *merely* endowing the tissue of the artery with excitabilities varying with
the intensity of the stimulus—because the stimulus we apply to
the nerve is the same in the two cases. Nor are we permitted
to ascribe different powers of conduction to the nerve-trunks
along which the stimuli travel. We seem therefore compelled
to say that in passing from the inhibitory nerves to the arteries
the stimulus is changed. And to effect such a change we must
assume a local mechanism interposed, and this will probably
take the form of an arrangement of ganglion cells or nuclei.

The rarity of the collections of nuclei on the nerve-trunks in
the frog's web mentioned above may possibly be connected with
the preponderance of the contractile over the inhibitory tendency
in that part of the body.

Finally I conclude:—

I. That local irritants do *not* cause dilatation by *direct* paralysis of the tissue of the arteries.

II. That Schiff's view is correct: viz. that local irritants
produce their effects on vessels by acting on the peripheral terminations of the vaso-motor nerves.

III. That when the vaso-motor nerves include both inhibitory and constrictor fibres, both are stimulated by local irritants,
and the resulting alteration in the calibre of the vessel is the
result of the victory of one set over the other.

ON THE ORIGIN AND HISTORY OF THE URINO-GENITAL ORGANS OF VERTEBRATES. By F. M. BALFOUR, B.A., *Fellow of Trinity College, Cambridge.*

RECENT discoveries[1] as to the mode of development and anatomy of the urinogenital system of Selachians, Amphibians and Cyclostome fishes, have greatly increased our knowledge of this system of organs, and have rendered more possible a comparison of the types on which it is formed in the various orders of vertebrates.

The following paper is an attempt to give a consecutive history of the origin of this system of organs in vertebrates and of the changes which it has undergone in the different orders.

For this purpose I have not made use of my own observations alone, but have had recourse to all the Memoirs with which I am acquainted, and to which I have access. I have commenced my account with the Selachians, both because my own investigations have been directed almost entirely to them, and because their urinogenital organs are, to my mind, the

---

[1] The more important of these are :—

Semper—Ueber die Stammverwandtschaft der Wirbelthiere u. Anneliden. *Centralblatt f. Med. Wiss.* 1874, No. 85.

Semper—Segmentalorgane bei ausgewachsenen Haien. *Centralblatt f. Med. Wiss.* 1874, No. 52.

Semper — Das Urogenitalsystem der höheren Wirbelthiere. *Centralblatt f. Med. Wiss.* 1874, No. 59.

Semper—Stammesverwandtschaft d. Wirbelthiere u. Wirbellosen. *Arbeiten aus Zool. Zootom. Inst.* Würzburg. II Band.

Semper—Bildung u. Wachstum der Keimdrüsen bei den Plagiostomen. *Centralblatt f. Med. Wiss.* 1875, No. 12.

Semper—Entw. d. Wolf. u. Müll. Gang. *Centralblatt f. Med. Wiss.* 1875, No. 29.

Alex. Schultz—Phylogenie d. Wirbelthiere. *Centralblatt f. Med. Wiss.* 1874, No. 51.

Spengel—Wimpertrichtern i. d. Amphibienniere. *Centralblatt f. Med. Wiss.* 1875, No. 23.

Meyer—Anat. des Urogenitalsystems der Selachier u. Amphibien. *Sitzb. Naturfor. Gesellschaft.* Leipzig, 30 April, 1875.

F. M. Balfour—Preliminary Account of development of Elasmobranch fishes. *Quart. Journ. of Micro. Science,* Oct. 1874.

W. Müller—Persistenz der Urniere bei Myxine glutinosa. *Jenaische Zeitschrift,* 1873.

W. Müller—Urinogenitalsystem d. Amphioxus u. d. Cyclostomen. *Jenaische Zeitschrift,* 1875.

Alex. Götte—*Entwicklungsgeschichte der Unke (Bombinator igneus).*

most convenient for comparison both with the more complicated and with the simpler types.

On many points the views put forward in this paper will be found to differ from those which I expressed in my paper (*loc. cit.*) which gives an account of my original[1] discovery of the segmental organs of Selachians, but the differences, with the exception of one important error as to the origin of the Wolffian duct, are rather fresh developments of my previous views from the consideration of fresh facts, than radical changes in them.

In Selachian embryos an intermediate cell-mass, or middle plate of mesoblast is formed, as in birds, from a partial fusion of the somatic and splanchnic layers of the mesoblast at the outer border of the protovertebræ. From this cell-mass the whole of the urinogenital system is developed.

At about the time when three visceral clefts have appeared, there arises from the intermediate cell-mass, opposite the fifth protovertebra, a solid knob, from which a column of cells grows backwards to opposite the position of the future anus (Fig. 1, *pd.*).

This knob projects outwards toward the epiblast, and the column lies at first between the mesoblast and epiblast. The knob and column do not long remain solid. The knob becoming hollow acquires a wide opening into the pleuroperitoneal or body cavity, and the column a lumen; so that by the time that five visceral clefts have appeared, the two together form a duct closed behind, but communicating in front by a wide opening with the pleuroperitoneal cavity.

Before these changes are accomplished, a series of *solid*[2] outgrowths of elements of the 'intermediate cell-mass' appear at the uppermost corner of the body-cavity. These soon become hollow and appear as involutions from the body-cavity, curling round the inner and dorsal side of the previously formed duct.

---

[1] These organs were discovered independently by Professor Semper and myself. Professor Semper's preliminary account appeared prior to my own which was published (with illustrations) in the *Quarterly Journal of Mic. Science*. Owing to my being in South America, I did not know of Professor Semper's investigations till several months after the publication of my paper.

[2] These outgrowths are at first solid in both Pristiurus Scyllium and Torpedo, but in Torpedo attain a considerable length before a lumen appears in them.

Fig. 1.

TWO SECTIONS OF A PRISTIURUS EMBRYO WITH THREE VISCERAL CLEFTS.

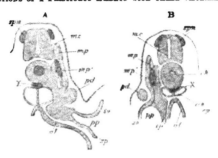

The sections are to show the development of the segmental duct (*pd.*) or primitive duct of the kidneys. In *A* (the anterior of the two sections) this appears as a solid knob projecting towards the epiblast. In *B* is seen a section of the column which has grown backwards from the knob in *A*.

spn. rudiment of a spinal nerve; *mc.* medullary canal; *ch.* notochord; *X.* string of cells below the notochord; *mp.* muscle-plate; *mp'.* specially developed portion of muscle-plate; *ao.* dorsal aorta; *pd.* segmental duct. *so.* somatopleura; *sp.* splanchnopleura; *pp.* pleuroperitoneal or body cavity; *ep.* epiblast; *al.* alimentary canal.

One involution of this kind makes its appearance for each protovertebra, and the first belongs to the protovertebra immediately behind the anterior end of the duct whose development has just been described. In Pristiurus there are in all 29 of these at this period. The last two or three arise from that portion of the body-cavity, which at this stage still exists behind the anus. The first-formed duct and the subsequent involutions are the rudiments of the whole of the urinary system. The duct is the primitive duct of the kidney[1]; I shall call it in future *the segmental duct;* and the involutions are the commencements of the segmental tubes which constitute the body of the kidney. I shall call them in future *segmental tubes.*

Soon after their formation the segmental tubes become convoluted, and their blind ends become connected with the

[1] This duct is often called either Müller's duct, the oviduct, or the duct of the primitive kidneys 'Urnierengang.' None of these terms are very suitable. A justification of the name I have given it will appear from the facts given in the later parts of this paper. In my previous paper I have always called it oviduct, a name which is very inappropriate.

2—2

segmental duct of the kidney. At the same time, or rather before this, the blind posterior termination of each of the segmental ducts of the kidneys unites with and opens into one of the horns of the cloaca. At this period the condition of affairs is represented in Fig. 2.

There is at *pd*, the segmental duct of the kidneys, opening in front (*o*) into the body cavity, and behind into the cloaca, and there are a series of convoluted segmental tubes (*st*), each opening at one end into the body cavity, and at the other into the duct (*pd*.).

Fig. 2.

DIAGRAM OF THE PRIMITIVE CONDITION OF THE KIDNEY IN A SELACHIAN EMBRYO.

pd. segmental duct. It opens at *o* into the body cavity and at its other extremity into the cloaca;   *x.* line along which the division appears which separates the segmental duct into the Wolffian duct above and the Müllerian duct below;   *st.* segmental tubes. They open at one end into the body-cavity, and at the other into the segmental duct.

The next important change which occurs is the longitudinal division of the segmental duct of the kidneys into Müller's duct, or the oviduct, and the duct of the Wolffian bodies or Leydig's duct. The splitting[1] is effected by the growth of a wall of cells which divides the duct into two parts (Fig. 3, *wd.* and *md.*). It takes place in such a way that the front end of the segmental duct, anterior to the entrance of the first segmental tube, together with the ventral half of the rest of the duct, is split off from its dorsal half as an independent duct (vide Fig. 2, *x*).

The dorsal portion also forms an independent duct, and into

---

[1] This splitting was first of all discovered and an account of it published by Semper (*Centralblatt f. Med. Wiss.* 1875, No. 29). I had independently made it out for the female a few weeks before the publication of Semper's account—but have not yet made observations about the point for the male.

My own previous account of the origin of the Wolffian duct (*Quart. Journ. of Micros. Science,* Oct. 1874), is completely false, and was due to my not having had access to a complete series of my sections when I wrote the paper.

Fig. 3.

TRANSVERSE SECTION OF A SELACHIAN EMBRYO ILLUSTRATING THE FORMATION OF THE WOLFFIAN AND MÜLLERIAN DUCTS BY THE LONGITUDINAL SPLITTING OF THE SEGMENTAL DUCT.

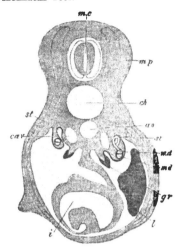

mc. medullary canal; *mp.* muscle-plate; *ch.* notochord; *ao.* aorta; *cav.* cardinal vein; *st.* segmental tube. On the one side the section passes through the opening of a segmental tube into the body cavity. On the other this opening is represented by dotted lines, and the opening of the segmental tube into the Wolffian duct has been cut through; *wd.* Wolffian duct; *md.* Müllerian duct. The Müllerian duct and the Wolffian duct together constitute the primitive segmental duct; *gr.* The germinal ridge with the thickened germinal epithelium; *l.* liver; *i.* intestine with spiral valve.

it the segmental tubes continue to open. Such at least is the method of splitting for the female—for the male the splitting is, according to Professor Semper, of a more partial character, and consists for the most part in the front end of the duct only being separated off from the rest. The result of these changes is the formation in both sexes of a fresh duct which carries off the excretions of the segmental involutions, and which I shall call the Wolffian duct—while in the female there is formed another complete and independent duct, which I shall call the Müllerian duct, or oviduct, and in the male portions only of such a duct.

The next change which takes place is the formation of

another duct from the hinder portion of the Wolffian duct, which receives the secretion of the posterior segmental tubes. This secondary duct unites with the primary or Wolffian duct near its termination, and the primary ducts of the two sides unite together to open to the exterior by a common papilla.

Slight modifications of the posterior terminations of these ducts are found in different genera of Selachians (vide Semper, *Centralblatt für Med. Wiss.* 1874, No. 59), but they are of no fundamental importance.

These constitute the main changes undergone by the segmental duct of the kidneys and the ducts derived from it; but the segmental tubes also undergo important changes. In the majority of Selachians their openings into the body-cavity, or, at any rate, the openings of a large number of them, persist through life; but the investigations of Dr Meyer[1] render it very probable that the small portion of each segmental tube adjoining the opening becomes separated from the rest and becomes converted into a sort of lymph organ, so that the openings of the segmental tubes in the adult merely lead into lymph organs and not into the gland of the kidneys.

These constitute the whole changes undergone in the female, but in the male the open ends of a varying number (according to the species) of the segmental tubes become connected with the testis and, uniting with the testicular follicles, serve to carry away the seminal fluid[2]. The spermatozoa have therefore to pass through a glandular portion of the kidneys before they enter the Wolffian duct, by which they are finally carried away to the exterior.

In the adult female, then, there are the following parts of the urinogenital system (Fig. 4):

(1) The oviduct, or Müller's duct (Fig. 4, *md.*), split off from the segmental duct of the kidneys. Each oviduct opens at its upper end into the body-cavity, and behind the two oviducts have independent communications with the cloaca. The oviducts serve simply to carry to the exterior the ova, and have no communication with the glandular portion of the kidneys.

---

[1] *Sitzen. der Naturfor. Gesellschaft*, Leipzig, 30 April, 1875.
[2] We owe to Professor Semper the discovery of the arrangement of the seminal ducts. *Centralblatt f. Med. Wiss.* 1875, No. 12.

Fig. 4.

DIAGRAM OF THE ARRANGEMENT OF THE URINOGENITAL ORGANS IN AN ADULT
FEMALE SELACHIAN.

md. Müllerian duct; *wd.* Wolffian duct; *st.* segmental tubes;
d. duct of the posterior segmental tubes; *ov.* ovary.

(2) The Wolffian ducts (Fig. 4, *wd.*) or the remainder of the segmental ducts of the kidneys. Each Wolffian duct ends blindly in front, and the two unite behind to open by a common papilla into the cloaca.

This duct receives the secretion of the whole anterior end of the kidneys[1], that is to say, of all the anterior segmental tubes.

(3) The secondary duct (Fig. 4, *d.*) belonging to the lower portion of the kidneys opening into the former duct near its termination.

(4) The segmental tubes (Fig. 4, *st.*) from whose convolutions and outgrowths the kidney is formed. They may be divided into two parts, according to the duct by which their secretion is carried off.

In the male the following parts are present:

(1) The Müllerian duct (Fig. 5, *md.*), consisting of a small remnant, attached to the liver, which represents the foremost end of the oviduct of the female.

(2) The Wolffian duct (Fig. 5, *wd*), which precisely corresponds to the Wolffian duct of the female, except that, in addition to functioning as the duct of the anterior part of the kidneys, it also serves to carry away the semen. In the female it is straight, but has in the adult male a very tortuous course (vide Fig. 5).

---

[1] This upper portion of the kidneys is called Leydig's gland by Semper. It would be better to call it the Wolffian body, for I shall attempt to show that it is homologous with the gland so named in Sauropsida and Mammalia.

Fig. 5.

DIAGRAM OF THE ARRANGEMENT OF THE URINOGENITAL ORGANS IN AN ADULT MALE SELACHIAN.

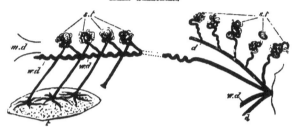

md. rudiment of Müllerian duct; *wd.* Wolffian duct, which also serves as vas deferens; *st.* segmental tubes. The ends of three of those which in the female open into the body-cavity, have in the male united with the testicular follicles, and serve to carry away the products of the testis; *d.* duct of the posterior segmental tubes; *t.* testis.

(3) The duct (Fig. 5, *d.*) of the posterior portion of the kidneys, which has the same relations as in the female.

(4) The segmental tubes (Fig. 5, *st.*). These have the same relations as in the female, except that the most anterior two, three or more, unite with the testicular follicles, and carry away the semen into the Wolffian duct.

The mode of arrangement and the development of these parts suggest a number of considerations.

In the first place it is important to notice that the segmental tubes develope primitively as completely independent organs[1], one of which appears in each segment. If embryology is in any way a repetition of ancestral history, it necessarily follows that these tubes were primitively independent of each other. Ancestral history, as recorded in development, is often, it is true, abridged; but it is clear that though abridgement might prevent a series of primitively separate organs from appearing as such, yet it would hardly be possible for a primitively compound organ, which always retained this condition, to appear

---

[1] Further study of my sections has shown me that the initial independence of these organs is even more complete than might be gathered from the description in my paper (*loc. cit.*). I now find, as I before conjectured, that they at first correspond exactly with the muscle-plates, there being one for each muscle-plate. This can be seen in the fresh embryos, but longitudinal sections shew it in an absolutely demonstrable manner.

during development as a series of separate ones. These considerations appear to me to prove that the segmented ancestors of vertebrates possessed a series of independent and segmental excretory organs.

Both Professor Semper and myself, on discovering these organs, were led to compare them and state our belief in their identity with the so-called segmental organs of Annelids.

This view has since been fairly generally accepted. The segmental organs of annelids agree with those of vertebrates in opening at one end into the body cavity, but differ in the fact that each also communicates with the exterior by an independent opening, and that they are never connected with each other.

On the hypothesis of the identity of the vertebrate segmental tubes with the annelid segmental organs, it becomes essential to explain how the external openings of the former may have become lost.

This brings us at once to the origin of the segmental duct of the kidneys, by which the secretion of all the segmental tubes was carried to the exterior, and it appears to me that a right understanding of the vertebrate urinogenital system depends greatly upon a correct view of the origin of this duct. I would venture to repeat the suggestion which I made in my original paper (*loc. cit.*), that this duct is to be looked upon as the most anterior of the segmental tubes which persist in vertebrates. In favour of this view are the following anatomical and embryological facts. (1) It developes in nearly the same manner as the other segmental tubes, viz. in Selachians as a solid outgrowth from the intermediate cell-mass, which subsequently becomes hollowed so as to open into the body-cavity: and in Amphibians and Osseous and Cyclostome fishes as a direct involution from the body cavity. (2) In Amphibians, Cyclostomes and Osseous fishes its upper end develops a glandular portion, by becoming convoluted in a manner similar to the other segmental tubes. This glandular portion is often called either the head-kidney or the primitive kidney. It is only an embryonic structure, but is important as demonstrating the true nature of the primitive duct of the kidneys.

We may suppose that some of the segmental tubes first united, possibly in pairs, and that then by a continuation of this process the whole of them coalesced into a common gland. One external opening sufficed to carry off the entire secretion of the gland, and the other openings therefore atrophied.

This history is represented in the development of the dog-fish in an abbreviated form, by the elongation of the first segmental tube (segmental duct of the kidney) and its junction with each of the posterior segmental tubes. Professor Semper looks upon the primitive duct of the kidneys as a duct which arose independently, and was not derived from metamorphosis of the segmental organs. Against this view I would on the one hand urge the consideration, that it is far easier to conceive of the transformation by change of function (comp. Dohrn, *Functions-wechsel*, Leipzig, 1875) of a segmental organ into a segmental duct, than to understand the physiological cause which should lead, in the presence of so many already formed ducts, to the appearance of a totally new one. By its very nature a duct is a structure which can hardly arise *de novo*. We must even suppose that the segmental organs of Annelids were themselves transformations of still simpler structures. On the other hand I would point to the development in this very duct amongst Amphibians and Osseous fishes of a glandular portion similar to that of a segmental tube, as an *a posteriori* proof of its being a metamorphosed segmental tube. The development in insects of a longitudinal tracheal duct by the coalescence of a series of transverse tracheal tubes affords a parallel to the formation of a duct from the coalescence of a series of segmental tubes.

Though it must be admitted that the loss of the external openings of the segmental organs requires further working out, yet the difficulties involved in their disappearance are not so great as to render it improbable that the vertebral segmental organs are descended from typical annelidan ones.

The primitive vertebrate condition, then, is probably that of an early stage of Selachian development while there is as yet a segmental duct,—the original foremost segmental tube opening in front into the body cavity and behind into the cloaca; with which duct all the segmental tubes communicate. Vide Fig. 2.

The next condition is to be looked upon as an indirect

result of the segmental duct serving as well for the products of the generative organs as the secretions of the segmental tubes.

As a consequence of this, the segmental duct became split into a ventral portion, which served alone for the ova, and a dorsal portion which received the secretion of the segmental tubes. The lower portion, which we have called the oviduct, in some cases may also have received the semen as well as the ova. This is very possibly the case with Ceratodus (vide Günther, *Trans. of R. Society*, 1871), and the majority of Ganoids (Hyrtl, *Denkschriften Wien*, Vol. VIII.). In the majority of other cases the oviduct exists in the male in a completely rudimentary form; and the semen is carried away by the same duct as the urine.

In Selachians the transportation of the semen from the testis to the Wolffian duct is effected by the junction of the open ends of two or three or more segmental tubes with the testicular follicles, and the modes in which this junction is effected in the higher vertebrates seem to be derivatives from this. If the views here expressed are correct it is by a complete change of function that the oviduct has come to perform its present office. And in the bird and higher vertebrates no trace, or only the very slightest, (vide p. 46) of the primitive urinary function is retained during embryonic or adult life.

The last feature in the anatomy of the Selachians which requires notice is the division of the kidney into two portions, an anterior and posterior. The anatomical similarity between this arrangement and that of higher vertebrates (birds, &c.) is very striking. The anterior one precisely corresponds, anatomically, to the *Wolffian body*, and the posterior one to the true permanent *kidney* of higher vertebrates : and when we find that in the Selachians the duct for the anterior serves also for the semen as does the Wolffian duct of higher vertebrates, this similarity seems almost to amount to identity. A discussion of the differences in development in the two cases will come conveniently with the account of the bird ; but there appear to me the strongest grounds for looking upon the kidneys of Selachians as equivalent to both the Wolffian bodies and the true kidneys of the higher vertebrates.

The condition of the urinogenital organs in Selachians is by no means the most primitive found amongst vertebrates.

The organs of both Cyclostomous and Osseous fishes, as well as those of Ganoids, are all more primitive; and in the majority of points the Amphibians exhibit a decidedly less differentiated condition of these organs than do the Selachians.

In Cyclostomous fishes the condition of the urinary system is very simple. In Myxine (vide Joh. Müller, *Myxinoid fishes*, and Wilhelm Müller, *Jenaische Zeitschrift*, 1875, *Das Urogenital-system des Amphioxus u. d. Cyclostomen*) there is a pair of ducts which communicate posteriorly by a common opening with the abdominal pore. From these ducts spring a series of transverse tubules, each terminating in a Malpighian corpuscle. These together constitute the mass of the kidneys. About opposite the gall-bladder the duct of the kidney (the segmental duct) narrows very much, and after a short course ends in a largish glandular mass (the head-kidney), which communicates with the pericardial cavity by a number of openings.

In Petromyzon the anatomy of the kidneys is fundamentally the same as in Myxine. They consist of the two segmental ducts, and a number of fine branches passing off from these, which become convoluted but do not form Malpighian tufts. The head-kidney is absent in the adult.

W. Müller (*loc. cit.*) has given a short but interesting account of the development of the urinary system of Petromyzon. He finds that the segmental ducts develop first of all as simple involutions from the body cavity. The anterior end of each then developes a glandular portion which comes to communicate by a number of openings with the body cavity. Subsequently to the development of this glandular portion the remainder of the kidneys appears in the posterior portion of the body cavity; and before the close of embryonic life the anterior glandular portion atrophies.

The comparison of this system with that of a Selachian is very simple. The first developed duct is the segmental duct of a Selachian, and the glandular portion developed at its anterior extremity, which is permanent in Myxine but embryonic in Petromyzon, is, as W. Müller has rightly recognized, equivalent to the head-kidney of Amphibians, which remains undeveloped in Selachians. It is, according to my previously stated view, the glandular portion of the first segmental organ or the segmental duct. The series of orifices by which this communicates with the body cavity are due to the division of the primary opening of the segmental duct. This is shown both by the facts of their development in Petromyzon given by Müller, as well as by the occurrence of a similar division of the primary orifice in Amphibians, which is mentioned later in this paper. In a note in my original paper (*loc. cit.*) I stated that these openings were equivalent to the segmental involutions of Selachians. This is erroneous, and was due to my not having understood the description given in a preliminary paper of Müller (*Jenaische Zeitschrift*, 1873). The large development of this glandular mass in the Cyclostome and Osseous fishes and in embryo Amphibians, implies that it must at one time have been important. Its earlier development than the remainder of the kidneys is probably a result of the specialized function of the first segmental organ.

The remainder of the kidney in Cyclostomes is equivalent to the kidney of Selachians. Its development from segmental involutions has not been recognized. If these segmental involutions are really absent it may perhaps imply that the simplicity of the Cyclostome kidneys, like that of so many other of their organs, is a result of degeneration rather than a primitive condition.

In Osseous fishes the segmental duct of the kidneys developes, as the observations of Rosenberg[1] ("Teleostierniere," *Inaug. Disser. Dorpat*, 1867) and Oellacher (*Zeitschrift für Wiss. Zool.* 1873) clearly prove, by an involution from the body cavity. This involution grows backwards in the form of a duct and opens into the cloaca. The upper end of this duct (the most anterior segmental tube) becomes convoluted, and forms a glandular body, which has no representative in the urinary apparatus of Selachians, but whose importance, as indicating the origin of the segmental duct of the kidneys, I have already insisted upon.

The rest of the kidney becomes developed at a later period,

[1] I am unfortunately only acquainted with Dr Rosenberg's paper from an abstract.

probably in the same way as in Selachians; but this, as far as I know, has not been made out.

The segmental duct of the kidneys forms the duct for this new gland, as in embryo Selachians (Fig. 2), but, unlike what happens in Selachians, undergoes no further changes, with the exception of a varying amount of retrogressive metamorphosis of its anterior end. The kidneys of Osseous fish usually extend from just behind the head to opposite the anus, or even further back than this. They consist for the most part of a broader anterior portion, an abdominal portion reaching from this to the anus, and, as in those cases in which the kidneys extend further back than the anus, of a caudal portion.

The two ducts (segmental ducts of the kidneys) lie, as a rule, in the lower part of the kidneys on their outer borders, and open almost invariably into a urinary bladder. In some cases they unite before opening into the bladder, but generally have independent openings.

This bladder, which is simply a dilatation of the united lower ends of the primitive kidney-ducts, and has no further importance, is almost invariably present, but in many cases lies unsymmetrically either to the right or the left. It opens to the exterior by a very minute opening in the genito-urinary papilla, immediately behind the genital pore. There are, however, a few cases in which the generative and urinary organs have a common opening. For further details vide Hyrtl, *Denk. der k. Akad. Wien*, Vol. II.

It is possible that the generative ducts of Osseous fishes are derived from a splitting from the primitive duct of the kidney, but this is discussed later in the paper.

In Osseous fishes we probably have an embryonic condition of the Selachian kidneys retained permanently through life.

In the majority of Ganoids the division of the segmental duct of the kidney into two would seem to occur, and the ventral duct of the two (Müllerian duct), which opens at its upper end into the body-cavity, is said to serve as an excretory duct for both male and female organs.

The following are the more important facts which are known about the generative and urinary ducts of Ganoids.

In Spatularia (vide Hyrtl, Geschlechts u. Harnwerkzeuge bei den Ganoiden, *Denkschriften der k. Akad. Wien*, Vol. VIII.) the following parts are found in the female.

(1) The ovaries stretching along the whole length of the abdominal cavity.

(2) The kidneys, which are separate and also extend along the greater part of the abdominal cavity.

(3) The ureters lying on the outer borders of the kidneys. Each ureter dilates at its lower end into an elongated wide tube, which continues to receive the ducts from the kidneys. The two ureters unite before terminating and open behind the anus.

(4) The two oviducts (Müllerian ducts). These open widely into the abdominal cavity, at about two-thirds of the distance from the anterior extremity of the body-cavity. Each opens by a narrow pore into the dilated ureter of its side.

In the male the same parts are found as in the female, but Hyrtl found that the Müllerian duct of the left side at its entrance into the ureter became split into two horns, one of which ended blindly. On the right side the opening of the Müllerian duct was normal.

In the Sturgeon (vide J. Müller, *Bau u. Grenzen d. Ganoiden*, Berlin Akad. 1844; Leydig, *Fischen u. Reptilien*, and Hyrtl, *Ganoiden*) the same parts are found as in Spatularia.

The kidneys extend along the whole length of the body cavity; and the ureter, which does not reach the whole length of the kidneys, is a thin-walled wide duct lying on their outer side. On laying it open the numerous apertures of the tubules for the kidney are exposed. The Müllerian duct, which opens in both sexes into the abdominal cavity, ends, according to Leydig, in the cases of some males, blindly behind without opening into the ureter, and Müller makes the same statement for both sexes. It was open on both sides in a female specimen I examined[1], and Hyrtl found it invariably so in both sexes in all the specimens he examined.

Both Rathke and Stannius (I have been unable to refer to the original papers) believed that the semen was carried off by transverse ducts directly into the ureter, and most other observers have left undecided the mechanism of the transportation

---

[1] For this specimen I am indebted to Dr Günther.

of the semen to the exterior. If we suppose that the ducts Rathke saw really exist they might perhaps be supposed to enter not directly into the ureter, but into the kidney, and be in fact homologous with the vasa efferentia of the Selachians. The frequent blind posterior termination of the Müllerian duct is in favour of the view that these ducts of Rathke are really present.

In Polypterus (vide Hyrtl, *Ganoiden*) there is, as in other Ganoids, a pair of Müllerian ducts. They unite at their lower ends. The ureters are also much narrower than in previously described Ganoids and, after coalescing, open into the united oviducts. The urinogenital canal, formed by the coalescence of the Müllerian ducts and ureters, has an opening to the exterior immediately behind the anus.

In Amia (vide Hyrtl) there is a pair of Müllerian ducts which, as well as the ureters, open into a dilated vesicle. This vesicle appears as a continuation of the Müllerian ducts, but receives a number of the efferent ductules of the kidneys. There is a single genito-urinary pore behind the anus.

In Ceratodus (Günther, *Phil. Trans.* 1871) the kidneys are small and confined to the posterior extremity of the abdomen. The generative organs extend however along the greater part of the length of the abdominal cavity. In both male and female there is a long Müllerian duct and the ducts of the two sides unite and open by a common pore into a urinogenital cloaca which communicates with the exterior by the same opening as the alimentary canal. In both sexes the Müllerian duct has a wide opening near the anterior extremity of the body cavity. The ureters coalesce and open together into the urinogenital cloaca dorsal to the Müllerian ducts. It is not absolutely certain that the semen is transported to the exterior by the Müllerian duct of the male, which is perhaps merely a rudiment as in Amphibia. Dr Günther failed however to find any other means by which it could be carried away.

The genital ducts of Lepidosteus differ in important particulars from those of the other Ganoids (vide Müller, *loc. cit.* and Hyrtl, *loc. cit.*).

In both sexes the genital ducts are continuous with the investments of the genital organs.

In the female the dilated posterior extremities of the ureters completely invest for some distance the generative ducts, whose extremities are divided into several processes, and end in a different way on the two sides. A similar division and asymmetry of the ducts is mentioned by Hyrtl as occurring in the male of Spatularia, and it seems not impossible that on the hypothesis of the genital ducts being segmental tubes these divisions may be remnants of primitive glandular convolutions. The ureters in both sexes dilate as in other Ganoids at their posterior extremities, and unite with one another. The unpaired urinogenital opening is situated behind the anus. In the male the dilated portion of the ureters is divided into a series of partitions which are not present in the female.

Till the embryology of the secretory system of Ganoids has been worked out, the homologies of their generative ducts are necessarily a matter of conjecture. It is even possible that what I have called the Müllerian duct in the male is functionless, as with Amphibians, but that, owing to the true ducts of the testis having been overlooked, it has been supposed to function as the vas deferens. Günther's (loc. cit.) injection experiments on Ceratodus militate against this view, but I do not think they can be considered as conclusive as long as the mechanism for the transportation of the semen to the exterior has not been completely made out. Analogy would certainly lead us to expect the ureter to serve in Ganoids as the vas deferens.

The position of the generative ducts might in some cases lead to the supposition that they are not Müllerian ducts, or, in other words, the most anterior pair of segmental organs but a pair of the posterior segmental tubes.

What are the true homologies of the generative ducts of Lepidosteus, which are continuous with the generative glands, is somewhat doubtful. It is very probable that they may represent the similarly functioning ducts of other Ganoids, but that they have undergone further changes as to their anterior extremities.

It is, on the other hand, possible that their generative ducts are the same structures as those ducts of Osseous fishes, which are continuous with the generative organs. These latter ducts are perhaps related to the abdominal pores, and had best be considered in connection with these; but a completely satisfactory answer to the questions which arise in reference to them can only be given by a study of their development.

In the Cyclostomes the generative products pass out by an abdominal pore, which communicates with the peritoneal cavity by two short tubes[1], and which also receives the ducts of the kidneys.

Gegenbaur suggests that these are to be looked upon as Müllerian ducts, and as therefore developed from the segmental ducts of the kidneys. Another possible view is that they are the primitive external openings of a pair of segmental organs. In Selachians there are usually stated to be a pair of abdominal pores. In Scyllium I have only been able to find, on each side, a large deep pocket opening to the exterior, but closed below towards the peritoneal cavity, so that in it there seem to be no abdominal pores[2]. In the Greenland Shark (*Læmargus Borealis*) Professor Turner (*Journal of Anat. and Phys.* Vol. VIII.) failed to find either oviduct or vas deferens, but found a pair of large open abdominal pores, which he believes serve to carry away the generative products of both sexes. Whether the so-called abdominal pores of Selachians usually end blindly as in Scyllium, or, as is commonly stated, open into the body cavity, there can be no question that they are homologous with true abdominal pores.

The blind pockets of Scyllium appear very much like the remains of primitive involutions from the exterior, which might easily be supposed to have formed the external opening of a pair of segmental organs, and this is probably the true meaning of abdominal pores. The presence of abdominal pores in all Ganoids in addition to true genital ducts and of these pockets or abdominal pores in Selachians, which are almost certainly

---

[1] According to Müller (*Myxinoiden*, 1845) there is in Myxine an abdominal pore with two short canals leading into it, and Vogt and Pappenheim (*An. Sci. Nat.* Part IV. Vol. XI.) state that in Petromyzon there are two such pores, each connected with a short canal.

[2] My own rough examination of preserved specimens was hardly sufficient to enable me to determine for certain the presence or absence of these pores. Mr Bridge, of Trinity College, has, however, since then commenced a series of investigations on this point, and informs me that these pores are certainly absent in Scyllium as well as in other genera.

homologous with the abdominal pores of Ganoids and Cyclostomes, and also occur in addition to true Müllerian ducts, speak strongly against the view that the abdominal pores have any relation to Müllerian ducts. Probably therefore the abdominal pores of the Cyclostomous fishes (which seem to be of the same character as other abdominal pores) are not to be looked on as rudimentary Müllerian ducts.

We next come to the question which I reserved while speaking of the kidneys of Osseous fishes, as to the meaning of their genital ducts.

In the female Salmon and the male and female Eel, the generative products are carried to the exterior by abdominal pores, and there are no true generative ducts. In the case of most other Osseous fish there are true generative ducts which are continuous with the investment of the generative organs[1] and have generally, though not always, an opening or openings independent of the ureter close behind the rectum, but no abdominal pores are present. It seems, therefore, that in Osseous fish the generative ducts are complementary to abdominal pores, which might lead to the view that the generative ducts were formed by a coalescence of the investment of the generative glands with the short duct of abdominal pore.

Against this view there are, however, the following facts :

(1) In the cases of the salmon and the eel it is perfectly true that the abdominal pore exactly corresponds with the opening of the genital duct in other Osseous fishes, but the absence of genital ducts in these cases must rather be viewed, as Vogt and Pappenheim (loc. cit.) have already insisted, as a case of degeneration than of a primitive condition. The presence of genital ducts in the near allies of the Salmonidæ, and even in the male salmon, are conclusive proofs of this. If we admit that the presence of an abdominal pore in Salmonidæ is merely

---

[1] The description of the attachment of the vas deferens to the testis in the Carp given by Vogt and Pappenheim (Ann. Scien. Nat. 1859) does not agree with what I found in the Perch (Perca fluvialis). The walls of the duct are in the Perch continuous with the investment of the testis, and the gland of the testis occupies, as it were, the greater part of the duct ; there is, however, a distinct cavity corresponding to what Vogt and P. call the duct, near the border of attachment of the testis into which the seminal tubules open. I could find at the posterior end of the testis no central cavity which could be distinguished from the cavity of this duct.

a result of degeneration, it obviously cannot be used as an argument for the complementary nature of abdominal pores and generative ducts.

(2) Hyrtl (*Denkschriften der k. Akad. Wien*, Vol. I.) states that in Mormyrus oxyrynchus there is a pair of abdominal pores in addition to true generative ducts. If his statements are correct, we have a strong argument against the generative ducts of Osseous fishes being related to abdominal pores. For though this is the solitary instance of the presence of both a genital opening and abdominal pores known to me in Osseous fishes, yet we have no right to assume that the abdominal pores of Mormyrus are not equivalent to those of Ganoids and Selachians. It must be admitted, with Gegenbaur, that embryology alone can elucidate the meaning of the genital ducts of Osseous fishes.

In Lepidosteus, as was before mentioned, the generative ducts, though continuous with the investment of the generative bodies, unite with the ureters, and in this differ from the generative ducts of Osseous fishes. The relation, indeed, of the generative ducts of Lepidosteus to the urinary ducts is very similar to that existing in other Ganoid fishes; and this, coupled with the fact that Lepidosteus possesses a pair of abdominal pores on each side of the anus[1], makes it most probable that its generative ducts are true Müllerian ducts.

In the Amphibians the urinary system is again more primitive than in the Selachians.

The segmental duct of the kidneys is formed[2] by an elongated fold arising from the outer wall of the body cavity, in the same position as in Selachians. This fold becomes constricted into a canal, closed except at its anterior end, which remains open to the body cavity. This anterior end dilates, and grows out into two horns, and at the same time its opening into the body cavity becomes partly constricted, and so divided into three separate

---

[1] This is mentioned by Müller (*Ganoid fishes*; Berlin Akad. 1844), Hyrtl (*loc. cit.*), and Günther (*loc. cit.*), and through the courtesy of Dr Günther I have had an opportunity of confirming the fact of the presence of the abdominal pores on two specimens of Lepidosteus in the British Museum.

[2] My account of the *development* of these parts in Amphibians is derived for the most part from Götte, *Die Entwickelungsgeschichte der Unke.*

orifices, one for each horn and a central one between the two. The horns become convoluted, blood channels appearing between their convolutions, and a special coil of vessels is formed arising from the aorta and projecting into the body cavity near the openings of the convolutions. These formations together constitute the glandular portion[1] of the original anterior segmental tube or segmental duct of the kidneys. I have already pointed out the similarity which this organ exhibits to the head-kidneys of Cyclostome fishes in its mode of formation, especially with reference to the division of the primitive opening. The lower end of the segmental duct unites with a horn of the cloaca.

After the formation of the gland just described the remainder of the kidney is formed.

This arises in the same way as in Selachians. A series of involutions from the body cavity are developed ; these soon form convoluted tubes, which become branched and interlaced with one another, and also unite with the primitive duct of the kidneys. Owing to the branching and interlacing of the primitive segmental tubes, the kidney is not divided into distinct segments in the same way as with the Selachians. The mode of development of these segmental tubes was discovered by Götte. Their openings are ciliated, and, as Spengel (*loc. cit.*) and Meyer (*loc. cit.*) have independently discovered, persist in most adult Amphibians. As both these investigators have pointed out, the segmental openings are in the adult kidneys of most Amphibians far more numerous than the vertebral segments to which they appertain. This is due to secondary changes, and is not to be looked upon as the primitive state of things. At this stage the Amphibian kidneys are nearly in the same condition as the Selachian, in the stage represented in Fig. 2. In both there is the segmental duct of the kidneys, which is open in front, communicates with the cloaca behind, and receives the whole secretion from the kidneys. The parallelism between the two is closely adhered to in the subsequent modifications of the Amphibian kidney, but the changes are not completed so far in

---

[1] It is called Kopfniere (head-kidney), or Urniere (primitive kidney), by German authors. Leydig correctly looks upon it as together with the permanent kidney constituting the Urniere of Amphibians. The term Urniere is one which has arisen in my opinion from a misconception; but certainly the Kopfniere has no greater right to the appellation than the remainder of the kidney.

Amphibians as in Selachians. The segmental duct of the Amphibian kidney becomes, as in Selachians, split into a Müllerian duct or oviduct, and a Wolffian duct or duct for the kidney.

The following points about this are noteworthy :

(1) The separation of the two ducts is never completed, so that they are united together behind, and, for a short distance, blend and form a common duct ; the ducts of the two sides so formed also unite before opening to the exterior.

(2) The separation of the two ducts does not occur in the form of a simple splitting, as in Selachians. But the efferent ductules from the kidney gradually alter their points of entrance into the primitive duct. Their points of entrance become carried backwards further and further, and since this process affects the anterior ducts proportionally more than the posterior, the efferent ducts finally all meet and form a common duct which unites with the Müllerian duct near its posterior extremity. This process is not always carried out with equal completeness. In the tailless Amphibians, however, the process is generally[1] completed, and the ureters (Wolffian ducts) are of considerable length. Bufo cinereus, in the male of which the Müllerian ducts are very conspicuous, serves as an excellent example of this.

In the Salamander (Salamandra maculosa), Figs. 6 and 7, the process is carried out with greater completeness in the female than in the male, and this is the general rule in Amphibians. In the male Proteus, the embryonic condition would seem to be retained almost in its completeness so that the ducts of the kidney open directly and separately into the still persisting primitive duct of the kidney. The upper end of the duct nevertheless extends some distance beyond the end of the kidney and opens into the abdominal cavity. In the female Proteus, on the other hand, the separation into a Müllerian duct and a ureter is quite complete. The Newt (Triton) also serves as an excellent example of the formation of distinct Müllerian and Wolffian ducts being much more complete in the female

---

[1] In Bombinator igneus, Von Wittich stated that the embryonic condition was retained. Leydig, *Anatom. d. Amphib. u. Reptilien*, shewed that this is not the case, but that in the male the Müllerian duct is very small, though distinct.

than the male. In the female Newt all the tubules from the kidney open into a duct of some length which unites with the Müllerian duct near its termination, but in the male the anterior segmental tubes, including those which, as will be afterwards seen, serve as vasa efferentia of the testis, enter the Müllerian duct directly, while the posterior unite as in the female into a common duct before joining the Müllerian duct. For further details as to the variations exhibited in the Amphibians, the reader is referred to Leydig, *Anat. Untersuchung, Fischen u. Reptilien.* Ditto, *Lehrbuch der Histologie, Menschen u. Thiere.* Von Wittich, *Siebold u. Kölliker, Zeitschrift,* Vol. IV. p. 125.

The different conditions of completeness of the Wolffian ducts observable amongst the Amphibians are instructive in reference to the manner of development of the Wolffian duct in Selachians. The *mode* of division in the Selachians of the segmental duct of the kidney into a Müllerian and Wolffian duct is probably to be looked upon as an embryonic abbreviation of the process by which these two ducts are formed in Amphibians. The fact that this separation into Müllerian and Wolffian ducts proceeds further in the females of most Amphibians than in the males, strikingly shows that it is the oviductal function of the Müllerian duct which is the indirect cause of its separation from the Wolffian duct. The Müllerian duct formed in the way described persists almost invariably in both sexes, and in the male sometimes functions as a sperm reservoir; *e.g.* Bufo cinereus. In the embryo it carries at its upper end the glandular mass described above (Kopfniere), but this generally atrophies though remnants of it persist in the males of some species (*e.g.* Salamandra). Its anterior end opens, in most cases by a single opening, into the perivisceral cavity in both sexes, and is usually ciliated. As the female reaches maturity, the oviduct dilates very much; but it remains thin and inconspicuous in the male.

The only other developmental change of importance is the connection of the testes with the kidneys. This probably occurs in the same manner as in Selachians, viz. from the junction of the open ends of the segmental tubes with the follicles of the testes. In any case the vessels which carry off the semen constitute part of the kidney, and the efferent duct of the testis is also that of the kidney. The vasa efferentia from the testis either pass through one or two nearly isolated anterior portions of the kidney (Proteus, Triton) or else no such special portion of the kidney becomes separated from the rest, and the vasa efferentia enter the general body of the kidney.

Fig. 6.

DIAGRAM OF THE URINOGENITAL ORGANS OF A MALE SALAMANDER.
(*Copied from Leydig's Histologie des Menschen u. der Thiere.*)

md. Müller's duct (rudimentary) ; *y.* remnant of the secretory portion of the segmental duct Kopfniere ; *Wd.* Wolffian duct ; a less complete structure in the male than in the female ; *st.* segmental tubes or kidney. The openings of these into the body-cavity are not inserted in the figure ; *t.* testis. Its efferent ducts form part of the kidney.

In the male Amphibian, then, the urinogenital system consists of the following parts (Fig. 6) :

(1) Rudimentary Müllerian ducts, opening anteriorly into the body cavity, which sometimes carry aborted *Kopfnieren.*

(2) The partially or completely formed Wolffian ducts (ureters) which also serve as the ducts for the testes.

(3) The kidneys, parts of which also serve as the vasa efferentia, and whose secretion, together with the testicular products, is carried off by the Wolffian ducts.

(4) The united lower parts of Wolffian and Müllerian ducts which are really the lower unsplit part of the segmental ducts of the kidneys.

In the female, there are (Fig. 7)

(1) The Müllerian ducts which function as the oviducts.

(2) The Wolffian ducts.

(3) The kidneys.

(4) The united Müllerian and Wolffian ducts as in the male.

Fig. 7.

DIAGRAM OF THE URINOGENITAL ORGANS OF A FEMALE SALAMANDER.

(Copied from Leydig's Histologie des Menschen u. der Thiere.)

Md. Müller's duct or oviduct; *Wd.* Wolffian duct or the duct of the kidneys; *st.* segmental tubes or kidney. The openings of these into the body-cavity are not inserted in the figure; *o.* ovary.

The urinogenital organs of the adult Amphibians agree in almost all essential particulars with those of Selachians. The ova are carried off in both by a specialized oviduct. The Wolffian duct, or ureter, is found both in Selachians and Amphibians, and the relations of the testis to it are the same in both, the vasa efferentia of the testes having in both the same anatomical peculiarities.

The following points are the main ones in which Selachians and Amphibians differ as to the anatomy of the urinogenital organs; and in all but one of these, the organs of the Amphibian exhibit a less differentiated condition than do those of the Selachian.

(1) A glandular portion (Kopfniere) belonging to the first segmental organ (segmental duct of the kidneys) is found in all embryo Amphibians, but usually disappears, or only leaves a remnant in the adult. It has not yet been found in any Selachian.

(2) The division of the primitive duct of the kidney into the Müllerian duct and the Wolffian duct is not completed so far in Amphibians as Selachians, and in the former the two ducts are confluent at their lower ends.

(3) The permanent kidney exhibits in Amphibians no distinction into two glands (foreshadowing the Wolffian bodies and true kidneys of higher vertebrates), as it does in the Selachians.

(4) The Müllerian duct persists in its entity in male Amphibians, but only its upper end remains in male Selachians.

(5) The openings of the segmental tubes into the body cavity correspond in number with the vertebral segments in most Selachians, but are far more numerous than these in Amphibians. This is the chief point in which the Amphibian kidney is more differentiated than the Selachian.

The modifications in development which the urinogenital system has suffered in higher vertebrates (Sauropsida and Mammalia) are very considerable; nevertheless it appears to me to be possible with fair certainty to trace out the relationship of its various parts in them to those found in the Ichthyopsida. The development of urinogenital organs has been far more fully worked out for the bird than for any other member of the amniotic vertebrates; but, as far as we know, there are no essential variations except in the later periods of development throughout the division. These later variations, concerning for the most part the external apertures of the various ducts, are so well known and have been so fully described as to require no notice here. The development of these parts in the bird will therefore serve as the most convenient basis for comparison.

In the bird the development of these parts begins by the appearance of a column of cells on the upper surface of the intermediate cell-mass (Fig. 8, *W.d.*). As in Selachians, the intermediate cell-mass is a group of cells between the outer edge of the protovertebræ and the upper end of the body cavity. The column of cells thus formed is the commencement of the duct of the Wolffian body. Its development is strikingly similar to that of the segmental duct of the kidney in Selachians. I shall attempt when I have given an account of the development of the Müllerian duct to speak of the relations between the Selachian duct and that of the bird.

Romiti (*Archiv f. Micr. Anat.* Vol. x.) has recently stated that the Wolffian duct developes as an involution from the body cavity. The fact that the specimens drawn by Romiti to support this view are too old to determine such a point, and the inspection of a number of specimens made by my friend Mr Adam Sedgwick of Trinity College, who, at my request has been examining the

Fig. 8.

TRANSVERSE SECTION THROUGH THE DORSAL REGION OF AN EMBRYO FOWL OF 45 h. TO SHOW THE MODE OF FORMATION OF THE WOLFFIAN DUCT.

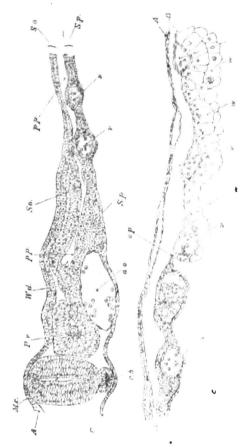

A. epiblast; B. mesoblast; C. hypoblast; Mc. medullary canal; Pv. Protovertebræ; Wd. Wolffian duct; So. Somatopleure; Sp. Splanchnopleure; pp. pleuroperitoneal cavity; ch. notochord; ao. dorsal aorta; v. blood-vessels.

urinogenital organs of the fowl, have led me to the conclusion that Romiti is in error in differing from his predecessors as to the development of the Wolffian duct. The solid string of cells to form the Wolffian duct lies at first close to the epiblast, but, by the alteration in shape which the protovertebræ undergo and the general growth of cells around it, becomes gradually carried downwards till it lies close to the germinal epithelium which lines the body cavity. While undergoing this change of position it also acquires a lumen, but ends blindly both in front and behind. Towards the end of the fourth day the Wolffian duct opens into a horn of the cloaca. The cells adjoining its inner border commence, as it passes down on the third day, to undergo histological changes, which, by the fourth day, result in the formation of a series of ducts and Malpighian tufts which form the mass of the Wolffian body[1].

The Müllerian duct arises in the form of an involution, whether at first solid or hollow, of the germinal epithelium, and, as I am satisfied, quite independently of the Wolffian duct. It is important to notice that its posterior end soon unites with the Wolffian duct, from which however it not long after becomes separated and opens independently into the cloaca. The upper end remains permanently open to the body-cavity, and is situated nearly opposite the extreme front end of the Wolffian body.

Between the 80th and 100th hour of incubation the ducts of the permanent kidneys begin to make their appearance. Near its posterior extremity each Wolffian duct becomes expanded, and from the dorsal side of this portion a diverticulum is constricted off, the blind end of which points forwards. This is the duct of the permanent kidneys and around its end the kidneys are found. It is usually stated that the tubules of the permanent kidneys arise as outgrowths from the duct, but this requires to be worked over again.

The condition of the urinogenital system in birds immediately after the formation of the permanent kidneys is strikingly similar to its permanent condition in adult Selachians. There is the Müllerian duct in both opening in front into the body

---

[1] This account of the origin of the Wolffian body differs from that given by Waldeyer, and by Dr Foster and myself (*Elements of Embryology*, Foster and Balfour), but I have been led to alter my view from an inspection of Mr Sedgwick's preparations, and I hope to show that theoretical considerations lead to the expectation that the Wolffian body would develop independently of the duct.

cavity and behind into the cloaca. In both the kidneys consist of two parts—an anterior and posterior—which have been called respectively Wolffian bodies and permanent kidneys in birds and Leydig's glands and the kidneys in Selachians.

The duct of the permanent kidney, which at first opens into that of the Wolffian body, subsequently becomes further split off from the Wolffian duct, and opens independently into the cloaca.

The subsequent changes of these parts are different in the two sexes.

In the female the Müllerian ducts[1] persist and become the oviducts. Their anterior ends remain open to the body-cavity. The changes in their lower ends in the various orders of Sauropsida and Mammalia are too well known to require repetition here. The Wolffian body and duct atrophy: there are left however in many cases slight remnants of the anterior extremity of the body forming the parovarium of the bird, and also frequently remnants of the posterior portion of the gland as well as of the duct. The permanent kidney and its duct remain unaltered.

In the male the Müllerian duct becomes almost completely obliterated. The Wolffian duct persists and forms the vas deferens, and the anterior so-called sexual portion of the Wolffian body also persists in an altered form. Its tubules unite with the seminiferous tubules and also form the epididymis. Unimportant remnants of the posterior part of the Wolffian body also persist, but are without function. In both sexes the so-called permanent kidneys form the sole portion of the primitive uriniferous system which persists in the adult.

In considering the relations between the modes of development of the urinogenital organs of the bird and of the Selachians, the first important point to notice is, that whereas in the Selachians the segmental duct of the kidneys is first developed and subsequently becomes split into the Müllerian and Wolffian ducts; in the bird these two ducts develope independently. This difference in development would be more accurately described by saying that in birds the segmental duct of the kidneys

---

[1] The right oviduct atrophies in birds, and the left alone persists in the adult.

developes as in Selachians, but that the Müllerian duct developes independently of it.

Since in Selachians the Wolffian duct is equivalent to the segmental duct of the kidneys with the Müllerian removed from it, when in birds the Müllerian duct developes independently of the segmental kidney duct, the latter becomes the same as the Wolffian duct.

The second mode of stating the difference in development in the two cases represents the embryological facts of the bird far better than the other method.

It explains why the Wolffian duct appears earlier than the Müllerian and not at the same time, as one might expect according to the other way of stating the case. If the Wolffian duct is equivalent to the segmental duct of Selachians, it must necessarily be the first duct to develope; and not improbably the development of the Müllerian duct would in birds be expected to occur at the time corresponding to that at which the primitive duct in Selachians became split into two ducts.

It probably also explains the similarity in the mode of development of the Wolffian duct in birds and the primitive duct of the kidneys in Selachians.

This way of stating the case is also in accordance with theoretical conclusions. As the egg-bearing function of the Müllerian duct became more and more confirmed we might expect that the adult condition would impress itself more and more upon the embryonic development, till finally the Müllerian duct ceased to be at any period connected with the kidneys, and the history of its origin ceased to be traceable in its development. This seems to have actually occurred in the higher vertebrates, so that the only persisting connection between the Müllerian duct and the urinary system is the brief but important junction of the two at their lower ends on the sixth or seventh day. This junction justly surprised Waldeyer (*Eierstock u. Ei*, p. 129), but receives a complete and satisfactory explanation on the hypothesis given above.

The original development of the segmental tubes is in the bird solely retained in the tubules of the Wolffian body arising independently of the Wolffian duct, and I have hitherto failed to find that there is a distinct division of the Wolffian bodies into segments corresponding with the vertebral segments.

I have compared the permanent kidneys to the lower portion of the kidneys of Selachians. The identity of the anatomical condition of the adult Selachian and embryonic bird which has been already pointed out speaks strongly in favour of this view; and when we further consider that the duct of the permanent kidneys is developed in nearly the same way as the supposed homologous duct in Selachians, the suggested identity gains further support. The only difficulty is the fact that in Selachians the tubules of the part of the kidneys under comparison develope as segmental involutions in point of time anteriorly to their duct, while in birds they develope in a manner not hitherto certainly made out but apparently in point of time posteriorly to their duct. But when the immense modifications in development which the whole of the gland of the excretory organ has undergone in the bird are considered, I do not think that the fact I have mentioned can be brought forward as a serious difficulty.

The further points of comparison between the Selachian and the bird are very simple. The Müllerian duct in its later stages behaves in the higher vertebrates precisely as in the lower. It becomes in fact the oviduct in the female and atrophies in the male. The behaviour of the Wolffian duct is also exactly that of the duct which I have called the Wolffian duct in Ichthyopsida, and in the tubules of the Wolffian body uniting with the tubuli seminiferi we have represented the junction of the segmental tubes with the testis in Selachians and Amphibians. It is probably this junction of two independent organs which led Waldeyer to the erroneous view that the tubuli seminiferi were developed from the tubules of the Wolffian body.

With the bird I conclude the history of the origin of the urinogenital system of vertebrates. I have attempted, and I hope succeeded, in tracing out by the aid of comparative anatomy and embryology the steps by which a series of independent and simple segmental organs like those of Annelids have become converted into the complicated series of glands and ducts which constitute the urinogenital system of . the higher vertebrates. There are no doubt some points which require further elucidation amongst the Ganoid and Osseous fishes. The most important points which appear to me still to need further research, both embryological and anatomical, are the abdominal pores of fishes, the generative ducts of Ganoids, especially Lepidosteus, and the generative ducts of Osseous fishes.

The only further point which requires discussion is the embryonic layer from which these organs are derived.

I have shown beyond a doubt (loc. cit.) that in Selachians these organs are formed from the mesoblast. The unanimous testimony of all the recent investigators of Amphibians leads to the same conclusion. In birds, on the other hand, various investigators have attempted to prove that these organs are derived from the epiblast. The proof they give is the following: the epiblast and mesoblast appear fused in the region of the axis cord. From this some investigators have been led to the conclusion that the whole of the mesoblast is derived from the upper of the two primitive embryonic layers. To these it may be replied that, even granting their view to be correct, it is no proof of the derivation of the urinogenital organs from the epiblast, since it is not till the complete formation of the three layers that any one of them can be said to exist. Others look upon the fusion of the two layers as a proof of the passage of cells from the epiblast into the mesoblast. An assumption in itself, which however is followed by the further assumption that it is from these epiblast cells that the urinogenital system is derived! Whatever may have been the primitive origin of the system, its mesoblastic origin in vertebrates cannot in my opinion be denied.

Kowalewsky (Embryo. Stud. an Vermes u. Anthropoda, Mem. Akad. St Petersbourg, 1871) finds that the segmental tubes of Annelids develope from the mesoblast. We must therefore look upon the mesoblastic origin of the excretory system as having an antiquity greater even than that of vertebrates.

ON ARTICULAR CARTILAGE. By ALEX. OGSTON, M.D.
Surgeon to the Aberdeen Royal Infirmary. (Pl. I. to VI.)

ARTICULAR Cartilage is generally looked upon as a structure which has few functions, other than the little that it has to perform. Its value as a buffer for diminishing concussion, and as a medium for preventing undue friction in joints, is universally admitted. But its situation beyond the reach of the vascular system, and its consequent slowness in responding to the stimulus of injury or of surrounding disease, have led to its being regarded as a tissue which, apart from the duties conceded above, participates only passively in physiological and pathological processes, and might, save for these functions, have been without damage omitted from the system.

It seems to me that this estimate of its nature and functions has been the cause why it occupies so subordinate a place in the accounts we possess of observations on, and investigations into, diseases of joints generally; and some of the conclusions at which I have arrived from a study of normal and pathological articular cartilage are so opposed to the generally received opinions, that I venture to believe they will have some little value in modifying these opinions, and enabling us to form a more correct estimate of the forces put forth by this structure both in its physiological and diseased conditions.

It will be conceded that a knowledge of the true functions which a tissue normally possesses, is of the utmost importance in every point of view. It enables us to assign it its just value among surrounding, and, it may be, dissimilar structures, to rank it in its proper place as one of the constituent elements of the living whole, and to follow with a clear comprehension its pathological alterations *per se*, as well as their reactions on surrounding parts and on the whole organism. We cannot afford to dispense with this knowledge in regard to any portion of the body, for we see that just in proportion to our comprehension of the various normal functions, characteristics, and tendencies of organs and tissues, is the clearness of our views of

their pathological conditions. Those parts with whose normal
characters we are best acquainted are those whose pathology is
best understood, while we comprehend little of the pathology of
such organs as the supra-renal capsules of whose physiological
design we are ignorant. Therefore our advances in physiological science, whatever be their intrinsic value, are eagerly applied
by the pathologist to his own researches, and always with profit
and enlightenment.

The contemptible *rôle* assigned to articular cartilage has
dissatisfied many. It has been remarked that, from our present
basis, we are incapable of comprehending what observation
teaches us of its behaviour, that we cannot, for instance, explain
why cartilage is not worn away, while the hardest and most
resisting structures, like the enamel of the teeth, show the
effects of friction in a short space of time. In fact it is plain
that, to comprehend cartilage at all, we must go back to the
very beginning, and acquire a correct notion of its physiology,
before we can progress with our present attainments in its
pathology.

In the first place, and bearing in mind the universal existence of incrusting cartilage in articulations, the question naturally suggests itself,—what would occur in a joint suppose it
were performing its usual functions unmutilated in every way
save in being deprived of its cartilage of incrustation? Such a
condition does not occur naturally under any circumstances, and
experimentally it would be impossible to produce it, for the
interference with other structures which would necessarily be
entailed, and the known results following such interference,
would vitiate or nullify any conclusions that might be sought to
be drawn from the experiment. Hence we are compelled to
fall back on other means of obtaining a clue to how the query
should be answered; and there is but one means, and that the
observation of a pathological process, which to any extent fulfils
the conditions required.

In Chronic Rheumatic Arthritis (*Arthritis Deformans*) we
have a disease where, while bones, ligaments and synovial membranes continue, at least in the earlier stages, to perform their
functions in something approaching to the normal manner,
considerable surfaces of the articular facets where bone rubs

against bone are deprived of the protective layer that incrusts them in the healthy joint. In this disease we find that, in the smaller articulations which are comparatively little exposed to pressure, the bones themselves are not perceptibly shortened, but retain nearly if not entirely their ordinary length, and, where apposed, are merely ground down to a smooth surface, the eburnation of which permits the continuance of motion to an extent and degree not very far off that they naturally enjoy. If, however, we go to the larger joints, such as the shoulder, hip, and knee, where we have in all the influence of large masses of muscle, and in some the necessity of supporting the weight of the body, intensifying the mutual pressure of the bones on each other, we find in addition a new phenomenon presenting itself. This consists in a wearing away of the bones by the pressure and attrition they are exposed to; and it goes to such an extent that the whole of the *cervix femoris* and nearly the whole of the head of the humerus are often worn away, and in time disappear in their respective joints, while similar, though less striking alterations of the same nature take place in the knee-joint. Even the eburnation (the so-called porcellanous deposit), occurring to a greater or less extent in the rubbed-down ends, is insufficient to arrest the destruction thus caused. The condition of the cartilage in Chronic Rheumatic Arthritis will be treated of further on, in the meantime it is sufficient to point out that the normal nutrition of the bone where it has lost its cartilage of incrustation is plainly insufficient to protect it against absorption from attrition and pressure.

This behaviour seems to suggest a clue to the function of articular cartilage, viz. that its purpose may be to resist the process of rubbing away continually going on. The next step is to examine how far the appearances of normal and abnormal cartilage support this suggestion.

I. NORMAL ARTICULAR CARTILAGE.

It is a striking fact that a section of healthy adult cartilage, made perpendicularly to the articular surface of a bone, reveals all the peculiarities of a structure in active growth. In its centre, somewhat nearer the joint-surface than the bone, there is seen (Fig. 1, b) a series of rounded groups of cells, one to three in number, imbedded in a finely granular hyaline matrix, which forms, round each cell or group of cells, the capsule external to the cell-wall characteristic of cartilage. The cells possess diameters varying from $\frac{1}{1800}$ to $\frac{1}{1300}$ of an inch, and present clear protoplasm, and a round central nucleus of about half the diameter of the cell. The groups are arranged at tolerably regular intervals about $\frac{1}{800}$ of an inch apart, and their general shape and distribution, while indicative of active growth, give no more clue to the direction towards which they tend to multiply than do the granulation cells in any piece of ordinary granulation tissue. Beyond this *focus of central growth*, as we may call it, matters assume quite another aspect. As they approach the surface of the bone, the cells evidently pass into a state of much greater activity, preparatory to some important end they are to serve. They proliferate so actively that each group now consists of five to ten cells, or even more, and the individual cells are usually though not invariably larger, often doubling or even trebling their former diameters. And still more marked is the arrangement of the groups into rows, or something approaching to rows, arranged perpendicularly to the surface of the bone (Fig. 1, c). The rows are about $\frac{1}{1300}$ of an inch broad, and $\frac{1}{400}$ to $\frac{1}{500}$ of an inch long. They are not always exactly perpendicular to the bone-surface (though they are always parallel to the direction in which the surface of the bone is growing), but they form to the surface at least something approaching to a right angle, and are never parallel to it. Corresponding to this alteration in the aspect and arrangement of the cells, the hyaline matrix becomes more distinctly granular as it approaches the surface of the bone.

On comparing these phenomena with those taking place in the ossification of epiphysal cartilages, the similarity is very striking. In the latter it is easy to observe that the stages and transformations are almost identical in quality. There exists the focus of central growth (Fig. 4, a), consisting of roundish or irregular cells, few in number, imbedded in a homogeneous matrix, and tolerably equidistant from one another, and in this focus the arrangement of the cells indicates no particular direction of growth. As they approach the diaphysis, however, the cells multiply into groups (Fig. 4, b) arranged perpendicularly to the bone, forming rows or *rouleaux*

(Fig. 4, g) $\frac{1}{300}$ to $\frac{1}{50}$ of an inch long and $\frac{1}{400}$ to $\frac{1}{300}$ of an inch broad, and acquire at the same time a much greater magnitude than they previously possessed. That the alterations in the two structures are identical in their nature is evident from the fact that in some places appearances exactly similar may be met with in each.

The next point to be examined is the boundary line where articular cartilage passes into bone. The determination of what happens here is the central point of the whole question. It is not easy to obtain sections which show satisfactorily the changes that take place. The process I have found most satisfactory is to decalcify the bones by soaking them some weeks in dilute chromic acid acidulated with nitric acid, then wash them from the acids, make sections perpendicular to the surface of the bone by the razor or Rutherford's microtome, and tint with logwood. Sections thus obtained show the existence, between the cartilage presenting the appearances above described and the bone, of a *zone of altered cartilage* (Fig. 6, c to i) $\frac{1}{500}$ to $\frac{1}{150}$ of an inch in thickness, marked off from the rest of the cartilage by a deeply-stained border forming an undulating line. In this zone the hyaline matrix is more pellucid and takes on a different, generally fainter, tint when acted on by staining solutions, than that further removed from the bone. The cells imbedded in it reach their maximum of proliferation, and often tend to form groups more rounded (Fig. 6, d) or less arranged in rows than before. The margin of the bone forms a slightly uneven line (Fig. 1, d; Fig. 2, g; Fig. 6, i) dotted at intervals by small rounded prominences (Fig. 2, d; Fig. 6, e and f) projecting into the zone of altered cartilage for a distance of from $\frac{1}{300}$ to $\frac{1}{550}$ of an inch, and exhibiting at their bases a corresponding breadth. The number of the prominences bears a distinct proportion to the number of groups of cartilage-cells in their vicinity; they seem to correspond in position with them and complete as it were the harmony of their arrangement, and point besides in a direction corresponding with that of the long groups of cartilage-cells. They vary much in distinctness in different sections, but are always present, and their arrangement is suggestive of their being somehow related to the cartilage-groups. This arrangement is not fortuitous. In favourable sections it can be seen that on the group of cartilage-cells reaching the margin of the bone, a most active proliferation and subdivision of its component cells commences at the end next the bone and quickly invades the rest of the group, which thus becomes converted into one of the prominences. In this proliferation the cells lose their capsule, and, ceasing to be recognisable as cartilage-cells, form by the process of multiplication a large mass of granulation tissue composed of small rounded cells, each about $\frac{1}{3500}$ of an inch in diameter, with granular protoplasm and large central nucleus whose diameter is about one half of that of the cell. It occasionally happens that one end of the cartilage-group still possesses its characteristic cells, while the other end is changed into granulation tissue (Fig. 2, c, c), but commonly the process once commenced is very rapid in involving the whole group. From this it is clear that the prominences are really groups of cartilage-cells which have undergone alteration. The alteration consists in a transmutation into a tissue not to be distinguished from medullary tissue such as occupies the cancelli of the spongy bones, in fact identical with it, and not unfrequently developing, like it, fat cells at various points.

It is not easy to see this process distinctly taking place. Out of a large number of sections there may be only one or two which exhibit it so clearly as in Fig. 2 of the accompanying plate, but once it has been observed in its entirety it is easy to recognize it in almost every section, and it seems to be a universal feature in the passage of the ono structure into the other.

The masses of granulation tissue thus formed are next transformed into true bony tissue, but the transformation shows appearances which vary a good deal according to the activity of growth. Sometimes the masses coalesce and form a layer (Fig. 2, e) between the cartilage and the bone, and on the surface of the layer next the bone the cells become transformed into bone-corpuscles (Fig. 2, f). This is attended by a considerable increase of their intercellular substance, so that each cell separates to a little distance from its fellows, the intercellular substance becoming hard by the deposition in it of calcareous salts, and the cells taking on the outlines of bone-corpuscles with imperfectly formed and not very numerous canaliculi. More frequently, on the other hand, the masses of granulation tissue remain separate and distinct (Fig. 3), and the peripheral cells of each mass become in the same manner transformed into layers of bone-tissue, so that a section presents islands of granulation tissue lying separated by a network of osseous substance. These variations are of interest as bearing on the pathological conditions which will fall to be considered afterwards.

Many observations have convinced me that these masses of medullary tissue, developed directly out of the cartilage cells, have at first no connection whatever with the medulla in the cancelli or with the vascular system. The observation of normal cartilage alone might be somewhat dubious, but as we shall afterwards find the process more distinctly marked in pathological states, this fact is, I believe, undeniable. I am aware that it is usual to describe prolongations of vascularized medullary tissue into the very margin of cartilage, and to assume that such prolongations are peculiar to inflammatory conditions and are the means by which cartilage is destroyed from its deeper aspect; but numberless observations made with this statement in view have convinced me of its inaccuracy, and I am satisfied that a fresh examination of the subject will be found to support in the clearest manner the statements just made as to the origin and nature of these prominences. It is true that in the insular development of bone from the prominences, examination shows that they do form an early communication with the neighbouring medulla and at the same time with the vascular system, as shown in Figure 6. The communication takes place by means of small channels (Fig. 6, h, h) eaten towards them through the already formed bone. The channels seem to be eaten by a tongue of medulla enclosing a loop or loops of blood-vessels, and which makes straight for the prominence almost as soon as it has formed. But in normal, and more especially, as will be afterwards seen, in pathological, cartilage (Fig. 3, c, d; Fig. 8, b, c, d) the prominences at an early stage contain no blood-vessel and have no communication with the neighbouring medulla, and yet are capable of performing the functions of medulla and developing new bone ere they have been provided with any vascular supply or any communication with the neighbouring medullary tissue.

At the line of transformation of cartilage into bone the hyaline matrix of the cartilage disappears. The alteration in its texture which can be brought out by staining is probably a process of softening, preceding and preparing for its absorption by the influence of the cell-groups round which it lies, and of the medullary tissue into which they become transformed. It is plain that the cartilage-cells are capable of modifying the condition of the matrix around them, and that each group possesses its surrounding "territory," as Virchow calls it, over which it bears rule. The staining usually shows, even at some distance from the bone, alterations in the "cell-territories" around the groups, becoming more marked as they approach the line of transformation. The alterations are probably of the same nature as the alteration at the line of transformation, and, like it, preparatory to the absorption of the hyaline matrix.

Where epiphysal cartilage passes into bone, changes are observable parallel to those described above. At the line of transformation the long *rouleaux* of cells can be followed with great ease across the line of ossification. Precisely at this line a great change (Fig. 4, c) in the behaviour of the cartilage-cells becomes manifest. Short of this line the cells have assumed the appearance of long sausage-shaped rows (Fig. 4, g), about $\frac{1}{800}$ of an inch in length and $\frac{1}{4000}$ of an inch in breadth. The cells composing them are flattened on their adjacent sides, and the groups which they form are not unlike the aspect of the well-known *rouleaux* of red blood disks; but they present all intermediate forms between those given in the plate and those described under articular cartilage, and in some places it would be impossible to tell from the appearances whether the section had been taken from epiphysal cartilage or articular cartilage. But the moment the boundary line is passed, the cells proliferate, lose their flat form, and change into a long sausage-shaped mass of medullary tissue (Fig. 4, d). Between these masses the hyaline matrix dips down, in processes $\frac{1}{150}$ to $\frac{1}{140}$ of an inch in length and $\frac{1}{2100}$ to $\frac{1}{800}$ of an inch in breadth, for a short distance into the bone (Fig. 4, e), and is finally absorbed by the medullary masses, which pass into bone, corpuscles just as in articular cartilage, the transformation commencing at the periphery of the sausage-shaped masses (Fig 4, f).

It is scarcely necessary to explain that the sausage-shaped masses of medullary tissue formed out of the *rouleaux* of cartilage-cells represent here the prominences spoken of under articular cartilage. Their tips form an even line, and the conversion of cartilage-cells into medulla progresses so harmoniously in the various groups that the line of advance is beautifully retained (Fig. 4, c). Some pathological conditions seem to modify this, and cause an unevenness of the line of transformation by permitting some groups to precede others in the change. These variations, however, it is foreign to our present purpose to enter into.

When the changes above described have been seen in their various stages, the observer cannot fail to be convinced that they amount to a complete demonstration of the fact that articular cartilage is continually producing new bone to supply the loss caused by the pressure sustained by the articulations. There is no other explanation possible, and, if any doubt remained, the comparison with the process of ossification occurring at the ends of the shafts by the development of the epiphysal cartilage into bone, showing as it does changes precisely similar in quality, though varying in activity, would, I submit, completely remove it. There is no doubt that cartilage is a structure admirably suited for resisting the pressure to which it is subjected. Hence the pressure is transmitted through it to the articular extremities of the bones, which are by their large size adapted to sustain and further to diffuse it. It seems probable that the spongy ends of the bones, by their size as well as by the elasticity (if the term be allowed) which their spongy structure confers upon them, succeed in reducing the pressure to its minimum and preventing its injurious action being concentrated on any one spot. They are however unable to neutralize the pressure completely, and the epiphyses would in time suffer absorption by the continual force thus acting upon them, were it not for the admirable provision of nature to replace what may be damaged. It seems probable that the spongy tissue of the articular ends of bones is being constantly renewed by a process of ossification proceeding from the articular cartilage.

Many facts in connection with diseases of joints become intelligible when they are viewed in the light of the processes described, and serve further to strengthen the conclusions arrived at; but some of these fall to be mentioned afterwards, and the others need not now be adverted to in detail.

The next point calling for attention is the condition of the cartilage where it is exposed to the attrition of the joint-movements.

Although the *pressure* of the articulating surfaces against each other seems to be transmitted through the cartilage to the bone, the *friction* to which the former is exposed cannot be similarly transmitted, but must be provided for by another arrangement. To understand this it will be necessary to return to the point whence we set out, viz. the focus of central growth. If, in examining a perpendicular section, we proceed from this point towards the joint-surface, we find that the clusters of cells lose their rounded form, and become elongated so as to lie parallel to the joint-surface (Fig. 1, *a*). As they approach it, they become flatter and flatter, and are composed of always fewer cells, so that in the immediate vicinity of the surface they are represented by isolated cells $\frac{1}{750}$ of an inch long and $\frac{1}{8000}$ of an inch broad, devoid of protoplasm and filled by a single elongated granular nucleus. Many such cells contain only a few fatty granules, and just before the surface is reached· the nuclei have all disappeared, and the cells are represented by thin clefts, containing at the most a few granules of fat (Fig. 1, *a* ; Fig. 5, *c*).

These changes are plainly preparatory to the rubbing down of the cartilage which takes place on its surface, and doubtless the infinitely fine *detritus* thus produced is absorbed by the lymph-spaces opening through the synovial membrane.

By this additional observation we recognise that *cartilage, from its focus of central growth, grows towards the joint-surface as well as towards the bone,* and that it is not a structure owing its permanence merely to its elastic consistence and smooth surface. Were this all, it would speedily, like the enamel of the teeth, be worn away; but nature has provided against this by conferring on it what I have termed a focus of central growth, whence it developes centrifugally, towards the bone to replace its loss from the forces acting on it, and towards the joint to renew continually a surface which is constantly being worn away.

The last point in connection with normal articular cartilage to be considered here is its relation to the synovial membrane, a fringe of which covers its margin for a short distance, and is supplied with loops of blood-vessels penetrating to its border. If a section be made perpendicularly through these structures

(Fig. 5), the following are the changes traceable in proceeding from cartilage to membrane. Starting from the focus of central growth (Fig. 5, a), the groups of cells show no rapid proliferation, they occur in groups of two, and further on as isolated cells (Fig. 5, e) $\frac{1}{2500}$ of an inch broad and $\frac{1}{1500}$ of an inch long, while they alter in shape, becoming elongated and fusiform, their long axis parallel to the synovial membrane. They do not appear to become effete, but consist of protoplasm and nucleus seemingly unaltered save in shape. Presently the hyaline matrix exhibits a fibrous transformation (Fig. 5, d), becoming arranged in fibres parallel to the long axis of the cells. At the same time the cells approach more and more in outline and resemblance to those of connective tissue, the fibrous matrix becomes indistinguishable from the fibrous intercellular substance of connective tissue, and the altered structure passes gradually into the synovial fringe (Fig. 5, b). Such a section too, if looked at with a low power, shows that the deeper layers at least of the fringe are not superimposed on the cartilage, but are directly continuous with it, so that if the plane in which they lie were prolonged towards the middle of the cartilage, the margin of the fringe would gradually end (Fig. 5, b to c) by passing into the cartilage-layer next the joint-surface, where it is being prepared for its function of being rubbed down.

What has already been discussed seems to warrant the following conclusions:

1st. That articular cartilage is continually renewing itself from a focus of central growth, and grows in two directions.

2nd. That articular cartilage developes in the direction of the joint an effete layer suitable for being worn away by the joint movements.

3rd. That, growing also towards the bone, it fulfils the important function of reproducing the spongy ends of bones, which would otherwise be destroyed by the pressure to which they are exposed.

4th. That it fulfils this function by means of its cell-groups.

5th. That the cell-groups develope into masses of medullary tissue, which ossify at their periphery.

6th. That these are at first unconnected with the vascular system.

7th. That the synovial fringe and the articular cartilage pass insensibly into one another.

8th. That the peculiar consistence of cartilage, whereby it transmits pressure to the underlying bone, is owing to its hyaline matrix, which becomes altered and absorbed where it is no longer required.

II. PATHOLOGICAL CHANGES IN ARTICULAR CARTILAGE.

If the preceding conclusions regarding the structure and functions of articular cartilage be correct, it may be assumed as probable that any additional changes produced by inflammation will be an intensifying of its normal processes. For in a structure shut out, like it, from the direct influence of the nervous, circulatory, and absorbent systems, their influence cannot, as in other structures, come into play in modifying and complicating the reactions due to tissue alone. Hence inflammation will show itself more in increased action and exaltation of its normal functions than in a perversion of them into a new groove. It has been stated that cartilage is a tissue insusceptible of inflammation, and observers have put themselves to great pains to refer the changes it displays in inflammation of the organs of which it forms part, to the action upon it of surrounding structures. This idea received a complete refutation at the hands of Redfern, but it has never lost altogether the hold it had acquired, and retains to the present day a prominent, though perhaps an unacknowledged place in the ideas that obtain regarding its inflammation. While freely admitting the insusceptibility of cartilage, from its very nature and position, to display its inflammation by some of the usual signs, such as redness and pain, or to participate in some of the usual results of inflammation, such as suppuration, I believe there will be no difficulty in showing that cartilage is not a passive structure in the presence of inflammation, but participates actively in it in its own peculiar way. I have as yet had no opportunities of studying the effect upon it of recent acute inflammations, such as might be sought for in Traumatic Arthritis, Pyaemic Arthritis, &c., but have carefully investigated its participation in Scrofulous Arthritis (pulpy synovitis), commencing in the joint proper, and in Chronic Rheumatic Arthritis (Arthritis Deformans). In them there are invariably found active changes which cannot be attributed to anything but inflammation. A word of caution is, however, not unnecessary on this point. At the present time we are too much carried away with the idea that inflammation is inseparably linked to the presence of blood-vessels and lymph-spaces. The observations of Cohnheim and Arnold, although of immense value in correcting the beliefs that previously existed, are apt to lead us somewhat astray from a sound appreciation of inflammation as a whole. The changes in the capillaries, the existence of stigmata and stomata, and the migration of the white blood-cells through them into the lymph-spaces to form pus or, it may be, to become a part of the fixed tissues, are so attractive and striking, that we are, at the present moment, rather inclined to forget the existence of the tissues themselves and of the changes going on in them. Important as the former are, they are tending to make us in the present day underrate or even misinterpret the proliferation of cells and other tissue-changes, the knowledge of which we owe to Goodsir and Virchow, and yet in cartilage we can find only the latter, for the very existence of the former is prevented by the peculiar isolation of the tissue. It must also be remembered that prolonged inflammation of any tissue of the connective series tends to manifest itself in a perverted increase of its normal function. Chronic Periostitis, for instance, leads to increased production of new bone, Chronic Osteo-myelitis to Osteo-sclerosis, and Chronic Inflammation of a serous or synovial surface to Hypertrophy and increased secretion. The non-vascularity of cartilage must render any inflammation it may partake in also chronic, so that changes analogous to those cited can alone be looked for as evidences of its existence.

1. *Articular Cartilage in Scrofulous Arthritis (Tumor Albus, Fungus Articuli, Strumous Arthritis).*

The earliest changes I have been able to find in this disease are such as result in an increased development of bone. The articular cartilage is increased in thickness, though only to a moderate extent, seldom attaining to more than double its normal measurement. This is sometimes present as a uniform thickening, sometimes in patches, in which latter case portions of it present a considerably greater depth than elsewhere, although everywhere the increased thickness is observable to some extent.

The group of cells between the focus of central growth and the surface of the bone take on a more tempestuous action, so to speak, and hurry more rapidly through their various stages. They reach a degree of development (Fig. 7, *a*) midway between those shown by normal and those shown by epiphysal cartilage. Instead of assuming, as in both of these, the form of *rouleaux* or sausages, the groups exhibit a more globular outline (Fig. 7, *a* ; Fig. 3, *b*), and consist of masses of proliferating cells, sometimes possessing the most irregular shapes. Since the acids employed must considerably have modified their true shapes, I would be cautious in drawing inferences from specimens prepared in the manner already stated, but, so far as can be inferred from them, the cells seem to present a variety of forms due to their mutual pressure, and to contain large granular masses of protoplasm in their interior, with a sometimes granular, sometimes transparent nucleus. The groups still preserve their vertical direction to the bone-surface, and continue enlarging from cell-multiplication as they approach it, until they come to contain from six to twenty separate cells. The hyaline matrix (Fig. 7, *b*) shows the granular alteration far more distinctly marked than does normal cartilage, faintly indeed in a zone immediately around the groups (Fig. 7, *c*), but beyond this, densely clouded as if from a fine punctuation. It also occasionally presents fatty transformations in the form of irregularly rounded oil-drops, varying from $\frac{1}{700}$ to $\frac{1}{500}$ of an inch in diameter, with clear transparent masses in the centre of each, the size of which is from $\frac{1}{850}$ to $\frac{1}{1500}$ of an inch. Save in these respects there is little difference between the normal and the inflamed cartilage at this part.

In the zone of altered cartilage lying immediately on the bone, its elements seem in a hurry to fulfil their functions. The number of cells in each group (Fig. 8, *a*) becomes even-greater than before, and the groups differ from those of the normal structure in often passing into medullary tissue before they have arrived at the junction line (Fig. 8, *d*). It is much easier to trace the changes here than in healthy cartilage. Some of the groups show their ends next the bone converted into medullary tissue, and their further ends still unaltered (Fig. 8, *b*). Other groups are entirely changed into medulla although they are still short of the boundary line (Fig. 8, *d*). And occasionally, though rarely, a group will not only have passed into medulla, but have commenced to produce bone at its periphery (Fig. 8, *c*) some time before it arrives at the line of ossification. In such a case the bone-production, like that of medulla, commences at the end next the bone by conversion of its peripheral cells into bone-corpuscles, with increase and ossification of their intercellular substance. In this way such a group may, at its end nearest the boundary line, be encased in a capsule of newly formed osseous tissue.

The haste to develope of which these are the expressions, further shows itself in the appearance of the boundary line between cartilage and bone. Instead of preserving on the whole an unbroken level, it shows an uneven condition (Fig. 8, *g*) to a remarkable degree, being comparatively advanced in some places, and receding in others, so as quite to have lost its normal regularity. Moreover, its ossifying prominences present an unusual aspect. Instead of rounded papillæ, not higher than they are broad, they may be seen as nipple-like or even tongue-like projections into the cartilage (Fig. 8, *e*; Fig. 3, *c*), or even project so far as to assume a long irregularly club-shaped appearance (Fig. 8, *f*). The tongues of hyaline matrix still dip between them (Fig. 3, *f*; Fig. 8, *k*), but the old regularity is lost, and the ossification of the prominences, hurried as it seems to be, takes place for the most part in the insular form already described, without there being the usual amount of confluence of the medullary masses.

In the direction towards the joint the behaviour of the cartilage in Scrofulous Arthritis differs even more remarkably from the normal condition. As, passing from the focus of central growth, the cells approach nearer the surface, they assume at first their usual elongated shape (Fig. 9, *d*), and,

diminishing in size, become fusiform and contain little if anything besides the nucleus. At a distance, however, of from
$\frac{1}{100}$ to $\frac{1}{150}$ of an inch from the surface, instead of passing into
mere clefts, they begin to increase again in size and activity
and become fusiform, subdivide so as to present two or even
three cells within the fusiform capsule (Fig. 9, c), and give
more distinct evidence of possessing an active protoplasm.
The nearer they approach the surface, the more marked do
these peculiarities become, and in addition the hyaline matrix
is seen to become fibrous, the fibres parallel to the surface, the
fibrous transformation more marked as it nears the surface,
until finally the cartilage is transformed into a fibrous tissue
with fusiform and oat-shaped cells, differing only from connective tissue in the firmness and cohesion of the intercellular
substance.

This change is most evident, and reaches its greatest development at the margin of the cartilage, where it is bordered
by the already inflamed synovial fringe, and at an early period
capillary loops are found to extend from the fringe into the
newly formed fibrous tissue. When it is thus provided with a
vascular supply, it soon shows its sympathy with the surrounding tissues by taking on, like them, the characters of granulation tissue (Fig. 9, b). The process of fibrous transformation
and vascularization spreads by degrees from the periphery over
the entire surface, and as it thus advances towards the centre it
gains at the same time the deeper peripheral layers of the
cartilage, until at last there remains no cartilage at all at the
periphery, but the bone is there covered by a structure which
it is impossible to distinguish from granulation tissue, and
whose origin from cartilage could hardly be inferred from its
appearance. Thus it comes about that a steadily diminishing
island of apparently unaltered cartilage often remains in the
centre of the articulation, surrounded by what appears a vascularized synovial fringe, while peripheral portions, and in the
knee-joint the semilunar cartilages, are already completely
transformed into pulpy granulation tissue.

Such are the minute changes present in the earlier stages of
Scrofulous Arthritis, and through them are produced the alterations commonly observed on the articular ends of the bones,

viz. pulpy degeneration of the cartilages, at first at the periphery, and subsequently even at the centre.

I would draw special attention to a peculiarity of this stage which is at once visible to the naked eye, but has been overlooked, and which is due to the combination of the processes described.

Of these one is continually, though at a slow rate, producing new bone, and so raising up the articular surface above its original level; the other is steadily effecting a centripetal diminution of the bone-producing cartilage. The sum of their action is that the new bone, which was at the beginning produced over the whole articular facet, soon ceases to be produced at the circumference where the belt of transformed cartilage lies. The cartilage remaining in the centre still goes on forming bone, but is constantly being destroyed at its edges by the ever-widening belt of transformation; and so it goes on till by the destruction of the last central piece of cartilage the process of bone-formation is for ever at an end. The raising up of the articular facet is, and must of necessity be, greatest in the centre, and diminishes as the periphery is approached.

This raising of the articular facet varies considerably in appearance, according to the form of the part affected by it. A flat articular surface like the head of the tibia is changed into a flat cone or pyramid. Very often the process is seen before its completion, and the head of the tibia is prolonged into a flattish truncated cone the apex of which is occupied by two smooth apparently little altered facets of cartilage, each bordered by a vascularized fringe like the normal synovial fringe, and articulating with two corresponding facets of the femur. The sides of the cone are covered with granulation tissue produced from the cartilage, and which is often studded with whitish points visible to the naked eye and not unlike tubercle, though the microscope shows that they have no such structure.

In a spherical articular surface the combined processes will result in its conversion into an egg-shaped surface. The head of the femur comes to present an ovoid elongation; and at its opposite extremity the surfaces of the condyles, which approach in form to the spherical shape, become elongated in the axis of the shaft. They may be found facetted at various parts according to the position in which the tibia has been retained, but usually the facets are at the posterior part of their articular surfaces.

On articular facets of irregular shape the deformity produced varies according to the peculiarities of the part, but the influence of the two factors is always traceable, and is usually very obvious.

This growth of the articular facets is of course often rendered less distinct, and even reabsorbed or obliterated by the destruction of the bones in the later stages of the disease. In the earlier stages its existence favours the occurrence of the subluxations so common in Scrofulous Arthritis, and it even happens that, when subluxation occurs before the cartilage has all been transformed, the growth of the articular surfaces becomes so great as to prevent the success of any other means than an operation in restoring the bones to their proper position.

I have hitherto been unable to observe the ulceration of cartilage described by Redfern. So far as I could judge, the bone in the later stages of the disease rarely becomes exposed at any place where the cartilage still remains tolerably healthy, but generally where the alterations have substituted for the resisting cartilage a tissue unable to withstand the mutual pressure of the bones. Even at points where the pressure is strongest and the tendency to *decubitus* greatest, such as between the patella and outer condyle of the femur, I cannot say that I have been able in any case to satisfy myself as to the actual production of an erosion of cartilage from *decubitus*. On the contrary, the erosion has, in the specimens at my disposal, seemed due to a centripetal extension of the pulpy transformation inwards to the point of strongest pressure.

In Scrofulous Arthritis the changes in articular cartilage may be summed up as follows :—

1. The cartilage is increased in thickness.

2. It is transformed from the periphery to the centre into vascularized granulation tissue.

3. This is effected by a proliferation of its cells, commencing on the surface at its border, extending inwards towards the centre, and gradually involving the deeper layers.

4. Its increased activity leads to an increased production of new bone, going on as long as any cartilage remains, and therefore greatest at the centre, where it remains longest unaltered.

5. Various alterations of form of the articular facets are the result of this process.

2. *Articular Cartilage in Chronic Rheumatic Arthritis (Arthritis Deformans, Arthritis nodosa, Arthritis sicca, Malum senile Articulorum, Rheumatic Gout).*

Ever since this disease was recognised as having an existence separate from ordinary rheumatic arthritis and from gout, its causation has been a mystery. Ranked by various authors at one time with the former, at another with the latter, or again held to be a disease *sui generis,* the utmost divergence of opinion has prevailed as to its nature, and this has found expression in the names which have been bestowed upon it. But mysterious though its origin may be, it cannot be denied that such a disease exists, owing its origin probably to constitutional tendencies, hereditary or acquired, more or less allied to those which produce Rheumatism and Gout, affecting chiefly persons past middle age and leading to destruction of articular cartilage on the middle of the articular facets of bones, and to the production at their margins of large portions of bone overhanging the ends in mushroom-shaped masses, as if forced like soft mortar out from between the articulating surfaces.

The characters revealed by the microscope in the cartilages in this disease differ very materially from any yet considered. On the whole there is doubtless the same exaggeration of the normal functions as in Scrofulous Arthritis, though marked by some very striking peculiarities, but the result is a characteristic combination of growth and destruction seemingly peculiar to this disease alone.

To commence with the first onset of the disease as it may be seen on the facets of the sternum and clavicle in the sterno-clavicular articulations of a person whose other joints, phalangeal for example, show well-marked evidences of the malady. In this joint, or in others little exposed, the incipient alterations can be most readily found and most easily traced.

Before being opened such a joint seems quite normal, or
the faintest feeling of knottiness of the margins of the cartilages
alone betrays the pathological condition. When opened, the
cartilages, articular and interarticular, have lost their trans-
lucency and are unusually yellowish, and, instead of having
smooth polished surfaces, are rough and even granular. A
section reveals the joint-surface of the articular cartilage proliferating and forming actively growing rounded cells, with
fibrillation of the hyaline matrix, identical in nature with the
changes described under Scrofulous Arthritis and depicted in
Figure 9.

Between the focus of central growth and the bone the
hyaline matrix is sometimes fibrous, sometimes, especially in less
aged subjects, more granular than usual, the granular alteration
most marked in the middle of the interspaces between the cell-
groups. The cell-groups proliferate more abundantly than
normally, but *show a tendency to pass more into groups of two
or three or even into single cells than into the large aggregations*
forming the groups in Scrofulous Arthritis. The zone of altered ·
hyaline matrix next the bone seems to be absent. The line of
ossification is quite irregular, just as in Scrofulous Arthritis,
and the islands of medullary substance produced by the transformed cartilage cell-groups appear, as in it, at the margin of
ossification or even beyond it. The islands of medullary substance are more sparing in number, and the newly-formed bone
between them seems to be unusually dense and disproportion-
ally abundant. This is caused by the *transformation of the
single cartilage-cells or groups of two or three cells directly into
bone-corpuscles without having passed through the intermediate
stage of medullary tissue.*

This forms a predominant feature in all stages of the disease
under consideration, and is most distinctly to be seen in a joint
where the malady has passed the incipient stage and attained
full development. Let us take, for example, a phalangeal joint
where the external knotting shows the fully-developed disease.
On opening it, the articular ends are found grooved from front
to back with deep ruts or grooves often nearly a quarter of an
inch in depth, each groove corresponding to a projecting ridge
on the opposing facet. The margins of the cartilages are pro-

longed into bony masses, not projecting into the joint nor into the normal tissues external to it, but forming as it were an enlarged base (Fig. 10, b, h), with which are connected the thickened ligaments and synovial capsule. The ridges running antero-posteriorly on the articular ends, are seen on section to be composed of dense bone, arranged in Haversian systems with vessels in their interior, forming a texture nearly as compact and little vascular as the compact tissue of the diaphyses of the long bones. The dense bone seen in these ridges by its hardness produces, owing to the movements of the articulation, the corresponding grooves in the opposed surfaces, and patches of still surviving cartilage are found here and there on the ridges and in the grooves.

The mode of formation of the compact bone may be seen by examining thin sections through any of the patches of cartilage which remain. The superficial layers of the cartilage will be absent as if scraped away (Fig. 11, a), but the remaining deeper part shows the cell-groups (Fig. 11, c, d) in active proliferation, tending however to form groups of two or three cells, or even scattered single cells. The single cells are most numerous as they approach the surface of the bone, and there they become transformed into bone-corpuscles, thus increasing the circumference of the Haversian systems by contributing additional layers to it.

The details of their conversion are: at the immediate margin of the bone such a thing as a group of cells is a rarity, they have broken up into isolated round cells, unusually transparent. When one of these comes so near the bone as to be in contact with it (Fig. 11, e), it becomes surrounded with a halo of ossified matter, commencing at the point of contact, running rapidly round the cell, and forming a ring round it $\frac{1}{5000}$ to $\frac{1}{3500}$ of an inch in breadth. It looks as if this zone of ossification were formed by deposition of calcareous salts in the hyaline matrix, but close investigation of this point shows it to proceed rather from an ossification of the capsule and wall of the cell spreading outwards and causing absorption of the matrix. The cell itself is now soldered on to the convex periphery of the Haversian system and forms a slight rounded prominence upon it. The same process occurring in other cells soon causes the appearance

of prominence at the part to be lost, and the cell passes gradually through the forms intermediate between the cartilage-cell
and the bone-corpuscle, till it finally becomes similar to the
other bone-corpuscles, with a distinct division into protoplasm
and nucleus.

Along the growing margin of the bone the newly-formed
bone-cells are large and rounded (Fig. 11, *h*), and to some
extent are crowded into clusters corresponding with the points
of most active cartilage-growth ; but, as they come to occupy
situations more remote from the margin and nearer the centre
of the Haversian system, they diminish in size (Fig. 11, *f*), and
the distinction between nucleus and protoplasm becomes less
marked.

There are thus formed at the margin of ossification a series
of Haversian systems (Fig. 11), each arranged in concentric
layers around its governing vessel, appearing to project like
rounded masses into the cartilage. The line of ossification
(Fig. 11) thus becomes irregular, the irregularity being due as
it were to a series of segments of bony circles projecting into
the cartilage. In time the masses of compact bone, favoured by
the joint movements and the alteration of the surface of the
cartilage, appear through the cartilage, and, scraping at every
movement on the parts opposite them, plough furrows in it,
and these furrows and ridges are the more distinct, the more
nearly the joint is a true ginglymus.

In time the cartilage is thus all scraped off from the middle
of the articular facets, and then the bone meets bone. The
friction which now ensues is too much for the bone to resist,
deprived as it is of any cartilage which might have replaced the
rubbed down portions, and accordingly the compact bone and
its Haversian systems are rubbed away (Fig. 10, *a*), just as one
stone might be rubbed down by another, the resulting facet
cutting off all projecting parts irrespective of their central
vessel, the vascular canals themselves being even laid open
(Fig. 10, *f*). Hence are formed the hard facets called "porcellanous deposit," being simply compact bone without any
special deposit. In large joints, where weight and pressure
must be borne in addition to the friction, the work of destruction goes on with irresistible force, till considerable portions of

the bones have disappeared, and nothing puts a stop to the process unless ankylosis occur, a termination which, in joints favourably situated, is occasionally met with.

At the periphery of the cartilage the process is identical in its nature though differing in its results. A section shows that the new bone formed at the peripheral part of the flat portion of the facet is both ground away and pressed outwards so as to form a flat collar (Fig. 10, *b*) within the attachment of the ligaments. The cartilage at the extreme outer margin, where the articular facet passes into the side of the bone (Fig. 10, *g*), and which is in intimate relation to the ligaments, is able to proliferate unchecked. It becomes altered so as to assume almost the characters of epiphysal cartilage. By rapid proliferation it comes to form a bulky outgrowth from the bone (Fig. 10, *g*). Its hyaline matrix (Fig. 12, *a*) appears less pellucid and more granular and striated, the striation for the most part perpendicular to the surface of the bone. Its cell-groups (Fig. 12, *b*) are changed into longitudinal rows, somewhat as in epiphysal cartilage, but more slender, composed of smaller and fewer cells ; these rows lie parallel to the fibrillation and at right angles to the bone. As they approach it, they tend to break up into single cells, so that by the time the bone is reached they exist no longer as groups but as solitary cells. These are, in their turn, on reaching the margin of ossification, transformed into bone-corpuscles (Fig. 12, *g*) by the process already described, without having undergone a transmutation into medulla. From a continuance of the process arise the fungous masses of bone, forming so marked a feature of the disease as to have gained for it the name of Arthritis Nodosa.

To sum up :—

(1) The cartilages proliferate in the same manner as in Scrofulous Arthritis, but with some distinguishing peculiarities, viz. :

(2) The cartilage-cells pass into bone-cells without having become medullary tissue.

(3) They form, by means of Haversian systems, a dense layer of compact bone, which, after the destruction of the cartilage, is worn into facets.

(4) The margins of the cartilage, similarly affected, form osseous masses overhanging the ends of the bones.

In conclusion, the facts briefly detailed above, seem to render necessary a revision of the present doctrines concerning many of the deformities and diseases connected with joints, for it cannot but be that they are considerably influenced by the process of new bone-formation, which has been seen to be the main function of articular cartilage.

It must be admitted that we at present understand little of the forces at work in modifying the forms and dimensions of articulations. In undertaking the cure of a congenital club-foot, for example, we know that we have to deal with bones still possessing their fœtal forms and differing greatly, both in size and shape, from those which are present in a well-formed limb. By properly directed force we succeed in giving the bones a shape not far removed from that which they ought to have had. It is evident that a power spontaneously to mould themselves must reside in the bones, and hitherto the perios-teum and medulla have been the only structures in a fully-formed bone that have been shown to possess this function. But if the functions claimed above be admitted as proven, it will be necessary to concede to articular cartilage a power of influencing the length of the bones similar to that possessed by periosteum, of influencing their breadth. In the changes pro-duced in the form of the tarsal bones during the cure of club-foot, therefore, it will in future be requisite to recognize the importance of the office performed by the joint-cartilages in supplementing the surgeon's efforts to restore the bones to their proper form and size.

Doubtless also the disease of the knee-joint known as "knock-knee" (genu valgum), owes its causation less to a subacute inflammatory condition of the bone as hitherto as-sumed, than to some alteration leading to an interference with the peculiar function of articular cartilage. I have as yet had no opportunity of examining such a case, or any case of similar deformity occurring in other articulations.

But it would be tedious to dwell upon the application, to almost every joint-affection in which cartilage is, however remotely, implicated, of an attempt, of which the preceding s a condensed *résumé*, to fill up what has long been a serious gap in our knowledge of the physiology and pathology of the articular surfaces of bones.

EXPLANATION OF PLATES.

PLATE I.

Figure 1. Perpendicular section of the articular cartilage on the upper surface of a healthy adult astragalus. *a.* Cartilage developing towards the joint into a non-cellular structure suitable for being destroyed by the friction movements of the articulation. The cells are replaced by mere clefts in the hyaline matrix; *b.* Focus of central growth whence the cartilage grows in all directions; *c.* Cartilage destined to form new bone, and growing towards the bone in elongated groups and rows of cells; *d.* Level surface of the bone marked by slight prominences, which, in the drawing, are not shown so distinctly as they should have been. Magnified 150 diameters.

Figure 2. Section through the boundary line between bone and articular cartilage from the upper surface of the tibia, in a slightly inflamed knee-joint (scrofulous synovitis). *a.* Granular hyaline matrix of cartilage; *b.* Elongated groups of proliferating cartilage-cells destined to form new bone; *c.* The same undergoing by proliferation a transformation into medulla; *d.* Prominences of medulla confluent with each other at their bases *e*; *f.* Development of the deeper surfaces of the prominences into bone; *g.* Wavy boundary line between articular cartilage and bone; *h.* Prolongations of the hyaline matrix of cartilage undergoing absorption. Magnified 320 diameters.

PLATE II.

Figure 3. Section through the boundary line between bone and articular cartilage of the tibia, in a case of scrofulous arthritis of the knee-joint. *a.* Granular hyaline matrix; *b.* Proliferating groups of cartilage-cells; *c.* Group of cartilage-cells transformed into medulla, and developing at the end next the bone into a cup of osseous tissue; *d.* Groups of cartilage-cells transformed into medulla, and passing at the periphery into osseous tissue; *e.* Osseous tissue forming an irregular boundary line between bone and cartilage; *f.* Prolongation of the hyaline matrix undergoing absorption; *g.* Fat-cells. Magnified 320 diameters.

Figure 4. Section through boundary line between diaphysis and epiphysal cartilage from the tibia of a boy of ten. *a.* Zone of central growth; *b.* Proliferation of the cartilage preparatory to the formation of bone; *d.* Boundary line between cartilage and bone; *d.* Sausage-shaped masses of medulla formed out of proliferating cartilage groups; *e.* Processes of hyaline matrix undergoing absorption; *f.* Bone developed by the periphery of the sausage-shaped masses of medulla; *g.* *Rouleaux* of cartilage-cells before their transformation into medulla; *h.* Portion of the cartilage next the epiphysis. Magnified 150 diameters.

PLATE III.

Figure 5. Perpendicular section through the joint-surface of healthy articular cartilage at its junction with the synovial fringe. *a.* Zone of central growth; *b.* Synovial fringe passing gradually into cartilage; *c.* Effete superficial layer of cartilage continuous with synovial fringe; *d.* Fibrillation of the hyaline matrix where the cartilage is passing into synovial fringe; e. Transformation of cartilage-cells into the oat-shaped cells of the synovial fringe. Magnified 200 diameters.

Figure 6. Section through articular cartilage and bone from the lower end of a femur expanded by a vascular tumour.   a. Hyaline matrix of the cartilage;   b. Proliferating cartilage cell-groups;   c. Boundary line between normal and altered cartilage;   d. Proliferating cartilage cell-groups;   e. Cartilage cell-groups transformed into medulla;   f. The same becoming transformed at the periphery into bone;   g. Bone;   h. Canals eaten through bone by a tongue of vascularized medulla going to the transformed cartilage cell-groups;   i. Border-line between bone and cartilage; c to i. Zone of altered cartilage. Magnified 200 diameters.

PLATE IV.

Figure 7. Articular cartilage in Scrofulous Arthritis.   a. Proliferating cartilage cell-groups.   b. Granular hyaline matrix.   c. Clear zone of hyaline matrix around the cell-groups. Magnified 500 diameters.

Figure 8. Section through articular cartilage and bone in Scrofulous Arthritis. From the head of the tibia.   a. Proliferating cartilage cell-groups;   b. The same transformed into medulla at the ends nearest the bone;   c. The same completely transformed into medulla, and developing into osseous tissue at their ends nearest the bone;   d. Cartilage cell-group transformed into medulla ;   e. Cartilage cell-groups transformed into medulla and producing osseous tissue at the boundary line : that to the left has become connected with the vascular system;   f. The same, club-shaped;   g. Irregular boundary line;   h. Fat-cells;   i. Bone;   k. Hyaline matrix. Magnified 150 diameters.

PLATE V.

Figure 9. Section of the joint-surface of articular cartilage in Scrofulous Arthritis. From the upper end of the tibia.   a. Joint surface;   b. Cartilage converted into granulation tissue;   c. Isolated cartilage-cells developing into rounded cells;   d. Isolated elongated cells. Magnified 300 diameters.

Figure 10. Section through the margin of the articular facet of a phalanx in Chronic Rheumatic Arthritis.   a. Articular facet formed of rubbed-down compact bone (porcellanous deposit);   b. Newly-formed compact bone pressed outwards;   c. Haversian canals;   d. Medulla;   e. Haversian systems ;   f. Haversian canals laid open by attrition;   g. Proliferating articular cartilage developing into h;   h. Compact bone;   k. Articular cartilage developing into bone b. Magnified 100 diameters.

PLATE VI.

Figure 11. Section through a patch of articular cartilage on a "porcellanous" facet in Chronic Rheumatic Arthritis.   a. Joint-surface. The superficial layers have been scraped away by the joint movements;   b. Granular hyaline matrix;   c. Proliferating cell-groups;   d. Isolated cartilage-cells;   e. The same soldered to the bone by a calcareous zone;   f. Haversian systems;   g. Haversian canals;   h. Clusters of bone-corpuscles formed out of cartilage-cells. Magnified 400 diameters.

Figure 12. Section through the boundary line between g and h in Figure 10, Plate V.   a. Fibrous transformation of hyaline matrix;   b. Cartilage cell-groups;   c. Bone;   d. Medulla;   e. Process of vascularized medulla, incipient Haversian canal;   f. Fat-cells;   g. Isolated cartilage-cells soldered to the bone and becoming transformed into bone-cells;   h. Groups of bone-corpuscles formed out of cartilage-cells;   i. An isolated cartilage-cell. Magnified 400 diameters.

The BRAIN AND CRANIAL NERVES OF ECHINORHINUS SPINOSUS, WITH NOTES ON THE OTHER VISCERA. By Wm. Hatchett Jackson, B.A., F.L.S., *New College,* and Wm. Bruce Clarke, B.A., *Burdett-Coutts Scholar, Pembroke College, Demonstrators of Anatomy, University Museum, Oxford* [1] [Pl. vii.].

Since the commencement of the present year, the Anatomical Department of the University Museum, Oxford, has received two specimens, both in a very good state of preservation, of the rare shark, Echinorhinus Spinosus. One specimen, a fine female, was captured on Feb. 15th, 1875, in Mount's Bay, Penzance, by a fisherman without any aid, a rather remarkable feat. It was received at the Museum, Feb. 20th, and dissected the same day, having kept exceedingly well; a circumstance probably due in part to the great cold then prevailing, in part, perhaps, to the deep-sea habitat of the fish. The other specimen, a male, was also caught at Penzance on Tuesday night, May 4th, and was dissected the following Saturday. The respective dimensions of the two sharks were as follows :—

		Male		Female.	
Tip of snout to end of tail.....................		5 ft.	1 in.	7 ft.	8 in.
Snout to nostril				0	6
„	eye	0	5¼	0	7
„	spiracle	0	9	1	0¼
„	1st gill-opening	0	11	1	6½
„	anterior edge of pectoral fin	1	3	1	11½
„	anterior edge of ventral fin	2	10½	4	3
„	base of 1st dorsal fin	3	0	4	7¾
„	base of 2nd dorsal fin..............	3	5¾	5	4½
Length of pectoral fin		0	8¼	0	10½
„	dorsal I...................	0	5	0	8
„	dorsal II	0	4½	0	7

[1] In this paper the description of the brain is by Mr Jackson. The dissection of the fifth group of nerves was made by Mr Clarke, of the eighth by Mr Jackson—and they are severally responsible for the accuracy of the descriptions of these groups. On the other hand, both authors concur in the homologies expressed. The notes on the viscera are extracted partly from the Museum Note-book, and are partly taken from the specimens preserved in the Biological Department.

	Male.		Female.	
Width at eyes	0ft.	7in.	0ft.	9in.
„ spiracle	0	$9\frac{1}{2}$	1	$1\frac{1}{4}$
Girth behind pectoral	2	0	2	$11\frac{1}{2}$
„ in front of pectoral	1	11		
Girth at root of tail	0	$8\frac{1}{2}$	1	$3\frac{1}{2}$
Length between tip of tail-lobes	1	$2\frac{1}{2}$	1	9

Weight of female 250 lbs.

There is no need to give any description of the animal, as Professor Turner has already done so with great care in the May number of this *Journal* (p. 295). In this paper we propose to treat first of the nervous system, which we have been able to work out to a very great extent, and then to append some notes on the visceral anatomy supplementary to those published by Professor Turner in his paper before mentioned. Professor Rolleston has kindly put all the material at our disposal, and has allowed us to extract from the Note-book of the Anatomical Department, aiding us in addition with his counsel and advice, for which we desire to thank him most heartily[1].

§ 1. The cranial cavity in both sharks was very large; and the brain occupied but a very small proportion of it. The cavity continued onwards, surrounding the olfactory tracts. This continuation, as well as the main cavity itself, was filled with a very clear and liquid arachnoid fluid, of a yellowish tint. . Strong connective-tissue fibres crossed it in every direction, supporting (apparently) numerous blood-vessels destined for the pia mater.

§ 2. The brains of both individuals disclosed on careful comparison no differences whatever. One—that of the female

---

[1] The following list gives the titles of the chief works to which we have occasion to refer repeatedly in the course of our paper, and the abbreviations employed for the sake of reference in each case.

Gegenbaur—Ueber die Kopfnerven von Hexanchus (mit Tafel XIII). *Jenaische Zeitschrift*, VI. 1871. (J. Z.)

Gegenbaur—*Untersuchungen zur Vergleichenden Anatomie der Wirbelthiere.* Heft III. 1872. (V. A.)

Von Miklucho-Maclay—*Beiträge zur Vergleichenden Neurologie der Wirbelthiere.* Leipzig, 1870. (M. M.)

Stannius—*Das peripherische Nerven-system der Fische.* Rostock, 1849. (St.)

Henle—*Handbuch der Systematischen Anatomie der Menschen.* Band III. Abtheilung, II. 1871. (H.)

Quain and Sharpey—*Elements of Anatomy*, Vol. II. Ed. 7, 1867. (Q. S.)

Stricker—*Manual of Histology.* Translated by H. Power, M.B., Vol. II. 1872. (Sr.)

Parker—On the Structure and Development of the Skull in the Salmon. *Phil. Trans.* 1873, p. 95. ('Salmon.')

Parker—On the Structure and Development of the Skull in the Pig. *Phil. Trans.* 1874, p. 289. ('Pig.')

—was removed and preserved as a permanent museum preparation, and it is from this specimen that the following description is mainly derived.

The brain of Echinorhinus belongs to the *first type* of Selachian brain described by Miklucho-Maclay. The distinctive marks of this type are according to him (*M. M.* pp. 9 and 10); (i) slight difference in external form between the adult and embryonic brains; (ii) the great permanent size of the cavities; (iii) the longitudinal extent of the adult brain, noticeable especially in the length of the olfactory tracts and the cerebral peduncles; and (iv) the unusual size of the fourth ventricle.

So·far as we have been able to ascertain, the brain of this particular shark appears never to have been figured or described. Accordingly the figure (Pl. VII., Fig. 1) has been given, and it is interesting to note the close resemblance existing between the brain of Echinorhinus and the brain of Hexanchus as drawn by M. Maclay (*M. M.* Taf. II. Figs. 8 and 9), and by Gegenbaur (*J. Z.* Taf. XIII. figs. 1 and 3).

§ 3. The olfactory tracts (*Olf.* Pl. VII., fig. 1) are separated by a slight constriction from the succeeding portion of brain. They are long, flattened, and are marked superficially by longitudinal striations. This striation clearly denotes a separation of the fibres into larger and smaller bundles. There is no terminal bulbous expansion: the bundles of fibres simply separate from one another, and spread out like a fan on the posterior surface of the nasal mucous membrane.

The olfactory tracts arise from the anterior angles of the still undifferentiated cerebral lobes and thalami optici. Maclay regards this part of the brain as representing the cerebral lobes alone (Vorderhirn, *M. M.*, p. 2), but there are good reasons for adhering to the ordinarily received nomenclature, as is here done. This portion of the brain is roughly triangular (*C.* Pl.VII., fig. 1) in shape, the truncated apex being turned backwards. The base of the triangle presents at its anterior edge in the middle line a slight furrow, which indicates a commencing differentiation into a right and left lobe. This furrow deepens as it extends on to the inferior surface, where the brain swells out anteriorly on either side into a perfectly distinct elevation. On the upper surface there is a slight indication of similar

swellings on either side. But by far the larger portion of the
upper surface is occupied by the third ventricle (*V.* 3). To the
roof of this ventricle which has been left on the right side is
hanging the long filamentary pineal gland (*Pi.*) attached at one
end to the roof of the cranium in the natural condition of the
parts, and traceable in the membranous roof of the third ventricle back to the *posterior edge* of the third ventricle constituted by the optic lobes. The sides of the third ventricle are
but slightly, if at all, thickened, and must be held to represent
the little developed optic thalami (*Th.*). On the inferior aspect,
and slightly in front of the posterior edge of the third ventricle,
rise the optic nerves (*Op.*), small and round, having a chiasma
of remarkable lateral extent. Immediately behind the chiasma
is the pituitary body—the front portion of which is light-
coloured like the rest of the brain—the hind portion (Hæmato-
sac) dark, vascular and thin-walled. The pituitary body was
closely attached to the floor of the cranium.

§ 4. The Mesencephalon (*C. B.*) has but a slight median
furrow, but bulges laterally on either side, thus showing the
division into a right and left optic lobe. It is partially covered
behind by the cerebellum. The inferior surface of this part of
the brain is smooth, broad anteriorly, and narrowing somewhat
posteriorly. It is slightly depressed in the middle, indicating
a separation of a right and left cerebral peduncle. The third
pair of nerves (III) arises from it about ⅛ in. behind the pituitary body by a single root on either side, situate very near the
middle line. M. Maclay regards this division of the brain as the
Thalamencephalon (= Zwischenhirn).

§ 5. The Cerebellum (Mittelhirn, *M. Maclay*) is lozenge-
shaped, and yields to the touch, having without doubt a large
internal cavity such as exists in Hexanchus. Anteriorly (*Ce.*)
it overlaps the optic lobes, posteriorly the fourth ventricle, both
to about the same proportionate extent. It is marked superiorly
by a central longitudinal depression, and there are two other
slight depressions, one on either side at the external or lateral
angles of the lozenge. The slender fourth nerve (IV) arises
between the cerebellum and the optic lobes on the upper or
dorsal surface.

§ 6. The medulla oblongata (*M. O.*) is of great extent—

making fully one half of the total length of the whole brain. This great length is however obscured partially by the over-hanging posterior angle of the cerebellum when the brain is viewed from above. Anteriorly the medulla is broad, about $\frac{3}{4}$ inch; it narrows posteriorly by degrees and passes into the spinal cord, which is barely $\frac{3}{8}$ inch in transverse diameter, and shews a somewhat elliptical figure in cross section.

Forming the floor and sides of the medulla oblongata are four perfectly distinct strands. The outermost ($s.$) is exceed-ingly slender at its commencement, gradually thickens, and be-coming convoluted forms at last the outermost part or free edge of the trigeminal lobes. Internal to this strand is a column (g) round and stout from the very beginning. It has five distinct nodular swellings—vagal lobes—ranged one in front of the other. Gegenbaur figures in Hexanchus six of these ganglionic masses. Anteriorly to the first swelling the tract appears somewhat enlarged and curves out of sight, probably joining the outermost tract, and entering into the formation of the trigemi-nal lobes. Both these tracts seem to spring from the posterior columns of the cord. ($V.\ A.$, pp. 266 and 267.) Internal to the nodular tract there is another (p') which takes a perfectly straight course—is broad below and narrow above. It appears to correspond chiefly to the lateral column of the cord, and is separated from the tract (g) by a well-marked furrow. On its inner side it is separated above and below by an equally well-marked furrow, from the most mesially situated tract of all (p), but at about the middle of the medulla this inner furrow is partially obscured by fibres which pass inwards from the tract (p') into the tract (p). This latter tract (p, anterior pyramid, Gegenbaur, fasciculi teretes, M. Maclay $loc.\ cit.$) seems to be derived from the anterior column of the cord, is yellowish in colour, and is separated from its fellow by a deep longitudinal fissure (sulcus centralis). It takes also a perfectly straight course fore and aft. In the absence of recent and accurate in-vestigations into the course of fibres from the cord into the medulla oblongata in fish and lower vertebrates, it is safer per-haps not to speculate confidently on the homologies of these several tracts. Judging superficially, the tracts (p) seem to correspond best with the anterior pyramids—the tracts (p') with the fasciculi teretes—the tracts (s) and (g) with the restiform bodies and the fasciculi graciles.

The inferior surface of the medulla is somewhat swollen on either side the middle line. From it rise anteriorly the fifth (v), seventh (VII), and auditory (au) nerves. About $\frac{1}{8}$ inch behind the last-named nerve and nearer to the middle line rises the abducens by three rootlets on each side. A quarter of an inch behind the auditory and more laterally placed is the origin of the glossopharyngeal (IX), and the upper (posterior) roots of the vagus (X) issuing from the sides of the medulla in a continuous line as far down as the posterior extremity of the fourth ventricle. On the inferior surface close to one another in the middle line are the lower (anterior) roots of the vagus (x'), four in number.

The roof of the fourth ventricle is exceedingly thin posteriorly. It is thicker in front, and there appears to contain a small quantity of nervous matter, which is at its thickest between the trigeminal lobes.

In discussing the homologies of the cranial nerves we have adopted Mr Parker's views of the visceral arches, embodied in his latest papers on the development of the skull of the Salmon, and Pig, and on the "Form and use of the facial arches" (*Monthly Micr. J.*, 1871, Vol. VI. p. 211).

Hence we consider that the trabeculæ cranii and the palato-pterygoid cartilages constitute two *præ-oral* visceral arches, with a cleft (the lacrymal cleft) between them. (Cp. 'Salmon,' p. 126.)

We have also adopted the view held by Gegenbaur, in his paper on Hexanchus, that each *individual* cranial nerve supplies the anterior and posterior edges of a visceral cleft. By applying these views to the anterior region of the skull, which has undergone considerable change from its simplest condition, it will be seen, in the sequel, that we have arrived at results different to those hitherto attained.

§ 7. The three motor nerves of the eye are all present. They present scarcely anything worthy of note save the fact that the third nerve sends a branch to the external rectus.

As to the homologies of these nerves, whether they are differentiations of the fifth, as held by Gegenbaur, or represent nerves belonging to three suppressed myotomes of Amphioxus, as suggested by Prof. Huxley (*R. Soc. Proc.*, Vol. 23, p. 130), we have no facts to guide us to any further elucidation. The only fact to which we wish to draw attention is this: Gegenbaur (*J. Z.* p. 513) considers it as an important point, that they invariably run below the ramus ophthalmicus: is it not equally important that they are invariably situated above the maxillary branches of the fifth?

§ 8. The roots of the fifth and the seventh are closely connected, as is commonly the case in fishes, therefore it will be as well to describe them together, though the nerves will be treated separately afterwards.

The references are all to Pl. VII. Fig. ii.

Three roots are plainly discernible.

V*a*. The anterior root arises from the inferior surface of the medulla oblongata, by two well-marked rootlets (V*a* 1 and 2) situated one above the other, which join together after about ½ inch. The trunk thus formed passes outwards for about an inch, until it reaches the exterior wall of the cartilaginous cranial wall, where it is joined by V*β*.

V*β*. This root has a well-marked superior rootlet, which arises from the lobus trigeminus, and after about ¼ inch joins the inferior rootlet, which arises below from the medulla oblongata in company with V*γ*. The nerve thus constituted then unites with V*a* as described above.

V*γ* and VII. This root is so closely united with V*β* inferiorly as to be scarcely distinguishable from it. Some of its fibres rise just above the auditory. It becomes closely connected with V*β*, and fibres appear to pass from one to the other.

Thus it will be seen that from the numerous rootlets, two main trunks are formed, which unite, and then differentiate as the Fifth and the Seventh nerves, the latter appearing to be chiefly derived from V*γ*.

§ 9. The distribution of the trigeminus is as follows:

Its branches may be divided into those, i. which pass above the orbit (R. ophthalmicus), and ii. those, Rr. maxillares—which pass transversely outwards below the eyeball.

i. The R. ophthalmicus (*Opth.*) divides into the R. ophth. superficialis and the R. ophth. profundus of Stannius. The former

(*Opth.*') of these runs, as is always the case in vertebrata, upwards, closely applied to the posterior wall of the orbit, giving
off on its way several small branches (dorsal branches, *d. d.*) to the
roof of the skull, which are distributed to the mucous canals and
skin.   It then pierces the cartilage at the antero-internal angle
of the orbit, and is distributed to the skin on the upper surface of the cartilaginous rostrum, from the anterior edge of the
orbit forward (*tr*).   It is here joined by small branches from the
R. ophth. profundus next to be described.

This nerve (*o.n.*) arises from the R. ophth. superficialis, and
gives off a long slender twig, the ciliary nerve (*ci.*), which runs
forward to the eyeball.   After giving off this ciliary nerve, the
main trunk passes on under the superior and internal recti and
the superior oblique as in man, and piercing the anterior wall
of the orbit runs in the substance of the cartilage underneath
the branches of the superficialis.   With this nerve it anastomoses, as has been mentioned above, and has a similar distribution (*tr.*).

Gegenbaur (*J. Z.* pp. 508, 545, et seq.) identifies the R.
ophthalmicus with the R. dorsalis of a spinal nerve, a view we
also put forward in the *Academy*, May 29, 1875, p. 561.   The
following considerations, however, have led us to change our
opinion.

In all vertebrata the Ophth. superficialis is constantly present, and invariably lies above the eye close to the ethmoidal
region, and it commonly sends branches forward to the pituitary
membrane of the nose, and even as far as the termination of
the face above the premaxillary.   In this shark it is distributed *above* the rostrum, and so also is the profundus.   The
rostrum, be it remembered, is a development of the trabeculæ
cranii.

It has been sufficiently proved that the trabeculæ cranii are
the first pair of præ-oral visceral arches, often perfectly independent of the investing mass with which they subsequently unite
(cf. 'Pig,' p. 327, Pl. XXVIII. Figs. ii. and v. *tr.*; 'Salmon,' p. 122,
Pl. I. Fig. v. Pl. II. Fig. v. *tr.*).   The direction of their long
axis, however, has undergone a change: the distal ends rotating upwards and forwards, consequently the primordial anterior surface has become continuous with the dorsal aspect of

the body, though homologically it is *not* a *dorsal* surface. Hence the R. ophthalmicus is a nerve distributed to the anterior edge of a visceral arch (viz. the trabeculæ cranii), as may be concluded from the description given above. Accordingly we regard this nerve as the *prætrabecular nerve;* a position strengthened by the fact that Prof. Huxley describes a nerve in Amphioxus running above the eyespot, and parallel with the chorda dorsalis, which he states " has the characteristic course and distribution of the orbito-nasal division of the trigeminal" (*loc. cit.* p. 130). If any branches are to be regarded as rami dorsales, they must be those we have lettered *d. d.*

As to the special homologies of these branches, the superficialis has in fish, reptiles and birds an extensive distribution, whereas in man it has become restricted to 'frontal' branches. The profundus, on the other hand, appears, from its relations to the muscles of the eye, and its distribution, to correspond with the nasal branch of the ophthalmic division of the fifth. The branches *d. d.* in Teleostei become intra-cranial branches, and in man probably correspond to the nervus recurrens ophthalmici of Arnold, which goes to the dura mater (*H.* p. 354).

ii. The branches below the eyeball are the superior and inferior maxillary nerves. After the R. ophthalmicus has been differentiated from the fifth, and before the trunk divides into these two nerves just mentioned, two or three small branches (*d′ d′*) arise, which pass upwards and outwards through the cartilage, and are distributed to the skin above and behind the orbit.

The nerve then divides into two trunks.

The upper of these, the superior maxillary (*mx.*), divides into three branches, an anterior branch (*n. p.*), which is by far the stoutest of the three, and turns over the anterior edge of the palato-pterygoid. It then courses along below the rostrum (trabeculæ cranii) on the inner side of the nose, supplying this part of the skin with sensory branches.

Of the two other branches, one (*I. O.*) arises from the upper part of the nerve, is exceedingly slender, and is distributed to the skin on the inner side and below the orbit, in front of the labial cartilages. The other (*s. pa.*), somewhat larger, is distributed along the outer or anterior edge of the palato-

pterygoid cartilage in its whole extent; it gives a twig (*m*) to
a small cylindrical muscle, which lies behind the labial cartilages, and extends between the inferior edge of the orbit and
the external angle of the mouth. All the other branches of this
nerve (*s. pa.*) are distributed to the skin between the labial
cartilages and the pterygo-palatine bar.

The second of the two trunks (*mn.*) into which the fifth
divides (*supra*), gives off first a slender branch (*b*), which is
distributed to the skin behind the orbit, and above the labial
cartilages—a spot to which pass also a few other fine twigs
from the main trunk. It then curves over the palato-pterygoid
cartilage, giving off (i.) a small fine twig (*au*) to a muscle connecting the palato-pterygoid cartilage to the auditory capsule;
(ii.) a stouter branch (*P. M.*) to a large muscle passing from the
outer end of the palato-pterygoid bar to the upper end of
Meckel's cartilage or the mandible. The main stem of the
nerve (*mn'*) curves round the angle of the mouth, and runs along
the outer surface of the mandible, distributing branches to the
skin and edge of the jaw.

Gegenbaur in his paper on Hexanchus, and his work on the
Vertebrata (*V. A.* p. 228, et seq.) regards the labial cartilages as
visceral arches. It is more than doubtful whether this is a
tenable view. On the other hand, he does not regard the
palato-pterygoid as a visceral arch at all, whereas it has undoubtedly been shown to be so by Mr Parker's researches (cf.
'Salmon,' pp. 141, 126, Pl. I. Figs. i. and vii. *pa*; Pl. II. Figs.
iii. iv. *p. pg.*; 'Pig,' p. 327, Pl. XXVIII. Figs. ii. v. *p. pg.*). He
regards the nerves accordingly as not related to the pterygo-
palatine, but to the labial cartilages.

A careful consideration of the superior maxillary in Echinorhinus has led us to regard it as the nerve supplying the
posterior margin of the trabecular arch, and the anterior edge
of the palato-pterygoid. The cleft between them, the lacrymal
cleft, is not persistent in fishes, save perhaps in the Dipnoi,
though it exists in the higher vertebrata.

The nerve (*n. p.*) runs along the inferior surface of the
trabeculæ, which it has been previously shown (*supra*) must be
homologous with the posterior margin of that arch. It is also
internally placed to the nose and the eye, and consequently

would be the nerve in front of the lacrymal cleft. This cleft is regarded by Mr Parker ('Pig,' p. 328) as representing externally the lacrymal duct, internally the posterior nares. The only nerve with which it can be homologous in the higher vertebrata is therefore the naso-palatine, which is distributed along the mesethmoidal cartilage, a development of the trabeculæ cranii.

The nerve (s. pa.) on the other hand is distributed along the whole of the edge of the pterygo-palatine. Its branches lie between the labial cartilages and the second præ-oral visceral arch, in relation with its anterior margin; they must therefore be post-lacrymal nerves, and consequently are to be regarded as supplying the anterior margin of the second præ-oral arch. Supposing the labial cartilages, which are most probably the foundations (so to speak) of the premaxillary and maxillary bones of the higher vertebrata (cf. Gegenbaur, Grunds. 1870, p. 645), were more developed, they would necessarily, from the mode in which the branches of this nerve are distributed, enclose it completely between themselves and the palatine arch. This is precisely the relation held in man by the spheno-palatine nerves, which are often independent, of the spheno-palatine ganglion (Q. S. p. 603 and H. p. 377), and which, where the premaxillaries, the maxillaries, and the palatine bones do not develop palatal processes, are situated externally to the internal nares,—the inner extremity of the lacrymal cleft.

The nerve (I. O.) appears to be a branch of (s. pa.) distributed below and to the inner side of the orbit; its inner fibres are distributed to the same spot as the inner fibres of the nerve (s. pa.): its outer fibres lie between the orbit and the labial cartilages, and would therefore lie externally to them, supposing they were more developed; it appears therefore to correspond best to the infra-orbital of man, and to the nerve in Hexanchus and Squatina, which is identified as infra-orbital by Gegenbaur (J. Z. pp. 508—511).

As to the inferior maxillary nerve, the prolongation (mn') along the Meckelian cartilage is no doubt homologous with the lingualis and the inferior dental. It is the nerve which runs behind the oral cleft, and supplies the anterior margin of

Meckel's cartilage. As Meckel's cartilage is the third visceral arch (the first post-oral), we ought to have a branch of the inferior maxillary nerve distributed in front of the oral cleft, and to the posterior margin of the palato-pterygoid cartilage. As a matter of fact there is a præ-oral branch of the infra-maxillary, the nerve (b), which inasmuch as it is distributed behind the eye, above the mouth and labial cartilages, and in front of the articulation of the mandible, corresponds most nearly, if to any, to the buccal branch of the infra-maxillary nerve in other vertebrata. But this nerve, though it is præ-oral, has lost all connection with the palato-pterygoid arch, and the posterior margin of that arch is supplied not by the fifth, but by the palatine branch of the facial (vide infra, § 10). This appears to be a somewhat startling difficulty, but it is a difficulty which disappears when we consider the changes of direction and growth which the palato-pterygoid arch undergoes. Instead of retaining its normal direction, more or less parallel with the course of the nerves, and of the arches in front and behind it, it has come to be placed more or less transversely to its original axis. Further, its posterior or upper extremity grows in the fish in such a manner as to become hooked on to the first post-oral arch, but as the direction of growth is *not external* to the præ-oral branch of the infra-maxillary, but *internal* to it, the palato-pterygoid comes to receive, what may be termed, for want of a better phrase, an adventitious nerve-supply from the seventh.

As to the dorsal branches of the nerve, before it divides into inframaxillary and supramaxillary, they seem most nearly to represent the nervi recurrentes of Arnold (*H*. pp. 367 and 381). In Teleostei they are represented by the intra-cranial branches (*St*. p. 47). Perhaps some of their branches, which stretch far out behind the orbit, at a level with the angle of the mouth, may be homologous with the auriculo-temporal. To sum up, the trigeminus appears to consist of three distinct nerves :

(i.) A prætrabecular nerve—the ophthalmic :

(ii.) A præpalatine nerve (supramaxillary), which divides into an anterior or præ-lacrymal branch (naso-palatine), and a posterior or post-lacrymal branch (spheno-palatine and infra-orbital).

(iii.) A præ-Meckelian nerve (inframaxillary), which divides into a præ-oral branch (buccal ?), and a post-oral (lingualis, and inf. dental).

§ 10. It remains now to discuss the distribution of the seventh nerve (VII). This nerve gives off one or two fine twigs (d''), which run upwards in the cartilage to the surface of the skull: these are Rami dorsales, and appear, so far as can be ascertained, never to have been discovered previously. The nerve then gives off a stout branch forwards, but just at its origin this branch gives off two or three fine nerves, which run between the rudimentary branchiostegal rays of the spiracle and the mucous membrane, which is thrown into folds: these are the præspiracular nerves (Sp.).

The rest of the branch runs on, and breaks up into numerous twigs (pt.), distributed to the mucous membrane of the mouth, and to the teeth of the palato-pterygoid cartilage. The main trunk of the facial running along the hyomandibular behind the spiracle distributes nerves (h. br.) to the muscles in front of the branchiostegal rays of the hyoidean arch, and to the skin; also a branch (an) forward to the skin, at the articulation of the mandible with the palato-pterygoid, a region also supplied by some of the dorsal branches of the trigeminus. It then divides into two branches, the R. mandibularis externus and the R. mandibularis internus of Stannius. The latter of these nerves (ch) runs above the ligament connecting the upper end of the mandible with the upper end of the hyoid; it is distributed to the mucous membrane between the hyoid and the mandible, as well as to the teeth on the posterior surface of that visceral arch. The former, or external mandibular branch, runs along the outer surface of the hyoid to the middle line, and distributes branches to the skin, and to the thin flat muscle which lies between the two diverging rami of the mandible, extending also far backwards (F').

It will thus be seen that the facial divides into a præ-spiracular and a post-spiracular nerve, but the branch from which the præ-spiracular nerves are given off is distributed also to the mucous membrane of the mouth. We here meet for the first time with a ramus pharyngeus destined for the walls of the digestive tract. This ramus pharyngeus, identified as such by

Gegenbaur (*J. Z.* p. 524), is the palatine nerve of Teleostei. In those osseous fishes it becomes connected with the superior maxillary nerve, supplies sensory branches to the palato-ptery-goid arch, and to muscles which draw the palatal apparatus to the base of the skull, as in Perca and Gadus. In Teleostei, therefore, it may be both sensory and motor: in the Sharks, on the other hand, it appears to be entirely sensory. On this ground Gegenbaur objects to its homology with the great super-ficial petrosal of Mammalia, because in Mammalia experimenta-tion shows that intracranial section of the facial paralyses the muscular branches to the soft palate, which are known to be derived from the great superficial petrosal. On the contrary, it may be urged, that the facts adduced by Hermann (*Grundriss,* Ed. 4, pp. 316, 317), as well as the arguments long ago urged by Mr Lewes (*Nat. Hist. Rev.* I. 176), clearly indicate that to determine the homology of a nerve by reference to its function is exceedingly dangerous.

In the frog it is stated that the facial nerve is entirely sensory, but no one for that reason doubts its homology with the facial of man, which is generally considered as entirely motor. In the Shark there are no muscles developed in the mucous membrane behind the palato-pterygoid, hence the nerve *cannot* be motor. In the Teleostei, already alluded to, there are such muscles, and they are supplied by the palatine nerve (cf. *St.* p. 56). Similarly the velum pendulum palati is muscular, and the branches of the seventh, which are distributed to it, are accordingly motor. Hence the purely sensory nature of the palatine in the Shark is *no* argument against its homo-logy with the petrosal nerves. On the other hand, it may be urged from morphology, that the relation of the great superficial petrosal to the inner end of the spiracle or Eustachian tube, is such as warrants us in considering its homology with the pala-tine nerve as all but certain. The pharyngeal branch of the spheno-palatine ganglion (*Q. S.* p. 605) is frequently derived altogether from the vidian, and it is a sensory nerve. Perhaps it may be represented by the præ-spiracular nerve in part[1]. Another argument in favour of the homology between the pala-

---

[1] A nerve which becomes greatly reduced when the spiracle is small or obsolete.

-tine nerve in Elasmobranchs and the great superficial petrosal, is that they both spring from a ganglion, the geniculate ganglion (cf. *St.* p. 55). It is possible that the palatine nerve represents both the petrosal nerves, as in the Sturgeon it is connected with the glosso-pharyngeal (vide infra § 12).

The Ramus mandibularis internus (*ch*), from the course it takes over the ligament connecting the mandible to the hyoid, is evidently homologous with the chorda tympani, while the Ramus mandibularis externus (*F.*) as evidently corresponds to the main trunk of the facial in Mammalia, from its close connection to the hyoid.

With reference to a point already discussed, viz. the curious relation subsisting between the palatine nerve and the posterior margin of the palato-pterygoid cartilage, it is as well to note the fact, that the post-spiracular nerve by means of its branch (*ch*) supplies the posterior margin of the Meckelian arch, whereas the nerve-supply ought to be derived from the præ-spiracular. This change is no doubt due to the secondary connections established between the first and second post-oral arches, just as in the former case it was argued that the change was due to a similar secondary connection established between the second præ-oral arch and the first post-oral.

§ 11.   The auditory nerve rises by a large root inferiorly placed to the root $V\gamma$ and VII, and it appears to have no connection whatever with that root. It divides into three primary branches :

(i.)   To anterior vertical semicircular canal (ampulla).

(ii.)   To vestibule, etc.

(iii.)   To ampulla of posterior-vertical semicircular canal.

Gegenbaur, in his paper on Hexanchus (*J. Z.* p. 545), and in his work on the Vertebrata (*V. A.* pp. 283 to 285), has broached the theory that the auditory is the long-missing Ramus dorsalis of the facial. In our shark, a Ramus dorsalis has been found, and hence it might be said that Gegenbaur's view is untenable. This is *not* necessarily the cause, because the auditory might still be a branch of the R. dorsalis. Meynert contends strongly for the view that the auditory "belongs to a different category from the other nerve-roots," from its anatomical connections with the cerebellum (*Sr.* Vol. II. p. 500). Perhaps investigations on the nuclei of origin of the auditory nerve in the lower Vertebrata might throw additional light on this obscure and puzzling question.

The fifth group of nerves quits the cranium in front of the auditory capsule. The next group, the eighth, consisting of the Glosso-pharyngeal and Vagus nerves, passes out below and behind it. The former is a single nerve, comparable to the seventh, as is shewn by its distribution to a single visceral arch, while the latter shows markedly its compound nature both in its roots and its distribution to more than one arch.

The references are to Pl. VII. Fig. iii.

§ 12. The Glosso-pharyngeal nerve (*Gp.*) arises from the side of the medulla oblongata. Its single root—a stout bundle of fibres—is situated about ¼ inch behind the root of the auditory nerve, and lies immediately below, or ventrally to, the first strand of the upper (*i.e.* posterior) roots of the vagus, and in consequence is completely concealed by that nerve when the brain is viewed from above. The two nerves here appear as in Teleostei to be more closely apposed than they are in Hexanchus (cf. *J. Z.* VI. Taf. XIII. Fig. iii. *Gp.*, and *M. M.* Taf. II. Figs. 8*a* and 9). A careful examination failed to detect any interchange of fibres at this point between the roots of these nerves.

The nerve-trunk runs obliquely backwards, at an angle with the medulla less acute than that formed by the vagus, and enters a canal in the cartilage. The course of this canal and its contained nerve is then as follows. It dips obliquely downward below the ampulla of the posterior semicircular canal of the ear, and then turns horizontally outwards and backwards. The external opening of the canal is somewhat trumpet-shaped, and lies at the level of the upper extremity of the first branchial slit. While in the canal the trunk of the glossopharyngeus is somewhat swollen, and when fresh of a yellowish colour, due probably to the presence of ganglionic elements. No communication of this nerve with the sympathetic was observed (cf. *St.* p. 76).

A small nerve (*d*), observed by Stannius in Spinax and Carcharias (p. 79), and by Gegenbaur in Hexanchus (p. 517), quits the glossopharyngeus while still in its canal. This *Ramus dorsalis* runs outwards and upwards through the cartilage in a special canal, and is lost on the superior surface of the cranium in the neighbourhood of the mucous canal. In man it is probably represented by the branch to the auricular nerve. Another small nerve (d^*) leaves the main trunk at the same point as the R. dorsalis, and running on in the main canal, is apparently lost in the upper part of a muscle connecting the suspensorium to the cranium.

The main trunk of the glossopharyngeus then divides into two branches, the Ramus anterior s. hyoideus, and the R. arcus branchialis primi (Stannius), which respectively supply, one the front, the other the hind, wall of the first branchial slit.

i. The anterior or hyoidean branch (br.), present in all Elasmobranchs and Ganoids (probably), variable in Teleostei, gives off close to its origin a visceral nerve, the R. pharyngeus (Ph.), homologous with the palatine branch of the facial and the Rr. pharyngei of the branchial nerves derived from the vagus.

This R. pharyngeus is rarely present in Teleostei. In the Carps it supplies their peculiar erectile organ. In Elasmobranchs it is distributed to the mucous membrane of the roof of the mouth, running forwards behind the spiracle, and in Accipenser it anastomoses with the palatine nerve, and then reaches as far forwards as the superior maxillary nerve. This course and anastomosis render its homology probable in the higher vertebrates with the tympanic or Jacobson's nerve, the connection of which with the small superficial petrosal is thus explained.

The *anterior* branch lies behind the branchiostegal rays of the hyoid, and divides into several twigs, the innermost supplying the mucous membrane of the inner surface of the hyoid, and in Spinax (St. p. 78) uniting with filaments of the facial. The outermost twigs are *here* in relation with the anterior gill-lamina of the anterior gill-pouch: in Accipenser they supply the opercular gill, and in Teleostei, which possess a pseudobranchia, a few filaments run to that organ. This anterior branch probably constitutes the main trunk of the nerve in the higher vertebrata, which is closely related to the anterior cornu of the hyoid.

ii. The *posterior* branch (br'.), or nerve of the first branchial arch, is apparently present in all fish. It is distributed to the musculature of the first branchial arch, to its mucous membrane, and the posterior gill-lamina of the first gill-pouch. According to Stannius (p. 79) some of its fibres are prolonged to the floor of the pharynx. In higher vertebrata, which have lost this gill-slit (second post-oral), this branch is possibly represented by the branches of the glossopharyngeal which unite with branches of the vagus to form the pharyngeal plexus.

The glossopharyngeal, of all the cranial nerves, best accords with the type of the spinal nerves. It is a mixed nerve, motor and sensory (cf. St. p. 75), and possesses a dorsal branch, a ventral branch, which gives off a nerve to the digestive tract, and divides into an anterior and a posterior branch in relation with the anterior and posterior margins of a gill-slit.

§ 13. The Vagus has, as in Hexanchus, Spinax, and Carcharias, both upper (posterior) and lower (anterior) roots. As the nerve-trunks formed by the union of these two sets of roots are in this shark distributed in very different ways, it will be as well to describe and discuss them separately.

The upper or posterior roots (*Vg*) appear to be connected with the ganglionic swellings mentioned (§ 6), in the medulla. They rise from the medulla in a continuous row—the hindmost roots as low down as the point at which the side walls of the fourth ventricle unite. They diminish as in Hexanchus (cf. the figures before alluded to) in size from before backwards, and the posterior strands turn obliquely forward to unite with the main trunk. The number of roots is not constant in both specimens, nor on both sides of the same specimen, varying from six to eight in number. They, however, unite more or less clearly into five trunks, which in their turn coalesce to form the vagus nerve. The roots show superficially fine longitudinal striations, indicating their component bundles of fibres.

The right vagus is slightly larger than the left. On each side the nerve turns obliquely backwards and enters a funnel-shaped canal in the cartilage which runs behind the arch of the posterior vertical semicircular canal. The internal opening of this canal is about one inch behind the internal opening of the canal for the glossopharyngeus. The trunk of the nerve while yet within the canal becomes of a yellowish hue, differing from the white tint of the rest of the trunk, and due probably to the presence of ganglionic elements; but it can scarcely be said to enlarge appreciably.

The nerve then divides into rami dorsales (and ramus lateralis) and the truncus branchio-intestinalis.

§ 14. The ramus dorsalis and the lateral line nerve are of special interest; the latter probably representing a number of rami dorsales united into a single trunk.

From the trunk of the vagus, still in its canal, there springs on each side a slender nerve, R. dorsalis (d'), which runs in a special canal of its own upward to the skin on the surface of the head. This nerve is present in Hexanchus, wanting apparently in other sharks, and probably corresponds (Gegenbaur) in the Teleostei to the branch which runs upwards inside the cranial cavity, and is present wherever a corresponding branch is developed from the trigeminus, or where a ramus lateralis trigemini is given off by the same nerve, with which this R. dorsalis unites. At the same spot in Echinorhinus, from which this R. dorsalis springs, there arise on the right side two, on the left three nerves (d''), which pierce the cartilage at short distances from one another, and then emerging run outwards along the fibrous septum which stretches down the sides of the body, separating the upper lateral from the lower lateral muscles. The nerves in question are distributed to the very edge of this septum, probably to the mucous canal, and the most anterior of them reaches that canal just *in front of the first branchial slit.* On the right side the most posterior of these branches anastomosed with a nerve given off from the lateral line nerve further down. These nerves probably correspond to the opercular branch of the lateral line nerve in Teleostei, which, as Stannius (p. 110) pointed out long ago, is homologous with the auricular nerve in the Mammalia.

All these nerves arise from the anterior margin of the trunk of the vagus. This portion of the trunk begins to show signs of separation at the spot where the vagus passes out of its canal, and finally separates entirely after the third branchial nerve has branched off, and becomes the lateral line nerve (L.). This important nerve runs along the fibrous septum (mentioned above), and from its outer margin pass off numerous twigs, distributed apparently to the mucous canal. No connection between this lateral line nerve of the vagus and the spinal nerves was observed—a point in which, according to Stannius (p. 96), it differs entirely from the ramus lateralis of the trigeminus, where such a connection always takes place.

Is the nerve of the lateral line serially homologous with a ramus dorsalis? Is it one R. dorsalis, or does it represent a number of rami dorsales fused? In his paper on Hexanchus Gegenbaur (p. 541) argues against the homology of the lateral line nerve with a R. dorsalis, on the ground that its distribution passes beyond the distribution of the rest of the vagus. The R. dorsalis, he says, must be restricted to the occipital region of the skull, into which are fused the vertebral segments to which the vagus corresponds. This argument must not count for much, because *no* similar objection can possibly lie against the homology of the R. lateralis of the trigeminus with a R. dorsalis (cf. *St.* p. 49, pp. 51—54). Yet certainly this nerve has also passed beyond its proper territory. Gegenbaur, however, in his work on the vertebrata before alluded to (p. 280), appears to regard the homology in question with a more favourable consideration—a conclusion to which he is led by the apparent correspondence of the R. intestinalis to a number of Rr. ventrales.

The arguments in favour of regarding the lateral line nerve as homologous with a R. dorsalis may be briefly summed as follows :

(i.) It shows in Echinorhinus early traces of a separation from the main trunk of the nerve. It is evidently continuous with that part of the vagus from which the R. dorsalis of Gegenbaur arises.

(ii.) There are present in the same shark two or three intermediate nerves, which are derived from the differentiating lateral line nerve close to the R. dorsalis, pierce the cranial cartilage, and are distributed to the edge of a fibrous septum that extends down the body, dividing the superior lateral masses of muscle from the inferior. A mucous canal exists at the edge of this fibrous septum.

(iii.) The lateral line nerve itself lies on this same septum, and its branches pass outwards to the edge in the same manner as do the nerves mentioned (ii. *supra*), and on one side of the body there exists an anastomosis with the hindermost of these nerves. Consequently we have here a gradation from the R. dorsalis into the lateral line nerve and its branches.

(iv.) Gegenbaur maintains, and there can be no reasonable doubt of his being right, that each branchial nerve derived from the vagus is the morphological equivalent of a single spinal nerve, or of such cranial nerves as the facial and glossopharyngeus. It is a striking fact that in Hexanchus (*J. Z.* vi. p. 524) the first branchial nerve is to all intents and purposes a distinct and separate nerve. The spinal nerves, the glossopharyngeus, and in this shark the facial, have each their Ramus dorsalis.

Now, unless we may regard the lateral line nerve as representing a R. dorsalis, we *must* suppose that the spinal nerves, which have fused to form the vagus, *retain* their Rr. ventrales (= Rr. branchiales), but have, on the contrary, *lost* every trace of their Rr. dorsales; a supposition at once unnecessary and improbable.

It is exceedingly likely, furthermore, that the lateral line nerve represents a number of Rr. dorsales, because:

(i.) It has so extensive a distribution (*V. A.* p. 280).

(ii.) There are in Hexanchus at least five branchial nerves (in Heptanchus six?) known to us which have lost their Rr. dorsales. Further, the R. intestinalis represents without doubt a number of Rr. ventrales—a number which is not known to us, and these Rr. ventrales imply in their turn the existence of corresponding Rr. dorsales.

Why the lateral line nerve with such an origin has acquired a distribution so extensive, and how this distribution originated, are difficult questions impossible to answer. But precisely similar difficulties attach to the R. lateralis nervi trigemini—an undoubted dorsal branch with a distribution greatly extended beyond the normal.

§ 15. The truncus branchio-intestinalis, the second great branch into which the vagus divides, is an exceedingly stout nerve. It gives off four branchial nerves before finally passing to the heart and stomach.

Of these branchial nerves the three first (*Br.* I. II. III.) are exact repetitions of the glossopharyngeus, and are related in a similar way to visceral arches and clefts. The only points to note are that the R. pharyngeus (ph') arises indifferently from the anterior (a) or posterior (p) division, but that the muscular branches appear to be derived invariably from the posterior division of the nerve.

The fourth branchial nerve (B. IV.) differs somewhat from the three foregoing, a difference due to morphological change in its arch. It rises from the vagal trunk beyond the spot where the lateral line nerve has become completely differentiated. The anterior (a) division is a stout branch distributed as usual, but it furnishes in addition a small twig (m') to the muscle at the top of the arch (IVth) and a slender pharyngeal branch (ph'). The posterior branchial division (p) or nerve of the Vth arch has become changed. At the spot where the main branch splits into the anterior and posterior nerves, the latter gives off a fine long nerve (m''), which is distributed to a muscle connected below to the front edge of the epicoracoid, above to the anterior margin of a large muscle arising from the exocci-pital region of the skull and attached to the supra- and præ-scapular regions of the shoulder-girdle. The posterior branchial division then passes on, distributing a few fine filaments (v) to mucous membrane and muscle clothing the anterior surface of Vth arch, and finally (ph'') turns round the posterior margin of the arch on its way to the pharynx. It has a branch of anastomosis (an.) with the first visceral or intestinal nerve.

Immediately after the fourth branchial nerve is given off, there arises from the trunk of the vagus a nerve (m'''), which is connected by a short and remarkable anastomosis with the cervical cord (vide infra, § 16), and is distributed to the outer part of the same muscle as m'' (supra).

The trunk of the vagus from which several fine nerves (œ) pass off inwards, apparently to the œsophagus, then resolves itself into the main intestinal branches (I), which subdivide several times, and turning over the margin of the Vth (aborted) arch, pass to the viscera (œsophagus, stomach, heart, &c.). These nerves, however, were not traced to their ultimate distribution.

The transition state of the fourth branchial nerve is interest-ing. It becomes the first R. pharyngeus inferior of Teleostei—a nerve which sends forward a twig, the R. anterior, to be distributed to the hind margin of the fourth branchial arch. This twig is small, or wanting, in Fish (*e.g.* Cyclopterus), where the IVth arch carries only a single row of branchial filaments. The main stem of the R. pharyngeus inferior encircles the œsophagus. Here the posterior ramus of the branchial nerve has a much reduced muscular branch, its main stem *is* pharyngeal. It is thus on the high road to becoming exclusively an intestinal branch. This is precisely what must have happened to the Vth branchial nerve of Hexanchus, which is not represented in Echinorhinus as a distinct branchial nerve at all. Gegenbaur (*V. A.* p. 279) considers it probable that the intestinal branches represent a number (unknown) of branchial nerves which have become restricted to a visceral distribution simply through the suppression or transitory nature of their visceral arches and clefts. This is exceedingly likely, for though the slit and arch disappear, the wall of the pharynx remains, but remains unbroken. It can therefore be scarcely supposed that the nerves of the arches, which are essentially only thickenings in the walls of the pharynx, have disappeared *in toto*. In the Abranchiate vertebrates the process of reduction has been carried to an extreme—the branchial nerves are represented *only* by intestinal branches, showing no trace of their original form. The change is due simply to changes of growth and development. When an air-bladder is present in Fishes it has its nervous supply from the vagus, because it is an outgrowth of that part of the tract which is supplied by the vagus. So, too, the trachea, larynx and lungs of pulmonate vertebrates are supplied by the vagus. Here nerves, which attain a great development, are engrafted on the original vagus, in obedience to a change of growth and development—a change of precisely the opposite character to that undergone by the branchial arches and slits.

§ 16. In Echinorhinus there appear to be four inferior or anterior roots (*Vg.*) of the vagus, increasing in size from before backwards. Hexanchus has "three to four pair," and in Spinax and Carcharias there are, according to Stannius (p. 83), a pair of anterior roots. But the distribution of these roots in Echinorhinus, as the sequel will show, is entirely different to what it is in the three other Sharks named. The number was the same in *both* our specimens.

The first (*Vg.* I.) is a very slender and fine nerve rising from

the anterior or inferior surface of the medulla oblongata, nearly
half an inch above the posterior limit of the fourth ventricle.
It runs backwards at a sharper angle with the medulla than do
the posterior vagal roots, and enters a long canal in the exocci-
pital (?) cartilage, divides into two or three fine filaments, and
is distributed to the upper part of the long muscle which,
springing from the exoccipital region of the skull, is inserted
into the supra- and præ-scapular regions of the shoulder-girdle.
This nerve is torn away from the brain, which we have figured,
but it was found in the skull.

The second root ($Vg.$ II.), a stouter nerve, rises about a
quarter of an inch behind the first, and enters a special canal in
the cartilage. The third nerve ($Vg.$ III.) rises nearly at the
same distance behind the second, and the fourth ($Vg.$ IV.), which
has a double root, at the same distance behind the third, and
both likewise enter special canals in the cartilage. These three
nerves each give a twig (d''') laterally into the muscle already
mentioned, but the chief part of their fibres (v') pass on and
join the cervical cord, which is also partly constituted by the
four first spinal nerves.

The first spinal nerve ($Sp.$ I.) rises about an inch behind the
posterior extremity of the fourth ventricle. The posterior root
(P) on each side is a fine slender nerve which passes obliquely
backwards for some distance in the spinal canal, then pierces
the cartilage, and forms a ganglion. The anterior root (A) has
two rootlets, and springs from the cord about ⅜ inch behind the
fourth (anterior) vagal nerve, and somewhat in front of its
dorsal root. It passes back similarly some way in the spinal
canal, and then through the cartilage about 4/10 inch behind
the canal for the last anterior root of the vagus. The nerve
then bifurcates—one branch (as) runs upwards to the ganglion
of the posterior root which it joins, the other branch (v'') joins
the anterior vagal nerves. From the ganglion there are branches
(ds) which pass into the muscle, and another fine twig (ps)
which passes down and joins the second of the strands into
which the anterior root divided, constituting the anterior primary branch. A kind of triangle is thus formed, very similar
to Stannius' figure of the ganglion and roots in the Sturgeon
($St.$ Taf. IV. Fig. 5).

All the following spinal nerves arise in the same manner,

only their roots pass out directly instead of running a long way within the spinal canal. The anterior root invariably has three or four rootlets, and is situated in front of its corresponding posterior root.

The anterior primary branches (*Sp.* II. III. IV.) of the three spinal nerves (after the first) join the first and the three last vagal roots at once, to constitute the cervical cord. This cervical cord gives off a small twig (*an'*) which unites with the muscular nerve of the vagus (cf. § 15 *m'''*), and then receives the anterior primary branch of the fifth spinal nerve (*Sp.* V.). It then divides into two branches which are distributed as follows.

The first, or more externally placed (*Cv*), is the stouter of the two branches. It runs down closely applied to the præcoracoidal region of the shoulder-girdle, gives off two small branches and then passes through a foramen at the base of the epicoracoid, and, coming out on the ventral surface, is distributed to the muscles of the pectoral fin. The second division (*Cv'*) runs behind the rudimentary Vth arch, and then breaks up, some of the twigs going to the muscular (?) membrane which connects the Vth branchial arch to the epicoracoid, and supplying it in conjunction with one of the small branches given off by the first division before entering the foramen. The second of the small branches derived from the first division, in conjunction with the remaining branches of the second division, are lost in the ventral muscles which have their attachment to the posterior edge of the epicoracoid.

In the distribution of the anterior vagal roots, and in the union of the three last with the five first spinal nerves, Echinorhinus differs completely from Hexanchus, Spinax and Carcharias, according to the descriptions given by Gegenbaur and Stannius respectively. It might be urged that because the anterior vagal roots in the three Sharks just named unite with the trunk formed by the union of the posterior vagal roots outside the skull, but here in Echinorhinus unite, not with the posterior roots of the vagus, but with spinal nerves; they are therefore not anterior vagal roots, but anterior roots of spinal nerves which have lost their posterior roots.

The answer to this objection, which is *primâ facie* a good one, lies apparently in the following facts.

i. The foremost of these anterior vagal roots arise from the
medulla oblongata some way above the posterior extremity
of the fourth ventricle. Consequently the posterior vagal roots
extending backwards to the very extremity of the fourth ventricle, extend also far behind the place of origin of the anterior
vagal roots mentioned. Hence we must either admit that the
roots in question are the anterior roots of spinal nerves *shifted*
far forwards and *deprived* of their posterior roots, or else allow
that they represent anterior roots corresponding to posterior
roots actually present and represented by the upper roots of the
vagus. The former supposition seems both gratuitous and ab-
surd; and the mode and place of origin of these nerves clearly
point to their connection with the vagus.

ii. The first anterior vagal root remains independent. It
does not unite with spinal nerves, and it enters a muscle which
connects the upper part of the shoulder-girdle to the skull. In
Spinax (*St.* p. 83) branches from the main trunk of the vagus
enter this same muscle, and in Cottus (*St.* p. 122) a branch
from a nerve, regarded by Stannius as the first spinal nerve
wanting the posterior root, enters the homologous muscle.

iii. Though it is true that the three posterior vagal roots
unite with spinal nerves, it must be remembered that each
sends a twig into the muscle into which the first anterior vagal
root enters exclusively, and also that the nerve-trunk, formed
by the conjoined anterior vagal roots and first spinal nerves, is
in close connection with a branch of the vagus by the branch of
anastomosis mentioned before (*supra*, § 15). Accordingly, in spite
of the union of the anterior vagal roots with spinal nerves, it is
possible that they have not lost their connection with the posterior roots.

It is also a remarkable fact, and one worth noting, that in
Teleostei branches derived from the Rr. pharyngei inferiores
are distributed to the muscular floor connecting the ventral
ends of the branchial arches and the ossa pharyngea inferiora
to the shoulder-girdle (*St.* p. 90). In the Sharks (*St.* p. 122)
the same parts are supplied by the two first spinal nerves.
Here in Echinorhinus they are supplied by conjoined vagus
and spinal nerves (see *ante*, *Cv'*).

An interesting point of inquiry in connection with the

vagus is the relation of the hypoglossus and accessorius Willisii of Mammalia and Sauropsida to these anterior roots of the vagus. Gegenbaur, in his paper on Hexanchus, so often referred to already, derives the hypoglossus from the anterior vagal roots. The spinal accessory, he says, "is obviously represented by the posterior part of the upper (*i. e.* posterior) row of roots of the vagus—it is still a part of the vagus itself" (*J. Z.* p. 530). This homology is somewhat doubtful.

In the first place, according to Meynert (*Sr.* pp. 524—5), the roots of both hypoglossus and spinal accessory are connected with masses of grey matter in the medulla oblongata, which in the cord correspond respectively to the inner portion of the anterior grey cornu, and to the processus lateralis of the same cornu. Below the level of the internal accessory olivary bodies the same grey masses give off anterior roots of the first pair of cervical nerves, which above these olivary bodies give off hypoglossus roots. The vagus of Man is, on the other hand, connected with grey masses homologous with the posterior grey cornu in part. The hypoglossus and spinal accessory of the Mammalia being thus so clearly connected anatomically with grey masses representing the anterior grey cornu in the spinal column, it is only reasonable to suppose that *both alike* are homologous anatomically with anterior nerve-roots. The spinal accessory will therefore have to be sought most probably in the anterior roots of the vagus.

In the second place, there is the fact to be taken into consideration, that the 'elevator' of the shoulder-girdle is supplied by an anterior vagal root in Echinorhinus and in Spinax by branches from the vagus *after it has been joined by the two anterior roots.* It is possible that this muscle corresponds more or less closely with the mass from which our trapezius and sternocleido-mastoid are differentiated. This fact would therefore also make for the derivation of the spinal accessory from the anterior vagal roots, as well as the hypoglossus, which is without doubt so derived.

The close union in Echinorhinus of the anterior roots of the vagus with the first spinal nerves certainly lends strength to the homology between these roots and the spinal accessory and hypoglossus. On the one hand in Mammalia the vagus is intimately connected with the spinal accessory. Indeed, evulsion of

this latter nerve is said to paralyse all the motor fibres contained
in the vagus (Hermann, *Grundriss*, p. 331). It is also connected
with the hypoglossus. It is this connection that we see exaggerated in Hexanchus, Spinax and Carcharias. On the other hand,
the vagus in Man is connected with the suboccipital nerve
(*Q. S.* p. 636), the spinal accessory is connected with the cervical plexus, and the hypoglossus receives the well-known commu-
nicans noni. It is this connection that we find exaggerated in
Echinorhinus. We may suppose it carried still further—even
to the suppression of the anterior roots as *visible external* roots.
Then we should be able to explain how it is in Amphibia that
the *first spinal* nerve represents the hypoglossus, being distributed to the hyoidean and tongue-muscles. The distribution in
Teleostei of branches derived from the first spinal nerve to the
homologous region led Cuvier and Büchner long ago to designate that nerve as hypoglossus, while Stannius pointed out
more correctly that it represented this nerve *in part.* It is of
course possible that the deep origin of the first spinal nerve
may in these cases embrace that of the anterior vagal roots, *i. e.*
extend far forwards into the medulla oblongata. This point,
however, is one which does not appear to have attracted observation; but that the supposition is far from improbable may be
surmised from the great variations both in the number of the
roots, and of the place of exit, displayed by the first spinal
nerves (*St.* p. 121). It is also possible where two distinct sets
of nerves come to supply identical regions from great morphological changes, that in given cases one set may disappear, the
other persist and develope, and *vice versâ.*

The accompanying table displays at one view the relations
of the cranial nerves to the visceral arches and clefts, as well as
the special homologies suggested in this paper.

The spinal column is not much differentiated[1]. The core is
composed of dense fibrous tissue: a complete fibrous septum is
developed at intervals, at a spot corresponding to the exit of
.the anterior roots of the spinal nerves. In this way a series of
oblong chambers is constituted internally, and these enclose the
gelatinous remains of the notochord. On the ventral surface of
this fibrous core, semi-rings of cartilage are developed, thickened

---

[1] In this description the present tense is used where the part described has
been preserved.

No.	Arch. Name.	Cleft.	Nerve.	R. dorsalis.	R. ventralis. a = anterior } division. p = posterior	R. pharyngeus.
I.	Trabecula cranii		Praetrabecular = Ophthalmic.	Nervus recurrens of Arnold.	Frontal branches, and orbito-nasal.	*Absent in these nerves.*
II.	Palato-pterygoid.	Lacrymal.	Praepalatine = Sup. Maxillary.	Several Br. dorsales = nervi recurrentes of Arnold and auriculo-temporal (?) = intracranial nerves of Teleostei.	a = naso-palatine. p = spheno-palatine and infra-orbital.	
III.	Meckel's Cartilage.	Mouth.	Praemeckelian = Inf. Maxillary.		a = buccal (?). p = Gustatory + Dental.	
IV.	Anterior Cornu of Hyoid	Spiracle Eustachian Tube.	Prehyoidean = Facial.	Auditory (Gegenbaur).	a = praespiracular nerve (= pharyngeal branch of Vidian?). p = Chorda Tympani + main stem in Man.	= Palatine nerve in Fishes. = Petrosal nerve in Man.
V.	Posterior Cornu of Hyoid = 1st Branchial Arch.	1st Branchial Slit.	1st Praebranchial = Glossopharyngeus.	Communicating branch to the auricular nerve.	a = main stem in Man. p = oesophageal branches.	Pharyngeal nerve = R. communicans in Sturgeon = Jacobson's nerve in Man.
VI.	2nd Branchial Arch.	2nd Branchial Slit.	2nd Praebranchial = 1st Branchial nerve of Vagus.	Auricular nerve.	a = R. anterior. p = R. posterior.	R. pharyngeus.

laterally, and forming short conical processes to which are attached the stunted cartilaginous ribs. The floor of the neural canal is fibrous, its sides are formed by cartilage. There is a series of crural cartilages, triangular in shape, and pierced by the anterior roots of the spinal nerves. Between these fit in large 'intercrurals,' also triangular in shape, the apices touching the fibrous core, as in Notidanus, Centrina, &c., and pierced by the posterior roots of the spinal nerves. At the summit of each crural cartilage and between each intercrural is a small cartilage, lozenge-shaped in transverse section.

There is little to say about the digestive organs after the description given by Professor Turner (IX. pp. 299—300 of this *Journal*). There is no duodenal dilatation, and duodenal cæca are absent. The spiral fold of the intestine, which is at first straight, but gradually becomes oblique, commences by a forked extremity, the fork embracing the pyloric aperture. The mucous membrane of the rectum is smooth. The length of the duodenum and valvular intestine in the female specimen was 23½ inches. The liver consisted of two equal lobes; in the female 4 ft. 3 in. long, weighing together 21 lbs. 12 oz., and of so low a specific gravity that they floated readily on water. There was a large gall-bladder containing some liquid yellow bile. The pancreas is a large bilobed gland; one lobe simple and thin, the other and larger lobe thick and bifid at the extremity. The longer of the two bifurcations curves round the pyloric extremity of the stomach just at the commencement of the duodenum. The bile and pancreatic ducts open into the duodenum below the commencement of the spiral valve.

In the digestive tract were found several examples of a large Distomum—species undetermined as yet.

The conus arteriosus gives rise to a long aortic stem, from which spring the branchial arteries, four in number, the anterior bifurcating, as is usual in Selachians. It contains four rows of valves. Of these the anterior or first row contains the largest valves, three in number. The second row has very small valves indeed, easily overlooked, one behind each of the valves in the first row. A fourth minute valve is intercalated in this row. The third and fourth rows each contain four valves, those in the third rather the larger. In shape they are broad, not deep, and they vary in size, two valves being larger than the other two. The ventricle has thick walls, with strongly marked internal muscular ridges. The auricle and sinus venosus have each their usual bivalved aperture.

There is a heart-shaped spleen attached to the curvature of the stomach. A second spleen, pyriform in shape, and in the female specimen about eight inches long, lies close to the pancreas.

Just below (in the natural position of the animal) the terminal division of the branchial artery, lying between the sterno-hyoid muscles (of Owen), there was an elongated slender structure, reddish in colour, and somewhat resembling atrophied thymus as it appeared on microscopical examination. It probably represents the thyroid.

A similar gland, reaching right across the body, lay close behind the branchial arches, thrusting itself between the œsophagus and the vertebral column. Its microscopic appearance was similar to that of the gland just mentioned. It is usually held to represent the thymus.

In the female the kidneys reached to within a foot of the apex of the heart. The two ureters dilate terminally and open separately into the cavity of the urethral papilla which projects from the dorsal wall of the cloaca. In the male the whole of the outer side of the vas deferens shows traces of kidney-substance, but it is only posteriorly that the characteristic colour of the kidney is evident, the forepart being colourless.

The ovaries are large, and consist of a loose stroma, embedding ova of every size, some quite microscopic, others of the size of ordinary pistol-bullets, and one fully as large as a hen's egg. They lie close behind the oviducts.

The oviducts were suspended each by a separate peritoneal fold, which coalesced with its fellow to form a single lamina before it was attached to the vertebral column. The mesogastrium commenced at the point of junction of the two oviducal folds. The upper part of the oviduct has a thin mucous membrane, elevated into numerous, close-set, and fine folds, somewhat sinuous. At the summit of each fold is a row of minute orifices (glandular no doubt) quite visible to the naked eye. A large nidamental gland was present, about 2 in. in transverse diameter, and 1½ in. broad. Professor Turner appears not to have found this gland in his specimen (IX. p. 299 of this *Journal*), though, according to Bruch (*Études sur l'appareil de la génération chez les Sélaciens,* Strasbourg, 1860), it is always present: *La glande de l'oviducte est constante,* p. 64. It is (p. 54) even present in viviparous species. The upper third of this gland is greyish in colour and apparently smooth, while the lower two-thirds is yellowish, and consists apparently of a number of transverse, thin, close-set, and parallel folds, between which, Bruch (p. 52) asserts, are situated the orifices of glands. The nidamental gland appears to vary much in size in different Selachians, and in the same species at different times of the year, being much larger at the period of sexual activity. It appears also that the relative arrangement of the grey and yellow zones is not always the same (cf. Bruch, pp. 51, 52). In our specimen the nidamental gland was situated 28 inches from the outlet of the oviduct.

The oviduct (uterus) below the nidamental gland has a thick mucous membrane produced into short wavy villous folds. These folds are aggregated more or less distinctly into four rows. In each row the central folds are well developed, while on each side they are stunted and feeble.

The mucous membrane at the termination of the oviducts is much more smooth, and has only a few papillæ. The apertures lie one on each side the cloaca. They are large, large enough indeed to admit two fingers easily: a point in which our specimen again differs from Professor Turner's, which from the minuteness of the ova, and the capillary size of the orifices of the oviducts, was not in a state of sexual activity. In the middle line between the oviducal apertures there lies below, the round rectal aperture; above, the long conical urethral papilla.

The abdominal pores in both specimens are large and slit-like, situated just without the rim of the cloaca at the base of the ventral fins and guarded by a fold of skin.

In the male the testes are thin, dark-coloured glands about 8 inches in length and much flattened. They are suspended by a well-developed peritoneal fold. The vas deferens is slender at its commencement from the epididymis, and much convoluted in the usual way. Lower down it becomes straight and of greater calibre, and terminally each has a well-marked and separate dilatation with thick muscular walls, and feebly marked internal rugæ. The vas deferens on each side opens separately into the cavity of the genito-urinary papilla, situated on the dorsal wall of the cloaca (cf. Bruch, *op. cit.* pp. 35, 36, and Pl. I. Figs. 1 and 2).

The claspers are large, about 5 inches in length. On their inner and posterior side is a slit about 2½ inches long, which leads into a deep conical cavity (cf. Owen, *Comp. Anat. Vert.* Vol. I. pp. 570, 1).

DESCRIPTION OF PLATE VII.

Fig. 1. The Brain of Echinorhinus Spinosus.

Olf. Rhinencephalon—Olfactory Tracts. *C.* Procencephalon. *Th.* Thalamencephalon. *Op.* Optic Nerve. *C. B.* Mesencephalon—Corpora Bigemina or Optic Lobes. *Ce.* Cerebellum. *M. O.* Medulla Oblongata. *L. Tr.* Lobi Nervi Trigemini. *Pi.* Pineal Gland. *V* 3. Third Ventricle. *V* 4. Fourth Ventricle: the roof thick in front, thin behind. *s.* Outermost strand of Medulla Oblongata. *g.* Ganglionated Tract. *p'.* Teretial (?) Tracts. *p.* Anterior pyramids. III. Oculomotor. IV. Trochlear. V. Trigeminus. V*a.* First root of Trigeminus. V*β* + V*γ*. Second and Third roots. VII. Facial. *Au.* Auditory Nerve. IX. Glossopharyngeus. X. Vagus (posterior roots). X'. Vagus (anterior roots—two last only represented). *Sp.* I. First Spinal Nerve.

Fig. 2. Fifth and Seventh Nerves (semi-diagrammatic).

V*a.* First root of Fifth; V*β.* Second root. V*γ.* VII. Third root and root of Seventh. V. Trigeminus. *Opth.* Ophthalmic. *d. d.* R. dorsalis. *Opth'.* Ophth. superficialis (frontal). *o. n.* Ophth. profundus (orbito-nasal). *Ci.* Ciliary Nerve. *tr.* Terminal twigs to upper (dorsal) surface of rostrum. *d'.* R. dorsalis. *m.x.* Superior Maxillary Nerve. *n. p.* Anterior division (= naso-palatine). *s. pa.* Spheno-palatine, and *I. o.* Infra-orbital, = posterior division. *Mn.* Inferior Maxillary Nerve. *b.* Buccal (?) Nerve. *au.* and *P. M.* Muscular Nerves. *Mn'.* Gustatory + Inferior dental. VII. Facial *d'.* R. dorsalis. *Sp.* Spiracular Nerve. *pt.* Palatine Nerve. *h. br.* Muscular Nerves. *an.* Nerve to angle of mouth. *ch.* R. mandibularis internus (chorda tympani). *F.* R. mand. externus (= main trunk).

Fig. 3. Glossopharyngeal, Vagus and First Spinal Nerves (semi-diagrammatic).

Gp. Glossopharyngeal; *d.* R. dorsalis; *d*.* Muscular (?) Nerve; *br.* Hyoidean (anterior) branch. *Ph.* Pharyngeal Nerve; *br'.* Posterior division (branchial). *Vg.* Posterior (dorsal) roots of Vagus; *d'. d".* Rr. dorsales. *L.* Lateral line Nerve. *Br.* I. II. III. IV. Branchial Nerves; *a.* and *p.* Anterior and posterior divisions; *m. m'.* Muscular branches; *ph'.* Pharyngeal branches. *ph".* Pharyngeal branch of IVth Branchial Nerve (= main portion). *v.* Twigs to mucous membrane of Vth arch. *an.* Anastomosis with I. *m".* Muscular Nerve of Posterior division of IVth Branchial Nerve. *m'".* Muscular Nerve from main trunk of Vagus. *oe.* Œsophageal branches. *I. I.* Visceral branches. *Vg.* I. II. III. IV. Anterior Vagal roots. *d".* Branches to Muscle. *v'.* Ventral branches. *Sp.* I. First Spinal Nerve; *A.* and *P.* Its anterior and posterior roots; *ds.* R. dorsalis. *as.* Branch to ganglion from anterior root. *ps.* Branch from ganglion to anterior root; *v":* R. ventralis. *Sp.* II. III. IV. V. Spinal Nerves—Rr. ventrales. *an'.* Anastomosis of Cervical Cord with Muscular Nerve of Vagus. *Cv.* External, *Cv'.* Internal, division of Cervical Cord.

NOTE ON CERTAIN PECULIAR· CELLS OF THE CORNEA DESCRIBED BY Dr THIN. By JOHN PRIESTLEY, *Platt Physiological Scholar, The Owens College.*

THE cornea, it is well known, has been made the object of investigation by great numbers of histologists and pathologists with a view to discover its intimate structure under both normal and pathological conditions, as is evidenced by the extent of the literature of the Normal and Inflamed Cornea. It is also well known that, in these investigations,. all the ordinary histological methods of treatment had been brought to bear on the organ. It was, therefore, a matter of astonishment and of very great interest to learn from a paper by Dr Thin published in the *Proceedings of the Royal Society* (vol. XXII., p. 515), that, by means of a new method, a multitude of peculiar cells had been discovered belonging to the proper corneal substance, and regarded by Dr Thin as the cellular elements of a sheath investing the fibrillary bundles. The results obtained by treating connective tissues according to the method referred to are stated by Dr Thin to be most satisfactory in the case of the cornea; and he properly regards his experiments on that structure as of fundamental importance.

The method is as follows (p. 515) : " If a cornea is placed in a saturated solution of caustic potash, at a temperature between 105° and 115° F., it is reduced, in a few minutes, to a white granulated mass of about a fourth of its previous bulk. In a small piece of the diminished.cornea, broken down with a needle and examined under the microscope in the same fluid, it is found that the only visible elements are a great number of cells." These cells, of which many are figured in Dr Thin's paper (Plate VIII.), are of very various forms, some being angular, or club-shaped with processes of various lengths, others resembling long, narrow rods, and oblong cells, united together in many ways. They are granular, and sharply outlined, and the nuclei are nearly equal in length in all the cells,

but in the narrower sort they are compressed. Those cells
which are long are also narrow; and the shorter kind are
noticed to be broader.

An inspection of the drawings of cells (see Plate VIII.)
obtained by the above method from the corneæ of oxen, sheep
and frogs, as well as certain statements in Dr Thin's description,
led to the suspicion that the cells he had seen were epithelial
cells belonging to the anterior epithelium of the cornea. Thus,
as he himself notices, many of them were sutured to undoubted
epithelial patches which were present in the preparations (p.
517); again, as has just been mentioned, the longer sort of cells
were very constantly narrow, while the shorter ones were wide,
a circumstance which might well indicate a relationship to some
one form, the originator of both. Moreover, the treatment of
tendon (which, of course, possesses no epithelium like that of
the anterior surface of the cornea) by the above method "seldom
led to satisfactory results" (p. 519); while in the cutis "the
demonstration of the cells was easy" (p. 520). The following
observations were made in the Laboratory of the Brown Institution, at the suggestion, and under the direction, of Dr Klein,
with a view to decide whether the suspicion were well founded
or not.

In the first place the fresh cornea of a frog was treated
according to the potash method, by being kept for four or five
minutes in a thoroughly saturated solution of caustic potash
within the prescribed limits of temperature, by means of a
water-bath. The cornea was mounted in the same fluid and
examined, when an abundance of cells, exactly resembling those
figured in Dr Thin's paper (Pl. VIII., Figs. 1, 2, 3), were seen,
along with others which were, from position and shape, undoubtedly anterior epithelial cells. A few sections were made
from the corneal surface of a sheep's eye (which had been immersed as a whole for a week in 0·5 p.c. solution of bichromate
of potash) and subjected to the potash method with similarly
favourable results.

Another fresh cornea of frog was then taken, and the anterior
epithelium scraped off with the edge of a sharp knife before it
was treated with potash as above. The cornea after treatment
was mounted as a whole, being pressed out firmly under the

cover-glass; far fewer cells could be seen than in the previous frog's cornea, and they were collected in patches here and there on the surface of the cornea, and especially where the knife had been less effectively used (e. g. near the edges). *When the scraping had been performed with exceeding care the whole cornea became transformed into an amorphous mass,* falling to pieces on being lifted from the solution after having been subjected to its action for the usual time. This is obviously a fact of very great importance, and it is the only point in which Dr Thin's experience differs from that recorded here; for he states (p. 516): " The smallest piece that can be removed by the needle from a cornea which, before being put into the solution, has had this (*i.e.* the anterior) epithelium scraped off and Descemet's membrane removed, shows under the microscope a multitude of cells."

The sheep's cornea used above, *i.e.* which had been kept for a week in 0·5 p.c. bichromate of potash, was cut into horizontal sections, and the sections separated into three sets, viz., 1. All those which bore anterior epithelium. 2. All those which were made through proper corneal substance alone. 3. The lowest layer of corneal substance proper still covered by Descemet's membrane and posterior endothelium. Several specimens of each of the three sets were then subjected successively to the caustic potash solution, with the following constant results : Those sections containing anterior epithelium exhibited satisfactorily the cells described by Dr Thin : those of the proper corneal substance alone *never exhibited any such cells, or anything indeed but a more or less granular amorphous mass,* which often fell to pieces while in the solution if allowed to remain there even slightly longer than two or three minutes: those including Descemet's membrane and posterior endothelium exhibited only cells of the quadrate and angular kind, and none of the long, rod-like, imbricated description; patches were sometimes noticed having a superficial resemblance to spindle-shaped imbricated structures (unlike, however, anything figured in Dr Thin's paper), but they were ascertained to be due to foldings of Descemet's membrane.

It seems, therefore, clear that when care is taken to remove entirely all traces of anterior epithelium, it is not possible, by the method devised by Dr Thin, to get indications of any cells whatever belonging to the proper tissue of the cornea.

The potash method is quite competent to modify epithelial cells so as to produce cells of the narrow, elongated, isolated and imbricated forms figured in Pl. VIII., as is shown by the following experiments: Vertical sections were made of a rabbit's cornea which had been hardened in spirits for some months. They were mounted as they were and examined, and, while under observation, were irrigated freely with hot saturated solution of caustic potash; it was seen that the proper corneal substance underwent considerable shrinking, the result of which was to contract laterally, and consequently to elongate, the cells of the anterior epithelium, which itself became more or less fanshaped. Where the epithelial covering had chanced to become detached from the subjacent tissue, at the end of a section, it remained sensibly of the same thickness and length. Oblique sections of the sheep's cornea above used, when treated in the same way, showed the same contraction of the proper substance, which was so powerful at certain spots as to force out individual elongated cells into the surrounding medium, much as pressure might squeeze out the interior of a soft fruit.

On p. 517 Dr Thin states, as to the method, that " there are conditions of success, as to the nature of which I have not yet come to a definite conclusion. Sometimes the same solution, applied at the same temperature to different corneæ, succeeds in one and fails in another, and sometimes a solution prepared with every precaution has failed to afford me any result. The two essential conditions to success are complete saturation and temperature. I have never succeeded with a temperature above 120°, nor with one below 102°" (F.). It may be here stated that when sections of the cornea of an ox, which had only been immersed in very dilute bichromate of potash for 16 hours, were treated in the usual manner, scarcely any cells were seen, and those that did appear were ill-defined and granular: this is evidently due to the easier destruction of the fresher tissue by potash. Dr Thin does not say in what condition as to hardness the corneæ were which he examined. Temperature had a marked influence on the modifying powers of the potash solution. The limits prescribed by Dr Thin, viz.,

102°-120° F., *i.e.* 39°-49° C., below and above which respectively he failed to get results, are perhaps more likely to be correct than those found most serviceable in these experiments, as being based on more extended series of observations. But for the cornea of sheep (hardened for two weeks in 0·5 p. c. solution bichromate of potash) temperatures of 48°-52° C. were found most useful. Temperatures of 43° C. and of 55° C. gave preparations exhibiting ordinary epithelial cells resembling those in successful preparations at 48°-52° C. only in being scattered about and aggregated in a similar manner.

No drawings have been inserted in this paper of the cells of the anterior epithelium of the cornea modified in the manner described above, as the excellent drawings given by Dr Thin in Figs. 1, 2 and 3 of Plate VIII. are very truthful, and agree exactly with the cells described in this paper.

A glance at Fig. 11, representing " cells of the cutis of the frog," and at Fig. 12, "cells isolated from the skin of the ox," prepared by the potash method, makes it difficult to doubt that the cells are analogous structures to those seen in the corneæ and figured in Pl. VIII., Figs. 1, 2 and 3; that is to say, that they also are merely modified cells from the epithelium covering the skin.

ON THE STRUCTURE OF HYALINE CARTILAGE.

By E. CRESSWELL BABER, M.B. Lond.: *late Demonstrator of Anatomy at St George's Hospital.* (Pl. VIII.)

IN a recent number of Max Schultze's *Archiv für Mikroscopische Anatomie*[1] appears an article by Dr Hermann Tillmanns, entitled "Contributions to the Histology of the Joints," in which occur the following remarks:—

"The fibrillation of the ground substance of hyaline cartilage, a process which occurs so frequently and is pathologically so well known, ought long ago to have led to trying whether *normal hyaline cartilage* could be artificially fibrillated by chemical reagents. The fact of the pathological fibrillation of cartilage might have led to the hypothesis, that the homogeneous hyaline matrix consisted of cartilage fibres, which were perhaps united by a cementing substance. The next step, therefore, was to attempt the artificial fibrillation of normal hyaline cartilage by the influence of chemical reagents. And I was in fact successful in resolving hyaline cartilage artificially into cartilage fibres. If fresh normal cartilage of a dog or rabbit just killed (*e.g.* from the patella or femoral condyles) be placed for several days into a moderately strong solution of permanganate of potash, the solution being changed several times daily, or into 10 per cent. solution of chloride of sodium, the homogeneous hyaline, previously non-fibrillated ground substance of the cartilage, becomes resolved into separate fibres and bundles of fibres, fibres that are so like connective-tissue fibrils, that for my own part I am unable to distinguish them. The process of fibrillation takes in from three to seven days. A necessary condition for the success of the experiment is that the cartilage be completely cleared of soft parts; bone-substance can remain attached to it, though the best and most reliable method is to lay only pieces of hyaline cartilage in the above solutions. The moderately strong solution of permanganate of potash in water, by which I understand a moderately dark violet-coloured (*mitteldunkelvioletgefärbte*) fluid, must then be changed daily four to six times, or oftener. Before the histological examination the cartilage is to be thoroughly washed in water. As regards the preference of the two solutions above named, the 10 per cent. solution of common salt is decidedly inferior to the permanganate of potash, the former acting more slowly and easily producing unreliable appearances."

Tillmanns goes on to say, that the cell-territories, into which the hyaline matrix of cartilage is resolved by the action of chlorate of potash and nitric acid, do not appear in the fibrillated matrix. And

[1] Volume x. Part IV. p. 401.

he further states, that by the action of permanganate of potash the cartilage-cells become free and lie isolated between the fibres and the bundles of fibres.

These extracts contain the substance of all Tillmanns says on this subject.

The importance of these assertions, if confirmed by other observers, is evident, and it is this that has led me to make the following observations with a view to either proving or disproving their accuracy. It will be seen that the only indefinite part in Tillmanns' directions is the strength of the permanganate solution. The colour of this of course varies according to the thickness of the layer of fluid looked at. When poured into a watch-glass ¼ p. c. solution gives a "moderately dark violet" colour. It was therefore with this strength that I commenced, following Tillmanns' directions, and with the following results.

Observation 1. June 14th.—Sections of the cartilage of the head of the tibia of a freshly-killed frog were placed in ¼ p. c. solution of permanganate of potash. Solution changed five or six times a day, and specimens microscopically examined daily. The cartilage soon became darkly stained, and broke up on the slightest pressure into irregular masses. *No sign of any fibrillation resembling that of connective tissue.* June 19th. It broke up entirely.

Observation 2. June 14th.—Sections of the hyaline cartilage of the condyles of the femur of freshly-killed kitten, put into ¼ p. c. solution of the permanganate. Solution changed five or six times a day, and the specimens examined almost daily. They broke up irregularly on pressure, as in the preceding instance. *No appearance of fibrillation.*

Observation 3. June 15th.—Sections of cartilage of condyles of femur of freshly-killed rabbit, put into ¼ p. c. solution of permanganate of potash. Fluid changed five or six times a day, and sections examined almost daily till June 22nd. Sections broke up irregularly as in the preceding instance, and *no signs of fibrillation appeared.*

Simultaneously I made the following observations with salt solution :—

Observation 4. June 14th.—Sections of cartilage of head of tibia of freshly-killed frog, put into 10 p. c. solution of chloride of sodium, and examined June 16th, 17th, 18th, 19th, 21st, 22nd, 26th, 28th, and 30th, and July 2nd and 5th, *without any appearance of fibrillation.*

Observation 5.   June 14th.—Sections of cartilage of condyles of femur of freshly-killed kitten, put into 10 p. c. solution of common salt, and examined June 16th, 17th, 18th, 21st, 22nd, 26th, 28th, and 30th, and July 2nd. *No fibrillation of the matrix apparent.*

Observation 6.   June 15th.—Sections of cartilage of condyles of femur of freshly-killed rabbit, put into 10 p. c. solution of common salt. Examined June 16th, 17th, 18th, 19th, 21st, 22nd, 26th, 28th, and 30th, and July 2nd and 5th, *with the same negative result.*

Up to this date (July 5th) I had therefore obtained no fibrillation resembling that described by Tillmanns that I could at all attribute to the action of the reagent. Wavy fibrillation was seen at the edge of some sections (joint-surface of the cartilage), but this did not increase in extent, nor did I consider it due to the reagent. I was therefore, as may be imagined, nearly despairing of success, and somewhat at a loss to reconcile my negative results with Tillmanns' statements; when, considerably to my surprise, I made the following observation :—

Observation 5 (continued).   July 5th (after 21 days maceration). Happening accidentally to make momentary pressure on the thin glass covering one of the sections before examining it, I was astonished to find the greater part of the small specimen beautifully marked out by a fine, but distinct, parallel striation. I produced a similar appearance in other sections by pressure. Whilst under examination, however, the striation disappeared. There was therefore no doubt *that the matrix of these sections which I had previously examined ten times* (see above) *and seen to be hyaline—not fibrillated—in appearance, had suddenly become distinctly fibrillated.   No fibrillation was seen without pressure.* The same absence of fibrillation without pressure, and its appearance on pressure, were observed on July 6th, 7th, 10th, and 13th. Observations 4 and 6 were proceeded with as follows :—

Observation 4 (continued).   July 6th.—Matrix hyaline, no

8—2

fibrillation produced by pressure. July 10th and 12th. A few cracks seen, but no distinct fibrillation, with or without pressure. July 13th. Distinct fibrillation seen without pressure having been applied knowingly. Slight movement of the cover-glass caused the fibrillation to disappear, nor was it reproduced by pressure. In the greater part of these sections the cartilage-cells were far apart, and some calcareous deposit had taken place. In one small piece, however, in which the cells were closer together and in which there was no deposit, pressure shewed (on July 14th) fine fibrillation, not visible before the pressure was applied. Short application of continuous pressure produced partial fibrillation in the part where the calcareous deposit had occurred. Pressure applied on July 22nd to a section produced partial fibrillation, but not more than on the last occasion.

Observation 6 (continued). July 5th, 6th, 10th, 13th and 16th.—Fibrillation was produced by pressure in the previously to all appearance hyaline matrix. July 24th. A section was examined and considerable pressure was required to produce fibrillation in it.

For convenience it may be advisable to give the rest of my observations with 10 p.c. salt solution at this place, before detailing those with other solutions.

Observation 7. June 26th. — Cartilaginous expansion of shoulder-girdle of freshly-killed newt (partly deprived of its perichondrium), put into 10 p.c. solution of common salt. Examined June 30th, July 1st, 2nd, 3rd and 5th without pressure, and on July 5th, 10th, 13th, 16th, 21st, 22nd, 23rd with pressure. On none of these occasions could fibrillation be detected with or without the application of pressure. But after a time firm pressure caused the matrix to break up irregularly, allowing the cells to escape into the surrounding fluid.

Thinking that this property of the matrix of hyaline cartilage of splitting up into fibres, might be confined to the cartilage of joints, and that here, as it were, the fibrils of connective tissue were continued into the matrix, I experimented with costal cartilage, with the following results :—

Observation 8. July 12th.—Longitudinal and transverse sections of costal cartilage of a freshly-killed kitten, placed in

10 p. c. salt solution. Examined at once, with and without pressure, and shewed a matrix slightly granular, but presenting no fibrillation. July 14th and 16th. The same appearances. July 20th. Doubtful fibrillation was seen in a longitudinal section. Jully 22nd. Fibrillation observed in a longitudinal section on pressure—none in a transverse section. July 24th. Fibrillation seen in both longitudinal and transverse sections on pressure.

The two following complete the observations with 10 p. c. salt solution.

Observation 9. July 12th.—The sections and fluid are the same as in Obs. 5. July 16th (after four days maceration). Partial fibrillation was plainly seen in one section on pressure. July 21st (nine days maceration). Perfect fibrillation appeared in one specimen on pressure.

Observation 10. July 12th.—Sections of cartilaginous head of femur of freshly-killed kitten, placed in 10 p. c. solution of chloride of sodium. Examined and seen to consist of hyaline cartilage, with a slightly granular matrix. July 16th. No fibrillation visible with or without pressure. July 23rd. Pressure produced well-marked fibrillation which could be seen running up among cartilage-cells. July 24th and 27th. Fibrillation again produced by pressure.

Above I have given merely the outlines of these observations with the results obtained, proposing now to give some particulars concerning the fibrillation observed and the manner in which it was obtained. These remarks refer only to the experiments with 10 p. c. salt solution.

First, with regard to the method. The time required for the maceration appears to vary somewhat; the earliest fibrillation I observed was after four days maceration (Obs. 9), and in this instance it was, although imperfect, distinctly recognizable. It would, however, appear that from a week to a fortnight is the average time required for maceration. The macerated section may be placed in a drop of the salt solution on a slide. Moderately firm pressure is then applied for a moment [1] (*e. g.*

---

[1] Continuous pressure with a spring-clip has been tried several times, but was of no special advantage. Throughout the rest of this paper the term "pressure" is used to signify pressure applied for a few moments, unless the contrary be specified.

with the point of a needle) on the cover-glass over the specimen
and the same immediately examined with a high power (e. g.
Vérick. No. 8). In some portion, often a large part, of the section,
fibrillation will probably be seen. If not, the pressure may
have been insufficient or too great; in the latter case slightly
raising the cover-glass with a needle frequently brings the
fibrillation into view. Whilst under examination, perhaps a
minute after the pressure was applied, the section may swell
up, present the appearance of a wave passing over it, and the
fibrillation entirely disappear, but may again be produced by
pressure. After this has been repeated several times, pressure
no longer brings the fibrillation into view. If the pressure have
been sufficient to prevent the cartilage from recovering, by its
natural elasticity, its hyaline condition, the appearance of fibrillation, of course remains, and the specimen may be permanently
preserved by irrigating, first with distilled water, and then
with a nearly saturated solution of acetate of potash, and
finally sealing with balsam solution or dammar varnish. During
this process it is important not to move the cover-glass at all,
as by a slight shifting of it I have several times entirely lost
the fibrillation. In one observation only (Obs. 4) did I observe
distinct fibrillation without pressure being applied. In all the
hyaline cartilages examined, with one exception, I have obtained the fibrillation of the matrix. Its non-appearance in
the shoulder-girdle of the newt is curious; though I can
scarcely regard this point as settled, for I macerated and
examined only one specimen. As regards the appearance presented by the fibrillation, the latter appears either complete or
incomplete, both forms commmonly occurring in the same
specimen. When complete it consists of a very fine but distinct, more or less parallel and wavy, striation, which can be
traced to pass above or beneath each cartilage-cell, as the case
may be. In fig. 1, taken from the femoral condyles of the
kitten, the fibrillation appears nearly complete. Not uncommonly incomplete fibrillation presents the appearance seen in
figs. 2 and 3, where the striæ run principally from cell to cell
or run out for a short distance on one or on both sides of a cell.
The striæ can in this case also always be followed either under
or over each cell. It appears to me not improbable that this

appearance (see figs. 2 and 3) is produced in the following manner: Assuming the cartilage-cells to be more resistant than the matrix, or at any rate to have become so by the maceration —a fact which seems to me probable from their comportment in Obs. 7, where the matrix broke up under pressure and allowed the cells to escape intact into the surrounding fluid—then those portions of the matrix which are situate under and over a cell will necessarily, on pressure being applied to the section, be subject to greater compression than the matrix in other parts, and it is here, therefore, that the fibrillation first appears. In incomplete fibrillation the striæ are not uncommonly seen running at an angle to one another. As regards the direction of the fibrillation, I have very commonly found it in the cartilage of joints run more or less at right angles to the edge of the section (i.e. the joint-surface of the cartilage), and have here seen it continuous with the fibrillation normally existing there. In longitudinal sections of costal cartilage I found the fibrillation running principally parallel to the long axis of the cartilage. In the transverse sections of this cartilage, on the other hand, the fibrillation ran in many different directions.

Having therefore obtained these results with salt solution I naturally supposed that my failures in Obs. 1, 2, and 3 were due to some error in the mode in which they were conducted, and concluded that by varying the strength of the solution of permanganate I should probably be more successful. I therefore made the following observations—as will be seen, however, with only very partial success:—

Observation 11. June 26th.—Cartilaginous expansion of shoulder-girdle of freshly-killed newt, put into $\frac{1}{8}$ p. c. solution of permanganate of potash, the perichondrium having been partly removed. Fluid changed five or six times daily, and specimen examined every day till it broke up and became too opaque to be examined (July 4th). *No fibrillation visible.*

Observation 12. June 26th.—Ditto, put into $\frac{1}{4}$ p. c. solution of the permanganate. Fluid changed five or six times a day, and specimen examined daily till July 2nd, when it became too opaque. *No fibrillation visible.*

Observation 13. June 26th.—A similar specimen put into $\frac{1}{2}$ p. c. solution of permanganate of potash. Fluid changed five or six times a day. Specimen examined daily till too opaque (June 29th). *No fibrillation visible.*

Observation 14. June 28th.—Sections of condyles of femur of freshly-killed rabbit, placed in ½ p. c. solution of permanganate of potash. Fluid changed five or six times a day, and specimens examined daily till they broke up and became too opaque for examination (July 4th). *No fibrillation observed.*

Observation 15. June 28th.—Similar sections, put into ⅛ p. c. permanganate solution, and examined daily till July 7th. Fluid changed five or six times daily. *No fibrillation visible.*

Observation 16. June 28th.—Similar sections, placed in $\frac{1}{16}$ p. c. solution of permanganate. Fluid changed five or six times a day, and specimens examined daily till July 7th, on which day *traces of fibrillation* were observed in one section. None could be seen in other sections. I could not be certain of any fibrillation at any other examination.

Observation 17. June 30th.—Some sections were removed from the preceding fluid and placed in $\frac{1}{32}$ p. c. solution of the permanganate. Fluid changed five or six times a day, and specimens examined daily till July 7th. *No distinct fibrillation visible.*

Having made these observations with this negative result I made the following, using still weaker solutions of the macerating fluid. In these I did not change the fluid as above, but thought the deoxidation of the solution to any extent might be prevented by putting the sections into bottles containing a considerable quantity of the fluid (about 100 cc.) and closing the same tightly.

Observation 18. July 12th.—Transverse and longitudinal sections of costal cartilage of freshly-killed kitten, put into $\frac{1}{60}$ p. c. solution of permanganate. Fluid changed on July 15th. Sections examined daily till July 16th, and no fibrillation observed. Pressure made them break up irregularly. Examined July 20th and 22nd, and each time a *trace of fibrillation* was seen in a longitudinal section.

Observation 19. July 12th.—Sections of femoral condyles of freshly-killed kitten, put into $\frac{1}{75}$ p. c. solution of permanganate, and examined daily till July 16th. On July 15th, pressure being applied, *distinct fibrillation* was seen in two sections, but it almost immediately disappeared, and could not be produced again. On July 16th a similar appearance presented itself in one section, but could not be observed in any other. July 20th. No fibrillation to be seen. July 22nd. The sections broke up without any fibrillation appearing.

Observation 20. July 12th.—Similar sections placed in $\frac{1}{80}$ p. c. permanganate solution and examined daily till July 16th. Fluid changed on the evening of the 12th (as it was turning colour slightly), and again on July 15th. July 15th. Slight fibrillation seen in one section on pressure. It, however, almost immediately disappeared, and could not be reproduced. None seen in other sections. July 22nd. Slight trace of fibrillation seen in one section.

This completes my experiments with permanganate of potash. The results (with solutions of the above strengths) are seen to be very uncertain and inferior to those obtained with 10 p. c. solution of common salt, though there is no doubt that permanganate of potash *can* produce a fibrillation similar to that above described. The results when the fluid was frequently changed—a point insisted on by Tillmanns—were no better than when this trouble was not taken.

Baryta-water and lime-water, being also commonly in use for macerating fibrous tissue, might naturally be expected to be serviceable in fibrillating hyaline cartilage. I therefore made the following observations with them :—

Observation 21. June 26th.—Two cartilaginous expansions of shoulder-girdle of freshly-killed newt placed in baryta-water. The one cleared of perichondrium, the other not. Examined June 27th and 28th. Much crystalline deposit on the specimens. Ground substance perfectly clear. *No fibrillation visible.* June 29th. Specimens lost in the deposit. The surface of the fluid became covered with a white scum, which fell to the bottom and produced white flakes, among which it was impossible to find the specimens. In the two following experiments this was prevented by putting the fluid in a test-tube and inserting a plug of calico into it under the surface of the liquid. When this was taken out the scum was removed with it, and the sections could thus be kept tolerably free from crystals.

Observation 22. July 12th.—Longitudinal and transverse

sections of costal cartilage of kitten placed in baryta-water.
July 16th. Matrix clear. *No fibrillation with or without
pressure.* July 20th. Same appearance; the sections became
soft and elastic, and much pressure broke them up irregularly.
July 27th. Same appearance. Sections scattered with crystals.

Observation 23. July 12th.—Sections of femoral condyles of
freshly-killed kitten placed in baryta-water. Examined July
16th, 20th, 27th. *No fibrillation visible.*

Observation 24. August 3rd. Sections of femoral condyles
of freshly-killed kitten put into lime-water. Examined August
4th (after 30 hours' maceration). Without pressure no fibrillation visible. With firm pressure *well-marked fibrillation* seen,
both partial and complete. Aug. 5th. Firm pressure produced
the same effect. Aug. 6th, 7th, and 9th. Without pressure
some partial fibrillation seen. Aug. 9th. Firm pressure produced considerable fibrillation, as on Aug. 5th. Matrix granular.

Observation 25. Aug. 3rd.—Longitudinal section of costal
cartilage of freshly-killed kitten placed in lime-water. Aug. 4th.
No distinct fibrillation observed with or without pressure.
Aug. 5th. *Extensive fibrillation*, often very perfect (see fig. 4),
brought out by pressure, which required to be rather firm.
Aug. 6th. One section shewed, without pressure, fibrillation
running transversely, and principally from cell to cell. Slight
pressure caused this to disappear, and firm pressure brought a
quantity of very fine longitudinal fibrillation into view. Another
section shewed no fibrillation without pressure, but it appeared
on pressure.

Thus, baryta-water, whether followed by pressure or not,
proved ineffective in producing fibrillation of the matrix of
hyaline cartilage.

Lime-water, as has been shewn, produces fibrillation in the
matrix when combined with pressure; without pressure, it occasionally does the same, though less constantly, and in a less
complete manner. Formation of crystals in the lime-water
may be prevented to a great extent by putting it into small
tightly-corked bottles; it is, however, advisable to wash the
sections in distilled water, before examining them, in a drop of
that fluid. A specimen in which fibrillation has been produced
by pressure and is seen to remain, may be mounted perma-

nently in acetate of potash solution, in the manner above described. Lime-water has the advantage over 10 p. c. salt solution in producing the required effect more rapidly, and the fibrillation when once produced seems to have less tendency to disappear. I may remark that costal cartilage appears to require longer maceration both in lime-water and in 10 p. c. salt solution (compare Obs. 8 and 9, and Obs. 24 and 25) than the cartilage of the femoral condyles. As regards the salt, the question might naturally suggest itself, whether it is necessary that the solution should be so strong as 10 p. c. This question is plainly answered in the negative by the following observation :—

Observation 26. Aug. 3rd.—Sections of the femoral condyles of a freshly-killed kitten, placed in ½ p. c. solution of chloride of sodium. Aug. 4th. In a section placed under continuous pressure for some hours, imperfect fibrillation was seen, but by prolonging this pressure it disappeared. Aug. 9th. Matrix appeared granular, and a few cracks only were visible in it. On pressure *abundant fibrillation* appeared, which could be preserved as above described in acetate of potash solution. There is therefore no necessity that the salt solution should be so strong, for fibrillation appears to be equally easily produced by a solution containing only ½ p. c. of the salt.

This observation seems to me to explain the occurrence of physiological fibrillation of the matrix on the free surface of the hyaline cartilage of joints. For the cartilage is here bathed by synovial fluid (which being "a transudation from the blood-vessels" according to Tillmanns (*l. c.* p. 437) must consequently contain Na Cl.), and pressure is applied to it by the contact of the opposing joint-surfaces.

To sum up shortly: From the above observations I feel justified in drawing the conclusion that *fibrillation of the previously homogeneous matrix of normal hyaline cartilage can be artificially produced in many instances.* I have shewn this to be the case in the cartilage of the head of the tibia of frog, the condyles and head of femur and costal cartilage of kitten, and the condyles of femur of rabbit. In fact, in all I have examined, with the solitary exception of the shoulder-girdle of the newt. These appearances lead me to conclude, with Tillmanns, that, in these cases at least, *the hyaline matrix is made up of fine fibrils, cartilage-fibres, held together by an interfibrillar cement-substance, which can be dissolved by certain reagents.* As regards these reagents, my results differ materially from those of Tillmanns. This observer finds 10 p. c. salt solution far inferior to solution of permanganate of potash, and it is the latter that he recommends. *I*, on the contrary, *find the effects of the permanganate solution (in the strengths that I have employed) very slight and uncertain, whereas I can readily produce the required effect by solution of common salt (both ½ and 10 p. c.) or by lime-water, in either case followed by pressure.*

Should all three methods be found by further experience to give equally good results, the one acting most expeditiously and involving least trouble (which seems to me to be the lime-water method) will probably be the one made use of to demonstrate this important fact in normal histology.

In conclusion, I would offer my best thanks to Dr Klein, under whose directions the above observations were made.

ADDENDUM.—I have recently had the opportunity of ascertaining that also in newts' cartilage fibrillation of the ground substance may be produced by lime-water.

The following are the observations :—

Observation 27. Aug. 5th.—Cartilaginous expansion of shoulder-girdle of freshly-killed newt (partly cleared of perichondrium) placed in lime-water. Aug. 6th. No fibrillation of matrix visible with or without pressure. Aug. 7th. No fibrillation observed. Aug. 9th. *Well-marked fibrillation* of the previously homogeneous matrix obtained by firm pressure.

Observation 28. Aug. 7th.—Similar specimens placed in lime-water. Aug. 9th. *Extensive fibrillation* was seen without pressure in several specimens. The fibrillation of the matrix in these specimens is fine and wavy, and appears to run principally from cell to cell. It can, as in other cartilages, be traced to pass either under or over each cell. Some parts of these specimens show cracks, which apparently mark out the matrix more or less completely into the well-known cell-territories.

I can therefore now say that I have been able to obtain the fibrillation of the matrix of hyaline cartilage in *all* the instances in which I have searched for it.

Since the above was written, I have also succeeded in obtaining fibrillation of the matrix of hyaline cartilage by means of baryta-water. The failure in the above observations is owing to the specimens having been macerated too long before microscopical examination. The observations shewing this were as follows :—

Observation 29. Cartilaginous expansions of shoulder-girdle of freshly-killed newt placed in baryta-water, after being partly cleared of perichondrium. Examined after four hours' maceration, they shewed no fibrillation of the matrix, with or without the application of pressure. The period of maceration was therefore still too long.

Observation 30. Similar specimens macerated in baryta-water for about one hour. They shewed fibrillation of the matrix resembling that in Obs. 27 and 28. Pressure was not always required to bring this into view.

Observation 31. Sections of the femoral condyles of a freshly-killed rabbit placed in baryta-water, and examined after maceration for periods of $\frac{1}{4}$, $\frac{3}{4}$, 1, 1$\frac{1}{2}$, 2, and 3 hours respectively.

In *all* these specimens well-marked fibrillation was seen after pressure had been applied. Occasionally, however, it was observed without the application of pressure. After maceration for only $\frac{1}{4}$ hour, followed by the application of pressure, some fibrillation was also seen.

Baryta-water therefore offers *the most rapid method* of resolving the matrix of hyaline cartilage into its constituent fibres. The method of procedure for preparing and mounting is the same as with lime-water (vide page 122), and the fibrillation obtained does not appear to differ from that produced by the latter reagent.

EXPLANATION OF PLATE VIII.

Figure 1. Hyaline cartilage of femoral condyles of kitten, shewing almost complete fibrillation of the matrix. After 22 days maceration in 10 p. c. solution of common salt and subsequent pressure. Vérick. No. 8. Oc. L.

Figure 2. Hyaline cartilage of femoral condyles of rabbit, shewing partial fibrillation of matrix. Treated by maceration in 10 p. c. salt solution and subsequent pressure. Obj. 8. Oc. L.

Figure 3. Another portion of the same specimen as preceding, presenting rather a different appearance. Obj. 8. Oc. L.

Figure 4. Longitudinal section of costal cartilage of kitten macerated for 48 hours in lime-water and subjected to pressure, shewing complete fibrillation of the matrix. The cartilage-cells are seen altered by the lime-water. Obj. 8. Oc. L.

EDITORIAL NOTE.—Many years ago Dr Leidy, an American histologist, described the fibrillar structure of the matrix of hyaline cartilage. The late Prof. Goodsir, about twenty years ago, repeated Leidy's observations, and gave to the writer of this note a demonstration of the structure as seen in a thin section of the encrusting cartilage of the human femur, which had been macerating for some weeks in water. (W.T.)

ON THE STRUCTURE OF THE DIFFUSED, THE POLYCOTYLEDONARY AND THE ZONARY FORMS OF PLACENTA[1]. By PROF. TURNER.

IN all inquiries into the anatomy of the Placenta, two series of structures have to be investigated, the one belonging to the fœtus, the fœtal placenta, the other to the mother, the maternal placenta. The placenta is therefore a compound organ, and the complexity of its structure in any given mammal is in proportion to the degree in which, in the course of its development and growth, the originally separable fœtal and maternal portions have become interlaced with each other.

The fœtal placenta and membranes, continuous with and derived from the germinal layers of the embryo, consist of the membranous sacs, known as the umbilical vesicle, the allantois, the amnion, and the secondary or persistent chorion. The persistent chorion, with the villi growing from it, forms the fœtal placenta properly so called. It is the outer envelope of the fœtus and is the medium of connection with the maternal placenta. There are four structures, of which one is developed at the periphery of the ovum, whilst the remaining three ultimately reach the periphery, which from their position may enter into the formation of the persistent chorion : viz. the zona pellucida, the subzonal membrane, the allantois, and the umbilical vesicle. The structureless zona, with its simple structureless villi, which together form the primitive chorion, very early disappears, either by becoming incorporated with the subzonal membrane, so as no longer to be recognised as an independent membrane, or by becoming absorbed. The subzonal membrane, or serous envelope of the ovum of Von Baer, originally continuous with the amniotic folds, and through them with the epiblast layer of the blastoderm, persists, and forms the external or epithelial non-vascular layer of the persistent chorion, and of the villi which grow from it. The allantois

[1] This Memoir contains the substance of a course of three Lectures, on the Comparative Anatomy of the Placenta, delivered June 14th, 15th, 16th, 1875, in the Theatre of the Royal College of Surgeons of England.

grows and expands so as to come into intimate relation with the whole or the greater part of the inner surface of the subzonal membrane, and conveys to it, from the mesoblast layer of the embryo, the connective tissue and blood-vessels, which form the inner or vascular layer of the persistent chorion, and the vascular matrix of the villi. In most mammals the umbilical vesicle takes no part in the formation of the permanent chorion, but in the *Rodentia* it expands, reaches and remains in contact with a limited portion of the subzonal membrane, to which it conveys connective tissue and blood-vessels. The persistent chorion, therefore, is a compound membrane, produced by the union of the subzonal membrane, from which its epithelial layer is derived, with the allantois from which it derives its blood-vessels and connective tissue.

The maternal placenta is formed by the mucous membrane lining the uterus, in which important changes occur during pregnancy. This membrane in the non-gravid state is covered on its free surface by a ciliated columnar epithelium. Beneath the epithelium is the sub-epithelial connective tissue, which contains a very large proportion of fusiform, ovoid and spherical corpuscles. In this tissue the blood-vessels, lymph-vessels, and nerves of the mucosa ramify, and in it lie the utricular glands. The glands of the mucosa are branching tubes, which lie, more or less, perpendicularly to the plane of the surface of the membrane, and are separated from each other by the interglandular corpusculated connective tissue, the proportion of which between any two glands is often not more than equal to the transverse diameter of a gland, though at others its amount is considerably greater. The epithelial lining of the glands, as was first demonstrated by Nylander and Leydig in the pig[1], and as has subsequently been shown by Lott[2] in various other mammals, is columnar and ciliated.

When the fertilized ovum is received into the cavity of the uterus the mucosa undergoes important changes. It swells up, becomes thicker, softer, and more vascular. Its epithelial covering usually, though not always, loses its columnar form; its glands enlarge throughout their entire length: the inter-

[1] Müller's *Archiv*, 1852.
[2] Stricker's *Handbuch*, article *Uterus*.

glandular tissue increases largely and rapidly in quantity, by a multiplication not only of the cells of the surface-epithelium, but by a proliferation of the corpuscles of the sub-epithelial connective tissue, so that the glands are separated from each other by a much greater amount of interglandular tissue than in the non-gravid state; the blood-vessels not only increase in numbers but in size. At the same time the free surface of the mucosa is perforated by multitudes of small openings, easily to be seen with a pocket-lens. Those openings lead into depressions in the swollen mucous membrane, which are usually regarded as the dilated mouths of the tubular glands, but which, as I shall show in this memoir, are the mouths of crypt-like depressions situated in the interglandular part of the mucous membrane. These crypts are for the lodgement of the villi, which project from the outer surface of the persistent chorion. As the area of distribution of the villi on the chorion varies very considerably in extent in different forms of placenta, the distribution of these crypts in the uterine mucosa necessarily also varies; for the crypts and the villi are correlated with each other.

The arrangement and structure of the placenta in those animals in which it is said to be diffused will first engage our attention.

THE DIFFUSED PLACENTA.—In the diffused placenta the villi are distributed over almost the entire outer surface of the chorion, and the uterine crypts exist in a corresponding area of the mucous membrane. This form of placenta is found in the Pigs, the Solipeds, the *Cetacea*, in *Manis*, the *Camelidæ*, the *Tragulidæ*, the Tapir, the Hippopotamus, and, in all probability, the Rhinoceros.

I shall commence by describing the structure of the placenta as I have myself observed it in the common Pig, and shall in the first place speak of the fœtal placenta. The surface of the chorion of a pig, where the embryo was 1·3 inch long, was traversed by multitudes of feeble ridges, visible under low powers of the microscope, but no true villi could be seen. A distinct and compact capillary plexus was present both in the ridges and intermediate parts of the chorion. In the ridges the plexus was elongated, but it formed a polygonal network in the intermediate areas. The polar ends of the chorion were smooth and free from ridges for about three inches from each pole. A uniform layer of squamous epithelial cells, the nuclei in which were distinct, covered the face of the chorion. In an older specimen, where the fœtus was six inches long, the ridges on the chorion were more strongly elevated, but still requiring a microscope for their examination. The summit of each ridge was broken up into numerous short, simple villi, just as a mountain-ridge may be broken up into short peaks and summits. In injected preparations these ridges and villi were seen to be very vascular[1]. Scattered irregularly over the surface of the chorion were quantities of circular or almost circular slightly elevated spots, which varied in number in a given area. Sometimes as many as 30 were seen in a square inch, in other places not more than 20. The spots varied in diameter from $\frac{1}{10}$th to $\frac{1}{16}$th inch, occasionally one $\frac{1}{8}$th inch in diameter was seen. In most of the red injected specimens these spots were white and free from colour, as if non-vascular, but when the injection was pushed further they became red also, though not so completely as the rest of the chorion. Examined microscopically each spot was observed to have a minute central depression surrounded by villi, which were the terminal villi of a group of ridges, and it was now seen that the villous ridges were arranged on the chorion with especial reference to these spots; for each spot was a centre from which the ridges radiated outwards as the spokes do from the centre of a wheel. After proceeding some distance the ridges not unfrequently branched, and adjacent branches joined together so as to form a network. Hence the villous surface of the chorion may be regarded as mapped out into a number of areas, the centre of each area being a circular spot, from which the villous ridges radiate. Each end of the chorion, for nearly three inches from the pole, had a smooth non-villous surface, and though possessing considerable vascularity was not so vascular as the villous part of the chorion. Hence the chorion of the pig is not uniformly

---

[1] For valuable aid in injecting this and the other injected placentæ described in this Memoir, I have to express my thanks to my Museum Assistant, Mr A. B. Stirling.

Fig. 1.

Portion of injected Chorion of Pig, as seen under a low magnifying power, to show the minute spot *b*, enclosed by a vascular ring, from which villous ridges, *r, r, r*, radiate. Figures 1, 2 and 3 were drawn on wood from my preparations by my assistant Mr A. H. Young.

villous, but the villi, as was indeed known to von Baer, are distributed over the middle and not the polar regions of the chorion. Though the structure of the placenta in the pig is simple like that of the mare and other animals with a diffused placenta, yet, as regards the distribution of the villi, they are arranged as a broadly zonular band which does not reach to within several inches from the poles of the chorion.

On examining the maternal placenta in the gravid uterus of a pig, where the fœtus weighed only 12 grains, the free surface of the mucosa was seen to present an undulating appearance, owing to numerous shallow furrows and fossæ separated by intervening ridgelets. Opening on the surface of the mucosa, the openings being marked by shallow depressions distinct from the furrows above referred to, were the mouths of the utricular glands, and not unfrequently a plug of epithelium projected through the orifice. Each gland-orifice was surrounded by a smooth portion of the mucosa. In a more advanced specimen, where the fœtus was 6 inches long, the mucosa was thrown here and there into transverse folds. When examined with a pocket-lens fine ridges and furrows were seen, which were adapted to furrows and ridges on the surface of the

9—2

chorion. When more highly magnified the furrows were seen to be subdivided into shallow crypts into which the villi of the chorion fitted. Scattered over the surface of the mucosa were numerous smooth almost circular depressed spots, free from ridges and crypts, corresponding in size and numbers to the circular, radiating spots already described on the surface of the chorion; and, as von Baer[1] and Eschricht[2] have described, the star-like elevations of the chorion are adapted to these smooth spots on the mucosa when the two surfaces are in contact. In the minutely-injected uterus the ridges and the walls of the crypts were seen to contain a compact capillary plexus, whilst the smooth spots possessed a feeble vascularity, so that they appeared as distinct white spots, surrounded by highly vascular ridges, on the injected mucous surface. Beneath the superficial crypt-layer of the mucous membrane was a well-defined glandular layer, the glands in which were tubular and branched repeatedly, so that each gland-stem or duct had connected with it numerous

Fig. 2.

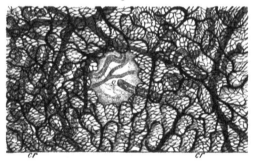

Surface-view of a portion of the injected Uterine Mucosa of a Pig to show a depressed circular spot in which the mouth of a gland g, opens. This spot is surrounded by numerous vascular crypts cr, cr. The branching glandsof the glandular layer and the larger vessels lie deeper than the crypts. Magnified same scale as Fig. 1.

branching tubes. The depressed circular spots had a special relation to the ducts of these glands, for opening either in the centre of each spot, or near its border, by an obliquely-directed

[1] *Ueber Entwickelungsgeschichte der Thiere*, p. 250, 1837.
[2] *De organis quæ respirationi et nutritioni fœtus mammalium inserviunt*. Hafniæ, 1837.

orifice, was a gland-stem, which could be seen running, some-what tortuously, from the deeper glandular layer of the mucosa to the spot. Eschricht had in 1837 described the mouths of the utricular glands in the pig as situated in small circular spots (*areolæ*), distinct from the surrounding crypts (*cellulæ*), which circular spots in injected specimens, owing to their feeble vascularity, appeared white, when compared with the highly vascular crypts. These observations of Eschricht seem to have been overlooked by most subsequent observers. In 1871[1] I described a similar arrangement in the uterus of a pig which I examined. In 1873 a description with a characteristic figure of a spot in an uninjected specimen was given by Ercolani[2]. It is clear therefore that in the pig the glands do not open into the crypts of the gravid mucosa, but into special depressions of the mucous membrane distinguished by a difference in form and in the degree of vascularity from the surrounding crypt-like surface. The crypts therefore are interglandular in position, and are produced by modifications in the interglandular part of the mucous membrane, and not by a dilatation of the gland-orifices themselves. The free surface of the mucosa was covered by an epithelium which also lined the crypts. The epithelial cells were columnar in form, finely attenuated at their deeper end; and not unfrequently I saw an appearance as if cilia projected from the broad end of the cell; but the animal had been too long dead to enable me to determine their presence by vibratile movements. Owing to the shallowness of the crypts and the very short villi of the chorion in the pig the uterine and chorionic surfaces separated from each other with great readiness. As the tubular glands did not open into the crypts their secretion did not come into immediate contact with the general villous surface of the chorion. The mucous membrane of the gravid uterus of the pig differs from that of the non-gravid animal in the following characters : in the presence of a layer of crypts, in the increased size and greater obliquity of the glands, and in the much greater vascularity of the mem-brane generally.

I have not had the opportunity of examining the fœtal

---

[1] *Trans. Roy. Soc. Edinb.* Vol. xxvi. p. 490.
[2] *Mem. dell' Accad. delle Scienze di Bologna*, Plate 2, Fig. 1.

placenta of the Mare in the early period of gestation, but two
specimens in advanced stages have come under my observation.
In both, the surface of the chorion presented a soft, velvety,
vascular appearance, due to its being almost uniformly covered
with vascular villi, even up to the poles. But at each pole,
where the chorion was in relation to the orifice of the Fallopian
tube, a spot, a fraction of an inch in diameter, smooth and bare
of villi, was present. Opposite the os uteri internum a well-
defined bare patch, which in one specimen was about an inch
in diameter, having in its centre a faint papillary elevation, was
present. Radiating outwards from this patch were five long
branching arms, also free from villi, and immediately beyond
these some irregularly-shaped bare patches were seen. The
surface of the uterine mucosa was also very vascular, except
opposite these bare patches on the chorion, and the folds and
depressions radiating from the os uteri corresponding to the
radiated patch on the chorion had a comparatively slight vascularity. In the chorion from another mare the bare spot oppo-
site the os uteri measured $2\frac{1}{2}$ inches long by from $\frac{1}{2}$ to $\frac{3}{4}$ inch
broad, and had distinct bare radii passing off from it. In each
specimen an irregular-shaped non-villous patch, about an inch
long, was found on a portion of the chorion not in relation to a
uterine orifice. Though in the mare the villi seem to the naked
eye to be closely set over the surface of the chorion, yet when
examined with low powers of the microscope they are seen to
be arranged in brush-like clusters or tufts, separated by narrow
non-villous intervals, an arrangement the importance of which
will appear when the corresponding surface of the mucosa is
described. The tufts are like minute fœtal cotyledons and the
villi in each tuft are filamentous in shape and contain a loop of
capillary blood-vessels.

I have not yet been able to procure the uterus of the
mare in an early period of gestation. But on inspecting with
a simple lens the surface of the uterine mucosa of a mare
which had reached an advanced stage of pregnancy[1], I found
it subdivided into multitudes of irregular polygonal areas, varying in diameter from $\frac{1}{12}$th to $\frac{1}{30}$th inch, by slender ridges,

[1] I am indebted to J. R. W. Dewar, Esq., V. S. of Midmar, Aberdeen, for
this specimen.

which anastomosed with each other so as to have a reticulated appearance. In injected preparations the ridges were seen to be less vascular than the areas which they enclosed, and consequently they were more readily recognised in injected than in non-injected portions of the mucosa. But in addition the ridges were smooth on the surface, whilst the enclosed areas possessed a delicate punctated appearance. When more highly magnified each . area was seen to be subdivided into multitudes of crypts, which passed more deeply into the mucosa than in the pig. The arteries and veins of the mucosa occupied the ridges, and broke up into small branches which ended in a

Fig. 3.

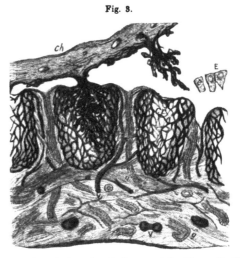

Vertical section through the injected placenta of the Mare. *Ch.* the chorion with its villi, partly *in situ*, and partly drawn out of the crypts, *cr. g, g,* the utricular glands. *V, V,* the blood-vessels of the mucosa imbedded in the connective tissue. *E* loose epithelial cells which formed the lining cells of a crypt.

compact capillary plexus situated in the walls of the crypts, the artery in each ridge giving off branches to the crypt-areas between which that ridge was situated.

The glandular layer of the mucosa contained numerous branched tubular glands. From each gland a stem or duct proceeded which ascended almost vertically into a ridge be-

tween the crypt-areas, and opened on the summit of the ridge
by a circular or oval aperture, which was usually situated at a
spot where convergent ridges became continuous with each
other. In the mare, therefore, as in the pig, the utricular
glands do not open into the crypts; but on definite surfaces of
mucous membrane between the crypts, so that the crypts are
interglandular in position, and produced by changes in the
interglandular part of the mucosa. The demonstration of the
want of any communication between the utricular glands and
the crypts in the mucosa of the gravid mare, and the consequent interglandular position of the crypts, was made a few
years ago by Prof. Ercolani[1] of Bologna.

The surface of the mucosa and of the wall of the crypts was
covered by an epithelium, which when examined *in situ* showed
a polygonal pattern, like the broad free ends of columnar epithelium-cells. When the cells were teased asunder, some were
seen to have the elongated form of ordinary columnar epithelium; others were so swollen out that their length but little
exceeded their breadth; whilst others were irregular in shape.
The protoplasm was distinctly granular, more especially in the
irregularly-shaped cells, which resembled in appearance the
cells of the serotina as seen in the higher mammals. Numerous cells exhibiting transitional forms between ordinary columnar epithelium and serotina-cells were seen, so that the large
granular cells of the serotina are to be regarded as a modified
epithelium.

The filiform villi of the tufts of the chorion occupied the
crypts in the mucosa, which represented therefore the maternal
cotyledons of a ruminant animal, and the size of the crypt-areas
correspond to the size of the tufts. The ridges between the
areas filled up the intervals between the tufts. The secretion
of the uterine glands was not poured into the crypts so as to
come into immediate relation with the villi, but opposite the
inter-villous surface of the chorion. The villi of the chorion
were so closely fitted into the uterine crypts, that, in the specimen of the gravid uterus near the full time, it required a
little force to draw the villi out of the crypts. As in the pig,

[1] See the French translation of his Memoir, and the figures in Plates
3 and 5. Algiers, 1869.

the gravid mucous membrane differed from the non-gravid in the presence of a layer of crypts, in the increased size and greater obliquity of the glands, and in the greater vascularity of the membrane generally.

The diffused distribution of the villi over the surface of the chorion in the *Cetacea*, and the velvety appearance due to this arrangement, have been recognised by several anatomists, from observations made more especially on the common porpoise. In 1869 I examined several square feet of the chorion of *Balænoptera Sibbaldii*[1], and showed that the whalebone whales agreed with the toothed whales in the diffused distribution of the villi. But in the *Cetacea*, as in the pig and mare, patches of chorion bare of villi are also present. Some years ago Prof. Rolleston pointed out[2] that in a specimen (species unknown), which he examined, a bare spot was situated at each pole. In a gravid *Orca gladiator*, which I examined in 1871[3], I found not only the polar bald spots, but a stellate non-villous surface, nearly the size of a crown piece, and with several bare lines radiating from it opposite the os uteri; the arrangement corresponding very closely with that just described

Fig. 4.

Stellate non-villous portion of the Chorion of *Orca*, opposite the os uteri.
About half the size of nature.

in the chorion of the mare in the same locality. These large radiating bare spots in the mare and cetacean are exaggerated

[1] *Trans. Roy. Soc. Edinburgh*, 1870.
[2] *Trans. Zool. Soc.* p. 307, 1866.
[3] *Trans. Roy. Soc. Edinb.* 1871.

representations of the small radiating spots, so abundantly distributed over the chorion of the pig.

When the chorion of *Orca* was examined microscopically the villi were seen to vary in number and in arrangement in different parts. Sometimes they were set in rows and formed parallel ridges: at others they were collected into tufts, irregular in form and size, which sometimes consisted of two, three or four villi, but frequently of a larger number. Solitary villi were also met with in the irregular intervals between the tufts and ridges; and it was not uncommon, as Eschricht had observed in *Phocœna*[1], to see short stunted simple villi projecting from the general plane of the chorion. The tufts not unfrequently swelled out into a branching crown, which, to adopt Eschricht's description of the shape of the villi in *Phocœna*, formed a miniature representation of the head of a cauliflower. The secondary villi of a tuft, as well as the simple villi, were club-shaped.

A layer of spherical or ovoid corpuscles was situated immediately within the free surface of each villus, and not unfrequently the periphery of the villus was slightly elevated immediately above the individual corpuscles, so that the outline of the villus had a gently undulating appearance. These corpuscles I have named from their position the sub-epithelial corpuscles of the villus. In their form and appearance they are not unlike the white corpuscles of the blood, and it is possible that they may have migrated out of the blood-vessels into the connective tissue of the villus. The villi contained a distinct capillary network. Not only in the *Orca*, but in the pig and mare, the capillaries of the chorion were not limited to the villi, but an extra-villous capillary network, which freely anastomosed with the intra-villous capillaries, was situated beneath the general plane of the chorion. The blood in its passage from the terminal twigs of the umbilical artery to the umbilical vein had to flow not only through the capillaries within the villi, but through the extra-villous network, from which the rootlets of the vein arose.

Eschricht in 1837[2] and Stannius in 1848[3] described the

[1] *De Organis*, &c., p. 6.   [2] *De Organis*, &c.
[3] Müller's *Archiv*, p. 402, 1848.

presence of numerous little recesses on the free surface of the uterine mucosa of the gravid porpoise, in which the villi of the chorion were lodged. In 1871[1] I had the opportunity of dissecting the gravid uterus of *Orca gladiator*, and of determining much more minutely, than had previously been done, the structure of the gravid uterine mucosa in this order of mammals. The free surface of the mucosa had a delicate reticulated appearance, due to an anastomosing arrangement of slender bands of the membrane. Sometimes a subdivision of the surface into irregular polygonal areas was seen, at others its surface was traversed by an elongated ridge and furrow

Fig. 5.

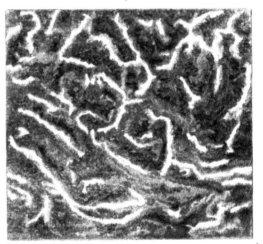

Surface-view, under a low power of the microscope, of a portion of the uninjected uterine mucous membrane of *Orca gladiator*.

arrangement. The polygonal areas and the furrows were subdivided by more delicate bands into small crypt-like compartments, and the intermediate ridges and bands of the mucous membrane were not unfrequently covered by similar crypts.

Beneath the crypt-layer was the glandular layer of the mucous membrane. The glands, as in the pig and mare, were

[1] *Trans. Roy. Soc. Edinburgh*, p. 467, 1871.

tortuous tubes and branched repeatedly. The mode of termination of the glands on the surface of the mucosa was more difficult to determine than in the pig and mare, but after repeated examinations, both of vertical sections through the membrane, as well as of surface-views, I came to the conclusion that the tubular glands opened into the bottom of some of the crypts[1]. But as the crypts were very much more numerous than the ducts of the glands, and as those crypts into which tubular glands opened were deeper and more funnel-shaped than those into which glands did not open, I was led to divide the crypts into two groups, non-glandular cup-shaped crypts, and glandular funnel-shaped crypts. The relation of the glands

Fig. 6.

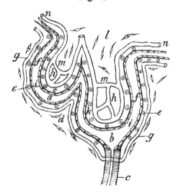

Diagrammatic section through the placenta of *Orca gladiator*. *a.* cup-shaped crypt. *b.* funnel-shaped crypt. *c.* tubular gland-stem with its epithelial lining. *d.* fusiform and *e.* spheroidal sub-epithelial connective-tissue corpuscles. *f. f.* epithelial lining of crypts. *g. g.* maternal capillaries in the walls of the crypts. *h. h.* chorionic villi occupying the crypts. *i. i.* epithelial covering of the villi. *k.* spheroidal and *l.* fusiform corpuscles of the villi. *m. m.* intra-villous fœtal capillaries continuous with *n. n.* extra-villous capillaries. The space represented between the fœtal epithelium *i. i.* and the maternal epithelium *f. f.* is to give distinctness to the diagram, for in the placenta itself the two epithelial surfaces are in close apposition.

[1] I am not prepared to say that on a surface so extensive as the mucous membrane of the gravid uterus in *Orca* there may not be here and there a spot free from crypts at which a tubular gland may open, but should this be so, it would have to be regarded as the exception and not the rule. For in a portion of mucosa about half an inch square I found several gland-openings at the bottom of a corresponding number of funnel-shaped crypts.

to the funnel-shaped crypts seemed to justify the inference that these pouch-like depressions in the mucosa were (as was stated, by Dr Sharpey, to be the case in the pits or "cells" on the surface of the uterine mucosa in the gravid bitch) the mouths of the glands somewhat enlarged and widened. But however this might be the case with the funnel-shaped crypts it obviously could not be so with the cup-shaped crypts, which were interglandular in position, and, as in the pig and mare, could only have been produced by changes in the interglandular part of the mucous membrane. I guarded myself however against too absolute an acceptance of the view that the funnel-shaped crypts were merely the widened mouths of the glands by stating (p. 501) that they, like the cup-shaped crypts, may be formed by a folding of the greatly hypertrophied mucous membrane: only in the one case the hypertrophy and folding take place between the glands, in the other at the mouth of the gland itself. The difference between the mode of opening of the glands in *Orca* on the one hand, and in the pig and mare on the other, seemed to be this: that in *Orca* the free surface of the mucosa was much more uniformly crypt-like than in the pig and mare, so that there were no intermediate surfaces destitute of crypts on which the glands could open, whilst in the pig and mare the crypts were collected into definite areas, with distinct smooth surfaces intermediate to them. This more compact arrangement of the crypts in *Orca* corresponds with the more crowded condition of the villi on the surface of the chorion. That the whole series of crypts however in the cetacean uterus, as in the pig and mare, are to be regarded as interglandular formations, is supported by the observations of Eschricht on the mucosa of the gravid porpoise. For he states (p. 35) that in this animal the glands open on the surface of the mucous membrane, not into the "cells" in which the villi are lodged, but into separate shallow areolæ; that in the porpoise, as in the pig, these *areolæ* are much less vascular than the surrounding crypts; and that for so great a multitude of gland-ramifications there are not more openings on the surface of the mucous coat than in the pig.

The walls of the crypts and the interglandular connective tissue in *Orca* contained numerous nucleated corpuscles. In

the interglandular tissue they were mostly fusiform, but in
the walls of the crypts a distinct layer of globular lymphoid-
looking corpuscles was seen close to the free surface, which
was not unfrequently elevated, in a sinuous outline, immediately
superficial to these corpuscles, which may, from their position,
be called the sub-epithelial corpuscles of the crypts. It is not
improbable that these corpuscles may have migrated through
the walls of the adjacent capillaries. The walls of the crypts
were very vascular and contained a compact capillary network.
Owing to the closer arrangement of the crypts, the surface of
the mucosa generally presented a more uniform vascularity
than in either the pig or mare. In all these animals indeed the great vascularity of the crypts was one of the most
striking features in the structure. The capillaries in the walls
of all the crypts belonging to the same group formed a continuous network, and in the *Orca*, owing to the more uniform
crypt-formation, the capillaries of one group freely anastomosed
with those of adjacent groups. The capillaries in the crypt-
walls in each animal formed a series of anastomosing festoons,
and usually a distinct capillary ring surrounded the mouth of
each crypt. Moreover there was a great difference between
the vascularity of the crypts and that of the deep layer of the
mucosa in which the glands were situated, for the vascularity
of the latter was not more than may be seen in connection with
the tubular glands of the intestine. The crypts in *Orca* were
lined by a well-defined layer of epithelium-cells, which closely
followed the various irregularities of the mucous surface. The
free ends of the cells were polygonal, often hexagons, or penta-
gons, though sometimes elongated into a pyriform shape. In
my original memoir on the placentation in *Orca*, I stated that
they had the appearance of a pavement-epithelium, though
they were not larger than the broad free ends of the cylindrical
epithelium lining the glands. I have since re-examined this
layer of cells, and have now come to the conclusion that they
cannot be associated with either the pavement-epithelium
(*i.e.* if we employ the term as equivalent to squamous), or with
the cylindrical epithelium. The cells are neither sufficiently
elongated for the one, nor flattened for the other, but have
an intermediate or transitional form.

The villi of the chorion fitted into the crypts, but were easily extracted from them. Only those villi which occupied the funnel-shaped crypts were brought into immediate contact with the secretion of the tubular glands, but in all the crypts the villi were in contact with the epithelial lining.

In *Manis*, as was pointed out by Dr Sharpey[1], the chorion was studded with villous ridges, but a bare band free from villi ran longitudinally along the concavity of the chorion, and there was a corresponding bald space on the surface of the uterine mucous membrane. The ridges of the chorion started from the margins of the bald stripe and ran round the ovum. The mucous membrane of the uterus possessed a finely reticulated appearance on the surface, and was punctated with the orifices of numerous shallow crypts in which the villi had been lodged. Branched cylindriform glands were very abundant in the deeper layer of the mucosa, but their mode of opening on the surface could not be satisfactorily ascertained, owing to the condition of the specimen.

The *Camelidæ*, unlike the ordinary Ruminant mammals, possess, as has for many years been recognised, a diffused placenta. I have recently, through the courtesy of Prof. Flower, had the opportunity of examining a considerable part of the chorion of a Dromedary, preserved in the Museum of the Royal College of Surgeons, London. The free surface of the chorion was thickly studded with short villi, but at one spot a bare patch about $1\frac{1}{2}$ inch long was seen; the relation of which to the wall of the uterus could not be ascertained. The villi were not arranged in tufts but arose singly from the chorion, having a somewhat constricted base, and expanding at the free end in a club-shaped manner. The villi varied in length from about the $\frac{1}{16}$th to $\frac{1}{12}$th inch. The larger proportion were unbranched, but some of the longer villi were divided into two or three short offshoots at the free end. The vessels of the chorion had been injected with size and vermilion, and a beautiful intra-villous network of capillaries was displayed. An extra-villous plexus of capillaries, not unlike that which I have described in

---

[1] Quoted in Huxley's *Elements of Comparative Anatomy*, p. 112, 1864; and with additional details in my Memoir on the Placentation of the Sloths in *Trans. Roy. Soc. Edinb.* 1873.

the mare and in *Orca*, was also seen. Although the gravid uterus itself was not in the museum for examination, yet there can be no doubt that its free surface must have been thickly studded with crypts for the reception of the villi.

In the *Tragulidæ* also, as has been described and figured by M. A. Milne-Edwards in *Tragulus Stanleyanus*[1], the villi are not collected into cotyledons, but are uniformly diffused over the surface of the chorion.

In the Tapir, as was shown by Sir Everard Home[2], the chorion is villous as in the mare. In *Tapirus Malayanus*, I am told by my friend Dr John Anderson, there is "a long bare area as in *Manis, Platanista* and *Orcella*, but proportionally of much greater size. The uterus also has the general characters of that organ in the gravid *Platanista*[3]." In the Hippopotamus M. H. Milne-Edwards has described[4] large villi disseminated over the whole surface of the chorion, except at the poles, where the membrane is smooth. Mr A. H. Garrod has also seen[5] the uniformly villous covering of the chorion in the placenta of this animal.

THE POLYCOTYLEDONARY PLACENTA.—This form of Placenta is characteristic of animals belonging to the order *Ruminantia*. It consists of a number of thick tuft-like masses of villi—the fœtal cotyledons, which are lodged in crypts situated in an equal number of thick, spongy elevations of the uterine mucous membrane—the maternal cotyledons. The fœtal cotyledons are separated from each other by considerable areas of smooth chorion, and the maternal cotyledons have equally large areas of smooth mucous membrane between them. Each cotyledon is complete when the fœtal are lodged within the maternal cotyledons, and each when complete forms a miniature placenta.

The first indication of the formation of a maternal cotyledon,

---

[1] *Ann. des Sciences Naturelles*, p. 101, Vol. II. 1864.
[2] *Lectures on Comp. Anatomy*, v. p. 328, and Plate 27.
[3] The characters of the gravid uterus in the rare Cetacean genera, *Platanista* and *Orcella*, have been specially studied by Dr Anderson, and will be described by him in a memoir to be shortly published.
[4] *Leçons sur la Physiologie*, IX. p. 562, 1870.
[5] *Proc. Zool. Soc.* Nov. 19, 1872.

as has been pointed out by Ercolani[1], is an elevation of the mucosa, which presents an irregular undulating surface. As the development proceeds this irregularity increases, until well-defined depressions or crypts are formed. The walls of the crypts continue to grow in length, and the crypts are not only deepened, but smaller compartments branch off from them. In the course of time they assume the appearance of deep pits, subdivided into numerous crypt-like compartments, into which the villi of the chorion closely fit. The shape of the fully formed cotyledons and the disposition of the pits vary in different Ruminants. I shall especially describe what I have seen in the Sheep and Cow.

In the Sheep the maternal cotyledons projected as cup-shaped mounds from the uterine wall. They were covered on the outer convex surface by the uterine mucosa, which was prolonged as far as the free inverted edge of the cup. The inner surface of the cotyledon was composed of a soft, spongy material, containing numerous pits, which extended almost

Fig. 7.

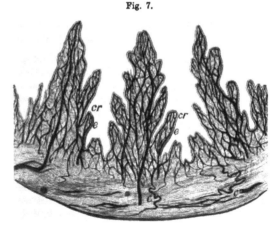

Semi-diagrammatic vertical section through a portion of a maternal cotyledon of a Sheep. *cr. cr.* pit-like crypts, with *e. e.* the epithelial lining. *v. v.* the veins, and *c. c.* the curling arteries of the sub-epithelial connective tissue.

[1] *Mem. dell' Accad. delle Scienze di Bologna*, 1870, Plate I. and 1878, Plate II.

vertically, and divided as they passed deeper into its substance into smaller crypt-like compartments, which radiated towards the outer wall of the cotyledon, without diverging much from each other. The pits were lined by well-marked cells, most of which were irregular in shape, polygonal, ovoid, or even somewhat caudate, and of considerable size, though some appeared like modified columnar cells. They consisted of granular protoplasm, in which one, two, or sometimes three, well-defined ovoid or elliptical nuclei were imbedded, but without a cell-wall. Not unfrequently the outline of the individual cells was very indistinct, and they seemed as if composed of a layer of protoplasm studded with nuclei.

The cells rested on a highly vascular sub-epithelial connective tissue, which formed the proper wall of the pits. The mucous membrane investing the cotyledon was continuous at the mouth of the cup with the walls of the pits in the spongy tissue, so that the cells lining the pits were in the same morphological plane as the epithelium covering the mucosa. The cotyledons were highly vascular. Some of the arteries in the sub-cotyledonary connective tissue were corkscrew-like; and in the deeper part of the cotyledon itself I have seen tortuous vessels. The greater number of the vessels within the cotyledon passed, however, vertically towards the surface, lying in the connective-tissue walls of the pits; branching repeatedly, as a rule, in a dichotomous manner, prior to forming a compact maternal capillary plexus,—not dilating into maternal blood-sinuses.

The mucous membrane of the uterus between the cotyledons contained numerous tortuous, branched, tubular glands. Some of these extended almost vertically to the surface, and could be seen in almost their entire length in vertical sections—others ran more obliquely, and, owing to their tortuosity, were repeatedly divided in vertical sections. The mouths of the glands could readily be seen with a pocket-lens opening on the surface, the orifice being partially surrounded by a minute elevation of the mucosa. In the mucosa around the base of the cotyledons a ring-like series of gland-openings was seen. In the mucosa covering the cotyledons, glands were also present, but their orifices were much stretched, as if by the pressure due to the great growth of the subjacent spongy tissue of the cotyledon. The sub-epithelial connective tissue, in which the glands lay, was not by any means so vascular as that which formed the walls of the pits within the cotyledons. In some sections through the cotyledons and adjacent mucosa no glands were to be seen in the connective tissue intervening between the cotyledon and muscular wall, but they were collected in considerable numbers around the cotyledon, as if pushed outwards by its rapid growth. In other sections, however, tubular glands were seen in the sub-cotyledonary connective tissue; but they seemed to be the deep ends of branching glands, the stems of which had inclined obliquely, so as to open on the surface of the mucous membrane covering the cotyledon. None of these subjacent glands, or those situated on the surface of the cotyledon, were seen to open into, or in any way to communicate with, the pits within the cotyledon itself.

The fœtal cotyledons consisted of numerous villi, which collectively formed a ball-like mass, occupying the concavity of the maternal cotyledon. Each villus consisted of a main stem, which gave off a tuft or cluster of spatulate branches. The villi entered the maternal pits and branched along with them, so that every compartment was occupied by a branch of the villus; but there was necessarily no great divergence of these branches from the main stem. At their deeper end these spatulate branches gave off slender terminal offshoots. The villi were formed of gelatinous connective tissue, in which very distinct fusiform and stellate corpuscles were arranged in an anastomosing network. At the periphery of the villus was a layer of flattened cells, with small but distinct nuclei arranged so as to form an epithelial-like investment. The umbilical vessels ramified within the villi and formed networks of capillaries. The villi were in close contact with the epithelial cells lining the maternal pits. Owing to the inversion of the free edge of the maternal cotyledon and the radiated arrangement of the pits, with their contained villi, it was impossible to disengage the maternal and fœtal cotyledons from each other without drawing away with the fœtal villi portions of the maternal cotyledon. I invariably found that, in drawing the fœtal villi out of their compartments, flakes of epithelial cells accompanied them, which showed how readily this element of the maternal issue is shed. During parturition, however, when the parts are relaxed, the disengagement of the two structures can necessarily be more easily accomplished.

In the Cow the maternal cotyledons differed in form from those in the sheep. They were fungiform or umbrella-shaped, and were connected to the uterine wall by a broad neck, around which the uterine mucosa was prolonged as far as the border of the umbrella. The whole convex surface of the cotyledon was riddled with pits, which passed vertically into its spongy substance, and divided into smaller compartments in the deeper part of the cotyledon. Projecting from the wall of each pit were delicate bands, visible to the naked eye, arranged as a rule in a vertical direction, and in the intervals between these bands the wall was perforated by numerous orifices, easily seen with a pocket-lens, which were the mouths of depressions or crypts in the wall of the pit, some lying almost at right angles, others obliquely to the wall of the pit itself. The pits, with their numerous crypts, were lined by cells, similar in character to those of the sheep. But I should state that a larger proportion of these cells had preserved the columnar form of the epithelium of the non-gravid uterine mucosa. They rested on a highly vascular connective tissue, in which the maternal capillaries formed a compact network.

The surface of the uterine mucosa between the cotyledons presented the mouths of the tubular, branched, utricular glands, which extended more obliquely to the surface than in the sheep, so that in vertical sections through the membrane they were frequently cut through and divided; segments of each gland were, as a rule seen, though sometimes the stem of a gland mounted to the surface to open by an obliquely-directed orifice. Glands were also present in the connective tissue forming the neck of the cotyledon, but none were seen to communicate with the pits.

The fœtal cotyledons were situated on the umbrella-shaped maternal cotyledons, and their numerous villi occupied the pits. The stems of the villi were comparatively large, and studded with multitudes of minute tufts, which, arising obliquely or almost at right angles to the main stem, entered and occupied the crypts. The minute villi forming these tufts were so slender and filiform that each terminal offshot contained only a single capillary loop. The villi were in contact with the epithelium-cells, and in drawing them out of the pits, more especially in drawing the tufts out of the crypts, multitudes of cells of the lining epithelium came away with them. From the differences in shape of the maternal cotyledon in the cow and in the sheep, there is not the same difficulty in unlocking the fœtal from the maternal placenta in the former animal as in the latter.

In the Red-deer the general form of the cotyledons is not unlike what I have described in the Cow. The form and to some extent the structure of the cotyledons in the Roe-deer have been described by Bischoff[1]. Prof. Owen has given some beautiful figures of the fœtal cotyledons of the Giraffe[2], and has pointed out that some of large size were arranged in longitudinal rows, whilst numerous smaller ones, of irregular form and unequal dimensions, projected from the outer surface of the chorion in the inter-space of the normal larger cotyledons. I have examined microscopically the villi of this specimen as preserved in the Museum of the Royal College of Surgeons, London, and have found them to possess some variations in form. Some were filiform and almost cylindrical, others broader and more flattened. Some were unbranched except at the free end, where they gave origin to two or three short bud-like off-shoots: others were much more deeply cleft, but none could be said to have an arborescent form. The cotyledons were very vascular, and each villus contained a compact capillary network. Although the uterine mucosa of the gravid Giraffe has not apparently been examined, there can be no doubt that maternal cotyledons containing pits for the reception of the fœtal villi must exist, and Owen has shown that, even in the non-gravid state, elevations are to be seen in the uterine wall, which correspond in position to the future cotyledons.

It is necessary that we should now consider whether the pits and crypts in the maternal cotyledons of the ruminant placenta, in which the villi of the chorion are lodged, are merely the greatly enlarged mouths of the utricular glands of the

---

[1] *Entwicklungsgeschichte des Rehes*, Giessen, 1854.
[2] *Trans. Zool. Soc.* Vol. III.

mucosa, or are structures specially formed during pregnancy, by great hypertrophy and folding of the inter-glandular part of the mucous membrane, as in the diffused forms of placenta. Prof. Spiegelberg[1] was of opinion, from some observations which he had made, that they were only remarkable dilatations of the utricular glands, and Bischoff was at one time disposed to regard them as the largely developed glands of the uterus. Subsequently however Bischoff figured in the cotyledons of the Roedeer[2], the utricular glands ascending to open on the surface of the uterus, not in the cotyledons, but around its circumference. Eschricht however had previously stated that in the cow the glands open, not into the cotyledons, but on the surface of the uterus between them. Ercolani also has·figured and described[3] both in the sheep and cow the glands as situated around the cotyledons, and not communicating with· the cavities within them. In the description which I have given of the cotyledons in the sheep and cow, I have stated that I was unable to detect any communications between the glands and crypts: the glands indeed appeared as if they had been pressed to the periphery of the cotyledons by the great development of its spongy substance. Hence it would appear that in· the· polycotyledonary, as in the diffused placenta, the crypts in which the fœtal villi are lodged are not produced by an enlargement and dilatation of the tubular glands of the mucosa; but are new structures formed, during pregnancy, by a great hypertrophy and folding of the interglandular part of the mucous membrane.

THE ZONARY PLACENTA.—The Zonary or Annular placenta is found in its most characteristic form in the *Carnivora* and *Pinnepedia*, though it is present also in *Hyrax* and in the Elephant.

The gravid uterus of the dog, cat, and other pluriparous *Carnivora* possesses a moniliform appearance. Each dilatation is a compartment of the uterus containing an embryo, with its membranes, and between adjacent compartments the uterine

---

[1] *Henle and Pfeuffer's Zeitschrift*, XXI. quoted by Ercolani, p. 17, of the French translation of his Memoir.
[2] *Entwieklungsgeschichte des Rehes*, Plate VIII. 1854.
[3] *Memoir of* 1878, Plate 2.

cavity forms a narrow tube. If one of these compartments be opened, in a well-advanced stage of development of the embryo, the chorion will be seen to be smooth and bare of villi, except in about its middle third, where the villi are arranged as a zonular band around the transverse circumference of the ovum. The uterine mucosa possesses a similar zone closely blended with the zonular band of the chorion. The mucous membrane on each side of the zone is smooth and vascular: it lies in apposition with the smooth part of the chorion, but has no attachment to it. Where the zonary and smooth parts of the mucosa are continuous with each other a narrow strip of mucous membrane is reflected on the margin of the zonular band of the chorion, and forms a rudimentary decidua reflexa. In the true carnivora the decidua reflexa is so very narrow that it has often been overlooked; but in the grey seal, where the placenta is large, the reflexa is from $\frac{3}{4}$ to $1\frac{1}{4}$ inch broad. As the zonary placenta is much more complex in structure than either the diffused or polycotyledonary forms, it is necessary, in order to understand the arrangements, that it should be examined in different stages of development. I shall first describe what I have seen in the domestic cat.

In the earliest impregnated Cat's uterus, which I have examined, the compartments were ovoid, and the long diameter of each, measured along the arc, did not exceed $\frac{8}{10}$th inch. When a compartment was opened the chorion readily separated from the mucous lining. At each pole of the compartment an area of mucosa $\frac{1}{10}$th inch in its long diameter was smooth; but the rest of the membrane was hypertrophied, spongy, swollen, and elevated above the smooth polar portions, and formed the placental area. The placental area possessed on its surface an extremely delicate reticulation, many of the strands of which had a sinuous direction. It was thickly studded with minute orifices barely visible to the naked eye, but easily seen with a pocket-lens. These orifices were the mouths of the pits or crypts in which the villi of the chorion had been lodged. A few of these openings were two or three times larger than the rest. The appearance which I saw in the cat is evidently similar to that figured by Dr Sharpey in the bitch (Fig. 211)[1], and

[1] Baly's Translation of *Müller's Physiology*, note p. 1576.

by Bischoff in the same animal (Fig. 48, A)[1], though, as will be seen further on, I interpret its mode of production in a different manner from those anatomists. The crypts passed vertically into the spongy substance, and when vertical sections were made through it, they were seen to be separated from each other by trabeculæ; the chief beams of which lay vertically, and when they reached the free surface formed the strands of the reticulum already described. The vertical trabeculæ were connected together by others directed obliquely or in a sinuous manner, and these lateral connections were especially seen about midway in their length. Hence not only on the surface, but when horizontal sections were made through the placental area, a reticulated arrangement was seen, and the crypts constituted the interstices of the reticulum. As these trabeculæ were formed of the thickened mucous membrane of the placental area, they were necessarily composed of the somewhat modified·tissues of that membrane. On the surface was a definite layer of epithelium, the cells of which were short columns, with distinct, circular or ovoid, brightly-refracting nuclei. These cells rested on a delicate sub-epithelial connective tissue in which the maternal capillaries ramified.

The trabeculæ and the sub-mucous connective tissue were carefully examined with the object of ascertaining their relations to the tubular glands. In vertical sections the glands were distinctly seen, transversely or obliquely divided, lying in a definite layer of connective tissue situated deeper than the crypts. Sometimes the divided glands were separated by comparatively broad bands of connective tissue from the crypts and trabecular structure, but in other places they were immediately subjacent. They were lined by a well-defined columnar epithelial layer. I looked for the stems of the glands to see if I could ascertain whether they opened into the crypts or passed along the trabeculæ to open on the free surface of the mucosa, but did not succeed in tracing them to their orifices.

As it was important however to ascertain if the crypts equalled in number in a given area the glands of the mucosa in the same area, or if the crypts much exceeded in number the glands, I submitted different parts of the mucosa of the

---

[1] Entwicklungsgeschichte des Hunde-Eies, 1845.

gravid uterus of this cat to microscopic examination, and compared the appearances seen with those presented by the mucosa of the non-impregnated uterus. In the non-gravid cat the stems of the glands were almost perpendicular to the free surface of the mucosa. They were so tortuous at their deeper ends as to be repeatedly cut across in a vertical section through the membrane. The interglandular connective tissue, containing numerous corpuscles, formed well-marked bands between the glands. Vertical sections made through the mucosa lining the constrictions between the compartments of the uterus of this gravid cat showed the tubular glands to be on the average $\frac{1}{4}$th wider than in the non-gravid condition, the interglandular connective tissue was much smaller in quantity, so that the glands were more closely crowded together; but in the placental area of the mucosa of the same cat the interglandular tissue was greatly increased in quantity, so that the glands were much further apart. The glands themselves were, as in the non-placental area, dilated, but the number of glands seen in the sections did not nearly equal the number of crypts.

In a cat's ovum, which had reached a somewhat more advanced stage of development, where the long diameter of the uterine compartment, measured along the arc, was $1\frac{1}{2}$ inch, I found that the villi of the chorion readily disengaged from the uterine crypts. By far the larger part of the chorion was still villous, not more than $\frac{3}{10}$ths inch at each pole being smooth. The line of demarcation between the placental and non-placental polar areas of the mucosa was very distinct. The placental area, or the hypertrophied and spongy mucosa, possessed a reticulated appearance, the principal strands of which were sinuous, and gave off numerous collateral branching offshoots, which joined adjacent branches to form the walls of the numerous pits or crypts which opened on the surface. The strands and branches were larger and the pits and crypts were more dilated than in the younger ovum already described, and on looking down the larger pits, their subdivision into smaller crypts could be seen. The crypts were lined by an epithelium, numbers of the cells of which possessed a columnar form, though others were swollen and otherwise altered in shape, so as to be irregularly polygonal. The cell-protoplasm was

granular and the nucleus was distinct. The sub-epithelial
connective tissue was vascular. When vertical sections were
made through the placental area the more dilated size of the
crypts and pits than in the younger specimen was distinctly
recognised, being thus in conformity with the larger size of the
chorionic villi. Between the deeper closed end of the crypts
and the muscular coat was a definite layer in which portions of
gland-tubes, lined by an epithelium, some of which were transversely, others obliquely divided, could be seen. The glands
were dilated as in the younger specimen, and not so numerous
as the crypts, neither could I obtain satisfactory evidence of the
communication of the mouths of the glands with the crypts.
I am led therefore to the conclusion that the crypts formed in
the early period of gestation in the placental area of the cat are
not due to a mere widening of the mouths of the tubular
glands; but are produced, as in the pig and mare, by a great
increase in the amount of the interglandular part of the
mucosa, which becomes folded so as to form the crypt-like
arrangement which I have just described. In this respect,
therefore, my observations agree with those of Ercolani on
the same animal[1]. The interpretation, therefore, which Ercolani and I have put on the appearances seen in the placental
area of the cat in the early stage of gestation differs from that
given by Dr Sharpey on the appearance seen in the uterine
mucosa of the bitch at a similar stage. As is so well known,
Dr Sharpey held that the pits and "cells" (crypts) seen on the
inner surface of the uterus, which receive the villi of the
chorion, are the mouths of the utricular glands enlarged and
widened. It is possible that in the cat, as in the *Orca*, the
utricular glands may open into some of the crypts, so as to
seem to justify the inference that they were formed by a
widening of the mouths of the pre-existing glands. But this
interpretation obviously cannot be given of the formation of
those crypts which are interglandular in position. Hence it
seems to be more in conformity with the structural arrangements of the organ to conclude, that the crypts which arise
in the uterine mucosa during pregnancy are new formations,

[1] *Mem. dell' Accad. delle Scienze di Bologna*, 1870, Plates 2, 3, 4.

produced by a great hypertrophy and folding of the surface of the mucous membrane.

When the ovum of a cat, which had completed about one-half the period of gestation was examined, a most important advance in placental formation was observed. The zonary villous band on the chorion was restricted to its middle third, and an equally large smooth surface was found at each pole. The zone on the chorion was now so completely interlocked with the corresponding zone in the uterine mucosa, that the two surfaces could not be detached from each other. The placenta could only be separated by rupturing the slender marginal band of decidua reflexa, and tearing through, or altogether pulling off, the placental area of the mucosa, which area was intermediate between the placenta proper and the muscular coat of the uterus, and formed a well-defined decidua serotina.

The villi of the chorion had the form of broad sinuous leaf-lets, which became attenuated at their uterine ends and gave off bud-like offsets from the free border. When vertical sections were made through the placenta the villi were seen to pass vertically through the organ up to its uterine aspect. The trabeculæ of maternal tissue, which formed the walls of the pits or crypts in which the villi were lodged, passed between the villi up to the chorion, and closely followed the sinuosities of the villi, so as to form an intimate investment for them, and in horizontal sections through the organ they were seen to be arranged as a series of laminæ, winding in a sinuous manner between the leaf-like villi. Between the placenta proper and the muscular coat was a well-defined layer of serotina, equal in thickness to the muscular coat itself. It was traversed by the numerous blood-vessels which passed into and out of the placenta, and which formed not unfrequent anastomoses with each other. The decidua serotina consisted not only of the vascular connective tissue, but of the epithelial cells of this part of the mucosa, which were similar in character to those described in the preceding stage of development. In thin sections, tubes, lined by an epithelium, were seen cut transversely or obliquely; they were about equal in diameter to the gland-tubes seen in the serotina in a less advanced stage of gestation, and were without doubt the dilated glands of this portion of the mucosa. It may here be stated, that in the non-placental area of the same uterus the tubular glands were distinctly seen separated from each other by comparatively wide intervals of interglandular tissue. The chorionic villi dipped into depressions in the decidua serotina, and were in contact with its epithelium. The trabeculæ and laminæ situated in the substance of the placenta were also continuous with the serotina and were invested by an epithelial layer, the cells of which were modified columns, like the cells of the decidua serotina. The blood-vessels of the serotina entered the laminæ and trabeculæ and ramified in them throughout the maternal part of the placenta. In the placenta of one of the embryos, where the maternal vessels were injected, they formed a network of capillaries of ordinary magnitude. In the other placentæ from the same uterus the maternal capillaries when injected with red gelatine were dilated to two or three times the size of the capillaries in the fœtal villi, and ascended almost vertically in the trabeculæ. Not unfrequently near the chorionic surface they dilated into sinus-like enlargements, which were crowded with blood-corpuscles. It is possible that these dilatations may, to some extent, have been due to the force employed in filling the maternal vessels with injection, but this will not, I think, account for the whole extent of the dilatation[1]. The vessels of the capillary network of the fœtal villi were injected with a blue colour and showed no dilatations; and the contrast between the two systems of vessels within the organ was well seen both in horizontal and vertical sections.

The placenta of a cat, shed in the ordinary course of parturition, was covered on its uterine surface by a layer of soft yellowish-white tissue, which was smooth and uniform in character, and was without any flocculent, ragged processes projecting from it. This layer was the deciduous serotina, and from it laminæ and trabeculæ passed into the substance of the placenta, which had a similar sinuous arrangement and relation to the fœtal villi as in the placenta at half time. Examined microscopically, the vascular connective tissue of the serotina

---

[1] The dilatation of the maternal vessels in the feline placenta has also been referred to by Eschricht and other observers.

with its epithelial investment was recognised, but as it was not possible in a detached placenta to inject the maternal blood-vessels their disposition could not be made out. I examined thin sections through the serotina for the presence of utricular glands. I saw indistinct appearances of tubes transversely or obliquely divided, which might be interpreted as tubular glands, but the aggregation of cells within and around them was so great that it was difficult to speak positively on this point. The chorionic system of fœtal blood-vessels was injected, and the leaf-like villi, with their remarkable compact capillary plexus, were readily seen. On examining with a pocket-lens the uterine surface of the serotina, many minute, rounded, scattered holes were seen in it, through each of which a ter-minal bud of a leaf-like villus projected so as to reach the uterine surface of the placenta. These buds were often clavate in form, and contained a capillary plexus, continuous with that of the body of the villus. It is clear, therefore, that when the placenta of the cat is shed at the time of parturition, a con-tinuous layer of serotina, interrupted only by these minute orifices, is shed along with it.

The presence of a layer investing the uterine surface of the cat's placenta, analogous to the caducous layer of the human placenta, was distinctly recognised by Eschricht; who also de-scribed the thin, perpendicular, flexuose laminæ of maternal structure passing through the entire thickness of the organ and investing the fœtal villi as if with sheaths[1]. Though Eschricht was at first inclined to the view that the layer investing the uterine surface of the placenta was nothing else than the mucous tissue of the uterus, further consideration led him to state that it altogether differed from that tunic. But he also came to the conclusion that the mucous tunic was left entire in the placental zone, exhibiting only torn and broken-off vessels.

There can be no doubt however, from its position and structure, that this layer is the mucosa of that part of the uterus which corresponds to the placental zone, for it and the intra-placental laminæ and trabeculæ are merely a more advanced condition of the crypt-like modification of the mucosa, which I have described in the earlier stages of placental formation in

[1] *De Organis*, &c., pp. 14, 18.

this animal. Is the whole thickness of the mucosa corresponding to the placental zone shed along with the placenta? or is this layer merely the superficial part of the membrane? are questions which may now be asked. These of course can only be satisfactorily answered after the uterus of a cat killed immediately after parturition has been examined. But I may state that, in the uterus of the cat in the mid-period of gestation, I found, on peeling off the placenta, that the serotina did not split into two layers, the one, a deciduous serotina attached to the placenta, the other, a non-deciduous serotina remaining connected to the uterine wall, but that the whole thickness of the serotina came away with the placenta, leaving the muscular coat exposed; moreover, the uterine surface of the placenta presented a smooth surface precisely similar to that exhibited by the organ when shed at the full time. A similar separation also took place more than once in the process of injecting the vessels of the gravid uterus.

Though the placenta in the Bitch, as in the cat, possesses the zonary form, yet its minute structure in the two animals presents sufficient differences to enable the anatomist readily to distinguish the one from the other. If the description and figures by Sharpey and Bischoff of the early stages of formation in the bitch be compared with the corresponding stages in the cat, a close resemblance is seen; but in the more advanced stages characteristic differences can be recognised.

In the Bitch, both at half and full time, when the placenta was stripped off the uterine zone, a distinct mucous membrane was left on the uterus, which was continuous at the margins of the zone with the narrow band of decidua reflexa and through it, with the mucosa covering the non-placental area. This zonary mucous membrane was subdivided into numerous, irregularly polygonal pits or trenches, bounded by folds of the mucous membrane. These folds had a ragged, flocculent appearance. The membrane was very vascular, and at the ragged edges of the fold numerous torn blood-vessels were seen. When examined microscopically the free surface not only of the pits and trenches, but of the folds, was seen to be covered by a layer of cells—the epithelium of the mucous membrane—which rested on the vascular sub-epithelial connective tissue. When this

epithelium was looked at from the surface, a pattern of polygonal cells was seen like the free ends of columnar epithelium;
but the cells were bigger than one usually finds this form of
epithelium to be, and had, more especially in the uterus at full
time, a distinct yellow colour, as if the cells were undergoing
fatty degeneration. When the cells were scraped off, so as to
be seen in profile, their columnar form was easily recognised.
As this mucous membrane was not detached from the uterus
along with the placenta it is to be regarded as a non-deciduous
serotina.

The uterine surface of the placenta also had a ragged appearance, for the numerous folds of the mucous membrane had
entered the placenta, and, when it was stripped off, their torn
ends were seen on its outer surface, but the flocculent appearance
was still further increased by the free ends of the chorionic
villi, which reached the surface. The prolongations of the
mucous folds entered the placenta at a multitude of points in
the interspaces between the villi, and as they ascended to the
chorion they branched repeatedly, so as to give investments to
the branches of the villi of the chorion. These intra-placental
prolongations of the mucosa consisted of sub-epithelial connective tissue, in which the maternal vessels ramified, and of an
epithelium composed partly of columnar cells, and partly of
cells the regular columnar form of which had been modified
into irregular polygons. These cells were larger and more distinct than the cells on the corresponding structures in the cat,
and their protoplasm was so very granular as in many cases to
obscure the nucleus. These prolongations of maternal tissue
constituted a deciduous serotina. The shed placenta of the
bitch, whilst possessing in its substance numerous prolongations
of maternal tissue not unlike those previously described in the
cat, yet differs from the latter animal, as has also been pointed
out by Prof. Rolleston[1], in the absence of a continuous layer of
deciduous serotina on its uterine aspect.

The chorionic villi in the bitch were arborescent and not
leaf-like as in the cat. They terminated in short villous tufts.
The umbilical arteries ended in a compact capillary plexus.

---

[1] *Trans. Zool. Soc.* v. 1863.

The villi were in close contact with the epithelial cells investing the intra-placental prolongations of the mucous membrane.

I may now relate some observations which I have made on the glands in the non-gravid uterine mucous membrane of the bitch. It is well known that two kinds of glands were described by Dr Sharpey[1] in the uterine mucous membrane of this animal, viz. short, simple, unbranched tubes, and compound tubes having a long duct dividing into convoluted branches, both kinds opening close together on the surface of the mucosa. These observations were supported by Weber and Bischoff, and generally accepted by anatomists and physiologists; but Prof. Ercolani of Bologna, in his first memoir on the Structure of the Placenta[2], stated his inability to distinguish more than one kind of gland, and concluded that only the long tubular glands were present. I have felt it necessary therefore carefully to examine the uterine mucous membrane of the unimpregnated bitch with reference to this question. On a surface view the mouths of the glands could be distinctly seen closely crowded together, as is so well represented in Dr Sharpey's figure (fig. 209), and in Bischoff's memoir (*Entwicklungsgeschichte des Hunde-Eies*, Plate XIV. Fig. 47). When horizontal sections were made through the membrane near its surface the glands were seen to be transversely divided, and so closely set together that the interval between any two adjacent glands was in some cases not equal to, in other cases about equal to, the transverse diameter of a gland-tube; further, all the gland-tubes in any given transverse section exhibited the same structural characters. When vertical sections through the membrane were examined, long compound tubular glands were readily seen passing into the deeper part of the mucosa, and between these, short and simple tubes were also recognised, so that, under low magnifying powers, at first sight these sections seemed to confirm the observations of Sharpey, Bischoff and Weber, which were made under magnifying powers of 10 and 12 diameters. When magnified more highly these apparently short simple glands were seen to vary considerably in length, some dipping for only a short distance

---

[1] Baly's Translation of *Müller's Physiology*, Note, p. 1576.
[2] *Mémoire sur les Glandes Utriculaires de l'Uterus*, p. 22, French Translation, Algiers, 1869.

from the surface of the mucosa, others for a greater distance, and exhibiting indeed every gradation in length up to the branched tubular glands themselves. But in the connective tissue, immediately below the short glands, portions of tubes were seen extending in line with the short tubes though apparently not continuous with them, but often with careful focussing a continuity could be traced, though obscured by overlying connective tissue. I am therefore of opinion that the utricular glands in the bitch, as in so many other mammals, lie in the mucosa, some almost vertically, others in various degrees of obliquity, so that, when vertical sections are made, some are cut short across, others longer, whilst others again may be seen in almost their entire length. I conclude therefore that all the glands belong to the type of compound tubular glands, that the apparent differences in length are simply due to the mode in which the glands are cut across in making the section, and that the physiological division proposed by Bischoff into simple mucous crypts and proper tubular glands cannot be supported.

From a dissection which I have made of the gravid uterus of a Fox at about the mid-period of gestation, I have satisfied myself that it corresponds in many respects with the bitch, though with specific differences. The uterine mucosa remained on the uterus when the placenta was stripped off, and possessed pits or trenches with intermediate ragged folds. The uterine face of the placenta was flocculent, owing to the prolongations of the folds into the substance of the placenta being torn across in the process of separation. These prolongations entered the placenta at a number of points, and passed with a sinuous course up to the chorion, and gave off many branches, which not unfrequently were arranged as an anastomosing reticulum, in the meshes of which the lateral offshoots of the villi were lodged. They were very vascular and their vessels were larger than ordinary capillaries. Compared with the capillaries of the foetal villi they were from twice to four times as big, so that they may be regarded as indicating the early stage of a dilatation into maternal sinuses, such as is still more clearly seen in the sloth, and reaches its maximum development in the human placenta. Many of these vessels ran vertically through the placenta, so that when horizontal sections were made through the organ, they were seen in transverse section. In many cases these transversely divided vessels were surrounded by a ring of cells—the epithelial investment of the process of maternal tissue in which the vessel lay—which showed that the process only contained a single dilated capillary. The epithelial cells investing the intra-placental prolongations of the decidua were remarkably large and distinct, and on the average about ¼th or even ⅓rd as large as the corresponding cells in the bitch. The fox therefore, like the bitch, has no continuous layer of modified mucosa, such as is seen in the cat, on the uterine face of the separated placenta. The villi of the chorion had an arborescent arrangement, and gave off both lateral and terminal offshoots in which a network of capillaries ramified.

I have studied the zonary placenta of the *Pinnepedia* in the Grey Seal, *Halichœrus gryphus*, a specimen of which, in the sixth month of gestation, I examined, in 1872. In this animal, as in the dog and fox, when the placenta was peeled off the uterus, a well-defined layer of mucous membrane was left on the muscular coat, which layer presented on its placental aspect numerous irregular pits and trenches, in which the convoluted folds of the placenta had fitted. From this layer numerous broad laminæ of the mucosa were prolonged into the placenta, not however, as in the dog and fox, irregularly over its uterine surface, but by means of a definite series of fissures. The laminæ dipped into the substance of the placenta, as the pia mater dips between the convolutions of the cerebrum, and the fissures, which they entered, may from their size be called primary. Each convoluted fold of the placenta was split up into elongated plates by secondary fissures, into which processes of the mucosa, derived not only from the broad laminæ, but from the mucosa in contact with the uterine face of the convolutions, penetrated. Each plate was again subdivided into polygonal lobules by tertiary fissures, into which more delicate processes of the mucosa entered, which could be traced through the thickness of the placenta up to the chorion. In drawing the placenta away from the uterus the laminæ were drawn out of the primary fissures, but the more delicate processes, which entered the secondary and tertiary fissures, were torn through, and remained in the substance of the placenta, entangled between the placental lobules and amidst the fœtal villi. The placenta of the seal when removed from the uterus had not on its uterine face a continuous layer of mucosa, as is seen in the cat.

The layer of membrane left on the surface of the uterine zone had all the structural characters of a mucous membrane. The free surface was covered by a layer of short columnar epithelial cells; the sub-epithelial connective tissue was very vascular and contained scattered, branched, tubular glands. The vascularity of the membrane was considerably greater than that of the mucosa of the non-placental area. The broad laminæ of this membrane had a similar structure. The more delicate secondary and tertiary processes consisted of a vascular connective tissue covered by a columnar epithelium, but without glands.

The villi of the chorion were long and very arborescent, and formed the polygonal lobules already described. The larger branches of a villus reached the periphery of a lobule, and instead of terminating in a cluster of bud-like off-shoots, the ends of many of the branches derived from the same parent stem were joined together so as to form a continuous layer of grey membrane, situated not only on the uterine surface of the lobule, but reaching for some distance down its sides. From the sides of the stem of the villus, as well as from its branches, multitudes of villous tufts arose. The prolongations of the maternal mucosa which passed into the lobules so as to come in contact with the villous tufts, were not derived directly from the non-deciduous layer of mucosa investing the muscular coat, for the greyish membrane situated on the uterine face of the lobule prevented a direct entrance. The intralobular maternal tissue arose from the processes which entered the secondary and tertiary fissures, which gave off lateral branches into the lobules. Within the lobules these branches subdivided into a reticulated lattice-like arrangement of sinuous trabeculæ, and the meshes of this reticulum were occupied by the villous buds. The trabeculæ had the same structure as the processes of the mucosa from which they were derived. The seal therefore in the reticulated arrangement of those portions of its mucosa, which are in direct contact with the terminal villi, presents a general correspondence with the fox, but the subdivision of the mucosa is more complete, and the cells of the epithelial investment are not so big as in the fox[1].

Although the zonary form of the placenta in *Hyrax capensis* was pointed out many years ago by Sir Everard Home[2], and although its structure has been examined by several anatomists, there is by no means an agreement on the exact relations of its fœtal and maternal portions. Prof. Huxley is convinced from his investigations[3] that the placenta in *Hyrax* has such an interblending of the fœtal and maternal portions that it is as truly deciduate as that of a Rodent. Prof. Owen states[4] that the villi are imbedded in a decidual substance, and the surface of attachment to the uterus is less limited than in the Elephant. On the other hand, M. H. Milne-Edwards describes[5] the placenta as only adhering very feebly to the walls of the uterus. Its villi, he says, are simple, very analogous to those of an ordinary pachyderm. In the midst of the zone there are vascular vegetations engaged in corresponding uterine cavities, but they adhere no more, than do the analogous prolongations in the ruminant, to the crypts in which they are included: they can be detached with the same facility without tearing through anything and without carrying away any portion of uterine tissue. There is nothing, he concludes, to indicate the presence of a caduca, and the allantois does not overstep the limits of the placental zone[6].

No observations have been recorded on the structure of the uterine mucosa in the gravid Elephant, but Prof. Owen has described and figured[7] the fœtal membranes. The chorion was encompassed at its middle by an annular placenta, 2 ft. 6 in. in circumference, varying from 3 to 5 in. in breadth, and from 1 to 2 in. in thickness:

[1] I have given a detailed description of the placenta of this Seal in the *Trans. Roy. Soc. Edinburgh*, 1875, and have figured not only its structure, but that of the cat and fox.

[2] *Lectures on Comparative Anatomy*, v. 325, Pl. 61.

[3] *Lectures on Comparative Anatomy*, 1864, p. 111.

[4] *Comparative Anatomy of Vertebrates*, III. 742.

[5] *Considérations sur la Classification des Mammifères*, Paris, 18 8.

[6] In the June number for 1875 of the *Annales des Sciences Naturelles*, M. George figures not only the placenta of Hyrax, but the gravid uterus. He says nothing however of its structure.

[7] *Phil. Trans.* 1857, p. 847.

"The placenta presents the same spongy texture and vascularity as does the annular placenta of the *Hyrax* and of the *Carnivora;* but the capillary filaments or villosities enclosing the fœtal vessels enter into its formation in a larger proportion, and are of a relatively coarser character. The greater part of the outer convex surface of the placenta is smooth: the rough surface, which had been torn from the maternal or uterine placenta, exposed the fœtal capillaries, and occupied chiefly a narrow tract near the middle line of the outer surface. A thin brown deciduous layer is continued from the borders of the placenta, for a distance varying from one to three inches, upon the outer surface of the chorion. In addition at each of the poles of the chorion was a villous and vascular subcircular patch, between two and three inches in diameter, the villi being short, ⅛th of a line in diameter, or less."

This specimen is preserved in the Museum of the College of Surgeons, London, and through the courtesy of Prof. Flower I have been permitted to obtain a slice for microscopic examination. Notwithstanding the number of years the placenta had been in spirit, I succeeded in passing some injection into the vessels of the chorion and the larger trunks in the stems of the villi, so that I was able to follow the villi more precisely into the substance of the placenta than I should otherwise have been able to do. The placenta was very compact and was clearly composed both of a fœtal and a maternal portion closely interlaced with each other. Many of the villi were of large size and passed through the entire thickness of the organ, branching repeatedly in an arborescent manner. Others again were of smaller size, and did not pass more than one-third through the organ, but, like the longer villi, branched repeatedly. The tissue of the villi was delicately fibrillated, and in it ran the branches of the umbilical vessels.

Interlocked between the villi was a tissue, which contained a very distinct network of minute tubes, obviously capillary blood-vessels, and on the surface of this tissue a layer of cells was seen with some difficulty. I succeeded more than once in isolating a few of these cells, and found them to be rounded or ovoid, with definite nuclei and with granular protoplasm. I believe these cells to be the epithelial covering of the laminæ of maternal mucosa, forming the walls of the highly-developed crypts in which the villi were lodged, whilst the capillary network subjacent to these cells, belonged to the intra-placental maternal vascular system. Several times I saw an appearance as if the intra-placental mucosa was split up into a reticulated arrangement of trabeculæ, similar to what I have described in the seal, but from the condition of the specimen it was difficult to speak positively on this point. There could be no doubt however that in this separated placenta of the elephant a large amount of uterine mucosa was inextricably locked in between the fœtal villi.

GENERAL MORPHOLOGY OF THE PLACENTA.—In the study of the morphology of the placenta in any mammal the presence of two parts, a fœtal and a maternal, originally quite distinct and separable from each other, must be clearly kept in view.

The morphology of the fœtal part presents no difficulty. It consists simply of a vascular villous membrane covered by an epithelium. The sub-epithelial part of the membrane is composed of a delicate connective tissue, containing numerous corpuscles in which the terminal branches of the umbilical vessels, with their capillary network, are distributed. The vascular villi may be either simple or branched, and in some of the mammals, whose placentation has just been described, e.g. the seal, the branching may assume a highly arborescent arrangement.

The morphology of the maternal part of the placenta presents greater difficulty, not only because the uterine mucous membrane, out of which it is produced, is more complex in structure than the chorion, but because this membrane becomes greatly modified in the course of placental development, and not unfrequently becomes so interlocked between the fœtal villi as to be separated from them with great difficulty.

In all the forms of placenta, along with the growth of the villi from the surface of the chorion, depressions or crypts arise in the uterine mucosa for their reception, and the walls of these crypts are formed by foldings of the hypertrophied mucosa.

In the diffused placenta the changes in the uterine mucosa are less complicated than in the other forms. The villi of the chorion are short, and branch but slightly. The crypts in the uterine mucosa are consequently shallow, so that the relations of the fœtal and maternal parts can be easily seen. Two free surfaces are in close apposition, the villi of the chorion fit into the crypts of the mucosa, but they can be drawn asunder without difficulty, so that the compound nature of the placenta can be at once demonstrated.

In the polycotyledonary placenta the villi are longer and more branched. The pits or crypts for their reception are consequently deeper and divided into smaller compartments, and the maternal mucosa in the site of the cotyledons is more hypertrophied, thicker, and more spongy. Two free surfaces are here also in apposition; but the length and branching of the villi, and the depth and subdivision of the crypts, render it somewhat more difficult to draw the two surfaces asunder than in the diffused placenta.

In the zonary placenta as seen in the *Carnivora, Pinnepedia* and *Elephas,* the villi are long and usually arborescent, though in the cat they are leaf-like and very sinuous. The foldings of the uterine mucosa, which have led to the production of the crypts, are more complicated, so much so indeed in the fox and seal, as to give rise to a remarkable subdivision of the membrane into a microscopic network. The two surfaces in apposition have become so interlocked that it is almost impossible to disengage them from each other. Hence in the process of parturition more or less of the uterine mucosa in the placental area is separated and shed in the substance of the placenta.

The morphological elements in the gravid mucosa of all mammals, are, as in the non-gravid membrane, epithelium, sub-epithelial connective tissue, blood- and lymph-vessels, glands and nerves. Of the arrangement of the lymph-vessels and nerves in the placenta we have no precise information. The epithelium, the sub-epithelial connective tissue, and the blood-vessels form the walls of the crypts in which the villi are lodged. The glands have no necessary relation to the crypts. In the pig, as has been shewn by Eschricht, myself, and Ercolani; in the mare, as has been pointed out by Ercolani and myself; and in the porpoise, as has been described by Eschricht, the mouths of the glands can be distinctly seen opening on the surface of the mucosa, in smooth areas intermediate to and quite distinct from the crypts. In *Orca gladiator,* though at first sight the funnel-shaped crypts seemed to be the dilated mouths of glands, yet further consideration has satisfied me that neither they nor the cup-shaped crypts are derived from the glands. In the *Ruminantia*, Eschricht, Bischoff, Ercolani and I have been unable to see any communication between the glands and the pit-like crypts of the cotyledons. In the *Carnivora*, though, as was interpreted by Sharpey, Weber, and Bischoff, the crypts seen in the placental area in the early stage of gestation may seem to be merely the mouths of the glands enlarged and widened, yet a more minute analysis of the structure shows that, though some of the glands may, as in *Orca*, open into crypts, yet that the crypts are much more numerous than the glands, and are consequently not derived from them. Hence in all these, and I believe in other placental mammals, the crypts are not modified glands, but are interglandular in position. The crypts do not exist in the non-gravid uterus, but, as was first definitely shown by Prof. Ercolani, are formed during pregnancy by a folding on itself of the mucous membrane.

The crypts are lined by an epithelium, which is derived from the epithelial lining of the uterus: the increase in the number of epithelial cells, owing to the greater magnitude of the mucous surface, being effected by proliferation of the pre-existing epithelial cells. In many mammals the cells lining the crypts have the columnar form, like the epithelium of the non-gravid mucosa, and in the pig the cells are apparently ciliated: but in some mammals the columnar form is not preserved, and the cells are rounded, or polygonal, and with granulated protoplasm. These cells form the cells of the decidua serotina, and they are homologous with the rounded, or polygonal, colossal, granulated cells of the decidua serotina in the human placenta.

The connective tissue in the walls of the crypts is derived from the sub-epithelial connective tissue of the non-gravid mucosa, through a rapid increase in the number of its corpuscles, though it is possible that there may also be a migration of white blood-corpuscles into it. The blood-vessels in the walls are continuous with the vessels of the mucosa, and are greatly increased in numbers. In the diffused and polycotyledonary forms of the placenta they are arranged as a capillary network, but in the zonary placenta they exhibit a tendency to dilate into colossal capillaries, which are the first indications of a maternal intra-placental blood-vascular sinus system, such as attains much greater development in the sloth, and acquires its maximum size in the quadrumana, and the human female. The vascular connective tissue forming the walls of the crypts constitutes the vascular part of the decidua serotina, by which term is signified the maternal mucous membrane situated between the fœtal placenta and the muscular wall of the uterus ; or, in other words, the maternal part of the placenta. In the diffused form of placenta the serotina consists of the whole of the mucous surface in which the crypts are met with. In the polycotyledonary it forms the maternal cotyledons. In the zonary placenta it consists of the annular band of mucosa, with the intra-placental laminæ and trabeculæ.

As is well known, the form of the placenta, the arrangement of the fœtal membranes, and the behaviour of the uterine mucosa at the time of parturition, have been taken by many zoologists as affording a basis of classification of the placental mammals. In 1828 von Baer[1] published a classification of animals based on their development. He divided the placental mammals into groups according to the size of the umbilical vesicle and allantois, and pointed out diversities in the form of the placenta in the different genera. In 1835 Prof. Weber communicated to a meeting of German Naturalists[2] a classification of the placental mammals based on the presence or absence of maternal parts in the separated placenta. Where the vascular folds or "cells" of the uterus are so closely attached to the vascular folds or villi of the chorion, that they fall away at the birth of the placenta they are, he says, "*hinfällig, organa caduca;*" whilst in mammals, where the uterine and fœtal parts are so loosely attached that the surface remains uninjured at birth, there are no "*zufällige organe.*" Eschricht in his essay published in 1837[3] employs a similar classification, and divides placental mammals into two families, in one of which

[1] *Ueber Entwickelungsgeschichte der Thiere*, p. 225, Königsberg.
[2] *Froriep's Notizen*, Oct. 1835, p. 90.
[3] *De Organis*, p. 80.

the uterine placenta is caducous, in the other non-caducous.
M. H. Milne-Edwards published a system of classification in
1844[1], in which he attached great weight to the size and disposition of the allantois and the form of the placenta. In a
subsequent memoir published in 1868[2] he lays stress upon the
presence of a caduca uterina in mammals with a zonary or
discoid placenta, and as these animals lose blood at the time
of birth he groups them together under the common term
Hématogénètes. In 1864 Prof. Huxley, in a Lecture on Classification[3], suggested that the terms deciduate and non-deciduate
were to be preferred to caducous and non-caducous, and arranged
the placental mammals into the groups Deciduata and Non-
deciduata; an arrangement which has been adopted by several
subsequent writers. By the term Deciduata is meant those
mammals which shed, along with the fœtal placenta, more or
less of the vascular constituents of the maternal mucosa in the
placental area, whilst the Non-deciduata do not part with any
of the mucosa in the act of parturition. A sharp line of demarcation therefore is drawn between these two groups of
mammals. In employing these terms it should be distinctly
kept in mind that the same anatomical elements exist in both
types of placenta, and that the shedding or non-shedding
of maternal tissue is determined by the degree of interlacement
of the fœtal and maternal parts of the organ, and not from the
presence in the deciduata of structures which do not exist in
the non-deciduata.

All anatomists agree in regarding the diffused placenta as
non-deciduate, for the uterine crypts are so shallow that the
chorionic villi can be drawn out of them with great ease; and
the fœtal membranes are shed in the act of parturition, without
entangling and drawing away maternal mucosa.

* The polycotyledonary placenta is also regarded as non-
deciduate. But from observations made on the shed membranes of the sheep and cow, I recently ascertained[4] that intermingled with the villi of the fœtal cotyledons were quantities

[1] *Ann. des Sciences Naturelles*, 1844, p. 92.
[2] *Considérations sur la Classification des Mammifères*, Paris, 1868, p. 22.
[3] *Elements of Comparative Anatomy*, London, 1864, p. 103.
[4] *Proc. Roy. Soc. Edinburgh*, May, 1875.

of cells, which possessed the characters of the epithelial cells of the pits and crypts of the maternal cotyledons; so that the fœtal cotyledons carried away with them during parturition portions of the epithelial lining of the crypts, and in so far therefore these animals are undoubtedly deciduate. But from the bloody state of the external parts of the ewe, for some hours after the birth of the lamb, I think it not improbable that the disruption of some, if not all, of the maternal cotyledons had been deeper than an epithelial shedding; that the maternal vessels had, in some places at least, been torn across, so as to have occasioned hæmorrhage.

There is no difference of opinion as to the deciduate nature of the zonary placenta. But it has not been sufficiently recognised that considerable variations occur in the relative proportion of maternal tissue which is shed along with the fœtal placenta. In the seal, the dog and the fox the decidua serotina, or mucous membrane of the placental zone, does not form a continuous layer on the uterine face of the separated organ. A definite layer is however left, when the placenta is shed, on the uterine zone itself, which is subdivided into pits or trenches by projecting folds. When the organ is *in situ* these folds dip into the substance of the placenta, but are torn through in the process of parturition, so that the only portions of maternal tissue which are shed are the intra-placental prolongations. That the membrane left on the uterus in the placental zone is the mucosa is proved by its vascularity, the layer of columnar epithelium on its free surface, and the utricular glands; which structures, the glands excepted, are also in the intra-placental prolongations. In the feline *Carnivora*, again, as illustrated by the common cat, the mucosa not only sends prolongations into the substance of the shed placenta, but forms a continuous layer on its uterine surface, so that there is a corresponding deficiency on the uterus itself. Hence though all the *Carnivora* part with a considerable portion of the maternal mucosa in the separation of the placenta, yet they exhibit differences in the degree in which the shedding takes place. The *Felidæ* have a higher grade of deciduation than the *Canidæ*, and with the latter the *Phocidæ* correspond. Hence the dogs and seals, in their placental affinities, are less removed from the *Cetacea*, the

Suidæ and the *Solipedia* than are the cats. The pits and trenches of the mucosa, which one sees on the uterine zone, after the separation of the placenta in a seal, a fox, or a dog, are obviously similar in their morphological characters to the crypts of the mucosa of a mare, a cetacean, or other animals with a diffused placenta. In the seal the pits and trenches possess a precision of form more than is seen in the dog and fox, a circumstance which is undoubtedly due to the subdivision of the placenta of the seal into definite minute lobules. The higher grade of deciduation in a cat may perhaps be accounted for by the broadly laminated villi, their very sinuous form, and the depth in the mucosa to which their terminal bud-like offshoots penetrate, giving to the fœtal part of the placenta a "grip," if I may so term it, over the maternal part, as to interlock the latter more firmly with the villi, and thus to cause the mucosa to be more completely shed in the process of parturition.

In the fox and seal the intra-placental prolongations of the mucosa are subdivided into a reticulated arrangement of slender trabeculæ, each bar of which contains only a single dilated capillary; but in the seal this subdivision is carried out to a greater extent than in the fox. In the seal occurs that very remarkable anastomosis of the distal ends of the primary branches of the chorionic villi, which gives to the placenta its precise lobular subdivision, and walls in each lobule at its uterine periphery with the greyish membrane. From a somewhat cursory examination of the placenta of the *Phoca vitulina*, in the Museum of the Royal College of Surgeons of England, it appeared to me that a similar membrane existed also in this animal; so that I am disposed to consider the arrangement as one which is of more than generic, indeed of ordinal value.

From the general correspondence in shape and structure between the placenta of the *Pinnepedia* and that of the true *Carnivora*, there can be no doubt, that, in both orders, the early stage of formation is marked by the production of crypts in the placental area of the uterine mucosa. In the grey seal the villi of the chorion, which are lodged in these crypts, acquire, not only a considerable length, but a highly arborescent form, and give origin to multitudes of villous tufts. As the branching and growth of the villi proceed in the course of development, the crypts will necessarily become divided into smaller compartments; and as the villous tufts increase in number and size, the walls of the crypts will become no doubt thinned, until at length they will lose their uniformly continuous surface, and become subdivided into the reticulated arrangement already described, in the meshes of the network of which the villous tufts are lodged. That the increased area of the uterine mucosa in the pregnant seal is due to a great increase in the interglandular part of the membrane, is proved by the much wider separation of the glands seen in both the non-placental and placental areas of the gravid as compared with the non-gravid uterus of *H. Gryphus.*

It has been customary to regard a placenta as deciduate only when the *vascular* constituents of the uterine mucosa are shed with the fœtal membranes. This acceptation of the term seems to me, however, to be too limited, and does not cover all the cases in which maternal tissue is shed in the separated placenta. I suggest therefore that the definition should be enlarged so as to embrace those cases in which epithelium alone is parted with, as well as those in which both the epithelium and the sub-epithelial vascular uterine tissue come away in the separated placenta. In studying the types of placenta which have formed the subject of this Memoir we have passed by successive gradations from the diffused placenta, which is apparently non-deciduate, to the polycotyledonary placenta in which the epithelial layer of the mucosa only has been found; then to the zonary placenta of the *Canidæ* and *Phocidæ*, where the entire constituents of the intra-placental prolongations of the mucosa are shed, but where a well-marked layer of mucous membrane is left on the uterine zone; and lastly to the *Felidæ*, where apparently the entire mucosa in the uterine zone is shed as a part of the placenta. It follows therefore that the line of demarcation between a diffused non-deciduate, and a zonary deciduate placenta, is not so sharp as has usually been supposed, but is graded over by the ruminant polycotyledonary placenta, in which the epithelial layer is the preponderating if not the only element of the mucosa which deciduates during parturition. But to prevent misconception it should be stated, as indeed has been already done by Owen[1], Ercolani[2], and myself, if not during parturition, at least afterwards, all placental mammals are deciduate, for in the pig, mare, and cetacean, "during the period of involution which follows parturition, it is obvious that great changes, either from actual shedding of portions of its substance, or from degeneration and interstitial absorption, must take place in the constituents of the crypt-layer before it can be restored to its proper non-gravid condition[3]." In the ruminants also, the thick, vascular, spongy tissue of the maternal cotyledon must disappear before the uterus can assume its normal unimpregnated aspect.

PHYSIOLOGICAL REMARKS.—The fœtal placenta possesses an absorbing surface; the maternal placenta a secreting surface. The fœtus is a parasite, which is nourished by the juices of the mother.

The absorbing structures of the fœtal placenta are the villi of the chorion, and the vessels they contain are the structures which transmit the materials absorbed to the fœtus. In the diffused placenta the smooth inter-villous part of the chorion is probably engaged in absorption as well as the villi themselves, for in both a capillary network is present. As the smooth intercotyledonary part of the ruminant chorion, and the smooth extra-zonary part of the zonary chorion are feebly vascular, they are probably little, if at all, engaged in absorption. Though the villi in the cotyledonary and zonary placenta are much fewer in relation to the extent of the chorion than in the diffused placenta, they are longer, more branched, or more sinuous, so that the surface for absorption is probably as great.

The illustrious Harvey distinctly recognised that the placenta prepared for the fœtus alimentary matters derived from the mother; that in the deer, for example, the pits in the cotyledons were

"filled with a muco-albuminous fluid (a circumstance already observed by Galen), and that from this source the ramifications of

[1] *The Anatomy of Vertebrates*, Vol. III. p. 727, 1868.
[2] *Sur les Glandes utriculaires de l'Uterus*, &c. Algiers, 1869.
[3] *Trans. Roy. Soc. Edinburgh*, 1871, and *Proceedings*, May, 1875.

the umbilical vessels absorbed the nutriment and carried it to the
fœtus; just as, in animals after their birth, the extremities of the
mesenteric vessels are spread over the coats of the intestines, and
thence take up chyle[1]."

Haller applied to this fluid the name of uterine milk; a term
which has been adopted by many subsequent writers, and the
cotyledons themselves have been regarded as uterine mammæ.

By what structures in the maternal placenta can this fluid
be secreted? Weber stated that in the ruminants it was in part
separated from the capillaries of the "cells" (crypts), but that
in addition there were uterine glands. Eschricht looked upon
the utricular glands as the sources of the secretion of this nutrient albuminous fluid, whilst he apparently regarded the
"cells" (crypts) as the places of formation of ordinary mucus,
and this conclusion, at least as regards the function of the
glands, has been adopted by various anatomists. Signor Ercolani, in his important memoirs on the placenta, has given a new
aspect to this question. He admits the presence of utricular
glands in the mucosa, and their increase in size during pregnancy, but conceives that their chief function is to furnish
nutritive materials during only the early stage of gestation;
and that as soon as the crypts (or follicles as he terms them)
are formed, a new glandular organ is produced, which prepares
a secretion that supersedes that of the utricular glands.

On this important subject a few words will now be said.
There can be no doubt that the utricular glands are secreting
structures, and that they enlarge during pregnancy. In the
diffused form of placenta they have the appearance of being
structurally perfect up to the completion of gestation; and in the
examples described in this memoir, *Orca* excepted, their secretion
is poured out so as to be brought into direct relation with the
inter-villous, and not with the villous portion of the chorion;
but as the whole free surface of the chorion is provided with
capillaries, the one is no doubt as capable of absorption as the
other, and the glands are presumably active throughout intra-
uterine life. In the polycotyledonary placenta the utricular
glands are not situated in the cotyledons, so that the uterine
milk cannot be formed by them. They exist abundantly in the

---

[1] *The Works of Harvey, translated by Dr Willis*, p. 562.

intercotyledonary parts of the mucosa, and their secretion is
brought in contact with non-villous, feebly vascular areas of the
chorion, where the power of absorption of the chorion is probably feeble. In the zonary placenta the glands are altered, and
degenerated in the placental zone. In the non-placental area
they also show a want of structural completeness, and if any
secretion is formed by them, it is poured out in relation to the
smooth feebly vascular parts of the chorion. Both in the cotyledonary and zonary forms this secretion is therefore of little or
no importance in fœtal nutrition, when the placenta itself is
formed.

The crypts possess the structural characters of secreting
organs. Each crypt is lined by an epithelium, descended from
the epithelial lining of the uterine mucosa; which from the size
and appearance of the cells is obviously endowed with great
functional activity. This epithelium rests upon a highly-vascu-
lar, sub-epithelial tissue, the vascularity of which is doubtless
proportioned to the amount of secretion formed by the epithelial
cells. The arrangements of the diffused and zonary forms of
placentæ are not such as to permit the secretion of the crypts
to be collected and examined free from mixture with the secretion of the utricular glands. In the polycotyledonary placenta,
where the spongy tissue of the maternal cotyledons consists exclusively of crypts, the secretion of uterine milk can not only be
shown to be derived from the crypts, but can be collected and
analysed. From the researches of Professors Prevost[1], Schloss-
berger[2] and Arthur Gamgee[3], it has been proved to contain fatty,
saline and albuminous materials dissolved in water, so that from
its composition it is well suited to act as a nutrient material.
The appearance of these crypts in the early stages of placental
formation, and their persistence throughout intra-uterine life,
though in the zonary form they may become somewhat difficult
to recognise, owing to complexities arising during growth, furnish evidence of their importance. The intimate relation which
they bear to the villi, which, in the whole series of placentæ

[1] *Ann. des Sc. Nat.*, 1829, xvi. p. 157, and in conjunction with M. Morin
in *Mém de la Soc. de physique de Genève*, 1841, ix. p. 285.
[2] *Ann. der Chemie und Pharm.*, 1855.
[3] *Brit. and For. Medico-Chir. Review*, 1864, xxxiii. p. 180.

described in this Memoir, are lodged within the crypts, shows that the secretion they form is in a position best fitted for being absorbed by the villi.

Under the stimulus imparted by the presence of the fertilized ovum the uterus undergoes enormous development and growth. The muscular coat increases so as to provide an arrangement capable of expelling by its contraction the fœtus, when its period of intra-uterine development is completed. In the mucous coat are developed multitudes of crypts, in which is produced a secretion capable of nourishing the fœtus during its intra-uterine life.

But the placenta is regarded as an organ, which not only provides nutriment for the fœtus, but serves as its respiratory apparatus, and it is believed that an interchange of gases takes place between the fœtal and maternal blood-vessels. Undoubtedly there are many facts on record which seem to show that the *fœtus in utero* needs to respire, and that the placenta is the organ where respiration goes on. But there is no evidence that the respiratory changes during intra-uterine life are actively carried on. The experiments of Wm. Edwards indeed show that new-born dogs and cats can resist asphyxia for more than half an hour. The interposition of a layer of secreting cells between the two systems of vessels necessarily throws difficulties in the way of the ready passage of gases from one set of vessels to the other.

The consideration of the structure of the human and the discoid form of placenta generally, will be deferred until a subsequent memoir.

ON THE FREEZING MICROTOME. *A Reply to Mr Lawson Tait.* By WILLIAM RUTHERFORD, M.D., F.R.S.E., *Professor of the Institutes of Medicine in the University of Edinburgh.*

WHEN I described in the pages of this *Journal* for May, 1871 (324), a method by which tissues might readily be frozen and cut for microscopical purposes by a modification of Stirling's microtome devised by me, I was aware that I had rendered the freezing process available to the histologist in a way which would certainly prove of great service. And I was also aware that the freezing method would in the end largely supersede all other methods of imbedding the tissues for the purpose of making microscopical sections. I stated that, with the aid of my freezing apparatus (p. 326), "beautiful sections of frozen fresh lung, liver, kidney, muscle, skin, brain, etc., may be made with an ordinary razor," and moreover that the capabilities of my apparatus might be readily appreciated from the following experiment (p. 328): "I killed a rabbit, and immediately removed a portion of lung, liver, intestine, muscle, and a whole kidney. I rapidly washed them in 0·75 p.c. salt solution, and put them, hot as they were, at once into the well of the machine and covered them with 0·75 p.c. salt solution. I put the freezing mixture into the box and covered the whole with cotton wadding. To freeze them thoroughly required sixteen minutes. I then made sections, as fine as any one could possibly desire, of all the several tissues at once, picked them off the knife with a camel-hair pencil, and put them into separate vessels for examination."

After a year's experience of my apparatus, I constructed another, in which the well for holding the tissue was smaller in proportion to the size of the box for holding the freezing mixture than in my first apparatus. Rotation of the brass plug at the bottom of the well was prevented.   The screw for elevating the plug was made steadier, and an indicator for giving sections of any thinness was attached, so that the tyro might at once use the apparatus successfully.   I described this second microtome in the *Lancet* (July, 1873).   Mr Lawson Tait makes the re-markable statement in the last number of this *Journal* (IX. 250), that "of Prof. Rutherford's apparatus I have always heard it said, that it either never completely freezes the tissue, or that the complete freezing lasts for so short a time as to be of little use."   I have only to say in reply to this that the identical microtome figured and described by me in the *Lancet* has been constantly used by my assistants and myself for more than two years without its ever having been discovered that it "*never completely* freezes the tissue, or that the complete freezing lasts for so short a time as to be of little use."   On the contrary, the appa-ratus is used with the greatest success for providing a great class of practical histology with something like two hundred sections of the retina, the same number of unhardened lymphatic gland, and other soft organs.   Moreover, at the recent meeting of the British Medical Association held in Edinburgh, I showed the very instrument described in the *Lancet*.   In the presence of Professor Struthers of Aberdeen, Dr Martyn of Bristol, and a number of other gentlemen, I froze a piece of intestine, and then sliced it.   The freezing was complete, and the sections were made within fifteen minutes from the commencement, although the air of the room was warm.   I then quoted Mr Lawson Tait's statement to the witnesses of my method and asked them to judge for themselves.

It is indeed amusing to read the first paragraph of that communication of this newly-fledged histologist (*lib. cit.*)   "If I may take the directions given in Foster and Balfour's *Elements of Embryology* as being an exposition of the most recent methods of section-cutting, *I can only feel that I have, by patient endeavours, been able to achieve a success which is not much known beyond the limited circle of those who have seen my work.   I therefore hasten to describe some improvements I have made in the methods of section-cutting, which seem to me to constitute as important an advance over the rude hardening and imbedding processes, as the results obtained by them were superior*

to the sections made by double knives[1]." It is instructive to see
how Mr Tait "hastens" to acquaint his learned friends with his
success so little "known beyond the limited circle of those who
have seen his work." I leave them to adopt what methods
they like best, but they will doubtless be as much surprised as
I have been to see this attempt of Mr Tait's to make it appear
that he is the inventor of the method which constitutes "so
important an advance."

Mr Tait is careful to tell us (p. 250) that "before the
appearance of Prof. Rutherford's instrument (July, 1873), I
had had one almost identical in use for many months, but I
have since had to modify it very much on account of repeated
failures." I will charitably suppose that Mr Tait is in ignorance
of my apparatus described in this *Journal* in May, 1871. He
can extend his "many months" beyond that date if he likes.
I will also charitably suppose that he is ignorant of the fact
that Mr Needham described and figured in the *Monthly Journal
of Microscopical Science* for June, 1873 (IX. 258), a modification of my microtome devised by Mr McCarthy, which in some
respects closely resembled that which Mr Tait described six
months ago. If Mr Tait continues his "patient endeavours,"
however, he will probably "achieve" a still greater "success,"
when he succeeds in comprehending why it is that as the result
of my experience I recommend the following to histologists as
worthy of their attention.

The freezing microtome described by me in the *Lancet*
consists (Fig. 1) of a plate of gun metal (*B*) with a circular
opening in its centre (*A*). The opening leads into a well closed
at the bottom by a brass plug (*K*, Fig. 2) capable of being
moved up or down by means of a screw (*D*). The tissue to be
frozen is placed in the well, and sections are made by gliding a
knife over its top. The thickness of the sections is easily
regulated by an indicator (*E*). The microtome is clamped to a
table by the screw (*F*). A mixture of salt and pounded ice is
placed in the box (*C*), the water from the melting ice flows
away by the tube (*H*). The tissue is seen in *A*, and the
freezing mixture in *C*, (Fig. 2). A section of the knife employed
is seen at *R* (Fig. 2).

---

[1] The italics are mine.

Fig. 1.

THE FREEZING MICROTOME.

Fig. 2.

VERTICAL SECTION OF THE FREEZING MICROTOME.

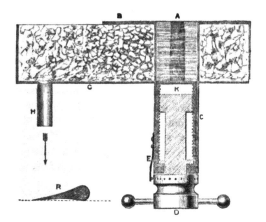

The instrument serves a double purpose : 1. For cutting tissues hardened in the ordinary way by chromic acid, or spirit, &c. 2. For cutting tissues hardened by freezing.

For the first purpose, a mixture of solid paraffin (five parts) and hog's lard (one part) is melted by a gentle heat and poured into the well, the tissue is then imbedded in the fluid and sections are made in the usual way. As every one knows, however, such substances as paraffin, wax and oil, are by no means perfect imbedding agents, for they cling to slices of structures of irregular shapes, and often necessitate much loss of time in getting rid of them, and if the section be of a delicate organ such as the retina, or an embryo, it may be altogether spoiled. These difficulties vanish when the freezing process is adopted. A slice of frozen tissue can be laid in any position on a cooled slide, and the moment the ice thaws the fluid flows, or may be washed away, leaving the section—be it retina or embryo, lung or kidney—*in situ* and perfectly clean. In fact when one discovers how great a luxury this microtome is when used for freezing, he is generally anxious to avoid most other methods of imbedding tissues. We would use it for everything, were it not that when the instrument is in almost daily use, as it is with us, the ice comes to be expensive.

The following is the mode of using the instrument for freezing : Unscrew the plug (*K*, Fig. 2), and cover its outer and inner surface with almond oil, so that it may not become fixed by freezing. This once done need not be repeated for months [1]. Encircle the freezing-box with a thick jacket of cotton wadding, felt, flannel, or cork. I always use the first, for it is always dry and clean, and only requires to be pushed between the microtome and the edge of the table in order to fix it. Pour into the well a solution of gum arabic made as follows : To 10 oz. cold water add 120 minims camphorated spirit and 5 oz. *clean* gum arabic. When the gum has dissolved, strain through calico and preserve for use in a corked bottle. Cover the well and the brass plate around it with a sheet of thin vulcanised india-rubber, in order to prevent the entrance of salt from the freezing

---

[1] In my former description, I stated that methylated spirit should be poured into the space under the plug *K* to prevent fixation during freezing, but I have long since abandoned this as unnecessary.

mixture and of heat into the well. Secure the india-rubber with a weight. Place in the freezing-box small quantities of finely-powdered ice (place the ice in a thick canvas bag and beat it with a wooden mallet) and salt alternately, by means of a bone spatula, *e.g.* an ordinary paper-knife. Constantly stir the freezing mixture around the well, and keep the tube (*H*) open to permit of the egress of water from the melting ice. When the solution of gum has begun to freeze and a film of ice has formed inside the well, introduce the tissue with a pair of forceps and hold it in any desired position against the advancing ice until it is fixed: then replace the india-rubber cap and complete the freezing. The process is exceedingly simple, and it may be completed within ten or fifteen minutes, even in summer. A cold room is of course always advantageous. The tissue can readily be kept frozen any length of time, by placing from time to time a little more of the freezing mixture in the box and stirring it round the well. It is possible, however, especially in the winter, to have the tissue frozen too hard. It splinters when it is too hard. This is prevented by discontinuing the further addition of the freezing mixture and by breathing upon the knife in the process of section. Always manipulate the sections with a brush to prevent injury to the face of the knife, and transfer them to slides, or to a vessel containing water, or a three-quarter p.c. salt solution to wash away the gum. For all ordinary purposes I employ a razor concave on both sides (*R*, Fig. 2). For large sections, however, it is advantageous to have a knife concave on the left and flat on the right side—(when the face is held downwards)—a half broader and a half longer than a razor. No wetting of the knife is needed, for the melting gum readily does this. I find it always most convenient to *push* and not to pull the blade through the tissue. I object to Mr McCarthy and Mr Tait's proposal to have the freezing-box projecting at the right as well as the left side of the cutting-table, for it does not appear to me to have any important advantage, and it certainly is calculated seriously to interfere with the hand in using the knife. I have avoided this arrangement from the first. It is advantageous, however, to have the table for the knife a little broader and the freezing box a little larger than in the instrument above figured. This was suggested by Mr Young, instrument maker, North Bridge, Edinburgh, last November, and from him the instrument of which I completely approve may be obtained. It has been lately proposed by Dr Fleming of Glasgow, to freeze the tissue by causing spirit cooled by passing through a tube—like that of a still—surrounded by a mixture of ice and salt, to flow round the well of the machine. An examination of Dr Fleming's apparatus has convinced me, however, that although its expense is more than twice as great, it possesses no real advantage over mine, and moreover its mechanical construction is even less perfect than my original instrument of four years ago.

There is another practical point worthy of mention. If it be desired to freeze a tissue that has been previously immersed in spirit, place it for a night in a large beaker of ordinary water. By the morning the amount of spirit contained in the tissue will be so small that it will not interfere with the freezing. Also this: at the end of a day wash the instrument in water, to get rid of the salt in the freezing-box. This obvious and simple expedient at once gets rid of what Mr Tait is pleased to term one of the "great faults" of my apparatus—to wit, the accumulation of salt in the freezing-box. His ingenuity in finding faults appears to considerably surpass his ability in devising improvements.

The process which I have described is really very simple, and can at any time be seen in my laboratory. There are two main points which should be noted. 1. The simplification of the process of freezing and cutting frozen tissues by means of the freezing microtome, so as to render this process readily available for histological purposes—that is my invention. 2. The use of an imbedding fluid which when frozen will cut like a piece of cheese instead of splintering to pieces, as frozen water does. That fluid is a solution of ordinary gum, and for this valuable suggestion I am indebted to my former assistant, Dr Pritchard, of King's College. Mr Tait places *water* in the well outside the tissue to be frozen. So did I four years ago, suggesting, however, that probably a better supporting fluid would be found. Mucilage is that fluid, and although I stated all this two years ago in the *Lancet* referred to by Mr Tait, he has evidently never tried it, otherwise he would scarcely have recommended the use

of water instead of it. So much for Mr Tait's "patient endeavours" by which he "has been able to achieve a success which
is not much known beyond the limited circle of those who have
seen his work," and, his "hastening to describe his improvements" (!).

There is no doubt whatever that physiologists, but especially
pathologists, will find the freezing microtome an instrument of
the greatest value in aiding histological study. Many indications for its use, which cannot be given here, will be found
in my *Outlines of Practical Histology* now published by
Churchill, New Burlington Street, London.

ON A NEW METHOD OF PREPARING THE SKIN FOR HISTOLOGICAL EXAMINATION. By W. STIRLING, D. Sc., M.D., *Demonstrator of Practical Physiology in the University of Edinburgh.*

DURING the winter of 1873, while working in Prof. Ludwig's
laboratory in Leipzig, I had occasion to make some investigations upon the structure of the skin, but the difficulty of the
subject rendered it imperative to have some method other than
those commonly in use for examining the structure and arrangement of its tissues. After trying a variety of methods I found
that the following process—viz. *digestion* of the tissue—yielded
excellent results. An artificial digestive fluid may be made
thus: 1 cc. of pure hydrochloric acid is mixed with 500 cc.
water at 38° C. and 1 gramme pepsine added. [The pepsina
porci prepared by Bullock and Reynolds, Hanover Street,
Hanover Square, London, should be employed, unless a glycerine extract of the stomach itself is used, which I found to
answer the purpose admirably.] After keeping the mixture for
three hours at 38° C. shake it thoroughly. The piece of skin
to be investigated is then stretched and tied over the mouth of
a glass dialysing jar. The skin is then digested in the above
fluid at 38° C. for a period varying from two to eight hours
according to the size and age of the skin. I always found that

the skin of young animals was digested more quickly than that of old ones. The digestion is completed more rapidly if the fluid is renewed from time to time, thus getting rid of the peptones which are formed and retard the digestive process. When digestion has been partially completed, the skin is then placed in water for twenty-four hours. In this it becomes swollen up and quite transparent. The small blood-vessels and nerves can be seen with the utmost clearness through a very thick layer of the tissue. The skin can now be cut with perfect ease, and may be hardened in the ordinary fluids, or stained with any of the ordinary staining reagents. The fibres of the white fibrous tissue composing the mass of the skin become swollen up, quite clear and transparent, and thus permit of a clear view of the structure and arrangement of the other tissues being obtained. By this method I have been able to make out the arrangement of the elastic fibres in the skin, and this with other points on the histology of the skin I hope shortly to publish *in extenso*. I have no doubt that this method will be found of value in the investigation of the structure of other organs and tissues.

THE ACTION OF JABORANDI ON THE HEART.

By J. N. LANGLEY, B.A., *St John's College, Cambridge.*

(From the Physiological Laboratory, Cambridge.)

JABORANDI has, in addition to its other remarkable properties, a distinct action on the heart, reducing in a marked manner the number of its beats, and ultimately bringing it to a complete standstill. In a preliminary account (*Brit. Med. Journ.*, Feb. 20th, 1875), I briefly stated this result, and pointed out the antagonistic action of atropia. In the present paper I wish to discuss somewhat more fully the exact manner in which this slowing or inhibition is brought about.

The preparations of jaborandi with which I have been principally working are : The alcoholic extract of jaborandi leaves, and the glycerine solution of that extract, kindly sent to me by Mr Martindale, of Cavendish Street. Since these preparations contained a considerable amount of alcohol and glycerine respectively, the results were complicated by effects due to these substances. Such effects have been as much as possible eliminated—especially with regard to frogs—by direct comparison with the results obtained by administering alcohol and glycerine alone.

Recently I have received from Mr H. B. Brady, of Newcastle, to whom my best thanks are due for the trouble he has taken in the matter, some much more concentrated alcoholic extract, and an aqueous extract of the alcoholic residue. All the experiments have not been repeated with these, but so far as they have, the results coincide with former results.

It is worth remarking that the aqueous appears to have all the properties of the alcoholic extract.

Mr Gerrard, in the *Chemist and Druggist* for June 15th, described the preparation of an alkaloid from jaborandi leaves, to which he gave the name of pilocarpine, and of the hydrochlorate and nitrate of this alkaloid. I have made some experiments with the nitrate, and the results are, I think, such as to justify the conclusion drawn from former experiments, that more than one active principle is present in the ordinary preparations. The quantity of pilocarpine obtained from jaborandi leaves is very small, about 0·75 per cent. This has to be borne in mind in comparing its effects with those of the extracts.

The main features of the action of jaborandi on the heart may be briefly described as follows:—If a few drops of the glycerine solution, or if one of the extracts be injected under the skin of a frog or toad, the heart will become red and dilated, and the beat slower. These characters increase till the heart is of a dark-red colour, very much distended in diastole. I have invariably found that the ventricle stops beating before the auricles. According to Vulpian (*Lond. Med. Record*, July, 1875) the auricles are the most readily affected; this certainly is not the case if it be taken to mean that the auricular beats are the first to cease. It does however not unfrequently happen that the auricles are the first to shew the effects distinctly, beating only in every alternate heart-cycle. This has been most frequently observed after the aqueous extract has been given. At a later stage the heart-beats generally become very irregular; at one time the auricles contracting only in every alternate heart-cycle, at another the ventricle; then perhaps a complete stoppage of both, followed by a few normal beats. Moreover, in point of energy the different parts of the heart beat seem to be quite independent of one another; at one time the auricles contracting feebly and the ventricle strongly, at another exactly the contrary. But with all these variations the ventricle has always in the many animals I have examined been the first to finally cease contracting.

The auricles as a rule cease contracting before the sinus, but I should hesitate to say that this was always the case; sometimes certainly if the sinus beats at all the beat is confined to the part closely adjoining the auricles, where it is very difficult to tell whether the movement is caused by a contraction nearly synchronous with the auricles, or by the auricular contraction alone.

Atropia sulphate quickly restores the heart-beat, but with less and less readiness as the heart has been stopped for a longer time. The restored beats are at first feeble, then get stronger till an almost or quite normal beat takes place. The auricles shew this effect most strikingly, and not unfrequently the auricular beat will be resumed on injection of atropia sulphate, but not the ventricular.

Thus it would appear that the auricles have a greater tendency to functional activity than the ventricle; their rhythmic beat is more difficult to stop, and when stopped is more easily set going again than that of the ventricle.

According to Vulpian (l. c.) jaborandi does not slow the heart, if urari be previously given; he therefore concludes that the slowing is caused by a stimulation of the peripheral ends of the pneumogastric inhibitory fibres in the heart. The contrary statement was made by me some time ago (l. c.), and I have made a fresh series of experiments to determine the matter.

(a) Urari was given to a frog in such quantity that the striated muscles did not contract on strong nerve-stimulation, e. g. a Daniell's cell with Du Bois-Reymond's Induction Coil, the secondary coil being at 0; yet jaborandi caused a slowing, though naturally not so rapidly as in an animal to which urari had not been given, since the absorption would not be so rapid.

(β) A frog was taken, and inhibition of the heart obtained by stimulation of the pneumogastric; then urari given till stimulation of the pneumogastric gave no inhibition. Still jaborandi caused a slowing of the heart.

(γ) The reverse of these experiments was also tried; by injection of jaborandi the heart has been nearly stopped, then urari given, but in no case has the rate of heart-beat been quickened.

(δ) The same is also true of the mammalian heart. The right carotid of a rabbit under chloral was connected with a mercurial manometer, and a tracing taken in the ordinary way. Urari was then given till stimulation of the peripheral end of the left pneumogastric had no effect whatever upon the heart (artificial respiration being kept up). Then the aqueous extract of jaborandi was injected into the left jugular vein. The heart-beat was slowed nearly one half.

But perhaps the most decisive experiment was the following :

(e) Urari was given to a rabbit, and when the respiration had all but ceased tracheotomy was performed and artificial respiration kept up. A needle was pushed into the heart and the beats counted ; these were from 270 to 290 in a minute. Stimulation of the pneumogastric had no effect on the heart. Then some of the aqueous extract of jaborandi was injected under the skin in four places, so that absorption should be more rapid. In ten minutes the beats had fallen to 120 in a minute.

I cannot then avoid the conclusion that jaborandi does slow the heart after urari has been given, and that therefore it produces this slowing by acting on something else than on the inhibitory nerve-fibres going to the heart.

I may here call attention to a remarkable result, which seems at first contradictory to the above statements, but about which I think I have but little doubt, viz. that an electric stimulus of definite strength sent into the pneumogastric of a toad soon after the injection of jaborandi when the heart is just a little slowed, produces a greater inhibitory effect on the heart than it does before the injection.

In the first set of experiments which showed this, a Daniell's cell with a Du Bois-Reymond's Induction Coil were used, the cell being in action the whole time. Thus any change in the strength of the current would be towards getting weaker, and in the nerve towards being less irritable. There was a possibility of the current becoming stronger by a gradual pushing up of the screw regulating the oscillating hammer, and thus making and breaking more times in a second.

Accordingly as a verifying experiment, three cells of remarkable constancy (kindly lent me by my friend Mr Dew-Smith, of Trinity College, for whom they were prepared by Mr Muirhead) were taken, and connected by long, rather thin wires, to a metronome, so that there was a continual current when the metronome was not working, and then by short thick wires to a key by which the current could be sent into the electrodes.

Thus when the metronome was set in action very little additional resistance was introduced into the circuit.

With this the same result was obtained; but perhaps the details of a particular experiment will make the matter clearer. The metronome was set so that the current was broken 110 times in a minute. The pneumogastric of a pithed toad being on the electrodes, the key was opened for 11 seconds. The beats which before were five in that time were brought just a little slower, being reduced to 4·5 in 11 seconds. This was repeated with a similar result. Then 18 small drops of the glycerine solution of jaborandi were injected under the skin. In 8 minutes the stimulus was again sent into the nerve for 9 seconds. This time the heart was completely stopped for 11 seconds. (The rate of heart-beat had by this time become a little slower than at first, viz. 4·5 in 11 seconds.) This was repeated with the same result. After another interval of 8 minutes the stimulus was once more applied to the nerve, and for a longer period, viz. for 48 seconds. Two beats of the heart took place after the beginning of stimulation, and then there was inhibition for 41 seconds. The rate of heart-beat had by this time come down to 2·5 in 11 seconds.

In the rabbit, I have not been able to determine any similar exaltation of the inhibitory function of the pneumogastric as the result of the jaborandi injections. Of this I can at present offer no explanation. I do not think however that there can be any error in my observations on the frog and toad. Though the results quoted above were obtained with the glycerine solution, they cannot be attributed to the glycerine itself, since they also appear when the aqueous extract is used; and I see no escape from regarding them as really due to the jaborandi.

Singularly enough this initial exalting influence ultimately gives way to exactly the contrary effect. Thus in the experiment related above, at a still later period when the beat of the heart had become very infrequent (about 5 a minute), stimulation of the pneumogastric nerve produced no effect whatever. Not only so, but even stimulation applied at the junction of the sinus with the auricles, which at first produces marked inhibitory results, at last, as the action of the drug progresses, is without any effect whatever.

Thus jaborandi, though, as we have seen, it will produce its slowing effect even when the inhibitory fibres of the pneumogastric have been rendered inert by urari, first exalts and then later on apparently paralyses these same inhibitory fibres.

A partial explanation of these contradictory results, is, I believe, to be sought for in the fact that the extracts of jaborandi contain other active principles besides the alkaloid pilocarpine. For pilocarpine, though in a general way behaving exceedingly like the extracts of jaborandi, slowing the heart, even (in sufficient doses) to complete standstill, producing a copious flow of saliva, &c. &c., differs from them in this particular action on the pneumogastric. In fact, if any experiment like that above described be undertaken to ascertain whether the irritability of the pneumogastric (using this phrase for convenience sake) be increased after pilocarpine, as it is after jaborandi, it will be found that instead of an increased effect no inhibition at all is obtained. Nor is there any stoppage if the strength of the shocks be increased. Two or three drops of a 5 per cent. solution of nitrate of pilocarpine are sufficient to produce this state. Further, it will be found that stimulation of the sino-auricular line will no longer stop the heart. Here then we have a substance which, as it is usually expressed, paralyses the inbibitory fibres of the pneumogastric. Now this substance is contained in jaborandi in very small quantity, and consequently when jaborandi is given the pilocarpine contained in it requires some length of time before it can produce its proper effects; hence it is not until a late stage that we find the inhibitory fibres paralysed.

The amount of pilocarpine needed to paralyse the inhibitory fibres produces comparatively little alteration in the rate of heart-beat, but if a larger dose be given, the heart becomes red and distended, and the beats slower until they finally cease. That a substance which prevents the heart being stopped by stimuli which ordinarily produce that effect, should itself stop the heart, appears to me a fact of some importance towards a knowledge of the heart's action. Pilocarpine I have invariably found to act as above stated, not only in frogs and toads, but also in rabbits and dogs. The following figure, I think, shows this very convincingly.

Fig. 1.

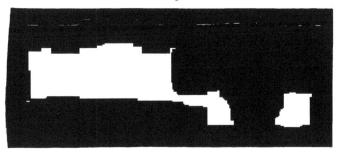

This is a tracing of the blood-pressure in the carotid of a rabbit under chloral taken in the ordinary manner with a mercurial manometer. The rise in the upper line shows the time during which the pneumogastric was stimulated. In the first part of the tracing, reading from right to left, the heart was beating very rapidly, more than 300 in a minute, the separate heart-beats being in the woodcut indistinguishable; the larger curves are the respiratory ones. A single Daniell's cell and a Du Bois-Reymond's Induction Coil were used, and when the first stimulus was thrown into the nerve, the secondary coil was at 10; it will be noticed that this produced a very marked inhibitory effect. Almost immediately afterwards 10 minims of a 5 per cent. solution of pilocarpine were injected into the jugular vein. The second part of the figure gives the tracing after an interval of two minutes. The pneumogastric was again stimulated for a longer time and with a much stronger stimulus, the secondary coil being at 5; yet no effect whatever was produced. It will also be noticed that the blood-pressure had fallen very considerably, and that the heart was beating more slowly, about 200 in a minute. Later, by a fresh injection, the rate of heart-beat was reduced to 126 in a minute. The stimulus was repeated with the same result, but after an interval of twenty minutes stimulation of the pneumogastric produced a slight slowing, which gradually became more and more evident as time went on, till at last distinct inhibition could be obtained; the action of the pilocarpine apparently passing off. Thus there seems to be present in the crude extracts of jaborandi, (1) Pilocarpine, the effect of which is in small doses to paralyse (sit venia verbo) the inhibitory fibres of the pneumogastric with a moderate slowing only, and in larger doses to increase the slowing up to complete standstill. (2) Some other substance (or substances) which exalts the inhibitory function of the pneumogastric, and at the same time has a powerful action in slowing the heart. For the slowing produced by the extracts, since it is already very manifest, while as yet the inhibitory function, so far from being depressed, is exalted, cannot be attributed to the pilocarpine alone.

I must now return to the relative actions of jaborandi and atropia. When the heart has been slowed or stopped by jaborandi the injection of atropia brings back the beat almost to its normal condition. Of this there can be no doubt; but the reverse action requires discussion.

In my preliminary account I stated that after atropia had been given jaborandi did not produce a slowing of the heart. This statement has since also been made by every one, I believe, who has paid attention to the subject. Recent experiments have however convinced me that it requires modification, and render it most probable that the condition of the heart, apart from other circumstances, depends on the relative amounts of jaborandi and atropia present. In these experiments, of which the following may serve as examples, the aqueous extract and the concentrated alcoholic preparations were used.

1. Six drops of a solution of atropia were injected under the skin of a frog. In ten minutes a piece of the semi-solid aqueous extract of jaborandi rather larger than a pea was placed under the skin of the back. In three quarters of an hour the heart was distended, the ventricle at rest, the auricles beating only four or five times a minute.

2. The heart of a pithed frog having been stopped by subcutaneous injection of the aqueous extract of jaborandi, atropia was given, and the heart-beat restored. The aqueous extract was again injected, and in an hour the heart was once more stopped in the characteristic jaborandi manner.

These experiments have been repeated with the other preparations, and they all lead to the conclusion that a definite quantity of atropia can only prevent a proportionate definite quantity of jaborandi from producing its effects on the heart.

This can be shown more easily, though possibly from the mode of experimenting less conclusively, by applying the substances directly to the heart. Thus take the following.

3. The brain and cord of a frog were destroyed and a little jaborandi put on the heart. In ten minutes the heart was stopped. Three drops of solution of atropia were put on the heart, and in seven minutes the heart was beating normally. Again, jaborandi was put on the heart, and in five minutes the heart again stopped. The excess of jaborandi was removed, and five drops of the atropia solution let fall on the heart. For more than half an hour the heart remained in diastole, the auricle very much distended, and not contracting when mechanically irritated. At the end of that time the upper part of the sinus and adjoining parts of the auricles began to contract feebly. In fifteen minutes more the auricles were freely contracting. In another fifteen minutes the ventricle contracted when mechanically irritated, and at the end of the next fifteen began to beat spontaneously; in fact, the heart-cycle was resumed, and so continued for a couple of hours.

This appears to me fairly conclusive with regard to the frog. Experiments, similar to those above described under (δ) and (ϵ), (pp. 189, 190), have been tried, substituting atropia for urari; but, although a slowing was in all cases obtained, it was not great, and I am not certain that it might not have been due to other causes.

Local applications of the extracts to the heart have enabled me to observe some facts which seem to me to possess interest as bearing on the theory of the heart's action.

If the apex of the heart be tilted over and a little of the extract cautiously smeared over the sinus venosus and roots of the venæ cavæ, these parts are arrested in their beats, while the auricles and ventricle continue beating for some time, provided that none of the material comes into contact with them. Of course, owing to the jaborandi getting into the blood and so being carried to all parts of the heart, this state of things cannot be maintained for any great length of time. But it may exist for a certain period, which period moreover may be distinctly prolonged by previously moistening the auricles and ventricle with a solution of atropia.

I have also succeeded, by local application of the extract, in causing the auricles as well as the sinus venosus to stop, while the ventricle moistened with atropia solution still remained unaffected. I was thus able to bring the heart into such a state that the ventricle continued to pulsate with great regularity while the auricles and sinus were perfectly quiescent, or at most showed nothing more than a feeble twitch in some part or other of the auricular walls. While the heart was in this condition I wiped the jaborandi away from the auricles and sinus, and carefully smeared these parts, by means of a camel's hair brush, with a solution of atropia. At the same time I smeared the ventricle with the extract of jaborandi. In consequence of the local action the ventricle ceased to beat, but became pale, conical, and contracted. This tonic contraction, which corresponds to the twitchings which are observed when the extract is applied directly to striated muscular fibre, is not due to the action of any alcohol in the extract, for it is witnessed when the aqueous extract is employed. It has been noticed by Vulpian. The auricles and sinus, which, owing to the contracted state of the ventricle, had become very much distended, now began gradually to resume their beat; and thus the heart, which a few minutes before exhibited quiescent auricles and a pulsating ventricle, now presented a quiescent ventricle and pulsating auricles. On wiping away the jaborandi from the ventricle, and sponging that organ with atropia solution, its pale contracted condition gave place to one of flaccidity; so that now the auricles were beating, and the ventricle flaccid and quiescent. Very soon, however, every third or fourth auricular beat was followed by a feeble ventricular contraction. These in time became more extensive, more forcible and more frequent, until at last the ventricular systole followed regularly upon that of the auricles, and the entire normal heart-cycle was resumed.

I will now briefly discuss the bearings of the facts described above on the theory of the action of jaborandi and atropia, and of the inhibitory mechanism of the heart.

The observations on urari and jaborandi have already enabled us to dismiss the view that jaborandi acts by stimulating the peripheral fibres of the pneumogastric.

The fact that after urari- and nicotin-poisoning stimulation of the sinus produces inhibition (though stimulation of the pneumogastric trunk is without effect), which inhibition is absent if atropia be previously given, and the facts of muscarin-poisoning, have led to the belief in the existence of a special inhibitory mechanism (inhibitory ganglia), with which the fibres of the pneumogastric are connected and which atropia is capable of paralysing.

According to this view we must suppose that jaborandi excites this inhibitory mechanism, which atropia paralyses. But if so, how is it then that jaborandi causes a slowing even after atropia has been given? If the mechanism has been previously *paralysed* how can it be *excited?* Evidently the word ' paralysis ' must in this case be used in a peculiar sense. The results can only be fairly stated by saying that atropia produces an unfavourable and jaborandi a favourable action on the inhibitory mechanism. But would it not be equally true in this case to say that atropia produces a favourable and jaborandi an unfavourable action on the automatic mechanism?

A similar argument may be used against the complex hypothesis that atropia paralyses or affects not the actual inhibitory mechanism itself but an intermediate apparatus connecting that mechanism with the terminations of the pneumogastric fibres.

Again, if we suppose that either jaborandi or atropia work solely on some nervous mechanism (inhibitory or automatic), this mechanism must be widely scattered over the heart, or rather must be composed of a group of mechanisms, each part of the heart having its own mechanism. Otherwise how can we explain the facts mentioned above, that either sinus, auricle, or ventricle, may be independently inhibited by jaborandi? Taking into account the speedy restoral by atropia, I cannot regard the local action of jaborandi as being essentially different

in kind from the general action which follows upon the introduction of the drug into the blood. The tonic contracted
state which is observable in the ventricle does not seem an
essential factor, and indeed is not witnessed in the sinus or
auricles.

Surely it is a far easier way out of the difficulty to suppose
that jaborandi (and therefore of course atropia also) acts directly
on the whole neuro-muscular cardiac tissue. In what exact
manner must be left for future inquiry.

Further, the effects of the local application of jaborandi on
the sinus militate strongly against the very generally accepted
theory of an automatic motor centre in the sinus and an inhibitory one in the auricles. Unless we suppose that the jaborandi thus applied acts on the muscular tissue alone and not
at all on any nerve-cells, a supposition very difficult to accept,
it ought, while stopping the sinus, to stop all the rest of the
heart also; but we have seen that it does not necessarily do
so. We may even go further and say that the same facts
argue against any localised automatic centres at all, inasmuch
as the ventricle will go on beating while both sinus and auricles
are quiescent.

I naturally feel hesitation in thus opposing such apparently
well-founded and so prevalent views, but at present I see no
way of escape out of my facts.

The particular stage of jaborandi-poisoning in which the
ventricle is arrested while the sinus and auricles continue
beating, has enabled me to make a few observations on the
rhythmic contractions of the bulbus arteriosus. Though this
organ is admitted by anatomists to be muscular and rhythmically contractile, its action has received little attention at the
hands of physiologists. Even in the normal heart the rhythmic
contractions of the bulbus can be readily observed, but they
become much more evident in the stage of jaborandi-poisoning
mentioned above. In such instances its beat takes place in
its proper rhythm, i.e. at a time just succeeding that at which
the ventricle would under normal conditions contract. It
seems to me not unworthy of notice that the fleshy massive
ventricle should be more easily affected by the drug when

introduced into the general circulation than the thinner-walled auricles, sinus or bulbus. And that it should þe thus possible to eliminate a median link (the ventricle) in the cardiac chain, while leaving the others, at either extreme, intact, supports the views expressed above.

In some hearts I have noticed the bulbus to continue contracting for some time after the ventricle has stopped, and later on the beats to assume an independent rhythm, not agreeing with that of the sinus venosus and auricles. Sometimes the expression of it is a regular rhythmic fibrillar twitching of a part only of the bulbus, generally of the strip bordering the ventricle.

Further, the bulbus may sometimes be seen contracting after the apex of the ventricle has been cut off, and I have occasionally been able to obtain it contracting alone, the ventricle and the greater part of the auricles having been removed.

I have spoken of most of the results of the experiments on the mammalian heart, but there are some points still left to notice.

The animals experimented on were the rabbit and dog, and in nearly every case the method adopted has been to use Ludwig's kymograph and a mercurial manometer, and so to obtain blood-pressure tracings of the right carotid, the substance being injected into the left jugular, and the left pneumogastric stimulated. By this means the variations in the rapidity of the heart's beat are most satisfactorily recorded.

Jaborandi in whatever form injected causes a slowing of the rate of heart-beat (whether the vagi have been cut or no) and a fall of blood-pressure. Of the former sufficient mention has already been made.

The fall of blood-pressure does not depend solely on the lessened rapidity of the heart's beat. By injecting slowly, often the first effect visible is a fall of the blood-pressure without an appreciable slowing of the pulse. The fall then resembles in a remarkable degree that caused by stimulation of the central end of the depressor. In Fig. 2 are two tracings from the same rabbit; the upper is that caused by stimulation of the central

end of the depressor, the lower by the injection of jaborandi.
Moreover, when atropia is given the blood-pressure does not

Fig. 2.

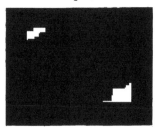

rise to its former level, although the rapidity of beat be restored.
Jaborandi given after atropia still causes a fall in the blood-
pressure, though not so considerable as in the absence of atropia.
Stimulation of the central end of the depressor still produces its
characteristic effect after injection of pilocarpine or any of the
preparations of jaborandi, but not to the normal extent, owing
to the mean pressure being already lowered by the drug.

I would here wish to take the opportunity to correct an
error in my preliminary account. I there stated that if a
glycerine preparation of jaborandi be injected under the skin of
a frog, tetanic movements and violent spasms follow. So indeed
they do: similar movements though less in extent being also
witnessed when a part only of the spinal cord is left; but later
experiments have led me to suspect that they are in some way
due to the glycerine, and not to the jaborandi. If a little of the
aqueous extract, or very strong alcoholic extract, forming a stiff
cohesive mass, be placed under the skin of a frog, the animal,
after the preliminary irritation has subsided, remains quiet, and
so continues, making few or no spontaneous movements, till it
dies. For some considerable period, depending on the amount
of jaborandi given and the strength of the frog, reflex action is
easily produced. Pinching the skin will make the frog leap
strongly; after which, it will as a rule remain in the place
where it alighted, till it receives a fresh stimulus. The eyes
on being touched are at once shut. If placed on its back it

will strive to regain its normal position. At this stage the heart beats very irregularly, with perhaps a pause now and then in the auricular or ventricular beats, or even a total stoppage of the heart for a short time. Later on, as the heart gets slower, the frog, if kept on its back for a little while, will remain so, and reflex action generally diminishes. In fact, beginning at the cerebral hemispheres, the functional activity of the various parts of the nervous system gradually disappears. These phenomena may be in part due to the gradual enfeeblement of the circulation, but to what extent I am not prepared to say.

NOTE TO MR BALFOUR'S PAPER ON THE ORIGIN AND HISTORY OF THE URINOGENITAL ORGANS OF VERTEBRATES.

SINCE writing this paper I have had an opportunity of examining a specimen of Acanthias and of Raia batis. In both of these I find that in the position corresponding with the blind pockets of Scyllium there is a pair of large and conspicuous abdominal pores.

F. M. B.

REPORT ON PHYSIOLOGY. By WILLIAM STIRLING, D.Sc.,
M.D., *Demonstrator of Practical Physiology in the University
of Edinburgh*[1].

Nervous System.

" Does the so-called ' facialis-centrum ' stand in relation to the
secretion of saliva ?"—E. Külz (*Centralblatt*, Nov. 26, 1875). The
author's results in two experiments on dogs are purely negative.

CO-ORDINATING CENTRA IN THE BEE.—Dönhoff (*Reichert und Du
Bois-Reymond's Arch.*, 1875, Heft I. 47) observes, that if the head is suddenly cut off a bee and honey applied to its proboscis sucking movements are made. The centra for co-ordinating the sucking movements
must lie in the head. If the body is separated from the head the bee
makes instinctive movements as if collecting pollen. If the body is
placed on its back, it is turned over, so that the bee comes to stand
on its feet. The centres for these movements lie in the thorax. If
the abdomen is cut off and pressed up, the sting is pushed out and
retracted, just as an intact bee does when it is touched on any part
of the body. The centre for these stinging movements lies in the
abdomen. The co-ordinating centres are therefore distributed over
the brain and the ventral cord of the thorax and abdomen.

CHEMICAL COMPOSITION OF THE BRAIN.—J. L. W. Thudicum contributes researches on the above subject. (*Privy Council Reports*,
1874.) Nerve-matter contains abundance of water which is chemically
combined, and termed by him "*water of colloidation.*" Three groups
of nerve-matter occur in the brain—phosphorised bodies, nitrogenised
bodies, oxygenised bodies. Phosphorised bodies contain P. in the
form of phosphoric acid combined proximately with glycerine, and may
be divided into the sub-groups, *kephalines, myelines, lecithines.* The
kephalines have free affinities for oxygen, but are not in a state of
atomic tension, and require for their decomposition the continued
influence of powerful extraneous affinities in the presence of water and
heat. The myelines have no apparent free affinities for oxygen, are
not affected by heat except to the extent of fusion, their atoms are
not in a state of chemical tension, but require for decomposition the
influence of strong external affinities, as water and heat. The lecithines are in a state of great atomic tension, and are easily decomposed. The nitrogenised bodies in many respects, but in a lesser
degree, imitate the properties of the phosphorised. They consist of
cerebrine, phrenosine and kerasine. Both the phosphorised and the
nitrogenised are strictly colloid, and do not pass through the septum

[1] To assist in rendering this Report more complete, authors are invited to
send copies of their papers to Dr Stirling, Physiological Laboratory, Edinburgh
University.

of a dialyser. The oxygenised principles consist mainly of alcohols with very slight combining powers. The most prominent member is cholesterine, which is not a fat, but a monodynamic alcohol.

The mode of isolating the different chemical principles, their compositions and properties, are minutely detailed in the report.

HITZIG ON THE SEAT OF THE CEREBRAL FLUID.—G. Hitzig (*Reichert und Du Bois-Reymond's Archiv*, 1874, *Centralblatt*, No. 19, 1875) remarks that the majority of authors generally found no fluid in the sac of the dura mater cerebralis on the vertex. The author, on the contrary, from operations upon dogs, has convinced himself of the existence of a not inconsiderable quantity of fluid in that sac. On passing a fine scalpel during life between the dura and pia either a fluid clear or mixed with blood always flowed out, whilst several hours after the death of the animal the operation was without result, thus coinciding with what is found in the human subject. On the contrary, the lateral ventricles were always full of fluid. Immediately after the death of the (poisoned) animal water trickled out of small openings in the dura, when here there was no question of blood-pressure. During life, therefore, there must be a higher secretion-pressure than the blood-pressure and elasticity of the tissues which press the brain against the skull. After death the water will be pressed into the brain from the sac of the dura by the remaining elasticity of the compressed brain.

ON THE INFLUENCE OF CHLORAL HYDRATE ON THE EXCITABILITY OF THE NERVOUS SYSTEM.—P. Rokitansky (*Wiener med. Jahrb.* 1874, 294,—*Centralblatt*, No. 29, 1875) finds that when large doses of a concentrated solution of chloral are injected directly into the heart from the jugular vein the heart at once ceases to beat, the respirations continuing. If the chloral acts less intensely and more slowly the heart resists longer than the nervous system, which becomes narcotised in the interval, specially however the respiratory central apparatus. In the latter case life can be prolonged by artificial respiration, the blood-pressure being low, and the pulse-beats medium in frequency however. Like Rajewski the author finds that the heart brought to a standstill by concentrated chloral solution is excitable by direct stimuli (electrical and mechanical), the cardiac muscle therefore is not paralysed. In that the heart with a proper dose does not begin again of itself to beat, the author does not accept the assumption of a stimulation of the peripheral cardiac fibres of the vagus (which ultimately must fatigue). The standstill must be caused by an action of the chloral upon the motor nervous apparatus of the heart. An animal gradually narcotised by chloral, where respiration begins to stop, again begins to breathe when artificial respiration has been kept up for some time. Such an animal conducts itself exactly like one whose medulla has been separated from its brain. The excitability of the respiratory nervous apparatus sinks in both cases in consequence of the injury, and is improved by the artificial respiration. Both animals die without spasms, the excitability of the motor centres of the muscles of the trunk is diminished by both injuries. That the excitability of the vaso-motor centre is also depressed is shewn by the author, apart from other data, thus: a direct electrical stimulus which on being applied to different parts of the central nervous system in the normal animal, was followed by marked increase of blood-pressure, was without effect in the chloralised animal, and on suspension of the artificial respiration the normal increase of pressure did not take place. The respiratory curves taken by means of a Marey's cardiograph shewed a superiority of (active) respiration over the inspiration in the chloral narcosis, just as is the case in animals whose medulla is separated from the brain.

Eye.

"Hyoscyamin and its importance in Ophthalmology." R. Simono-witsch, *Arch. f. Augen- u. Ohrenheilk.* IV. I. 1. (Abstract in *Central-blatt*, No. 20, 1875).——"Crossing of the Optic Nerves." M. Reich, *Militärarzt. Journ.* 1875 (Abst. in *Centralblatt*, No. 29, 1875).—— "On the binocular mixture of colours." W. v. Bezold, *Ann. d. Physik u. Chemie Jubelhand*, 1874, 585, and W. Dobrowolsky, *Pflüg. Arch.* x. 56 (Abstract of both in *Centralblatt*, No. 30, 1875).

CROSSING OF THE TROCHLEARES.—S. Exner (*Wien. Acad. Sitzungsb.* 1874, LXX. 151) remarks the difficulty of deciding from anatomical considerations whether there is an actual crossing of the two Nn. trochleares in the velum medullare (Stilling, Meynert), or whether only a commissure is present (Schröder, van d. Kolk). Exner stimu-lated the corresponding part of the velum electrically and found that only the eye of the stimulated side moved. There is therefore only a commissure, and not a crossing of the Nn. trochleares in the velum, else both must have moved simultaneously.

P. L. Panum, "Sur la détermination de la distance qui sépare les centres de rotation des Yeux." *Nordiskt med. Arkiv.* VII. 1875. The author in this research shews how the prismatic effects which arise from using the peripheral parts of concave and convex spectacles may be avoided. It is also shewn how the distance of the centres of rotation of the eyes is measured by determining the distance of the optic axes, when these are fixed in a parallel position, and also when the distance of the pupil is measured, the eye being accommodated for a distant object.

Blood and Circulating System.

ON OZONE AND ITS ACTION ON THE BLOOD. — Joh. Dogiel (*Centralblatt*, No. 30, 1875) sums up the results of his investigations thus :—1. By the action of electricity upon oxygen, according to the author, two different modifications of ozone are produced. 2. Ozone prepared from pure O. by electricity and collected over water, or pure H_2SO_4, did not decompose completely for a long time (seven days). 3. Water only absorbs a small quantity of undecomposed ozone. 4. The blood is very strongly changed by ozone, its action being chiefly upon the coloured corpuscles. 5. The colouring matter of the blood-corpuscles is excreted and the blood becomes darker even in the course of 5 to 15 min. after the passage of the ozone. 6. By a prolonged (one hour) action of ozone upon previously defibrinated blood, the blood becomes darker, and in thin layers transparent. 7. Blood changed in this way loses the property of excreting hæmoglobin crystals on the addition of ethylic alcohol, or ether, or chloroform. 8. As the blood becomes darker it becomes more sticky, and the bubbles of gas which arise during the process last very long. 9. After this stage a tough body is secreted from the blood in flakes, which under the microscope are seen to be composed of fibres : by repeated washings in water and when prepared pure this body in its physical characters is not to be distinguished from fibrin. 10. With prolonged conduction (three, four or more hours) of ozone through it the colour of the hæmoglobin of the blood-corpuscles changes, the red of the hæmoglobin and hæmatin becoming changed into a dirty yellowish green, similar to that which is produced by the action of H_2S upon blood. At last the change in the colour goes so far that the blood becomes quite colourless. 11. The discoloration by ozone proceeds specially rapidly when blood much diluted with water is taken instead of the undiluted defibrinated blood. 12. Defibrinated blood discoloured by ozone consists of a colourless fluid and an albuminous body which in its physical characters corresponds to fibrin. 13. The formation of this substance from defibrinated blood (dog and frog) is obviously caused by the change of the hæmoglobin of the blood-corpuscles. 14. A solution in acetic acid of hæmatin prepared after the method of Prof. Gwosdew and washed with ethylic ether, becomes colourless on passing ether through it. 15. Blood of a dog poisoned with carbonic oxide gas acquires in a comparatively short time the properties of normal blood; on passing ozone through it it becomes darker through the action of CO_2, and arterial by the taking up O. Ozone acting on blood poisoned with CO causes CO_2 to be given off. 16. The blood of a dog poisoned with CO does not become so rapidly dark on passing ozone through it, and also becomes more slowly discoloured than normal blood, and blood treated with ozone loses its property of excreting crystals of hæmoglobin more rapidly than blood poisoned with CO. 17. The change which the red and white blood-corpuscles undergo through the action of ozone must not be confounded with the change that CO_2 produces (as Dewar and McKendrick have done). 18. The bile (dog and frog) assumes at first a yellowish brown colour on the passage of ozone, but ultimately it becomes colourless, as has been already observed by v. Gorup-Besanez. 19. Chlorophyll loses its green colour and becomes yellow.

George Gulliver (*Proc. Zool. Soc.* June 15, 1875) gives copious observations on the SIZES AND SHAPES OF THE RED BLOOD-CORPUSCLES of vertebrates, with a plate of them all drawn to a uniform scale, and extended and revised tables of measurements. While presenting evidence of the hopelessness of absolute precision and agreement in such measurements, he insists on the practicability of a regular concordance in the estimates of the relative sizes of the corpuscles, notwithstanding their variations in this or that animal, and the occurrence of personal or instrumental errors. Thus in a single order or family the corpuscles are constantly seen to be smaller in some than in other species; *e. g.* in the Tragules than in other Ruminants, in Paradoxures than in Dogs, in Hippopotamus than in Elephas, in Mus than in Hydrochærus, in Dasypus villosus than in Orycteropus capensis, in Rhea americana than in Casuarius japonicus, in Zootica vivipara than in Anguis fragilis, in Bufo viridis than in Bufo vulgaris, in Osmerus than in Salmo. In like manner different orders are often easily distinguishable, such as the Edentates, by the largeness and the Ruminants by the smallness of the corpuscles. Still, though these two orders are thus characterised, there are ruminants in which the corpuscles are larger than in some Feræ, and other Feræ in which the corpuscles are larger than in a few Edentates. In this last-named order, in the pinnipid Feræ, and in the two elephants, the largest known mammalian corpuscles occur; in the Tragules the smallest. As to the important medico-legal question concerning the diagnosis, by micrometry, of human blood-corpuscles, the author decides, according to his old observations, that it would be difficult or impossible thus to distinguish human blood from that of dogs and monkeys; and from his recent measurements of the corpuscles from the musk-rat (Sorex indicus), it appears that the difficulty of distinction between the blood-discs of this Insectivore and those of the human subject might prove insuperable, as it did prove when the author's son originally made the examination; and, considering the frequency of this so-called musk-rat within the houses and elsewhere at the Mauritius, the likeness of its corpuscles to those of man may some day give rise to perplexity in the law-courts. In the camels only among mammals are the blood-corpuscles oval; but both in structure and size they conform to the apyrenæmatous type. In birds and reptiles there is no exception to the oval shape of the corpuscles, and so little difference in either class, that each whole class has in these respects less value than an order of Apyrenæmata. The breadth of the corpuscles in birds corresponds to the diameter of the corpuscles in mammals. The largest corpuscles of birds are found in the Cursores and Rapaces, and the smallest in the little Granivoræ, Insectivoræ and Anisodactyli. Neither in the class of Birds nor in that of Reptiles are the corpuscles ever more than twice larger in one than in another species; though in some mammals this difference is sometimes five-fold. While, as long since known, the largest corpuscles of vertebrates occur in the tailed Batrachians, the anourous species of this class have corpuscles not always larger than those of the sharks and rays and of a few reptiles. In the two cauducibranchiate Batrachians, Amphiuma and Sieboldia, the corpuscles are larger than in the perennibranchiate Siredon. In osseous fishes the largest corpuscles are those of the Salmonidæ, and of the river Eels of Britain and Roderiguez; and the smallest are found in the little Acanthopteri and Anacanthini, and in the Sprat and Herring. Lepidosiren has the corpuscles so large as to lose in this respect the character of any regular fish, and present a true saurobatrachian relation. Though the largest ichthyic corpuscles occur in Plagiostomi, the corpuscles are not larger in the cartilaginous sturgeon than in the osseous salmon. As to the relation between the size of the corpuscles and that of the species, this in Apyrenæmata by no means applies to the class, but only to orders or smaller sections of it; while in birds this relation extends throughout the class; but neither to reptiles nor fishes, save in some partial instances rather indeterminate or accidental than regular. The physiological import of the corpuscles is not developed in the present memoir; but the author insists on their taxonomic value, and on the propriety of always recognising this in the descriptions of the classes and orders, and sometimes of the species, in works of systematic zoology.

ON THE CONTRACTILE ELEMENTS IN THE BLOOD AND LYMPH-CAPILLARIES.—F. Tarchanoff (*Pflug. Arch.* IX. 407, *Centralblatt*, No. 30, 1875) has studied the properties of the contractile elements in the walls of the blood- and lymph-capillaries with regard to stimuli. He employed for the most part the transparent tail of the young larvæ of frogs, and as a stimulus the induced current of different intensities and duration. For narcotisation and to render the animals passive, he employed a three p.c. mixture of alcohol, in which, after about fifteen minutes, the animals became quite passive. The voluntary muscles no longer reacted to strong stimuli, whilst no disturbance of the circulation was to be observed. If a blood-capillary with lively circulation and clearly obvious spindle-shaped nuclei was placed under a very high power, and was then stimulated by a moderately strong current for 15—20 seconds, then a shortening and thickening of the spindles soon became obvious, by which the lumen was narrowed, and sometimes completely obliterated. If the stimulus acts for 2—3 minutes, the elements retain their contracted form and the capillaries remain impervious to the currents of blood, or if the stimulating current is soon interrupted however, the spindle-elements soon regain their original shape. The blood-current returns to its normal condition. The same experiment can be repeated several times on the same elements, until at last the cells no longer return to the normal state. By the closure of the capillaries a slowing of the current is produced in the corresponding artery with consecutive dilatation, whilst stasis begins in the veins.

Mechanical and chemical stimuli produce the same changes, and the freshly excised webs and membrana nictitans of a frog conduct themselves analogously. Further, in employing the electricity it was found that the spindles at the point of origin of the capillary from the artery contracted first, and generally very early. The lymph-capillaries shewed the same phenomena; if, however, very strong currents were employed, the spindle-elements of the walls of the lymphatic capillaries were not only simply thickened as with the blood-vessels, they rather became pale, their contours disappeared, and their nucleus appeared in a rounded form. Twice the author observed that the protoplasm of those cells which had become pale fell into fine particles, which fell into the passing lymph-current. The author further believes that the rounded nuclei of the disintegrated cells, which remained at first, are ultimately removed by the lymph-currents. The changes in the spindle-elements of the capillary-wall the author, in opposition to Goluben, regards as a vital process.

In the course of his investigations T. had an opportunity of proving that the body of the fixed connective-tissue corpuscles also becomes thicker after a long continued electrical stimulation; the change is clearly but slowly produced. With gentle inflammatory stimuli the spindles, after they have been shortened for a day, gradually resume their original form. The out-wandering of colourless blood-corpuscles, with the other phenomena, continue. The author concludes that the hypothesis of Goluben—that all inflammatory phenomena are caused by the change in the capillary wall—is false.

TARCHANOFF ON THE CANALS WHICH ARE SUPPOSED TO CONNECT THE BLOOD-VESSELS WITH THE LYMPHATICS.—J. Tarchanoff (*Journal de la Physiologie*, July, 1875) reviews the opinions of Kölliker, Virchow, Recklinhausen, and specially of Arnold, upon the direct communication of the blood-vessels with the lymphatics. He has injected the blood-vessels and lymphatics of the web of the frog's foot with Prussian-blue solution, either alone or mixed with gelatine. Even after ligature of the veins of the limbs, followed as it is by œdema, the author's uniform result was, the injection-mass was never found outside the walls of the vessels, unless there had been rupture of the walls of the vessels, when the mass spread itself between the interstices of the connective tissue outside. As to the injection of the corpuscles of the connective tissue by such a method, the author believes that such a view is quite untenable. He utterly denies the existence of a set of canals connecting the blood-vascular and lymphatic systems.

ROUGET ON THE MIGRATIONS AND METAMORPHOSES OF THE WHITE CORPUSCLES OF THE BLOOD.—Ch. Rouget's new investigations on the circulation of the larvæ of frogs have shewn (*Archiv. de Physiolgie Norm. et Pathol.*, 1874, 812, and *Centralblatt*, No. 21, 1875) that the red corpuscles in their diapedesis through the walls of the vessels remain perfectly passive. The intra-vascular pressure causes one corpuscle after another to pass through the cell-protoplasm and the structureless cuticula of which the walls of the young capillaries consist. In that the red corpuscles are incapable of self-movement, they cannot again regain their normal forms, which they have lost by being passed through the capillary walls. They therefore soon degenerate in their foreign surroundings outside the blood-vessels. The white blood-corpuscles arise from the fixed connective-tissue corpuscles, and are returned to the blood by the lymph. In virtue of their amœboid movements they are able, independent of the blood-pressure, to pass through the vascular wall. As soon as they meet a red blood-corpuscle outside the vessel they surround it with their processes. In the interior of the white corpuscle the red one is dissolved, it falls into pigment-granules, and thereby transforms the corpuscle into a pigment-cell. The latter, just like the original leucocytes, are capable of amœboid movements, they pass partly into the vessels, and their further fate is unknown ; and partly they form pigmented tunicæ adventitiæ of the vessels and nerves, together with the chromatogene layer of the subcutaneous tissue. The star-like subepidermal pigment-cells arise originally from white blood-corpuscles. On the application of mechanical stimuli the pigment-cells which arise from the white corpuscles collect around the scar and form neoplasms whose structure is similar to the caro luxurians of the wounds of mammalia.

LEWIN ON THE ACTION OF ACONITIN ON THE HEART.—L. Lewin (*Centralblatt*, No. 25, 1875), working under Liebreich's direction, finds the following results :

1. The action of aconitin in doses of 0·025—0·015 grammes in frogs diminishes very pronouncedly the frequency of the heart's action, and this diminution is most rapid when the drug is injected into the veins ; it occurs more slowly when given subcutaneously, and slowest of all when it is injected into the stomach. The mean duration of the time when it is injected into the veins is one hour, and by subcutaneous injection two hours.

2. In some cases this slowness was followed by an increase of the frequency of the heart's action, lasting for a very short time, the heart's action soon becoming irregular, or it passed into standstill, always in diastole.

3. At the end of almost all experiments, an arhythmical action of the heart occurred, mostly caused by a more frequent pulsation of the auricles, sometimes the auricles alone contracting.

4. The peripheral nerves suffered with medium doses (0·005—0·008 grammes) a marked diminution in their excitability, with larger doses paralysis.

5. *Post-mortem* electrical stimulation did not always cause the heart to contract.

In warm-blooded animals the phenomena are more marked, the most pronounced symptom being the strong dyspnœa. In opposition to Achscharumow, the author finds that rabbits, even after an absolutely fatal dose, can be kept alive by artificial respiration. The author assigns a direct action on the respiratory centre in the medulla as the cause of the dyspnœa.

As constant as the dyspnœa is a quantitative and qualitative change in the heart's action, similar to the effects already observed by Böhm and Wartmann. The author tabulates his views thus :

1. The anomalies in the heart's action, occurring in poisoning with aconitin, are not caused by an affection of the medulla oblongata.

2. The observed and apparently contradictory results of the experiment can be united into two groups ; both comprise a lesion of the ganglionic centres in the heart, and are thus distinguished, the

one coincides with the integrity of the vagi, the other with their paralysis.

3. The integrity or paralysis of the vagi depends upon whether the intracardial endings of the vagus are stimulated for some time or are paralysed at once.

4. The difference in the action is quite an individual one, and does not belong to the poison.

The often observed arhythmical pulse is explained by the non-simultaneous and unequal effect of the aconitin on the one or other cardiac centre, perhaps caused by its unequal distribution in the blood.

EXPERIMENTAL CONTRIBUTIONS TO THE DOCTRINE OF TRANSFU-SION.—Ponfick (*Virch. Arch.*, 1874, LXII. 273, *Centralblatt*, No. 26, 1875) investigated exclusively on dogs, the two factors, viz. the mechanical and chemical effects of transfusion. As to the mechanical effects, the author's results coincide with those of Worm Müller (*Jour. of Anat. and Phys.* IX. 222). Large quantities of artificial or natural serum, which increased the original quantities of blood about one-half, were only able to produce quite temporary oppression.

With regard to the chemical effects, the author shews the complete harmlessness of large doses of similar blood, whether defibrinated or transfused directly. It is quite different, however, with other kinds of blood. On transfusing the blood of lambs into a dog death occurred, as well with defibrinated as with non-defibrinated blood, and this when the dose was twelve per thousand. Experiments with different kinds of blood shewed that a scale of different kinds of blood could be constructed for the dog when death took place. It begins with 2 per 1000 with the blood of the calf and pig, and with the blood of hens 20—25 per 1000. The other kinds of blood stand between these extremes.

The most pronounced post-mortem appearance from transfusion with heterogeneous blood is in the kidney. They are greatly swollen, and mostly pale. The microscope shews that the convoluted -and straight tubules are filled with solid plugs of a brownish colour, the colour not arising from the presence of blood-corpuscles, but from a uniform imbibition with a substance like hæmoglobin. Red-coloured transudations were only observed in the large cavities when death occurred very rapidly. Clots were not found.

The changes described in the kidneys form the anatomical sub-stratum of the often-remarked bloody colour of the urine in man and animals. The condition should not be termed hæmaturia, but hæmoglo-binuria. This condition begins 30—60 minutes after the operation, as was shewn by introducing a catheter into the bladder of bitches and examining the urine from time to time spectroscopically. The quantity of colouring matter gradually reaches a maximum and then falls. The quantity of blood necessary to produce this result varies with the different kinds of blood.

In the dog the lowest limit was with the blood of the pig and calf, which produced hæmoglobinuria with 1·25—1·60 per thousand,

and the blood of hens with 4·20—5·50 per thousand. In the rabbit the values are much smaller. It can be shewn with the greatest certainty that this symptom is due to the solution of the transfused coloured blood-corpuscles, which after several hours are recognisable as specks in the blood, and the same is true if lake-coloured blood, in which the red blood-corpuscles have been destroyed by repeated freezing, is taken from the same animal and injected instead of the fresh blood. The organism does not appear to react to small doses of the lake-coloured as well as of the fresh heterogeneous blood, there is a condition of latency for a single transfusion of a small dose. If the small doses are repeated hæmoglobinuria also occurs.

The injurious property of heterogeneous blood therefore appears to depend upon the destruction of the blood-corpuscles introduced. The kidneys have the chief share in eliminating the hæmoglobin set free ; they may therefore be thrown into a condition of inflammation which may be so intense that copious exudations into the renal tubules may lead to complete secretory insufficiency of the kidneys, and so cause death.

From this Ponfick concludes that if heterogeneous blood is at all useful, it is only so in virtue of its plasma and its colourless corpuscles, whilst the value of similar blood is doubtful.

JAKOWICKI ON THE PHYSIOLOGICAL ACTION OF TRANSFUSION OF BLOOD.—A. Jakowicki (*Gazeta Lekarska*, 1875, No. 1—11, and *Centralblatt*, No. 23, 1875) confirms the results of W. Müller as to defibrinated blood when injected into an animal not producing blood-extravasions, as stated by Magendie, and also the results of Landois, that injection, or direct transfusion of foreign blood, or of defibrinated blood, is followed by a change in the entire blood of the animal, as manifested in the excretion of hæmoglobin in the urine. If the quantity injected is not too large, the urine clears up again and resumes its normal condition. In these experiments the blood of the cat, horse, and calf were transfused into dogs.

In injection of blood where the corpuscles had been dissolved, no matter the method employed to produce solution or injection of a solution of hæmoglobin, the author, like other observers (Francke and Naunyn), found that coagula were formed—specially pulmonary infarcti—which caused death. In dogs hæmaturia was constant, and often convulsions were present.

According to Alex. Schmidt the fibrin-ferment is formed after the blood escapes from the vessel. On injecting defibrinated blood a considerable quantity of this ferment is introduced into the blood. The author prepared this ferment after the method of Alex. Schmidt, and injected large quantities of it after having tested it with the plasma of dogs' and horses' blood. The author concludes (1) that the fibrin-ferment is a normal constituent of the blood circulating in the organism (in opposition to Alex. Schmidt). (2) The organism can destroy the ferment injected into the vessels so that the surplus disappears after a time, and the quantity of the ferment returns to the normal. (3) The organism presents conditions which are able to

14—2

limit the action as well of the ferments introduced from without as that circulating normally in the blood.

Respiratory System.

RESPIRATORY MOVEMENTS.—S. Mayer (*Wiener Acad. Sitzungsber.* LXIX. 3 Abth., *Centralblatt*, No. 20, 1875) remarks that if the vagus is stimulated until complete standstill of the heart is produced, during the complete cessation of the heart's action the respirations become more rapid and deeper. If the stimulation is now suddenly interrupted so that with the returning contractions of the heart the blood-pressure tends to rise rapidly to the height it had before the stimulation, then there follows the previously noted rapid and deep respirations, a complete standstill of the respiration, which under the most favourable circumstances may last half a minute. The standstill of the respiration occurs with expiration. If the stimulation of the vagus is interrupted after a short sandstill of the heart, so that a series of rapid beats may follow, and then a new standstill be again produced, then a long pause in the respiratory movements follows each period that the heart begins to beat again. These phenomena are explained by the author upon the basis of Rosenthal's theory of respiration. By the standstill of the heart we stop the supply of blood to the respiratory centre in the medulla, and as blood which has stagnated by the progressive diminution of its oxygen excites this centre more strongly, so more rapid and deeper respirations result (Dyspnœa). These deep and rapid respirations will obviously assist the quantity of blood in the lungs, and render it highly arterial. The succeeding standstill in the respiration is nothing more or less than a condition of apnœa. If during the standstill of the heart the animal is allowed to respire in a close space by means of a narrow connection which renders the exchange of gases difficult, the standstill in the respiration does not appear, but the symptoms of dyspnœa continue.

EXCRETION OF ALCOHOL BY RESPIRATION.—A. Schmidt (*Central-blatt*, No. 23, 1875), working in Binz's laboratory, finds in his experiments when 50 cc. of absolute alcohol were taken that at most traces were excreted by the lungs.

Spleen.

TARCHANOFF AND SWAEN ON THE WHITE CORPUSCLES OF THE BLOOD OF THE SPLEEN.—The spleen is, as is well known, regarded as a former of white blood-corpuscles, and the splenic venous blood is generally admitted to contain a disproportionately large number of white blood-corpuscles. J. Tarchanoff and A. Swaen (*Journal de Physiologie*, July, 1875) were led to test the accuracy of this statement. Their experiments were made upon dogs and rabbits, which had had no food for sixteen or eighteen hours previous. The method adopted for counting the blood-corpuscles was that of Melassez. They find that in the dog, under conditions as normal as possible, there is never such an enormous quantity of white blood-corpuscles in the venous blood of the spleen as is usually attributed to it. Dilatation of the spleen (such as is produced by section of its nerves) is immediately followed by a diminution of the number of white corpuscles in the splenic veins.

Immediately after section of the nerves there is a considerable diminution of the number of white blood-corpuscles in the splenic veins, but the longer the time after section of the nerves the more pronounced is the diminution of their number. This diminution of the white blood-corpuscles in the total mass of the blood can only be accounted for by a mechanical accumulation of the corpuscles in the splenic pulp, by their destruction in the interior of the spleen, or by their transformation into red corpuscles. The experiments were made in Cl. Bernard's laboratory.

BLOOD OF THE SPLEEN.—L. Malassez and P. Picard, *Comptes Rendus*, LXXIX. 1511. The blood of the splenic vein is of a different colour according as the splenic veins are stimulated or paralysed. The latter condition is accompanied by an increase in the blood-corpuscles and the capacity for 0. The difference between the blood of the splenic vein and artery is only pronounced after section of the splenic nerves ; the blood of the vein then exceeds that of the artery in its quantity of corpuscles and capacity for 0. This difference is a function of the spleen, for it is not manifested by other venous bloods, *e.g.* from the jugular vein after section of the sympathetic. The authors then investigated the effect of section of the splenic nerves upon the number of blood-corpuscles in the entire blood, and found their number more or less pronouncedly increased in the arterial blood some time after the operation, which then again disappeared. Under normal conditions the blood of the splenic vein is richer in corpuscles than the arterial blood of the body. As the mean of four observations 1 ccm. of the former contained 5,352,500 blood-corpuscles, 1 ccm. of the latter 5,092,500.

Digestive System.

ON PEPTONES, AND FEEDING THEREWITH.—P. Plosz, *Pflüg. Arch.* IX. 325, and ON THE CHEMICAL COMPOSITION AND PHYSIOLOGCIAL IMPORTANCE OF PEPTONES. R. Maly, *Ebendas.* 585 (Abstract of both papers in *Centralblatt*, No. 29, 1875). Plosz fed animals with peptones to test whether these are materials fit for building up cells, *i.e.* are again transformed into albumen within the animal body. He prepared an artificial nutrient fluid consisting of 5·0 peptone, 5·0 grape-sugar, 3·0 fat (butter free from albumen), and 1·2—1·5 salts in 100 ccm., and fed therewith a dog ten weeks old, the dog originally weighing 1335 grms. The dog received of this solution, which latterly was made more concentrated, 360—440 ccm. in single portions, intro-duced into the stomach artificially for 18 days. The quantity of nutrient fluid corresponded in N. to about the quantity of milk that would have been taken spontaneously. The weight of the animal increased during this time to 1836 grm., *i.e.* 500 grm. The newly-formed tissue could only have proceeded from the supplied peptones, and it is thereby proved that peptones in fact become retransformed into albumen.

These same questions have been investigated independently by Maly. He investigated the elementary composition of fibrin and the peptones prepared by digesting the same. The fibrin employed after having been purified in the ordinary way was treated most thoroughly with ether. By this process it did not lose any of its properties, but acted still upon hydric peroxide. As a mean the analysis gave per cent. C. 52·51, H. 6·98, N. 17·34. Various old analyses, especially those of Dumas and Cahours, correspond well with this. For the preparation of the peptones fibrin was employed together with artificial gastric juice. (The gastric mucous membrane—after Brücke—digested with phosphoric acid, and precipitated with lime-water; the precipitate dissolved in HCl, and the solution subjected to dialysis, until it no longer gave the chlorine reaction.) The peptone-solution was freed by dialysis from the NaCl which might enter with carbonate of soda for the neutralisation of the acid solution. The solution was then evaporated and precipitated with alcohol. Evaporating the alcoholic solution and precipitating with alcohol, a new precipitate was obtained. All these precipitates upon analysis gave nearly coinciding results, of which this is the mean: C. 51·4, H. 6·95, N. 17·13 per cent. On comparing these numbers with those of the fibrin there is a small diminution in the quantity of C. and N. and an increase of the O. The analyses of Thiry gave exactly the same results for albumen and albumen-peptones; the peptones vary very little from albumen, and are probably hydrates of albumen, and are certainly the only product that arises from albumen during digestion. The varying results of Möhlenfeld on the composition of peptones the author thinks are to be explained by changes in the peptones through the method of preparation. The question whether peptones can supply the place of albumen in digestion was tested by Maly upon pigeons. They were fed upon wheat, and the quantity of wheat necessary to keep up the body-weight was determined; they then received an artificial diet in the form of pills, which resembled the wheat, only peptones were substituted for the albumen. The composition of the pills was the following: water 12·6, starch 66·1, ash 1·6, fat 2·0, cellulose 3·5, peptones 10·2, gum 4·0 (the peptone was not purified by dialysis). As a rule the pigeons received wheat and peptone-pills together in varying proportions, and sometimes peptone-pills alone. The body-weight was estimated daily. In all there are 16 experiments, lasting from 4 to 30 days. In all experiments it is shewn that the weight stationary with feeding on wheat increases on employing the pills, and in the later series of experiments it is proved that the body-weight fell again on feeding with wheat. The cause of this phenomenon probably lies in this, that the peptone is better absorbed than the albumen of the wheat. Certainly however there is here again the result that the peptone passes again into albumen.

PANCREAS PEPTONE. — B. Kistiakowsky (*Pflüg. Arch.* IX. 438) wished to compare the elementary composition of fibrin with the peptone which results from it. The fibrin was purified by treating it with a 3 per cent. sol. of NaCl, washed for a long time in water and extracted with alcohol and ether, and yielded 52·32 C., 7·07 H., 16·23 N., 1·35 S., 23·03 0. The digestion was produced by the glycerine extract of the pancreas, and was continued as long as was necessary to dissolve the fibrin, the peptone was precipitated by alcohol, and freed from leucine by washing with alcohol, and then treated with silver oxide. The solution was strongly fluorescent. The analysis yielded 42·72 C., 7·13 H., 15·92 N., 1·03 S., 33·2 O : shewing a marked difference from the fibrin employed. The peptone treated with metallic salts prepared from vegetable casein had almost the same composition, viz. C. 43·4, H. 7·02, N. 16·16, S. 0·78, O. 32·74, and the same physiological properties. Peptones prepared from vegetable caseine (almonds) by the action of pepsin and HCl gave almost the same composition. The formation of peptones by the gastric digestion is not, according to the author, accompanied by the formation of gas ; and even in the pancreatic digestion the formation of gas was not observed.

Liver.

ON THE BILIARY SECRETION, by Prof. Rutherford and M. Vignal, *Brit. Med. Journ.*, Aug. 14, 1875. It has been shewn by Professor Hughes Bennett's Committee appointed by the British Medical Association, that in dogs with permanent biliary fistulæ, and living upon a fixed diet, that "spontaneous diarrhœa, dysentery, and purgation produced by blue pill, calomel, corrosive sublimate, and podophylline, always diminished the solid constituents of the bile, and, with one exception, the fluid portion of the bile also."

More recently, Röhrig performed experiments on the action of cholagogues in fasting curarised animals with temporary fistulæ, and found that large doses of croton-oil greatly increased the secretion of bile, and that a similar effect, though to a less extent, was produced by colocynth, jalap, aloes, rhubarb, senna, and sulphate of magnesia, the potency of these agents as hepatic stimulants being in the order mentioned. He found also that castor-oil had little effect, and that calomel, while it seldom recalled the biliary secretion, nevertheless somewhat augmented it when it was taking place slowly.

In Dr Rutherford's and M. Vignal's experiments a modification of Röhrig's method was adopted. Dogs which had fasted for eighteen hours were curarised, and artificial respiration maintained. A cannula was tied in the common bile-duct; the cystic duct was clamped. The bile flowed from the cannula into a finely graduated cubic centimeter measure, and the quantity secreted was recorded every fifteen minutes. It was shewn that this method of continuous observation yielded results far more reliable and instructive than that adopted by Röhrig.

Two experiments on the secretion of bile in dogs that had fasted for eighteen hours, and which received nothing more than the doses of curara used in all the experiments for the purpose of keeping the animals at rest, shewed that the biliary secretion was not affected by the doses of curara given; that the biliary secretion, on the whole, somewhat diminishes in the course of an experiment lasting from six to eight hours, but that the chemical composition of the bile remains almost exactly the same.

The curara was always injected into a vein; the various substances hereafter mentioned were injected directly into the duodenum; for this purpose, the wound in the abdominal wall was opened, and the substances injected through the wall of the viscus.

Three experiments with croton-oil shewed that, although it produced violent irritation in the alimentary mucous membrane in all cases, it increased the biliary secretion in only one instance. A high place is, therefore, not assigned to this substance as a stimulant of the liver.

Six experiments with podophylline proved that this substance greatly increases biliary secretion. A definite statement regarding the composition of the bile before and after podophylline will be given in the report.

Röhrig's statement that aloes deserve a high place as a hepatic stimulant was confirmed by three experiments, in which the extract of socotrine aloes was employed. The analysis of the bile (not hitherto given), however, shewed that after aloes the bile is more watery; nevertheless, the velocity of secretion is so much increased, that it certainly causes the liver to excrete more biliary matter.

Three experiments with rhubarb proved that it is a far more important hepatic stimulant than Röhrig has stated it to be. Doses of rhubarb were given nine times in the course of the experiments, and they never failed to excite the liver within half an hour after they were given. Analysis of the bile before and after rhubarb in all the three experiments proved the remarkable fact that, notwithstanding the greatly increased velocity of secretion after rhubarb, the bile-solids secreted by the hepatic cells are not diminished. The rhubarb apparently calls forth an increased secretion of normal bile.

Three experiments with senna proved that its power as a cholagogue is far below that of rhubarb. The bile is rendered more watery.

Four experiments with the aqueous extract of colchicum proved that it is a very decided cholagogue. The bile was rendered more watery, but the increase in the velocity of secretion was such that the amount of biliary matter excreted by the liver was certainly increased.

Two experiments with the solid extract of taraxacum proved it to be a cholagogue, though not a powerful one.

Two experiments with scammony proved that it has a slight cholagogue action.

Of four experiments with calomel, the secretion of bile was slightly increased in one, but there was nothing but diminution of the secretion in the other three. Purgative action was produced in all. The bile was rendered more watery.

Two experiments with gamboge gave no evidence that this substance is a cholagogue.

One experiment with castor-oil confirmed Röhrig's statement that this substance has scarcely any cholagogue power.

Two experiments with dilute alcohol injected into the stomach shewed that, after the alcohol was given, the secretion of bile slightly diminished.

Post-mortem examinations were always made.

It was shewn that the increased biliary flow from podophylline, rhubarb, etc., in these experiments could not be ascribed to reflex contraction of the gall-bladder; for this had been previously well-nigh emptied by digital compression, and the cystic duct had been clamped; nor could it be ascribed to reflex spasm of the larger bile-ducts, for the exaggeration of the biliary flow was far too great and far too prolonged to be explained in this way. Reasons were adduced for regarding it as probable that the agents are absorbed, and act on the liver directly. It was not professed, however, that their *mode of action* was definitely settled, the experiments having had for their primary object a determination of the facts of the case.

The opinion was expressed that powerful purgative action tends to diminish the biliary secretion.

The diminished secretion of bile in *non-fasting* animals after podophylline, observed in the experiments of Dr Bennett's Committee, probably resulted from a diminished absorption of food from the alimentary canal, in consequence of the purgative action.

When a hepatic and intestinal stimulant, such as podophylline, is administered to an animal that is not fasting, it is probable that (1) the liver is excited to secrete more bile; (2) the absorption of bile and food from the small intestine is diminished on account of the purgative effect.

ACTION OF THE BILE-SALTS ON THE ANIMAL ECONOMY.—J. G. Brown, (*Proc. Roy. Soc. Edin.*, 1874—75, 52, 3) concludes from his experiments:—

1. That a mixture of glycocholate and taurocholate of soda when injected hypodermically, in rabbits, in doses of 40 grains and under, does not cause any immediate disturbance, but is almost always fatal (unless the dose be small) in a period varying from 30 hours to 3 or 4 days.

2. That such injections often cause an increased nitrogenous excretion by the urine.

3. That they frequently cause diarrhœa.

4. That they are followed by an excretion of a small amount of bile-acid by the urine invariably, and in some cases that bile-pigment is also so excreted.

5. That they are followed by a fatty degeneration of the hepatic secreting cells, and of the renal epithelium.

6. That they cause a destruction of red blood-corpuscles, and consequently an apparent increase of the white.

7. That they are frequently followed by drowsiness and somnolence.

ON THE COMPOSITION OF HUMAN BILE.—D. Trifanowsky, *Pflüg. Arch.* IX. 493. (Abst. in *Centralblatt*, No. 28, 1875.)

Genito-Urinary System.

"Origin of calculi from the presence of foreign bodies in the bladder." F. Hoffmann, *Arch. d. Heilkunde*, 1874, 477.

SEEGAR ON THE REDUCING ACTION OF SUGAR AND URIC ACID IN THE COLD.—J. Seegar (*Centralblatt*, No. 21, 1875) observes that it is well-known that uric acid, like sugar, can reduce cupric oxide when heated. In order to ascertain the worth of the usually cited control-experiment, he tested the solution with Fehling's solution, and allowed the mixture to stand in the cold from six to twenty-four hours. It is said that if sugar is present reduction follows in the cold, whilst uric acid alone in the cold has not this effect. The author, however, finds that when the sugar is in small amount it loses the property of reducing cupric oxide in the cold. A 0·1 per cent. watery solution of sugar produced a scarcely perceptible reduction in the cold, whilst a 0·5 per cent. solution had no reducing action at all in the cold. A 0·5 per cent. solution of uric acid acts powerfully as a reducing agent in the cold (such a quantity, of course, never occurs in the urine). The usually cited control-experiment is, therefore, valueless, when we have to deal with small quantities of sugar.

Mammary Glands.

DE SINÉTY ON THE MAMMÆ OF NEW-BORN INFANTS.—De Sinéty (*Journal de Physiologie*, July, 1875) confirms the well-known fact that the mammæ of new-born children of both sexes, specially at the third and fourth day after birth, secrete milk. The author finds that the milk which is obtained from the mammæ of new-born children, several days after birth, is the result of a true secretion, and that the anatomical and physiological state of the mammary gland corresponding to that period is in many respects comparable to that which is to be observed during lactation in the adult female.

Skin.

ABSORPTION BY THE SKIN.—A. V. Wolkenstein (*Centralblatt*, No. 26, 1875) employed the skin of frogs stripped from off the legs to test its permeability to solutions of various substances, perchloride of iron, ferrocyanide of potassium, iodide of potassium ext. He finds (1) that the skin is permeable for watery solutions, but not for concentrated ones. (2) An increase of temperature of the solution increases the absorptive power of the skin; the absorption is in direct relation to the temperature of the fluid. (3) In young animals (rabbit, cat, mouse) the skin absorbs better than in old animals of the same species. (4) Hair and wool hinder the absorption. (5) Some alkaloids (atropine, strychnine, cyanide of potassium, curara) are also absorbed by the skin and are followed by the phenomena of intoxication.——"Influence of the time of the year upon the skin of embryos." Dönhoff, *Reichert und Du Bois' Archiv*, 1875, 46.

Muscle.

ON THE EXCITABILITY OF FUNCTIONALLY DIFFERENT MUSCULAR APPARATUS.—A. Rollet, *Centralblatt*, No. 22, 1875, points out that the flexors of the foot reply to much weaker stimuli applied to the nerves than do the extensors. The author proposes shortly to reply to a paper done in Fick's *Laboratory*, by J. P. Bour, 1875.

REACTION OF THE MUSCLES OF BEETLES TO STIMULI.—E. Fleischl (*Centralblatt*, No. 29) amputated the limb of a water-beetle (Hydrophilus piceus), and placed pins to act as electrodes through its upper end, and then stimulated the limb with a weak constant current. The character of the resulting movement of course varied with the strength of the stimulus, but it was found that the individual joints came to rest at very different times after one and the same stimulation. Neither is the beginning of the movement after *one* stimulation simultaneous in all the joints. The effect of one stimulation is as it were decomposed into several acts separated from each other by intervals of time. The same result was obtained with non-polarizable electrodes.

CONTRACTION OF MUSCULAR FIBRE.—C. Kaufmann, (*Reichert u. du Bois' Archiv*, 1874, 273. *Centralblatt*, No. 23, 1875), working under Krause's direction, employed for his investigation the muscles of the leg of beetles (Amara apricaria, Carabus nemoralis, Pygära bucephala). The animals were placed in toto in alcohol, and dissected at once under water. The muscular bundles so obtained were tinged with logwood and mounted in dammar. The results obtained from fibres in most conditions of their activity are the following: the contracted muscular fibre diminishes in length, and increases in breadth: during a contraction, only the isotropic substance becomes less in the long axis of the fibre, whilst the anisotropous does not lose anything in the same direction, or only to such a small degree as cannot be measured by the present apparatus. The anisotropous transverse bundles (and following them of course the isotropous) appear during a contraction not as straight but as bent stripes, whose convexity is directed towards the contracted end. The folding in of the sarcolemma described by Englemann in the contracted fibre could not again be found by K., on the contrary, the' sarcolemma appeared to be nearly a straight, *i. e.* not mathematically, line.—At the end of his paper the author communicates the following observation of Krause. An injection of chloroform into the A. femoralis of a living animal was practised, and then there occurred after a time the characteristic signs of waxy muscular degeneration, viz. rigidity and weakness. On investigation spots were found in the muscle-spindles, on which were present shining discs of coagulated muscular substance. By the double-tinging method with watery aniline green and carmine these spots were coloured green (and this is the interesting result) whilst the remaining muscle-substance shewed a pale red colour.

Miscellanea.

STEINER ON THE ACTION OF CURARA.—J. Steiner (*Reichert und Du Bois-Reymond's Archiv*, 1875, Heft 2, p. 145), during his residence at Naples, has investigated the action of curara upon some of the lower animals. His results may be summed up thus. 1. In fishes curara paralyses—*a.* the central organ of voluntary movement; *b.* the respiratory centre; *c.* the motor nerves. 2. This action varies with the time, and corresponds to the letters *a*, *b* and *c*, where *a* represents the earliest effect. 3. Paralysis of the motor nerves occurs much later than in the highly-organised amphibians, birds and mammals; still the passage to the fishes is not abrupt, but occurs gradually through animal species on both sides. 4. The late occurrence of the paralysis of the motor nerves increases with the size of the fish in spite of a larger dose. 5. In the electric rays the paralysis of the electrical nerves occurs much later than that of the motor nerves. 6. The other rays and hag-fish are affected in the same way by the poison as the freshwater fish. 7. The cause of the phenomena of 3 and 4 is quite unknown. 8. Paralysis also takes place in crabs, which, however, occurs relatively later than in the fish. 9. In snails, starfish, and holothuria only paralysis of the central organ of voluntary movement occurs. 10. In the medusæ it appears, from very few observations however, that the poison exerts no influence.

ON THE ACTION OF AMMONIA ON THE ANIMAL ORGANISM.—O. Funke (in conjunctin with A. Deahna, *Pflüger's Archiv.* IX. 416, *Centralblatt*, No. 29, 1875) takes up the old dispute between himself and Kühne as to the action of ammonia, and holds to his old view, that ammonia, if a weaker stimulus, stimulates the peripheral motor nervous elements. Frogs and rabbits served for the new experiments, caustic ammonia in dilute solution being injected into the veins or subcutaneously.

Following Kühne's view one might suppose that the powerful spasms, one of the most pronounced features in poisoning with NH_3, arose from direct stimulation of the muscles, but the posterior extremities, from which the blood is cut off by ligature of the iliac arteries, also participate in the spasms, and remain at rest when the corresponding nerve-paths are cut across. If a weak current of solution of ammonia is propelled in the frog from the aorta into the extremities, only weak fibrillar contractions occur, but no tetanus. There seems rather to be an enormously increased excitability of the central nervous system implicating the brain as well as the spinal cord. In opposition to Rosenstein the author finds that in frogs with divided cervical spinal cord the spasms occur to the same extent as before.

Probably, just as with strychnine, there is a pronounced increase in the excitability of the reflex centres, which is more pronounced in frogs than rabbits. That in the case of ammonia there is only a pronounced opisthotonous depends upon the exhaustion of the necessary apparatus. The action of the poison is always clearly manifested on the circulatory system. In slightly curarised frogs the vessels of the web of the foot contract some time after the subcutaneous injection of the ammonia—the lumen of the vessel even disappearing. The same is true of the ear of the rabbit immediately after the injection of the ammonia into the blood, whilst the pulmonary vessels of the frog only react slightly. This contraction in frogs becomes very much smaller after section of the medulla or the plexus ischiadicus; its cause must be sought for essentially in an excitation (direct or reflex) of the chief centre in the medulla. This indirect assumption was confirmed by manometric observations. At once after injection the blood-pressure and pulse frequently sink; after a few seconds the former begins to rise rapidly even to twice and more than its original value, and then to sink again gradually after a few minutes. At the same time the rare large beats become more frequent and smaller, still always remaining behind the original number. If the vagi are divided the original depression of the pulse-frequency and blood-pressure disappear. The author therefore concludes that ammonia excites the vagus as well as the vaso-motor centre, the latter more slowly, but more powerfully than the former. There is no support given to Lange's view that ammonia stimulates the heart directly.

As to the effect of ammonia on the respiration the author's results differ somewhat. With intact vagi the injection is generally followed by acceleration and flattening of the respiration, which, in two or three seconds, passes into a standstill. After this pause there follows as a second period, when no spasms occur, very deep and accelerated respirations with speedy return to the normal. If tetanus occurs, the respiration stops in expiration. The disturbances of the respiration are more pronounced with divided vagi.

The well-known type of respiration after section of the vagus occurs, characterised by rare, deep respirations, and is not disturbed by the ammonia, as asserted by Lange.

The further effects are the occurrence for 8—10 secs. of a pause in the respiration, generally at once after the injection without previous acceleration, which either passes into a dyspnœic period or into an attack of tetanus, which passes into the continually constant standstill of the respiration. The above-mentioned dyspnœic period is of very varying duration—a few seconds to several minutes—united with enormous respiratory excursions and active expiratory effects, much greater than occur with intact vagi. The author does not yet feel himself in a position to give a satisfactory explanation of his results.

MORIGGIA ON SUGAR IN THE FŒTUS AND ADULT.—Moriggia (*Reale Accademia dei Lincei, Estr. de Sess.*, Feb., 1873, *Centralblatt*, No. 10, 1875) investigated the quantity of sugar and partly also of glycogen in numerous organs and fluids in the fœtus and adult of many classes in the animal kingdom. His chief results are: that the blood of carnivora and herbivora at all times of life and in all stages with similar food contains sugar. The fresh bile of adult animals generally contains sugar, the acid or sometimes faintly alkaline reaction which

sometimes appears, arises from the transference of the sugar into
lactic acid. Sugar is always found in the muscles, heart and lungs,
sometimes in the spleen, but never in the kidneys, urine, brain,
salivary glands, or the pancreas of the adult or newly-born fœtus.
Sugar was found in the amniotic fluid up to the end of intra-uterine
life, but the quantity seems to diminish towards the end of pregnancy.
In very young fœtuses, whose whole body was examined at once,
only traces of sugar were to be found, whilst at all later periods of
fœtal life it is to be found plentifully in the urine, in the bile, and
peritoneal fluid, so that the fœtus may be regarded as truly diabetic
[as Cl. Bernard had already shewn]. Sugar was found plentifully
in the muscles, lungs and heart, in traces in the spleen, pancreas,
parotid, both placentæ, and the skin from the earliest fœtal period to
maturity. The latter is also true of the kidneys, and M. supposes
that their sugar passes through the urachus with the amniotic fluid.
Sugar was never found in the brain. Sugar and glycogen were
found plentifully at all times in the liver, just as in the adult. In
the white and the yellow of eggs, fresh, old and hatched, there is
much sugar.

The author regards the maternal blood as the source of the sugar
in the fœtus, at least in the earliest period of intra-uterine life.

The obvious constancy in the presence of sugar at all times of
animal life shews the immense importance and necessity of a certain
quantity of sugar for the development and sustenance of the organism.

INLUENCE OF LIGHT ON THE WEIGHT OF THE BODY.—Fubini
(*Influenza della luce sal peso degli animali*, Torino, 1875, 8°, pp. 19)
placed intact and blinded frogs, in other respects as nearly as possible
alike, alternately in darkness and in diffuse light—generally 10 to 15
hours—and gave them no food. From numerous tables it is
shewn that in the light the intact frogs lost more in weight than the
blinded ones in the proportion of 2·22 per cent. to 1. In the dark
both increased in weight; but the intact ones more than the blinded
ones. The great loss in weight the author ascribes to the lively
respiration, the increase in weight to the accumulation of O and H_2O.

Fig. 1

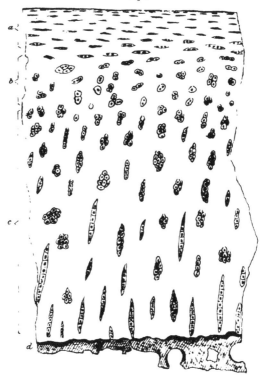

Fig. 2.

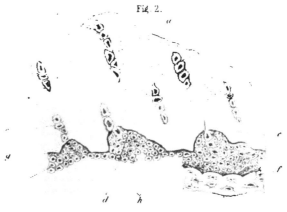

Plate II

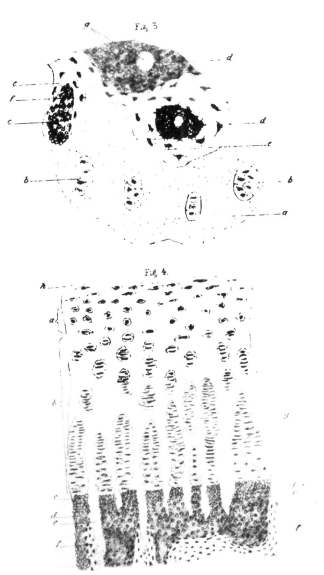

Fig. 3

Fig. 4.

ARTICULAR CARTILAGE.

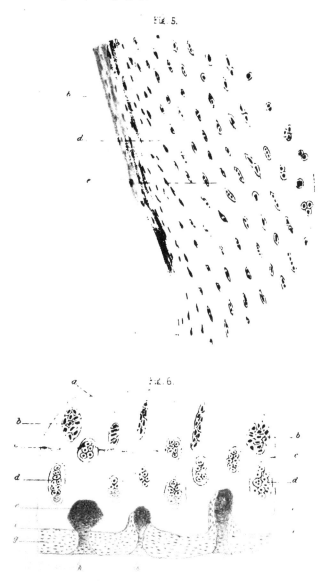

FIG. 5.

FIG. 6.

ARTICULAR CARTILAGE.

Alex. Ogston M.D. Del.

Fig 7.

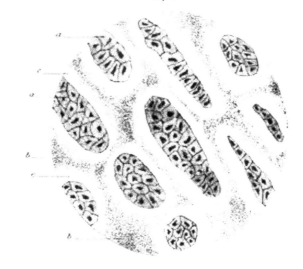

Fig 8.

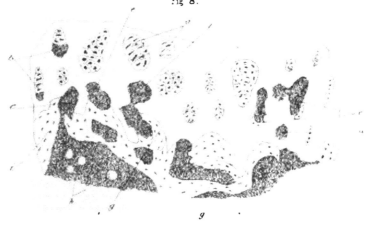

ARTICULAR CARTILAGE.

A del.d Cestan ML Del.

Fig. 9.

Fig 10

Fig. 11

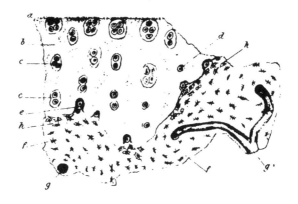

Fig. 12.

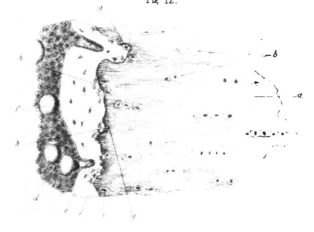

ARTICULAR CARTILAGE.

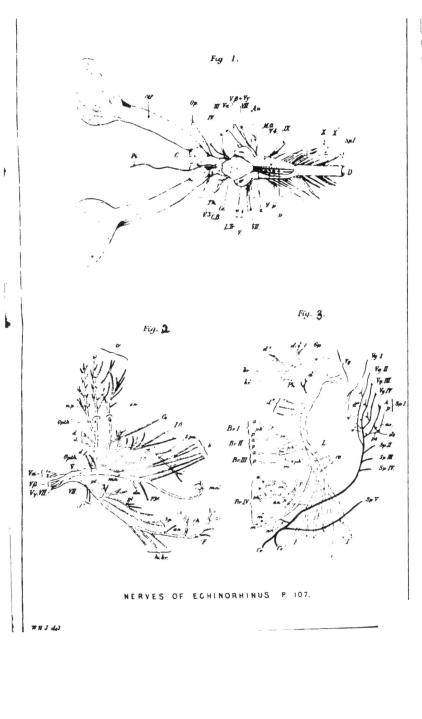

Fig. 1.

Fig. 2.

Fig. 3.

NERVES OF ECHINORHINUS P. 107.

W H J del

Fig. 1.

Fig. 2.

Fig. 3.

Fig. 4.

FIBRILLATION OF CARTILAGE. P 125

Journal of Anatomy and Physiology.

A CONTRIBUTION TO THE ANATOMY OF THE LENS. By G. THIN, M.D., *London*, and J. C. EWART, M.B., *University College, London.* (Plate IX.)

PREPARATIONS of the lens obtained by the usual methods of treating the tissues exhibit when observed under the microscope a series of parallel fibres. We have sometimes seen in spaces between the fibres cells similar to, if not in reality, white blood-corpuscles, indicating that the spaces are channels which play an important part in the nutrition of the lens. The size of these spaces in the lens of the toad is indicated in Fig. 1. They are well seen in the lens of the common fowl. Our investigations into the nature of the lens-fibres have yielded us the following results. We have found that the fibre, generally so called, is composed of a definite number of narrow flattened bands, which, to distinguish them from the fibre proper, we shall term in this paper primary fibres. The fibres are covered by a delicate structureless membrane, and the membrane is covered by elongated flat cells, the broadest part of which approaches that of the fibre. The primary fibres are covered by very narrow elongated cells, whose breadth is equal to that of the primary fibre. There are layers of oval cells in the lens, the breadth of which exceeds that of the fibres. A layer of polygonal cells covers the surface of the lens.

The preparations from which we have learned these important facts were obtained by the following methods. It has been known for some time to histologists that the large quadrangular cells of tendon can be occasionally demonstrated by exposing to sunlight in acidulated water pieces of tendon that had been previously in solution of chloride of gold. One of us has previously published[1] an account of flat cells seen in muscular fibre treated by a similar method.

This action of gold is essentially different from that by which nerves and the protoplasm of branched cells are stained a deep purple, and requires for its success somewhat different conditions. The tissue operated on must be so manipulated that the solution can penetrate the interstices with as little disturbance as possible of the natural relations of its component parts; and, as a rule, the solution must be allowed to act for a length of time varying from half-an-hour to three hours. The cells are by this process made visible by being what is technically called "fixed". When the tissue has been subsequently exposed to light in acidulated water they are seen to be of a uniform light mahogany colour, the nucleus being generally invisible. After the preparations have been kept some time, the cell becomes darker, and the nucleus, which does not undergo this change, is then more distinct. By frequent applications of this method we were able to obtain preparations shewing the cells of the surface, the large oval cells, and the cells which lie on the fibres of the lens.

In order to bring the gold solution into freer contact with the spaces we injected it by the arteries, and found abundant evidence that it penetrated the interfascicular spaces of all the tissues. In small animals we injected from the aorta, and in larger animals from the carotid. Immediately after death, when blood no longer escaped from the left ventricle of an animal which had been killed by bleeding, a canula was introduced into the aorta, and a tepid $\frac{1}{4}$ per cent. solution of chloride of gold injected until the eyes were greatly distended. A ligature was then passed round the aorta and the canula removed. Ten minutes afterwards the eyes were cut out, placed in a $\frac{1}{2}$ per cent. gold solution for 15 minutes, and then preserved in a mixture of glycerine and water. After 24 hours the lens was removed and placed for a few minutes in a strong solution of extract of logwood and alum. It was then broken

[1] *Ed. Med. Journal*, Sept. 1874.

up by needles in glycerine and examined. In successful preparations the following facts were observed. The lens-fibres were seen to be composed of smaller fibres hitherto not described, and which we name primary fibres. The cells seen on the fibres by the previous method were visible, and also much smaller and narrower cells lying on the primary fibres. The cells were seen both where the logwood-dye had stained and where there was no staining. Where it had stained, the whole cell was coloured uniformly, no predilection being shewn for the nucleus. The cells of the surface and a sheath covering the fibres were also well seen in these preparations.

Another valuable method in studying the structure of the lens is the employment of caustic potash after the manner previously indicated by one of us, and which consists in dissolving caustic potash in an equal weight of distilled water, and placing tissues in the solution when the temperature has fallen to between 107° and 105° Fahrenheit. Small portions of the lens are then removed by needles and examined in a drop of the solution. In such preparations are to be seen isolated portions of the sheath of the fibre, the fibres, and primary fibres.

We have not succeeded in isolating the cells by this method, although it is right to mention that we have applied it only in a few instances. We have, however, seen the nuclei of the cells which lie on the fibres, and the small narrow nuclei of the cells of the primary fibres adherent to isolated portions, the other parts of the cell having disappeared. The fibres of the lens as seen in Fig. 1 have been generally described as lying parallel to each other, with narrow spaces between them, and serrations along their edges; but in a lens injected with chloride of gold these fibres are seen to be more or less covered with narrow flat cells (Fig. 3). These cells have a granular appearance, sometimes lie in single rows upon the fibres, at others surround the fibres, so that besides the single rows represented in woodcut II., other cells are seen lying along the sides of the fibres (Fig. 3), and on changing the focus another row is seen adhering to the under surface. The fibres are seldom seen completely enveloped by the cells, but in all cases the well-marked rows on the upper and lower surfaces are evident. The ends of the cells fit closely to each other, thus forming a distinct nucleated band. This band, which runs along the centre of the lens-fibre, is somewhat narrower than the fibre itself, thus leaving a clear space at each side (woodcut II.). The cells, which at first might be mistaken for nuclei, on careful examination are generally seen to have a well-marked nucleus and nucleolus. The cells adhere firmly to the fibres and are not often found lying detached from them except in very carefully teased preparations. Though quite evident in unstained preparations, they are sometimes more apparent after staining with solution of extract of logwood. The cells as seen in Fig. 3, are about the breadth of a human red blood-corpuscle, and about four times longer than they are broad.

Though these cells are best obtained by injecting a solution of gold chloride from the aorta, we have also found them by treating the lens with gold in the usual way. Fig. 2 represents the same cells from the lens of a toad. After lying in a half per cent. solution for three hours and a half it was exposed to a bright sun-light for two days in a two per cent. solution of acetic acid. The preparations obtained by injection seldom shew the fibres completely enveloped by the cells, but seem to indicate that there are only two rows of cells, one along the under and another along the upper surface. When, however, preparations are allowed to lie for three or four hours in the gold solution, they clearly shew that the cells more or less completely invest the fibres.

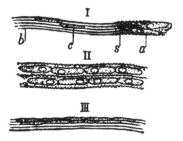

I. Fibre from lens of *Bufo Vulgaris*. *a*, cell lying on fibre; *b*, 4 primary fibres: *c*, cell lying on primary fibre: *s*, sheath of the fibre.
II. Two lens-fibres (*Bufo Vulg.*) with a row of cells on each.
III. Two rows of cells of primary fibres.

Though these cells seem to lie directly on the fibre, there is here, as in other organs in which flat cells are seen, a delicate structureless membrane on which the cells lie. This membrane invests the fibres, forming a sheath for them, and probably by the irregularities of its surface gives rise to the serrations seen in lens preparations obtained by the usual methods (Fig. 1). It is sometimes seen partly detached as in Fig. 4, sometimes as ending abruptly as in Fig. 3, and it is sometimes almost destroyed as in woodcut I., in which however at *a* the remains are still visible. In teased preparations detached pieces are occasionally seen in lenses from animals that have been injected with chloride of gold, and also in preparations obtained by treating a fresh lens with a warm saturated solution of potash. In both cases the serrations are still seen, though the cells are no longer visible. When this sheath has been separated, the fibre is found not to be a single homogeneous band, but to be made up of a definite number of very narrow straight parallel bands. These, in order to distinguish them from the ordinary lens-fibres, we have called primary fibres. We have always found four of these primary fibres uniting to form one lens-fibre (woodcut I.). Like the sheath they have only been seen in lenses of animals injected with solution of gold chloride and in potash preparations. In the gold preparations we sometimes found the sheath completely investing the fibre at one part but detached at another. In woodcut I. the sheath with its serrated edges remains at *a*, but is absent at *b*. Where the sheath is absent at *b* the four primary fibres are seen. In many gold-injected preparations large parts are seen to be composed of layers of the parallel primary fibres, the disappearance of the sheath of the larger ordinary lens-fibres rendering the outlines of the latter invisible. This appearance is represented in Fig. 5, where, though no division into the ordinary fibres is visible, some of the cells remain.

These primary fibres, as represented in woodcut I., though only about one-half the breadth of a human red blood-corpuscle, have also a single row of exceedingly narrow cells lying on their surface. Woodcut III. shews two rows of these cells, and woodcut I. shews two cells lying on two adjacent primary fibres. In several preparations obtained by injecting the gold solution, we have seen several rows of these delicate cells lying on the primary fibres, and above and below them are the larger cells (a) which lie on the sheath of the ordinary lens-fibre. We have not seen more than one row of these very narrow cells in connection with one primary fibre. Each cell lies immediately over the fibre and sometimes appears to be slightly broader than the fibre it covers.

In addition to the cells which cover the fibres and the smaller cells of the primary fibres, layers of oval cells are sometimes found by gently breaking up a fresh lens in gold solution, and after 1—3 hours placing the portions in acidulated water. These cells have a very large nucleus, and from their oval shape are in direct contact with each other only at certain points. Between two such layers the narrower cells covering the fibres can be sometimes seen. From the layers which they form being found in planes separated from each other by the lens-fibres, it may be inferred that they correspond to a division of the lens into large tracts or bundles, of which at present nothing further is known.

The appearance presented by these oval cells is seen in Figs. 8 and 9. Both in preparations teased in gold solution and in injected gold preparations, portions of a layer of polygonal cells are seen, sometimes detached from the lens, but usually in close apposition to it. Occasionally they can be seen in a very large unbroken layer, much larger than the part of one represented in Fig. 6. It forms the surface of an investing membrane which covers the lens proper.

In order to give a good idea of the relative size and form of these different cells, the figures representing them have all been selected from preparations of the lens of one animal, the common toad (*Bufo Vulgaris*), and have all been drawn to one scale. The cells which cover the primary fibres are not represented in Pl. IX. but are shewn in the woodcuts I. and III.

In portions of the lens sealed with Brunswick Black on an object-glass in half per cent. gold solution, and also occasionally in preparations obtained by gold and acidulated water in the ordinary way, we have seen the fibres intersected by straight transverse lines. These divide the fibres into portions of a definite and uniform length, but the lines which cut one fibre are not opposite those which cut the adjoining ones. The remarkable appearance which is thus produced is represented in Figs. 10 and 11.

We are not able to give a sufficient explanation of this appearance. Probably it is not due to any peculiarity of structure inherent in the fibres themselves, because in potash preparations isolated portions of fibres of the most varied length are seen, which have no analogy to the size of the portions marked off by the lines in question. On the other hand, in some of the gold preparations, a purple-gold staining in rosette-form in the position of the lines, suggestive of the presence of a branched cell, points to a possible action of the branches of such a cell on the fibre.

We have seen isolated in a frog's lens which had been long in gold solution, a small stellate cell with very fine processes.

The injection of gold solution from the arteries is a method that promises great results in histological investigation, as will be understood by a short resumé of what we have seen during the very short time that we have used it. Its advantages in the examination of the lens have been described above. In connective tissues generally it gives preparations of great beauty and instructiveness, and subsequent staining with logwood, though not essential, is advisable for purposes of demonstration. In the tongue of a rat, injected from the carotid, we have seen layers of flat cells investing the larger bundles of fibrillary tissue, like an epithelium, and contrasting with them small irregular nucleated clumps of protoplasm (stellate cells), with long slender glistening processes, indistinguishable from elastic fibres. Distinct from the stellate cells in such preparations are to be seen spindle cells with fine processes and very scant protoplasm, the varying size of the oval nucleus indicating the nature of the cell to which it belongs.

The elastic fibres in tendon are, by methods hitherto employed, difficult of demonstration, and one competent observer has lately confessed his inability to see them at all. Injection of gold solution by the femoral artery is an easy and simple method of demonstrating the elastic fibres in tendon. Longitudinal sections, made either directly or after hardening in alcohol, shew a number of straight strong fibres, bifurcating and anastomising as elastic fibres do elsewhere.

In the cornea, gold solution injected from the carotid, gives results, the importance of which it is impossible to overestimate. In a successful preparation, the stellate cells are seen to be the central points of a rich system of long, even, bifurcating and anastomosing fibres, similar to the elastic fibres just described in tendon, and separate and distinct from them parallel chains of spindle-cells are arranged in planes parallel to the layers of fibrillary tissue.

For a special study of spindle cells we recommend the sclerotic of a frog injected with gold.

In gold-injected tissues the solution frequently penetrates the nerve-sheaths, and hardens the individual medullated nerve-fibres, which, after subsequent logwood staining, are well fitted for study in regard to some of the points recently raised in regard to their structure.

In conclusion, we would remark, that the demonstration in the lens of layers of flat cells, possessing all the characters that are comprised in the definition of epithelium, by the use of reagents that demonstrate similar layers possessing equally the characters of epithelium in other tissues, is of great importance in forming a generalisation in regard to the plan on which animal structures are developed.

We have found the structures described above in the lens of the toad, frog, rat, and rabbit.

EXPLANATION OF THE PLATE.

The Figures have been all drawn as seen by Hartnack's Obj. 7, Oc. 3. Tube out.

Fig. 1.   Lens-fibres of *Bufo Vulgaris* in aqueous humour.

Fig. 2.   Cells of lens-fibres (*Bufo Vulg.*).   Gold.

Fig. 3.   Cells of lens-fibres (*Bufo Vulg.*).   Gold injection from the aorta.

Fig. 4.   Sheath of lens-fibre, from same lens as *Fig.* 3.

Fig. 5.   Primary fibres of same lens as *Fig.* 3.   Some of the cells of the fibres are seen.

Fig. 6.   Part of surface layer (*Bufo Vulg.*).   Gold.

Fig. 7.   *a.* Isolated cell from surface layer.   *b.* Oval cell isolated from a layer.   *c.* Cell isolated from a fibre.

Figs. 8 and 9.   Parts of layers of oval cells (*Bufo Vulg.*).   Gold.

Fig. 10.   From a sheep's lens.   Sealed in gold solution.

Fig. 11.   From a rabbit's lens.   Gold and acetic acid (ordinary method).

ON THE CENTRAL NERVOUS SYSTEM, THE CEPHALIC SACS, AND OTHER POINTS IN THE ANATOMY OF THE LINEIDÆ. By W. C. M'INTOSH, M.D., &c. *Murthly.* (Plates x, xi, xii, xiii.)

THE group of Nemerteans to which the Lineidæ belong is distinguished amongst other things by the long and narrow shape of the ganglia, and by the size, prominence and position of the cephalic sacs, which abut on the posterior border of the superior ganglionic lobes, above the origin of the great lateral nerve-trunk on each side. In the previous observations on these structures[1], it was shown that in some forms, for example *Lineus gesserensis,* O. F. Müller, the cephalic sacs were much more differentiated from the ganglia than in *Lineus sanguineus,* Jens Rathke, *L. lacteus,* Montagu, and others, though no doubt was expressed as to their being in every form structures *sui generis,* intimately related to the ganglia, but performing a special function. That function was unknown, though it was hinted it might be excretory. The revival of a similar view to that of the late Professor Keferstein, viz. that the sacs are mere continuations of the ganglia, by Dr Hubrecht, Conservator of the Museum of Zoology at Leyden, gives an opportunity of making some further remarks on the minute anatomy of the central nervous system of the group, which a more detailed study of the preparations warrants. The majority of the animals had likewise been investigated formerly in the living condition, and though there has been no chance of supplementing the information from this source, the figures and descriptions already made are sufficiently numerous to afford a fair counterpoise to the appearances observed in the mounted specimens.

The general form and structure of the ganglia and cephalic sacs have been indicated in the foregoing work, so that on the present occasion the remarks shall be confined to their minute anatomy in favourable preparations in the typical forms.

[1] *Ray Society,* Nemerteans, p. 112, Pl. 18, f. 10, Pl. 19, f. 1, 2, 3, Pl. 22, f. 1, 10, and Pl. 23, f. 9; and *Jour. Micros. Soc.* July, 1875.

In horizontal sections from above downwards it is found
that in *Lineus gesserensis* the superior sections present a uniformly cellular appearance (Pl. x. Fig. 1), except a circular
area (*a*) of fibres situated rather to the exterior of the middle
line, and somewhat posterior in position. The fibrous area, or
superior cornu, is bounded externally by a distinct band of
muscular fibres (*b*), and, in sections superior to that represented,
nearly reaches the outer border of the ganglion; while there is
a considerable fibro-cellular area behind the muscular band (*c*)
at the posterior margin. In *Lineus marinus* the superior
fibrous cornu is larger than in *Lineus gesserensis*, but the
ganglionic nerve-cells are less distinct. In the next slice the
fibrous area just mentioned in the latter form is enlarged, and
a considerable increase has taken place in the anterior ganglionic region. The upper wall of the cephalic sac is likewise cut,
and is readily distinguished by its somewhat coarse cellular
appearance. In *Lineus marinus* (Pl. xii. Fig. 8) the fibrous
area (*a*) is less rounded—being transversely elongated in the
form of a club—towards the level of the upper part of the
cephalic sac, *m*. The area is proportionally larger and is nearer
the posterior border of the ganglion. The muscular investment
of the ganglion posteriorly is thicker than in *Lineus gesserensis*.
In the next section in the latter species (Pl. x. Fig. 2) the
general superficies of the ganglion is greatly enlarged, but the
fibrous area (*a*) is still posterior and external in position. The
cephalic sac (*m*) now forms a large semicircular body behind,
separated distinctly from the ganglion by the investment of the
latter, and its own capsule. The cells in the centre of the sac
closely resemble those of the ganglion, while those near the
posterior border are larger and coarser. The cephalic sac in
Lineus marinus is developed within the dense posterior investment of the ganglion.

A great increase next occurs in the fibrous area, which at first
assumes the form of a curved sausage. Slanting from the old
(superior) area, the fibrous band curves inward and backward
and then outward and backward—projecting beyond the cellular posterior region of the ganglion, and indenting the cephalic sac (*hb*, Pl. x. Fig. 3). Moreover, while the inner and outer
portions of the sac are well defined from the ganglion, the

central fibrous part presents no such clear line of demarcation (Pl. x. Fig. 3). The majority of the cells in the sac, as already mentioned, very much resemble those of the ganglion, but towards the posterior and inner curve there are larger and coarser clear regions, some of which, however, may be due to folds. The latter area of the sac therefore presents a somewhat glandular appearance. In *Lineus marinus* (Pl. x. Fig. 8) the fibrous posterior cornu is much less developed, while the area (*ha*) in front occupies the greater part of the ganglion. The cornu, *hb*, also lies nearer the inner border of the ganglion. The cells of the sac, *m*, like those of the ganglion, are less distinct than in *Lineus gesserensis*. In sections beneath the foregoing (e.g. Pl. x. Fig. 9), the same continuity of the posterior fibrous cornu, *hb*, with the cellular contents of the sac, obtains. In this species the great fibrous area, *ha*, goes quite to the anterior border of the ganglion.

Immediately beneath the foregoing in *Lineus gesserensis* the fibrous area rapidly increases anteriorly, but projects less into the cephalic sac posteriorly, though there is no clear line of demarcation at its termination. It touches the circumference of the ganglion at the anterior third—externally. The coarse (gland-like) area of the cephalic sac is somewhat larger, and the capsule of the organ is very evident. Then a slight diminution in the general area of the ganglion takes place (Pl. x. Fig. 4). The fibrous area, *ha*, stretches nearly from front to back, but the posterior cornu, *ho*, is now directed inward, so that it abuts against the capsule of the ganglion ; and the investment of the cephalic sac is distinctly differentiated from the ganglionic border. The posterior end of the fibrous area has thus moved considerably inward. Another fibrous area, *hc*, further appears at the anterior and inner region, and fibres proceed from the front of this apparently to the opposite ganglion. The nerve-cells occur in greatest numbers behind the latter area, for externally they form a comparatively narrow belt. The cephalic sac, *m*, is nearly as broad (transversely) as the ganglion, and presents externally the opening, *m''*, of the cephalic duct, *m'*. The coarse cellular (rugose or gland-like) area is very conspicuous at the inner curve posteriorly.

The area of the ganglion still decreases as we proceed down-

wards, (Pl. x. Fig. 5), while the separation of the cephalic sac
becomes more pronounced.   In front of the well-defined posterior border of the ganglion there is now a considerable group of
cells, the great fibrous area, *ha*, having moved forward.   The
anterior fibrous area (*hc*) is larger, and approaches the internal
border of the ganglion.   The cellular region behind the foregoing shows three or four prominent transverse bands of fibres
(probably interfascicular), from the great fibrous area to the internal border; indeed, this is a common feature in the organ,
the cellular parts being traversed by such bands, as represented
in the various figures.   The cephalic sac, *m*, in this section, is
relatively large, and exhibits a gradation of cells from the
smaller series near the entrance (*m″*) of the cephalic duct (*m′*)
to the larger and somewhat rugose area at the inner border
posteriorly.   It will also be observed that the latter projects
much inward (towards the middle line), while a section of the
great nerve-trunk, *n*, in this and the previous figure appears at
its external border.

The next section (Pl. x. Fig. 6) shows three fibrous divisions,
ha, *hc*, and a smaller intermediate area.   The anterior (*hc*)
reaches the internal border of the ganglion anteriorly, and
covers the largest area; the two posterior being situated in the
substance of the cellular region, which is continued without interruption into the nerve-trunk, *n*; while in vertical sections,
e.g. (Pl. xi., Fig. 6), the fibres of the great nerve-trunk can be
followed anteriorly into the superior fibrous area.   The remnant
of the cephalic sac, *m*, has passed to the inner border of the
sheath of the great nerve-trunk, and is thus wholly posterior to
the ganglion.   In this preparation numerous fibres pass from
side to side between the sacs, and probably indicate the obliteration of the vascular spaces of the region.   In sections immediately beneath the foregoing, only a single fibrous area occurs at
the anterior and inner region of the ganglion, the rest being
cellular, with transverse fibrous streaks.   It is also worthy of
note that outside the ganglionic investment there are many
spaces (generally marked *e* in the figures), enclosed by the
longitudinal muscular and other fibres, filled with cells which
resemble those in the ganglion.

In the next section, Pl. x. Fig. 7, a larger fibrous area, *ha*, is found anteriorly, and, moreover, it almost reaches the internal and anterior borders, an arrangement due to the presence of the fibres of the inferior commissure (which is soon observed). The interfascicular streaks are well-marked in this region, and are directed for the most part backward and outward. The inferior part of the ganglion consists solely of cells resting on the investment.

In longitudinal vertical sections from without inward we find externally a somewhat rounded large fibrous area of the great nerve-cord near the ventral region, with cells above, in front of, and beneath it. The fibrous area abuts on the posterior boundary, then diminishes in size, and assumes an ovoid appearance. The cells also show a tendency to accumulate behind it inferiorly. The section of the cephalic duct lies in front of and above the fibrous area.

The following slice (Pl. XI. Fig. 1) has the fibrous area of its nerve-trunk, n, surrounded by cells, the posterior and upper part alone being somewhat thinly covered. The cells, h, of the external surface of the ganglion are also visible. A fibrous band, b, passes downward behind the latter to the front of the nerve-trunk, cutting off the cephalic sac, m (which now appears) and the trunk from the ganglion; while the cephalic duct, m', is in front of and rather above the level of the nerve. A groove, m'', is observed in this section of the cephalic sac, as if the duct were prolonged backward into the area behind, and in other preparations an appearance is presented as if the duct entered inferiorly, and ran upward and backward to the posterior border.

In the next section, Pl. XI. Fig. 2, the cephalic sac, m, comes out very distinctly. The band, b, behind the ganglion is less prominent, though still evident enough from the summit to the inferior part of the cephalic sac, where its fibres become obscure. This line of separation slopes from above downward and forward, the ganglionic section in front of it being somewhat pear-shaped, with the apex at the top posteriorly. The whole circumference of this pear-shaped area is cellular (h, h) except the upper angle, the centre being occupied by a fibrous area, a, shaped somewhat like a leg of mutton, with the shank directed towards the narrow end of the pear. The superior or pointed end (shank of the leg of mutton) evidently corresponds to the small fibrous area observed superiorly in horizontal sections, and this appears almost to touch the ganglionic circumference. This area is less characteristic in *Lineus marinus*, though still present. In contrasting similar sections of the latter and *Lineus gesserensis*, the great distinctness of the cephalic sac in the latter is apparent. The cephalic sac, *m*, Pl. XI. Fig. 2, is semicircular, and held in position by various vertical bands, *ma*, which pass from the roof of the vascular space to the investment. The inferior curve is rounded, the superior more pointed, and in the section the scar (*m''*) of the entrance of the cephalic duct is evident above the inferior border. The cephalic sac has a deeper yellowish tinge than the ganglionic cells in the preparation, and is filled with granular cells, the surface being somewhat corrugated as if from folds. At the upper border anteriorly is a dense granular deposit, and a similar arrangement is noticed at the inferior border posteriorly. The latter abuts on the nerve-trunk, which has its somewhat ovoid fibrous area surrounded by the cellular coating, the latter obtaining its greatest bulk inferiorly and anteriorly, the thinnest part being, as before, at its upper and posterior boundary.

In the next slice a considerable change takes place, Pl. XI. Fig. 3. The boundary-line, *b*, of the ganglion posteriorly is now less oblique, though it still inclines forward inferiorly. There is little distinction between the nerve-area and the ganglion, the whole having somewhat the appearance of a kidney, with the cephalic sac attached rather on one side of the hilum. The great fibrous area, *a*, has an irregular outline, pointed above, *a'*, where it nearly reaches the circumference of the ganglion. The upper margin is botryoidal, while the inferior presents a slight fibrous process, *ab*, near the posterior margin, which process passes downward and slightly forward. Towards the upper third of the great fibrous area is appended a large posterior process, *hb*, which, in the preparation, penetrates (by about $\frac{1}{3}$ of its bulk) the cephalic sac. A distinct (fibrous) separation is visible between this and the main area. The investment of the sac limits it for a short distance at the edge, and then it seems to be lost in the general mass of the sac, one band being apparent at the upper, and another at the lower end.   Inferiorly the ganglion merges into the cellular coating of the nerve-trunk, the fibrous area of which (n) occupies a similar point to that in the foregoing section.   The cellular area of the ganglion throughout presents a somewhat streaked appearance, the streaks being chiefly vertical.

The cephalic sac shows a distinct investment all round, with the exception of the hiatus for the entrance of the posterior fibrous cornu, hb.   This investment is contractile and probably muscular.   Bands (ma) pass from the upper region downward over the investment, as before.   In *Lineus marinus* (Pl. XI. Fig. 9) the arrangement somewhat differs.   The great fibrous area, a, forms a longer median mass, and its long axis is directed horizontally from the front to the back of the ganglion.   A long spur, ad, comes off anteriorly and inferiorly, and passes downward and forward.   The posterior cornu, hb, is very distinct. The small fibrous area, a', probably is in connection with the superior fibrous region (or cornu) seen in upper horizontal sections.   The fibrous area of the nerve-trunk (n) is likewise horizontally flattened.   Still further outward (for the sections are in this case from within outward) than in Fig. 9, in *Lineus marinus*, the arrangement, as shown in Pl. XI. Fig. 10, is as follows: The great area, a, has the form of a curved sausage, (with the concavity in front) stretching from the upper region of the ganglion downward and forward quite to its anterior margin.   Just below its upper third a spur, hb, passes upward and backward to the cephalic sac, which it enters and assumes a slightly bulbous form.   The great elastic outer investment (mc) of the sac is well shown in this preparation, and appears to be independent of the proper sheath of the organ.   The investment (md) presents a finely-streaked appearance in longitudinal, horizontal, and transverse sections, so that it forms a very efficient elastic and probably muscular compressor.   The elastic investment is very much more developed than in *Lineus gesserensis*.   The disproportion between the size of the posterior cornu is very distinct in the two species, *Lineus gesserensis* having the large process (actually), while its ganglion is much smaller.

The following slice (Pl. XI. Fig. 4) exhibits the great fibrous area (a) shorn of the superior limb, and somewhat oblong in outline. The superior region is convex. The inferior process, ab, is now larger, while only a trace of the posterior cornu, hb, exists at the sac. No indication of a separation between the cellular region around the area of the nerve-trunk (hn) exists, a feature to be borne in mind in connection with the complete disjunction between the cellular substance of the cephalic sac and the ganglion-cells. The fibrous area of the nerve-trunk has no cells outside its upper and posterior region, as before mentioned. The cephalic sac is also somewhat diminished.

In the succeeding section (Pl. XI. Fig. 5) the posterior cornu has retired quite within the cellular mass of the ganglion, the posterior investment (b) of the latter passing continuously from above down to the base of the fibrous area of the nerve-trunk. A great cellular region, h, occurs above the fibrous area superiorly, but the latter, at ac, now touches the anterior margin of the ganglion, and appears to send a process forward. There is a bulbous part below, and then the chief feature is the increase in the inferior cornu, ad. The cephalic sac, m, presents a uniformly cellular aspect, more opaque, however, than the cellular regions of the ganglion. Thus the fibrous band proceeding to the cephalic sac is greatly larger and more closely connected with the great area than in *Lineus marinus*, and the cephalic sac itself is somewhat more differentiated (in *Lineus gesserensis*). The fibrous area of the nerve-trunk shows a furrow superiorly.

A considerable change is apparent in the next section (Pl. XI. Fig. 6), for, instead of being chiefly superior, the fibrous area is now in the lower part of the ganglion. In the upper half a limited double area occurs near the anterior margin. Inferiorly the area of the nerve-trunk is connected by a curved band which passes beneath the former, but does not quite reach the anterior margin of the ganglion. In the concavity beneath is another ovoid area, a', somewhat larger than any of those above, and probably in connection with the inferior commissure. The great nerve-trunk thus derives its fibrous area from the inner and anterior part of the ganglion, and probably from the spur seen there in horizontal sections. In the preparation there is a small rounded process, ob, cellulo-granular in structure, projecting from the upper and posterior margin of the ganglion above the proboscidian sheath. The cephalic sac is almost separated, and has now descended to a lower level than formerly, that is its border projects downward and inward so as to be below the proboscidian sheath (o) in this section. It is still uniformly cellular. The shape of this inferior area quite differs in *Lineus marinus* (Pl. xi. Fig. 8). The investment of the cephalic sac, *md*, in this form springs superiorly from the capsule of the ganglion, indeed it seems to be a continuation thereof all round. In sections, such as that just mentioned, there are curved lamellæ of cells (often like folds) *e*, *e*, bounded by muscular fibres above and below the ganglion. These may account for the appearance seen in horizontal sections, where bands cut off certain parts of the area. A similar condition has been mentioned in *Lineus gesserensis*, and at first sight, in superficial horizontal sections, it appears that the fibrous area abuts on the border of the ganglion, whereas it only touches one of these muscular bands.

In the succeeding section the ganglion is divided into two portions by the proboscidian sheath and the proboscis. The upper is small and—with the exception of the fibrous area of the superior commissure in front—cellular. The inferior is fully twice the size of the superior. The fibrous area is superior —in the shape of a broad, curved mass, the anterior region joining the commissure, and the rest forming the communication with the great nerve-trunk. The fibrous area touches the dorsal margin of the ganglionic division throughout the greater part of its extent. The cells are inferior. A fragment of the cephalic sac still lies behind the region.

The cellular area in the next slice is much diminished in the upper division, the small rounded fibrous area being quite in front. The inferior division is separated by a wide interval, and is somewhat anterior in position, that is, its anterior border is nearer the snout. The fibrous area is much diminished, and lies just beneath the thick investment of the division dorsally.

Proceeding still further inward, Pl. xi. Fig. 7, we find the commissures nearly at a minimum. Superiorly there is the small rounded superior commissure, *f*, with a few cells behind it. Inferiorly the division shows only the section of the fibres of the inferior commissure, the rest being cellular. The fibrous area (g) is irregularly rounded, and occupies the upper and anterior region.

It will easily be seen from the foregoing that the inferior commissure is not a mere band of fibres, but that it has a series of cells forming a continuous investment inferiorly, but which, dorsally, are confined to the centre, the fibrous area reaching the circumference at the sides of the commissure. It is also, as already indicated, continuous with the great fibrous area. It is well, further, to bear in mind that in the living animal the parts present a very different appearance, for then the cephalic sac forms a large globular organ, and the fibrous peduncle of connection with the ganglion is probably considerably elongated. In the young animal the condition is still more divergent, since the cephalic sacs are situated considerably behind the ganglia.

As the minute structure of the snout of *Lineus marinus* was not specially detailed in the 'British Annelids' (*Lineus gesserensis* having been selected as the type), a few remarks on the subject may be made in the course of the description of the ganglia.

The chief point of interest in the anterior sections is the structure of the eyes (*oc*, Pl. xii. Figs. 1 and 2), which are composed of masses of black pigment somewhat irregular in shape but smooth in outline, as if in a capsule, and having a pale patch at one side (generally the dorsal)—probably the rudimentary condition of a lens. The stroma consists, immediately behind the cutaneous structures—in transverse vertical section, of a series of longitudinal fibres scattered generally over the area, but chiefly congregated in the region above a line drawn between the cephalic fissures. A somewhat radiating series of fibres converge to a point above the proboscidian channel (towards the dorsal border). As the channel falls (proceeding backward) into the centre of the snout, it is found that the radiating fibres in the line between the cephalic fissures assume the form of a powerful band above the proboscidian channel. Beneath the latter there are also many isolated transverse muscular bands, some of which decussate in the middle line. The great muscularity of this supra-

proboscidian arch becomes more marked in the succeeding
sections.   The proboscidian canal, *ao*, is not separated from
the stroma, but shows internally its mucous layer, *ab*, and is
surrounded by a vast series of longitudinal muscular fibres,
with various radiating and interlacing bands.   A considerable
increase in the size of the proboscidian channel, Pl. XII.
Fig. 1, next occurs.   The chief development of pigment is
dorsal, and the eyes, *oc*, lie in the region above the cephalic
fissure on each side.   The great muscular band, *em*, above the
proboscidian channel is very evident—being composed of
fibres which converge dorsally, ventrally, and laterally, with
longitudinal bundles in the meshes.   Around the proboscidian
canal is an investment (*ac*) of strong longitudinal bundles.
Beyond this, and a little above the level of the canal, a space
soon appears (Pl. XII. Fig. 2, *as*) on each side, the inferior
margin thereof being formed of fibres which slant downward
and inward to the bottom of the canal.   The vascular channel
is first formed by a splitting of certain of the diverging (or
converging) fibres from the band above the canal.   The rest of
the section inside the cutaneous textures may best and most
briefly be described as a great muscular plexus of longitudinal,
transverse, and oblique fibres, densest towards the centre,
though the longitudinal and other fibres are well developed
at the inner part of the cephalic fissure, *b*.   A more distinct
differentiation of the central region next occurs.   By the
rapid increase of the vascular channel (Pl. XI. Fig. 11, *as*)
on each side of the proboscidian canal, the latter is now free
laterally, but still has a dorsal and ventral attachment—the
latter being the broader.   The great dorsal band of fibres, *em*,
remains, but by the differentiation of its lateral and ventral
connections a distinct investment, *e,e,* for the combined vascular
and proboscidian region is formed.   From this investment
fibres, *er*, radiate in all directions, especially superiorly and
inferiorly, and often cross each other; while numerous longitudinal bundles, *lm*, occur in dense series over the greater
part of the area.   Within the general muscular sheath above
mentioned a thin series of longitudinal bundles, *am*, form a
kind of inner coating all round, with the exception of the
dorsal and ventral connection of the proboscidian canal.   This

layer in transverse section has a somewhat crenate aspect. Its inner surface is covered throughout by a granular coating, *mo*, which indeed lines the vascular canal all round. The canal is half-moon-shaped in ordinary sections, and is often filled with a granular fluid. In the centre of the united area is the proboscidian canal, *ao*, having externally its portion of the granular layer, *mo*, of the vascular chamber on each side, and then a thick investment of longitudinal muscular fibres (*ac*), the centre being occupied by the streaked, glandular or mucous layer (*ab*) of the canal.

A little behind the foregoing the proboscidian canal becomes attached by a broad base inferiorly, while superiorly the connection is narrower and less powerful. Around the united area are the strong radiating muscles, which form a comparatively smooth inner margin, but send off branches in all directions externally, so as to constitute a most elaborate plexus of crossed fibres all round, with the longitudinal muscles and nerves in the meshes. Superiorly, fibres converge downward from each side of the middle line and pass amongst the longitudinal series enveloping the proboscidian canal—a coating of granular tissue being likewise on each side of this isthmus. As formerly, the external lateral wall of the proboscidian canal is somewhat firm and granular, the region between this and the glandular layer being composed of interlaced circular and longitudinal fibres. A considerable change has thus taken place in the investment, which anteriorly was chiefly composed of longitudinal muscular fibres. Internally the canal is lined by its streaked glandular layer. Inferiorly it has a broad connection with the tissues, and by the crossing of bands from the circular fibres into the latter a very firm attachment is obtained. The layer of longitudinal fibres—covered internally by a glandular layer—still lines the common muscular investment at the vascular canals.

The chief change that takes place between the last-mentioned region and the front of the ganglia is the great increase in the lateral vascular spaces (on each side of the proboscidian channel) and the appearance of a new vessel in that portion which we have called the broad base. This new channel, Pl. XIII. Fig. 1, *vm*, has an ovoid form, a muscular investment resting on the general muscular boundary of the united area and arched over superiorly by the walls of the proboscidian channel. The longitudinal muscular lining of the circular or radiating investment of the common area is much diminished, and evidently is disappearing. The wall of the proboscidian channel is thick. Externally the somewhat fibro-granular outer part (not differentiated as a distinct layer), then a somewhat thick coat of longitudinal fibres, a c, finely fibrillated, and enveloped in a very distinct intermuscular substance, with numerous fibres passing downward from the arch above. This layer goes to the inferior common investment of the area, on each side of the new channel mentioned above. Within is a thick circular layer (ad), upon which the glandular internal lining (ab) rests. The latter quite differs in appearance from the glandular lining of the proboscis, so that there can be no doubt we are still describing the proboscidian channel, which thus has obtained its maximum degree of complexity. The cephalic fissures have now passed into large ducts. The hypertrophy of the circular layer outside the glandular internal lining inaugurates the great change which soon takes place. At the anterior border of the ganglia externally (next the ganglionic sheath) is a layer of longitudinal fibres with distinct fasciculi, then a narrow glandular layer with traces of a translucent basement-layer externally (next the longitudinal layer). Still further backward what appeared to be the ganglionic sheath now assumes the form of a special coat of circular fibres (c, Pl. XII. Fig. 3), having within it, ac, the well-marked longitudinal layer all round. Inferiorly, however, a separation has occurred in the longitudinal coat (and in this is the channel, vn), the somewhat rigid-looking glandular layer, ab, lying within a very thin basement-layer. The free surface of the channel, vn, has no special coat, though it is somewhat granular throughout. This view quite agrees with that of *Lineus gesserensis* given in Pl. XXII. Fig. 1, Ray Society, though the inner coats somewhat diverge, because in the former figure the cut has been a little further back.

The next noteworthy change is the appearance of a series of circular fibres (c', Pl. XII. Fig. 4) which separate the outer

longitudinal coat into two in the superior area (or area of the
tube), and soon assume considerable strength. Inferiorly
some of the circular fibres pass straight down to be attached
to the transverse fibres which support the arch over the channel
(vn). Moreover, a definite canal, p, appears between the
basement-layer of the glandular lining and the inner longitudinal muscular layer.

A further alteration occurs at the posterior part of the
ganglia, and consists of an analogous condition to the pre-
ganglionic arrangement, for the united ganglionic and proboscidian area now gets enveloped by strong circular fibres, which
permit the great increase of the infra-proboscidian channel, and
the appearance of other spaces at the cephalic sacs—very well
seen in all its features in Pl. XIII. Fig. 2. The important points
are the dense investment, md, of the cephalic sacs, the great
increase in the circular layer, c', around the inner longitudinal
muscular coat of the canal of the proboscis, and the absorption
of the outer longitudinal layer chiefly into the former, with the
exception of small isolated fasciculi, and at the sides superiorly
a few larger ones (ak), which become continuous with a longitudinal layer, ec, lining the circular general investment of the
united area (formerly mentioned). The rest of the area of the
section of the head is formed by longitudinal muscular fibres
clasped in interfascicular substance and radiating fibres, and
externally by the cutaneous tissues.

The occurrence of the infra-proboscidian canal in this region
is interesting, especially as its subsequent course shows it to be
continuous with the dorsal vessel, and, before describing the
ganglia, this will be disposed of.

The small sub-proboscidian canal (or dorsal vessel) presents,
at the above region, Pl. XIII. Fig. 2, the following appearance :
a strong arch of basement-tissue, vb, surmounted by a glandular
layer, g, better developed than the rest of the sheath, forms
the superior boundary of the vessel, while a distinct but thin
band, vb', from the edges of the foregoing, completes the circuit.
This thin inferior ring rests on the longitudinal muscular coat,
ac. Internally the canal is lined by a series of small glands.

About the region of the mouth a change occurs in the

sheath. Its circular fibres, Pl. XII. Fig 5, c', are now largely developed, and abut superiorly on the great circular body-layer which has, exactly in the central line, a solid granular (nervous?) band, be, probably analogous with that seen in *Amphiporus spectabilis* [1]. In this form, however, a granular belt proceeds from each side of the central mass between the longitudinal and circular layers. The inferior third of the proboscidian circle is made up of regularly interlaced circular and longitudinal fibres, ci, this mixed layer forming the sole coat underneath the dorsal vessel, p, and it is further intimately connected with the fibres of the region. Within is a well-defined longitudinal coat, ac, narrow above, increasing in size laterally, and again diminishing abruptly near the dorsal vessel, beneath which it ceases as an independent layer. The fasciculi of this coat are arranged in narrow rows—well seen at the thick inferior region. Next comes the translucent basement-layer, vb, having the glandular lining, g, internally. The arch of basement-membrane over the dorsal vessel seems to be denser than the rest, and the glandular lining, g', is still prominently developed. The inferior arch has now immediately beneath it a layer of lax granular tissue. The superior arch projects considerably into the sheath-cavity. In the interior of the dorsal vessel the glandular lining is still evident, and the circular fibres merge inferiorly into the decussating layer, which bounds internally the vascular meshes surrounding the œsophageal region, the central arrangement of the proboscidian sheath inferiorly being, indeed, in keeping with the structure of this system.

The chief changes behind the foregoing are (Pl. XIII. Fig. 3): the narrowing of the interlaced region (ci) of the proboscidian sheath inferiorly, and its greater differentiation from the surrounding tissues, with which, however, some of its fibres still mix. Another marked feature is the projection upward of the greatly enlarged dorsal vessel, p, so that in transverse section it forms a prominent fold in the cavity. The longitudinal layer, ac, has advanced downward, and between it and the first isolated group of longitudinal fibres, ac', is a strong band of circular fibres, c''. The wall of the anterior dorsal vessel is still thicker

[1] *Jour. Micros. Soc.* Vol. XV. N.S. Pl. XIV. fig. 1, b e.

superiorly, but the glandular tissue on the surface of the sheath immediately above forms a less continuous coating. On the internal surface of the vessel there appears to be a granular glandular lining. The characteristic granular cells occur in the sheath-cavity. The longitudinal coat of the sheath, ac, soon passes so far down as to be separated only by the vessel (Pl. XIII. Fig. 4), and the inferior decussating region (ci) is limited in area.

It is next noticed that the vessel projects less into the cavity of the proboscidian sheath, and that the longitudinal muscular layer sends processes along it in transverse section; and as soon as the layer becomes continuous over the vessel there is a distinct invagination of the sheath into the vessel (Pl. XIII. Fig. 5) instead of the opposite arrangement—hitherto characteristic. The strong band, c'', formerly alluded to as bounding the decussating region, has greatly increased in size and is continuous from side to side underneath the longitudinal layer of the sheath. The calibre of the dorsal vessel, p, has become much less, and its long diameter is transverse in the preparation instead of vertical. A continuous area, ac''', of longitudinal fibres bounds the vessel beneath and reaches up as a broad layer to the strong circular band above mentioned. The decussating region beneath is therefore narrowed. Then the longitudinal fibres, ac''', of the sheath increase in thickness over the vessel, Pl. XIII. Fig. 6, and the transverse inferior band, c'', becomes so strong that the wall now forms only a slight downward curve over the dorsal vessel, which is supported on a finely interlaced meshwork of longitudinal and circular fibres.

The dorsal vessel gradually increases in size (Pl. XIII. Fig. 7), and has the greater part of the circular coat in a continuous stratum above it; some bands, however, still pass downward beneath the latter, and in the central line a few longitudinal bundles also occur.

Finally (Pl. XIII. Fig. 8) the sheath, ao, is quite separated from the dorsal vessel, p, which is inferior and of large size. The vessel rests on the œsophageal wall, and is surrounded by its sheath and the stroma of the region.

The origin and relations of the dorsal vessel in the *Lineidæ* afford additional proof of the important connections subsisting

between the proboscidian sheath and its contents on the one
hand, and the circulatory system on the other. The occurrence
of distinct vessels from the chamber in *Amphiporus spectabilis*,
and of two lateral vessels on each side internally in that of
Valencinia Armandi, have been formerly mentioned. In no
feature is the distinction between the *Lineidæ* and the *Carin-
ellidæ* better shown than in the dorsal trunk. The latter vessel
does not appear to have any connection with the pre-ganglionic
meshes at the sides of the proboscidian canal, communication
between the latter and the posterior ganglionic meshes being
completed by the sub-proboscidian channel (*vn*).

In transverse vertical section the ganglia of *Lineus marinus*
present the following appearances :—Behind the section of the
anterior nerves, Pl. XII. Fig. 6 (left side), are the superior commissure, and the cut ends of several fibrous spaces, *a''*. In this
section there are also traces of the vascular spaces at *as* and
as'. A little further backward (right side of section) is superiorly the ganglionic commissure (*f*)—somewhat indistinctly seen,
a circular fibrous area, *a'*, immediately beneath, and probably
continuous therewith, and certain muscular bands obscuring the
parts above and below this area (two diverging transversely).
The rest of the ganglion consists of a large central fibrous area,
a, and an indistinct smaller area inferior and internal, the
commencement of the inferior commissure being shrouded in
fibres. A somewhat elliptical cellular region lies beneath the
great fibrous area. Then (Pl. X. Fig. 10) a cellular region,
hi, comes out clearly superiorly, having a large and somewhat irregular fibrous area, *ha*, bounded by a cellular belt, *hi*,
inferiorly. The inferior horn of the fibrous area is next seen to.
be continuous with the inferior commissure, *g*. A simplification
ensues behind this, the ganglion having superiorly a large cellular area, centrally an external fibrous area of a somewhat
rounded form, another smaller fibrous area, also rounded,
internally and below, and inferiorly a long cellular belt, which
quite passes from the proboscidian sheath to the limit of the
ganglion inferiorly. In Pl. XII. Fig. 3 (right) the changes are:—
The passage of the inferior fibrous area (*n*) towards the inner
border, an alteration of its long axis, and the great development of the cellular area (*hn*, commencement of the nerve-cord)
around this inferiorly and internally. The fibrous area, *n*,

touches the ganglionic investment at the proboscidian sheath.
The great external fibrous area (a) passes somewhat further
upward and has assumed a different shape, and a separate
fibrous area internally is still evident, but the smaller one above
the nerve-trunk has now disappeared. Thereafter two fibrous
regions only remain, the superior being diminished and surrounded on all· sides by the cellular.

The next great alteration is the decided separation which
occurs between the superior and inferior regions of the ganglion.
In a section immediately behind the foregoing the capsule of
the ganglion is somewhat better differentiated, especially on the
internal, superior, and inferior borders. There is a large and
distinct fibrous area towards the inferior part of the upper
division, but also an indication of a superior and outer one in
connection with the former. At the outer and inferior part of
this division is a slice of the cephalic sac (representing the
anterior inferior part of the organ), the aperture for the cephalic
duct passing clearly into it; and also the track of the duct
leading to the cephalic fissure. The outer border of the sac is
guarded by a strong investment, passing downward from the
capsule of the ganglion. Some bands beneath the ganglion
(superior division) seem to pass inward to the wall of the
sheath, and this attachment gives the area in section a somewhat triangular appearance, since there is a strain put upon the
inferior angles. The inferior region now shows the nerve-
trunk connected with the fibrous area, and surrounded on three
sides by the cellular investment. There is a small isolated
fibrous area at the base, close to the proboscidian sheath. In
a section very slightly further back (Pl. XII. Fig. 4), two distinct fibrous regions exist—one, a, at the inner and inferior
border of the ganglion, and the other, a', at the upper and outer.
A somewhat long slice (md) of the sheath of the cephalic sac
occurs, and the cephalic duct, m', can be traced a considerable
distance outward from its opening into the former at the external
border. The fibres (cm) from the roof of the vascular inter-gan-
glionic area pass at their outer angle into the capsule of the sac.
Then (Pl. XIII. Fig. 2, left) only a single fibrous area, hb (posterior cornu) occurs in the centre of the ovoid outline of the
ganglion and cephalic sac, and the latter is differentiated inferiorly. In such a section part of the ganglionic tissue appears,

since the cephalic sac is applied like a cap to its prominent posterior border (see longitudinal sections). The dense capsule, *md*, round this area is characteristic. Above the area is a portion, *h*, apparently of the posterior and upper part of the ganglion, which is probably continuous with the rest in the upper part of the oval area, and externally (outside the capsule of the sac) is a great vascular space, *s*, from summit to base. In the next section (same figure—right), the fibrous area has disappeared, and the section of the diminished sac, *m*, is purely cellular, while the investment, *md*, has considerably increased in thickness. The great longitudinal muscular layer, *e*, with interlaced intermuscular fibres and substance, surrounds the entire central area. In the median line dorsally is a granular trunk, *be* (longer than the posterior continuation of the same), which appears to proceed from the superior com-missure, and therefore is probably nervous. Within is the circular layer, *e′*, surrounding the common area of the vascular meshes, proboscidian sheath, and cephalic sacs. This at first passes between the posterior part of the ganglion and the great nerve-trunk, *n*, but by-and-by the latter pierces it, and the fibres form a great muscular belt all round. Superiorly it is lined by a longitudinal layer, *ec*, which is also continued inferiorly, beyond the bridge over the infra-proboscidian vascular space. The internal meshes of the circular layer at the inferior margin of the vascular space, *s′*, are occupied by a granular substance, *bu*, which probably represents the divided ends of nerve-cords.

In the ganglionic region a special arrangement of the longi-tudinal and circular muscles occur, for the former are split by the latter into numerous laminæ superiorly and inferiorly (Pl. XII. Fig. 7, *e*, *e*). This disposition is confined to the gan-glionic region proper, and gradually disappears anteriorly and posteriorly.

Thus in the *Nemerteans* the nervous system is highly developed and of large size, a feature in which they strongly contrast with the parasitic worms, with which, indeed, as formerly mentioned, their resemblances are rather remote. The cephalic sacs are evidently special organs of sense, their internal surface being in direct communication with the surrounding element (sea-water) by the ciliated duct, while the fibrous peduncle places their cells in continuity with the central nervous system. In the sections most of the ganglion-cells present a simple outline, but this is due less to their actual relation to the fibres than to the mode of preparation. In the living and fresh animals such connection is easily demonstrated, both in the present group and in the *Enopla*.

EXPLANATION OF THE PLATES.

THE following letters have for the most part been used to designate the same structures throughout the figures. *a.* Great fibrous area of the ganglion; *ac.* Longitudinal muscular layer of the proboscidian sheath; *ao.* Channel for the proboscis in the snout; *as.* Pre-ganglionic vascular channels; *b.* Posterior investment of the ganglia; *be.* Nervous band—probably from the superior commissure; *c'.* Circular layer of the proboscidian sheath; *ci.* Region of interlaced circular and longitudinal fibres; *e.* External longitudinal muscular layer of the body-wall; *e'.* Circular muscular coat of the body-wall; *e''.* Internal longitudinal layer of the same; *f.* Superior commissure of the ganglia; *g.* Inferior commissure; *h.* Cellular region of the ganglia; *hb.* Posterior fibrous cornu connected with the cephalic sac; *hn.* Cellular region at the origin of the great nerve-trunk; *m.* Cephalic sac; *m'.* Duct of the foregoing; *m''.* Entrance of the duct into the sac; *n.* Great lateral nerve-trunk; *o.* Proboscidian sheath; *p.* Dorsal blood-vessel; *s.* Vascular lacunæ behind the ganglia; *vn.* Infra-proboscidian vascular channel.

PLATE X.[1]

Fig. 1. Horizontal section of the right ganglion in *Lineus gesserensis* towards its summit. *a.* Fibrous area; *b.* Longitudinal muscular band; *c.* Muscular boundary posteriorly. × 350 diam.

Fig. 2. Horizontal section of the same ganglion somewhat deeper than the former. The letters are similar. × 210 diam.

Fig. 3. Horizontal section of the left ganglion at a lower level than the foregoing, with the cephalic sac (*m*) well-developed. *a.* Fragment of the proboscis; *ab.* Union of the proboscis with the sheath; *e.* Space bounded by fibres and filled with cells; *ha.* Great fibrous area; *hc.* Anterior fibrous area. × 210 diam.

Fig. 4. Horizontal section of the right ganglion below the former preparation. *ho.* The posterior fibrous cornu in its altered position. × 210 diam.

Fig. 5. Similar section of the right ganglion still lower. × 210 diam.

Fig. 6. Horizontal section of the right ganglion near the inferior part of the cephalic sac, which has now passed inwards. × 210 diam.

Fig. 7. Horizontal section of the right ganglion, showing the continuity of the ganglion and nerve-trunk. × 210 diam.

Fig. 8. Horizontal section of the right ganglion in *Lineus marinus*. The great strength of the investment of both ganglion and sac is apparent. × 55 diam.

Fig. 9. Horizontal section of the same at a lower level than the former, showing the relations of the posterior cornu to the cephalic sac. × 55 diam.

Fig. 10. Vertical section of the right ganglion in the same species. *hi.* Cellular region; *ha.* Fibrous area. × 55 diam.

PLATE XI.[2]

Fig. 1. Longitudinal vertical section of the ganglion and cephalic sac

---

[1] To understand the position of the ganglia in the animal, it is only necessary to remember that in this Plate all the figures have the anterior end directed upwards.

[2] In all the figures, with the exception of the last (*Fig.* 11), the anterior extremity is directed to the left.

of *Lineus gesserensis* near the external border.		*a.* Trace of the great fibrous area.		× 90 diam.

Fig. 2.		Similar section just within the former, showing the leg-of-mutton shape of the fibrous area, and the position of the cephalic sac. *e'.*		Muscular coats bounding the vascular spaces behind the ganglia; *ma.*		Muscular meshes passing from the former to the investment of the cephalic sac.		× 90 diam.

Fig. 3.		Section succeeding the foregoing, and (in conjunction with Plate X. *Fig* 3) exhibiting the connection of the posterior cornu with the cephalic sac.		*a'.*		Superior process of the fibrous area—that portion corresponding with *a*, Plate X. *Fig.* 1;		*ab.*		Inferior process of the great fibrous area;		*e.*		Muscular lamellæ with intervening cells.		× 90 diam.

Fig. 4.		Section next the preceding.		*ab.*		Fibrous process from the great area *a*.		× 90 diam.

Fig. 5.		Section following *Fig.* 4.		The cephalic sac is now diminishing. *ac.*		Anterior spur of the great fibrous area *a*;		*ad.*		Inferior process of the same;		*an.*		Granular structure resembling a section of a nerve-trunk.		The anterior boundary of the alimentary chamber is observed above the latter.		× 90 diam.

Fig. 6.		Section somewhat within the foregoing, and just as the cephalic sac is disappearing.		The connection between the fibres of the nerve-trunk (*n*) and the great fibrous area (*a*) is indicated.		*o.*		Proboscidian sheath; *ob.*		Small cellulo-granular processes attached to the posterior border of the ganglion.		× 90 diam.

Fig. 7.		Section of the same specimen near the middle line (and therefore internal to that represented in the previous figure), showing the commissures.		*ao.*		Fold uniting the proboscis and the proboscidian sheath; *vs.*		Vascular meshes behind the ganglia.		× 90 diam.

Fig. 8.		The first of a series of longitudinal vertical sections of a ganglion of *Lineus marinus* from within outwards (from near the middle line).		*e.*		Cellular spaces clasped by longitudinal fibres;		*ha.*		Probably the inner edge of the knuckle in front of the posterior cornu—seen in Plate X. *Fig.* 9, with which compare.		× 55 diam.

Fig. 9.		Section external to the foregoing, showing the ganglion in complete development.		*a'.*		Small fibrous area, corresponding with *a* in Plate XII. *Fig.* 8;		*ad.*		Inferior spur of the great fibrous area;		*mc.*		Special capsule of the cephalic sac;		*md.*		External investment of the region.		Other letters as before.		× 55 diam.

Fig. 10.		Section succeeding the previous.		× 55 diam.

Fig. 11.		Transverse vertical section of the central region of the snout in front of the ganglia in the same species (*L. marinus*), exhibiting the arrangement of the muscular fibres around the area containing the proboscidian and vascular channels.		× 90 diam.

PLATE XII.

Fig. 1.		Outline of a transverse vertical section of the anterior part of the snout in *Lineus marinus.*		*em.*		Great muscular bands above the proboscidian channel;		*oc.*		Eyes.		× about 20 diam.

Fig. 2.		Similar section somewhat posterior to the foregoing, showing the commencement of the vascular channels (*as*).		× about 20 diam.

Fig. 3.		Section through the ganglia in the same species in the region of the great infra-proboscidian channel (*vn*);		*c.*		Special coat of circular fibres surrounding the united area.		× 90 diam.

Fig. 4.		Transverse vertical section through the posterior part of the left ganglion in the same species.		*a.*		Posterior fibrous cornu; *a'.*		Indication of a superior fibrous area;		*cm.*		Point of junction of the fibres from the arch over the infra-proboscidian vascular space, with the investment of the cephalic sac. × 90 diam.

Fig. 5. Similar section of the proboscidian sheath in *Lineus marinus.* *g.* Internal or glandular coat of the organ ;   *g'.* The same layer specially thickened over the dorsal vessel ;   *vb.* Basement-layer.   × 90 diam.

Fig. 6. Transverse vertical section of the anterior ganglionic region in the same form just before the termination of the pre-ganglionic vascular spaces (*as* and *as'*).   *a'.* Fibrous area probably connected with the superior commissure ;   *a''.* Sections apparently of large nerve-trunks proceeding from the left ganglion anteriorly.   × 55 diam.

Fig. 7. Similar section of the same specimen in the line of the inferior commissure, exhibiting the great development of the laminated muscular layers *e, e.* On the right the cephalic duct is observed at *m'*.   × 55 diam.

Fig. 8. Horizontal section of the superior region of the right ganglion in *Lineus marinus.*   × 55 diam.

PLATE XIII.

Fig. 1. Transverse vertical section of the channel in the snout for the proboscis (in front of the ganglia) in *Lineus marinus;* *ab.* Glandular inner coat ;   *ac.* Longitudinal external layer ;   *ad.* Intermediate circular coat ;   *vn.* Commencement of the infra-proboscidian channel. × 55 diam.

Fig. 2. Similar section at the posterior border of the ganglia in the same form. *ak.* Longitudinal muscular bundles clasped by vertical and other fibres ;   *ec.* Mixed longitudinal layer ;   *g.* Glandular lining of the proboscidian sheath ;   *g¹.* Thicker portion of the same over the vessel (*p*);   *vb.* Basement-layer ;   *vb'.* Portion of the latter, proceeding under the vessel (*p*);   *bu.* Granular spaces—probably sections of nerve-trunks ;   *s'.* Great vascular space behind the ganglia. The elevation in the centre of the latter inferiorly is probably due to the approach of the mouth.   × 90 diam.

Fig. 3. Transverse vertical section of the proboscidian sheath of the same form posterior to that represented in *Fig. 5* of the previous Plate. *ac'.* Upper group of longitudinal fibres ;   *c''.* Strong band of circular fibres passing to join a similar one from the opposite side. The vascular meshes are observed at *s.* Other letters as before.   × 55 diam.

Fig. 4. Section posterior to the foregoing. The vessel *p* now projects less into the proboscidian sheath.   × 55 diam.

Fig. 5. Section of the same structures after a posterior interval. The wall of the proboscidian tunnel is now everted into the vessel inferiorly ; the separate circular band, *c''*, has now become larger, and the longitudinal bands, *ac'''*, better developed.   *j.* Wall of the alimentary chamber. × 55 diam.

Fig. 6. Similar section behind the foregoing, showing the great increase in the separate fasciculus, *c''*, and the grouping of the longitudinal fibres, *ac'''*, around the sides and ventral surface of the vessel *p*. The interlaced region (*ci*) beneath is still extensive.   × 55 diam.

Fig. 7. Section somewhat behind the previous, exhibiting the band *c''*, almost in full development. The interlaced region, *ci*, is now very limited. The dense walls of the vessel (*p*) prevent its collapse.   × 55 diam.

Fig. 8. Section of the proboscidian sheath posterior to the former, and in the condition in which it remains throughout the greater part of the body. The dorsal vessel (*p*) is now large and patent.   *g.* Glandular internal lining ;   *vb.* Basement-layer.   × 55 diam.

EXPERIMENTS ON THE BILIARY SECRETION OF THE DOG. By WILLIAM RUTHERFORD, M.D., F.R.S.E. *Professor of the Institutes of Medicine in the University of Edinburgh; and* M. VIGNAL.

INTRODUCTION.

THE influence of mercury, podophylline, and taraxacum upon the biliary secretion of the dog was some years ago investigated by a committee, of which Professor Hughes Bennett was chairman and reporter. Dr Arthur Gamgee and Dr Rutherford were the two junior members of the committee, upon whom devolved the task of performing the experiments. On the recommendation of the committee, they resorted to the method of establishing permanent biliary fistulæ in dogs; they then gave the animals a fixed diet, and analysed the bile secreted daily before and after the administration of various substances, and they observed that "spontaneous diarrhœa, dysentery, and purgation produced by pilula hydrargyri, calomel, corrosive sublimate, and podophylline, always diminished the solid constituents of the bile, and, with one exception, the fluid portion of the bile also" (*British Association Report*, 1868, p. 229). The observations were made with such laborious carefulness, that the truth of these facts need not be called in question. It is not intended to reopen here the controversy raised by opinions expressed regarding the interpretation of these and other facts, it being the object of the present report to give the results of experiments performed by a different method.

Two years ago, Röhrig ("Experimentelle Untersuchungen über die Physiologie der Gallenabsonderungen"—Stricker's *Jahrbücher*, 1873, p. 240) performed a number of experiments on the effect of various substances on the biliary secretion. He observed the rate of the biliary flow from temporary fistulæ in fasting curarised animals before and after the injection of purgative agents into the stomach and intestine. He found that large doses of croton oil greatly increased the secretion of bile, and that a similar effect, though to a less extent, was produced by colocynth, jalap, aloes, rhubarb, and senna, and sulphate of magnesia—the potency of these agents as hepatic stimulants being in the order mentioned. He found, moreover, that castor-oil had little effect, and that calomel, while it seldom recalled the biliary secretion after it had ceased, nevertheless somewhat augmented it when it was taking place slowly.

Röhrig's statement with regard to calomel does not much differ from that made by Hughes Bennett's committee, and he made no experiments with podophylline and taraxacum; nevertheless, he did find that certain purgative agents, when given to animals that are *fasting*, increased the biliary secretion, while the committee found that in *non-fasting* animals, purgative action induced by podophylline, calomel, etc., diminished the amount of water and solids of the bile secreted in the twenty-four hours.

It appeared to the reporter that the subject should not be allowed to remain in its present position: accordingly, he undertook the following research, and invited the co-operation of Monsieur Vignal, who had come to do some work in the physiological laboratory of the University of Edinburgh.

METHOD OF EXPERIMENT.

All our experiments were performed on dogs that had in nearly every instance fasted about eighteen hours. After paralysing the animal with curara, and establishing artificial respiration, we opened the abdomen in the linea alba, and tied a glass cannula in the common bile-duct, near its junction with the duodenum. To the end of the cannula which projected from the abdomen we attached a short India-rubber tube, and to the end of this again a short tube of glass, drawn to a narrow aperture, so that the bile might drop from it. The gall-bladder was then compressed, in order to fill the whole tubing with bile, and the cystic duct was clamped to prevent the return of the bile to the gall-bladder, and so compel all the bile secreted by the liver to flow through the cannula. The wound in the abdominal wall was then carefully closed; and in all our later experiments the animal was thoroughly covered with cotton-wool, in order to quickly restore it to its normal temperature.

Röhrig estimated the velocity of the biliary secretion by counting the seconds that elapse between the fall of the drops from the orifice of the tube. A single trial convinced us that this method is extremely laborious, and by no means accurate, seeing that it does not permit continuous observation for any length of time. Moreover, we saw that the degree of viscosity of the bile caused a variation in the size of the drops, and, therefore, in the intervals between their fall. We therefore abandoned this for the more accurate method of allowing the bile to flow into a fine cubic centimeter measure, and recording the quantity secreted every quarter of an hour. In addition to constant collection of the bile, this method has the great advantage of permitting a graphic representation of the results. We, moreover, took the trouble to analyse the bile in many cases, and to make *post mortem* examinations of the alimentary canal; points entirely omitted by Röhrig, but which, nevertheless, have yielded valuable results.

Until it is attempted, one might suppose that this mode of experiment is extremely simple, but it is by no means so simple as it appears. It is needful to manipulate the abdominal viscera with great care, and to avoid all dragging at the bile-duct, otherwise the secretion of bile becomes so irregular that the experiment may be useless. The cannula must be very carefully retained in a position which will permit its moving with the diaphragm, but will prevent it from twisting the duct, and thus impeding the exit of the bile by forming a valve at its orifice.

The respiration requires to be maintained with great regularity, otherwise the biliary flow is rendered unequal by irregular diaphragmatic compression of the liver. Moreover, if the respiration be deficient, the secretion of bile is always diminished (Röhrig). In the cases indicating the biliary secretion in these experiments, some of the oscillations are probably owing to variations in the respiration, especially to variations in the amount of diaphragmatic compression of the liver; for we were obliged to have the respiration kept up by the hand, and this is never so regular as a machine. Notwithstanding this, however, the main results of the experiments are perfectly clear.

As is well known, curara is of great value in such experi-

ments, for, by paralysing voluntary movement, it prevents the
irregular outflow of the bile, which certainly follows irregular
contraction of the abdominal muscles; and if care be taken to
give doses just sufficient to produce this paralysis, the biliary
secretion is not apparently affected; but if too much be given,
the heart is rendered weak and irregular, and the secretion of
bile diminishes.

Lastly, it is essential that the fall of temperature which
follows the opening of the abdominal cavity be speedily recovered, otherwise the biliary secretion may be abnormally low at the
outset of the experiment.

SECRETION OF BILE IN A CURARISED FASTING DOG.

In all the illustrations, the numbers under the horizontal
line indicate the hours during which the secretion of bile was
observed, while those to the left of the vertical line indicate in
cubic centimeters the amount of bile which flowed from the
cannula; the dots in the curve indicate the quantities of bile
collected every quarter of an hour. The vertical dotted lines
that cross the curves in the illustrations indicate that something
was given to the animal. In all such experiments the amount
of bile first collected is usually much greater than that at subsequent periods. This apparently results from the sudden diminution in the resistance to the exit of the bile which follows the
opening of the duct. The first one or two collections are therefore not reliable indices of secretion, and they are consequently
omitted from some of the curves altogether.

The solution of curara employed in all the experiments was
a filtered aqueous solution, every minim of which contained one
milligramme of the poison. The solution was always injected
into the jugular vein.

Experiment 1. Dog weighing 7.6 kilogrammes[1].—Twenty
milligrammes of curara were injected into the jugular vein (at
a, Fig. 1). The abdomen was then opened, and the cannula
placed in the common bile-duct, as above indicated. The wound
in the abdomen was closed, the animal enveloped in cotton-

---

[1] A *kilogramme* is 2.2 pounds.

wadding, and the bile collected. As the experiment proceeded, the effect of the curara gradually wore off, owing to its elimination, and it was necessary to inject from two to four milligrammes from time to time (*b, c, d, e, f, g,* Fig. 1). If the curve be

Fig. 1.

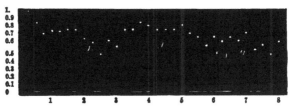

Secretion of bile in a dog that had fasted eighteen hours. Curara only given. *a*, 20 *milligrammes; b*, 2 *mill.; c* and *d*, 4 *mill.; e, f, g*, 8 *mill.* of curara injected into vein.

examined, it will be observed that these doses had no apparent effect on the biliary secretion. Large doses weaken the heart, and diminish the secretion possibly on that account; but doses so small as those given in these experiments have apparently no effect, and therefore their administration is not indicated in subsequent curves.

The secretion of bile in this case was fairly regular. After falling until the middle of the third hour, it increased for a time and then fell somewhat. At the eighth hour it was slightly below what it had been at the close of the first.

Experiment 2. Dog weighing 15 kilogrammes.—In this experiment the secretion of bile with doses of curara similar to the above was again observed. The biliary flow was not so regular in this case. It will be noticed (Fig. 2), that a high and a low reading succeed each other more than once. It is just possible that this may have been owing to a variation in the facility of exit of the bile, owing, perhaps, to some slight shifting of the cannula. The real amount of biliary secretion in such cases would probably be more nearly represented by taking the mean of the high and the low reading, as indicated by the dotted line and triangles.

Fig. 2.

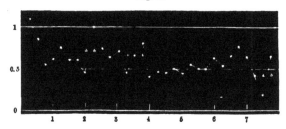

Secretion of bile in a dog that had fasted nineteen hours. Curara only given.
The unbroken curve joins the readings of the bile, that were actually taken
every quarter of an hour. The broken line and triangles indicate what was
probably the real secretion of bile when the outflow was irregular. (See
text.)

Composition of Bile in a Fasting Dog.—Analyses were made
of the bile secreted by the second dog during the first, fourth,
and last hours of the experiment. The following are the results.

TABLE ·I.

*Composition of Bile secreted by a Dog paralysed by Curara after fasting
eighteen hours.*

Experiment 2.				Bile secreted during		
				First Hour.	Fourth Hour.	Last Hour.
Water				89.53	89.58	89.55
Bile-acids, pigments, cholesterine						
fats				8.73	8.68	8.71
Mucus				0.71	0.72	0.72
Ash...				1.03	1.02	1.02
Total				100.00	100.00	100.00

It therefore appears that in the progress of the experiment
the composition of the bile remained almost precisely the same.
This is remarkable, seeing that the animal had been deprived of
water for so long a time, and, moreover, seeing that the entrance
of the bile into the intestine had been cut off. It should be
mentioned that in taking the bile secreted near the beginning
of such experiments for analysis, we were always careful to
eliminate that which had been expressed from the gall-bladder
into the cannula.

ACTION OF CROTON-OIL.

Röhrig has placed croton-oil at the head of his list of hepatic stimulants, with the statement that in doses from eighteen drops to a "teaspoonful" it has an exciting effect on the biliary secretion even under the most unfavourable circumstances (*lib. cit.*, p. 250). The substance was therefore made the subject of our first experiments with cholagogues.

Experiment 3. Dog weighing 7.3 kilogrammes.—Considering the small size of this dog, the secretion of bile was unusually great. This probably resulted from digestion being incomplete; for, although the animal was fed seventeen hours before the experiment, at death a quantity of elastic tissue, and a greyish fluid resembling chyme, were found in the stomach. After the secretion had fallen very low, fifteen grains (about thirty drops) of croton-oil, in sixty minims of almond-oil, were injected directly into the duodenum (at *c*, Fig. 3). The dose was a large one, but not so large nor yet so small as the quantities given by Röhrig. After half-an-hour, the fall in

Fig. 3.

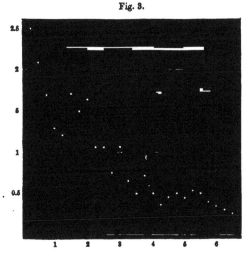

Secretion of bile in a dog. Digestion incomplete. Fifteen grains of croton-oil injected into duodenum at *c*.

the bile-secretion was arrested, and a slight rise took place. Towards the close of the experiment the pulse became extremely weak.

AUTOPSY[1].—The mucous membrane of the upper three-fourths of the small intestine was intensely red, especially in the duodenum, the colour of which resembled that of claret. There was evidence of impending purgation in the small intestine. The weak pulse at the close of this experiment, together with the violent intestinal irritation, suggested that the collapse had been occasioned by the drug, and that a smaller dose should be given in the next experiment.

Experiment 4. Dog weighing 5.9 kilogrammes.—This animal had refused almost all food for nearly two days. Six grains of croton-oil in sixty minims of almond-oil were injected into the duodenum (*c*, Fig. 4). No increase of the biliary

Fig. 4.

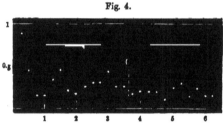

Secretion of bile in a fasting dog. Six grains of croton-oil injected into duodenum at *c*.

secretion followed. The pulse became so weak that the experiment was ended two hours and a half after the oil was given.

AUTOPSY.—The oil had found its way into the stomach. The gastric mucous membrane was of a claret colour. There was slight redness of the duodenum, but no evidence of purgative action.

Experiment 5. Dog weighing 3.1 kilogrammes.—In this experiment, only three grains of croton-oil in sixty minims of

[1] In all cases, unless otherwise stated, the autopsy was performed immediately at the close of the experiment.

almond-oil were injected into the duodenum. A decided increase in the biliary secretion began within an hour after the injection. The secretion soon reached a maximum, and then fell in the course of two hours to the same level as before the injection. (Fig. 5.)

Fig. 5.

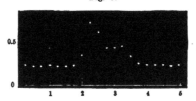

Secretion of bile in a dog that had fasted eighteen hours. Three grains of croton-oil injected into duodenum at *c*.

AUTOPSY.—A portion of the oil was found in the stomach, and another portion half way down the small intestine. The gastric mucous membrane was intensely red. There were patches of slight redness here and there in the duodenum. No evidence of purgative action.

These experiments led us to doubt the great potency of croton-oil as a cholagogue; and, seeing that probably no one would think of giving this irritant for the purpose of stimulating the liver, we laid it aside.

ACTION OF PODOPHYLLINE.

Experiment 6. Dog weighing 15.3 kilogrammes.—The secretion of bile fell very gradually (Fig. 6). Ten cubic centimeters of water were injected into the duodenum at *w*. There being no apparent effect, 100 cc. were injected at *w'*. The slight rise in secretion that ensued at the end of an hour may have been owing to this; but it is not likely, seeing that water is absorbed with rapidity. At *p*, ten grains of podophylline, suspended in 10 cc. water, were injected into the duodenum; and it is probable that the rise in secretion two hours afterwards was due to the podophylline.

AUTOPSY.—The mucous membrane of the duodenum, and to a slight extent below it, was very vascular, and this part

Fig. 6.

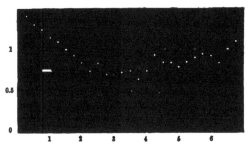

Secretion of bile in a dog that had fasted nineteen hours. w, 16 cc. of water; w', 100 cc. of water; p, ten grains of resina podophylli in 10 cc. of water injected into duodenum.

of the intestine contained a considerable quantity of a slightly brown fluid, thereby affording evidence of a purgative effect.

Experiment 7. Small Dog.—Eight grains of podophylline in 10 cc. of water were injected into the duodenum (p, Fig. 7).

Fig. 7.

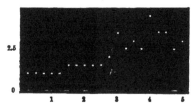

Secretion of bile in a fasting dog before and after injection into the duodenum of eight grains of podophylline in 10 cc. of water at p.

Before this, the bile-secretion was low, and remarkably regular. The distinct rise in the secretion an hour after the injection was probably owing to the podophylline, but the experiment was not continued sufficiently long to show the full effect. At death, there was increased redness of the duodenal mucous membrane, but no distinct evidence of purgative action.

Experiment 8. Small Dog.—Eight grains of podophylline in 25 cc. of water were injected into the duodenum (p, Fig. 8). The subsequent increase in the biliary secretion was most

Fig. 8.

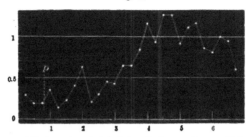

Secretion of bile in a dog that had fasted nineteen hours. Eight grains of podophylline in 25 cc. of water injected into duodenum at *p*.

marked about four hours after administration, but by the end of the sixth hour the effect had greatly diminished.

AUTOPSY.—The upper part of the small intestine contained a viscous brownish fluid. As the small quantity of water injected had probably been absorbed, the intestinal contents were regarded as distinct, though not abundant, evidence of purgative action. The mucous membrane, to the extent of about eighteen inches below the pylorus, was extremely vascular. The remainder of the intestine was pale. A small quantity of mucus was found in the stomach, the mucous membrane of which was pale.

Experiment 9. Dog weighing 6.6 kilogrammes.—Six grains of podophylline in 9 cc. of water injected into the duodenum (*p*, Fig. 9). The subsequent rise in the bile-secretion is very

Fig. 9.

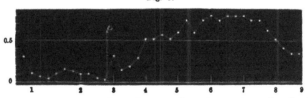

Secretion of bile in a dog that had fasted eighteen hours. Six grains of podophylline in 9 cc. of water injected into duodenum at *p*.

evident. The secretion attained its maximum between three and four hours after the administration of the podophylline.

As in the previous case, the effect on the liver had very greatly diminished by the end of the sixth hour after administration.

AUTOPSY.—There was distinct, though not abundant, evidence of purgative action in the small intestine, and decidedly increased vascularity of the mucous membrane in its upper two-thirds; nothing remarkable in the stomach or large intestine.

Probably every one will be struck by the slowness and the small extent of the purgative action in these experiments, notwithstanding the large doses of podophylline. That this was owing to the insolubility of podophylline in water is probable, from the two following experiments. Zwicke, Hagentorn, and Köhler having shewn (Fraser's Report, *Journal of Anatomy and Physiology*, v. p. 393) that convolvulin, elaterin, and some other substances have no purgative action unless they come in contact with bile—which, therefore, appears to be a solvent for them—it occurred to us that the tardy action of the podophylline might be owing to the non-entrance of the bile into the intestine. Accordingly, in the next experiment, the podophylline was suspended in bile.

Experiment 10.   Dog weighing 11 kilogrammes.—12.2 cc. bile injected into duodenum (*b*, Fig. 10).   Unfortunately, there is a hiatus in the curve immediately before the injection, owing to a loss of the bile ; nevertheless, it is evident that increased bile-secretion followed the injection when the biliary flow had become fairly constant.   Nine grains of podophylline, triturated in a mortar with 12 cc. bile, were injected into the duodenum (*p*). A rapid increase in the bile-secretion ensued ; but soon it diminished, and three hours after the injection it was lower than it had ever been.   In this remarkable experiment, therefore, the *diminution* of bile-secretion after podophylline was far more remarkable than its increase ; indeed, the increase might possibly have been owing to the injected bile, and not to the podophylline.   Towards the close of the experiment, the pulse became weak, but not excessively so.

AUTOPSY.—Mucous membrane of stomach and whole length of small intestine intensely red.   The small intestine contained a large quantity of fluid.   The large intestine contained a con-

Fig. 10.

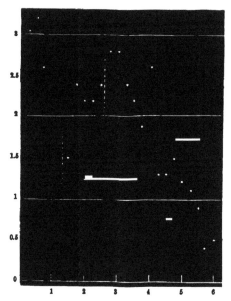

Secretion of bile in a dog that had fasted eighteen hours. *b*, 12.2 cc. bile; *p*, nine grains podophylline in 12 cc. bile injected into duodenum.

siderable quantity of liquid fæcal matters. There was, therefore, abundant evidence that excessive purgation was imminent.

In this experiment, the intestinal irritation and the purgative effect were far greater than they were in any of the previous experiments with podophylline, and it is evident that the principal change in the bile-secretion was *diminution*. It therefore appeared that, with a powerful solvent such as the bile, nine grains of podophylline produced a too violent effect upon the alimentary canal. The previous experiments having shown that, with a slighter action on the intestine, there was a more powerful action on the liver, suggested that with a smaller dose of podophylline given in the biliary solvent, an action on the liver would be evident, and that this would follow the injection more speedily than it had done in the experiments where the podophylline was not given in a state of solution.

Fig 11.

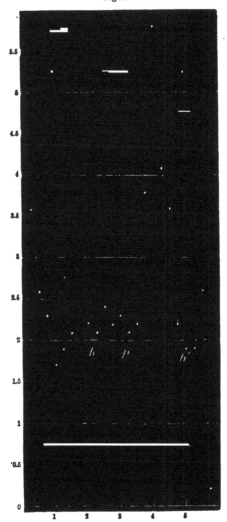

Secretion of bile in a dog that had fasted nineteen hours. 6 cc. bile and 6 cc. water injected into duodenum at *b* and *b'*. Four grains podophylline in 6 cc. bile and 6 cc. water injected at *p*.

The next experiment realised this anticipation in a very striking manner.

Experiment 11. Dog weighing 17.1 kilogrammes.—The bile-secretion was about 2 cc. per fifteen minutes before injection into the duodenum of 6 cc. bile and 6 cc. water (*b*, Fig. 11). The subsequent increase of secretion was trivial. An hour after this, four grains of podophylline, in the same quantity of bile and water, were injected (*p*). About half an hour afterwards, a great acceleration of the biliary flow began, and lasted about an hour. In one of the periods of fifteen minutes, no less than 5.8 cc. bile were secreted; a quantity never noticed in any other experiments even on larger dogs. When this great hepatic excitement had disappeared, 6 cc. bile and 6 cc. water were again injected (*b'*), as in the first instance. The fall in the secretion was for a time arrested; but within three hours after the administration of the podophylline, the action of the liver had almost entirely ceased. The pulse was weak, but not extremely so.

AUTOPSY.—The mucous membrane of the duodenum was intensely vascular, but that of the remainder of the small intestine did not show an increased vascularity nearly so great as in the previous experiment. The upper three-fourths of the small intestine contained very decided evidence of purgative effect. The gastric mucous membrane had a dull red appearance.

Experiment 12. Dog weighing 7.9 kilogrammes.—In this case, 0.5 cc. bile and 2.5 cc. water were injected into the duodenum (*b*, Fig. 12), without producing any noteworthy effect. The same quantity of bile and water, containing one grain of podophylline, was then injected, and the dose was twice repeated (*p, p', p''*). After the first dose, an increase in the biliary secretion was perceptible in half-an-hour; it never became very marked, and it lasted only about an hour. The second, but especially the third dose, was followed by a fall in secretion, probably owing to purgative action taking place.

AUTOPSY.—Evidence of severe irritation of mucous membrane of upper half of small intestine. Decided evidence of

Fig. 12.

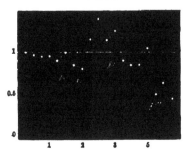

Secretion of bile in a dog that had fasted eighteen hours.   0.5 cc. bile and
2.5 cc. water injected into duodenum at *b*; the same, together with one
grain podophylline, injected at *p*, *p′*, and *p″*.

purgation in this portion of intestine. Lower part of small
intestine almost quite empty and dry.

COMPOSITION OF BILE BEFORE AND AFTER PODOPHYLLINE.

The next question to be answered was evidently this: Is the
increase in the quantity of bile, after podophylline, merely due
to an increase of water, or are the bile-solids also increased?
The bile secreted by dog 11, between the second hour and a half
and the third hour, and that secreted an hour and a quarter
after the administration of podophylline, were analysed with
the following results.   (Table II.)

TABLE II.—*Podophylline.*

Experiment 11.						Before.	After.
Water	...	...	...	...	...	90.88	91.07
Bile-acids, pigments, cholesterine, fats					...	7.75	7.84
Mucus	...	...	...	...	...	1.00	0.60
Ash ...	...	...	...	...	...	0.42	0.49
						100.00	100.00
Velocity of secretion per half-hour			...	...		4.6 cc.	9.6 cc.

It thus appears that, notwithstanding the great velocity of
bile-formation, the special bile-solids were not diminished; the

only noteworthy diminution being in the amount of mucus. This remarkable result was confirmed by the following analysis of the bile in Experiment 12. The table shows the composition of the bile secreted before and during the increase of the secretion after the podophylline was given.

TABLE III.—*Podophylline.*

Experiment 12.						Before.	After.
Water	...	...	...	...	...	94.26	94.28
Bile-acids, pigments, fats, cholesterine					...	4.66	4.68
Mucus	...	...	...	...	...	0.73	0.70
Ash ...	...	...	...	...	...	0.35	0.34
						100.00	100.00
Velocity of secretion per half-hour				...	...	1.86 cc.	2.47 cc.

Results of the Experiments with Podophylline.—1. Podophylline, when injected into the duodenum of a fasting dog, increases the *secretion* of bile. It is inferred that the increased biliary flow in the preceding experiments was due to increased *secretion,* and not merely to *expulsion,* because the gall-bladder had been wellnigh emptied by compression, and the cystic duct had been clamped: moreover, the increased flow was far too prolonged in some of the experiments to be attributable to spasm of the larger bile-ducts; therefore an increase in secretion must have been the cause. 2. When the bile is prevented from entering the intestine, podophylline acts less powerfully and less quickly than when bile is introduced. 3. Augmentation of the biliary secretion is most marked when the purgative effect is not severe; indeed, if the purgative effect be very decided (*Experiment* 10), diminution and not augmentation of the biliary secretion may be the chief result. 4. Podophylline purgation is apparently due to a local action, for the irritation of the intestinal mucous membrane extends gradually from above downwards. 5. The bile secreted under the influence of podophylline, although it may be in increased quantity, contains as much of the special biliary matter as bile secreted under normal conditions.

ACTION OF ALOES.

Although aloes has been found by Röhrig to accelerate the biliary secretion, we were anxious to compare its action, as ascertained by our method, with that of other substances; and we also desired to know the composition of the bile secreted before and after its administration.

Experiment 13. Dog weighing 8.6 kilogrammes.—Sixty grains aqueous extract of Socotrine aloes in 12 cc. water were injected into the duodenum (*a*, Fig. 13). A decided increase in the biliary secretion was perceptible within half-an-hour thereafter. After attaining a maximum about an hour and a half

Fig. 13.

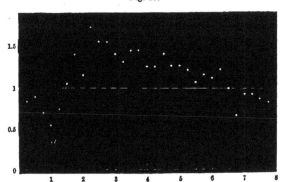

Secretion of bile in a dog that had fasted eighteen hours. Sixty grains ext. aloes Soc. in 12 cc. water injected into duodenum at *a*.

after the administration of the drug, the secretion gradually fell; but although the experiment was continued for seven hours after the aloes was given, the effect had not disappeared.

AUTOPSY.—The aloes had extended along two-thirds of the small intestine, which contained about an ounce and a half of viscous fluid as the only evidence of purgation. There was a decided increase in the vascularity of the mucous membrane in this part of the intestine. The stomach contained a little mucus. Its mucous membrane was pale.

Experiment 14. Dog weighing 5 kilogrammes.—Sixty grains extract of Socotrine aloes in 12 cc. water were injected into the duodenum (*a*, Fig. 14). As in the previous experi-

Fig. 14.

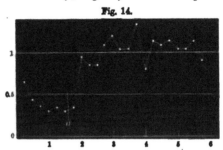

Secretion of bile in a dog that had fasted eighteen hours. Sixty grains ext. aloes Soc. in 12 cc. water injected into duodenum at *a*.

ment, the subsequent increase in the biliary secretion was decided within half-an-hour, and became very strongly marked.

AUTOPSY.—The aloes had extended half way down the small intestine. This portion of intestine contained about two ounces of viscous fluid; and its mucous membrane, together with that of the stomach, was intensely red.

Experiment 15. Dog weighing 9.5 kilogrammes.—In this experiment, it was proposed to test the effect of aloes on the

Fig. 15.

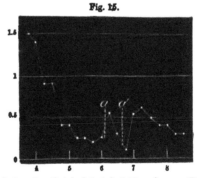

Secretion of bile in a dog that had fasted eighteen hours. Twenty grains ext. aloes Soc. in 5 cc. water injected at *a* and at *a'*. (The curve during the first three hours is omitted.)

liver when well-nigh exhausted: accordingly, at the sixth hour of an experiment on a dog that had fasted the usual period of eighteen hours, twenty grains extract of Socotrine aloes in 5 cc. water were injected into the duodenum (a, Fig. 15), and this dose was repeated in half-an-hour. The secretion of bile was increased, but the effect was not very marked; nevertheless, the result is noteworthy, seeing that in this case there was a great secretion of bile during the first four hours of the experiment. (Fig. 15.)

Composition of the Bile before and after Aloes.—Table IV shows the results of the analysis of the bile in Experiment 13, secreted before and during the first two hours after the administration of the aloes.

TABLE IV.—*Aloes.*

Experiment 13.		Before.	After.
Water		84.11	91.44
Bile-acids, pigments, cholesterine, fat		12.45	7.53
Mucus		1.77	0.88
Ash		1.67	0.65
		100.00	100.00
Velocity of secretion per half-hour		1.5 cc.	2.65 cc.

Table v gives the result of the analysis of the bile before and after the administration of aloes in Experiment 14.

TABLE V.—*Aloes.*

Experiment 14.		Before.	After.
Water		83.98	86.75
Bile-acids, pigments, cholesterine, fat		12.30	10.79
Mucus		2.74	1.49
Ash		1.08	0.97
		100.00	100.00
Velocity of secretion per half-hour		0.66 cc.	2.2 cc.

It is evident from Tables IV and V, that, under the influence of aloes, the bile became more watery; nevertheless, the amount of bile-solids secreted per unit of time increased.

Results of Experiments with Aloes.—1. Sixty grains of the extract of Socotrine aloes, when placed in the duodenum, powerfully stimulated the liver. 2. Under its influence the liver excreted a greater quantity of biliary matter in a given time, although the bile was rendered more watery. 3. Coincident with the marked action on the liver, there was only a slight purgative action.

ACTION OF RHUBARB.

The following experiments show that rhubarb is also a very remarkable hepatic stimulant. The ordinary infusion of the *British Pharmacopœia* was made with Indian rhubarb; it was then filtered and concentrated until 5 cc. contained the active part of seventeen grains of rhubarb. This was the dose employed.

Experiment 16. Dog weighing 22.2 kilogrammes.—5 cc. of the above infusion of rhubarb were injected into the duodenum four times in succession (r, r', r'', r''', Fig. 16). Within half-an-hour after every dose there was an increase in the biliary secretion.

Fig. 16.

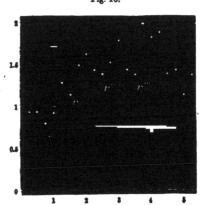

Secretion of bile in a dog that had fasted fifteen hours. 5 cc. of a concentrated infusion of rhubarb injected into duodenum at r, r', r'', and r'''.

. AUTOPSY.—The rhubarb had extended along about a third of the small intestine. There was no unusual redness of the mucous membrane, and there was only slight evidence of purgative action.

Experiment 17. Dog weighing 13.4 kilogrammes.—The artificial respiration, which was deficient at the commencement of this experiment, was improved at *a*, Fig. 17. This was followed by an increase in the secretion of short duration: 5 cc. of the same infusion of rhubarb as that used in the previous experiment were injected into the duodenum three times in succession (*r, r', r''*, Fig. 17). The biliary secretion was augmented within half-an-hour after each injection.

Fig. 17.

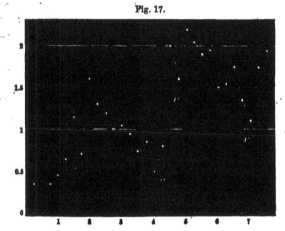

Secretion of bile in a dog that had fasted eighteen hours. Respiration improved at *a*. 5 cc. concentrated infusion of rhubarb injected into duodenum at *r, r',* and *r''*.

AUTOPSY.—The rhubarb had extended along four-fifths of the small intestine. There was no unusual redness of the mucous membrane. The portion of intestine through which the rhubarb had extended contained 120 cc. of a thick yellowish fluid: there was, therefore, decided evidence of purgative action.

Experiment 18. Dog weighing 22.7 kilogrammes.—In this experiment it was proposed to test the action of rhubarb upon the exhausted liver. The artificial respiration being defective at the commencement of the experiment, was improved at *a*, Fig. 18. At the middle of the fourth hour a remarkable in-

Fig. 18.

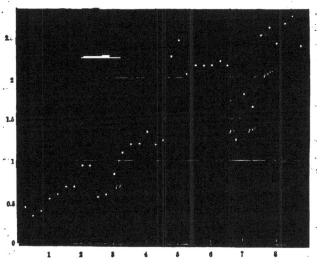

Secretion of bile in a dog that had fasted eighteen hours. Respiration improved at *a*. 10 cc. of the concentrated infusion of rhubarb injected into duodenum at *r*; and 5 cc. at *r′* and *r″*.

crease of secretion took place, and lasted for two hours. The cause of this was not apparent until it was observed that a warm July sun shone full upon the abdomen, from the middle of the fourth to the middle of the sixth hours of the experiment: the increase in the secretion was, therefore, probably due to a rise of temperature: 5 cc. of the concentrated infusion of rhubarb, given in the previous experiment, were injected into the duodenum three times in succession (*r, r′, r″*), and, notwithstanding the very high secretion shortly beforehand, the rhubarb increased it still further.

AUTOPSY.—The rhubarb had extended along two-thirds of the small intestine. There was no unusual vascularity of their mucous membrane. The upper portion of the intestine contained 87 cc. of a yellowish-green liquid, whereas only 20 cc. of water had been injected. Purgation had, therefore, evidently been produced.

In these three experiments, it appears that an infusion of seventeen grains of rhubarb was given nine times, and on one occasion twice this quantity, and that none of these doses ever failed to increase the biliary secretion within half-an-hour after administration. The amount of water given was so trivial that its effect may be entirely disregarded. It is to be noted, that the excitement of the liver produced by rhubarb was accompanied by far less intestinal irritation than was the case with podophylline and aloes.

Composition of the Bile before and after Rhubarb.—The bile secreted before and after the administration of rhubarb in all the three experiments was analysed. Tables VI, VII, and VIII show the result.

TABLE VI.—*Rhubarb.*

Experiment 16.	Before.	After the second dose.	At the close of the Experiment.
Water	88.80	89.28	88.98
Bile-acids, pigments, cholesterine, and fat	9.60	9.60	9.60
Mucus	1.00	0.60	0.80
Ash	0.60	0.52	0.62
	100.00	100.00	100.00
Velocity of bile-secretion per half-hour	1.9 cc.	2.95 cc.	2.55 cc.

From the foregoing analysis, it appears that the percentage amount of the special biliary matter was not diminished by the action of rhubarb, although there was so great an increase in the amount of bile. This result, which corresponds to that observed in the case of podophylline, is confirmed by the following analyses.

TABLE VII.—*Rhubarb.*

Experiment 17.		Before.	After.
Water		85.47	86.23
Bile-acids, pigments, cholesterine, fat		11.59	11.03
Mucus		1.87	1.72
Ash		1.07	1.02
		100.00	100.00
Velocity of bile-secretion per half-hour · ...		1.45 cc.	3.95 cc.

TABLE VIII.—*Rhubarb.*

Experiment 18.		Before.	After.
Water		85.35	85.52
Bile-acids, pigments, cholesterine, fat		12.07	11.98
Mucus		1.53	1.49
Ash		1.05	1.01
		100.00	100.0)
Velocity of bile-secretion per half-hour ...		4.32 cc.	5.87 cc.

It therefore appears that rhubarb, like podophylline, excites the liver-to secrete bile having a composition similar to that secreted under normal conditions.

Results of Experiments with Rhubarb.—1. An infusion of seventeen grains of Indian rhubarb, when placed in the duodenum, never failed to increase the secretion of bile. 2. The bile, although secreted in increased quantity, had the composition of normal bile as regards the biliary constituents proper. 3. The doses which so powerfully excited the liver had in one case no marked purgative effect, but in other two cases the purgative effect was considerable.

ACTION OF SENNA.

Senna excites the liver, but not so powerfully as rhubarb. The ordinary infusion of senna of the *British Pharmacopœia* was prepared and concentrated until 5 cc. contained the active part of forty-five grains of senna; a small dose for a man.

Experiment 19. Dog weighing 22.3 *kilogrammes.*—5 cc. of the concentrated infusion of senna, prepared as above, were in-

Fig. 19.

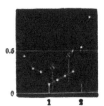

Secretion of bile in a dog that had fasted fifteen hours. 5 cc. concentrated
infusion of senna injected into duodenum at *s* and *s'*.

jected into the duodenum twice in succession (*s* and *s'*, Fig. 19).
After the second dose the secretion rose rapidly, but, unhappily,
the animal died from asphyxia, owing to the accidental closure
of the expiration aperture of the bellows-tube.

Experiment 20. Dog weighing 8 kilogrammes.—5 cc. of the
above-mentioned concentrated infusion of senna were injected
into the duodenum five times in succession (*s, s', s'', s''', s''''*, Fig.
20). The secretion of bile rose rapidly after the second dose, as

Fig. 20.

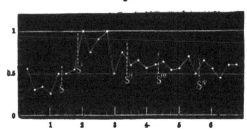

Secretion of bile in a dog that had fasted nineteen hours. 5 cc. concentrated
infusion of senna injected into duodenum at *s, s', s'', s''', s''''*.

in the previous experiment, but it soon fell again; and the
third, fourth, and fifth doses did not increase it.

AUTOPSY.—The senna had extended along three-fourths of
the small intestine, which contained 80 cc. of liquid. Seeing
that the amount of fluid injected was 25 cc., considerable purgation had been produced. There was a considerable increase in
the vascularity of the duodenal mucous membrane, but elsewhere there was no unusual redness.

It appears from this experiment that, although the amount of senna given was sufficient to distinctly act on the intestine, it did not produce any very marked effect on the liver. The purgative effect was greater than in the case of rhubarb, in Experiment 16, but less than in Experiments 17 and 18. However, the effect on the liver was less than in all the rhubarb experiments.

Experiment 21. Dog weighing 5 kilogrammes.—In this case it was proposed to test the action of senna on the exhausted liver. Accordingly, at the middle of the eighth hour of an experiment on an animal that had fasted for the usual period of eighteen hours, 7 cc. of a concentrated infusion, containing the active part of ninety grains of senna, were injected into the duodenum twice in succession (*s* and *s'*, Fig. 21). No noteworthy effect on the biliary secretion ensued.

Fig. 21.

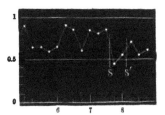

Secretion of bile in a dog that had fasted eighteen hours. 7 cc. concentrated infusion of senna injected into duodenum at *s* and *s'*.

AUTOPSY.—The senna had extended half-way along the small intestine, which contained about an ounce of viscous fluid, thus affording slight evidence of purgative action.

Röhrig, from some experiments with rhubarb and senna (*lib. cit.*, p. 253), concluded that in their activity as cholagogues they stand much on the same level; and, although far inferior in power to aloes, their infusions are nevertheless more powerful than water. The defective method of experiment employed by him is the probable cause of this statement. The estimate of the action of senna is indeed correct; but the foregoing experiments show that rhubarb holds a much more important rank as a hepatic stimulant than senna, and this is still further seen when we compare the analysis of the bile before and after senna with that of the bile before and after rhubarb.

TABLE IX.—*Composition of Bile before and after Senna.*

Experiment 20.	Before.	After.
Water	90.63	91.31
Bile-acids, pigments, cholesterine, fat	7.90	6.75
Mucus	1.80	1.90
Ash	0.87	0.74
	100.00	100.00
Velocity of bile-secretion per half-hour ...	0.82 cc.	1.186 cc.

It appears from this analysis, and from the velocity of secretion, that although senna causes the liver to excrete more biliary matter, its power is far below that of rhubarb.

Results of the Experiments with senna.—1. Senna is a hepatic stimulant of feeble power. 2. It renders the bile more watery.

ACTION OF COLCHICUM.

Colchicum has been recommended as a cholagogue in cases of gout, but its action on the liver has not hitherto been tested by direct experiment.

Experiment 22. Dog weighing 23.5 kilogrammes.—Sixty grains of the aqueous extract of colchicum of the *British Pharmacopœia* in 10 cc. of water were injected into the duodenum (c, Fig, 22). In an hour the biliary secretion began to increase, and five hours after the injection it was nearly five times more than before the drug was given. The secretion then fell, and just at the close of the experiment a large quantity of liquid fæces was discharged.

AUTOPSY.—Great vascularity of upper four-fifths of the mucous membrane of the small intestine. The vascularity of the duodenum was intense. The mucous membrane of the large intestine was also unusually vascular. The gastric mucous membrane was pale. There was evidence of considerable hy-

Fig. 22.

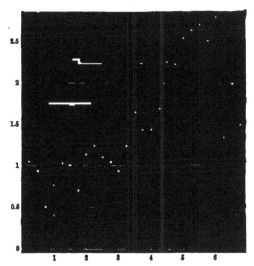

Secretion of bile in a dog that had fasted sixteen hours. Sixty grains of aqueous extract of colchicum in 10 cc. of water injected into duodenum at c.

drocatharsis in the small intestine. The large intestine was empty, owing to the recent discharge of fæcal matter.

Experiment 23. A somewhat weak dog, weighing 7 kilogrammes.—Twenty grains of aqueous extract of colchicum in 5 cc. of water injected into duodenum (c, Fig. 23). The effect

Fig. 23.

Secretion of bile in a dog that had fasted twenty hours. Twenty grains of aqueous extract of colchicum in 5 cc. of water injected into duodenum at c.

on the biliary secretion was evident at the end of an hour; but throughout the whole experiment the animal secreted only a small quantity of bile, and by the fifth hour the pulse was so weak that the experiment was closed.

AUTOPSY.—There was somewhat increased vascularity of the mucous membrane in the upper fourth of the small intestine. There was no distinct evidence of purgative action.

Experiment 24. Dog weighing 9.9 kilogrammes.—Forty grains of aqueous extract of colchicum in 8 cc. of water injected into duodenum (c, Fig. 24). The effect on the bile-secretion

Fig. 24.

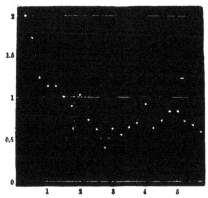

Secretion of bile in a dog fed on liver fifteen hours before the experiment began, but food was still found in the stomach at the close. Forty grains of aqueous extract of colchicum in 8 cc. of water injected into duodenum at c.

was not very marked; for, although after the lapse of an hour, the falling of the secretion was arrested, it did not, under the influence of the colchicum, rise to any very notable height. At the close of the experiment, a quantity of partially digested food was found in the stomach. On this account it was impossible to be sure whether or not a small amount of fluid in the upper part of the small intestine was due to cathartic action. The mucous membrane of the duodenum was intensely red.

Experiment 25. Dog weighing 23.6 kilogrammes.—Sixty grains of aqueous extract of colchicum in 10 cc. water injected into duodenum (c, Fig. 25). Although the biliary flow thereafter varied so much, a decided increase was evident an hour and a half after the administration of the drug. The increase lasted about four hours, after which the secretion gradually fell

Fig. 25.

Secretion of bile in a dog that had fasted eighteen hours. Sixty grains of aqueous extract of colchicum in 10 cc. of water injected into duodenum at c.

AUTOPSY.—There was increased vascularity of the mucous membrane of the upper three-fourths of the small intestine. The whole small intestine contained evidence of powerful cathartic action.

These experiments show that the aqueous extract of colchicum in large doses increases the biliary secretion.

TABLE X.—*Composition of the Bile before and after Colchicum.*

Experiment 25.						Before.	After.
Water	...	...	...	...	...	88.434	90.63
Organic Bile-solids		...	...	...	...	10.616	8.75
Ash ...	...	...	...	...	...	0.950	0.62
						100.00	100.00
Velocity of bile-secretion per half-hour...					...	1.2 cc.	2.24 cc.

It appears from the above analysis, that colchicum rendered the bile more watery; nevertheless, owing to the increased velocity of secretion, more biliary matter was excreted by the liver under its influence.

Results of Experiments with Colchicum.—1. Sixty grains of the aqueous extract of colchicum powerfully excited the liver, and produced hydrocatharsis; forty, and even twenty grains, also increased the biliary secretion, though less powerfully. 2. Colchicum, while it increases the amount of biliary matter excreted by the liver, renders the bile more watery.

ACTION OF TARAXACUM. ...

Experiment 26. Middle-sized dog.—One hundred and eighty grains solid extract of taraxacum in 25 cc. water injected into duodenum (*t*, Fig. 26), and two hours after this, one hun-

Fig. 26.

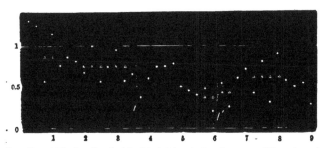

Secretion of bile in a dog that had fasted twenty-four hours. *t*, 180 grains, *t'*, 120 grains solid extract of taraxacum in 25 cc. water injected into duodenum.

dred and twenty grains in the same quantity of water were injected (*t'*). After both doses there was a greater increase in the biliary secretion than was at all likely to have been occasioned by the same quantity of water.

AUTOPSY.—The taraxacum had passed along nearly the whole length of the small intestine. Most of the fluid had been absorbed. There was no evidence of purgative action.

Experiment 27. Small dog.—One hundred and twenty grains solid extract of taraxacum in 15 cc. water were injected into duodenum (*t*, Fig. 27), and this dose was repeated in two

Fig. 27.

Secretion of bile in a dog that had fasted eighteen hours. 120 grains solid extract of taraxacum in 15 cc. water injected into duodenum at *t* and *t'*.

and a half hours. The increase of the biliary secretion after the second dose was trivial, but after the first it was considerable, although of short duration. Examination of the intestine at death revealed no purgative action.

From these experiments it therefore appears that taraxacum is a very feeble hepatic stimulant.

ACTION OF SCAMMONY.

The resin of scammony being insoluble in water was dissolved in dilute alcohol, and some bile was added in order still further to promote its absorption from the alimentary canal.

Experiment 28. Dog weighing 9.5 kilogrammes.—2.5 cc. bile were injected into the duodenum (*b*, Fig. 28). This pro-

Fig. 28.

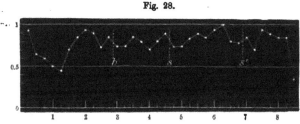

Secretion of bile in a dog that had fasted eighteen hours. Bile injected into duodenum at *b*, and scammony with bile and alcohol at *s* and *s'*.

duced no notable effect. Twenty grains scammony resin dissolved in 3.5 cc. rectified spirit, 3 cc. water, and 3 cc. bile were then injected (*s*), and this dose was afterwards repeated (*s'*). There was a slight increase in the biliary secretion.

AUTOPSY.—Greatly increased vascularity of the mucous membrane of the whole length of the small intestine. Vascularity of gastric mucous membrane also somewhat increased. Evidence of severe purgative action in the whole extent both of the small and large intestine.

Experiment 29. Dog weighing 6.8 kilogrammes.—In this experiment it was determined to give scammony in smaller doses, 1 cc. bile and 2 cc. water (*b*, Fig. 29). The exact effect of

Fig. 29.

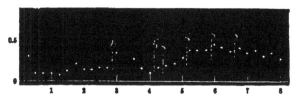

Secretion of bile in a dog that had fasted nineteen hours. Bile given at *b* and
at *b'*. Scammony, &c., at *s*, *s'*, *s''* and *s'''*. (See text.)

this was not ascertained owing to the loss of the bile secreted
during one of the periods. About an hour after this 0.25 cc.
bile 0.5 cc. rectified spirit and 1.25 cc. water were injected at *b'*.
This having scarcely any effect, it was given with four grains of
scammony resin at *s*, and again at *s'*. The amount of scammony
was doubled and this dose was given at *s''* and *s'''*. There was
an increase in the biliary secretion after the first two doses of
scammony, but after the third and fourth the secretion diminished.

As two experiments, not here reported, in which alcohol only
was given, proved that it certainly does not augment the biliary
secretion, the foregoing experiment shows that scammony is a
hepatic stimulant, although not a powerful one.

AUTOPSY.—The scammony had passed along two thirds of
the small intestine. There was decided evidence of purgation,
but no remarkable increase in the vascularity of the mucous
membrane.

From the two foregoing experiments it appears that scammony is a cholagogue of feeble power.

ACTION OF CALOMEL.

As is well known, Scott (*Beale's Archives*, Vol. i) and Hughes
Bennett's Committee (*lib. cit.*) found that purgative doses of
calomel when given to non-fasting dogs diminished the biliary
secretion. Röhrig (*lib. cit.* p. 254) investigated the action
of calomel by his method, and he states, that with large
doses (twenty grains for a dog) it rarely happened that the
secretion of bile was recalled after it had come to a standstill,

although this agent can increase the secretion when it is only diminishing. As the method adopted in our experiments is best calculated to afford accurate data, we performed the following four experiments.

Experiment 30. Dog weighing 19.1 kilogrammes.—Ten grains calomel in 7 cc. water injected into duodenum at *c* and at *c'*. An increase in the biliary secretion ensued: thus, during a period of two hours before the administration of calomel, the amount of bile secreted was 3.35 cc.: during the two hours immediately following the first dose the secretion rose to 4.25 cc., and during the two hours immediately following the second dose, it rose to 4.72 cc.

Fig. 30.

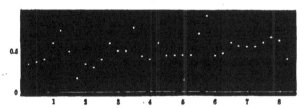

Secretion of bile in a dog that had fasted eighteen hours. 10 grains calomel in 7 cc. water injected into duodenum at *c* and *c'*.

AUTOPSY.—The autopsy in this case was made fourteen hours after death. The small intestine contained a large quantity of thick greyish liquid with green flakes here and there on the mucous membrane, thus affording evidence of profuse purgative action. The mucous membrane of the small intestine was pale throughout the greater part of its extent, but at intervals in the upper portion there was a limited area of redness.

Experiment 31. Dog weighing 7 kilogrammes.—Ten grains calomel in 4 cc. water injected into duodenum at *c* and again at *c'*. No increase of the biliary secretion ensued, but on the contrary, there was a gradual diminution.

·· AUTOPSY.—The autopsy in this case was made fourteen hours after death. The gastric mucous membrane was pale, and the stomach contained some viscous fluid of a brownish colour, with a patch of green matter clinging to the mucous

Fig. 31.

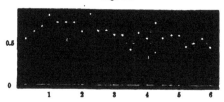

Secretion of bile in a dog that had fasted eighteen hours. 10 grains calomel given at *c* and at *c'*.

membrane near the pylorus. The green appearance was evidently due to mercury, for a little unchanged calomel was perceptible at the margin of the patch. The upper third of the small intestine was semidistended with a brown somewhat clear viscous fluid containing patches of green, thus affording evidence of purgative action.

Experiment 32. Dog weighing 13 kilogrammes.— Ten grains calomel in 9 cc. water injected into duodenum at *c* and again at *c'* (Fig. 32). No augmentation of the biliary secretion followed.

Fig. 32.

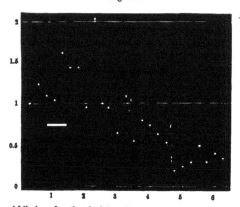

Secretion of bile in a dog that had fasted eighteen hours. 10 grains calomel in 9 cc. water injected into duodenum at *c*, and again at *c'*.

AUTOPSY.—Mucous membrane of upper half of small intestine extremely vascular. .This portion of bowel contained a large quantity of a greyish fluid with green patches. There was therefore evidence of profuse purgative action.

Experiment 33. Dog weighing 11 kilogrammes.—In this case we resolved to give smaller doses of calomel: accordingly, three grains of calomel in 3 cc. water were given four times in succession (*c. c′. c″. c‴*. Fig. 33). Three quarters of an hour

Fig. 33.

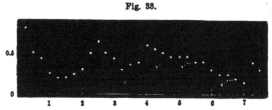

Secretion of bile in a dog that had fasted eighteen hours. 3 grains calomel in 3 cc. water injected into duodenum at *c, c′, c″, c‴*.

after the first dose there was an increase in the biliary secretion,—of doubtful import however; for an increase quite as great had occurred previous to the administration of the calomel: the last three doses were followed only by diminution.

AUTOPSY.—The autopsy was made twenty-four hours after death. There was evidence of powerful purgation in the whole length of the intestine. No increased vascularity of the mucous membrane.

Composition of the Bile before and after Calomel.

TABLE XI.—*Calomel.*

Experiment 33.						Before it was given.	After the third dose.
Water	...	...	...	...	...	86.00	87.78
Bile-acids, pigments, cholesterine, fat			...		...	8.40	7.85
Mucus	...	...	...	...	...	4.57	3.40
Ash	...	...	...	...	...	1.03	0.97
						100.00	100.00
Velocity of secretion per half-hour			...		...	0.78 cc.	0.4 cc.

It therefore appears that notwithstanding the smaller amount of bile secreted after the third dose of calomel in the

19—2

last experiment, its consistence was rendered more watery. There was therefore a decided diminution in the amount of biliary matter excreted by the liver during the calomel purgation in this case.

Results of Experiments with Calomel. 1. An increase of the biliary secretion followed the administration of two successive doses of ten grains of calomel in one case (*Exp.* 30). Diminution of the secretion was the only result of the same doses given under similar circumstances in other two cases (*Exps.* 31 and 32), and it was the most definite result of the administration of four successive doses of three grains in another case (*Exp.* 33). 2. In all the four experiments the calomel had a purgative effect. 3. Analysis of the bile secreted during the calomel purgation in *Exp.* 33 showed that notwithstanding a diminution in the quantity of bile secreted, the percentage amount of solids had become less.

The fact that three of the four experiments with calomel indicated no stimulating effect of calomel upon the liver suggests a careful scrutiny of the circumstances under which a different result was obtained in the first case (*Exp.* 30). It is a noteworthy fact that although this dog weighed 19.1 kilogrammes, while the others weighed respectively 7, 13, and 11 kilogrammes, it secreted an amount of bile that was unusually small for a dog of this size. This question therefore arises: In consideration of the negative results as regards the cholagogue action of calomel in *Exps.* 31, 32 and 33, is it not possible that the increase in the biliary secretion observed in *Exp.* 30 might have taken place during the course of the experiment had no calomel been given?

ACTION OF GAMBOGE.

Experiment 34. Dog weighing 4.8 kilogrammes.—3 cc. bile and 3 cc. water injected into duodenum at *b* (Fig. 34). Previous to this there was, considering the small size of the dog, a large secretion of bile. The increased secretion which followed the injection was probably owing to the action of the bile. Twenty grains of gamboge in the same quantity of bile and water were given at *g*, and forty grains in the same fluid at *g'*. Half an

Fig. 34.

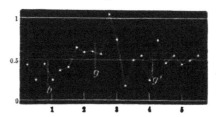

Secretion of bile in a dog that had fasted eighteen hours. 3 cc. bile and 3 cc. water injected into duodenum at *b*. 20 grains gamboge at *g*, and 40 grains at *g'* in the same fluid as at *b*, injected into the duodenum.

hour after the first dose, there was a decided acceleration of the biliary flow, but in an hour afterwards it had temporarily sunk nearly to zero. If the mean be taken, it will be found that the increase of secretion was so slight, that it might have been due to the bile that was given with the gamboge. On the whole, therefore, it can scarcely be said that the amount of bile secreted was increased by the gamboge, and certainly the next experiment lent no support to such a view of the matter.

AUTOPSY.—Great redness of the mucous membrane in the upper half of the small intestine. Evidence of profuse hydrocatharsis in this portion of the gut. Some of the gamboge had passed into the stomach, the mucous membrane of which was somewhat reddened.

Experiment 35. Dog weighing 8 kilogrammes.—1 cc. bile and 2 cc. water injected into duodenum at *b* (Fig. 35). Four grains gamboge in 0.2 cc. bile and 2 cc. water injected at *g*, *g'*, and *g''*. At *g'''*, *g^{iv}*, *g^v*, twice the amount of gamboge was given

Fig. 35.

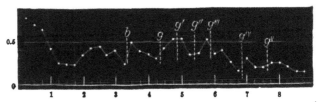

Secretion of bile in a dog that had fasted nineteen hours. Gamboge given. (See text.)

in the same fluid. The increase of the bile secreted after the first dose was trifling. The chief result of the experiment was *diminution* of the secretion.

AUTOPSY.—Profuse hydrocatharsis in the small intestine. No very noteworthy increase in the vascularity of the mucous membrane.

In *Exp.* 35 a smaller quantity of bile was given than in *Exp.* 34—in order to eliminate as far as possible its stimulating effect on the liver; 35 is therefore a better experiment than 34, and it affords no sufficient evidence that gamboge is a hepatic stimulant. It is interesting to contrast the negative effect on the liver of this hydrocathartic with the positive effect of colchicum, also a hydrocathartic.

ACTION OF CASTOR OIL.

Röhrig found that castor oil has scarcely any effect on the hepatic secretion. It appeared however desirable to emulsify the oil with bile, so that its condition in the intestine might more closely resemble that in any normal case.

Experiment 36. Dog weighing 7.7 kilogrammes.—3 cc. bile injected into duodenum at *b* (Fig. 36). One ounce castor oil emulsified with 3 cc. bile injected into duodenum at *c* and again at *c'*. A slight increase in the bile secretion followed the

Fig. 36.

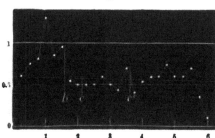

Secretion of bile by a dog that had fasted eighteen hours. 3 cc. bile injected into duodenum at *b*. The same with one ounce castor oil given at *c* and at *c'*.

second dose, but as its extent was trifling, it should probably be disregarded. There was a great diminution towards the close of the experiment.

AUTOPSY.—There was no unusual redness of the mucous membrane of the intestine, save at the lower part of the duodenum, where the vascularity was somewhat increased.

As this experiment, notwithstanding the difference in the method, gave a result similar to that obtained by Röhrig, it did not appear necessary to repeat it. It was therefore concluded that castor oil can scarcely be regarded as a hepatic stimulant.

CONCLUDING OBSERVATIONS.

Experiments with colocynth, jalap, and other substances will form the subject of a future Report, for although Röhrig by his method found that the substances just mentioned increase the biliary secretion, we as yet know nothing regarding their effect on the composition of the bile. It would be interesting to know this, and also to have curves of the effect on the secretion to compare with those above given.

Of the substances examined by us, podophylline, rhubarb, aloes and colchicum are the most notable stimulants of the dog's liver. In considering their *modus operandi* in these experiments we repeat what has been already mentioned in reference to the action of podophylline, that the increased discharge of bile was not owing to contraction of the gall-bladder, for this in all cases had been well-nigh emptied by digital compression after which the cystic duct was clamped at the beginning of the experiment: moreover the long continuance of the effect of these substances negatives the idea that reflex spasm of the larger bile-ducts occasioned the biliary flow.

The remaining points appear to be these. 1. Did these substances increase the activity of the hepatic cells by reflex excitement proceeding from the intestinal mucous membrane? 2. Was the effect due to dilatation of mesenteric vessels, and to a consequent increase in the stream of blood through the liver? 3. Did the substances stimulate the hepatic cells directly? These questions are difficult to answer, indeed, it is not professed that the data for a perfectly definite answer

.are given in the foregoing, for we have been more concerned in ascertaining the *facts* regarding the action of these substances, than in searching for the *explanations* of these facts. Nevertheless the experiments as far as they go are eminently suggestive regarding the *modus operandi*. Thus, the evidence of intestinal irritation was great after podophylline and colchicum, and very considerable after aloes, but on the other hand very slight after rhubarb, and yet it has been proved that rhubarb is a decided hepatic stimulant; apparently a much more powerful one than croton oil, although this is a far greater irritant of the intestinal mucous membrane. We are, therefore, disposed to think that reflex excitement of the liver was not a chief cause of the result. Nor can we believe that an increased stream of blood through the liver, owing to dilatation of intestinal vessels, was a principal cause, for after croton oil, there was great vascular dilatation in the alimentary mucous membrane, while the effect on the liver was slight; moreover, after calomel, castor oil, and gamboge, the vascular dilatation was very considerable, and the effect on the liver scarcely perceptible. It therefore appears most probable that the rhubarb, colchicum, aloes, and podophylline were absorbed and excited the liver directly. At the same time, however, we do not here profess to do more than offer suggestions touching the explanation of these results.

The question will, doubtless, be asked, How is it that the foregoing experiments prove that podophylline *increases* the biliary secretion of the dog, while Hughes Bennett, from experiments with this substance also performed on dogs, arrived at the opposite conclusion? The only apparent explanation of the two series of results is, that in the present experiments the dogs were in a *fasting* condition, whereas, in those employed by Hughes Bennett's Committee they had their usual food. The absorption of food is undoubtedly followed by greatly increased biliary secretion. The purgative probably diminished the amount of food absorbed, and the effect of this probably overbalanced in the course of the day the stimulation of the liver[1].

[1] The diminished absorption of biliary constituents from the intestine, which probably accompanies the action of a purgative in a normal case, is not here taken into account, for in the dogs which were made the subjects of experiment by Hughes Bennett's Committee, a permanent biliary fistula had been established, so that the entrance of bile into the intestine was entirely prevented.

Although for the purpose of shewing whether or not a substance excites the liver, experiments with *fasting* are of far greater value than experiments with *non-fasting* animals, at the same time both sets of experiments are of value, and the general conclusions deducible from them appear to be, that when such substances as podophylline, rhubarb, aloes, and colchicum are administered : 1. The liver is excited to secrete more bile. 2. If purgation result, the absorption of biliary matter, and of food—if digestion be taking place—from the intestine is probably diminished. Thus, by the twofold operation of increased hepatic action, and diminished absorption of biliary matter from the intestine, the composition of the blood as it passes through the portal system is probably rendered more pure.

The experiments with podophylline more especially suggest a point of great practical importance, as well as of much physiological interest; to wit, *that powerful purgative action tends to diminish the biliary secretion.* This conclusion was also deduced from the experiments of Hughes Bennett's Committee, but it was scarcely warranted, because in those experiments the diminished absorption of food from the alimentary canal, resulting from the purgation, was doubtless an important cause of the result. In the foregoing experiments, however, this disturbing factor has been eliminated; nevertheless, the podophylline experiments 10 and 11, the calomel experiments 32 and 33 more especially show that diminished biliary secretion accompanied the excessive purgation resulting from these substances. Whether or not this depressing effect on the liver is due to the tendency towards collapse that results from violent purgation, or to abundant abstraction of certain matters from the portal blood, is doubtful. The point, however, is one of such importance, that we shall investigate it further.

It must be borne in mind that the foregoing experiments directly apply only to the healthy dog. They are simply a contribution to comparative physiological pharmacology. To the clinical observer is left the task of comparing these results with those observed in man. He will probably accept them all as not opposed to clinical experience in the case of man, with the exception of the results of the action of calomel. It must be observed that we are not offering evidence regarding the action of calomel in man, nor do we profess to take the side either of those who regard calomel as a cholagogue in man, or with those who deny this and maintain that calomel produces bilious discharges merely by rapidly hurrying the bile through the intestines; we express no opinion on the subject, but restrict ourselves simply to showing that, although there are a number of substances supposed to be cholagogues in man, which excite the liver of the dog, calomel, judging from the chief results of our experiments, does not appear to be one of these, although, in the dog, as in man, it produces purgation; and although the researches of Hughes Bennett's Committee conclusively showed that mercury can produce salivation, ulceration of the gums and other signs of its constitutional action, in the dog as well as in man. These facts suggest that the clinical observer should be very cautious in coming to a conclusion regarding the explanation of the bilious discharges after the administration of calomel in the case of man.

It will, doubtless, be admitted that the foregoing experiments supply facts of no little importance regarding the physiological actions of these substances. The clinical observer may obtain from them *suggestions* which he will probably not discard. He will probably recognise the fact that this kind of research, when not strained to furnish conclusions that lie outside its proper sphere, may yield evidence of much value in clinical investigation, for it can—in some instances, as in the present—eliminate various factors which may so complicate the case in man, that no definite conclusion can be drawn; moreover, it may, as it has often done, show actions of the greatest importance, which have in many instances also been found when the experiment came to be tried in man.

Conclusion of Part I.

ON THE TRANSFORMATIONS OF THE PULSE-WAVE IN THE DIFFERENT ARTERIES OF THE BODY.

By A. L. GALABIN, M.A., M.D., *Late Fellow of Trinity College, Cambridge, Assistant Obstetric Physician to Guy's Hospital, and Assistant Physician to the Hospital for Sick Children.* Pl. XIV.

IN a paper published in the *Journal of Anatomy and Physiology*, Vol. VIII., I discussed the causation of the secondary waves seen in the sphygmographic tracing of the pulse. In the present communication I propose to give the results of a series of observations on the changes which these waves undergo at different distances from the heart; the study of which appears to throw further light on the question as to what is the cause, or combination of causes, by which each of them is originated. The investigation has been conducted partly by taking tracings from the different arteries of the body, and partly by the aid of the same schema representing the arterial system, which was described in the former paper. This consists of a series of bifurcating elastic tubes of india-rubber, and differs from that used in any similar experiments which have been published chiefly in the means used to avoid the occurrence of waves reflected from the periphery. Such reflection in the body is rendered in the highest degree improbable from the complexity and varying length of the arteries, but it is very apt to occur in experimental tubes. It has been successfully prevented by making the lengths of the numerous terminal tubes very un-equal, and by applying pressure to each of them, when it is desired to increase the resistance, not at one, but at several points widely apart. The heart is one of a single cavity, having two valves of india-rubber, and the contraction of the natural heart is imitated by the hand.

In order to render the schema as mechanically simple as possible, no attempt is made to imitate the arrangement of the individual arteries of the body. It is found however that, by a suitable adjustment of the pressure in the tubes and their distensibility, all the varieties of human pulse can be closely imitated. From this it may be inferred that the cause of the

secondary waves is not to be found in the curvature of the
aorta, or in the mode in which the large arteries arise from it,
but that this arrangement can exercise at most only a subordinate influence on the form of the pulse-wave.

' In the comparison of pulse-waves in different arteries of the
human body, the observations have been extended over the
widest possible range. For in investigating experimentally the
cause of any phenomenon it is essential to secure the greatest
possible variation of the conditions under which it occurs.
Hence even to ascertain the physiological explanation of the
healthy pulse-curve it is not sufficient to take tracings from
healthy persons. For in them only a limited variation in
arterial tension and pulse-rate is possible, but to observe the
effects of altered elasticity of arterial walls, of imperfection in
the aortic valves, of cardiac hypertrophy, or of a departure from
the normal relation between arterial tension and pulse-rate,
it is necessary to have recourse to the diseased. The aortic
tracing again can only be obtained when the aorta has undergone aneurismal dilatation.

In the transmission of any wave along an artery there is
one influence always at work, namely the effect of the elasticity
of the arterial walls, by which the intermittent flow of blood is
at length converted into a continuous stream. The tendency
of this is to broaden the base of each wave, to make both the
up-stroke and down-stroke more gradual, and at length to obliterate it altogether. This change is the most marked in the case
of abrupt and narrow waves having no great volume, as may be
seen especially by a comparison of the simultaneous tracings
Figs. 16 and 17, of which the first was taken close to the aortic
valves of the schema, the second at a distance of 9 feet. All
the minor oscillations are entirely lost in the latter tracing, and
the dicrotic wave C has lost its high and narrow summit, and
become broad and rounded. It will be seen however hereafter.
that there is some cause in operation by which the effect of this
change in rendering the up-stroke of the principal wave more
gradual, is, in some cases, counteracted up to a certain point.

Before describing in detail the changes of the pulse-curve, it
is necessary to revert to the nature of the tidal or first secondary wave (marked B in the figures) which, in most pulses, pre-

cedes the dicrotic wave (C). In my former paper I endeavoured to show that the relative position of this wave at different distances from the heart disproves the theory hitherto proposed as to its causation, namely that it has the same origin as either the primary or the dicrotic wave, but is transmitted with a different velocity, and so appears in the radial pulse-curve as a distinct elevation. I there explained the separation of the primary or so-called "percussion" and tidal waves as not really existing in the artery, but as produced in the trace by the velocity acquired by the sphygmograph in the sudden primary upstroke. Further observation has convinced me that, although this explanation applies to many cases, it yet does not express the whole truth, and that, in some instances at least, there is a real first secondary wave or oscillatory expansion in the artery. For this I shall use the term predicrotic in preference to that of tidal wave.

The most striking example of this which I have ever met with was in a patient suffering from aortic regurgitation, tracings from whose carotid, radial, and dorsalis pedis arteries are shown in Figs. 11, 12 and 13. In this case an actual double expansion in the carotid artery, whose pulsation was violent, was manifest not only to the finger but to the eye. On listening to the chest while the finger was placed upon the pulse, it was found that the second beat in the carotid preceded by a decided interval the second sound of the heart, and even in the radial artery the second beat was felt a little before the occurrence of the second sound. This second or predicrotic expansion had therefore nothing to do with the closure of the aortic valves, or with the dicrotic wave. On comparing the relative position of the predicrotic wave in the different arteries, it will be seen that while the interval is not precisely constant, yet there is no progressive difference. In the radial artery (Fig. 12) it is somewhat nearer to the primary wave than in the carotid (Fig. 11), the primary up-stroke being at the same time more sudden, but in the dorsalis pedis (Fig. 13) it is somewhat further off. This instance therefore, while it proves the existence of a true predicrotic wave, demonstrates also that it is not derived from the same origin as either the primary or the dicrotic, and that both primary and predicrotic waves, as they exist in the artery, are equally waves of expansion, and therefore also of increased pressure, and forward movement in the blood.

Is it then possible to find any criterion by which it may be determined whether the tidal wave, as seen in any ordinary pulse-curve, is made up of two components, and what is the relative proportion of each? To a great extent this may be done by observing the change of shape in the trace produced by an increase of the pressure applied. For the fact being established that the primary and predicrotic waves are equally waves of expansion, there is clearly no reason why one should be affected more than the other by increased pressure upon the artery. So far therefore as the relative height of the primary wave is diminished by this means, it may be inferred that its summit, as seen in the trace, is due to the velocity acquired in the sudden up-stroke. It is found that increase of pressure almost always decreases the relative height of the primary wave, and sometimes abolishes the distinction between primary and tidal. In the latter case therefore the distinction is produced only by the sphygmograph. I have introduced examples of the effect on the trace of increase of the pressure applied in two typical classes of cases. The first is that of acute Bright's disease, in which arterial tension is very high, but the arteries healthy. The tidal wave is here very distinctly separated, and, since but little change is produced by increase of pressure, it seems to be the effect of a great predicrotic wave in the artery. A series of tracings from such a patient at different pressures is shewn in Figs. 21, 22 and 23. In Figs. 24 and 25, on the other hand, is shewn the effect of increased pressure on the pulse of atheroma, and it will be seen that the primary summit is much diminished, and a rounded 'tidal wave' brought out. In that case therefore the pointed summit was mainly due to acquired velocity.

I think that an example may be found of the production of a tidal wave solely by the oscillation of the sphygmograph in the simultaneous tracings given by Marey[1], which were taken from the carotid artery of a horse whose pulse was unusually dicrotic, one with the sphygmograph, the other with Chauveau's hæmodromometer, representing the velocity of blood-

[1] *Physiologie Médicale de la Circulation du Sang*, p. 278.

current. In the former, there appears a small pointed tidal wave, in a position similar to that in Fig. 41 (*B*), but in the curve of blood-velocity this does not appear, and its position corresponds to the bottom of the notch which precedes the dicrotic wave, and here descends below the zero line. Now it is clear that no wave of expansion in the artery can occur without a wave of forward movement in the blood, although there need not be a correspondence in the shape of the two waves, or the position of their maxima. A true predicrotic wave of expansion should therefore appear in the trace of the hæmodromometer, and, since it is not found there, it may be inferred that, in this instance, the elevation seen in the sphygmographic trace is a recoil due to acquired velocity. The lever of the hæmodromometer, being immersed in the blood-current, would not be liable to such an oscillation.

The oscillation of the sphygmograph cannot be prevented by the use of a small spring to depress the long recording lever, since it is not caused, except in rare cases, by any separation of the lever from the knife-edge, but by a variation in the compression of the elastic tissues covering the artery, which occurs most when the pressure applied is small. This may be seen by a comparison of Figs. 41, 38, and 39, taken from the same full and dicrotic pulse. Comparing Fig. 38, taken at a pressure of 1 oz. with Fig. 41 taken at a pressure of 2 oz., it is clear that the deep notch which in the former trace immediately follows the high primary summit, and precedes the tidal wave *B*, descending far below the base line of the curve, is the effect of oscillation, since it is not seen in Fig. 41, taken at a somewhat higher pressure. Fig. 39 was taken with a small spring to depress the long lever at the same pressure of 1 oz. The amplitude of the whole trace is somewhat diminished by the increased friction, but the shape of the curve is almost exactly the same as that in Fig. 38, thus shewing that the oscillation occurred as much as before, and much more than it did when the pressure was increased to two ounces, and the small spring dispensed with. It may be readily tested by experiment that, if the sphygmograph be fixed upon the arm, and the pressure applied to the artery be moderate, the recording lever may be pushed down by a slight pressure, and that, if lifted to a height and allowed to drop, it will oscillate. If however the instrument be fixed upon a hard flat surface, this scarcely occurs at all, however small the tension of its spring may be.

The small secondary spring was used in Marey's original sphygmograph, but is omitted in the best instruments now constructed in England. If delicate enough, it scarcely alters the shape of the trace, but if somewhat stronger, it may break the tidal wave into two, as shewn in Fig. 21 of my former paper. The chief objection to its use is the impossibility of securing that it shall be of equal strength in different instruments.

If any further evidence be necessary to confirm the proof given on a former occasion[1], that the primary or so-called percussion wave is not the effect of any instantaneous shock or vibration, it may be found, I think, in the experiment of compressing an artery a few inches beyond the point at which the sphygmograph is applied. The effect of this is shewn in Figs. 31 and 32. Tracing 31 represents a full, soft pulse taken in the usual way at a pressure of 3 oz. from the radial artery rather high up; tracing 32 was taken at the same pressure while the artery was compressed at a distance of three inches below the pad. In the latter case the primary wave is much increased in height and suddenness. This implies that the front of the wave in the artery was reflected from the point compressed, and returned to the pad of the sphygmograph, having traversed a distance of six inches, in time to find the ascent of the primary wave still only commencing, and to add its own effect in enhancing that primary wave. Precisely the same effect was produced by compression of the artery upon the dicrotic wave (c) as upon the primary wave, thus indicating the analogous nature of the two waves.

The relative position of the predicrotic wave at different distances from the heart, not only in the case depicted in Figs. 11, 12, and 13, but in all other tracings, whether taken from the human body or the schema, shews that it can only be explained as some form of oscillation in the blood-column, which arises near the aortic valves, and is propagated to the periphery with a velocity not very widely differing from that

---

[1] *Journal of Anatomy and Physiology*, Vol. VIII.

of the primary or dicrotic waves. It remains to be considered whether any mechanical explanation can be found for it. There are two ways in which it is possible that such an oscillation might arise, the statement of which will perhaps be rendered more intelligible if I first briefly recapitulate the mode of causation of the dicrotic wave.

The first cause of the dicrotic wave is that which has been very generally accepted as depending upon the aortic valves. For let us consider a section of artery close to the valves. When the influx from the heart suddenly ceases at the end of systole, the fluid for an instant continues to flow away out of the section on account of its acquired velocity, and the pressure in the section therefore rapidly falls and the artery contracts. As soon as the velocity of the fluid is checked by the pressure in front, a reflux takes place, which, being stopped by the valves, causes a second increase of pressure and second expansion. This is propagated as the dicrotic wave to the periphery, and may itself again call out a second similar oscillation or tricrotic wave, which is not unfrequently seen in the pulse (vide Fig. 35, *D*). Even in the total absence of aortic valves the reflux, meeting with the current entering the ventricle, may cause a second increase of pressure or dicrotic wave, although this will be much less than in the former case. If the fluid in the tubes be air instead of blood or water, its momentum is so small that its velocity is checked instantly at the end of systole, and there is no perceptible dicrotic wave (vide Fig. 8). If on the contrary mercury be taken, both the dicrotic and succeeding waves become enormous, on account of the great momentum of the fluid, as was shewn by Marey. The fluid remaining the same, the oscillation will be more ample the greater its initial velocity, and the more slowly that velocity is checked. Thus dicrotism is promoted by a sudden action of the heart, and also by distensibility of arteries, by lowness of arterial pressure, and by freedom of outflow. I think that in considering this origin of the dicrotic wave sufficient attention has not generally been paid to the important part played in it by the inertia of the fluid, and to the fact that the aortic valves, although extremely important, are not absolutely essential.

. So far as it depends on this cause the dicrotic wave will be greatest near its origin at the valves, and will tend to diminish as it is transmitted, from the influence of the elastic arterial walls. This may be seen experimentally by a comparison of Figs. 16 and 29, taken close to the valves of the schema, with the simultaneous tracings 17 and 30, taken at a greater distance. It is found however that in many cases dicrotism increases up to a certain point, as the pulse-wave recedes from the heart. This is true to some extent of the healthy pulse (compare Fig. 35 from the radial artery with Fig. 33 from the carotid and Fig. 34 from the facial : compare also Fig. 14 with Fig. 15, and Fig. 26 with Figs. 27 and 28); but in some instances, especially when arteries are much relaxed, it occurs in a very remarkable degree. Thus the carotid pulse shewn in Fig. 40 is but little dicrotic, but in the radial pulse of the same person (Fig. 41) dicrotism is very great. It seems then that a second cause must contribute to the dicrotic wave, and this, in my former paper already referred to, I endeavoured to shew to be the effect of the inertia of the arterial walls, owing to which the expansion of the tube is somewhat delayed, is then by acquired velocity carried beyond the point which it would otherwise have reached, and is therefore followed by an increased elastic contraction, and a consequent slight second expansion. The effect of this cause would be a progressive one. For suppose the tube divided into a number of small transverse sections, A, B, C, &c. Any very minute recoil expansion arising in section A will be in part transmitted as a wave to section B, and the transmitted wave will be added to the recoil originating in section B itself. This effect will go on increasing until a kind of equilibrium is reached between this process and the tendency of the elastic walls to obliterate all waves.

I think that this gradual production of secondary waves may be illustrated by the series of simultaneous tracings in Figs. 8, 9, and 10. These were taken from different points of the schema, while its tubes were filled not with water but with air. There is therefore no dicrotic wave originating near the valves, or having anything to do with their closure, but for a considerable distance from the heart no secondary waves at all were visible (Fig. 8). At the distance of 4½ feet however two small oscillations have appeared (Fig. 9), and at a distance of 9 feet the first oscillatory wave seems to have become slower and more ample, and thus is separated by a greater interval from the primary wave (Fig. 10).

Let us now return to the consideration of the predicrotic wave. The two possible modes of origin of such an oscillation correspond to the two causes of the dicrotic wave. For just as the sudden cessation of the influx at the end of systole causes a rapid fall of pressure in the section nearest the valves owing to the acquired velocity of the fluid, and a consequent reflux and second expansion, so, if the velocity of the blood is greatly diminished before the end of systole, this diminution of velocity may give rise to an oscillation of a similar kind, but one less in degree, and occurring during the systolic period. In this case the second rise of pressure would be due, not to an actual reflux of fluid, but only to a diminution of onward velocity, and would therefore not depend upon any closure of valves. It may be doubted however whether the velocity of the blood propelled through the aortic orifice is much diminished before the end of systole, and in the case of the schema, where the heart was not emptied completely at each stroke, the velocity appeared to be pretty uniform. I am disposed therefore to believe that in most cases the predicrotic wave is, in the main, to be referred to the same oscillation, set up by the inertia of the arterial walls, which often contributes to the dicrotic wave. For it is probable that the recoil which in dicrotic pulses, when tension is low, helps to augment the dicrotic wave, will, when tension is high, or the arterial walls less distensible, occur more rapidly and produce a separate wave. This probability is confirmed by the position of the first oscillation in Fig. 9, obtained in the experiment in which no dicrotic wave was produced near the valves, because the fluid contained in the tubes was air. When air is used the oscillation is not perceptible until the wave has travelled some distance from the valves; but it does not follow that this is so if the tubes contain liquid. For in the former case the inertia of the tube-wall alone comes into play, that of the air being too small to have any perceptible effect, as shewn

by the absence of waves in Fig. 8 ; but in the latter the inertia
of the liquid is also involved in every movement, and the
oscillation would from the first have a more important character.

In those pulses in which the first oscillation occurs before
the end of systole, and forms the predicrotic wave, it would
be not the first but the second oscillation of the same series
which would be combined with and tend to augment the
dicrotic wave. The occurrence of this second oscillation is
shewn in Figs. 9 and 10.

The same qualities in the heart's action, namely a powerful
commencement and at the same time prolonged duration of
systole, which would promote the occurrence of this predicrotic
wave, tend also to cause the separate appearance of the tidal
wave in the trace, so far as that depends on the sphygmograph.
There is no practical disadvantage therefore in the combination
of two elements in the tidal wave, as it appears in the curve, so
far as regards the inferences to be drawn from the tracing as to
the state of the vascular system ; but on the contrary, the distinctive characters of each variety of pulse are in this way
magnified, and rendered more manifest. A practical lesson
may however be drawn, that, if the tracing of any pulse taken
at a low pressure shew a very marked primary summit, whose
proportionate magnitude is modified by increase of pressure,
then the tracing taken at the higher pressure more closely
represents the true pulse-wave. The form of trace at the lower
pressure may however have much significance, and in these
cases the whole of the information to be derived from the
sphygmograph cannot be compressed into any one curve, but
requires at least two for its expression, namely that trace which
has the greatest amplitude, and another taken at a higher
pressure.

Comparing the different waves in the pulse-curve, it may be
said that the dicrotic wave is an oscillation due primarily to the
inertia of the blood, and that the distinction of primary and
tidal waves is generally the joint effect of one oscillation set up
by the inertia of the arterial walls and another due to that of
the sphygmograph. When the tidal wave is absent, these oscillations contribute to the dicrotic wave.

The systolic part of the pulse-curve may be regarded as

made up of a simple uniform curve, or fundamental wave, such as would be described if no oscillation were set up by the inertia of arterial walls or sphygmograph, upon which fundamental wave are superposed first a pointed primary summit, and then one or sometimes even two (vide Figs. 16, 45 and 46) recoil waves. The recoil wave is the reflection, not of the whole primary ascent, but only of the height by which the primary summit rises above the fundamental wave, and it therefore deviates from the fundamental curve much less than does the primary summit. Hence it is often combined with and almost lost upon the rounded top of the fundamental wave (vide Figs. 14, 25, 26 and 46). It is in this case only, that the title of tidal wave appears to be truly appropriate, as applied to this rounded summit of the pulse-curve. It seems better however to retain the term tidal wave as denoting the first secondary wave, as seen in the trace, in all its forms. For in such a case as that just referred to, in which the part of the curve denominated tidal wave almost coincides with the corresponding portion of the fundamental wave (Fig. 25), it would be incorrect to call it the predicrotic wave, and yet every gradation may be found between this form of curve and that shewn in Fig. 21, in which the tidal wave B seems to be mainly due to the oscillatory predicrotic wave in the artery. The term predicrotic should then be used as a synonym for tidal or first secondary wave only when the elevation in the curve seems to be wholly or chiefly due to this component.

It remains to trace out more in detail the changes of the several waves as they recede from the heart. Both of the instruments used in taking the simultaneous tracings from the schema had an adjustment for altering the amplification given to the trace, and the magnifying power employed was much less than that of the sphygmograph. The pressure used was also much greater than can be applied to an artery, and there were no elastic tissues intervening between the pad and the tube, by which oscillation might be allowed. The chance of deviation in the trace was therefore reduced to a minimum, and moreover it was found that, when the tubes contained air, no secondary wave whatever was seen in a trace taken near the valves (Fig. 8), however sudden the up-stroke might be made.

It may be concluded that no oscillation occurred in the instrument, and that the curves are a very close, if not an absolute, representation of the waves in the fluid, and more strictly faithful than most of those obtained from arteries.

In many of the curves measuring lines are drawn to shew the relative distances of the waves. These are arcs of a circle having a radius equal to the length of the recording lever, care being taken that the centre of the arc corresponds precisely in relative position to the centre of motion of the lever at the time when the tracing was taken.

In tracings taken quite close to the valves of the schema (Figs. 16, 29 and 45), the dicrotic wave is found to be very narrow, sudden, and lofty, when the heart's contraction is short and forcible, but this quality is lost at a distance. (Compare Figs. 17 and 30 with 16 and 29.) These examples shew a fact which at first sight might appear improbable, namely that a recoil wave may be higher than its primary. The total mechanical energy in the reflected wave must of course be less, but, by losing in volume, it may gain in height, somewhat as a wave on the surface of water may, after striking against a wall, be tossed up into the air. For the production of such a lofty dicrotic wave close to the valves, it is necessary that the valves should be of not too yielding a material, and should be capable of closing quickly. In the tracings from a dilated aorta (Figs. 14 and 46) the dicrotic wave C has the same character of suddenness, but is not so lofty, and more nearly resembles the dicrotic wave in Fig. 1, taken 10 inches from the valves of the schema.

In these, as in former experiments, I found that, other things being equal, dicrotism is increased if arterial tension be lowered. A still greater increase however is produced by making the heart's action more short and sudden, and this condition can easily be made to preponderate over the effect of changed arterial pressure. In Fig. 45 is shewn the effect of a gradual modification of heart's action on the dicrotic wave C, in a tracing taken close to the valves of the schema. In the first beat the contraction was short and sudden, in the third steady and prolonged, and in the second of an intermediate character. We may hence understand how it is that in fever the first effect of the relaxed arteries and freedom of outflow is to make the pulse dicrotic, but that in a later stage, if the heart greatly fails, and its action becomes feeble and wavy, dicrotism may diminish, and the dicrotic wave even altogether disappear, although arterial tension remains as low as before. Hence also the bad prognosis to be derived from such a pulse under such circumstances.

In all curves taken from the schema it is found that, at a greater distance from the heart, the tidal wave is less distinctly separate, commences lower down in the descent, and at a somewhat greater distance from the primary wave. The interval however is always considerable, even near to the valves, and it is not therefore the fact that the two waves start together, and become separated as they proceed. The effect of a progressive diminution of arterial tension is shewn in Figs. 6, 7 and 43, 44, as compared with Figs. 3, 4. The tidal wave in the primary trunk is less marked, and in the distal tube it is almost lost in Fig. 7, and entirely so in Fig. 44. The curve becomes also more dicrotic than with the higher tension. The tracings 6 and 7 from the schema may be compared with Figs. 33 and 35, taken from the carotid and radial arteries of a healthy person.

The changes of the tidal wave in the different arteries of the body correspond very closely with those observed in tracings from the schema. In the healthy carotid trace, Fig. 33, it is much wider and more marked than in the radial or dorsalis pedis (Figs. 35 and 36). A similar relation may be seen also in the tracings of carotid, radial, and dorsalis pedis arteries taken from a case of atheroma (Figs. 26, 27 and 28). In small arteries nearer to the heart the pulse-curve has a similar character to that of the carotid, as will be seen in Fig. 34, obtained from the facial artery. Thus we see that the form of the curve depends on the distance from the heart, rather than on the size of the artery from which the tracing is taken. In tracings from an aneurismal dilatation of the aorta (Figs. 14 and 46), the rounded prominence of the tidal wave is still more marked as compared with the radial pulse of the same person (Figs. 15 and 47). This may be partly the effect of the dilatation, but the form of the curve resembles that obtained by Marey by means of his polygraph, representing the pressure

within the aorta of a horse[1]. When the aneurism is sacculated
and communication with the artery not free, all secondary
waves may be lost, and only a rounded swelling appear. But
when there are any secondary waves at all, I have always found
the dicrotic wave visible, and in the aortic trace, Fig. 14, it is
more marked than in the radial pulse tracing of atheroma, Figs.
24 and 25. It is seen also in Marey's aortic tracing. This
seems to me to be an experimental disproof of the theory of the
causation of dicrotism proposed by Dr Mahomed, and adopted
also by Dr Broadbent in his lectures on the pulse, recently published in the Lancet, namely, that there is no second expansion
in the aorta itself, but that the simple elastic recoil or contraction of the aorta produces a second expansion in peripheral
arteries. The same thing is shewn still more strongly by the
dicrotic pulse from the aorta of a dog (Fig. 37). It is confirmed
by the fact shewn by Marey, from the curve of the hæmodromometer (op. cit. p. 273), that in the dicrotic pulse of a horse
there is a marked retrograde current before the dicrotic wave
even in the carotid artery. So extensive a reflux would not be
required for the closure of the aortic valves, but it must be
expended in causing a second expansion in the arteries near to
the heart. When the pulse is less dicrotic, the notch in the
curve of blood-velocity does not fall so low as the zero line.

Marey attributed the dicrotic wave, in the abrupt and
pointed form in which it appears in the aorta (vide Figs. 14 and
46), and frequently also in the carotid artery, to the closure of
the aortic valves, but the dicrotic wave in the radial pulse he
considered to be of a different character, and to be an oscillation
produced in the peripheral arteries. In the schema, however,
the wave may be traced by a uniform gradation from one shape
to the other, and the same thing may be done, although of necessity not so perfectly, in the arteries of the body.

The aortic tracing, Fig. 46, shews a small rise Z just preceding the main up-stroke, and this appears also in Marey's
curve representing the aortic pressure. It is attributed by him
to a wave transmitted through the still closed aortic valves, and
caused by the commencement of ventricular contraction. His

---

[1] *Physiologie Médicale de la Circulation du Sang*, p. 189.

simultaneous tracings shew that the two events are synchronous, and this is probably the true explanation. Fig. 46 shews also another rounded swelling X, which is not unfrequently seen in tracings from arteries near the heart. It is somewhat difficult to explain, since from its shape and position it does not seem likely to be the reflection of any preceding wave. It may be suggested as possible that it may be a wave due to the auricular systole, transmitted through the unopened aortic valves. This explanation would agree with its position in point of time, although it would hardly be thought likely that such an effect would be perceptible.

The form of curve usual in arteries near the heart may, under some circumstances, be entirely altered. This is shewn in the carotid tracing Fig. 18, taken from a case of inflammatory fever, in which the curve is almost as dicrotic as the radial pulse of the same person (Fig. 19). The same thing appears in Fig. 37, taken from the aorta of a dog, exposed within the thorax, while artificial respiration was maintained. The tracing was taken very soon after division of the spinal cord below the medulla oblongata, the effect of which was suddenly to render both aortic and femoral pulses (Figs. 37 and 37, A) highly dicrotic. Previously to this the dicrotic wave had been scarcely visible.

These two instances would seem to shew that when dicrotism is increased, the element of the dicrotic wave which is augmented is generally that which originates close to the valves. An example of an opposite kind is shewn in the series of tracings (Figs. 40, 41, and 42) from the carotid, radial, and dorsalis pedis arteries of a patient having a healthy vascular system, but great relaxation of arteries. The carotid pulse (Fig. 40) is here moderately dicrotic, the radial (Fig. 41) highly so, while in the dorsalis pedis (Fig. 42) the dicrotic wave has altogether disappeared, and the tidal-wave has become more marked. My present belief, however, is, that the first of the two causes to which I have attributed the dicrotic wave, namely, the oscillation near to the valves, is generally the most important, and that the great variation of dicrotism which occurred in this patient is exceptional.

It is possible that if, in a section of artery at some distance

from the aortic valves, the systolic wave had become more lofty,
and more sudden in its conclusion, than it was near the valves,
it might be followed in that section by a greater proportionate
fall of pressure and consequent increased reflux, which meeting
with the advancing dicrotic wave might cause an increased
second expansion. The dicrotic wave would then be progressively increased by an increasing oscillation in the several sections, of a kind analogous to that which first produced it at the
aortic valves, and apart from any oscillation set up by inertia of
arterial walls. Almost invariably, however, the down-stroke of
the systolic wave becomes progressively more gradual, although
its up-stroke in some cases may retain or even increase its steepness. The great transformation of the pulse-wave in the patient just referred to (Figs. 40, 41, 42) was associated with a
very slow rate of transmission of the wave, the velocity of which
must have been as low as 16 feet per second. Thus is inciden-
tally shewn the fallacy of the view of Dr Sanderson, that the
apparent retardation of the pulse in distant arteries is due to
the tidal and not the primary wave being noticed by the finger,
for in this case the tidal wave in the radial pulse is but slightly
marked, and far smaller than the dicrotic. It is generally to be
observed, that there is a much more variable, and, on the average, a much lower velocity of the wave in arteries, than was
found to be the case in the schema, where the velocity varied
only between about 60 and 90 feet per second. This seems due
to the greater thickness of the india-rubber tube-walls, and is
probably to be associated with the fact that it requires a greater
length of india-rubber tube than of artery to produce the same
amount of change in the pulse-wave. Professor A. H. Garrod
has estimated the velocity of the primary wave seen in the
pulse-curve by the very exact method of taking simultaneous
tracings upon the same stage from the radial and posterior tibial
arteries[1]. If the rate of transmission be supposed uniform in
the two arteries, his results give a velocity varying between 25
and 30 feet per second.

In the carotid trace Fig. 40, the dicrotic wave C is broken
into two, and the same thing is often seen in tracings taken

[1] *Proceedings of the Royal Society*, No. 157, 1857.

from the schema not very far from the valves. It may be seen in Fig. 3, and is just visible also in Fig. 6. This is analogous to the division of the systolic wave into primary and predicrotic, and thus indicates suddenness of commencement. As the wave recedes from the heart, the two parts become blended into one, as will be seen by comparing the corresponding tracings taken at a distance, namely, Figs. 4, 7, and 41.

There is one respect in which the tracings at different points of arteries differ from those taken from the schema, namely, that there is not the same constant alteration of the relative position of the tidal wave. In tracings from the schema, we have seen that the interval between the commencements of primary and tidal waves is slightly greater at an increased distance from the heart. This holds true also in a considerable number of tracings taken from arteries (compare Fig. 14 with 15, 26 with 27 and 28, 40 with 41), but exceptions are numerous. Thus the interval is less in the radial pulse Fig. 12, than in the carotid Fig. 11, less in the dorsalis pedis Fig. 36, than in the radial Fig. 35, and much less in the dorsalis pedis Fig. 42, than in either the carotid Fig. 40, or the radial Fig. 41. On measuring the traces it is found that, when this is so, the time occupied in the main up-stroke is, in the distant artery, not greater than, or is even less than, that at the point nearer the heart; that is to say, that the front of the wave has retained its steepness, or has grown even steeper during its progress. The facts then do not admit of the supposition that the tidal wave is always transmitted with either a less or a greater velocity than the primary. They are rather to be explained on the ground that, in general, the primary up-stroke becomes more gradual, and attains its summit more slowly, after it has traversed a greater distance, and then the oscillatory predicrotic wave, or the oscillation due to the sphygmograph, occurs also after a longer interval. In some cases, however, the primary up-stroke does not undergo this change, and then there is no comparative retardation of the tidal wave in the more distant artery.

The fact just alluded to, that in some cases the primary wave does not, for a considerable distance, tend to lose any of its suddenness, brings us to the consideration of a general pro-

gressive change of shape, noticeable in the systolic part of the
pulse-wave. This consists in a shaving off of the right-hand portion of the systolic eminence, made up of the rounded tidal wave,
while at the same time the primary summit becomes narrow,
and comparatively loftier than before.   It may be described
as a change in the shape of the fundamental wave, by which
more and more of its volume is piled up in the front of the
wave, that is to say, at its left-hand side, as it appears in the
trace.   This transformation may be seen by comparing tracings
1 with 2, 3 with 4, 6 with 7, 43 with 44, all of which were
taken from the schema.   It is still more manifest in many
cases after transmission of the pulse-wave through a much
shorter length of artery (compare Fig. 11 with 12 and 13, 14
with 15, 26 with 27 and 28, 33 with 35 and 36, 46 with 47).
The result of this is that the tidal wave, which near the heart
is made up in great measure of the rounded summit or shoulder
of the fundamental wave (vide Figs. 14 and 33), becomes of
less and less importance in approaching the distal arteries,
where it may consist of little more than the oscillatory predicrotic wave, or the oscillatory wave set up by the sphygmograph,
both of which are comparatively small (vide Figs. 2 and 35,
compared with 1 and 33).   The type of healthy pulse therefore
differs widely in different arteries, and a curve which in the
carotid is healthy would be a proof of atheroma if found in the
radial artery, while on the other hand the subject of atheroma
moderate in degree may have a pulse-curve in the dorsalis
pedis like that in the radial of a healthy person.   When the
tidal wave is but small originally, it may entirely disappear at
a distance.   Thus it has almost vanished in Fig. 7 and entirely
so in Fig. 44, both tracings taken from the schema; in Fig. 36
from the dorsalis pedis it is but slight, and it is generally but
little marked in this artery when the vascular system is healthy.

The change of shape in the whole wave appears to include
not only the systolic but also the diastolic portion, so that this
becomes flatter, and of less volume, and the point of commencement of the dicrotic wave is nearer the base-line of the
curve (compare Fig. 11 with 13, 14 with 15, 26 with 28, 33
with 36).   When the change is more marked than usual, the
suddenness of the up-stroke may become actually greater as it

proceeds (compare Figs. 11 and 12), in opposition to the usual tendency of elastic tubes to render all waves more gradual. And in all cases the diminution of steepness, when it does occur, is much less in the primary than it is in the dicrotic wave (compare Figs. 14 and 15, 35 and 36).

Upon the discussion of the mechanical explanation of the transformation undergone by the pulse-wave I shall not attempt to enter in the present paper. The results of my observations are in accordance with the large number of corresponding tracings obtained by Wolff from the radial and dorsalis pedis arteries[1]. He found that in the normal curve from the foot the main up-stroke was slightly more slanting than in the radial artery, and the first secondary wave less marked. But when aortic regurgitation or any other cause of cardiac hypertrophy existed, the first secondary wave was often more marked than in the normal radial curve, and in about the same relative position (op. cit. Fig. 117), although less than in the radial pulse of the same person (op. cit. Figs. 33 and 34). An instance of a similar kind is shewn in Figs. 15 and 18 of my former paper[2]. Again, in senile degeneration of arteries the tracings of Wolff shew a first secondary or tidal wave, commencing high up, close to the primary summit (op. cit. Figs. 109, 110, 112), such as may be found under similar circumstances in the radial pulse. His observations shew another interesting result, which well illustrates the fact, that the form of curve depends on the distance traversed, and not on any local peculiarity in the artery. In tracings from the dorsalis pedis artery of children, and in one from a dwarf (op. cit. Figs. 115, 116, 119), a first secondary wave appeared even more marked than in the normal radial pulse of adults, and greater than was seen in them in the pulse-curve of the foot, except in cases of great cardiac hypertrophy. From this it would seem that we ought to take into account the length of a person's limbs, in estimating what should be the characters of his normal pulse.

It remains to notice the changes which take place in the position of the dicrotic wave. The interval between the commencements of primary and dicrotic wave does not vary much,

[1] *Charakteristik des Arterienpulses*, 1865.
[2] *Journal of Anatomy and Physiology*, Vol. VIII.

but I have not found it to be absolutely constant. Thus in Figs. 1 and 2, 3 and 4, from the schema, and in Figs. 26 and 28 from arteries, in all which cases tension was high, there is scarcely any perceptible difference. On the other hand, in Figs. 16 and 17 from the schema, and Figs. 40 and 41 from the human body, in both which cases tension was very low, the interval is decidedly greater at the point more distant from the heart. This seems to be true also with the dicrotic pulses from the dorsalis pedis shewn by Wolff (*op. cit.* Fig. 120). The point is of some interest, since Professor A. H. Garrod has considered the interval to be constant in all cases, and on this assumption, by superposing the cardiac upon the radial pulse-tracing, has founded equations for calculating the "syspasis," or interval between the commencement of ventricular contraction, and the opening of the aortic valves[1]. His results seem indeed to be somewhat improbable in themselves, as applied to extreme cases, for according to the equations the syspasis vanishes altogether if the pulse-rate reaches 170, a rate which in rare cases may be exceeded in the human body, and actually assumes a negative value if it passes beyond that point. In these cases of very low tension I have found the length of the systolic portion of the pulse-curve to deviate somewhat considerably from that deduced from the equation given by Professor Garrod as connecting it with the pulse-rate. This equation is $xy' = 47 \sqrt[3]{x}$, where $x =$ the pulse-rate, and $y' =$ the ratio borne by the systolic part to the whole beat. It appears to be approximately true in normal pulses, although I have not found the approximation to be quite so close as that observed by Professor Garrod himself. But in Fig. 41 the observed value of the ratio was 2.15, the calculated value 2.38, pulse-rate 88. The same deviation may be found in exactly the opposite condition, namely, when tension is very high. Thus for the patient whose pulse is shewn in Fig. 21, the ratio as observed was 2.33, as calculated 2.50, pulse-rate 82. This retardation of the dicrotic wave may depend in the former class of cases upon a slower transmission of its commencement, in the latter upon an undue prolongation of systole in proportion to the pulse-rate.

With regard to the summit of the dicrotic wave my obser-

[1] *Proceedings of the Royal Society*, No. 157, 1875.

vations agree with those of Wolff and of Professor Garrod, in shewing that its up-stroke grows rapidly more slanting, and thus the distance of its summit from that of the primary wave is progressively increased (compare Fig. 14 with 15, 26 with 28, 33 with 36). If therefore we estimate the velocity of the wave by that of its summit, the dicrotic wave is transmitted much more slowly than the primary. The same principle seems to apply still more to the tricrotic and other succeeding waves. Thus in an artery near the heart we may see several faint waves after the dicrotic; in the radial pulse we rarely see more than at most the tricrotic (Fig. 35, D); but in the dorsalis pedis (Fig. 36) all waves beyond the dicrotic are lost in the succeeding beat.

I have introduced one tracing, that in Fig. 5, which appears to me to furnish a refutation both of the more recent view of Dr Sanderson, that the dicrotic wave has nothing whatever to do with the closure of the aortic valves, and also of the theory which has been held by some, that the dicrotic wave in each segment of an artery originates in that very segment, simply from the effect of the elasticity of the arterial walls. It represents the radial pulse of a young man, having healthy arteries, who was suffering from very free aortic regurgitation, the result of rheumatism. In this instance the artery was elastic, the arterial tension was very low during the intervals of the pulse, and the primary wave in the radial artery extremely lofty and sudden, although it had quite a different character in the carotid. So far therefore as regarded that particular artery the conditions known to favour dicrotism were present in the highest possible degree, but on referring to the figure it will be seen that there is little or no dicrotic wave present in the pulse. The early position of the wave B shews that it is the tidal or predicrotic and not the dicrotic wave, and the small wave which follows it is probably the second oscillation belonging to the same series. It seems that the absence of a considerable dicrotic wave can only be explained by a failure of the wave which originates near the aortic valves.

I think that the general principle is established by a comparison of all the tracings, namely, that each small section of artery transmits to the section following a wave differing very

slightly in shape from that received from the section preceding, but that the cumulative effect of a considerable length
of artery causes a wide transformation in the general shape of
the pulse-wave, as well as in its minor features.

EXPLANATION OF PLATE XIV.

The tracings have been copied by photo-lithography, and are represented of the original size. In tracings taken from the smaller arteries
the magnifying power used was that usual in the sphygmograph, namely, an
amplification of the original motion about 90 times in vertical height. In
tracings from the schema, or from large arteries, a less magnifying power
was used, and in such cases the amount of amplification is specified
in the description of each trace. In the simultaneous tracings taken from
the schema the clockwork movement of the instruments used was not
precisely uniform, being generally somewhat slower in that placed at the
greater distance from the heart. The letter A denotes the primary, B the
tidal or first secondary, C the dicrotic, D the tricrotic wave.

Fig. 1. From main trunk of schema, 10 inches from aortic valves:
arterial tension somewhat high. The main trunk used in this experiment was of not very distensible tubing. Amplified 20 times in vertical
height.

Fig. 2. Tracing simultaneous with Fig. 1, from a small tube at a
distance of 6 feet. Amplified 40 times.

Fig. 3. From main trunk of schema 18 inches from valves: arterial
tension somewhat high. In this and subsequent cases (except Figs. 29 and
30) the main trunk was of more distensible tubing. Amplified 20 times.

Fig. 4. Tracing simultaneous with Fig. 3, taken from a small tube at a
distance of 4½ feet. Amplified 40 times.

Fig. 5. The radial pulse of a man æt. 22, suffering from free aortic
regurgitation. Pressure, 4 oz.

Fig. 6. From main trunk of schema, 18 inches from aortic valves.
Arterial tension somewhat low. Amplified 20 times.

Fig. 7. Tracing simultaneous with fig. 6, from a small tube at a distance of 4½ feet. Amplified 40 times.

Fig. 8. From schema close to aortic valves: tubes filled with air.
Amplified 20 times.

Fig. 9. Tracing simultaneous with Fig. 8, from a small tube at a
distance of 4½ feet. Amplified 40 times.

Fig. 10. Tracing simultaneous with Fig. 8, from a small tube at a
distance of 9 feet. Amplified 90 times.

Fig. 11. From the carotid artery of a woman æt. 22, suffering from
free aortic regurgitation. Amplified 40 times.

Fig. 12. From the radial artery of the same patient. Pressure,
3½ oz.

Fig. 13. From the dorsalis pedis artery of the same patient. Pressure,
2 oz.

Fig. 14. From an aneurismal dilatation of the ascending aorta, taken
in the second right intercostal space. Amplified 40 times.

Fig. 15. From the radial artery of the same patient. Pressure, 3 oz.

Fig. 16. From schema, close to aortic valves: arterial tension low.
Amplified 20 times.

Fig. 17. Tracing simultaneous with Fig. 16, from a small tube at a
distance of 9 feet. Amplified 80 times.

Fig. 18. From the carotid artery of a girl æt. 17, suffering from a
lumbar abscess. T. 102°. 6. Amplified 40 times.

Fig 19. From the radial artery of the same patient. Pressure, 2 oz.

Fig. 21. Radial pulse of a woman æt. 20, the subject of acute Bright's disease. Pressure, 4 oz.

Fig. 22. From the same pulse at a pressure of 5 oz.

Fig. 23. From the same pulse at a pressure of 6 oz.

Fig. 24. Radial pulse of a man æt. 62, having atheromatous arteries. Pressure, 3 oz.

Fig. 25. From the same pulse at a pressure of 6 oz.

Fig. 26. From the carotid artery of a man æt. 66, having somewhat atheromatous vessels. Amplified 35 times. In this and the two following tracings an instrument was used having a slower clockwork movement.

Fig. 27. From the radial artery of the same person. Pressure, 3 oz. Amplified 60 times.

Fig 28. From the dorsalis pedis artery of the same person. Pressure, 4 oz. Amplified 60 times.

Fig. 29. From schema close to aortic valves; arterial tension somewhat high, heart's action sudden. In this experiment, as in that depicted in figs. 1 and 2, the main trunk of the schema was composed of not very distensible tubing. Amplified 20 times.

Fig. 30. Tracing simultaneous with fig. 29, from the main trunk at a distance of 18 inches. Amplified 20 times.

Fig. 31. From the radial artery of a healthy man having a full soft pulse. Pressure, 3 oz.

Fig. 32. Taken from the same pulse while the artery was compressed at a distance of 3 inches. Pressure, 3 oz.

Fig. 33. From the carotid artery of a healthy man æt. 26. Amplified 30 times.

Fig. 34. From the facial artery of the same person.

Fig. 35. From the radial artery of the same person. Pressure, 4 oz.

Fig. 36. From the dorsalis pedis artery of the same person. Pressure, 3 oz.

Fig. 37. From the aorta of a dog, exposed within the thorax, while artificial respiration was maintained. Taken shortly after division of the spinal cord below the medulla oblongata. Amplified 25 times.

Fig. 37, *A.* From the femoral artery of the dog under the same circumstances.

Fig. 38. From the radial artery of a man æt. 30, the subject of Anæmia Lymphatica, having very relaxed arteries and a temperature of 100°. Pressure, 1 oz.

Fig. 39. Taken from the same pulse, while the small secondary spring was applied to the long lever of the sphygmograph. Pressure, 1 oz.

Fig. 40. From the carotid artery of the same patient. Amplified 30 times.

Fig. 41. From the radial artery of the same patient. Pressure, 2 oz.

Fig. 42. From the dorsalis pedis artery of the same patient. Pressure, 3 oz.

Fig. 43. From main trunk of schema, 18 inches from valves. Arterial tension somewhat low, as in Figs. 6, 7, but heart's contraction shorter. Amplified 20 times.

Fig. 44. Tracing simultaneous with Fig. 43, from a small tube at a distance of 4½ feet. Amplified 40 times.

Fig. 45. Taken from schema close to aortic valves. Heart's contraction sudden in first beat ; afterwards made progressively more prolonged and steady. Amplified 20 times.

Fig. 46. From an aneurismal dilatation of the ascending aorta, taken in the second right intercostal space. Amplified 20 times.

Fig. 47. From the radial artery of the same patient. Pressure, 3 oz.

NOTES ON THE BRONCHO-ŒSOPHAGEAL AND PLEURO-ŒSOPHAGEAL MUSCLES. By D. J. CUN-NINGHAM, M.B., *late Demonstrator of Anatomy, University of Edinburgh.*

To Prof. Hyrtl the credit is due of first having described these muscular slips (*Zeitschrift der Gesellschaft der Aertze zu Wien*, 1844, S. 155). Two years later, Sir J. Paget[1] when taking notice of his paper on this subject, confirmed his observations, but since then the subject seems to have remained quiescent in this country, for in most of our text-books of anatomy they are not mentioned at all, and in those, in which they are, the description is brief, and they are spoken of as occurring 'sometimes' or 'rarely.' Indeed, I was ignorant of their existence until, about nine months ago, my attention was called by a student, who was dissecting the posterior mediastinum, to a thick fleshy band crossing the aorta from the œsophagus to the left pleura. On consulting the seventh edition of Quain's *Anatomy*, I found a note, p. 822, stating that the longitudinal fibres of the œsophagus are sometimes reinforced by two bands, proceeding, the one from the left bronchus and the other from the left pleura. I then proceeded to make a series of special dissections of the thoracic portion of the œsophagus, with the view of throwing, if possible, a little more light upon the subject. The following are my results.

I have now made fourteen dissections, and in all of these, with one exception, the pleuro-œsophageal slip or slips were present, whilst in ten cases both the pleuro- and broncho-œsophageal muscles were found. The exceptional case in which I could discover neither the one nor the other, was that of a decrepit old man, whose lower limbs and one of his upper extremities had been paralysed for many years, and whose muscular system in consequence was much wasted and fattily degenerated. Whether this had anything to do with the absence of the œsophageal slips it is hard to say. The muscular fibres of the œsophagus were extremely pale, and the con-

---

[1] Report in *Medico-Chirurgical Review*, 1846.

nective tissue around it plentiful, so it is possible that they may have been overlooked in the dissection.

Pleuro-œsophageal Muscle.—To expose this it is best to make the dissection from the front by removing the heart. A longitudinal incision should then be made through the pericardium, where it forms the anterior wall of the posterior mediastinum, in a line corresponding to the centre of the œsophagus. The pericardium should then be raised and carefully turned outwards from right to left to the point where it is joined by the left pleura. With a little careful dissection the pleuro-œsophageal slip will be seen arising from the pleura, where it passes over the thoracic aorta, and forms the left boundary of the posterior mediastinum. From this origin it arches over the aorta to the left margin of the œsophagus, into which its muscular fibres enter and diverge—a few taking an upward direction, but the great majority passing down towards the stomach.

In size it varies much. In one case it was $1\frac{1}{4}$ inches broad and $1\frac{1}{2}$ inches long, and constituted a thin sheet of muscular fibres connecting the œsophagus to the pleura very effectually. In the majority of cases, however, it did not exceed $\frac{1}{4}$ to $\frac{1}{2}$ an inch in breadth, but in these cases it was generally more fleshy, and was, perhaps, accompanied by a second or even a third slip, very much smaller, but having the same connections. In some cases, indeed, no distinct band was to be found, simply a great number of muscular fasciculi or bundles, which passing outwards from the anterior surface of the œsophagus, as well as from its left margin, joined with each other to form a series of muscular arcades. Of these some were lost in the loose cellular tissue, whilst others had a distinct connection with the left pleura. I have seen as many as six of these œsophageal slips. When one band alone is present, its usual position is at a point corresponding to the eighth dorsal vertebra.

Broncho-œsophageal Muscle.—To expose this a further dissection is necessary. The left bronchus must be cut through close to where it enters the lung and the latter removed; then fixing a hook into the cut end of the bronchus it should be

drawn gently forwards.   In the loose tissue between it and the
œsophagus the muscle is found.

In no case is this so well marked as the pleuro-œsophageal
muscle, and, as we have seen, it was altogether absent in four out
of the fourteen cases.   It is generally double and represented
by two slips springing from the posterior aspect of the bronchus,
and, expanding as they pass downwards and backwards, their
constituent fibres mingle with the longitudinal muscular fibres
on the anterior surface of the œsophagus.   In size these slips
vary but little, being usually nothing more than two muscular
fasciculi, with a circumference not greater than a piece of whip-
cord.   In one case, however, the highest slip was fully ¼ inch
in breadth and flattened out like a ribbon.   They enter the
œsophagus at a point a little above the level of the pleuro-
œsophageal muscle.

But these are not all the muscular connections of the œsophagus in the thorax.   I have in one or two cases seen muscular
fasciculi pass from the anterior surface of the œsophagus to be
inserted into the pericardium, and also sometimes into the
angle formed by the point of contact between the pleura and
pericardium, in other words, into the left anterior corner of the
posterior mediastinum.   In Henle's work upon *Systematic Anatomy*, I notice that Treitz has not only described these but also
others which pass to the right wall of the mediastinum, to the
trachea, to the thoracic aorta and to the left subclavian artery.
I have on several occasions looked for these latter, but have
never been able to find them.

It is very difficult to ascribe any definite action to these
muscular slips.   Various theories have been advanced, but all
are, more or less, unsatisfactory.   Hyrtl considered that the two
muscles acted together—the broncho-œsophageal pulling the
posterior wall of the bronchus outwards, whilst the pleuro-
œsophageal held the œsophagus in a downward direction.   He
was obliged to depart from this view, however, by the occasional
absence of the broncho-œsophageal slip.   Henle, on the other
hand[1], believes that they protect the arteries from pressure and
friction.   According to him the broncho-œsophageal, during

---

[1] *Handbuch der systematischen Anatomie*, Vol. II. p. 152.

swallowing, draws the œsophagus forwards, and in this way moderates the pressure of the bolus of food upon the bronchial artery. The pleuro-œsophageal, on the other hand, protects the œsophageal vessels. This explanation must strike every one as being very far-fetched, and with regard to the œsophageal arteries altogether erroneous, because these vessels, where they have any connection at all with the muscular slips, lie behind them, and must, consequently, be compressed —not protected— by them when they contract. I have thought much upon this point, but the only reasonable explanation which I am able to offer is that these muscles by their attachments serve to give the œsophagus fixed points upon which it can contract the more readily in the process of swallowing. Probably also they may have some influence in restoring the gullet to its right position after each descent it makes with the diaphragm in inspiration.

ON THE SUMMATION OF ELECTRICAL STIMULI APPLIED TO THE SKIN. By WILLIAM STIRLING, D.Sc., M.D., *Demonstrator of Practical Physiology in the University of Edinburgh*[1].

THE experiments which form the basis of the present communication were performed in the Physiological Laboratory of Prof. Ludwig of Leipzig, to whom and also to Prof. Kronecker I am deeply indebted for many useful hints and much valuable assistance rendered me during the course of the investigation.

In the older papers upon reflex action, the time which elapsed between the application of a stimulus to a sensory nerve and the beginning of the resulting movement remained quite unnoticed.

Prochaska[2] thought that the nerves "took up the external and internal impressions of the stimuli and conducted them with a light-ning-like speed to their destination." Even Joh. Müller, who was acquainted with the peculiarity of the "individual error," like Bessel, ascribed the difference in time which elapses between the perception of an impression upon the eye (by observing an audible pendulum) and the marking of the same, to this, that we cannot perceive two sensations simultaneously (through eye and ear). Müller adds, "The time required for the propagation of a sensation from the surface of the brain and spinal marrow, and for the reflected action producing contractions of the muscles, is immeasurably short. Frogs poisoned with opium or nux vomica are, at first, in a state of such excessive sensibility that the slightest touch on the skin excites convulsions of the whole body. Here the impression made on the skin is propa-gated first to the spinal cord, and from the cord to all the muscles, and, nevertheless, I have been unable to detect the slightest interval between the moment when the skin was touched and the occurrence of the muscular spasms[3]."

In the fundamental treatise of Edward Weber, "Muskelbewegung[4]," there is the first notice, as far as I am aware, of the pause which

---

[1] This paper has already been published *in extenso* in German (Ueber die Summation elektrischer Hautreize, *Berichte der kön. sächs. Gesellsch. d. Wis-sensch. math.-phys. Classe*, December, 1874), and was also presented to the University of Edinburgh as a prize Thesis for graduation in medicine.

[2] *Physiologie d. Menschen*, 1840, Bd. I. p. 186.

[3] Joh. Müller, *Handbuch der Physiologie d. Menschen* (Elements of Physi-ology), 1848, Bd. I. p. 588. Translated by W. Baly, Vol. I. p. 781.

[4] *Wagner's Handwörterbuch d. Physiologie*, Bd. III. Abth. II. 19.

exists between the application of the stimulus and the reflex movement. He records that in a beheaded frog, one of whose sciatic nerves was isolated and stimulated with the electro-magnetic rotation apparatus, that "the movements did not occur at the same moment in which the galvanic current began, as is the case when the spinal cord or nerves were stimulated directly, but nevertheless, in spite of the powerful action of the apparatus, a considerable time always elapsed before the movements followed."

The first measurements of this time "of latent stimulation" we owe to Helmholtz[1]. The time necessary for the conveyance of an impression from the sensory to the motor root in the spinal cord he estimated at $\frac{1}{80}$ to $\frac{1}{10}$ of a second and more, i.e. twelve times as long as the time necessary for the excitation to pass along the sensory and motor nerves. The duration is specially long when the frog, poisoned with strychnine, possesses a high reflex excitability.

8. Exner[2] "measured the sensory conduction in the spinal cord in man." He reckons the rapidity of propagation in the spinal cord (taking the rapidity of propagation in the nerves at 62 meters per second) at 8 meters per second. The "time of reaction ('Reactionszeit') increases with fatigue, diminishes with increasing intensity of the stimulus and in consequence of exercise."

Wundt[3] also found that the time of latent stimulation diminished with increase in the strength of the current. At the same time he remarks (what Volkmann and Weber had concluded from simple observation) that the curve of the reflex contraction is of much longer duration than that which is obtained by direct stimulation. After poisoning with strychnine the minimal contraction is also longer than the one obtained under normal conditions. "With increased action of the poison it very soon passes into a tetanic contraction, so that the time of latent stimulation can be increased to more than twice its usual duration. Sometimes the differences in the time of latent stimulation with strong and weak stimuli increase enormously."

Rosenthal[4] found that the reflex time ("Reflexzeit"), i.e. "the time necessary for the transference of a sensory stimulus to a motor-nerve," on stimulating (with simple electrical shocks?) the intact skin or the exposed nerves, was dependent on the strength of the stimulus, "so that with very strong stimuli it may become imperceptibly small." He did not employ such stimuli as gave the maximum of reflex action. "The time also of transverse conduction ("Querleitung")—i.e. the difference in time between the reflex time from one point on the skin to a muscle situated on the same side, and the reflex time from the corresponding point on the skin to the muscle of the opposite side—with sufficient stimuli reaches a maximum, which becomes smaller with over-maximal stimuli,

[1] Berichte d. k. Akademie d. Wissensch. zu Berlin, 1854, p. 328.
[2] Die experimentelle Untersuchung der einfachsten physiologischen Processe, Pflüg. Arch., 1873. Bd. vii. 632 and 638.
[3] Physiologische Psychologie, Leipzig, 1874, p. 262.
[4] Berichte der Akademie der Wissenschaften zu Berlin, 1873.

and quite inappreciable with very strong ones." Both latent times increase with the fatigue. "The reflex time is greater for a point further removed from the spinal cord, than for a nearer one; with strong stimuli the difference is small." S. Exner[1] also found, that the time of reaction like the reflex time, (closure of the eyelid after electrical stimulation of the cornea) with strong stimuli is smaller than with weaker stimuli (0.05—0.06 secs.).

Schiff[2] asserts that "the time which is necessary for the occurrence of reflex movements is longer, the more the thickness of the gray substance at one point which lies between the point stimulated and the part to be moved is diminished by incisions into the spinal cord (only tried at one point in the length of the spinal cord in *frogs*)."

When several single stimuli are applied to sensory nerves with a certain frequency their effects become summated in the organ which produces the reflex act. Methodical investigations upon this point have up to this time been wanting. Of the investigations upon this subject those of Setschenow are the most extensive. In his monograph "Ueber die elektrische und chemische Reizung der sensiblen Rückenmarknerven des Frosches[3]" he gives as the result of one series of observations, "The reflex and locomotor centra possess in a high degree the property of summing up individual impulses communicated to them" (p. 25). The central end of the sciatic nerve of a frog was stimulated by the closing and opening of a constant current. "The limit of the summation of the effects of single shocks in the time (*i.e.* how often shocks of a given strength and direction must succeed each other in order that a summation of their effects may take place) could not be estimated by my apparatus" (p. 11).

He employed frogs from which only the cerebral hemispheres were removed, and cites one or two experiments; but remarks that "these experiments must be further investigated." "Increase of the current or increase in the number of interruptions causes the result under similar conditions to follow more quickly" (p. 12). By means of induction-shocks also, each single shock having very little effect, very considerable effects can be obtained when the Wagner's hammer is allowed to vibrate[4]. A similar increase of the effect with the strength and duration of the stimulus as has been shown for electrical stimulation, has already been shown (Türck[5]) to be true for chemical (dilute sulphuric acid) stimulation, and this point has been methodically investigated by W. Baxt[6]. The latent effect of very dilute acid (0.0006 per cent.) may last even as long as two minutes. Baxt found that the times of action (from the dipping of the limb into the dilute acid until it was again withdrawn) increased in a geometrical progression, whilst the degree of acidity followed an arithmetical

[1] *Pflüg. Arch.*, 1874, p. 531.
[2] *Lehrbuch d. Physiologie*, Jahr. 1858—59, p. 228.
[3] Graz, *Universitäts-Buchhandlung*, 1868.
[4] P. 14. Compare Fick, *Pflüg. Arch.*, 1873. Bd. III. p. 329, Wundt. l. c. p. 262.
[5] *Zeitschrift der Ges. d. Aertze*, Wien, 1850, Heft III.
[6] *Arbeiten aus d. phys. Anstalt zu Leipzig*, 1871.

one (p. 86). "The stimulating action of the acid therefore, does not depend alone upon the quantity which penetrates to the nerve, but also upon the rapidity with which the diffusion occurs. While the quantity of acid which penetrates to the nerve is equal to the product of the density into the time of diffusion, and while with its diminishing density, the time of diffusion necessary for a contraction increases more rapidly than the acid has diminished in density, it therefore follows, that for the production of a contraction, more acid must have passed in, the slower it has penetrated."

This condition is easily explained if we assume that each particle of acid coming to the nerve produces therein a periodic process, a kind of momentary impulse, e.g. of induction-shock. According to this view of the question, what causes the stimulation is not the number of particles of acid present in the nerve, but the number of particles of acid which penetrate to it in the unit of time; or otherwise expressed, it is not the condition remaining after the entrance of the acid which produces the excitation, but only the act of decomposition, or the changes in the nerve-mass. "The possibility of producing contractions would come to the acid when its masses followed more rapidly than the course of the nervous excitations, so that each successive particle of acid strengthened the movement induced in the nerve-mass by its predecessor[1]."

To test and prove this view by means of analytical electrical experiments was the task set me by Professor Ludwig.

The animal experimented on was *Rana esculenta*. The brain and spinal cord were destroyed as far down as the origin of the brachial plexus, and to prevent bleeding the upper part of the canal was plugged. The animal was suspended by the very practical support of Sanders-Ezn[2], the head being fixed in a small clamp. The part stimulated was one foot. "The effect of stimuli which produce reflex movements is modified and increased by the peripheral expansions of the nerves." This was expressed by Volkmann[3] even in 1838, who showed how much more sensitive the skin is, than are the nerves, which run from the skin to the coverings of the muscles.

Setschenow placed his electrodes on the sciatic nerve of the frog after amputating the limb as high up as possible, following the procedure of Marianini[4] and Pflüger[5]. Fick also chose the small cutaneous nerve-trunks instead of the skin. He found (like Volkmann)

[1] W. Baxt, l. c. p. 87.
[2] *Arbeiten aus d. physiol. Anstalt zu Leipzig*, 1867, p. 16.
[3] *Müller's Arch.*, p. 25.
[4] E. du Bois Reymond's *Untersuchungen üb. thierische Electricität*, Bd. I. 1848, p. 859.
[5] *Ueb. d. elekt. Empfindungen, Untersuch. aus den physiol. Laborat. zu Bonn.* Berlin, 1865, p. 155.

that stimulation of the skin produced washing-movements, and that extraordinarily strong induction-shocks were necessary to obtain reflex movements from the trunk of the sciatic, whilst an extremely weak stimulus applied to the skin produced energetic movements (p. 328). Nevertheless he did not employ cutaneous stimuli. Rosenthal (l. c.) seems to have stimulated the skin electrically, though in his preliminary communication he does not give his method. Wundt only speaks of stimulation of the central end of a sensory nerve-root (p. 261).

Meihüizen[1] rejects the skin as the point of application of the stimulus, "because the contractions by direct stimulation of the muscles render it impossible to employ the skin of an extremity for electrical stimulation without further preparation."

I have tried to free stimulation of the skin from all these imperfections, and have found that the results obtained were by no means less precise than by electrical stimulation of a nerve-trunk. The foot must always be kept moist with ordinary water, and the electrodes must be so adjusted that they do not stimulate mechanically nor change their place.

The arrangement of the experiment most employed is rendered plain by the following schema Fig. 1.

Fig. 1.

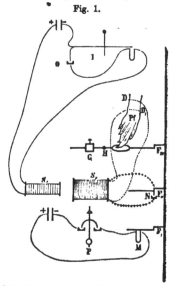

---

[1] *Pflüger's Arch.*, 1873. Bd. vii. p. 202.

Three pens F', F'', F''', wrote with anilin blue—horizontally above each other on an endless strip of paper rolled on a kymographion which moved with a regular velocity. As long as the pens were at rest the lines drawn were horizontal, parallel, and straight. The lowest pen F' was by means of an electro-magnet drawn down several mm. as long as the seconds-pendulum P at the end of each vibration completed the time-marking circuit (......). The middle pen F'' was fixed to a secondary closure N, which was introduced into the course of the induced current (+++++). As long as the key was depressed and the pen F'' kept down, it conducted away the current from the preparation, which therefore was stimulated as long as the line drawn by F'' ran high. The upper pen F''' was elevated by the counterpoise G, as soon as the foot of the frog was raised from the plate fixed to the lever H.

The electrical stimuli which caused the reflex movements were applied to the skin of the foot by two loops D, D' of fine gold thread, which surrounded the ankle-joint and removed from each other about 5 mm. The threads were well isolated by sheet gutta percha and ran to the secondary spiral S'' of a du Bois Reymond's inductorium which was graduated into units[1]. The primary spiral S' supplied by two Grove's elements was interrupted and closed at the mercury contact O at the requisite intervals by a metronome or a Ruhmkorff's interruptor. The wires l and m (Fig. 2) conducted the primary current to and from the interruptor, whose essential parts are here figured. The transverse bar g sent a platinum style to the Hg. contact o, from which it was pulled out when the magnet p was attracted by the electro-magnet i. The contact thus broken interrupted the current, rendered the electro-magnet ineffective, and the bar g was permitted to spring back. Thus the vibration of the pendulum is induced and maintained just as in the Wagner's hammer of an induction-machine. The moveable ball h on the pendulum permits the complete duration of a vibration of the pendulum-system being varied between the limits $\frac{2}{5}''$—$\frac{1}{5}''$. By the vessel o, which may be raised or depressed, the contact can be so regulated that it only takes place in the

[1] Fick's *Untersuch. aus d. phys. Labor. d. Zürich. Hochschule.* Wien, 1869, p. 38.

quiescent condition of the pendulum; *i. e.* an opening induction-current is produced as soon as the magnet is set in action. After $\frac{3}{4}''$ vibrations the contact is again completed (closing-shock), again after $\frac{3}{4}''$ vibrations (whilst the style dips in and out of the Hg.) an opening-shock, and so on, *i.e.* in equal intervals of half the duration of a vibration alternately one opening and one closing shock. By an arrangement similar to that of Helmholtz on the induction-machine the oppositely directed induction-shocks can be made of very nearly equal intensity. The opening spark, however, never completely disappears.

THE WASHING-APPARATUS. (Fig. 2.)

This apparatus is meant to keep the Hg. contact (O in Fig. 1, *o* in Fig. 2) free from the isolated particles which are formed at the point of breaking the primary current, and which influence the constant completeness of the resistance and therewith the intensity of the current.

The layer of alcohol which, following Poggendorf[1], stands over the Hg. to weaken the spark at the point of interruption, is constantly being made dirty by particles of oxydised mercury, and must be continually renewed. This is done by a fine stream (regulated by a glass stop-cock *s*) which flows from a small funnel *b* filled with dilute alcohol, through the tube *a* into the contact-glass. The overflow through the syphon-tube *r* into the glass *f* is so arranged that the circulating alcohol stands at a level of 1 centimeter above the Hg. In order that the small filter may remain constantly full, a Mariott's flask is placed above it. Through the bottom of this, a glass tube 5 mm. in diameter is fitted water-tight. It projects for 2 ctm. into the funnel. A conical glass rod is placed in the tube as a moveable valve. The pointed end of the valve projects below somewhat beyond the tube, so that it is raised by the wall of the funnel as soon as the flask is placed thereon. An ascending tube *b* (1 ctm. diam.) fills up a second hole in the bottom of the flask, and ends in an obliquely ground opening 1 ctm. under the bottom of the flask, whilst its upper end reaches into the space filled

[1] *Annalen*, Bd. 94. p. 389.

Fig. 2.

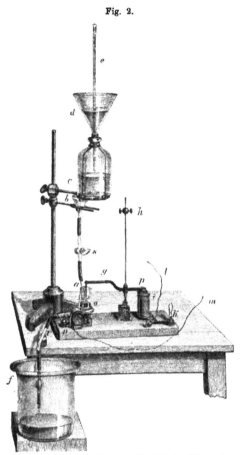

with air in the Mariott's flask. If this is filled, its contents
run into the funnel until the mouth of the ascending tube is
closed by the level of the fluid.   Then the fluid is prevented
by the pressure of the outer air, which cannot be equalised
with that within the flask, from passing out through the valve-
tube, till the level, by being caused to sink, allows air-bubbles to
rise through the ascending tube.   Thus the level of the fluid
under the wall of the funnel is kept at a constant height.

At first I applied the most frequent possible stimuli to the skin of the frogs prepared as above described. The primary current was interrupted by the Wagner's hammer of du Bois Reymond's inductorium or by a König's tuning-fork of 100 complete vibrations.

Minimal stimuli acted even after very short time of latent stimulation, and soon lost their effect, if at least half a minute's rest was not given between the periods of stimulation. Increased currents produced reflex movements after very short latency. If such discharges, however, were often produced in pauses of five to ten seconds, the times of latent stimulation increased to three and exceptionally to five seconds. Simultaneously the reflex contractions become weaker. The duration of the latency could not be essentially diminished by strengthening the frequent stimuli. When ineffective stimuli were *rapidly* strengthened, the reflex act occurred with an intensity of the current which remained powerless when the current was *gradually* strengthened to the same degree. The often proved observation, that the reflex minimal contraction produced by frequent stimulation after a short latent period, and that strengthened stimuli are not able essentially to shorten the duration of the latency, seem to indicate, that the great differences in the duration of latent stimulation of the limb dipped in acid, are not caused by varying intensity of impulses of equally high frequency. I therefore tried to observe the effect of a change in the frequency of the stimuli. Slow vibration of Wagner's hammer (about 40 per second) appeared sometimes to cause the limb to move reflexly after somewhat longer latency than the rapidly following shocks of the high-sounding spring. This result was however highly doubtful, and was probably caused by irregularities in the contact. Nor did I observe any great difference in the latency whether the stimuli were weak or strong, when I gave the limb 48 induction-shocks per second. Only the amount of contraction increased with the strength of the current, and the latency only where the fatigue became apparent; in that for equally strong stimuli it increased the duration of the latency. This observation may be explained by the following example. At the beginning of the corresponding experiment, stimuli of different intensities following each other at large intervals were compared.   In such a case, the latency was essentially lengthened, when the intensity of the current diminished.

For the convenience of tabular arrangement the amount or extent of contraction is divided into four degrees, thus: degree I. means elevation of the foot alone, degree II. flexion of the leg at the knee, degree III. flexion of the thigh and leg (at knee and hip-joint), degree IV. violent repeated spasms of one or both lower extremities.

A 0 under "Degree of contraction" indicates that in the corresponding period of stimulation no reflex movement occurred.   The sign ∞ under "Latent time" also shows that the expected contraction did not occur.

TABLE I.

Shows, how with small interval of stimulation the times of latent stimulation are independent of the strength of the stimulus.

Periods of rest between the periods of stimulation.	Interval of stimulation.	Strength of stimulus in Units (E).	Latent time in seconds.
30 seconds	½"	200	0,8
,,	,,	150	0,4
,,	,,	100	4,4
,,	,,	125	1,5
1 minute	,,	125	2,3
30 seconds	,,	125	2,0
,,	,,	150	1,5
,,	,,	175	1,0
..	,,	200	1,0
,,	,,	150	2,0
,,	1⅛"	150	1,2
,,	,,	100	∞
2 minutes	,,	125	1,2
30 seconds	,,	125	1,6
,,	,,	125	1,3
,,	,,	150	1,5
,,	,,	175	1,0
..	,,	150	1,4
,,	,,	175	1,8
,,	,,	150	1,8
,,	,,	175	1,5
,,	,,	200	1,5
,,	,,	200	∞

TABLE II.

Shows, how with small interval of stimulation the times of latent stimulation are independent of the strength of the stimulus, but the degree of contraction however varying therewith.

Periods of rest between the periods of stimulation.	Interval of stimulation.	Strength of stimulus in Units (*S*).	Latent time in seconds.	Degree of contraction.
30 seconds	¼"	50	0,14	I (after 0,5" III)
"	"	25	0,5	I (after 1,0" III)
"	"	20	0,5	I (after 2,0" III)
"	"	15	1,3	I (after 2,5" III)
"	"	10 ·	2,0	I (after 3,0" III)
10 minutes	¼"	10	0,5	II
30 seconds	"	30	0,3	II
"	"	40	0,3	I
"	"	40	0,3	II
"	"	37	0,3	I
"	"	70	0,3	III
"	"	60	0,3	II
"	"	55	0,3	I
"	"	45	0,4	minimal
"	"	45		0
"	"	50		0
"	"	55	0,5	I
"	"	60	0,5	I
"	"	65	0,7	I
"	"	70	0,7	I
"	"	80	1,0	I
"	"	90		0
"	"	100	1,0	I
"	"	125		0

These examples are selected because in the one case the latent times, in spite of comparatively strong and frequent stimuli, soon become large, whilst in the other case only with long intervals and the weak stimuli does the latency become considerably lengthened, while with a rapid succession of shocks, until the preparation is nearly dead, the latent times remain small. In both examples, no dependence of the latent times upon the intensity of *frequent* stimulation is to be observed. The stimuli must be gradually strengthened to obtain any effect whatever, and must also act *longer* before they produce a contraction. Already however, before the time of latent stimulation had reached two seconds, the excitability is generally extinquished. At the same time the contractions become small, *i.e.* only the foot is slightly raised. This agrees with the observations of Volkmann[1] "that the *extension* of the reflex movement is pronouncedly dependent on the strength of the stimuli and the degree of excitability."

The above-cited variations in the excitability of different frogs led me to investigate whether the excitability of the spinal cord changed with the general conditions under which the animal is placed. Calling to mind the observation of Leube[2] made under Rosenthal's direction, and its extension by Uspensy[3], that "apnœa" (produced by strong artificial respiration) extinguished the occurrence of reflex exchange in the spinal cord, I tried if powerful ventilation of the lungs influenced the reflex processes. Although, on account of the important cutaneous respiration in frogs a smaller effect was to be expected in them than in mammalia, still a long series of experiments convinced me, that artificial respiration of different depths and frequency exercises no influence on the reflex excitability of frogs. On the contrary, most of the preparations which lasted for a long time showed an increase of excitability at the beginning of the experiment (which was generally begun from a quarter to half an hour after section of the spinal cord), and often it was to be observed on fresh (winter) preparations, that weak stimuli were more effective after stronger ones than before them;— modifications in the excitability, such as have been observed in the motor-nerves by Wundt[4], Türck[5] and W. Baxt[6] on reflex preparations with chemical stimulation. In general, the excitability diminishes with the time, but in very different degrees, according to the individuality of the preparations placed under similar external conditions. Generally the limb with the piece of spinal cord from which it is supplied with nerves hangs motionless until a stimulus of sufficient strength is applied to it. The oftener it has been stimulated, the stronger must the impulses be to give a reflex action. The smaller the periods of rest between the stimulations, the sooner is the excitability lost. This may be kept constant for a long time when the periods of rest vary from three to ten minutes.

[1] *Müller's Arch.*, 1838, p. 23.
[2] *Reichert and du Bois-Reymond's Arch.*, 1867.
[3] *Ibid.* 1868.
[4] *Ibid.* 1859, p. 537.
[5] *Sitzungsb. d. Gesell. d. Aertze zu Wien*, 1850, Nov.　　[6] l. c. p. 74.

After having found in the first series of experiments that with a *rapid* succession of shocks, the strength of the reflex action varies with the intensity of the stimuli, but that the latent times do not, I tested the influence of varying strength of current with *medium* succession of stimuli. On employing a moderate interval of stimulation ($\frac{1}{2}''$ to $\frac{1}{15}''$) the duration of the latent stimulation diminished, when the sub-maximal intensity of current was increased.

The following table will serve to show how the latency varies with the intensity of the current.

What was to be observed from the first half of the previous table is very obvious in this one, viz. that the latent time diminishes whilst the strength of the stimulus increases. To render clearer the relation of the increase of stimulus and the times of latent stimulation, I have calculated the quotients of each two neighbouring current-intensities, and the quotients of each two successive latent times, and placed their values between the times of the two corresponding factors. On comparing now the proportions of the strength of stimulus with the reciprocal relative numbers of the latent times of similar height, then in general it is to be observed, similar neighbouring stimulation-values have also tolerably similar latent values. In a few cases variations are to be found, which however only become marked when the stimulating tempo is changed; a point which will be alluded to later. The inconstancy sometimes apparent in the results is not to be ascribed alone to imperfections in my apparatus, but also to the variability of the nervous structures.

Between each two observations where nothing is noted there is always a pause of half a minute.

In spite of some irregularities, it is to be observed that after several periods of stimulation the excitability rises somewhat, so that 80 units (E) produced a reflex action first after 1.8—2.0″ and then after 1.4—1.5″.

TABLE III.

Shows, the dependence of the times of latent stimulation, and the degree of contraction on the intensity of the current used to stimulate at moderate intervals.

Running number of the experiment.	Interval of stimulation.	Strength of stimulus in Units (E).	Relation of the		Latent time in seconds.	Degree of contraction.
			Strength of the stimuli.	Latent times.		
1	$\frac{1}{1}''(\frac{1}{1}'')$	100			1,0	II
			1,0 : 1	1 : 1,0		
2		100			1,0	II
			1,0 : 1	1 : 1,0		
8		100			1,0	III
			1,25 : 1	1 : 1,8		
4		80			1,8	Y
			1,0 : 1	1 : 1,11		
5		80			2,0	ī
			1 : 1,12	1,66 : 1		
6		90			1,2	II
			1 : 1,11	1,09 : 1		
7		100			1,1	II
			1 : 1,25	1,1 : 1		
8		125			1,0	II
			1,89 : 1	1 : 1,2		
9		90			1,2	Y
			1,12 : 1	1 : 1,17		
10		80			1,4	·
			1,14 : 1	1,08 : 1		
11		70			1,8	ī
			1,16 : 1			
12		60				0
			1 : 1,16			
18		70			1,8	Y
			1,0 : 1	1 : 1,05		
14		70			1,9	▾
			1 : 1,14	1,27 : 1		
15		80			1,5	–
			1 : 1,12	1 : 1,0		
16		90			1,5	ī
			1 : 1,11	1,25 : 1		
17		100			1,2	II
			1 : 1,25	1,7 : 1		
18		125			0,7	IV
			1,25 : 1	1 : 1,7		
19		100			1,2	IV
			1 : 1,25	1,5 : 1		
20		125			0,8	III
			1,25 : 1	1 : 2,12		
21		100			1,7	II
			1 : 1,0	1,42 : 1		
22	$\frac{1}{15}''(\frac{1}{15}'')$	100			1,2	Y
			1 : 1,25	1,5 : 1		
28		125			0,8	ī
			1 : 1,20	2,66 : 1		
24		150			0,8	III
			1,20 : 1	1 : 2,66		

Running number of the experiment.	Interval of stimulation.	Strength of stimulus in Units (N)	Relation of the		Latent time in seconds.	Degree of contraction.
			Strength of the stimuli	Latent times.		
25	¹⁄₈″ (¹⁄₁₃″)	125			0,8	II
			1 : 1,0	1,0 : 1		
26		125			0,8	II
			1 : 1,20	1,14 : 1		
27		150			0,7	I
			1,20 : 1	1 : 1,14		
28		125			0,8	I
			1 : 1,20	1,14 : 1		
29		150			0,7	II
			1,0 : 1	1 : 1,28		
30	¹⁄₁″ (¹⁄₁₃″)	150			0,9	II
			1,20 : 1	1 : 1,0		
31		125			0,9	I
			1 : 1,20	1 : 1,22		
32		150			1,1	II
			1,20 : 1	1 : 1,09		
33		125			1,2	V
			1 : 1,20	1,2 : 1		
34	¹⁄₁″ (¹⁄₁₃″)	150			1,0	–
			1,20 : 1	1 : 1,7		
35		125			1,7	I
			1 : 1,20	2,12 : 1		
36		150			0,8	II
			1 : 1,17	3,12 : 1		
37		175			0,25	III (?)
			1,17 : 1	1 : 3,12		
38		150			0,8	II
			1 : 1,17	3,12 : 1		
39	2 M. Rest	175			0,25	III
			1,17 : 1	1 : 5,0		
40		150			1,25	V
			1 : 1,17	5,0 : 1		
41	¹⁄₈″ (¹⁄₄₄″)	175			0,25	III
			1,17 : 1	1 : 5,0		
42		150			1,25	I
			1 : 1,17	5,0 : 1		
43		175			0,25	III
			1,17 : 1	1 : 6,8		
44		150			1,7	I
			1,0 : 1	1,18 : 1		
45		150			1,5	I
			1 : 1,07	5,0 : 1		
46		160			1,5	I
			1 : 1,09	1,5 : 1		
47		175			0,8	III
			1,09 : 1	1 : 1,5		
48		160			0,2	III
			1 : 1,09	1,5 : 1		
49		175			0,8	III
			1,09 : 1	1 : 2,0		
50		160			0,2	III
			1 : 1,0	1,33 : 1		
51		160			0,4	II
			1 : 1,09	1,5 : 1		
52		175			0,8	II

Running number of the experiment.	Interval of stimulation.	Strength of stimulus in Units (N).	Relation of the — Strength of the stimuli.	Relation of the — Latent times.	Latent time in seconds.	Degree of contraction.
			1,09 : 1	1 : 1,25	0,2	II
53	¹⁄₁₆″ (¹⁄₁₆″)	160				
			1 : 1,0	1 : 1,2	0,25	II
54		160				
			1 : 1,09	1,5 : 1	0,8	II
55		175				
			1,09 : 1	1 : 2,0	0,2	II
56		160				
			1 : 1,09	2,0 : 1	0,4	II
57		175				
			1,09 : 1	1 : 1,5	0,2	II
58		160				
			1 : 1,09	1,5 : 1	0,8	Iİ
59		175				
			1,09 : 1	1 : 8,5	0,2	II
60		160				
			1 : 1,09	2,88 : 1	0,7	II
61		175				
			1,09 : 1	1 : 2,88	0,8	II
62		160				
			1 : 1,09	2,88 : 1	0,7	I
63		175				
			1,09 : 1	1 : 2,88	0,8	II
64		160				
			1 : 1,09	2,88 : 1	0,7	II
65		175				
			1,09 : 1	1 : 8,88	0,8	II
66		160				
			1 : 1,09	4,0 : 1	1,0	II
67		175				
			1,09 : 1	1 : 2,8	0,25	II
68		160				
			1 : 1,09	8,5 : 1	0,7	II
69		175				
			1,09 : 1	1 : 2,0	0,2	II
70		160				
			1 : 1,09	2,0 : 1	0,4	II
71		175				
			1,09 : 1	1 : 4,0	0,2	II
72		160				
			1 : 1,09	8,2 : 1	0,8	I
73		175				
			1,09 : 1	1 : 8,2	0,25	II
74		160				
			1 : 1,0		0,8	II / 0
75		160				
			1 : 1,09		0,2	II
76		175				
			1,09 : 1	1 : 6,0	1,2	II
77		160				
			1 : 1,0	1,09 : 1	1,1	II
78		160				

End of the Experiment.

Apart from this and other modifications in the excitability, we find that where the values of the quotients of the stimuli vary from 1, the latency-quotients also vary, and this in the opposite direction, so that diminishing latent times correspond to increasing stimuli. But we also see that the quotients of the stimuli, and the latency-quotients of the same series, are not exactly reciprocal values (as might have been expected).

It is evident from this table, that the times of latent stimulation are by no means simple functions of the strength of the stimuli, as one sees the height of a muscular contraction dependent on the stimuli to motor-nerves; but equal increments to the stimulus under different conditions of excitability in the same preparation correspond to quite different amounts of excitation. The results of this observation may be thus formulated. There is only a small limit between maximal stimuli (small latent time) and minimal ones (large latency). If the two stimuli which are being compared are beyond the limit (i.e. both maximal), then the corresponding values of the latent times differ little from each other, no matter how different the absolute values of the intensity of the stimuli may be. If the stimulus however is below the limit, or in consequence of fatigue, comes into the limit of the minimal, whilst the other one therewith compared is above the limit, then the effects of the two stimulations vary considerably, although their intensities, estimated according to absolute values, may possibly show only a small difference.

As I thought over the cause of this peculiar phenomenon it occurred to me that the frequency of the stimulation had also a great influence on the latent time; and that perhaps with the intensity, the frequency also of the stimuli employed varied. The vibrating rod which caused the rapid interruptions (22 and 30 per sec. in this experiment) gave in each second 11 and 15 closings and as many openings of the primary circuit. As the opening induction-shocks are stronger stimuli than the closing ones, the former will be active with a position of the spirals when the latter may still be without effect. If however by approximating the spirals the closing shocks also become effective, then the frequency of the effective stimuli becomes doubled. Under this assumption (confirmed later) I have placed in the second column of the table in brackets double the actual interval of stimulation.

To prove this view I tried to make the alternating induction currents as equal as possible by means of a special arrangement of the Ruhmkorff's coil (p. 330).

The following curve (Fig. 3) gives the result. It is so constructed, that above every initial point of the abscissa divided into units, each of which corresponds to a period of the experiment, together with the pause (of one minute) the corresponding times of latent stimulation are drawn as ordinates, each unit in the first ordinate indicating one second. The lines connecting the ends of the ordinates give the curve of latent times. The vertical thin lines which run from the broken ends of the curves to the upper line indicate that at this period no reflex act was produced, i.e. the latency may be indicated as endlessly large.

The curve shows, that the duration of the latency of a fresh preparation is not essentially lengthened when the strength of stimulus is diminished. 125 E (E = unit) 112, and 100 units all act after nearly the same duration of the stimulation, whilst at the 10th period with unchanged strength of stimulus the latent time gradually lengthens from 3" to 4". Here it remains even with 80 E. When diminished still 20 E more, the latent time increases pronouncedly, and the stimulus (60 E) soon becomes ineffective (latency ∞). The further variations in the strength of the stimulus and the latency are indicated in the curve.

This experiment, confirmed of course by many others, shows that the duration of the latent stimulation by no means so essentially depends on the intensity of the stimulation as has up to this time been imagined. When moderate stimuli, with diminishing excitability, becomes almost minimal, the latent times become considerably lengthened, and can again be considerably diminished by stronger stimulation. We see however, at the same time, that strong (sub-maximal) stimuli are not able, as in the nerve-muscle preparations, to compensate the weakening influence of the resulting actions. The power of energising, which the reflex-frog, apart from the intensity of the contractions, shows through the rapidity of the reaction,

Fig. 3.

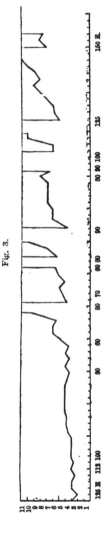

Curve of latent stimulation from a reflex preparation of a frog. One foot was stimulated at intervals of ¼″ by alternating induction-currents whose two directions were of equal intensity. The variations in the strength of the stimulus between 150 E (units) and 60 E is noted where these took place. Between each reflex act a pause of one minute. The ordinate units indicate seconds of latent-time, the abscissa-units periods of stimulation.

diminishes in a much more irregular manner than in muscle; its course is not, like that of the fatigue of muscle, to be obtained very gently by large pauses, but towards the end becomes more rapid and abrupt, and then ceases quite suddenly.

Very great variations occur in individual frogs in the length of such fatigue curves, and this is true also of currents in different directions. There is no generally valid law of contraction for reflex preparations just as for frog's muscles stimulated with the induced current[1]. The inconstancy is even greater in the former.

In Fig. 4 are given the results of a series of experiments in which the effects of differently directed induction-currents in the preparation were investigated. The shocks employed for each period of stimulation were of the same sort, the opening-shocks being conducted away from the preparation after Pflüger's method[2]. Closing induction-shocks passing in the same direction were reversed after every period of stimulation, so that the foot was stimulated alternately by ascending (↑) and descending

Fig. 4.

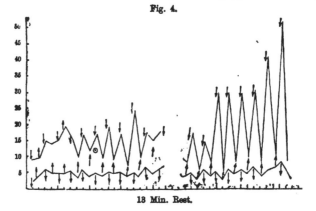

13 Min. Rest.

Reflex preparation stimulated by closing induction-shocks of 20 E intensity at intervals of ¼″. Periods of stimulation follow each other at intervals of 3 mins. alternately in descending (↓) and ascending (↑) direction. The abscissa-units mark periods of stimulation. Each of the ordinate-unit 5 secs. of latent stimulation. The upper curve represents the time of strong reflex contraction.

[1] H. Kronecker, Monatsber. d. Akad. d. Wissensch. zu Berlin, 1870, p. 640.
[2] E. Pflüger, Untersuch. üb. d. Physiol. d. Elektrotonus, Berlin, 1858, p. 189.

($\downarrow$) currents. In Fig. 4 we will at present only consider the upper curve.

At the beginning of the (upper) curve we observe, that the latent times gradually increase independent of the direction of the current. From the 8th period, the currents in an ascending direction ($\uparrow$) begin to be distinguished by their short latent times, whilst the descending ($\downarrow$) induction-shocks require somewhat longer time than formerly, and once (17th period) 24″ to produce a strong reflex act. After thirteen minutes rest very great regular differences in the action of the currents in both directions are to be noticed, the latency therefore increases very considerably (to more than 51″) only for currents in *one* direction; a peculiarity analogous to what is seen in Table III. where we see that the latencies only slightly increase with the fatigue for strong stimuli, whilst for weak stimuli largely, and that therefore the quotients of the latent times also exceed considerably the reciprocal proportions of the stimuli. In Fig. 4 the difference between the actions of the ascending and descending currents disappear. Later also the latent times increase with the fatigue for both directions of the current. With currents of greater intensity contractions with very short latencies can be again obtained.

Only with infrequent stimuli do the times of latent stimulation reach the pronounced value of one minute and more. Only in these cases therefore can we, by change in the intensity of the stimulus, observe great differences in the latent time.

It was therefore to be expected that the duration of the latent stimulation would be essentially influenced by the frequency of the stimuli. The three tables already given also show the influence of varying frequency of stimulation.

In Table I. (p. 333) in the last line which contains the interval of stimulation $\frac{1}{4}$″ the latent time 2.0″ is noted, which corresponds to the strength of stimulus 150 units (E), whilst the next period of stimulation of equal intensity with an interval $\frac{1}{50}$″ only requires 1.2″ for a contraction. More pronounced is the difference in Table II. (p. 334) between two neighbouring latency-values 2.0″ and 0.5″. In both cases induction-shocks of equal intensities were applied to the frog's foot, but in the first case the interval was $\frac{1}{4}$″, and in the second $\frac{1}{18}$″. In Table III.

(p. 337) there were small changes in the interval of stimulation ($\frac{1}{11}''$ and $\frac{1}{15}''$), but here the time of latency varies in a distinct manner with the change in the interval.

1.	$\begin{cases} 11 \text{ stimuli} \\ 15 \quad\text{,,} \end{cases}$	$\begin{matrix}(100\ E) \\ (100\ E)\end{matrix}$	per sec. acted after ,,	,, ,,	,, ,,	1,7″ latency. 1,2″ ,,
2.	$\begin{cases} 15 \quad\text{,,} \\ 11 \quad\text{,,} \end{cases}$	$\begin{matrix}(100\ E) \\ (100\ E)\end{matrix}$	,, ,,	,, ,,	,, ,,	0,7″ ,, 0,9″ ,,
3.	$\begin{cases} 11 \quad\text{,,} \\ 15 \quad\text{,,} \end{cases}$	$\begin{matrix}(100\ E) \\ (100\ E)\end{matrix}$	,, ,,	,, ,,	,, ,,	1,0″ ,, 0,8″ ,,

A view of this table might lead one to the hypothesis that in every stage of the excitability of the preparation a certain number of stimuli of a certain intensity was necessary to produce a reflex act. In the two first parallel cases 18.7 and 18.0 stimuli respectively were applied to the limb, before it was raised; in the second pair 10.5 and 9.9 shocks were effective. In the third pair the limb reacted to the eleventh and twelfth shock. Such an assumption is only true in very few cases.

A tabular arrangement of the number of shocks which were necessary to produce a reflex act in neighbouring periods of stimulation of equal intensity, but different intervals, will not only serve to enfeeble this hypothesis, but yield us other indications.

Even a passing glance at the following table (Table IV.) shows, that in the same preparation, in the same stage of excitability, with unchanged intensity of stimulus, the shorter latency belongs to the smaller interval of stimulation.

At the same time it is quite inadmissable to compare the results of several periods in the experiment with each other.

Notwithstanding all individual and experimental differences, this much is noticed throughout, that frequent stimuli of from $\frac{1}{4}''$ interval downwards, even under unfavourable circumstances (diminished excitability and small strength of current), do not reach the long latent times such as belong to infrequent stimuli ($\frac{1}{2}''$ to $2'$).

The greater the difference of the corresponding interval of stimulation, the greater is generally the difference in the times of latent stimulation.

The number of shocks required to produce a reflex action are given in the fifth column.

TABLE IV.

Shows, the influence of different intervals of stimulation on the duration of the latent stimulation.

Running number of the Experiment.	Strength of Stimulus in Units (*E*).	Interval of Stimulation in Seconds.	Necessary for the reflex movement.	
			Duration of Stimulus in Seconds.	Number of Shocks.
1	150 150	$\frac{1}{16}$ $\frac{1}{10}$	2,0 1,2	16,0 60,0
2	10 10	$\frac{1}{4}$ $\frac{1}{48}$	2,0 0,5	8,0 24,0
8a	100 100	$\frac{1}{8}$ $\frac{1}{11}$	2,2 0,8	17,6 6,8
8b	100 100	$\frac{1}{21}$ $\frac{1}{8}$	0,7 2,4	14,7 19,2
4	100 100	$\frac{1}{8}$ $\frac{1}{21}$	2,8 0,7	18,4 14,7
5a	100 100	$\frac{1}{8}$ $\frac{1}{21}$	1,7 0,7	18,6 14,7
5b	175 175	$\frac{1}{31}$ $\frac{1}{8}$	0,25 1,0	5,25 8,0
6	20 20	$\frac{1}{4}$ $\frac{1}{8}$	7,5 2,0	80,0 16,0
7	15 15	$\frac{1}{4}$ $\frac{1}{8}$	5,0 1,0	20,0 8,0
8a	11 11	$\frac{1}{4}$ $\frac{1}{8}$	4,0 2,0	16,0 16,0
8b	80 80	$\frac{1}{4}$ $\frac{1}{8}$	1,0 0,5	4,0 4,0
9	8 8	$\frac{1}{4}$ $\frac{1}{8}$	7,0 8,0	28,0 24,0
10	10 10	$\frac{1}{8}$ $\frac{1}{4}$	5,0 15,0	40,0 60,0
11	15 15	$\frac{1}{8}$ $\frac{1}{4}$	1,5 4,0	12,0 16,0
12a	12 12	$\frac{1}{5}$ $\frac{1}{2}$	2,5 7,0	12,5 15,5
12b	20 20	$\frac{2}{5}$ $\frac{1}{5}$	80,0 5,0	75,0 25,0
12c	20 20	$\frac{1}{5}$ $\frac{2}{5}$	10,0 45,0	50,0 112,5
18a	10 10	$\frac{1}{2}$ $\frac{1}{4}$	6,0 2,5	12,0 10,0
18b	10 10	$\frac{1}{2}$ 1	4,0 15,0	8,0 15,0
18c	10 10	$\frac{1}{4}$ $\frac{1}{2}$	19,0 50,0	19,0 88,0
14	15 15	$\frac{1}{2}$ $\frac{1}{4}$	5,0 8,0	10,0 12,0

Running number of the Experiment.	Strength of Stimulus in Units (*E*).	Interval of Stimulation in Seconds.	Necessary for the reflex movement. Duration of Stimulus in Seconds.	Number of Shocks.
15	8	1/4	1,0	4,0
	8	1/2	2,5	5,0
16a	500	1/4	1,0	4,0
	500	1/2	3,0	6,0
16b	400	1/3	18,0	52,0
	400	1/2	30,0	60,0
17	15	1/2	46,0	92,0
	15	1/4	5,0	20,0
18	900	1/2	25,0	50,0
	900	1/4	5,0	20,0
19	400	1/2	4,5	9,0
	400	1	34,0	34,0
20	1000	1	9,0	9,0
	1000	1/2	5,5	11,0
21a	1000	2	35,5	17,75
	1000	3/2	8,5	5,6
21b	1000	3/2	16,5	14,3
	1000	2	55,0	27,5
22	500	3/2	6,7	4,36
	500	2	35,5	17,75
23	1000	2	80,0	40,0
	1000	3/2	68,0	42,0

From these numbers it is obvious that one dare not assume that with unchanged excitability of the limb, in order to produce a reflex act, a certain amount of stimulation is required, which with a varying interval always requires a similar number of equally strong shocks. The corresponding number of beats in a pair of experiments generally vary considerably from each other. *In by far the most cases a large number of beats belongs to the large intervals, i. e. the duration of the latent stimulation is with infrequent stimuli not only absolutely longer than with frequent ones, but also more than is necessary in order to complete the similar number of stimuli.* This phenomenon has a certain analogy to the result of W. Baxt (p. 326), who showed that the latent times increased more rapidly than the stimulating acid increased in strength. The cause of this peculiar phenomenon I have not yet had time to investigate. Further from (5a, b, and 8a, b) is to be observed, how when the intensity of the stimulus is great the time of latency diminishes

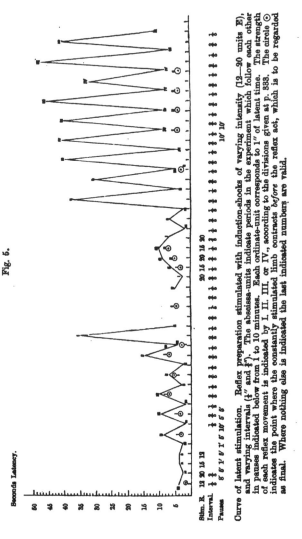

Fig. 5.

Curve of latent stimulation. Reflex preparation stimulated with induction-shocks of varying intensity (12—20 units E), and varying intervals ($\frac{1}{4}''$ and $\frac{3}{4}''$). The abscissa-periods in the experiment which follow each other in pauses indicated below from 1 to 10 minutes. Each ordinate-unit corresponds to 1″ of latent time. The strength of each reflex movement is indicated by I. II. III, or IV., according to the divisions given at p. 333. The circle ⊙ indicates the point where the constantly stimulated limb contracts *before* the reflex act, which is to be regarded as final. Where nothing else is indicated the last indicated numbers are valid.

somewhat; fatigue operating, as was shewn (p. 342, Fig. 3), in the opposite direction.   The lengthening influence of fatigue upon the latency can be compensated by increasing the stimulus; but this is not valid in the same degree for all intervals.

Fig. 5 shows the increase of the latency by stimulating with shocks of different intervals.

This series of experiments shows how the fresh preparation reacts equally rapidly to stimuli of tolerably different intensity, (20 E and 12 E) and different intervals ($\frac{1}{4}''$ and $\frac{2}{3}''$); the extent of the second contraction however is somewhat greater than that of the next following one, which only reaches the II. degree.

For the second half of the experiment stimuli of 15 E and 20 E were required.   20 E $\frac{2}{3}''$ remained almost as long without effect as 15 E $\frac{1}{4}''$.   On comparing both frequencies the same strength of current being employed, the inferiority of the infrequent stimuli is seen in a very pronounced manner.   The latency with $\frac{2}{3}''$ is about ten times as long as for an interval of $\frac{1}{4}''$, i.e. for an interval of double the frequency.   In the first case, five times as many stimuli were applied to the nerves as in the latter. At the same time it is to be observed that the reflex movement which occurs late is by no means weaker than that which occurs early.   If, however, only a weak discharge (I.) takes place when a strong reflex movement was to be expected, then, either the stimulus employed is ineffective, or if strong currents have been employed we may conclude that the preparation will soon die. In passing I would remark that at the beginning of the curve where the large indentations begin, the increase in the latency for infrequent stimuli does not occur directly, but a weak contraction ($\odot'$) instead of the expected strong one occurs, which follows several seconds later.   Such sub-maximal reflex movements we will call "preliminary" in opposition to the "final" one.

Fig. 6 also is of interest in consequence of the distribution of the "preliminary reflex", whose importance will be considered later in connection with other phenomena.   This curve also shows the temporary modification in the excitability, such as I have mentioned in Table III.

This curve shews that the latencies of infrequent stimuli ($\frac{1}{2}''$) are longer than those of more frequent ($\frac{1}{4}''$) ones.   On trying to increase the latency of the frequent stimuli by diminishing the

Fig. 6.

Seconds Latency.

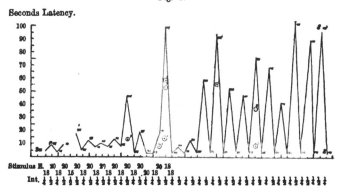

Curve of latent stimulation. Reflex preparation stimulated with induction-currents of varying intensity (18 and 20 units), and varying intervals ($\frac{1}{4}''$ and $\frac{1}{2}''$). The abscissa-units indicate periods in the experiment which follow each other in pauses of three minutes. Each ordinate-unit is equal to latencies of 5 seconds. The signs I. II. III. indicate (p. 333) degree of contraction, the circle ⊙ preliminary reflex.

intensity of the current (to 15 E) the reflex did not occur. Alternating stimulations with 20 E $\frac{1}{4}''$ int. and 18 E $\frac{1}{4}''$ int. required latencies, represented graphically by the slightly indented piece of curve. Then there occurred a preliminary reflex (⊙) at the time where the final one was to be expected, and this was delayed till 45″ after the beginning of the stimulation. The further variations are indicated by the curve.

While the two former curves for comparing the action of infrequent and frequent stimuli presented absolutely long latencies, the following curve (Fig. 7) gives a case in which the infrequent stimuli ($\frac{1}{2}$) even at the maximum only required 11 seconds, whilst the frequent ones ($\frac{1}{8}$) acted even after 0.5 seconds, and never required more than 2.5″ to produce a reflex act. The cause of this condition is to be sought for in the tolerably high frequency of both successions of stimuli which we are here comparing, and also in individual variations of the excitability. As a mean, however, the latent times which belong to the interval $\frac{1}{4}''$ are confined between the limits 2″ and 10″, whilst for $\frac{1}{8}''$ they extend between 0.5″ and 4″.

In order that the small variations in the latent times may

appear more clearly, in the construction of the following curve, I have made the space between each two lines equal 1″.

Fig. 7.

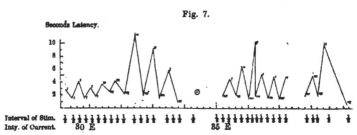

Curve of times of latent stimulation. Reflex frog preparation stimulated with induction-shocks whose intensity remained constant in each half of the experiment (first 30 E and then 35 E). The intervals (¼″ and ½″) alternate. The abscissa-units indicate minutes, the ordinate-units seconds. The degree of contraction is indicated by I. II. and III. Where these are omitted it is of the normal final strength (III.).

With varying interval of stimulation the latency of the *fresh* preparation varies only slightly. Suddenly however, the times of latent stimulation for infrequent shocks rise to a high value, whilst for frequent stimuli the small values remain. In this experiment only the reflex contraction which occurs *first* is noticed. The stimulation was interrupted as soon as the preparation had reacted. Preliminary reflexes are therefore always noted as final. Hence is to be partly explained, on taking into account the degree of contraction, the relatively great variations in the latent times. We will return to this point.

Up to this time I have only examined the changes which the duration of the latent stimulation undergoes when the intensity of the stimulating current is changed, and when the interval varies. Only occasionally has the effects of simultaneous change in both variables been noticed.

The next curve (Fig. 8) shows how the reflex times conduct themselves when the interval of stimulation and the strength of the stimulus are varied in opposite directions.

The absolute values of the latent times are here also small, varying between 1″ and 8″. We have to deal with tolerably rapid shocks. The first part of the toothed curve shows latencies varying with the same strength of stimulus, but with the intervals (¼″ and ½″). When once the same interval

Fig. 8.

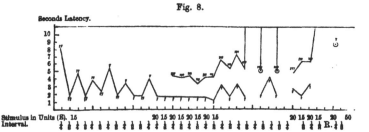

Curve of times of latent stimulation. Reflex preparation stimulated with induction-shocks whose intensity was at first kept constant while the interval of
stimulation varied, but later was also changed in the opposite direction.
Then the intervals with constant strength of stimulus again alternate ; lastly,
both variables again. The abscissa marks periods in the experiment which
are separated from each other by intervals of 8 minutes. R indicates rest
for 10 minutes. Each ordinate-unit corresponds to 1 second. The signs I.
II. III. indicate the degree of contraction.

was retained the latent time remained unchanged. This could
however be kept at the same level by combining a stronger
strength of current with a smaller interval; *e.g.* induction-
shocks of 20 E int. $\frac{1}{4}''$ remained equivalent to currents of
15 E int. $\frac{1}{8}''$.

From this experiment it is obvious that it is possible to
compensate the diverging effects of stimuli of different frequency and intensity, but there is no indication as to how
the two variables must be chosen in order to obtain constant
latencies.

One can certainly, with varying medium frequency of the
stimulus, obtain latent times of constant amount, by making
the stimuli immoderately strong; but then there always occurs
the minimum of latency. We may therefore say, that, *by
variation in the intensity of the individual shocks, the latent
time can only be changed within narrow limits; on the contrary,
however, the graduation of the interval of stimulation affords
a means of obtaining extraordinarily large variations in the
duration of the latent stimulation.* Hence small differences
in the latency, caused by a change in the interval, can be
compensated by a change in the intensity of the current,
whilst large ones cannot.

Further, the time of latency cannot be varied indefinitely

by a change in the interval. The intensities of the current which are sufficient for frequent stimuli are not so for infrequent ones. This we have already seen in Figs. 5 and 7. The infrequent stimuli (at first of very effective intensity) even there were soon without effect, without maximal latencies having been obtained. We must strengthen the intensity of the current in order to obtain results from the infrequent stimuli. The more the interval of stimulation is increased, the more powerful must be the strength of the current which is necessary to produce a reflex act.

Table V.

Shows, how the times of latent stimulation increase with the interval of stimulation; infrequent stimuli, to be effective, soon require very strong currents, which rapidly extinguish the excitability.

Periods of rest between the periods of stimulation.	Interval of stimulation in seconds.	Strength of stimulus in Units (E).	Latent time in seconds.	Degree of contraction.
24 minutes	0,5	10	6	II
11 ,,	0,25	10	2,5	II
5 ,,	0,5	10	5,0	I
			6	II
5 ,,	0,25	10	2,5	II
5 ,,	0,5	10	8	I
			4	II
9 ,,	1,0	10	8	II
5 ,,	1,0	10	9	I
			15	II
5 ,,	1,0	10	7	I
			14	II
5 ,,	1,0	10	8	II
			19	II
5 ,,	1,5	10	50	II
5 ,,	1,5	10	68	II
7 ,,	1,5	10	68	II
5 ,,	1,5	10	∞	
5 ,,	1,5	15	80	II
			90	0
5 ,,	1,5	15	∞	
5 ,,	1,5	15	∞	
5 ,,	0,25	10	40	I
5 ,,	1,5	80	45	III
5 ,,	0,25	15	14	II
5 ,,	1,5	80	∞	
5 ,,	0,25	80	6	II
7 ,,	1,5	40	∞	

In the table the bracketed numbers indicate the latent times of successive reflex contractions ("preliminary" and

23—2

TABLE VI.

Shows, how very infrequent stimuli only producing reflex movements from the fresh preparation when the strength of current is extraordinarily powerful, how they soon become ineffective; and how with sinking excitability refuse to act with intervals of stimulation always becoming smaller.

Pauses for Rest. Minutes.	Interval of Stimulation.	Strength of Stimulus in Units.	Latent time in Seconds.	Degree of Contraction.
38	2,5	1000	5,0	III
1	2,5	1000	∞	
1	2,0	1000	∞	
2	2,0	1000	⎰25,5	I
			⎱48	I
			58	II
1	1,5	1000	10,0	III
1	1,5	1000	36,0	II
1	1,5	1000	12,0	III
1	1,5	1000	18,0	III
1	2,0	1000	⎰46,0	I
			⎱66,0	II
1	2,0	1000	85,5	I
1	1,5	1000	7,5	III
1	1,5	1000	18,0	II
1	1,5	1000	16,5	III
1	2,0	1000	64,0	III
1	2,0	1000	42,5	III
1	2,0	1000	49,0	III
8	1,5	1000	19,0	III
1	1,5	1000	18,75	III
1	1,5	1000	21,0	III
1	1,5	1000	86,0	II
1	1,5	1000	40,5	II
1	1,5	1000	26,0	III
1	1,5	1000	53,75	I
1	1,5	1000	∞	
8	1,5	1000	∞	
1	1,0	1000	10,0	II
1	1,0	1000	11,0	II
1	1,0	1000	14,0	I
1	1,0	1000	16,0	II
1	1,0	1000	∞	
10	0,5	1000	8,5	II
1	0,5	1000	8,0	II
1	0,5	1000	8,5	II
1	0,5	1000	9,0	I
1	0,5	1000	10,0	I
1	0,5	1000 pinching	∞ —	II
2	0,5	1000	11,5	
1	0,5	1000 pinching	∞	0

"final"). It is noticeable that the latent times for small- intervals (0.25") which are at first short, become extraordinarily long after the periods of strong stimulation with large intervals, even with the same strength of stimulus; and can only be diminished to the ordinary amount by relatively strong currents, whilst we have generally seen that as a rule they remain at the same level, in spite of larger latencies going between, the intensity of the current being little changed.

The more distant are the stimuli chosen, the earlier in the stage of fatigue do they become completely inactive. Induction shocks (closing and opening) following at intervals of 2.5" I have only several times seen effective, and that on employing a very powerful stimulus with fresh preparations. The duration of the latent stimulation in such a case was very short (Table VI.). Soon the excitability was extinguished for frequent stimuli, and then gradually for those becoming always more frequent. The extraordinarily intense shocks, even when the preparation was about to die, had not long latent times. Suddenly, the excitability was completely extinguished, even to the generally infallible pinching.

Very strong stimuli therefore act very rapidly in a deleterious manner without producing any apparent reflex action. We shall soon see that sub-maximal shocks, i.e. those which are just sufficient to produce a reflex act with a moderate interval of stimulation, but do not do so by a large interval, also strongly diminish the excitability. There is here a fundamental difference from the condition of the voluntary muscles, in which Kronecker observed "that weak stimuli, when they had no further important effect, do not fatigue, as maximal stimuli do, without producing any mechanical action, that the muscle treated with ineffective stimuli conducts itself almost as a passive one[1]."

Thus we see that successive stimuli tend to produce two opposite effects in the preparation; to produce movement and to extinguish life. This injurious property seems to belong in a high degree only to the intense stimuli: movement, as we have seen, is better assisted by repetition of the shocks, than by

[1] *Arb. aus d. physiol. Anstalt zu Leipzig*, 1871, p. 261.

increasing their strength. With infrequent stimuli, which require strong currents, to be effective at all, the injurious influence is more obvious than with frequent stimuli.

How then are we to set before ourselves the summation of movements?

We may, I believe, obtain some important indications by considering the already mentioned phenomenon of the " preliminary reflex."

Türck says in his already often cited paper[1]: " Very frequently a powerful movement and extension of the extremity (frog) dipped in acid, does not follow suddenly, but it is preceded by gentle, slow movements." " For each hind foot, therefore, two numbers are obtained, e.g. 12—17 seconds of latency in the normal, and 7—12 in the hyper-æsthetic."

Sanders-Ezn[2] describes " the repetition of the same movement with continued (chemical) stimulation" in a special chapter. He finds that between the first and second appearance of the same movement there is always a clearly recognisable pause which is shorter the greater the excitability. " This alternate disappearance and appearance of a movement shows that the stimulus, although it remains continually, only expresses its discharging force periodically." He often observed, that when two similar weak stimuli were applied shortly after each other, that the movement following upon the last one was more energetic than that which occurred after the first stimulus. As the most probable explanation, he holds therefore, "that the forces rendered disposable by the first stimulus were only partly employed by the consequent movement, whilst the remaining part adds itself to those forces which are rendered disposable by the second stimulus."

A similar phenomenon was observed by Edward Weber when he passed the alternating currents of the electro-magnetic rotation-apparatus through the lower limb of a frog, the sciatic nerve being the only connection between the limbs and the body. Weber found (as is noticed at the beginning of this paper) that a considerable time elapsed ere the movements followed; that "further, in spite of the action of the rotation-apparatus continuing uninterruptedly, the muscular movements passed off, being separated by pauses, and

[1] l. c. p. 8.
[2] Arb. aus d. physiol. Anstalt zu Leipzig, 1867, p. 29.

again returned, as if the animal made actual efforts ; that, lastly, always the same muscles were not thrown into movement[1]." Similar peculiarities were observed by Setschenow and Nothnagel[2] on stimulating the central end of the sciatic nerve electrically.—On stimulation, with weak tetanising currents of the induction-machine, Setschenow observed, that, at once after beginning only a single temporary movement of the anterior extremities, or afterwards a series of tremulous movements lasting for several seconds took place, then rest. Moderately strong stimulation caused at once strong movement in both anterior extremities, which then assumed a tetanic character, to give place in a few seconds, as in the former case, to absolute rest. "Nevertheless if the stimulation is continued, after some time (sometimes after one to two minutes), a tetanic wave is observed to pass over the body of the frog. This wave generally begins in the remainder of the femoral muscles of the side stimulated, whence it is propagated to the abdominal muscles and the other extremities, and ends in the form of a strong, continuing extensor-tetanus. Later, there follow irregular movements." "With strong stimulation the first phase of movement is omitted. Instead of it quite an inconsiderable movement is noticed in the extremities, followed by rest for several seconds, or the occurrence of the second phase of movement in the form of a tetanic elevation of the extremities followed by extensor-tetanus." Setschenow explains the rest after the first discharge as an inhibition phenomenon, produced by the strong stimulus, for when the stimulation of the nerve was interrupted at the beginning of the period of rest, tetanic extension of the arms (sometimes very strong) was always obtained as an after-effect.

A step in advance, in the analysis of the process of summation was made by Setschenow, by stimulation of the central end of the sciatic nerve with single electrical shocks. He says[3] "If one interrupts constant galvanic currents (which are so weak, that the closing and opening of the current singly have no effect) e. g. 60 times per min., then after several interruptions the first contraction, still weak, occurs in a limited number of muscles, the second, third, set become always stronger and more extended, till at last a movement of the whole extremity occurs, but nevertheless, even now, every variation in the current is often replied to by a contraction in the extremity already in motion.

In none of the above papers are any precise results given as to the relations in time in which the reflex movements, produced during any one period of stimulation, stand to each other.

Table II. (p. 334) of this paper gives some values which show that with moderate frequency of the stimulus ($\frac{1}{4}''$) a tolerably

[1] Art. Muskelbewegung, *Wagner's Handwörterb. d. Physiologie*, Bd. III. Abth. II. 1846, p. 19.

[2] Zur Lehre vom klonischen Kramp, *Virchow's Arch*. Bd. 49, p. 276.

[3] l. c. p. 11.

rapid increase, in the latent times as well of the preliminary as of the final reflex actions, occurs. In Fig. 4 a series of experiments are given, where in each period of stimulation the preliminary reflexes are represented.

Fig. 4.

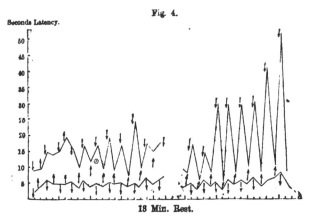

The reflex preparation was stimulated by closing induction-shocks of 20 E (intensity in ¼" interval). The periods of stimulation follow each other after pauses of 3 minutes, alternately in descending (↓) and ascending (↑) direction of the current. The lines on the abscissa indicate periods of stimulation, each of those on the ordinate 5 seconds of latent stimulation. The lower curve indicates the latent times of the preliminary (weak) reflex contraction, the upper connects the points of final (strong) excitation.

The course of the lower curve varies in a pronounced manner from that of the upper one. Whilst the latter by its large notches shows that the latencies for the descending currents soon increase to very large values, the latencies of the preliminary reflexes remain within narrow limits (3"—8"), although here also a slight action of the descending current is obvious.

Further, in this series the preliminary reflexes have a peculiar form. They were not rapid temporary contractions, but slight continued elevations of the foot, which increased quite gradually, then weak movements of the knee set in, till suddenly the flexors of the foot, leg, and thigh became strongly contracted, when the stimulation was generally interrupted and the limb sank to its position of rest.

Generally, however, the preliminary reflexes are distinctly

separated from the final ones. Their course is sometimes flat, mostly steeply ascending and descending.

According to their succession in time, two different kinds of reflex movements occur: (1) in large regular or irregular pauses, mostly, so that all the reflex contractions are small (Degree I.), or that the later ones are larger than the first; (2) (after shorter or longer latency) at intervals, which are isorhythmical with all stimuli; or when these are unequally strong, with the stronger. These preliminary contractions are almost always of the first degree.

The facsimile, Fig. 9, gives an example of the first sort.

Fig. 9.

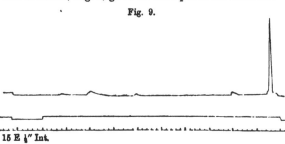

15 E ¼″ Int.

The limb of a reflex frog weighted with 5 grms. and fixed to the writing lever, which writes its elevations on the upper line of the endless paper, was stimulated with alternating induction-shocks of 15 E intensity at intervals of ¼″. The middle line is elevated by a secondary circuit as soon as the stimulus is applied to the limb, and depressed as soon as the stimulus is interrupted. Seconds are marked on the lowest line.

Here the first small elevation of the foot follows 3.5″ after the beginning of the stimulation, one less strong after 6″, then a single weak one after pauses of 9″ and 20″, lastly, after 48″ latency there is a strong elevation of the whole limb (Degree III.).

To what extent however in excitable preparations a succession of infrequent weak stimuli may increase, is shown by the facsimile curve Fig. 10, which, on account of the length of the original, is given in pieces.

We see that 57″ of stimulation are required till the first preliminary contraction (Degree I.) appears; from there 33″ till the second, then till the third 18″, and from here till the fourth 43″. After 27″ of further stimulation the limb makes a series of violent elevations and depressions, and does not come at once to rest, even after the stimulation has ceased.

If after such a violent discharge the stimulation is continued,
an excitable limb with moderate intensity and frequency of the
stimulus, remains for a considerable time at rest. Then it can

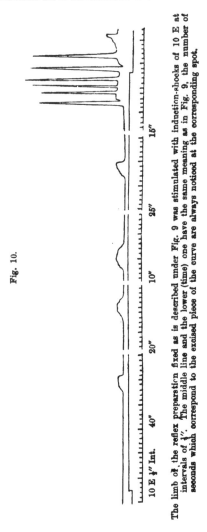

Fig. 10.

The limb of the reflex preparation fixed as is described under Fig. 9 was stimulated with induction-shocks of 10 E at intervals of ½″. The middle line and the lower (time) one have the same meaning as in Fig. 9, the number of seconds which correspond to the excised piece of the curve are always noticed at the corresponding spot.

be brought to a new, perhaps to a strong, but scarcely again to a spasmodic discharge. Thus, *e. g.* the limb which wrote the original for Fig. 10, was stimulated at a later period by currents of 10 E ¼″ int. and produced a strong reflex spasm after 2.5″ latency, then the continually acting stimulus remained 35″ latent, then produced a contraction of Degree III., and then no more, though stimulated for two minutes. The limb was by no means killed thereby, for slight mechanical stimulation produced a strong reflex towards the end of this period of stimulation. After a rest of a few minutes the same stimulus was capable of producing within 1.5″ repeated strong contractions (Degree III.), herewith however, the limb was exhausted for the remaining three minutes of stimulation. A rest of six minutes promptly restored the limb to a high degree of excitability. Minutes are not even required. Even 10″ gave back to the preparation a high degree of excitability. Thus is proved what I have already indicated (p. 355), that sub-maximal stimuli fatigue the reflex preparation. Only a very short time however is required for recovery.

What has already been shown for those reflexes which are confined to the limb stimulated electrically, is also valid for those movements which occur in the second limb after intense or long stimulation of a spot of skin.

If a limb is feebly stimulated, whether only weak stimuli are applied to it, or whether its excitability has become diminished, then not only does the excitation remain longer latent, but is no longer propagated to the second limb.

Can the stimulus be applied so that its motor forces, but not its injurious effects, become summated ? This postulate appears to involve a *contradictio in adjecto.* For if the movement from stimulus to stimulus is summated, and only the movement produced, but not the discharging moment fatigues, then the exhaustion ought to increase only in the degree in which the internal activity increases. If, however, one regards the reflex discharges as the resulting effects of independent oppositely acting forces—accelerating and inhibitory—then there is nothing to prevent one's conceiving that with a certain succession of stimuli the inhibitory ones were only able to sum themselves up in an unimportant manner, whilst the motor impulses survive-

the intervals; by well-appointed pauses the small amount of
time necessary for the recovery of the exciting nerves might be
obtained, which might not be sufficient for the inhibitory ones.
From the former explanations, it may be doubted whether a
succession of stimuli exists by which the after effects of the
shocks last above a very short time. The largest possible interval of stimulation I have indicated as 2.5″, and expressly
noticed that stimuli, which are applied at such long intervals,
in order not to lose their after-effect, must be very strong, and
in consequence soon extinguish the excitability. We have
however seen, from my own investigations and those of others,
that violent reflex spasms last longer than the stimulus, sometimes even in a very pronounced manner. It appears, therefore,
that internal movements increased to a considerable intensity
have a longer after-effect than individual shocks.

It must be to a certain extent possible, therefore, to obtain the summation of sums, such however must require large
intervals of stimulation. In fact, I have succeeded in making
the after-effect of the reflex excitation visible for more than
twelve seconds of a pause. The sluggish preparation required strong stimuli, 1000 E with the interval ¼″. The small
experimental series is again most conveniently arranged in a
tabular form.

Pause for rest.	Duration of			The contraction remaining after the termination of the stimulation.	Maximal height of the curve.
	Latent stimulation.	Period of stimulation.	Complete contraction.		
	1,0″	10,5″	12,0″	2,5″	12,0 Millm.
12,5″	2,0″	8,0″	8,0″	2,0″	9,0 ,,
10,0″	2,2″	6,0″	6,2″	2,4″	5,0 ,,
11,0″	2,3″	6,0″	6,0″	2,7″	6,0 ,,
12,5″	2,5″	6,0″	6,0″	2,5″	5,5 ,,
10,3″	2,0″	7,0″	4,2″	0″	57,0 ,,
12,0″	8,0″	18,5″	10,0″	0″	2,0 ,,

Thus the excitations become heaped up from period to
period, without the height of the preliminary contractions having
increased, or without the length of the visible after-effect having been at all importantly augmented. After the *strong* dis-

charge however only a few small reflexes were to be obtained. One can here scarcely assume that the excitability and not the excitation was increased by the stimuli. It is scarcely to be believed that so extraordinarily strong stimuli, which were able to kill the preparation so soon, should have previously increased the excitability. It is most probable that neighbouring parts were set in motion, in unison with the nearest excited central parts of the nervous system, and this movement, when the external impression continues, always goes further (hence the reflex movements became more extended).

When a large mass is once set in motion, the primary impulse can be withheld for a short time without the whole coming to rest at once. The powerful new impulses which, after several seconds of rest meet the somewhat recovered ends of the nerves, suffice to restore to the whole system its former amplitude, and then to increase it. Single strong stimuli therefore, following at regular, long $(2''-2.5'')$ intervals, expend their forces. The vibrations, which were produced by the individual impulses did not long preserve their intensity, and remained local. Hence it is with difficulty that one succeeds with infrequent stimuli so to increase the small remainder, that extensive movements result; whilst not unfrequently with frequent stimuli $(\frac{1}{4}'')$, of only moderate strength, the most complicated apparently purposive washing-movements, such as are so striking with chemical stimuli, may be produced.

The strong stimuli specially fatigue the peripheral nerves, so the impulse proceeding from these becomes weaker. Hence it comes that preliminary reflexes are often wanting, where, from the analogy of previous and following periods of stimulation, they ought to have occurred (compare Fig. 5 and 6, pp. 348, 350). A small diminution of the induction-current or of the sensibility of the nerves is sufficient to retard the result. On the contrary, often quite a weak help, sometimes even slight contact (with a camel-hair pencil or the finger) during the electrical stimulation suffices to produce a reflex. The inhibitory hypothesis therefore appears to me to be quite superfluous for the processes here considered.

As the second category of preliminary reflex movements I have cited the regular contractions isorhythmical with the stronger stimuli. They occur very often, sometimes when the skin is stimulated with shocks of moderate frequency and intensity. Their size is mostly small; sometimes, however, very considerable. One has the impression as if the foot was moved by direct stimulation of its muscles, but by careful observation one may convince himself that these contractions are also reflex; for in fatigued preparations they occur after considerable, sometimes after long latency, and frequently change their strength ; a phenomenon which we will presently consider more fully. Hence the (already mentioned, p. 328) apprehension of Herr Meihuizen, that with electrical stimulation of the skin one would be unable to distinguish between direct and reflex contractions—is also quite unnecessary for this case of the rhythmical preliminary reflex movements. Very strong stimuli of course act upon the deeper muscles.

We have observed the preliminary rhythmical reflex movements with the apparatus already described (p. 328), but in order to obtain a better view I also used the cylinder of the kymograph. Thus we could have a series of 100 reflex experiments on one sheet of paper, and in order to spare room the movements of the writing lever were limited in their extent. The movements of the foot are here essentially distinguished from those of the leg by their evanescence. The degree of contraction was always noted at the corresponding point during the experiment. The following Fig. 11a and 11b (which to save space is divided into two) gives an example of a curve so registered.

The small dimensions of the contraction-curves require exact observation in order to ascertain the details. A cursory glance shows that the rhythmical reflex movements began after a latency of 0.2 to 0.6 seconds, and passed into the final elevation of the limb after a total latency of 1.3″ to 1.8″.

In the two pieces of curve before us, we see that the latent times of the preliminary reflexes remained tolerably equal (0.2″—0.4″), that the final reflexes, however, generally require 1.9″ to 2.2″. At the same time, the second experiment shows that with long-continued stimulation the final reflex tetanus again ceases, and now rhythmical contractions of the foot reappear.

Fig. 11a.

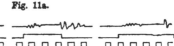

35 E ¼″

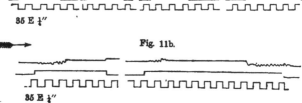

Fig. 11b.

35 E ¼″

The foot of the suspended reflex preparation rested on the writing lever, which was capable of moving within narrow limits. This writes the upper line (from left to right) straight as long as it is at rest, wavy when it is rhyth. mically depressed and raised by the foot and counterpoise. The middle line indicates, by its high course, that the skin was stimulated with induction. shocks of 35 E units at intervals of ¼″, by its lower course, that the stimu. lation was interrupted. The lowest line indicates seconds divided into halves.

Fig. 11a represents neighbouring parts of the curve which belong to the earlier stages of fatigue. Fig. 11b, the same from a later period.

Fig. 12, gives an example of very regular movement of the foot.

Fig. 12.

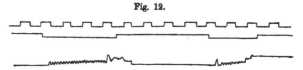

Reflex preparation stimulated with induction-shocks of 100 E and 125 E at intervals of ¼″, current interrupted by vibrating spring. Apparatus as in last figure, only here the lines indicating time and stimulation are placed *above* the contraction-curve, and the middle line by its *descent* indicates the beginning of the stimulation.

In this case the foot only replied to each second stimulus (opening-shock), and did not visibly react to the closing-shock, exactly like an over-weighted, weakly stimulated, or very fatigued muscle. The reflex nature of this contraction is here also proved by the length of the latency. Further, the later reflex periods showed an increase of the virtually invisible rhythmical movements, what I have never noticed in muscle with stimulation of such frequency. The neighbouring curve, produced by 125 E, has a much shorter stage before the proliminary reflex, *i. e.* the latency to the final reflex is shorter than with stronger stimulation.

I also employed another very sensitive method for registering the course of the slightest reflex contractions. A thin plate of cork upon which the toes of the preparation rested was fixed on one of the ends of a Schortmann's relais[1], the platinum style at the opposite end of the delicately balanced beam dipped into Hg as soon as the foot was raised from the double lever.

If a current is hereby closed, the beam of a small electro-magnet provided with a style writes on a kymographion-cylinder the moment of elevation, and with a rhythmical succession of contractions a finely indented curve is obtained.

The question now arises, are these rhythmical reflex movements fundamentally different from the single preliminary ones already noticed?

I have already advanced the view that every stimulus adds an increase or increment to the movement of the reflex centres, that several single shocks sum themselves up, until a preliminary reflex appears, that then the summation goes further until a final contraction or a spasm closes the reaction. Every small contraction therefore does not cause the cessation of the movement, this is only done by strong discharges. When by the summation of several stimuli the limit has once been reached, why is not *every* stimulus followed by its small discharge, whose remainder increases till a final reflex movement, similar to that in Table VI., is produced, the primary small reflex-sums heaping themselves up to large secondary ones? Only great equality of the stimulus and excitability are necessary for this.—Thus this process would be represented as a special case of the general. In order to establish this view, I present various forms of discharge which show the passage of the single reflexes into the rhythmical.

The first curve shows a whole period of stimulation with the syn-arhythmical, tolerably strong reflexes, which arise' after short (almost 1″) latency, then increase after 5″ to a spasm, in whose curve the impulses composing it are to be recognised.

[1] Vide Bowditch, *Arbeiten aus d. physiol. Anstalt su Leipzig*, 1871, p. 142.

Fig. 13.

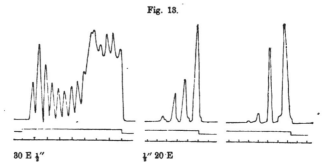

30 E ¼″   ¼″ 20 E

The originals of this curve were written on endless paper by two reflex preparations whose ankle-joint was fixed by thread to the writing-lever. The
one preparation was stimulated by induction-shocks of 30 E at intervals of
¼″, the other with 20 E at intervals of ¼″. The middle line by its high
course shows the time of stimulation. The under line marks seconds.

The second curve represents a group of reflex contractions which
appear after 5″ latency. The preliminary ones follow each other
at intervals of about 1″, and rapidly increase to the final contraction. The third group, lastly, gives the final piece of a long period
of stimulation (20 E ¼″ int.), in which before this discharge single
preliminary ones appeared after 13″ (degree I.), 14″ (degree II.),
20″ (degree II.), 28″ (degree II.), 31″ (degree II.) respectively;
then there occurred the first small elevation of the leg of this
group. After 38″ and at intervals of 1″—1·5″ a full discharge
took place. Here *elevations of the leg proper*, singly or in
rhythmical succession, preceded the final reflex.

Fig. 14 gives an example of a rhythmical succession of small
reflexes which were written by the leg proper. I have in my
curves a number of pictures which may serve as intermediate
forms of the varieties here presented. In Fig. 8, an example
is given in which from the middle of the series onwards, in
every experiment, a weak (I.) initial reflex occurred, which
was followed after a pause by the strong (III.) one. In the la-
tency-curve of Fig. 6 also, several preliminary reflexes indicated
by circles ⊙ at very unequal distances are indicated, together
with those where only simple discharges were obtained. If one
compares these modifications, the bridge between the simple
final reflex discharges to the isolated reflexes of Figs. 9 and 10,
and from these to the rhythmical ones of Figs. 11 and

12 will be found, and will come to the conclusion that between the boundary forms a number of combinations determined by the number of stimuli is possible; when, however, one takes into consideration the continually variable heights of the contractions, an endless number.

For the sake of simplicity we have till now called the first strong reflex movement "final." With justice too; for we have already said that generally no other movement follows the first strong simple or spasmodic reflex in the same period, by stimulation with moderate or weak stimuli, the excitability being moderate. It is otherwise, however, with very excitable preparations, or those treated with intense stimuli. There often many very high contractions follow at long or short intervals. After some time the discharges become smaller, less frequent and irregular, and disappear completely. The penultimate stadium demands our especial interest. When the preparation is only capable of medium or weak reflexes, which contraction have we then to regard as final, and when can we interrupt the stimulation as objectless?

Rosenthal[1] in estimating the "Reflex-time" departs entirely from such stimuli which do not give the maximum of reflex action, and compares only such stimuli which are just sufficient to produce this maximum ("adequate stimuli"), with strong ones ("uber-maxima"). In my experiments, I could not have followed this system without omitting many valuable data. If in the curves I had omitted the periods marked I. and II., numerous disturbing imperfections would have remained. Compare, for example, Fig. 5. How often there exactly at the expected place, instead of the maximal contractions (III.), are those of degree II. and I. on the one hand, and on the other, strong and spasmodic ones (IV).

Similar results are to be found in other examples. It therefore appears that it is not the strength or extent of the reflex contraction which is of importance for the principle, but the circumstance that a complete discharge has taken place.

Which process have we to regard as a complete discharge? This is the most important question which we have as yet met with in our investigation.

[1] *Sitzungsb. d. phys.-med. Societ. zu Erlangen*, 1873, p. 18.

When we observe a preparation which requires tolerably strong electrical cutaneous stimuli to set it in motion, and we treat it with moderately frequent shocks, we often observe, that after a short time the foot begins to move rhythmically. This beating time in whole or half *tempo* gradually increases till a large reflex results.

When the preparation is stronger, the leg also begins to move at the same time, and writes curves of which Fig. 4 is a fac-simile. If the excitation is weaker (because stimulus or excitability have been diminished), we now no longer have a large reflex (degree III.), but the vibrations of the foot or small vibrations of the leg increase only to feeble elevation of the leg.

Fig. 14.

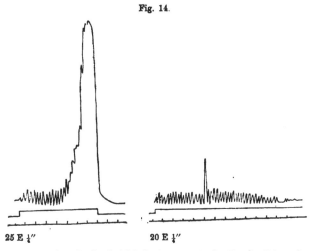

25 E ¼″   20 E ¼″

Reflex preparation stimulated with induction currents of 25 E and 20 E intensity, at intervals of ¼″. The contractions were written upon the endless paper of a kymographion by means of a lever fixed to the ankle-joint. The middle line indicates the duration of the stimulation, the lower one seconds.

Such conditions are represented in the second curve of Fig. 14. Or, lastly, one sees that the rhythmical beating of the foot increases or diminishes periodically, without the leg being also moved. Like the preliminary reflex contraction, such groups of beats during a period of stimulation often return, and it often happens that the first group no longer causes elevation of the leg, but after several seconds of complete rest the movements of the foot commence, and may increase to elevation of the leg or also of the thigh. If one has gained some experience in such observations, one can see during a long period of stimulation whether with the stimulus employed a reflex is to be expected. If the beats of the foot commence late, and only increase slightly, and soon diminish again, the return the second time not stronger, but weaker, then we may well despair of a good reflex.

As Edward Weber (already cited) compared the intermittent movements of the reflexly-excited limbs to those of an animal "that makes actual exertions," so I would call the above-described process a Sisyphus-task. Near to the result (the discharge or liberation of a powerful reflex) the force diminishes and *the work already done becomes fruitless, it must be begun again anew.* This seems to me to indicate the moment of the discharge. *A simple or complex discharge exhausts the store present.* This can in a fresh preparation be very quickly regained. The energy becomes disposable through the stimuli, fed by the regenerating exchange of material. Even the minimal excitations which discharge the limited reflexes in the foot appear to fatigue, for we have seen that these alone may increase or disappear rhythmically, but this diminution in the excitability appears to be very trifling, for such vibrations often last for minutes.

We may perhaps make the process clearer by a simple comparison. A pendulum requires minimal impulses to be raised a little from the position of rest; returning however it is only able to overcome very small resistances. With a greater swing it gains on the return also greater velocity, and is able, with corresponding mass, to convey a larger amount of kinetic energy to a resisting body. If the resistance is not great, it expends the remainder of its energy in rising in the opposite direction; on its return it can again overcome obstacles, and so on until its energy is used up, its vibrations from the point of equilibrium become invisible, and it at last stands still. Let us further imagine a second pendulum of the same length, and so placed with regard to the first one that the planes of vibration of both are parallel. An easily moveable process (e.g. a piece of watch-spring with little elasticity) is fixed to the second pendulum, so that first the pendulum in passing the position of rest strikes against it. With small excursions the spring will receive a feeble shock, not sufficient to bend it. The process conveys the shock to the second pendulum. This moves more feebly than the first, which was hindered by the resistance but not brought to a standstill.

With greater elongation the first pendulum falls with greater velocity against the spring-process, presses it aside, and completes the other quarter of its vibration. In the mean time, however, the second pendulum has taken up the part of the force for which the spring was still inflexible. It makes a small vibration; but as its duration of vibration is equal to that of the first, also a quarter vibration, it arrives simultaneously with the strongly moved one at the position of equilibrium. There it receives from the rapidly falling one a new impulse, which the spring again partly conveys to the second one. In this way the impulses of the first pendulum become summated in their action on the second, and at last it may acquire the complete amplitude of the first one.

The application of this picture to the reflex process is simple. We require to conceive that the nerve-motion of the sensory nerves impinges upon the first pendulum, that the next motor nerve receives an impulse from the pendulum, as soon as this on its return passes the position of rest. From the secondarily moved pendulum the ends of other nerves are set in motion, as soon as the kinetic energy of the shock has exceeded a certain limit. A third similarly coupled pendulum would represent a vibrating complex arrangement of the third order. To explain the increased consumption of material, arising from secondary and tertiary discharges, one must introduce a certain complication into our picture, that at the moment where the forces of the secondary, or else of the tertiary pendulum, have become sufficiently increased to introduce a movement in the ends of the nerves, a sort of explosion takes there, which sends the motor impulse towards the periphery but also reacts upon the centre. The production of heat and the production of acid may be partly to blame for the temporary diminution of the central excitability. By violent discharges the whole system becomes more difficult to move, hence the conveyance from one vibrating system to another must last very long before the force sufficient for a discharge reaches the second or third system.

But the exhaustion of the central parts need not necessarily be blamed for the inaction of the cutaneous stimulation. The peripheral nerves may refuse to act and render the bridge to the centre impassable to the electrical impulse.

This is easily ascertained by testing the reflex excitability of the limb experimented on by that of the other. There still remains to be explained a weighty point which is of fundamental importance for the principle of summation.

Up to this time we have regarded the reflexes in as far as they are the effects of summated electrical cutaneous stimuli; we have pointed out the circumstances which render the effect desired possible, favour or retard it. The question now arises, Can a reflex action be produced without the summation of stimuli? This appears unquestionable. A puncture is sufficient to cause an animal to fly, and pressure on the foot of a reflex frog to cause violent contraction. The discharges of a Leyden jar of *very short duration* have already been found by Cavendish and Volta to act on the sensory nerves of fishes and man. Setschenow, Fick, Rosenthal have investigated, in the works already cited, the effects of single electrical discharges (interruptions of the current and induction-shocks)[1]. I have already remarked in passing (p. 351) that with very strong infrequent stimuli applied to the fresh preparation the reflex occurred sooner than the second shock. All investigators are at one upon this point, that for reflex excitation by single induction-shocks of incomparably greater strength of the current are required than for stimulation of motor nerves, whilst for the summation-reflexes comparatively weak stimuli are sufficient.

To ascertain the cause of this remarkable difference I have in several series of experiments compared the effects of single induction-shocks with the effects of repeated stimuli.

The following was the result of the experiments with *single induction-shocks.*

---

[1] E. du Bois-Reymond, *Untersuchungen üb. thierische Electicität.* Bd. 1. p. 200, 1848.

Reflex movements were liberated by simple (always applied to the skin of the leg) induction-shocks only when these were very strong. In most cases, the limb only contracted a few times to such stimuli, even when long rest was allowed between the shocks. Only very hardy preparations could survive more than fifty of such stimuli. These had at last to be increased to quite an abnormal extent. The latent time may increase to 3″ or more. How it increases the following fac-simile curve (Fig. 15 a and b) may serve to show.

We observe at all parts of the curve, with the exception of the first (P_1), two or three differently formed parts, a pointed short elevation and a long variously formed double-tooth, and between or behind these here and there a small simple or double-tooth. The indication is plain. The first point represents the compressed curve of contraction which follows the direct stimulation of the motor elements of the limbs without obvious latency.

Fig. 15a.

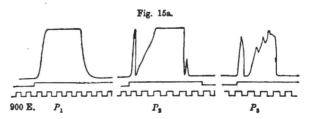

900 E. P_1 P_2 P_3

Fig. 15b.

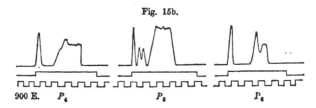

900 E. P_4 P_5 P_6

The left foot of a large frog rested upon the already described writing-lever. An opening induction-shock of 900 units (greatest intensity of an induction-machine armed with two fresh Grove's elements) stimulated it in each of the six periods (P_1 to P_6). After every stimulation there was a minute's rest. The stimulating pen makes a notch upwards in the middle line, when the primary current is opened. The time-marker indicates seconds (½″ above, ¼″ below).

In the first period (P$_1$) this remains invisible, because it is united to the long reflex curve, which rose after a short latency before the first one had time to fall from the maximum. In the other periods represented, which in the original follow each other almost continuously, we see that the contraction-curves of the reflexes become always more widely separated from the one produced by direct stimulation. They have always a tetanic character; sometimes with indications of clonic spasms. The rapid fall of the writing-lever between the two contractions shows that sluggish vibrations of the lever have not lengthened the contraction. The reflex latent time in P$_1$ reaches 1·7″.

The increase of the latency with the fatigue is also a common phenomenon in reflex single contractions. The dependence of the duration of the latent stimulation on the intensity of the exciting current is not easy to detect, because one must rapidly increase the initial strengths of the current, already high enough without this, in order to obtain a long series of reflexes. In some cases where it was possible to compare the increasing with the diminishing strength of current, the intensity appeared to be without influence on the latent time.

The closing induction-shocks are only in rare cases capable of liberating a reflex movement, but they are always surpassed in their action by that of the opening shocks. As closing shocks of 800 units (E) liberated reflex contractions after latent times of from 1·25″ to 1·75″, 900 E could only diminish it to 1·7″. Opening shocks of the same intensity liberated a strong contraction after 0·7″ to 0·8″ latent time. Now, as a specially strong oscillating discharge has been proved for opening induction-shocks[1], the thought arises, of regarding its reflex contractions also as the effects of summation.—In fact, as single induction-shocks of great intensity (700—900 E) acted after long latency (1·7″), and then had no further effect, tetanising stimuli of moderate strength (100 E) gave reflex contractions after inappreciably short latency. This result was striking in the highest degree, because, according to the usual interpretation, it is im-

---

[1] Donder's *Proces-verbaal van de Academie te Amsterdam*, 1868. 30 Mai, No. 1. Helmholtz *Verhandl. des Naturhist. Vereins zu Heidelberg*, 1869. Both referred to in Wiedemann's *Lehre vom Galvanismus und Elektromagnetismus.* 2 Auflage, Braunschweig, 1874. Bd. II. Abth. 2, p. 360 and 128.

possible to understand how one single strong stimulus should require a longer velocity of propagation than many weak ones. Lastly, to explain this result, experiments were made in which the proved effect of a single shock was compared with that of several following each other at different intervals.

At first a closing shock acted like an opening shock (obtained by introducing a good secondary circuit) of 600—700 E, after latent times of 0·7—0·6 seconds. Shocks following each other at an interval of 1″ were able to sum up their actions to reflexes much stronger than the simple. By employing the ordinary opening and closing shocks this succeeded still at intervals of 3″. When however two stimuli followed each other at an interval, which was shorter than the duration of the latent stimulation of a simple shock (0·2″), the reflex followed the second stimulus *directly*, i.e. after a much shorter latent time than after an equally strong simple induction-shock. Thus it is proved, that the observed time of latent stimulation does not give the velocity of propagation of the excitation from peripheral sensory nerves to motor peripheral ones, for if it did, one cannot see how a second stimulus should arrive sooner than the first one equally strong, but that the estimation of the latent time also includes the time which is necessary to initiate the movement in the spinal cord. This view would be in unison with that of S. Exner who defines "reduced reflex time" as that time "which the nervous centres require to convert a sensory into a motor impression[1]."

As to the question, how it arises that demonstrably simple stimuli can at all produce summation-effects, the indication has been given by Engelmann[2]; very strong induction-shocks act by producing thermal (and chemical) changes in the nerves. In fact we have seen in our experiments that a sensitive limb, treated with the strongest single shock, passes *directly* into a tetanus, which must be liberated from the motor structures. Many observers have long ago proved that a single strong induction-shock applied to the sciatic nerve is able to produce a tetanus in the muscles of the leg.

Thus we have brought the apparently single stimuli into the

[1] *Pflüg. Arch.*, 1874, Bd. viii. p. 530.
[2] *Pflüger's Arch.*, 1872, Bd. v. p. 37.

category of the summated ones, and can therefore say that
*Reflex movements can only be liberated by repeated impulses com-
municated to the nervous centres.*

The results of the foregoing experiments find their confirmation in many observations, which we are able to make upon
ourselves. A puncture of the nasal mucous membrance is pain-
ful, tickling excites laughing.   Coughing gives a clear example
of a reflex discharge following summated stimuli ; we feel a
particle or a fibre which may have come in contact with the
mucous membrane of the larynx.   This slight sensation increases, without any new cause, to a stimulus; gradually we feel
obliged to clear the throat, then follow infrequent, short coughs.
If this is not sufficient to remove the offending body, the expiratory efforts becoming more violent, and lastly increase to a spasmodic attack.   A stage of fatigue which gives rest follows the
powerful discharges till the excitation is able again to become
effective.   But even when the offending body has been expelled,
the feeling still lasts for a time, and is expressed in gentle
coughs and hawking, which give a certain amount of relief, like
the movements produced after swallowing, and after hiccough.

The sensation without a reflex movement can also be increased by repeated small stimuli.   Intermittent gentle contact
of the skin produces tickling, which, when long continued
becomes unbearable.   The never tiring, constantly returning
fly can, by its infrequent, weak punctures, set nervous people
into a high state of excitement; a strong blow, or pressure, or
cut is easily tolerated.

ON THE DEVELOPMENT OF ELASMOBRANCH FISHES. By F. M. BALFOUR, B.A., *Fellow of Trinity College, Cambridge.*

INTRODUCTION.

DURING the past two years, I have devoted the greater part of my spare time to the investigation of the development of the Elasmobranch Fishes.

I commenced my observations at the Zoological Station in Naples in the early months of 1874; and a short resumé of the results arrived at during that year was read by me before the British Association in August, and subsequently published in the October number of the *Quarterly Journal of Microscopical Science.*

Since that date I have continued my investigations during a second residence at Naples in the spring of this year; and again during the past summer I have been able to carry on my work partly with specimens brought from Naples, and partly with Scyllium embryos reared in the Brighton Aquarium.

In addition to my original preliminary notice, I have published two papers bearing upon the development of these Fishes. The first of these was entitled, " A Comparison of the Early Stages in the Development of Vertebrates," and appeared in the July number of the *Quarterly Journal of Microscopical Science*, while the second, "On the Origin and History of the Urinogenital Organs of Vertebrates," was published in the last number of this *Journal.* Also a paper by me "On the Development of the Spinal Nerves of Elasmobranchs," has been communicated to the Royal Society.

These various papers contain in a more or less complete form many of the results at which I have arrived. But while some of these are incompletely stated, of a still larger number no mention is made.

These considerations have determined me to put together my observations in the more complete form of a monograph, which I propose to publish as a series of papers in this *Journal.* The more complete my study of the development of these

Fishes becomes, the more is the conclusion forced upon me that they retain in their development ancestral features which have become lost or modified in other vertebrates, and which are of great value in understanding both the origin and history of many of the organs of the higher members of that sub-kingdom, as well as in pointing backwards to the characters which were possessed by their early ancestors.

I cannot omit this opportunity of expressing my thanks to the various friends who have assisted me in the course of my work.

To Dr Dohrn and Dr Hugo Eisig my thanks are especially due for their unremitting kindness and continual assistance during my two visits to Naples.

To the same gentlemen as well as to Dr Michael Foster, Professor Huxley, Professor Lankester, and Dr Kleinenberg, I am indebted for many valuable criticisms and suggestions. To Mr Henry Lee and the Directors of the Brighton Aquarium I am under special obligations for the more than liberal manner in which they have supplied me with Scyllium embryos reared in the Aquarium—an Aquarium which seems likely to be as great a boon to biologists as it has proved an attraction to the world at large.

I. THE RIPE OVARIAN OVUM.

Each ripe ovarian ovum is enclosed in a capsule of fibrous tissue. This capsule is a prolongation of the general connective tissue framework of the ovary, and serves, not only to support the ovum, but also to attach it to the ovary[1].

The ovum itself is nearly spherical, and, after the removal of its capsule, is found to be unprovided with any form of protecting membrane.

My investigations on the histology of the ripe ovarian ovum have been made with the ova of the Gray Skate (*Raja batis*) only, and owing to a deficiency of material are somewhat imperfect.

The bulk of the ovum is composed of yolk spherules, imbedded in a protoplasmic matrix. Dr Alexander Schultz[2],

---

[1] A satisfactory description of this capsule can only be given in connection with its development, and will be given subsequently.
[2] *Archiv für Micro. Anat.* Vol. XI. 1875.

who has studied with great care the constitution of the yolk, finds, near the centre of the ovum, a kernel of small yolk spherules, which is succeeded by a zone of spherules which gradually increase in size as they approach the surface. But, near the surface, he finds a layer in which they again diminish in size and exhibit numerous transitional forms on the way to molecular yolk-granules. These Dr Schultz regards as in a retrogressive condition.

Another interesting feature about the yolk is the presence in it of a protoplasmic network. Dr Schultz has completely confirmed, and on some points enlarged, my previous observations on this subject[1]. Dr Schultz's confirmation is the more important, since he appears to be unacquainted with my previous investigations. In my paper (loc. cit.), after giving a description of the network I make the following statement as to its distribution.

"A specimen of this kind is represented in Plate XIII. Fig. 2, n. y, where the meshes of the network are seen to be finer immediately around the nuclei, and coarser in the intervals. The specimen further shews, in the clearest manner, that this network is not divided into areas, each representing a cell and each containing a nucleus. I do not know to what extent this network extends into the yolk. I have never yet seen the limits of it, though it is very common to see the coarsest yolk-granules lying in its meshes. Some of these are shewn in Plate XIII. Fig. 2, y. k."

Dr Schultz, by employing special methods of hardening and cutting sections of the whole egg, has been able to shew that this network extends, in the form of fine radial lines, from the centre to the circumference; and he rightly states, that it exhibits no cell-like structures. I have detected this network extending throughout the whole yolk in young eggs, but have failed to see it with the distinctness which Dr Schultz attributes to it in the ripe ovum. Since it is my intention to enter fully both into the structure and meaning of this network in my account of a later stage, I say no more about it here.

At one pole of the ripe ovum a slight examination demonstrates the presence of a small circular spot, sharply distin-

[1] Quart. Journ. Micro. Science, Oct. 1874.

guished from the remainder of the yolk by its lighter colour. Around this spot is an area which is also of a lighter colour than the yolk, and the outer border of which gradually shades into the normal tint of the yolk. If a section be made through this part (vide Pl. xv. fig. 1) the circular spot will be found to be the germinal vesicle, and the area around it a disc of yolk containing smaller spherules than the surrounding parts. The germinal vesicle possessed the same structure in both the ripe eggs examined by me; and, in both, it was situated quite on the external surface of the yolk.

In one of my specimens it was flat above, but convex below; in the other and, on the whole, the better preserved of the two, it had the somewhat quadrangular but rather irregular section represented in (Pl. xv. fig. 1). It consisted of a thick-ish membrane and its primitive contents. The membrane surrounded the upper part of the contents and exhibited numerous folds and creases (vide fig. 1). As it extended down-wards it became thinner, and completely disappeared at some little distance from the lower end of the contents. These, therefore, rested below on the yolk. At its circumference the membrane of the disc was produced into a kind of fold, forming a rim which rested on the surface of the yolk.

In neither of my specimens is the cavity in the upper part of the membrane filled by the contents; and the upper part of the membrane is so folded and creased that sections through almost any portion of it pass through the folds. The regularity of the surface of the yolk is not broken by the germinal vesicle, and the yolk around exhibits not the slightest signs of displacement. In the germinal vesicle figured the contents are somewhat irregular in shape; but in my other specimen they form a regular mass concave above and convex below. In both cases they rest on the yolk, and the floor of the yolk is exactly moulded to suit the surface of the contents of the germinal vesicle. The contents have a granular aspect, but differ in constitution from the surrounding yolk. Each germi-nal vesicle measured about one-fiftieth of an inch in diameter.

It does not appear to me possible to suppose that the pecu-liar appearances which I have drawn and described are to be looked upon as artificial products either of the chromic acid, in which the ova were hardened, or of the instrument with which sections of them were made. It is hardly conceivable that chromic acid could cause a rupture of the membrane and the ejection of the contents of the vesicle. At the same time the uniformity of the appearances in the different sections, the regularity of the whole outline of the egg, and the absence of any signs of disturbance in the yolk, render it impossible to believe that the structures described are due to faults of manipulation during or before the cutting of the sections.

We can only therefore conclude that they represent the real state of the germinal vesicle at this period. No doubt they alone do not supply a sufficient basis for any firm conclusions as to the fate of the germinal vesicle. Still, if they cannot sustain, they unquestionably support certain views. The natural interpretation of them is that the membrane of the germinal vesicle is in the act of commencing to atrophy, preparatory to being extruded from the egg, while the contents of the germinal vesicle are about to be absorbed.

In favour of the extrusion of the membrane rather than its absorption are the following features,

(1) The thickness of its upper surface.

(2) The extension of its edge over the yolk.

(3) Its position external to the yolk.

In favour of the view that the contents will be left behind and absorbed when the membrane is pushed out, are the following features of my sections:

(1) The rupture of the membrane of the germinal vesicle on its lower surface.

(2) The position of the contents almost completely below the membrane of the vesicle and surrounded by yolk.

In connection with this subject, Oellacher's valuable observations upon the behaviour of the germinal vesicle in Osseous Fishes and in Birds at once suggest themselves [1]. Oellacher sums up his results upon the behaviour of the germinal vesicle in Osseous Fishes in the following way (p. 12):

[1] *Archiv für Micr. Anat.* Vol. VIII. p. 1.

"The germinal vesicle of the Trout's egg, at a period when the egg is very nearly ripe, lies near the surface of the germinal disc which is aggregated together in a hollow of the yolk....... After this a hole appears in the membrane of the germinal vesicle, which opens into the space between the egg-membrane and the germinal disc. The hole widens more and more, and the membrane frees itself little by little from the contents of the germinal vesicle, which remain behind in the form of a ball on the floor of the cavity formed in this way. The cavity becomes flatter and flatter and the contents are pushed up further and further from the germinal disc. When the hollow, in which lie the contents of the original germinal vesicle, completely vanishes, the covering membrane becomes inverted ......and the membrane is spread out on the convex surface of the germinal disc as a circular, investing structure. It is clear that by the removal of the membrane the contents of the germinal vesicle become lost."

These very definite statements of Oellacher tell strongly against my interpretation of the appearance presented by the germinal vesicle of the ripe Skate's egg. Oellacher's account is so precise, and his drawings so fully bear out his interpretations, that it is very difficult to see where any error can have crept in.

On the other hand, with the exception of those which Oellacher has made, there cannot be said to be any satisfactory observations demonstrating the extrusion of the germinal vesicle from the ovum. Oellacher has observed this definitely for the Trout, but his observations upon the same point in the Bird would quite as well bear the interpretation that the membrane alone became pushed out, as that this occurred to the germinal vesicle, contents and all.

While, then, there are on the one hand Oellacher's observations on a single animal, hitherto unconfirmed, there are on the other very definite observations tending to shew that the germinal vesicle has in many cases an altogether different fate. Götte[1], not to mention other observers before him, has in the case of Batrachian's eggs traced out with great precision the gradual atrophy of the germinal vesicle, and its final absorption into the matter of the ovum.

Götte distinguishes three stages in the degeneration of the germinal vesicle of Bombinator's egg. In the first stage the germinal vesicle has begun to travel up towards the surface of the egg. It retains nearly its primitive condition, but its contents

[1] *Entwicklungsgeschichte der Unke.*

have become more opaque and have partly withdrawn themselves from the thin membrane. The germinal spots are still circular, but in some cases have increased in size. The most important feature of this stage is the smaller size of the germinal vesicle than that of the cavity of the yolk in which it lies, a condition which appears to demonstrate the commencing atrophy of the vesicle.

In the next stage the cavity containing the germinal vesicle has vanished without leaving a trace. The germinal vesicle itself has assumed a lenslike form, and its borders are irregular and pressed in here and there by yolk. Of the membrane of the germinal vesicle, and of the germinal spots, only scanty remnants are to be seen, many of which lie in the immediately adjoining yolk.

In the last stage no further trace of a distinct germinal vesicle is present. In its place is a mass of very finely granular matter, which is without a distinct border and graduates into the surrounding yolk and is to be looked on as a remnant of the germinal vesicle.

This careful investigation of Götte proves beyond a doubt that in Batrachians neither the membrane, nor the contents of the germinal vesicle, are extruded from the egg.

In Mammalia, Van Beneden[1] finds that the germinal vesicle becomes invisible, though he does not consider that it absolutely ceases to exist. He has not traced the steps of the process with the same care as Götte, but it is difficult to believe that an extrusion of the vesicle in the way described by Oellacher would have escaped his notice.

Passing from Vertebrates to Invertebrates, we find that almost every careful investigator has observed the disappearance, apparent or otherwise, of the germinal vesicle, but that very few have watched with care the steps of the process.

The so-called Richtungskörper has been supposed to be the extruded remnant of the germinal vesicle. This view has been especially adopted and supported by Oellacher (loc. cit.), and Flemming[2].

---

[1] Recherches sur la Composition et la Signification de l'Œuf.
[2] Studien in der Entwicklungsgeschichte der Najaden, Sitz. d. k. Akad. Wien, Bd. LXXI. 1875.

The latter author regards the constant presence of this body, and the facility with which it can be stained, as proofs of its connection with the germinal vesicle, which has, however, according to his observations, disappeared before the appearance of the Richtungskörper.

Kleinenberg[1], to whom we are indebted for the most precise observations we possess on the disappearance of the germinal vesicle, gives the following account of it, pp. 41 and 42.

" We left the germinal vesicle as a vesicle with a distinct doubly contoured membrane, and equally distributed granular contents, in which the germinal spot had appeared......The germinal vesicle reaches 0·06 mm. in diameter, and at the same time its contents undergo a separation. The greater part withdraws itself from the membrane and collects as a dense mass around the germinal spot, while closely adjoining the membrane there remains only a very thin but unbroken lining of the plasmoid material. The intermediate space is filled with a clear fluid, but the layer which lines the membrane retains its connection with the mass around the germinal vesicle by means of numerous fine threads which traverse the space filled with fluid. ......At about the time when the formation of the pseudocells in the egg is completed the germinal spot undergoes a retrogressive metamorphosis, it loses its circular outline and it now appears as if coagulated; then it breaks up into small fragments, and I am fairly confident that these become dissolved. The germinal vesicle...... becomes, on the egg assuming a spherical form, drawn into an eccentric position towards the pole of the egg directed outwards, where it lies close to the surface and only covered by a very thin layer of plasma. In this situation its degeneration now begins, and ends in its complete disappearance. The granular contents become more and more fluid; at the same time part of them pass out through the membrane. This, which so far was firmly stretched, next collapses to a somewhat egg-like sac, whose wall is thickened and in places folded.

"The inner mass which up to this time has remained compact now breaks up into separate highly refractive bodies, of spherical or angular form and of very different sizes; between them, here and there, are scattered drops of a fluid fat......I am very much inclined to regard the solid bodies in question as fat or as that peculiar modification of albuminoid bodies which we recognise as the certain forerunner of the formation of fat in so many pathologically altered tissues; and therefore to refer the disappearance of the germinal vesicle to a fatty degeneration. On one occasion I believe that I observed an opening in the membrane at this stage; if this is a normal condition it would be possible to believe that its solid contents passed out and were taken up in the surrounding plasma.

[1] *Hydra.* Leipzig, 1872.

What becomes of the membrane I am unable to say ; in any case the germinal vesicle has vanished to the very last trace before impregnation occurs."

Kleinenberg clearly finds that the germinal vesicle disappears completely before the appearance of the Richtungskörper, in which he states a pseudocell or yolk-sphere is usually found.

The connection between the Richtungskörper and the germinal vesicle is not a result of strict observation, and there can be no question that the evidence in the case of invertebrates tends to prove that the germinal vesicle in no case disappears owing to its extrusion from the egg, but that if part of it is extruded from the egg as Richtungskörper this occurs when its constituents can no longer be distinguished from the remainder of the yolk. This is clearly the case in Hydra, where, as stated above, one of the pseudocells or yolk-spheres is usually found imbedded in the Richtungskörper.

My observations on the Skate tend to shew that, in its case, the membrane of the germinal vesicle is extruded from the egg, though they do not certainly prove this. That conclusion is however supported by the observations of Schenk[1]. He found in the impregnated, but not yet segmented, germinal disc a cavity which, as he suggests, might well have been occupied by the germinal vesicle. It is not unreasonable to suppose that the membrane, being composed of formed matter and able only to take a passive share in vital functions, could, without thereby influencing the constitution of the ovum, be ejected.

If we suppose, and this is not contradicted by observation, that the Richtungskörper is either only the metamorphosed membrane of the germinal vesicle with parts of the yolk, or part of the yolk alone, and assume that in Oellacher's observations only the membrane and not the contents were extruded from the egg, it would be possible to frame a consistent account of the behaviour of the germinal vesicle throughout the animal kingdom, which may be stated in the following way.

The germinal vesicle usually before, but sometimes immediately after impregnation undergoes atrophy and its *contents* become indistinguishable from the remainder of the egg. In

[1] Die Eier von Raja quadrimaculata, *Sitz. der k. Akad. Wien*, Bd. LXVIII.

those cases in which its membrane is very thick and resistent,
e.g. Osseous and Elasmobranch Fishes, Birds, etc., this may be
incapable of complete resorption, and be extruded bodily from
the egg. In the case of most ova, it is completely absorbed,
though at a subsequent period it may be extruded from the egg
as the Richtungskörper. In all cases the contents of the
germinal vesicle remain in the ovum.

In some cases the germinal vesicle is stated to persist and
to undergo division during the process of segmentation. The
observations on this point stand in need of confirmation, but
even if true would not be altogether inexplicable on the supposition that whether the germinal vesicle vanishes or divides, the
matter of the first segmentation-nuclei is derived from it.

My investigations shew that the germinal vesicle atrophies
in the Skate before impregnation, and in this respect accord
with very many recent observations. Of these the following
may be mentioned,

(1) Oellacher (Bird, Osseous Fish).
(2) Götte (Bombinator igneus).
(3) Kupffer (Ascidia Canina).
(4) Strasburger (Phallusia Mamillata).
(5) Kleinenberg (Hydra).
(6) Metschnikoff (Geryonia, Polyzenia leucostyla, Epibulia
aurantiaca, and other Hydrozoa).

This list is sufficient to shew that the disappearance of the
germinal vesicle before impregnation is very common, and I
am unacquainted with any observations tending to shew that
its disappearance is due to impregnation.

In many cases, e. g. Asterocanthion[1], the germinal vesicle
vanishes after the spermatozoa have begun to surround the egg;
but I do not know that its disappearance in these cases has
been shewn to be due to impregnation. To do so it would be
necessary to prove that in ripe eggs let loose from the ovary, but
not fertilized, the germinal vesicle did not undergo the same
changes as in the case of fertilized eggs; and this, as far as I
know, has not been done. After the disappearance of the
germinal vesicle, and before the first act of division, a fresh

[1] Agassiz, *Embryology of the Star-Fish.*

nucleus frequently appears [—vide—Auerbach (Ascaris nigro-venosa), Fol (Geryonia), Kupffer (Ascidia canina), Strasburger (Phallusia mamillata), Fleming (Anodon), Götte (Bombinator igneus)], which is generally stated to vanish before the appearance of the first furrow; but in some cases (Kupffer and Götte, and as studied with especial care Strasburger) it is stated to divide. Upon the second nucleus, or upon its relation to the germinal vesicle, I have no observations; but it appears to me of great importance to determine whether this fresh nucleus arises absolutely de novo, or is formed out of the matter of the germinal vesicle.

The germinal vesicle is situated in a bed of finely divided yolk-particles. These graduate insensibly into the coarser yolk-spherules around them, though the band of passage between the coarse and the finer yolk-particles is rather narrow. The mass of fine yolk-granules may be called the germinal disc. It is not to be looked upon as diverging in any essential particular from the remainder of the yolk, for the difference between the two is one of degree only. It contains in fact a larger bulk of active protoplasm, as compared with yolk-granules, than does the remainder of the ovum. The existence of this agreement in kind has been already strongly insisted on in my preliminary paper; and Schultz (*loc. cit.*) has arrived at an entirely similar conclusion, from his own independent observations.

One interesting feature about the germinal disc at this period is its size.

My observations upon it have been made with the eggs of the Skate (Raja) alone; but I think that it is not probable that its size in the Skate is greater than in Scyllium or Pristiurus. If its size is the same in all these genera, then the germinal disc of the unimpregnated ovum is very much greater than that portion of the ovum which undergoes segmentation, and which is usually spoken of as the germinal disc in impregnated ova.

I have no further observation on the ripe ovarian ovum; and my next observations concern an ovum in which two furrows have already appeared.

II. THE SEGMENTATION.

I have not been fortunate enough to obtain an absolutely complete series of eggs during segmentation.

In the cases of Pristiurus[1] and Scyllium only have I had any considerable number of eggs in this condition, though one or two eggs of Raja in which the process was not completed have come into my hands.

In the youngest impregnated Pristiurus eggs, which I have obtained, the germinal disc was already divided into four segments.

The external appearance of the blastoderm, which remains nearly constant during segmentation, has been already well described by Leydig[2].

The yolk has a pale greenish tinge which, on exposure to the air, acquires a yellower hue. The true germinal disc appears as a circular spot of a bright orange colour, and is, according to Leydig's measurements, $1\frac{1}{2}$ m. in diameter. Its colour renders it very conspicuous, a feature which is further increased by its being surrounded by a narrow dark line (Pl. xv. fig. 2), the indication of a shallow groove. Surrounding this line is a concentric space which is lighter in colour than the remainder of the yolk, but whose outer border passes by insensible gradations into the yolk. As was mentioned in my preliminary paper (*loc. cit.*), and as Leydig (*loc. cit.*) had before noticed, the germinal disc is always situated at the pole of the yolk which is near the rounded end of the Pristiurus egg. It occupies a corresponding position in the eggs of both species of Scyllium (stellare and canicula) near the narrower end of the egg to which the shorter pair of strings is attached. The germinal disc in the youngest egg examined, exhibited two furrows which

---

[1] In my preliminary paper (*Quart. Journal of Micr. Science* 1874), I stated that I had chiefly dealt with the eggs of a species of Mustelus. Pristiurus however, and not Mustelus, is the genus I have been working on. I was led into this unfortunate mistake partly by the mendacity of the Neapolitan fisherman who brought me a species of Mustelus as the fish from which they derived the eggs they procured for me, and partly by my own ignorance in not being aware that neither species of Mustelus is oviparous. I was able to correct my error on returning to Naples in the spring of this year, when I actually removed eggs, similar to those I had been working with, from the oviduct of a Pristiurus.

[2] *Rochen und Haie.*

crossed each other at right angles in the centre of the disc, but neither of which reached its edge. These furrows accordingly divided the disc into four segments, completely separated from each other at the centre of the disc, but united near its circumference.

I made sections, though not very satisfactorily, of this germinal disc. The sections shewed that the disc was composed of a protoplasmic basis, in which were imbedded innumerable minute spherical yolk-globules so closely packed as to constitute nearly the whole mass of the germinal disc.

In passing from the coarsest yolk-spheres to the fine spherules of the germinal disc, three bands of different sized yolk-particles have to be traversed. These bands graduate into one another and are without sharp lines of demarcation. The outer of the three is composed of the largest sized yolk-spherules which constitute the greater part of the ovum. The middle band forms a concentric layer around the germinal disc, and is composed of yolk-spheres considerably smaller than those outside it. Where it cuts the surface it forms the circle of lighter colour immediately surrounding the germinal disc. The innermost band is formed by the germinal disc itself and is composed of spherules of the smallest size. These features are shewn in Pl. xv. fig. 6, which is the section of a germinal disc with twenty-one segments; in it however the outermost band of spherules is not present.

From this description it is clear, as has already been mentioned in the description of the ripe unimpregnated ovum, that the germinal disc is not to be looked upon as a body entirely distinct from the remainder of the ovum, but merely as a part of the ovum in which the protoplasm is more concentrated and the yolk-spherules smaller than elsewhere. Sections shew that the furrows visible on the surface end below, as indeed they do on the surface, before they reach the external limit of the finely granular matter of the germinal disc. There are therefore at this stage no distinct segments: the otherwise intact germinal disc is merely grooved by two furrows.

I failed to observe any nuclei in the germinal disc just described, but it by no means follows that they were not present.

In the next youngest of the eggs[1] examined the germinal disc was already divided into twenty-one segments. When viewed from the surface (Pl. xv. fig. 3), the segments appeared divided into two distinct groups—an inner group of eleven smaller segments, and an outer group of segments surrounding the former. The segments of both the inner and the outer group were very irregular in shape and varied considerably in size. The amount of irregularity is far from constant and many germinal discs are more regular than the one figured.

In this case the situation and relations of the germinal disc to the yolk were precisely the same as in the earlier stage.

In sections of this germinal disc (Pl. xv. fig. 6), the groove which separates it from the yolk is well marked on one side, but hardly visible at the other extremity of the section.

Passing from the external features of this stage to those which are displayed by sections, the striking point to be noticed is the persisting continuity of the segments, marked out on the surface, with the floor of the germinal disc.

The furrows which are visible on the surface merely form a pattern, but do not isolate a series of distinct segments. They do not even extend to the limit of the finely granular matter of the germinal disc.

The section represented Pl. xv. fig. 6, bears out the statements about the segments as seen on the surface. There are three smaller segments in the middle of the section, and two larger at the two ends. These latter are continuous with the coarser yolk-spheres surrounding the germinal disc and are not separated from them by a segmentation furrow.

In a slightly older embryo than the one figured I met with a few completely isolated segments at the surface. These segments were formed by the apparent bifurcation of furrows as they neared the surface of the germinal disc. The segments thus produced are triangular in form. They probably owe their origin to the meeting of two oblique furrows. The last formed of these furrows apparently ceases to be prolonged after meeting the first-formed furrow. I have not in any case observed an example of two furrows crossing one another at this stage.

[1] The germinal disc figured was from the egg of a Scyllium stellare and not Pristiurus, but I have also sections of a Pristiurus egg of the same age, which do not differ materially from the Scyllium sections.

The furrows themselves for the most part are by no means simple slits with parallel sides. They exhibit a beaded structure, shewn imperfectly in Pl. xv. fig. 6, but better in Pl. xv. fig. 6 a, which is executed on a larger scale. They present intervals of dilatations where the protoplasms of the segments on the two sides of the furrow are widely separated, alternating with intervals where the protoplasms of the two segments are almost in contact and are only separated from one another by a very narrow space.

. A closer study of the germinal disc at this period shews that the cavities which cause the beaded structure of the furrows are not only present along the lines of the furrows but are also found scattered generally through the germinal disc, though far more thickly in the neighbourhood of the furrows. Their appearance is that of vacuoles, and with these they are probably to be compared. There can be little question that in the living germinal disc they are filled with fluid. In some cases, they are collected in very large numbers in the region of a furrow. Such a case as this is shewn in Pl. xv. fig. 6 b. In numerous other cases they occur, roughly speaking, alternately on each side of a furrow. Some furrows, though not many, are entirely destitute of these structures. The character of their distribution renders it impossible to overlook the fact that these vacuole-like bodies have important relations with the formation of the segmentation furrows, but to this subject I intend to return.

Lining the two sides of the segmentation furrows there is present in sections a layer which stains deeply with colouring re-agents; and the surface of the blastoderm is stained in the same manner. In neither case is it permissible to suppose that any membrane-like structure is present. In many cases a similar very delicate, but deeply-stained line, invests the vacuolar cavities, but the fluid filling these remains quite unstained. When distinct segments are formed, each of these is surrounded by a similarly stained line.

The yolk-spherules are so numerous, and render even the thinnest section so opaque, that I have failed to make satisfactory observations on the behaviour of the nucleus. I find nuclei in many of the segments, though it is very difficult even to see them, and only in very favourable specimens can their structure be studied. In some cases, two of them lie one on each side of a furrow; and in one case at the extreme end of a furrow I could see two peculiar aggregations of yolk-spherules united by a band through which the furrow, had it been con¬ tinued, would have passed. The connection (if any exists) between this appearance, and the formation of the fresh nuclei in the segments, I have been unable to elucidate.

The peculiar appearances attending the formation of fresh nuclei in connection with cell-division, which have recently been described by so many observers, have hitherto escaped my observation at this stage of the segmentation, though I shall describe them in a later stage. A nucleus of this stage is shewn on Pl. xv. fig. 6 c. It is lobate in form and is divided by lines into areas in each of which a deeply-stained granule is situated.

The succeeding stages of segmentation present from the surface no fresh features of great interest. The somewhat irregular (Pl. xv. figs. 4 and 5) circular line, which divides the peripheral larger from the central smaller segments, remains for a long time conspicuous. It appears to be the representative of the horizontal furrow which, in the Batrachian ovum, separates the smaller pigmented spheres from the larger spheres of the lower pole of the egg.

As the segments become smaller and smaller, the distinction between the peripheral and the central segments becomes less and less marked; but it has not disappeared by the time that the segments become too small to be seen with the simple lens. When the spheres become smaller than in the germinal disc represented on Pl. xv. fig. 5, the features of segmentation can be more easily and more satisfactorily studied by means of sections.

To the features presented in sections, both of the latter and of the earlier blastoderms, I now return. A section of one of the earlier germinal discs, of about the age of the one represented on Pl. xv. fig. 4, is shewn in Pl. xvi. fig. 7.

It is clear at a glance that we are now dealing with true segments completely circumscribed on all sides. The peripheral segments are, as a rule, larger than the more central ones, though in this respect there is considerable irregularity.

The segments are becoming smaller by repeated division; but, in addition to this mode of increase, there is now going on outside the germinal disc a segmentation of the yolk, by which fresh segments are being formed from the yolk and added to those which already exist in the germinal disc. One or two such segments are seen in the act of being formed (Pl. XVI. fig. 7 *f*) ; and it is to be noticed that the furrows which will eventually mark out the segments, do so at first in a partial manner only, and do not circumscribe the whole circumference of the segment in the act of being formed. These fresh furrows are thus repetitions on a small scale of the earliest segmentation furrows.

It deserves to be noticed that the portion of the germinal disc which has already undergone segmentation, is still surrounded by a broad band of small-sized yolk-spherules. It appears to me probable that owing to changes taking place in the spherules of the yolk, which result in the formation of fresh spherules of a small size, this band undergoes a continuous renovation.

The uppermost row of segmentation spheres is now commencing to be distinguished from the remainder as a separate layer which becomes progressively more distinct as segmentation proceeds.

The largest segments in this section measure about the $\frac{1}{100}$th of an inch in diameter, and the smallest about $\frac{1}{500}$th of an inch.

The nuclei at this stage present points of rather a special interest. In the first place, though visible in many, and certainly present in all the segments[1], they are not confined to these : they are also to be seen, in small numbers, in the band of fine spherules which surrounded the already segmented part of the germinal disc. Those found outside the germinal disc are not confined to the spots where fresh segments are appearing, but are also to be seen in places where there are no traces of fresh segments.

This fact, especially when taken in connection with the forma-

---

[1] In the figure of this stage, I have inserted nuclei in all the segments. In the section from which the figure was taken, nuclei were not to be seen in many of the segments, but I have not a question that they were present in all of them. The difficulty of seeing them is, in part, due to the yolk-spherules and in part to the thinness of the section as compared with the diameter of a segmentation sphere.

tion of fresh segments outside the germinal disc and with other facts which I shall mention hereafter, is of great morphological interest as bearing upon the nature and homologies of the food-yolk. It also throws light upon the behaviour and mode of increase of the nuclei. All the nuclei, both those of the segments and those of the yolk, have the peculiar structure I described in the last stage.

In specimens of this stage I have been able to observe certain points which have an important bearing upon the behaviour of the nucleus during cell-division.

Three figures, illustrating the behaviour of the nucleus, as I have seen it in sections of blastoderms hardened in chromic acid, are shewn in Pl. XVI. figs. 7 *a*, 7 *b* and 7 *c*.

In the place of the nucleus is to be seen a sharply defined figure (Fig. 7 *a*) stained in the same way as the nucleus or more deeply. It has the shape of two cones placed base to base. From the apex of each cone there diverge towards the base a series of excessively fine striæ. At the junction between the two cones is an irregular linear series of small deeply stained granules which form an apparent break between the two. The line of this break is continued very indistinctly beyond the edge of the figure on each side.

From the apex of each cone there diverge outwards into the protoplasm of the cell a series of indistinct markings. They are rendered obscure by the presence of yolk-spherules, which completely surround the body just described, but which are not arranged with any reference to these markings. These latter striæ, diverging from the apex of the cone, are more distinctly seen when the apex points to the observer (Fig. 7 *b*), than when a side of the cone is in view.

The striæ diverging outwards from the apices of the cones must be carefully distinguished from the striæ of the cones themselves. The cones are bodies quite as distinctly differentiated from the protoplasm of the cell as nuclei, while the striæ which diverge from their apices are merely structures in the general protoplasm of the cell.

In some cells, which contain these bodies, no trace of a commencing line of division is visible. In other cases (Fig. 7 *c*), such a line of division does appear and passes through the junction of the two cones. In one case of this kind I fancied I could see (and have represented) a coloured circular body in each cone. I do not feel any confidence that these two bodies are constantly present; and even where visible they are very indistinct.

Instead of an ordinary nucleus a very indistinctly marked vesicular body sometimes appears in a segment; but whether it is to be looked on as a nucleus not satisfactorily stained, or as a nucleus in the act of being formed, I cannot decide.

With reference to the situation of the cone-like bodies I have described I have made an observation which appears to me to be of some interest. I find that bodies of this kind are found in the yolk *completely outside* the germinal disc. I have made this observation, in at least two cases which admitted of no doubt (vide Fig. 7 *nx'*).

We have therefore the remarkable fact, that whatever connection these bodies may have with cell-division, they can occur in cases where this is altogether out of the question and where an increase in the number of nuclei can be their only product.

These are the main facts which I have been able to determine with reference to the nuclei of this stage; but it will conduce to clearness if I now finish what I have to say upon this subject.

At a still later stage of segmentation the same peculiar bodies are to be seen as during the stage just described, but they are rarer; and, in addition to them, other bodies are to be seen of a character intermediate between ordinary nuclei and the former bodies.

Three such are represented in Pl. XVI. figs. 8*a*, 8*b*, 8*c*. In all of these there can be traced out the two cones, which are however very irregular. The striation of the cones is still present, but is not nearly so clear as it was in the earlier stage.

In addition to this, there are numerous deeply stained granules scattered about the two figures which resemble exactly the granules of typical nuclei.

All these bodies occupy the place of an ordinary nucleus, they stain like an ordinary nucleus and are as sharply defined as an ordinary nucleus.

There is present around some of these, especially those

situated in the yolk, the network of lines of the yolk described by me in a preliminary paper[1], and I feel satisfied that
there is in some cases an actual connection between the network and the nuclei. This network I shall describe more fully
hereafter.

Further points about these figures and the nuclei of this
stage I should like to have been able to observe more completely than I have done, but they are so small that with the
highest powers I possess (Zeiss, Immersion No. 2 $= \frac{1}{18}$ in.) their
complete and satisfactory investigation is not possible.

Most of the true nuclei of the cells of the germinal disc
are regularly rounded; those however of the yolk are frequently irregular in shape and often provided with knob-like
processes. The gradations are so complete between typical
nuclei and bodies like that shewn (Pl. XVI. fig. 8 c) that it
is impossible to refuse the name of nucleus to the latter.

In many cases *two nuclei* are present in one cell.

In later stages knob-like nuclei of various sizes are scattered in very great numbers in the yolk around the blastoderm. In some cases it appears to me that several of these are
in close juxta-position, as if they had been produced by the
division of one primitive nucleus. I do not feel absolutely
confident that this is the case, owing to the fact that in the
investigation of a knobbed body there is great difficulty in
ascertaining that the knobs, which appear separate in one plane,
are not in reality united in another.

I have, in spite of careful search, hitherto failed to find
amongst these later nuclei cone-like figures, similar to those I
found in the yolk during segmentation. This is the more remarkable since in the early stages of segmentation, when very few
nuclei are present in the yolk, the cone-like figures are not uncommon; whereas, in the latter stages of development when
the nuclei of the yolk are very common and obviously increasing rapidly, such figures are not to be met with.

In no case have I been able to see a distinct membrane
round any of the nuclei.

I have hitherto attempted to describe the appearances

---

[1] *Loc. cit.*

bearing on the behaviour of the nuclei in as objective a manner as possible.

My observations are not as complete as could be desired; but, taken in conjunction with those of other investigators, they appear to me to point towards certain definite conclusions with reference to the behaviour of the nucleus in cell-division.

The most important of these conclusions may be stated as follows. In the act of cell-division the nuclei of the resulting cells are formed from the nucleus of the primitive cell.

This may occur;—

(1)   By the complete solution of the old nucleus within the protoplasm of the mother cell and the subsequent reaggregation of its matter to form the nuclei of the freshly formed daughter cells,

(2)   By the simple division of the nucleus,

(3)   Or by a process intermediate between these two where part of the old nucleus passes into the general protoplasm and part remains always distinguishable and divides; the fresh nucleus being in this case formed from the divided parts as well as from the dissolved parts of the old nucleus.

Included in this third process it is permissible to suppose that we may have a series of all possible gradations between the extreme processes 1 and 2. If it be admitted, and the evidence we have is certainly in favour of it, that in some cases, both in animal and vegetable cells, the nucleus itself divides during cell-division, and in others the nucleus completely vanishes during the cell-division, it is more reasonable to suspect the existence of some connection between the two processes, than to suppose that they are entirely different in kind. Such a connection is given by the hypothesis I have just proposed.

The evidence for this view, derived both from my own observations and those of other investigators, may be put as follows.

The absolute division of the nucleus has been stated to occur in animal cells, but the number of instances where the evidence is quite conclusive are not very numerous. Recently F. E. Schultze [1] appears to have observed it in the case of an Amœba in an altogether satisfactory manner. The instance is

[1] *Archiv f. Micr. Anat.* xi. p. 592.

quoted by Flemming[1]. Schultze saw the nucleus assume a dumb-bell shape, divide, and the two halves collect themselves together. The whole process occupied a minute and a half and was shortly followed by the division of the Amœba, which occupied eight minutes. Amongst vegetable cells the division of the nucleus seems to be still rarer than with animal cells. Sachs[2] admits the division of the nucleus in the case of the parenchyma cells of certain Dicotyledons (Sambucus, Helianthus, Lysimachia, Polygonum, Silene) on the authority of Hanstein.

The division of the nucleus during cell-division, though seemingly not very common, must therefore be considered as a thoroughly well authenticated occurrence.

The frequent disappearance of the nucleus during cell-division is now so thoroughly recognised, both for animal and vegetable cells, as to require no further mention.

In many cases the partial or complete disappearance of the nucleus is accompanied by the formation of two peculiar star-like figures. Appearances of the kind have been described by Fol[3], Flemming[4], Auerbach[5] and Strasburger[6] and possibly also Oellacher[7] as well as other observers.

These figures[8] are very probably due to the streaming out of the protoplasm of the nucleus into that of the cell[9]. The

[1] *Entwicklungsgeschichte der Najaden*, LXXI. Bd. der *Sitz. der k. Akad. Wien*, 1875.

[2] *Text-Book of Botany*, English trans. p. 19.

[3] *Entw. d. Geryonideneies. Jenaische Zeitschrift*, Bd. VII.

[4] *Loc. cit.*

[5] *Organologische Studien*, Zweites Heft.

[6] *Zellbildung u. Zelltheilung.*

[7] *Beiträge z. Entwicklungsgeschichte der Knochenfishen: Zeit. für Wiss. Zoologie.* Bd. XXII. 1872.

[8] The memoirs of Auerbach and Strasburger have unfortunately come into my hands too late for me to take advantage of them. Especially in the magnificent monograph of Strasburger I find drawings precisely resembling those from my specimens already in the hands of the engraver. Strasburger comes to the conclusion from his investigations that the modified nucleus always divides and never vanishes as is usually stated. If his views on this point are correct part of the hypothesis I have suggested above is rendered unnecessary. The striæ of the protoplasm, which in accordance with Auerbach's view I have considered as being due to a streaming out of the matter of the nucleus, he regards as resulting from a polarity of the particles in the cell and the attraction of the nucleus. My own investigations though, as far as they go, quite in accordance with those of Strasburger, do not supply any grounds for deciding on the meaning of these striæ; and in some respects they support Strasburger's views against those of other observers, since they demonstrate that in Elasmobranchs the modified nucleus does actually divide.

[9] This is the view which has been taken by Auerbach (*Organologische Studien*).

appearance of striation may on this hypothesis be explained as due to the presence of granules in the protoplasm. When the streaming out of the protoplasm of a nucleus into that of a cell takes place, any large granule which cannot be moved by the stream will leave behind it a slack area where there is no movement of the fluid. Any granules which are carried into this area will remain there, and by the continuation of a process of this kind a row of granules may be formed, and a series of such rows would produce an appearance of striation. In many cases, e. g. Anodon, vide Flemming[1], even the larger yolk-spherules are arranged in this fashion.

On the supposition that the striation of these figures is due to the outflow from the nucleus, the appearances presented in Elasmobranchs admit of the following explanation.

The central body consisting of two cones (Figs. 7a, 7c) is almost without question the remnant of the primitive nucleus. This is shewn by its occupying the same position as the primitive nucleus, staining in the same way, and by there being a series of insensible gradations between it and a typical nucleus. The contents must be supposed to be streaming out from the two apices of the cones, as appears from the striæ in the body converging on each side towards the apex, and then diverging again from it. In my specimens the yolk-spherules are not arranged with any reference to the radiating striation.

It is very likely that in the cases of the disappearance of the nucleus, its protoplasm streams out in two directions, towards the two parts of the cell which will eventually become separated from each other; and probably, after the division, the matter of the old nucleus is again collected to form two fresh nuclei.

In some cases of cell-division a remnant of the old nucleus is stated to be visible after the fresh nuclei have appeared. These cases, of which I have not seen full accounts, are perhaps analogous to what occasionally happens with the germinal vesicle of an ovum. The whole of the contents of the germinal vesicle become at its disappearance mingled with the protoplasm of the ovum, but the resistent membrane remains and is eventually ejected from the egg, vide p. 381 et seq. If the

[1] Loc. cit.

remnant of the old nucleus in the cases described is nothing more than its membrane, no difficulty is offered to the view that the constituents of the old nucleus may help to form the new ones.

In many cases the total bulk of the new nuclei is greater than that of the old one; in such instances part of the protoplasm of the cell necessarily has a share in forming the new nuclei.

Although, in instances where the nucleus vanishes, an absolute demonstration of the formation of the fresh nuclei from the matter of the old one is not possible; yet, if cases of the division of the old nucleus to form the new ones be admitted to exist, the derivation in the first process of the fresh nuclei from the old ones must be postulated in order to maintain a continuity between the two processes of formation; and, as I have attempted to shew, all the circumstantial evidence is in favour of it.

Admitting the existence of the two extreme processes of nuclear formation, I wish to shew that my results in Elasmobranchs tend to demonstrate the existence of intermediate steps between them. The first figures I described of two opposed cones, appear to me almost certainly to represent nuclei in the act of dissolution; but though a large portion of the nucleus may stream out into the yolk, I question whether the whole of it does[1].

I described these bodies in two states. An earlier one, in which the two cones were separated by an irregular row of deeply stained granules; and a later one in which a furrow had already appeared dividing the cones as well as the cell. In neither of these conditions could I see any signs of the body vanishing completely. It was as clearly defined and as deeply stained as an ordinary nucleus, and in its later condition the signs of the streaming out of material from its pointed extremities were less marked than in the earlier stage.

All these facts, to my mind, point to the view that these cone-like bodies do not completely disappear, but form the basis for the new nuclei. Possibly the body visible in

---

[1] After Strasburger's observation it must be considered very doubtful whether the streaming out of the contents of the nucleus, in the manner implied in the text, really takes place.

each cone in the later stage, was the commencement of this new nucleus. Götte[1] has figured structures somewhat similar to these bodies, but I hardly understand either his figure or his account sufficiently clearly to be able to pronounce upon the identity of the two. In case they are identical, Götte gives a very different explanation of them from my own[2].

A second of my results, which points to a series of inter-mediate steps between division and solution of the nucleus, is the distribution in time of the peculiar cone-like bodies. These are present in fair abundance at an early period of segmen-tation, when there are but few nuclei either in the blastoderm or the yolk. But at later periods, when there are both more nuclei, especially in the yolk, and they are also increasing in numbers more rapidly than before, no bodies of this kind are to be seen. This fact becomes the more striking from the lobate appearance of the later nuclei of the yolk, an appearance which exactly suits the hypothesis of the rapid budding off of fresh nuclei.

The observations of R. Hertwig[3] on the gemmation of *Podo-phrya Gemmipara*, support my interpretation of the knobbed condition of the nuclei. Hertwig finds (p. 47) that

The horse-shoe shaped nucleus grows out into numerous anastomosing projections. Over the free ends of the projections little knobs appear on the surface of the body, into which the lengthening ends of the processes of the nucleus grow up. Here they bend themselves into a horse-shoe form. The newly-formed nucleus then separates from the original nucleus, and afterwards the bud contain-ing it from the body.

From the peculiar arrangement of the net-work of lines of the yolk around these knobbed nuclei, it is reasonable to con-clude that interchange of material between the protoplasm of the yolk and the nuclei is still taking place, even during the later periods.

These facts about the distribution in time of the cone-like

[1] *Entwicklungsgeschichte der Unke*, Pl. i. fig. 18.
[2] As I before mentioned, Strasburger (*Zellbildung u. Zelltheilung*) has repre-sented bodies precisely similar to those I have described, which appear during the segmentation in the egg of *Phallusia Mammillata* as well as similar figures observed by Butschli in eggs of *Cucullanus elegans* and *Blatta Germanica*. The figures in this monograph are the only ones I have seen, which are identical with my own.
[3] *Morphologisches Jahrbuch*, Bd i. pp. 46, 47.

bodies afford a strong presumptive evidence of a change in the manner of nuclear increase.

The last argument I propose urging on this head is derived from the bodies (Pl. XVI. fig. 8 *a, b, c*) which I have described as intermediate between the true cone-like bodies and typical nuclei. They appear to afford evidence of less and less of the matter of the nucleus streaming out into the yolk and of a large proportion of it becoming divided.

The conclusion to be derived from all these facts is that for Elasmobranchs in the earlier stages of segmentation, and during the formation of fresh segments, a partial solution of the old nucleus takes place, but all its constituents serve for the reconstruction of the fresh nuclei.

In later periods of development a still smaller part of the nucleus becomes dissolved, and the rest divides; but the two fresh nuclei are still derived from the two sources. After the close of segmentation the fresh nuclei are formed by a simple division of the older ones.

The appearance of the cone-like bodies in the yolk outside the germinal disc is a point of some interest. It demonstrates in a conclusive manner that whatever influence (if any) the nucleus may have in ordinary cases of cell division, yet it may undergo changes of a precisely similar character to those which it experiences during cell division, without exerting any influence on the surrounding protoplasm. If the lobate nuclei are also nuclei undergoing division, we have in the egg of an Elasmobranch examples of all the known forms of nuclear increase unaccompanied by cell division[1].

The next stage in the segmentation does not present so many features of interest as the last one. The segments are now so small, as to be barely visible from the surface with a simple lens. A section of an embryo of this stage is represented in Pl. XVI. fig. 8. The section, which is drawn on the

---

[1] Strasburger's (*loc. cit.*) arguments about the influence of the nucleus in cell division are not to my mind conclusive; though not without importance. It is difficult to reconcile his views with the facts of cell division observable during the Elasmobranch segmentation; but even if their truth be admitted they do not bring us much nearer to a satisfactory understanding of cell division, unless accompanied (and at present they are not so) by a rational explanation of the forces which produce the division of the nucleus.

same scale as the section belonging to the last stage, serves to show the relative size of the segments in the two cases.

The epiblast is now more distinct than it was. The segments composing it are markedly smaller than the remainder of the cells of the germinal disc, but possess nuclei of an absolutely larger size than do the other cells. They are irregular in shape, with a slight tendency to be columnar. An average segment of this layer measures about $\frac{1}{700}$ inch.

The cells of the lower layer are more polygonal than those of the epiblast, and are decidedly larger. An average specimen of the larger cells of the lower layer measures about $\frac{1}{400}$ in. in diameter, and is therefore considerably smaller than one of the smallest cells of the last stage. The formation of fresh segments from the yolk still continues with fair rapidity, but nearly comes to an end shortly after this.

Of the nuclei of the lower layer cells, there is not much to add to what has already been said. Not infrequently two nuclei may be observed in a single cell.

The nuclei in the yolk which surrounds the germinal disc are more numerous than in the earlier periods, and are now to be met with in fair numbers in every section (Fig. 8 n').

These are the main features which characterise the present stage, they are in all essential points similar to those of the last stage, and the two germinal discs hardly differ except in the size of the segments of which they are composed.

In the last stage which I consider as belonging to the segmentation, the cells of the whole blastoderm have become far smaller (Pl. XVI. fig. 9).

The epiblast (ep) now consists of a very marked layer of columnar cells. It is, as far as I have been able to observe, never more than one cell deep. The cells of the lower layer are of an approximately uniform size, though a few of those at the circumference of the blastoderm considerably exceed the remainder in the bulk.

There are two fresh features of importance in germinal discs of this age.

Instead of being but indistinctly separated from the surrounding yolk, the blastoderm has now very clearly defined limits.

This is an especially marked feature of preparations made with osmic acid. In these there may frequently be seen a deeply stained doubly coloured line, which forms the limit of the yolk, where it surrounds the germinal disc. Lines of this kind are often to be seen on the surface of the yolk, or even of the blastoderm, but are probably to be regarded as products of reagents, rather than as organised structures. The outline of the germinal disc is well rounded, though it is occasionally broken, from the presence of a larger cell in the act of being formed from the yolk.

It is not probable that any great importance is to be attached to the comparative distinctness of the outline of the germinal disc at this stage, which is in a great measure due to a cessation in the formation of fresh cells in the surrounding yolk, and in part to the small and comparatively uniform size of the cells of the germinal disc.

The formation of fresh cells from the yolk nearly comes to an end during this period, but it still continues on a small scale.

The number of the nuclei around the germinal disc has increased.

Another feature of interest which first becomes apparent during this stage is the asymmetry of the germinal disc. If a section were made through the germinal disc, as it lay *in situ* in the egg capsule, parallel or nearly so to the long axis of the capsule, one end of the section would be found to be much thicker than the other. There would in fact be a far larger collection of cells at one extremity of the germinal disc than at the other. The end at which this collection of cells is formed points towards the end of the egg capsule opposite to that near which the yolk is situated. This collection of cells is the first trace of the embryo; and with its appearance the segmentation may be supposed to terminate.

The section I have represented, though not quite parallel to the long axis of the egg, is sufficiently nearly so to shew the greater mass of cells at the embryonic end of the germinal disc.

This very early appearance of a distinction in the germinal disc between the extremity at which the embryo appears and the non-embryonic part of the disc, besides its inherent interest, has a further importance from the fact that in Osseous Fishes a similar occurrence takes place. Oellacher[1] and Götte[2] both agree as to the very early period at which a thickening of one extremity of the blastoderm in Osseous Fishes is formed, which serves to indicate the position at which the embryo will appear. There are many details of development in which Osseous Fish and Elasmobranchs agree, which, although if taken individually are without any great importance, yet serve to show how long even insignificant features in development may be retained.

The segmentation of the Elasmobranch egg presents in most of its features great regularity, and exhibits in its mode of occurrence the closest resemblance to that in other meroblastic vertebrate ova.

There is, nevertheless, one point with reference to which a slight irregularity may be observed. In almost all eggs segmentation commences by, what for convenience may be called, a vertical furrow which is followed by a second vertical furrow at right angles to the first. The third furrow however is a horizontal one, and cuts the other two at right angles. This method of segmentation must be looked on as the normal one, in almost all the important groups of the animal kingdom, both for the so-called holoblastic and meroblastic eggs, and the gradations intermediate between the two. The Frog amongst vertebrates exhibits a most typical instance of this form of segmentation.

In Elasmobranchs the first two furrows are formed in a perfectly normal manner, but though I have not observed the actual formation of the next furrow, yet from the later stages, which I have observed, I conclude that it is parallel to one of the first formed furrows; and it is fairly certain that, not till a considerably later period, is a furrow homologous with the horizontal furrow of the Batrachian egg formed. This furrow appears to be represented in the Elasmobranch segmentation by the irregular circumscription of a body of central smaller spheres from a ring of peripheral larger ones (vide Pl. xv. figs. 3, 4 and 5).

In the Bird the representative of the horizontal furrow

[1] *Zeitschrift für Wiss. Zoologie*, Bd xxiii. 1878.
[2] *Archiv. für Micr. Anat*. Bd ix. 1873.

appears relatively much earlier. It is formed when there are eight segments marked out on the surface of the germinal disc[1]. From Oellacher's[2] account of the segmentation in the fowl[3] it seems certain, as might be anticipated, that this furrow is nearly parallel to the surface of the disc, so that it cuts the earlier formed vertical furrows and causes the segments of the germinal disc to be completely circumscribed below as well as at the surface. In the Elasmobranch egg this is not the case; so that, even after the smaller central segments have become separated from the outer ring of larger ones, none of the segments of the disc are completely circumscribed, and only appear to be so in surface views (vide Pl. xv. fig. 6). Segmentation in the Elasmobranch egg differs in the following particulars from that in the Bird's egg:

(1) The equivalent of the horizontal furrow of the Batrachian egg appears much later than in the Bird.

(2) When it has appeared it travels inwards much more slowly.

As a result of these differences, the segments of the germinal disc of the Birds' eggs are much earlier circumscribed on all sides than those of the Elasmobranch egg.

As might be expected, the segmentation of the Elasmobranch egg resembles in many points that of Osseous Fishes, (vide Oellacher[4] and Klein[5]). It may be noticed, that with Osseous as with Elasmobranch Fishes, the furrow corresponding with the horizontal furrow of the Amphibian's egg does not appear at as early a period as is normal. The third furrow of an Osseous Fish egg is parallel to one of the first formed pair.

In Oellacher's[6] figures, Pl. xxiii. fig. 19—21, peculiar beadings of the sides of the earlier formed furrows are distinctly shown. No mention of these is made in the text, but they are unquestionably similar to those I have described in the Elasmobranch furrows. In the case of Elasmobranchs I

[1] Vide *Elements of Embryology*, p. 23.
[2] *Stricker's Studien*, 1869, Pt. i, Pl. ii. fig. 4.
[3] Unfortunately Professor Oellacher gives no account of the surface appearances of the germinal discs of which he describes the sections. It is therefore uncertain to what period his sections belong.
[4] *Zeitschrift für Wiss. Zool.* Bd. xxii. 1872.
[5] *Monthly Microscopical Journal*, March, 1872.
[6] *Loc. cit.*

pointed out that not only were the sides of the furrow beaded, but that there appeared in the protoplasm, close to the furrows, peculiar vacuole-like cavities, precisely similar to the cavities which were the cause of the beadings of the furrows.

The presence of these seems to show that the molecular cohesion of the protoplasm becomes, as compared with other parts, much diminished in the region where a furrow is about to appear, so that before the protoplasm finally gives way along a particular line to form a furrow, its cohesion is broken at numerous points in this region, and thus a series of vacuole-like spaces is formed.

If this is the true explanation of the formation of these spaces, their presence gives considerable support to the views of Dr Kleinenberg upon the causes of segmentation, so clearly and precisely stated in his monograph upon Hydra.

I have not observed the peculiar threads of protoplasm which Oellacher[1] describes as crossing the commencing segmentation furrows. I have also failed to discover any signs of a concentration of the yolk-spherules, round one or two centres, in the segmentation spheres, similar to that observed by Oellacher in the segmenting eggs of Osseous Fish. The appearances observed by him are probably connected with the behaviour of the nucleus during segmentation, and are related to the curious bodies I have already described.

With reference to the nuclei which Oellacher[2] has described as occurring in the eggs of Osseous Fish during segmentation, there can, I think, be little doubt that they are identical with the peculiar nuclei in the Elasmobranch eggs.

He[3] says:

In an unsegmented germ there occurred at a certain point in the section......a small aggregation of round bodies. I do not feel satisfied whether these aggregations represent one or more nuclei.

Fig. 29 shews such aggregation; by focusing at its optical section eleven unequally large rounded bodies measuring from 0.004 —0.009 *Mm.* may be distinguished. They lay as if in a multi-locular gap in the germ mass, which however they did not quite fill. In each of these bodies there appeared another but far smaller body.

[1] *Loc. cit.*
[2] *Loc. cit.*
[3] *Loc. cit.* p. 410, 411, &c.

These aggregations were distinguished from the germ by an especially beautiful intense violet gold chloride colouration of their elements. The smaller elements contained in the larger were still more intensely coloured than the larger.

He further states that these aggregations equal the segments in number, and that the small bodies within the elements are not always to be seen with the same distinctness.

Oellacher's description as well as his figures of these bodies leaves no doubt in my mind that they are exactly similar bodies to those which I have already spoken of as nuclei, and the characteristic features of which I have shortly mentioned, and shall describe more fully at a later stage. A moderately full description of them is to be found in my preliminary paper[1].

Their division into a series of separate areas each with a deeply-stained body, as well as the staining of the whole of them, exactly corresponds to what I have found. That each is a single nucleus is quite certain, though their knobbed form might occasionally lead to the view of their being divided. This knobbed condition, observed by Oellacher as well as myself, certainly supports the view, that they are in the act of budding off fresh nuclei. Oellacher conceives, that the areas into which these nuclei are divided represent a series of separate bodies—this according to my observations is not the case. Nuclei of the same form have already been described in Nephelis, and are probably not very rare. They pass by insensible gradations into ordinary nuclei with numerous granules.

One marked feature of the segmentation of the Elasmobranch egg is the continuous advance of the process of segmentation into the yolk and the assimilation of this into the germ by the direct formation of fresh segments out of it. Into the significance of this feature I intend to enter fully hereafter; but it is interesting to notice that Oellacher's descriptions point to a similar feature in the segmentation of Osseous Fish. This however consists chiefly in the formation of fresh segments from the lower parts of the germinal disc which in Osseous Fish is more distinctly marked off from the food-yolk than in Elasmobranchs.

[1] *Loc. cit.* p. 415.

I conclude my description of the segmentation by a short account of what other investigators have written about its features in these fishes. So far as I know, the earliest description of this process was given by Leydig[1]. To his description of the germinal disc, I have already done full justice.

In the first stage of segmentation which he observed 20—30 segments were already visible on the surface. In each of these he recognised a nucleus but no nucleolus.

He rightly states that the segments have no membrane, and describes the yolk-spherules which fill them.

The next investigator is Gèrbe[2]. I have unfortunately been unable to refer to this elaborate paper, but I gather from an abstract that M. Gerbe has given a careful description of the external features of segmentation.

Schenk[3] has also made important investigations on the subject. He considers that the ovum is invested with a very delicate membrane. This membrane I have failed to find a trace of, and agree with Leydig[4] in denying its existence. Schenk further found that after impregnation, but before segmentation, the germinal disc divided itself into two layers, an upper and a lower. Between the two a cavity made its appearance which Schenk looks upon as the segmentation cavity. Segmentation commences in the upper of the two layers, but Schenk does not give a precise account of the fate of the lower. I have had no opportunity of investigating the impregnated ovum before the commencement of segmentation, but my observations upon the early stages of this process render it clear that no division of the germinal disc exists subsequently to the commencement of segmentation, and that the cavity discovered by Schenk can have no connection whatever with the segmentation cavity. I am indeed inclined to look upon this cavity as an artificial product. I have myself met with somewhat similar appearances, after the completion

---

[1] *Rochen u. Haie.*

[2] *Recherches sur la segmentation des products adventifs de l'œuf des Plagiostomes et particuliement des Raies.* Robin, *Journal de l'Anatomie et de la Physiologie,* p. 609, 1872.

[3] *Die Eier von Raja quadrimaculata innerhalb der Eileiter. Sitz. der k. Akad. Wien.* Vol. LXXIII. 1873.

[4] *Loc. cit.* My denial of the existence of this membrane naturally applies only to the egg after impregnation, and to the genera Scyllium and Pristiurus.

of segmentation, which were caused by the non-penetration of my hardening reagent beyond a certain point.

Without attempting absolutely to explain the appearances described by Professor Schenk, I think that his observations ought to be repeated, either by himself or some other competent observer.

Several further facts are recorded by Professor Schenk in his interesting paper. He states that immediately after impregnation, the germinal disc presents towards the yolk a strongly convex surface, and that at a later period, but still before the commencement of segmentation, this becomes flattened out. He has further detected amœboid movements in the disc at the same period. As to the changes of the germinal disc during segmentation, his paper contains no facts of importance.

Next in point of time to the paper of Schenk, is my own preliminary account of the development of the Elasmobranch Fishes[1]. In this a large number of the facts here described in full are briefly alluded to.

The last author who has investigated the segmentation in Elasmobranchs, is Dr Alexander Schultz[2]. He merely states that he has observed the segmentation, and confirms Professor Schenk's statements about the amœboid movements of the germinal disc.

[1] Loc. cit.
[2] Die Embryonal Anlage der Selachier. Vorlaufige Mittheilung, Centralblatt f. Med. Wiss. No. 33, 1875.

EXPLANATION OF PLATES XV. AND XVI.

Illustrating Mr Balfour's Paper on the Development of Elasmobranch Fishes.

PLATE XV.

Fig. 1. Section through the germinal disc of a ripe ovarian ovum of the Skate. *gv.* germinal vesicle.

Fig. 2. Surface-view of a germinal disc with two furrows.

Figs. 3, 4, 5. Surface-views of three germinal discs in different stages of segmentation.

Fig. 6. Section through the germinal disc represented in *Fig.* 3. *n.* nucleus; *x.* edge of germinal disc. The engraver has in this figure not accurately copied my original drawings in respect to the structure of the segmentation-furrows.

Figs. 6 *a.* and 6 *b.* Two furrows of the same germinal disc more highly magnified.

Fig. 6 *c.* A nucleus from the same germinal disc highly magnified.

PLATE XVI.

Fig. 7. Section through a germinal disc of the same age as that represented in *Fig.* 4. *n.* nucleus; *nx.* modified nucleus; *nx'.* modified nucleus of the yolk; *f.* furrow appearing in the yolk around the germinal disc.

Figs. 7 *a,* 7 *b,* 7 *c.* Three segments with modified nuclei from the same germinal disc.

Fig. 8. Section through a somewhat older germinal disc. *ep.* epiblast; *n'.* nucleus of the yolk.

Figs. 8 *a,* 8 *b,* 8 *c.* Modified nuclei from the yolk from the same germinal disc.

Fig. 8 *d.* Segment in the act of division from the same germinal disc.

Fig. 9. Section through a germinal disc in which the segmentation is completed. It shews the larger collection of cells at the embryonic end of the germinal disc than at the non-embryonic. *ep.* epiblast.

ON THE NATURE OF THE CRANIOFACIAL APPA-
RATUS OF PETROMYZON, by T. H. HUXLEY, *Sec. R. S.*

IN the first part of the 'Vergleichende Anatomie der Myxinoiden,' published in 1835, Johannes Müller gave an exact and
exhaustive account of the form and arrangement of the various
parts which make up the cartilaginous skeleton of the skull and
face in the Lampreys. He distinguishes a 'Hirn-capsel' or *brain
case;* two 'Gehör-capseln' or *auditory capsules;* a 'Nasen-capsel'
or *olfactory capsule;* and 'Gesichts-knochen' or *facial cartilages.* The latter are, in front, a 'Ringförmige knorpel-stuck' or
annular cartilage; two 'Griffel-förmige Knochen' or *styliform
cartilages,* connected by their anterior ends to the annular cartilage and by their posterior ends giving attachment to lateral
muscles; a 'vordere grosse Mundschild' or *anterior dorsal cartilage;* a 'hintere Mundschild' or *posterior dorsal cartilage;* two
vordere Seiten platten' or *antero-lateral cartilages;* and two
'zweite Seiten platten' or *postero-lateral cartilages.*

The *Brain case* consists of a *basilar plate,* the centre of
which is traversed by the anterior end of the notochord. This
is continued superiorly into the narrow *occipital arch,* which
forms the only cartilaginous part of the roof of the skull; while
laterally it passes into two cartilaginous bars, which bound the
lower lateral regions of the skull. The brain rests upon a
fibrous membrane stretched between these, which may be
termed the *sub-cerebral membrane.* The inner and ventral edge
of each of these *lateral bars* of the skull is continued into a
solid cartilaginous floor, which lies between the naso-palatine
canal and the mucous membrane of the mouth. Müller terms
this the *hard palate.* It terminates behind by an excavated
edge between the *auditory capsules;* while, in front, it is closely
united with the hinder edge of the posterior dorsal cartilage.
The olfactory capsule rests on this cartilage, and is united with
the lateral bars of the skull, and the naso-palatine canal extends backwards, between the sub-cerebral membrane and the
hard palate, to terminate in a cæcal dilatation behind the posterior edge of the latter.

The anterior end of the hard palate is prolonged outwards and downwards on each side into an *anterior lateral process* (Vorder-seiten Fortsatz); this meets at an acute angle with a *posterior lateral process* (Hinter-seiten Fortsatz), which at its dorsal end is closely connected with the auditory capsule. Diverging from the dorsal or attached extremity of this, at an acute angle, is the downwardly directed *styliform process* (stiel-förmige Fortsatz, *i*).   With the ventral extremity of this, the elongated horizontally directed *cornual cartilage* ("Knorpel-platte, am Fortsatz *i* befestigt ; dient zur Befestigung der Zungenmuskeln") is connected.

In the middle ventral line is the long *lingual cartilage* ("knorpeliger Stiel der Zunge"), which is pointed behind, but bifurcates in front, the two short branches supporting the lobes of the tongue.

On the ventral side of this is another, elongated, but much shorter cartilage, the *median ventral cartilage* ("Zungenbein"), the anterior end of which is transversely enlarged, and lies immediately on the ventral side and in front of the anterior end of the lingual cartilage.   The extremities of the anterior end of this cartilage are connected with the antero-lateral cartilages, which again are united by ligamentous fibres with the inner surface of the anterior dorsal cartilage close to its anterior edges.

Müller considers that the annular cartilages answer to the labial cartilages of the Elasmobranchs; that the styliform process and the cornual cartilages, with the median ventral cartilage, correspond with parts of the hyoidean arch in other Vertebrata; that the lateral bars of the skull are the homologues of the palatine bones of the latter ; and that the inverted subocular arch formed by the anterior and posterior lateral processes corresponds with the "temporale, tympanicum, · jugale, transversum, pterygoideum (†) of Cuvier" in Osseous Fishes (l. c. p. 163).   He is further of opinion that the posterior dorsal cartilage has nothing to do with the skull, and that, together with the anterior dorsal cartilage and the lateral cartilages, it forms a series of structures special to the Lampreys and not represented in other Vertebrates (l. c. p. 164).

Agassiz, in the "Recherches sur les Poissons fossiles" (Tome 1,

1835—43), availing himself of Müller's description and figures, and of the embryological investigations of Vogt, makes a most important rectification in the nomenclature of the parts.

·   In describing the skull of *Ammocœtes branchialis* (p. 113) he justly states that it corresponds with that of the embryo at the moment of the first appearance of cartilage. "The point of the notochord advances freely into the space comprised between the 'anses latérales' absolutely as in the embryo of Salmonidæ."

In *Petromyzon* he recognizes the "anses latérales" in the lateral cranial bars described by Müller and regarded by him as palatines. The subocular arch is the "*arc ptérygoïdien*," which, in the Cyclostomes, as in embryos, is not yet separated from the cranium by articular faces[1]. The styliform cartilage is considered to be the hyoid, while the other cartilages are reckoned as labial.

In 1844—6, Professor Owen[2] gave the following account of the structure of the lamprey's skull.

"In the lamprey (*Petromyzon*, fig. 26) the occipital cartilage is continued backwards in the form of two slender processes (*o*) upon the under part of the chorda dorsalis (*ch*) into the cervical region. The hypophysial space (*hy*) in front of the occipital cartilage remains permanently open, but has been converted into the posterior aperture of the naso-palatine canal. The sphenoidal arches (5) are very short and approximated towards the middle line and the presphenoid and vomerine cartilage (13) is brought back closer to the sphenoidal arches. Two cartilaginous arches (24) circumscribe elliptical spaces outside the presphenoid plate: these appear to represent the pterygoid arches, but, as in the embryo of higher fishes, are not separated from the base of the skull by distinct joints. The basal cartilages after forming the ear capsules (16) extend upwards upon the sides of the cranium (fig. 11), arch over its back part, and leave only its upper and middle part membranous, as in the human embryo, when ossification of the cranium commences. Two broad cartilages (*ib.* 20, 21) may represent upon the roof of the infundibular suctorial mouth the palatine and maxillary bones, and anterior to these there is a labial cartilage (*ib.* 22). There are likewise cartilaginous processes (*ib. r. s.*) for the support of the large dentigerous tongue and the attachment of its muscles." *l. c.* pp. 72, 3.

It appears from the context (p. 71) that by "sphenoidal arches" Professor Owen means the *trabeculæ cranii* of Rathke,

[1] l. c. p. 114. At p. 132, what is here termed "*arc ptérygoïdien*," is named "*arc palatin.*"
[2] Lectures on the Comparative Anatomy and Physiology of the Vertebrate Animals, 1846.

as he applies the term to the *anses latérales* described by Vogt in *Coregonus*, which are nothing but these *trabeculæ*, and the homology of which with the lateral bars of the cranium in the lampreys, had already been pointed out by Agassiz.

In 1858 I stated that "the skull of the lamprey is readily reducible to the same plan of structure as that which is exhibited by the tadpole, while its gills are still external and its blood colourless[1]," having been led to that conviction by a careful study of the early stages of development of the frog's skull.

In 1863[2] I expressed this view more fully, and compared the margin of the oval space upon the base of the skull to the divergent *trabeculæ cranii*, as Agassiz had originally done, and the posterior dorsal plate to the ethmo-vomerine cartilage. I expressed the opinion that the inverted cartilaginous arch which gives attachment to the hyoidean and mandibular apparatuses of a tadpole is "strictly comparable" to the subocular arch of the lamprey; but I added a doubt "whether the accessory buccal cartilages can be strictly compared to anything in other fishes, though some of them are doubtless, as Müller has suggested, the analogues of labial cartilages." I gave a figure (Fig. 75) shewing the true relation of the skeleton to the enclosed soft parts in *Petromyzon marinus*.

In his masterly monograph upon the cephalic skeleton of the Selachians[3], Gegenbaur explains that he has taken the skull in the Selachians and not that of the more lowly organized Cyclostome for the starting-point of his investigations, because the latter are in many respects abnormal and so much less directly affiliated with the other Vertebrata, that it is not wise to attempt to begin with them. He approves of the sharp line of demarcation which Haeckel has drawn between the Lampreys and Hags, as *Monorrhina*, on the one hand, and all the higher Vertebrata, as *Amphirrhina*, on the other; and though he admits the possible correctness of the interpretation of some of the parts which I have given, he doubts whether it really tends to bridge over the hiatus.

[1] Croonian Lecture, *Proceedings of the Royal Society*, 1858.
[2] Lectures on the Elements of Comparative Anatomy, on the Classification of Animals, and on the Vertebrate Skull, 1864, p. 194.
[3] *Das Kopfskelet der Selachier*, 1872.

"Da mir das Knorpelcranium der Selachier zum Ausgangspunkt
so wichtig erschien, weil es einen tiefer stehenden Zustand repräsen-
tirt, bedarf es noch der Rechtfertigung wegen des Ausschlusses der
Cyclostomen, die gleichfalls mit einem Knorpelskelete versehen und
zudem noch in der ganzen übrigen Organisation eine tiefere Stufe
einnehmend, jenen Anförderungen noch besser hätten entsprechen
müssen. Darauf kann erwiedert werden, dass sowohl in dem Cranium
wie in vielen Punkten ihrer übrigen Organisation die Cyclostomen
bedeutend abweichende Verhältnisse darbieten und keinen so directen
Anschluss an die übrigen Wirbelthiere bieten. Sie würden daher
von Haeckel mit allem Recht als Monorrhina den Amphirrhinen
gegenüber gestellt. Die darin ausgesprochene Auffassung kann
kaum schärfer präcisirt werden. Von so abweichenden keine
stricten Vergleichungen zulassenden Formen auszugehen, wäre kein
glücklicher Gedanke. Wenn auch die Deutungen, welche Huxley
einzelnen Theilen des Craniums gab, dasselbe dem Cranium der
Amphirrhina näher gerückt scheint, so besteht darüber doch keines-
wegs Sicherheit. Die gewiss vorhandene Verbindung mit den Amphirrhinen mag noch zu weisen sein, aber die Entfernung, welche
zwischen diesen und den Cyclostomen liegt, wird dadurch nicht
vermindert." [1]

In describing the skull of *Menobranchus lateralis*[2], I have
remarked that

"No known Elasmobranch, Ganoid, or Teleostean fish presents so
incompletely developed a chondrocranium as that of *Menobranchus*.
On the other hand, the latter is much like that of a Lamprey if we
leave the ossification of the *Menobranchus* skull and the accessory
cartilages of the *Petromyzon* out of consideration. And this fact,
taken together with the curious resemblance in development between
the Lampreys and the Amphibia (which are much closer than those
between any of the higher Fishes and the Amphibia[3]) suggest to my
mind the supposition that, in the series of modifications by which the
Marsipobranch type has been converted into that of the higher fishes,
the most important terms must have been forms intermediate in
character between the *Dipnoi* and the Marsipobranchii." (*l. c.* pp.
197, 8[4]).

Finally, in a 'Preliminary Note upon the Brain and Skull of
Amphioxus lanceolatus,' read before the Royal Society last year,

[1] Gegenbaur, *Das Kopfskelet der Selachier*, Einleitung, p. 9, 1872.
[2] *Proceedings of the Zoological Society of London*, 1874.
[3] Unfortunately we know nothing of the development of the *Dipnoi*.
[4] I have not yet had time to study Goette's large and elaborate work on the
development of Bombinator igneus ("Entwickelungsgeschichte der Unke"), published this year, with the attention it deserves, but I notice at p. 692, the remark,
that "der vollendete Zustand des Kopfes, wenigstens der Neunaugen ganz
entschieden auf die Anuren-larven hinweist." I am further glad to find that
Dr Goette takes the same view as I have done respecting the relations of the
Anura with the Cyclostomes (p. 744).

I compared the skull of the Lamprey in its *Ammocœtes* stage
with that of *Amphioxus*, with the view of shewing (1) that
numerous anterior proto-vertebræ, answering to those which
in the higher Vertebrata give rise to vertebræ (among other
products of their metamorphosis), but which neither in the
head, nor in any other part of the body of *Amphioxus* develope
vertebræ, correspond with the region in which the chondrocranium is developed in the Ammocœte, and (2) that neither
in the Ammocœte do these proto-vertebræ give rise to vertebræ,
but that the brain-case, as in *Menobranchus*, is formed partly
by the parachordal cartilages (which are chondrifications of the
investing mass of the notochord, comparable to that which
precedes the development of the ·vertebræ in the spinal column
of the Frog); and partly by the trabeculæ which, in my view,
are homologous with branchial arches[1]. In the Croonian
Lecture to which I have already referred, the following passage occurs :—

" The cranium never becomes segmented into somatomes: distinct
centra and intercentra like those of the spinal column are never
developed in it.   Much of the basis cranii lies beyond the notochord.
In the process of ossification there is a certain analogy between the
spinal column and the cranium, but that analogy becomes weaker
and weaker as we proceed towards the anterior end of the skull.
    Thus it may be right to say that there is a primitive-identity of
structure between the spinal or vertebral column and the skull; but
it is no more true that the adult skull is a modified vertebral column,

---

[1] Goette has put forward a very different view of the nature of the tra-
beculae (l. c. p. 629).   "The original foundation of the whole skull consists
firstly, of the posterior basis cranii, a cartilaginous plate, which incloses the
notochord, in which I distinguish, as in the trunk, an axial part, the notochord
with its external sheath, and lateral plates homologous with the arches of the
vertebræ."   (This answers to what I have named the parachordal portion of the
skull.)   "To this are added the two pair of arches, which, as continuations of
these lateral plates, at their anterior and posterior ends, embrace laterally, and
eventually arch over the anatomical base of the fore-brain and a part of the
hind-brain.   This anterior pair of arches belongs therefore to the first segment
of the body under discussion; it forms the first vertebral arch, which in agreement with the general position of this segment is horizontal.   But it has no
corresponding centrum—inasmuch as the axial structure by which the latter
should be produced, the notochord, was drawn back from the fore-head (Vorder-
kopf) to the anterior margin of the hind-head."   According to this interpretation, which is worthy of serious consideration, though I entertain grave doubts
whether it can be sustained, the trabeculæ represent not the most anterior
pair of visceral arches, as I have supposed, but the most anterior pair of neural
arches.   How this view is to be reconciled with the relations of the trabeculæ
to the trigeminal nerve and to the organs of the higher senses, is not clear
to me.

than it would be to affirm that the vertebral column is a modified skull."

The immense extension of our knowledge of the minute details of the structure and development of the vertebrate skull in the seventeen years that have elapsed since those words were written, and which is largely due to the investigations of Mr Parker, whose elaborate and numerous contributions to this difficult branch of anatomy seem hardly to be as well known as they ought to be abroad[1], has, so far as I know, revealed no fact inconsistent with their fullest signification.

The segments of the *Amphioxus* head are not vertebræ, and in the Lamprey, as in every known vertebrate animal, the cranial region takes on the characters of a skull without passing through any stage of vertebration[2].

At the end of my brief paper I stated that I proposed at some future time to shew "in what manner the skull of the Marsipobranch is related to that of the higher vertebrata, and more especially to the skull of the frog in its young tadpole state."

For this purpose it was needful to go over the structure of the lamprey's skull afresh, and unfortunately my supply of these fish was small, and I have been unable to procure fresh ones until recently. I have examined them by the ordinary way of dissection, and by making transverse sections which form admirable Canada-balsam preparations.

The dorsal wall of the circular lip which surrounds the mouth of the lamprey (Pl. XVII. Fig. 1) is longer than the ventral wall, and hence has the form of a hemispherical bell, set obliquely on to the body. On the ventral side, the lip is separated by a deep transverse constriction from the rest of the head, but this constriction dies away laterally, and hardly any trace of it remains on the dorsal aspect. The inner surface

---

[1] For example, I do not find Mr Parker's name in the long 'Autoren Verzeichniss' appended to Goette's work, though the memoir *On the Structure and Development of the Skull of the Common Frog*, the best piece of work of its kind which has appeared since Dugès' *Recherches*, was published in 1871.

[2] It is to be hoped that this statement will prevent persons of even the largest powers of misunderstanding from imagining that the demonstration of the multi-segmentation of the head of *Amphioxus*, is a relapse on my part into archetypal fancies such as those of which I endeavoured to shew the futility a score of years ago.

of the lip is beset with the well-known horny teeth, and its margins are provided with numerous short lamellar papillæ, of which those on the ventral side are the longest and largest.

The opening into the buccal chamber, in the posterior and dorsal region of the bell-shaped cavity of the lip, is small. Immediately below it is a projection (Pl. XVII. Fig. 1, *o*) convex forwards, flat or concave on its dorsal aspect, and having a horny envelope, the edge of which is produced into a number of denticulations, the longest of which is median. I shall term this the *mandibular tooth*. On the ventral side of it is a transverse fold which bears two papillary eminences. The *annular cartilage* (*a*) is lodged in the posterior part of the lip, and its ventral part lies just below this fold.

The *inferior median cartilage* (Pl. XVII. Fig. 1, *l*) lies beneath the integument, and separated from it by a large subcutaneous sinus[1], amidst the ventral muscles. The anterior edge of the cross-piece in which it terminates is connected by fibrous tissue with the support of the mandibular tooth. Behind the latter, the tongue (*p*) rises from the floor of the buccal cavity and nearly closes it. It is divided by a deep longitudinal groove into two lobes, united for a short distance in front. The opposed surfaces of these lobes present minute horny denticulations arranged in a curved anteroposterior series.

The *buccal cavity* itself has the form of an elongated tube with delicate and transparent walls. Its roof is folded longitudinally so as to form a groove (Pl. XVIII. Fig. 1, *g*) which is much deeper in front than behind. A small papillose elevation (*t*) bounds the anterior end of this groove, and for a short distance from its commencement its sidewalls are obliquely folded. This buccal portion of the alimentary canal terminates behind by dividing into two tubes, one of which lies on the dorsal side of the other, both occupying the median plane. The upper tube (Pl. XVII. Fig. 1, *œ*) is the very slender *œsophagus* which traverses the whole length of the branchial region to pass into the gastrointestinal division of the alimentary canal. The lower, much larger tube is the so-called 'respiratory bronchus,' or *branchial*

---

[1] This communicates with the system of cavities described as Lymphatic by Langerhans. It contained blood in the specimens I examined. See Milne-Edwards' "Leçons," III. p. 369.

canal (*Br.*). The wall of the buccal cavity between these two tubes is produced into a sort of horizontal shelf, the free edge of which is directed forwards and is divided into five tentacular processes, of which the median is the shortest (Pl. XVII. Fig. 1, *f*; Pl. XVIII. Fig. 1, *u*). These overhang the entrance of the branchial canal, and doubtless serve to prevent the entrance of solid particles into it. Two small, flat pieces of cartilage which are wide in front and narrow behind, and are similar in colour to the other cartilages, support the horizontal shelf from which the tentacles spring. They diverge outwards and backwards towards the styliform processes. The axis of each tentacle and the middle of the 'shelf' between the two cartilages just described, are occupied by a colourless cartilaginous tissue. Behind and below this tentaculated shelf the entrance of the branchial canal is further protected by two folds of the lining membrane (Pl. XVII. Fig. 1, *r*; Pl. XVIII. Fig. 1, *v l*), the free edges of which are directed backwards and towards one another. The dorsal half of each of these valves is nearly vertical; the ventral half slopes backwards until it becomes nearly horizontal. These valvular folds constitute the *pharyngeal velum*, and are, doubtless, the metamorphosed velum of the Ammocœte. They must readily allow of the passage of water into the branchial canal, but must obstruct its exit[1]. Behind these, is another smaller and much more delicate pair of valvular folds (Pl. XVIII. Fig. 1, *vl'*), which when they flap back, cover the first internal branchial aperture. There is a depression behind each of the pharyngeal vela, and a bristle could sometimes be passed through the wall into a small space outside it. This I conceive to be the remains of the hyoidean cleft which opens externally in the Ammocœte[2].

When the lining wall of the buccal cavity is removed, two large muscles are seen to lie between it and the lateral skeletal parts. The upper is attached by a long tendon to the middle of the hinder edge of the anterior dorsal cartilage, on each side of the papillose elevation at the anterior end of the dorsal groove already described. The lower, fleshy throughout

---

[1] See Stannius, *Handbuch der Zootomie*, I. p. 240, and Rathke's description of these parts in his *Bemerkungen ueber den innern Bau der Pricke*, 1825.
[2] *Proceedings of the Royal Society*, 1875, p. 128, Fig. D. 1.

its length, is inserted into the tongue. Each of these muscles arises from the styliform process, the position of which exactly corresponds with that of the outer attached edge of the velum. Immediately subjacent to the ventral wall of the buccal cavity lies the strong aponeurotic sheath of the lingual muscles with the lingual cartilage (Pl. XVII. Fig. 1, *k*), which they ensheath. The two *cornual cartilages* (*i*) are imbedded in this sheath, their long edges being close to, and nearly parallel with, one another (Pl. XVIII. fig. 3).

Thus it is obvious that the *lingual cartilage* has the same relations as the median ventral element of the hyoidean arch in the higher Vertebrata, and that the *cornual cartilages* and the *styliform processes* represent lateral elements of the same arch. Müller attached the same signification to the cornual cartilages, but not to the lingual cartilage.

In this case what is the *median ventral cartilage* which Müller regarded as the body of the hyoid? The hyoidean arch is complete without it and has no special connexion with it, the bent up anterior end of the lingual cartilage simply playing over it. I conceive it to be a median ventral element of the mandibular arch; notwithstanding that, in the higher vertebrates, such an element, though the analogy of the other arches would lead us to expect its presence, is not known to occur.

The third division of the trigeminal nerve passes over the expanded anterior end of this cartilage, traverses the ventral half of the annular cartilage, and runs along the anterior edge of the latter to its dorsal extremity[1]. Thus although its halves are united dorsally, the annular cartilage would seem to be essentially a post-oral structure. The inverted subocular arch formed by the posterior (*f*) and anterior (*f'*) lateral processes lies at the sides of the posterior part of the buccal cavity. The second and third divisions of the trigeminal nerve (Pl. XVIII. fig. 1, V^2, V^3) perforate the membrane which connects them, the third running obliquely downwards and forwards to its distribution (Pl. XVII. fig. 1); the second turning outwards and passing to the sides of the head.

[1] See Born, *Ueber den inneren Bau der Lamprete*, Heusinger's *Zeitschrift*, 1827, Tab. VI. fig. 7.

The posterior lateral process therefore answers, in all essential respects, to the suspensorial cartilage or proximal division of the mandibular arch, though possibly the dorsal end of the hyoidean arch may be united with it.

Having proceeded thus far, the further study of the cranio-facial apparatus of the Lampreys will be facilitated by comparison with that of the tadpole (of *Rana temporaria*) in the stage in which the right opercular cleft is closed, while the hind-limbs are still in the condition of mere buds (Pl. XVII. fig. 2; Pl. XVIII. figs. 2, 4 and 6).

Dugès has given an admirably clear and accurate account of the structure of the skulls of the tadpoles of *Pelobates fuscus*, *Hyla viridis*, and *Rana esculenta* in this stage; and it has subsequently been treated of by Reichert, by myself and by Mr Parker (*Phil. Trans.*, 1871). Figures of a somewhat more advanced condition of the skull of *Bombinator igneus* are given by Goette in the work already cited.

The skull of the tadpole, at this stage, consists of a cartilaginous parachordal basilar plate, the middle of which contains the anterior end of the notochord. On each side are the auditory capsules, while, in front, the basilar plate extends forwards on each side of the pituitary space to form the trabeculæ. These unite in front, but speedily diverge again to terminate abruptly, close to the upper labial cartilages (Pl. XVIII. fig., 2 *U. lb*). The nasal sacs are situated on the dorsal aspect of the head, one on each side of the origin of these ethmoidal processes of the skull (*e*), and they open directly into the anterior part of the buccal cavity. On comparing this skull with that of the young Ammocœte (Pl. XVIII. fig. 5) the justice of the comparison instituted by Agassiz, between the 'anses' of the latter (*Tr*) and the trabeculæ of vertebrate embryos in general becomes manifest. Only, in the Ammocœte, there is, as yet, nothing answering to the ethmoidal processes of the Frog's skull.

In the young Ammocœte again the median nasal sac (*Na*) is conical and its brief ventral prolongation merely overlies the commissure of the trabeculæ. If the sac could be divided into two by a median constriction it would answer to the nasal sacs of the Frog before they communicate with the buccal cavity. The only representative of any part of the subocular arch of the adult, in the Ammocœte stage of the Lamprey, is a slender cartilaginous process (g) which corresponds with the dorsal end of the posterior lateral process of the adult[1].

It is obvious from the position of the lateral bars of the cranium in the Lamprey that they are, as Agassiz determined them to be, homologous with the 'anses' of the Ammocœte and therefore are modified trabeculæ—while the 'hard palate' is a chondrification of the tissue which lies between them, corresponding with that process of chondrification by which the floor of the skull in the Frog becomes completely converted into cartilage.

Nor in my judgment, can it be doubted that the posterior dorsal cartilage of the Lamprey answers to the ethmoidal processes of the tadpole's skull, the interspace between them being similarly chondrified.

Again, at the sides of the Frog's skull there is a subocular arch (Pl. XVIII. fig. 2, p, f, g), the posterior limb of which (g) is in all respects comparable to the corresponding part of the Lamprey's subocular arch; except so far as the styliform cartilage of the Lamprey may possibly represent the upper end of the hyoidean arch. For in the Frog, as I pointed out in my paper on *Menobranchus*, the hyoidean arch is simply articulated with the suspensorium and does not coalesce with it.

I have formerly assumed that the anterior pillar of the subocular arch in the tadpole (Pl. XVII. fig. 2; Pl. XVIII. fig. 2, p.) answers to the anterior lateral process in the Lamprey, to which it is indeed extraordinarily similar. But further consideration shews that there is a difficulty in the identification of the two. Both the second and the third divisions of the trigeminal nerve pass through the subocular membrane, and therefore on the ventral side of the arch; whereas in the Frog, as in all other Vertebrata, they run on the dorsal aspect of the palatine arcade. This is a singular anomaly (which occurs also in the Myxinoids), and it leads to the suspicion that the anterior lateral process

---

[1] When this passage was written I had not seen the valuable paper of Langerhans, *Untersuchungen über Petromyzon Planeri*, Freiburg, 1873. I find that he has described and figured the process here mentioned, as well as the cartilaginous olfactory capsule known to Rathke, but overlooked by Müller.

may be represented, not by the palatine process of the suspensorium of the tadpole (*p.* fig. 2), but by the orbital process, which, as is well known, arches over the jaw-muscles and nerves until it reaches the skull, with which it becomes united by fibrous tissue.

The otic process of the tadpole's suspensorium has no representative in the Lamprey, and therefore the posterior division of the seventh nerve, which takes the same course as in the Frog, does not pass through a foramen.

The oral aperture of the tadpole, in this stage, is surrounded by a deep, fringe-like, transversely-oval lip, produced on each side into a fold (Pl. XVIII. fig. 6). Dorsally, this lip, as in the Lamprey, passes evenly into the integument of the head; ventrally, it is marked off from the rest of the integument, just as in the Lamprey, by a deep transverse constriction. The inner surface of the lip is raised into ridge-like linear elevations, the free edges of which are beset with the singular spoon-shaped serrated hooks, which result from the modification of the epithelium.

There are four rows of these denticles in the lower division of the lip, and two in the upper. Behind these come the horny jaws, which are structures of a very similar nature, moulded upon the edges of two pairs of labial cartilages—an upper and a lower (*U.lb; L,lb*).

Each pair of these labial cartilages are so closely united in the middle line in this stage that it is not always easy to discern the traces of their primitive distinctness. The upper pair overlap the lower, which last form a half circle. In *Rana temporaria* the second small upper labial cartilage attached to the outer angle of each of the principal pair, which is described and figured by Dugès in *Pelobates,* appears to be absent.

The angles of the upper and lower labial cartilages are united by fibrous tissue. The outer part of the dorsal face of each lower labial cartilage articulates with the outer end of the posterior face of the short Meckelian cartilages (*Mk*). These are therefore separated by a considerable interval, occupied by the floor of the mouth (Pl. XVIII. fig. 4). Their proximal ends articulate with the ends of the suspensoria, and the long axis of each Meckelian cartilage is inclined downwards and backwards. These cartilages therefore lie at the sides of, as well as beneath the buccal cavity (Pl. XVII. fig. 2).

The larger hyoidean cornu (Pl. XVIII. fig. 4, Hy^1), articulated with the posterior face of the suspensorium by a part of its anterior edge, ends dorsally in a free point; ventrally, it narrows and passes into the anterior, transversely expanded, end of a median cartilage (Hy^2), which tapers posteriorly, and is received between the coalesced ventral ends of the branchial arches. Of these there are four. The ventral part of the most anterior is still distinctly marked off from the coalesced ventral ends of the three posterior arches. The dorsal moieties of the branchial arches, when they are separated by the branchial clefts, are bent so as to be strongly convex outwards and concave inwards.

The cornu of the hyoid in the tadpole obviously answers to the *cornual cartilages* in the Lamprey. The median cartilage, the anterior expanded end of which raises the mucous membrane of the anterior part of the floor of the mouth into a rudimentary tongue, no less closely resembles the *lingual cartilage* of *Petromyzon*, while the branchial arches represent the four anterior branchial arches of the Lamprey. Not only so, but in the present stage, the branchiæ of the tadpole are, as is well known, pouches, which present no merely superficial likeness to the branchial sacs of the Lamprey. A septum extends inwards from the concave face of each branchial arch, and the septa of the two middle arches (Pl. XVIII. fig. 4) terminate in free edges in the branchial dilatation of the pharynx. Vascular branchial tufts beset the whole convex outer edge of the branchial arch, and are continued inwards in parallel transverse series of elevations, which become smaller and smaller towards the free edge of each septum, near which they cease[1].

In the young Ammocœte the septa of the branchial chambers similarly bear vascular processes, which are first developed close to the external branchial aperture, and thence extend inwards transversely[2].

The recesses at the sides of the floor of the pharynx into

---

[1] Dugès (*l. c.* p. 97) has carefully described the branchiæ of *Pelobates*.

[2] If these first-formed long branchial filaments of the Ammocœte projected through the small gill-clefts outwards instead of inwards, they would resemble the first-formed 'external gills' of Elasmobranchs. And this difference of direction seems to indicate the solution of the difficulty, that external gills, which are so generally developed at first in *Elasmobranchii*, *Ganoidei* and *Dipnoi*, are apparently wanting in *Marsipobranchii*.

which the interseptal clefts, or internal branchial clefts open, answer, taken together, to the branchial canal of the Lamprey, which is not shut off from the œsophagus in the Ammocœte. The anterior boundary of each of these recesses is marked by a fold of the mucous membrane, the free edge of which projects backwards and is produced into papilliform angulations so as to appear scalloped (Pl. XVIII. fig. 4). The anterior face of this fold is convex, its posterior face is concave. The inner angle of each fold passes into its fellow by a ridge, produced into one or two papillæ, which is closely adherent to the median part of the floor of the mouth. The outer angle is continued into a more delicate fold of the mucous membrane lining the roof of the mouth, the free edge of which also projects backwards. It is plain that these structures answer to the pharyngeal velum of the Lamprey[1].

Thus I think there can be no doubt that the cornua of the hyoid in the Frog, and the median cartilage which connects them, are the homologues of the cornual cartilages and the lingual cartilage of the Lamprey. Whether the styliform process of the Lamprey is really the upper end of the hyoidean arch, or whether it simply answers to the part of the mandibular arch of the tadpole which articulates with the hyoidean cornu elongated into a process, is more than I can, at present, venture to decide. The analogy of the frog and of *Chimæra* however is against the hyoidean nature of the styliform process. The lower labial cartilages in the tadpole (*l. lb*) occupy just the same position in the lip, in front of the ventral constriction, as the ventral half of the annular cartilage does in the Lamprey. Considering that the corresponding structure in *Amphioxus* is an incomplete ring open on the dorsal median line; and considering, further, the distribution of the third

---

[1] Some very singular tentacular structures are arranged in definite order in the roof and on the floor of the tadpole's mouth. Three or four are situated in a transverse row upon the rudimentary tongue: two over the junction of Meckel's cartilage with the lower labial cartilage: one large one immediately behind the inner or posterior nostril. Between these the roof of the mouth presents a triangular papillose elevation with its apex directed backwards—which is comparable to the papilla at the anterior end of the dorsal buccal groove in *Petromyzon*—and two parallel rows, one on each side of the roof, and one on each side of the floor of the mouth. The extremities, and sometimes the sides of these tentacula, are more or less papillose, and the central axis is in structure very similar to developing cartilage.

division of the trigeminal nerve, I incline to think that the annular cartilage of the Lamprey represents only the lower labial cartilages of the Frog. A knowledge of the development of the ring in question would decide this point, but I have not yet been able to obtain young Lampreys in which the buccal cartilages are just making their appearance. If the annular cartilage of the Lamprey answers to the lower labial cartilages of the Frog, then the upper labial cartilages will correspond in form and position to the anterior dorsal cartilage, and the small antero-lateral cartilages will perhaps have a parallel in the upper 'adrostral' cartilages in *Pelobates*.

The *posterior lateral cartilages* are directly connected with that end of the suborbital arch which answers to the articular end of the suspensorium in the frog (Pls. XVII., XVIII. figs. 1 and 2), and, in their position, exaggerate the peculiar arrangement of the tadpole's Meckelian cartilage. That they are parts of the mandibular arch I believe to be certain, but in the absence of any knowledge of their mode of development, I leave the question as to their exact homology open. Finally, the median ventral cartilage appears to have no representative in the tadpole; and, as I have already said, I take it to be an inferior median piece of the mandibular arch—not represented, so far as our present knowledge goes, in the higher vertebrata.

Thus the craniofacial apparatus of the Lamprey can be reduced to the same type as that of the higher Vertebrata, by means of the intermediate terms afforded by the Tadpole's skull; and there appears to me to be no sufficient foundation, in the present state of knowledge, for regarding the Marsipobranch skull as one which departs in any important respect from the general vertebrate type.

To what extent all the identifications here made will stand the test of the study of the development of the Lamprey's facial cartilages remains to be seen; but the only doubt which exists in my mind is with regard to the anterior dorsal and the postero-lateral cartilages. If the annular cartilage is developed by the confluence of primitively distinct upper and lower labial cartilages, the homologues of the anterior dorsal cartilage will have to be sought in some of the anomalous palatal cartilages of the Rays; among which it might not be impossible to find representatives of the postero-lateral cartilages. But that the parts of the face of the Lamprey present no structures, which are not to be found in one shape or another among the higher Vertebrata, appears to me to be clear.

In the Myxinoid *Marsipobranchii* there is even less difficulty in reducing the skull to the ordinary vertebrate type. Three pairs of cartilaginous rods here spring from the anterior end of the parachordal region, one pair passing forwards as the trabeculæ, and two curving backwards and to the ventral side as the mandibular and hyoidean arches respectively. The two latter have swung backwards until they take a position unlike that which they have in the Tadpole and the Lamprey, and like that which they have in the adult Frog. Not only in this respect, but in the structure of the circulatory and respiratory apparatuses the Myxinoid fishes exhibit a higher stage of organization than the Lampreys. But, at present, I can only indicate the outlines of a comparison which requires fuller discussion than can be given to it on the present occasion.

In conclusion, I will only advert to the singular resemblance in structure and mode of working between the tongue of the Marsipobranch and the odontophore of a Mollusk, as a point worthy of attention.

DESCRIPTION OF THE PLATES.

PLATE XVII

Fig. 1. Vertical and longitudinal section of the anterior part of the body of Lamprey (*P. fluviatilis*) × 3.   *a.* annular cartilage ;   *b.* anterior dorsal cartilage ;   *c.* antero-lateral ;   *d.* postero-lateral cartilage ;   *e.* posterior dorsal cartilage ;   *e'.* hinder margin of the hard palate ;   *f.* anterior lateral process ;   *g.* posterior lateral process ;   *f'.* angle formed by the junction of these ;   *h.* styliform process ;   *i.* cornual cartilage ;   *k.* lingual cartilage ;   *l.* median ventral cartilage ;   *m.* occipital arch ;   *n.* posterior wall of the nasal capsule ;   *o.* lip-like fold with two papillæ ;   *p.* tongue ;   *q.* tentaculate branchial valve ;   *r.* pharyngeal velum. *N.* nasal aperture ;   *Ol.* olfactory sac ;   *Hm.* the cerebral hemispheres ; *M.* midbrain ;   *Cs.* cerebellum ;   *Md.* medulla oblongata ;   *My.* myelon ; *Ch.* notochord ;   *Oc.* nasal canal.
Fig. 2. Vertical and longitudinal section of the anterior part of the body of a Tadpole of *Rana temporaria* (× 10) *Ch., My., Md., Cc., Ms., Hm., œ., N.* as before ;   *N'.* posterior nasal aperture ;   *Mk.* Meckel's cartilage ;   *U.l b, L.l b.* upper and lower labial cartilages ;   *e.* eth-

moidal cartilage;   *f'.* angle of junction of the palatine cartilage, *p.* with the suspensorium, *g.* and the articular surface for Meckel's cartilage;   *k.* median cartilage connecting the hyoid and branchial arches ;   *Br. Vl. Vl.* pharyngeal velum ;   *H.* heart.

PLATE XVIII.

Fig. 1.   The anterior part of the roof of the branchial canal and of the buccal cavity of a Lamprey (× 3) ;   *b.* The two anterior branchial openings ; B. The right anterior branchial sac ;   *x.* a bristle passed into the opening behind the right half of the *velum vl.*;   *vl².* the second valvular fold in front of the first branchial aperture, more distinctly shewn than in nature;   *u.* the 'shelf' with its tentacles;   *g.* the median groove ;   *t.* its anterior termination ;   *e.* the anterior end of the posterior dorsal cartilage ;   *v²v³.* the second and third divisions of the trigeminal nerve ; *Au.* the auditory capsules.

Fig. 2.   The roof of the mouth of a Tadpole of *R. temporaria* (× 10) *Ch., Au.* as before—*vl.* the superior vela ;   *t.* a median large triangular papilla, between the two which lie behind the posterior nares ;   *Tr.* trabeculæ ;   *g.f.p.* the sub-ocular arch ;   *e.* the ethmoidal processes ; *U. lb.* the upper labial cartilage.

Fig. 3.   The lingual cartilage *Hy²*, the cornual cartilages *Hy'* and the styliform processes *Hy* (?) of a Lamprey magnified three times;   *Lt.* the mandibular tooth ;   *l.* the anterior end of the median ventral cartilage, the posterior prolongation of which is supposed to be seen through the lingual cartilage ;   *Vl.* the lower halves of the pharyngeal vela—between which a small portion of the mucous membrane of the floor of the mouth is seen.   Beneath this is the fibrous aponeurosis of the muscular sheath of the tongue and the tendon of the long retractor muscle.

Fig. 4.   The floor of the mouth of the Tadpole of *Rana temporaria;* *Llb.* lower labial cartilage ;   *Mk.* Meckel's cartilage;   *Hy'.* hyoidean cornu ;   *Hy³.* median inferior piece of the hyoid ;   *Vl.* inferior velum ; *s.* 1. 2. 3. 4. walls of the branchial sacs.

Fig. 5.   The skull of a young Ammocœte, or larval Petromyzon, with the brain in situ ;   *Ch.* the notochord;   *Au.* the auditory capsules ; *g.* the lateral process ;   *Tr.* the trabeculæ ;   *w.* the lateral walls of the cranium in the region of the cerebral hemispheres ;   *Na.* the nasal opening embraced by the crescentic cartilage.

Fig. 6.   Lips and horny upper jaw of the Tadpole (× 10).

THE SECONDARY ARCHES OF THE FOOT.

By S. Messenger Bradley, F.R.C.S.

The construction and mechanism of the three principal arches of the foot, viz the antero-posterior vertical, the transverse vertical, and the horizontal arches, have been carefully described by Ward, who also gives an interesting account of the peculiar features of the various intertarsal joints. He draws attention chiefly, 1st, to the great strength of the short posterior pier of the antero-posterior arch reaching from the ankle to the heel for the purposes of support, and to the much greater elasticity of the anterior pier of the same arch, which is chiefly concerned in distributing concussion ; 2nd, to the fact that in the rotation of the os calcis beneath the astragalus, the body of the calcaneum advances a quarter of an inch and descends one-eighth of an inch, while the lesser process ascends, so that in thus rotating the outer side of the foot advances and the toes point inwards, while the outer border is at the same time depressed ; 3rd, that the astragalo-scaphoidean and calcaneo-cuboidean articular surfaces glide on each other ; and that when this gliding takes place cotemporaneously with the preceding movements the toes move inwards 25°, and the outer border of the foot is depressed an inch and a half : and 4th, he gives a very careful and accurate description of the means provided for elasticity and the general distribution of concussion in the angles at which the various tarsal bones articulate with each other and with the metatarsal bones. There is, however, one point which appears to have escaped his notice, as it has also escaped the notice of others : I refer to the mechanical shapes of the articular surfaces of the various intertarsal joints. This may be most easily seen and studied when a vertical section is made from before backwards through the foot with the tibia still *in situ* in the direction which the line of pressure takes when standing erect. I have lately made several such sections in well-formed feet ; the one which the following measurements were taken from was a somewhat small foot belonging to a young woman, and the section in this instance fell through the tibia, astragalus, os calcis, scaphoid, cuboid, external cuneiform, and third metatarsal, as shown in the annexed woodcut. In walking, and even in standing, it is of course true that the foot is slightly everted, and the line of greatest pressure falls upon the first and not upon *the third toe*, as here represented, but, from other sections, I find that the result is the same, whether the cut traverses the first or third metatarsal bone. In the section which is here represented every articular surface divided by the saw proved to be a segment of a circle. Other sections of other feet have shown that the joint between the scaphoid and cuneiform is not always an arc of a circle, but with this exception all the articular surfaces appear to be of this nature. The

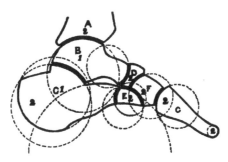

A, Tibia. *B*, Astragalus. *C*, Calcaneum. *D*, Scaphoid. *E*, Cuboid. *F*, External Cuneiform. *G*, Third Metatarsal. 1, 1, Spongy cancellous tissue out of line of pressure. 2, 2, 2, 2, Firm curves of cancellous tissue in line of pressure. The various circles are indicated by dotted lines.

circles differ as do the segments in size, the larger are situated in the short posterior pier of the antero-posterior arch, *i.e.* between the tibia and the heel, the smaller and more numerous in the long anterior pier, *i.e.* between the astragalus and the toes. The line of pressure, though deflected by the astragalus, chiefly passes of course to the heel, and in this passage tiaverses two segments of comparatively large circles, viz. one between the tibia and the astragalus, and the other between the astragalus and os calcis. The circles serve to transmit the pressure chiefly to the dense periphery of these bones, and to the strong cancelli, leaving the spongy centres fully equal to supporting their lesser share in the burden[1]. The other line passes downwards and forwards to the toes and traverses more numerous but smaller segments of smaller circles, viz., one between astragalus and scaphoid, one between os calcis and cuboid (vertically beneath the preceding), one between scaphoid and (in this particular section) external cuneiform, and one between external cuneiform and third metatarsal.

Finding the centres of these segments they proved to have the following dimensions :—

				Inches
Upper articular surface of astragalus, segment of circle				2·5 in diam.
Under ditto	ditto	ditto	...	2·8 ,,
Under surface of os calcis		ditto	...	2·5 ,,
Anterior surface of astragalus		ditto	...	1· ,,
Anterior surface of os calcis	...	...	...	·84 ,,
Anterior surface of scaphoid	...	...	...	1· ,,
Anterior surface of external cuneiform		...	...	1· ,,
Anterior end third metatarsal	...	...	...	·5 ,,

[1] Mr Wagstaffe has shown that the cancellous tissue of the bones of the foot form a series of curves of various sizes, which are also important elements in the mechanical construction of the pedal arches. Vide *St Thomas's Hospital*

As Professor Humphry remarks in a letter which I have received from him upon this subject, when one bone revolves upon another the surfaces must be segments of circles, and, as he further states, there are probably only a very limited number of pure gliding joints in the body. This appears to be precisely the case in the foot, for, with the exception of the joint between the scaphoid and cuneiform, which I have before stated is occasionally a non-circular, and therefore a gliding surface, every tarsal movement must be one of rotation. It is further interesting to note a sort of compensatory movement which takes place among these tarsal joints, by means of which the length of the foot remains unaltered during its various changes of position; thus when the medio-tarsal joint is concerned in extreme adduction of the foot, the os calcis is at the same time swinging beneath the astragalus so as to preserve an exact equality of length. As far as I have been able to see in the movements of the living body, while making careful measurement, whatever motion is performed, whether of abduction or adduction, there is by this principle of compensatory movement no alteration in the distance between the toes and the heel.

One other point is noteworthy in reference to this subject: besides preserving a uniformity of length, the fact that the intertarsal movements are rotatory instead of gliding, preserves a uniformity of pressure upon the various articular surfaces which must be preventive in a great degree of extreme and injurious pressure falling upon any one point, and which would almost of necessity be the case if the curves were not circular, and the movements of rotation were exchanged for simple sliding motions.

Reports, vol. v. See also Professor Humphry's remarks on the disposition of the cancelli in his *Treatise on the Human Skeleton*, 1858, pp. 474, 498, and elsewhere.

NOTE ON THE PLACENTAL AREA IN THE CAT'S UTERUS AFTER DELIVERY. By Professor Turner.

In my description of the placentation of the Cat in the last number of this *Journal* I stated (p. 158) that, on peeling off the placenta from the uterus of a cat killed in the mid-period of gestation, the whole thickness of the mucosa in the placental area came away with the placenta and left the muscular coat exposed. I have since the publication of that number obtained the uterus of a cat killed five hours after giving birth to four kittens, so that I am now able to supplement my description of the placentation of this animal by stating what is the *post partum* condition of its uterus.

The uterus was contracted and the mucous lining thrown into well-defined rugæ. Each placental area was a narrow zonular trench, bounded at each margin of the zone by a fold of the mucosa. The surface of the non-placental part of the mucosa was unbroken and covered by epithelium. The surface of the placental zone was blood-stained, and with a number of shreds of membrane hanging from it, so that it had a torn and flocculent appearance. When thin flakes were removed from the surface of the placental zone and examined microscopically, they were seen to consist of multitudes of free red blood-corpuscles, of very delicate fibres of connective tissue, intermingled with which were fusiform and lymph-like corpuscles, and here and there a patch of cells, evidently epithelium. A series of vertical sections was then made through the placental area and adjacent non-placental part of the mucosa, and examined with low and high magnifying objectives. The free edge of the section in the non-placental area was covered by a well-defined layer of columnar epithelium, deeper than which was a thick layer of sub-epithelial connective tissue, intervening between the epithelium and the muscular coat. Lying vertically in this connective tissue were numerous utricular glands, which opened on the free surface of the mucosa, and were lined by columnar epithelium. In the placental area itself the surface epithelium was absent, and the free edge of the section had not a smooth outline, but was irregular and with slender filaments of connective tissue projecting from it. The thickness of the connective tissue layer on the surface of the muscular coat was appreciably less (on the average about one-third) than in the non-placental area. In this connective tissue sections through utricular glands were seen. Some of these sections were transverse to the tube of the gland, others oblique, others almost longitudinal. The epithelial lining of the glands was present, and it is not unlikely that the occasional patch of cells found on the surface of the placental area may have belonged to the glands and not to the surface epithelium. In more than one of the sections I saw in the placental area gland-structures which had not the form of cylindrical tubes, but were somewhat irregularly dilated.   Numerous blood-vessels, which were the vascular trunks going to the placenta, were also seen plugged with collections of blood-corpuscles.

From this description it will be seen that in the normal separation of the placenta at the time of parturition so complete a shedding of the mucosa of the placental zone does not take place as was effected by artificially tearing off the placenta in an earlier period of gestation.   The muscular coat is not exposed, but is covered by a layer of connective tissue, forming the deeper part of the sub-epithelial connective tissue of the placental area, in which portions of the gland-tubes and blood-vessels may be seen.   This layer forms a non-deciduous serotina.   The surface epithelium of the area and the more immediately subjacent portion of the connective tissue are however shed along with the placenta, and form a well-defined deciduous serotina, so that, although the entire thickness of the serotina is not shed during parturition, yet a larger proportion falls off with the placenta than in the bitch, fox and seal.

NOTICES OF BOOKS.

The Anatomy of the Lymphatic System. Part II. The Lung.
By E. KLEIN. London, 1875.

Dr KLEIN has now published the second part of his work on the
Lymphatic System, which like the first part is based on researches
undertaken for the Medical Department of the Privy Council. In
the first chapter he describes the endothelium of the pulmonary
pleura, and points out that during inspiration the endothelial cells
are flattened, because the lung is then distended, whilst during expiration, when the lung occupies a much smaller volume, the cells
are polyhedral or even columnar, and their contents are distinctly
granular. The endothelium of the pulmonary pleura rests on connective tissue, which contains elastic fibres and cellular elements.
The groups of bundles of fine connective-tissue fibres are separated by
spaces, which contain flattened connective-tissue corpuscles and correspond to the lymph-canalicular system of other serous membranes.
The connective tissue of the pleura is continuous with the septa that
separate the groups of the superficial alveoli of the lung. In guinea-
pigs bundles of unstriped muscular fibres form a distinct coat beneath
the proper connective tissue of the pleura ; these bundles are more
widely separated in inspiration than in expiration. A network of
superficial lymphatics lies in the grooves between the lobules. Their
wall is formed of a single layer of elongated endothelial plates, and
some of the branches have lateral blind dilatations. The sub-pleural
lymphatics are continuous with a deeper lymphatic network. He
believes that there is satisfactory evidence of their communication
with the pleural cavity through stomata among the endothelium.
These stomata are, he believes, opened widely during inspiration ; the
lymph-vessels and inter-muscular spaces become distended, so that
the lymph-system is filled. During expiration again they become
compressed. A network of lymphatic vessels lies in the connective-
tissue adventitia of the bronchial tubes. These Dr Klein calls *peri-
bronchial lymphatics.* Spherical, oblong, or even cord-like accumu-
lations of adenoid tissue are situated in the wall of the lymphatic
which lies next the bronchus, which resemble those previously described by the author (Part I.) as *peri-lymphangeal follicles,* and are
analogous to the lymph-follicles found in other mucous membranes.
The peri-bronchial lymphatics receive rootlets from the inner layers
of the bronchial wall, which rootlets are as a rule merely lymph-
spaces, *i. e.* inter-fascicular spaces between the bundles of the connective-tissue matrix. He describes nucleated cells, apparently
branched, between the epithelial cells of the bronchial mucosa, and
he regards them as inter-epithelial connective-tissue corpuscles similar
to the cells described by Mr H. Watney in the alimentary mucosa.
A system of *peri-vascular lymphatics* is described accompanying the

pulmonary artery and vein with their branches, which are in some parts replaced by a system of lymph-spaces between the connective-tissue fasciculi, lined by connective-tissue corpuscles, arranged like a true endothelium. Dr Klein then describes various pathological changes of the lymphatic system of the lungs. The volume is beautifully illustrated with six large plates, provided by the scientific grant committee of the Royal Society.

Traité Technique d'Histologie. Par L. RANVIER. Fasciculi 1—3. Paris, 1875.

AMONGST living histologists no one stands in more estimation than M. Ranvier, both from his ingenuity in devising new methods of research, and from his acuteness in interpreting the appearances produced on the tissues by the several agents he employs. The publication of a systematic Treatise on Histology by so accomplished an observer is therefore an event of some importance and interest. The fasciculi already published form a large octavo volume of 480 pages. The first book contains an account of the instruments, reagents, and general methods employed. In the second book the simple tissues are described and illustrated by numerous wood-cuts. The description comprises an account of the corpuscles of the lymph and blood ; the endothelium and epithelium ; the cartilaginous, osseous, and connective tissues, and the commencement of the chapter on the muscular tissue.

Jahresberichte über die Fortschritte der Anatomie und Physiologie, herausgegeben von Prof. HOFMANN und Prof. SCHWALBE. Leipzig.

WHEN in the year 1872 Professors Henle and Meissner of Göttingen discontinued the publication of their annual Report on the Progress of Anatomy and Physiology, which for so many years had been highly esteemed by students of those sciences, the need was felt in Germany of supplying its place by a new annual record of their progress. The well-known house of Vogel of Leipzig has accordingly undertaken the publication of the *Jahresberichte,* and the labour of editing has been entrusted to Prof. Hofmann of Leipzig and Prof. Schwalbe of Jena. These gentlemen have associated with them several eminent investigators, not only in Germany, but in Switzerland, Bohemia, Poland, Denmark and Sweden, so that arrangements have been made for furnishing an extended analysis of the results of research in both anatomical and physiological science. Three volumes of Reports are now before us, giving an analysis of the numerous memoirs published during the years 1872, 1873, 1874. Not only is the literary part of each volume carefully and methodically executed, but it is printed on good paper in a clear and readable type. We cordially recommend the Reports to all who take an interest in the literature of these sciences.

Lehrbuch der Vergleichenden Anatomie, von Dr A. NUHN. Ersten
Theil. Heidelberg, 1875.

PROFESSOR NUHN, of the University of Heidelberg, has published
the first part of a new work on Comparative Anatomy, in which he
describes the modifications in the form and arrangement of the
digestive, respiratory, circulatory, urinary and generative organs in
both the Vertebrata and Invertebrata. The descriptions, though
brief, are pointed, and bring out the salient features of structure.
The part is profusely illustrated with numerous well-executed wood-
cuts, many of which give diagrammatic representations of the objects,
which will be found of service not only by the student but the teacher.

---

*Leçons sur la Physiologie et l'Anatomie Comparée de l'Homme et des
Animaux.* Par H. MILNE EDWARDS. Part II. Vol. XI.
Paris, 1875.

THIS part of Professor Milne Edwards's most comprehensive Treatise
cóntains the conclusion of his description of the anatomy of the
Nervous System in the Vertebrata. Also four lectures on its physiology, in which he treats not only of sensation in general, but
specially the senses of touch, taste, and smell in the animal kingdom.

Lectures and Essays on the Science and Practice of Surgery. By
ROBERT MACDONNELL, M.D., F.R.S., Surgeon to Steevens'
Hospital, Dublin. Part II. The Physiology and Pathology
of the Spinal Cord. Fannin and Co., Dublin.

DR ROBERT MACDONNELL is one of the few practical surgeons who
not only keeps himself *au courant* with modern physiological research
and makes contribution to it, but also applies the knowledge derived
from it to the investigation of disease, as is evinced by the interesting
'Lectures and Essays' before us. In them he gives a succinct account
of recent views on the Physiology of the Spinal Cord, the effect of
nervous influence on the circulation, and the contraction of muscle,
and draws from them explanations of various clinical phenomena,
such as metastasis of disease, sympathetic irritation, &c. He also
gives the following ingenious 'New Theory of Nervous Action' to
explain the phenomena of varied impressions of temperature, pain,
tickling, &c., being transmitted by the same channel instead of by
different kinds of nerve-fibres, as has been supposed by Brown-
Séquard and others.

"I conceive that the various peripheral expansions of sensitive
nerves take up undulations or vibrations, and convert them into
waves capable of being propagated along nervous tissue (neurility, as
it has been well named by Lewes). Thus the same nerve-tubule
may be able to transmit along it vibrations differing in character,
and hence giving rise to different sensations ; and consequently the
same nerve-tubule may, in its normal condition, transmit the wave

which produces the idea of simple contact, or that which produces the idea of heat; or, again, the same nerve-tubules in the optic nerve which propagate the undulations of red may also propagate, in normal vision, those which excite the idea of yellow or blue, and so for the other senses."

Elements of Physiology. By D. L. HERMANN, Professor of Physiology in the University of Zürich. Translated from the German by ARTHUR GAMGEE, M.D., F.R.S., Professor of Physiology in the Owens College, Manchester. London: Smith and Elder.

THIS excellent translation by an accomplished Physiologist of the best systematic treatise on Physiology which has recently appeared in Germany will prove a great boon to English students, and, we doubt not, will be freely used by them.

Quain's Elements of Anatomy. Edited by Dr SHARPEY, Professor ALLEN THOMSON and EDWARD ALBERT SCHÄFER. Eighth Edition. Vol. I. London: Longmans.

THIS volume, containing the descriptive anatomy of the bones, joints, muscles, vessels and nerves, has been edited by Professor Thomson, and bears the evidence of his wide reading and good judgment. With so eminent an editor it is no wonder that the work holds its rank as the best handbook of descriptive anatomy in the language.

The Student's Guide to Human Osteology. By W. W. WAGSTAFFE, Assistant Surgeon to and Lecturer at St Thomas's Hospital. London: Churchill.

WE are glad to welcome a book which makes some attempt to set forth the mechanical features of the skeleton. To this subject Mr Wagstaffe has already shewn that he has paid attention, and we trust that this effort "to interest the student in the mechanical wonders of his framework" will prove as successful as it deserves to be.

Nouveaux Éléments de Physiologie Humaine. Par H. BEAUNIS. Professeur de Physiologie de la Faculté de Médecine de Paris. 8vo. p. 1140. Paris. J. B. Baillière.

M. BEAUNIS has in this book professedly broken through the customary methods of writing physiological treatises; it is mainly devoted to general physiology; but whilst histology is almost entirely neglected, there are short abstracts of other subjects connected with physiology though not usually included in works on the subject, such as The Correlation of Forces, The Characteristics of living bodies, and The Origin of Species. The attempt at comprehensiveness has led M. Beaunis to give a description of a physiological laboratory, and at the beginning of each chapter a description of the apparatus and experiments especially belonging to the text of the chapter. This is certainly a mistake; for the accounts are too concise to be of any practical value to a beginner, and the advanced student would not need them. The chemical properties of the tissues and their products are separated from the physiology, and are collected together, as they deserve to be, in a special chapter.

The main part of the book is however on general physiology, and here M. Beaunis has in many cases brought the subject, in outline at least, up to the present day; but there are some curious instances to the contrary. Thus in the article on the coagulation of the blood, although several theories are discussed, deserving only to be forgotten, yet the experiments of Alexander Schmidt are not even mentioned.

There is one character of the older text-books which might well have been omitted, viz. that of giving at intervals certain lists and statistics of little physiological value, as for instance, the percentage of alcohol in the various French light wines; but as a make-weight to this, at the end of each chapter references are given to the original memoirs of the subject of the chapter.

Although one can without much difficulty find points to object to in this book of M. Beaunis, yet on the whole it is likely to prove of considerable value to physiological students.

J. N. L.

REPORT ON THE PROGRESS OF ANATOMY[1]. By Prof. TURNER and D. J. CUNNINGHAM, M.B., C.M.

OSSEOUS SYSTEM.—Emil Rosenberg (*Morphologisches Jahrbuch*, 1875, 83) writes on the DEVELOPMENT OF THE VERTEBRAL COLUMN and the Os CENTRALE CARPI IN MAN. This memoir occupies the greater part (115 pages) of the new Morphological Journal, edited by Gegenbaur. The object of the author is to see what evidence may be found in the development of the spine and carpus, of the descent of man, along with other Primates, from a common stock. The investigation into the development of the spine is of an elaborate nature, and should be read by all who may be working at this interesting subject. From it he concludes that the existing dorsal vertebræ in man are to be regarded as the most conservative section of the spine; the 20th to the 24th vertebræ have undergone a transformation into lumbar vertebræ; the 25th to the 29th vertebræ have passed through a second transformation, and become sacral vertebræ; whilst the 30th to the 35th have gone through a third metamorphosis, and become caudal vertebræ.——W. Gruber records the following VARIETIES IN BONES (*Reichert u. du Bois-Reymond's Archiv*, 1875): a case in which the *lateral tuberosity* of the tarsal end of the *5th metatarsal* was a distinct *epiphysis;* also a case in which the *anterior tuberosity* of the *trapezium* formed a distinct epiphysis; also on p. 194, an additional case to those previously recorded (Report, VIII. 159) of a malar bone divided into two parts: also in *Virchow's Archiv*, LXIII. the following: an *os triquetrum* in the *anterior fontanelle;* enormously deep *fossæ maxillares;* ossifications in the wall of a subcutaneous trochanteric bursa, in the external intermuscular ligament of the femur, in the fibrillar head of the soleus; also an unusual canal for a deep temporal artery arising from the middle meningeal artery; and in *Virchow's Archiv*, LXV. the following cases: an enormously wide *canalis mastoideus;* a crista galli containing a cavity; an articulation between the shafts of the 1st and 2nd ribs; a costal cartilage giving off a branch laterally; an exostosis in the sigmoid sulcus of the mastoid process, and an exostosis in the external auditory meatus.——H. von Jhering enquires (*Reichert u. du Bois-Reymond's Archiv*, 1875) into the TEMPORAL RIDGE on the side of the SKULL. As a rule, he says, it is a double line; the lower division corresponds to the border of origin of the temporal muscle, the upper to the attachment of the temporal fascia.——C. Aeby investigates the occurrence of SESAMOID BONES IN THE HUMAN HAND (*Reichert u. du Bois-Reymond's Archiv*, 1875) with reference to their occurrence in other digits than the thumb. Of seventy-one extremities examined he found a single sesamoid present in the index in thirty cases, in the little finger in fifty cases,

[1] To assist in preparing the Report Professor Turner will be glad to receive separate copies of original memoirs and other contributions to Anatomy.

whilst in twenty they were absent from both fingers.——W. W. Wagstaffe (*St Thomas's Hospital Reports*) describes the mechanical structure of the CANCELLOUS TISSUE OF BONE—and endeavours to shew that this has a definite arrangement insuring the greatest strength and elasticity along the lines of greatest pressure. He states that if an antero-posterior vertical section be made through the body of a vertebra, the osseous columns, or lamellæ, are seen to be arranged both vertically and horizontally. In both cases they are curved—the concavity of the vertical columns being directed towards the centre of the body, whilst that of the horizontal columns is turned towards its upper and lower surfaces. The curve, whilst it adds to their elasticity, does not detract from the strength of the columns. A similar section through the lamina shews a series of curves springing from the lower edge and radiating into the body and articular process, where they terminate at right-angles to the surface of pressure. A horizontal section of a vertebra shews a series of diverging fibres springing from the walls of the pedicles, and arching into the body, where a definite arrangement is difficult to make out. In the dorsal transverse processes, the fibres spring from the posterior wall at the root, and diverging, end perpendicularly to the articular surface. In a vertical section made from one side of the *sacrum* to the other, he describes five sets of fibres. (1) A more or less vertically directed set, slightly curved, and having their concavities turned towards the middle line. (2) A set springing from the upper surface and passing obliquely outwards to the iliac articulation—their curve increasing as they pass downwards from the first foramen until they run nearly vertically to the second foramen. (3) A set passing down from the second foramen, and ending partly upon the outer and partly upon the lower surfaces. (4) A set at first parallel with the iliac surface, and curving inwards below. (5) A small set running from the lower part of the iliac surface, and passing inwards in a diverging manner. A vertical section through the glenoid cavity of the *scapula* shews one set of fibres springing from the articular surface at right-angles, and then diverging with their concavities directed downwards. Another series crosses these, having their concavities directed towards the articular surface. He next describes the arrangement of the osseous lamellæ in the head of the *humerus*. The fibres take their origin from the walls of the surgical neck. Those from the inner side are arranged in two series, one of which curves outwards, interlacing with a similar set springing from the outer wall, and forming with them a number of gothic arches. The other fibres springing from the inner wall curve towards the articular surface, where they end at right-angles to it. Of the fibres springing from the outer wall, some, as we have seen, go to form the arches, others curve vertically upwards to end perpendicularly to the upper surface of the great tuberosity. In the lower extremity of the humerus, the fibres curve from the walls of the shaft to the opposite side, their concavities being directed towards the centre of the bone, and thus a series of inverted arches is formed. The mass of the cancellous tissue is

composed of columns running down from the shaft with their concavities directed towards the middle line, and finally ending at right-
angles to the articular surface. A set is also visible running parallel
to the articular surface. He describes in detail the structure of the
cancellous tissue of all the other bones of the upper extremity,
which, generally speaking, he states to be composed of diverging and
curving columns, some of which end at right-angles to the articular
surfaces, whilst others curve obliquely across the bone, forming in
many cases a series of arches, the convexities of which are directed
towards the extremities. He then passes to the bones of the lower
extremity, and deals with them with equal minuteness. The same
principles are exhibited here as in the bones of the upper extremity.
In the neck and head of the femur, fibres spring from the inner and
outer walls of the shaft, and interlace with each other so as to form
arches. The uppermost fibres from the inner wall run into the
head, where they end at right-angles to the articular surface. The
intermediate fibres from the inner wall, less definite in form, constitute the cancellous tissue of the neck. The second series of fibres
from the outer wall strengthen the upper end of the shaft, but some
can be traced into the head, where they are parallel to the articular
surface. The fibres which run into the great trochanter are very
indefinite.——Renaut (Arch. de Phys. Aug. et Sept. 1875) contributes a long article upon THE ELASTIC TISSUE OF BONES. In this
he first establishes the presence of this tissue in a particular peripheral
zone of the bone, and points out that the long bones of birds are
the best to use for this purpose. In the second part of the article
he describes the regular distribution of this tissue in the long bones
of birds, and he then concludes by endeavouring to determine the
origin and the morphological significance of the elastic networks in
the midst of the osseous tissue.

MUSCULAR AND ARTICULATORY SYSTEMS.—K. Bardeleben records
(Centralblatt, No. 27, 1875) observations on the MUSCULUS STERNALIS.
From an anlysis of the numerous cases described by authors he concludes that it cannot be considered as a true homologue of the pan-
niculus carnosus.——C. Gegenbaur (Morphologisches Jahrbuch, 1875,
243) discusses the OMO-HYOID MUSCLE and its clavicular attachment. His conclusions are that the omo-hyoid belongs to the group
of muscles which are also represented in man by the sterno-hyoid
and thyroid : in lower animals the origin of this group of muscles
extends continuously from the region of the sternum along the clavicle,
and is continued on to the scapula : through a splitting up into distinct portions the sterno-cleido- and omo-hyoideus arise as distinct
muscles : the most frequent variety of the omo-hyoid is an attachment
to the clavicle : the fascia fixing the omo-hyoid to the clavicle is a
return to the cleido-hyoid muscle.——H. Welcker considers the movements of PRONATION and SUPINATION of the FORE-ARM (Reichert u.
du Bois Reymond's Archiv. 1875), also the connections and use of
the ilio-tibial band of the fascia lata or "tractus ileo-tibialis" as he
calls it.——Welcker also communicates to His and Braune's new

Zeitschrift für Anatomie und Entwicklungsgeschichte, Leipzig, 1875, 41, an essay on the HIP-JOINT and SHOULDER-JOINT. He describes the capsular ligament of the hip as containing four longitudinal bands of fibres: *a*, a *superior ileo-femoral band* at the upper and outer part of the capsule; *b*, an *anterior ileo-femoral band* from the anterior inferior iliac spine to the lower end of the anterior inter-trochanteric line; *c*, a *pubo-femoral band* from the pecten of the pubes to the lower end of the anterior inter-trochanteric line; *d*, an *ischio-femoral band*, which extends from the groove in the ischium, through which the tendon of the obturator externus passes, horizontally outwards to the digital fossa of the great trochanter: *a* limits extension, external rotation and abduction; *b* limits extension; *c* limits abduction; *d* limits rotation inwards. He agrees essentially with Henle in describing the *zona orbicularis* of the capsule as a fibrous ring-like arrangement not directly attached to the bones, but only through its blending with the longitudinal fibres of the capsule; it is thinnest behind and below and acts as an orbicular ligament. He thinks it very improbable that the *lig. teres* limits the movements of the joint as long as the capsule of the joint is intact; neither does he consider it to be of importance in conveying blood-vessels to the head of the femur, but believes that its object is to promote the circulation of the synovia over the articular surfaces during the movements of the joint. In the shoulder-joint the tendon of the biceps has, he considers, the same function, and in the atlo-odontoid, knee and other joints, arrangements exist which have the same object. He describes the coraco-humeral ligament of the shoulder-joint as consisting of two bands, an anterior and a posterior, between which is a recess for the lodgement of the synovia, and he regards the lig. teres as a further development in the hip-joint of the anterior band of the coraco-humeral.—— Hesse by means of sections (*His and Braune's Zeitsch. für Anat. u. Entwick.* 1875, 80), made transversely, longitudinally, and parallel to the surface, studies the arrangement of the MUSCLES of the HUMAN TONGUE. In the same *Zeitschrift*, 107, W. Henke describes the MUSCLES of the UPPER and LOWER LIPS.——W. Gruber records (*Reichert u. du Bois Reymond's Archiv.* 1875) the following MUSCULAR VARIATIONS: a *musculus pisohamatus* from the unciform to the pisiform; four cases of *m. ext. digt. com. manus* with five tendons to all the fingers; a case of *m. ext. digit. longus pedis* with five tendons to all the toes; a case in which the *m. flexor pollicis longus* acted as a tensor bursæ tendinum, or as a head of the flexor profundus digitorum.——W. Gruber also records in *Virchow's Archiv.* LXV. two additional cases of *rudimentary external oblique muscles;* a *m. scapulo-clavicularis;* a new case of *m. tensor semivaginæ articulationis humero-scapularis;* a case of *m. ext. digiti sec. pedis longus.*——E. Meyer makes some observations (*Reichert u. du Bois Reymond's Archiv.* 1875, 217) on PALE AND DARK-COLOURED TRANSVERSELY-STRIPED MUSCLES.——W. W. Keen (*Trans. Coll. Phys. Phila.* 1875) gives the results of some EXPERIMENTS ON THE MUSCLES OF RESPIRATION AND LARYNGEAL NERVES in a criminal executed by hanging. He states that after the body had been suspended for half an hour it was cut down, and the left vagus and recurrent laryngeal nerves dissected out as low down in the neck as possible. A small piece of india-rubber cloth was then introduced beneath each as an insulator, and an opening made between the thyroid cartilage and hyoid bone so as to admit the mirror for the purpose of examining the effect produced upon the vocal cords by stimulation of the nerves. He found that repeated faradization and galvanization both weak and strong of the recurrent and vagus produced marked movements of the left cord only. From this he concludes that no chiasma of the inferior laryngeal nerves exists. In less than two hours after death he dissected out and insulated the phrenic nerve, but neither the constant nor interrupted current applied to this produced the slightest change in the capacity of the chest. Lastly, he proceeded to apply stimuli to the intercostal muscles. The interrupted current applied to the inter-cartilaginous portions of the internal intercostals produced an elevation of the cartilage below them. When the external intercostals were faradized the upper rib was pulled down very markedly, whilst the lower was perceptibly raised. The depression of the first rib was but slight, but as each external intercostal was stimulated in turn it became more and more decided from the first downwards. He now reflected and removed the external intercostals, and faradized the internal muscle from between the cartilages to the axillary line, and this was followed by depression of the upper rib and a marked elevation of the lower rib. He concludes therefore that the internal are inspirators and the external expirators.

NERVOUS SYSTEM.—J. Marshall gives (*Proc. Roy. Soc. London,* June 17, 1875) a preliminary note on the PROPORTIONS of the several LOBES of the CEREBRUM in Man and some higher Vertebrata, with an attempt to explain some of the ASYMMETRY OF THE CEREBRAL CONVOLUTIONS IN MAN ; also a note on the influence of STATURE ON THE WEIGHT OF THE ENCEPHALON. He has been led to believe that some of the asymmetry is due to the right-handedness of Man, and that the immediate neighbourhood of the left fissure of Rolando and the right parietal lobule furnish stronger evidences of essential as distinguished from non-essential asymmetry. He analysed Dr Boyd's Tables, and found that the total increase in the weight of the encephalon is in the male $2\frac{3}{4}$ oz. with a mean range of 7 inches between the tallest and shortest individuals; and in the female $1\frac{1}{4}$ oz. with a mean range of 6 inches. The cerebrum increases more than the cerebellum. The increase in the weight of the encephalon is not *pari passu* with the stature ; on the contrary there is a gradual and progressive relative diminution in the proportion of encephalon to the stature as the latter itself increases. Hence shorter persons of either sex have proportionately to their height a larger amount of brain than taller persons, and the proportion is larger in the male than the female, bearing out the well-known sexual difference that the weight of the male brain overrides the influence of stature, which has a tendency to diminish his proportionate amount of brain.——A report on the BRAIN AND SKULL, in cases of Microcephalic Idiocy, by

W. W. Ireland, is in *Edinb. Med. Jour.* August to October, 1875. The cases are described in *Rivista Clinica di Bologna* by Lombroso and Valenti.——Ad. Pansch communicates (*Centralblatt*, No. 38, 1875) observations on the CORRESPONDING REGIONS in the CEREBRUM of the PRIMATES and CARNIVORA. He considers that he has established three primary fissures on the outer surface of the cerebrum of all mammals with convoluted brains. In the *Primates* they lie radially above the Sylvian fissure, but in other mammals the first or most anterior is more vertical, whilst the second and third have a sagittal direction; the convexity of the first is directed forwards, that of the others upwards. The first primary fissure is the *præcentral* of Ecker in the *Primates*, the *vordere Bogenfurche* in the dog; the second primary fissure is that of *Rolando* in the *Primates*, the *obere Bogenfurche* in the dog; the third primary fissure is the *intraparietal* of Turner in the *Primates*, the anterior part of the *mittlere Bogenfurche* in the dog.——D. N. Knox (*Glasgow Med. Jour.* April, 1875) records a case of DEFECTIVE CORPUS CALLOSUM in the brain of a female idiot. He states that when this brain was placed upon its base, after being stripped of its membranes, the hemispheres fell outwards, shewing no appearance of the corpus callosum and fornix, and exposing the floor of the lateral ventricles. These cavities were small—about 4¾ in.; the posterior horns were dilated and the lining membrane thickened. The third ventricle measured from the anterior to the posterior commissure 1⅛ in. The corpus callosum, apparently only represented by a slight ridge, "began above the lamina cinerea, and passed upwards and backwards, attached to the side of the general cavity of the ventricle, forming the upper border of a layer of white matter, the lower border of which was part of the fornix." Separating from the fornix it ended in the anterior and lower part of the hippocampal convolution. The fornix was cleft into two in the middle line, and between the anterior part of this and the callosal ridge was the septum lucidum. The anterior and posterior commissures were present, but the gyrus fornicatus was wholly and the calloso-marginal fissure partially absent. In consequence of the absence of the former, the occipital and calcarine fissures passed independently down into the ventricular cavity. He next gives a summary of fourteen cases recorded by various authors, and from a study of these he divides them into two classes according to the anatomical lesion. First those in which the corpus callosum is absent or very rudimentary, and in which the gyrus fornicatus, the posterior commissure, and usually the middle commissure, are also wanting; second, those in which the corpus callosum is present in considerable part, and in which the other structures are also present. He points out that the most frequent cause of this deformity is to be found in dropsy of the ventricles, and that its degree is influenced by the period at which the dropsy supervenes. Thus if it appears before the formation of the commissural system, the result is the total absence of the latter; and if it appears later, then the deformity is less in proportion to the amount of the corpus callosum developed.—— L. Gustave Richelot (*Arch. de Phys.*, March, April, 1875) describes the distribution of ·THE COLLATERAL NERVES OF THE FINGERS. He states that the *Median* supplies the palmar collateral branches of the thumb, of the index, of the middle, and of half of the ring-finger. From these come off twigs to the dorsum of the first phalanx of the thumb, and the collateral dorsal branches of the index, middle, and half the ring-finger. They end in the digital pulp, and give sub-ungual branches with which the terminal twigs of the collateral dorsals unite. The *Ulnar* furnishes, by its palmar branch, the collateral palmars of the little finger, and of half the ring-finger, and the collateral dorsal of the same half of the same finger. By its dorsal branch it gives off the collateral dorsals of the little finger, and also twigs to the first phalanx of the ring-finger, and of the middle finger. As in the other fingers, the collateral palmars of the little finger give off a sub-ungual twig with which the terminal branch of the collateral dorsals unite. The *Radial* furnishes collateral dorsals to the thumb, and also twigs to the dorsal surface of the first phalanx of the index and middle finger. Finally, the author gives the clinical deductions which he draws from this arrangement as well as from sections of the nerves of the upper extremity.——W. Gruber records (*Virchow's Archiv.* LXV.) another variety additional to those previously described by him in the arrangement of the MUSCULO-CUTANEOUS NERVE.——R. Clement Lucas, (*Guy's Hospital Reports*, 1875) writes upon the normal arrangement of the BRACHIAL PLEXUS OF NERVES. He begins by referring to the unsatisfactory manner in which this plexus is described in our standard text-books of Anatomy, and he then gives the arrangement which he believes to be normal. Out of thirty consecutive observations the arrangement he describes was present in twenty-seven cases. He states that at first the plexus consists of *an upper cord* formed by the junction of the 5th and 6th cervial nerves, *a middle cord* consisting of the 7th cervical which runs alone, and *a lower cord* formed by the junction of the 8th cervical and the first dorsal. The upper and middle cords next bifurcate into anterior and posterior divisions. Of these the two anterior join to form the outer cord, whilst the two posterior join to constitute the posterior cord. The lower cord passes down as the inner cord, but in addition gives a slender fasciculus of reinforcement to the posterior cord. His paper is accompanied by a diagram illustrating this arrangement.——H. Eichorst writes (*Virchow's Archiv.* LXIV.) on the DEVELOPMENT of the human SPINAL CORD and its constituent tissues.——A. Willigk describes and figures (*Virchow's Archiv.* LXIV.) some examples of the ANASTOMOSIS OF MULTIPOLAR NERVE-CELLS in the spinal cord.——M. Nesterowsky describes (*Virchow's Archiv.* LXIII.) the arrangement of the NERVES OF THE LIVER. He figures a network of very fine nerve-fibres surrounding the inter-lobular branches of the portal vein, the intra-lobular of the hepatic, and the intra-lobular capillaries.——J. M⁰Carthy, (*Quart. Journ. of Micros. Science*, Oct. 1875) publishes a memoir upon SPINAL GANGLIA AND NERVE-FIBRES. He first describes his method of preparing the ganglia for microscopical observation. Hardening them by means of mono-chromate of ammonia, he stains the sections with hæmatoxylon, and mounts the preparations in Canada balsam. He states that the ganglion is surrounded by a thick layer of connective tissue containing numerous blood-vessels, and sending processes into the ganglion. These processes ramify so as to form a delicate meshwork, which supports the nerve-fibres, or joins with the capsules of the ganglion-cells. These cells, very variable as regards size, are arranged in clusters, or spindle-shaped groups, the long axis of which corresponds with the long axis of the ganglion. Each cell possesses a sharply defined margin in which nuclei of connective-tissue corpuscles may be recognized. Inside this is a layer of hyaline substance, faintly fibrillated when examined by a high power, and containing round or oval nuclei varying greatly in number and arrangement. The hæmatoxylon gives them a pale violet hue, which contrasts markedly with the deep tint of the nuclei of the connective-tissue corpuscles. Internally this hyaline layer has no definite boundary, but shades into a cloudy substance which in turn is continuous with molecular matter enclosing a large nucleus. Around the nucleus the molecular mass appears to be concentrically fibrillated, whilst outside this the granules also seem to possess more or less of a concentric arrangement, and here a group of pigment-granules is generally observed. The nucleus consists of a distinct membrane, enclosing granular matter and a nucleolus. The nuclei of different cells present differences. Sometimes they resemble closely the nuclei of the hyaline layer; at other times they are vesicular in appearance, colourless, and enclose a shrivelled nucleolus like a pigment-granule. From this the author draws the inference that the ganglion-cells are regenerated from the nucleated hyaline substance. As regards the nerve-fibres, the author states that they pass irregularly through the ganglion so that a section cuts some transversely, some obliquely, and some longitudinally. The constrictions of Ranvier can be well studied on these fibres, and the nuclei often appear more within than in the sheath of Schwann. The medullary sheath has the appearance of being formed of minute rods, radiating from the axis-cylinder to the sheath of Schwann, and arranged differently in different fibres. Sometimes they occupy the whole space between the central axis and the sheath, whilst in other cases they are shortened, and seem to be embedded in some homogeneous substance.

CONNECTIVE TISSUE AND EPITHELIUM.—G. B. Ercolani communicates (*Mem. dell' Accad. delle Scienze di Bologna,* 1875) a memoir on the minute structure of TENDINOUS TISSUE.——G. Thin (*Proc. of the Royal Soc.,* London, No. 160, 1875) gives an account of his investigations into TRAUMATIC INFLAMMATION OF CONNECTIVE TISSUE. In addition to the primary bundles of fibrillary tissue covered by the two kinds of flat cells and the stellate cells which he previously described, he now gives a description of parallel chains of spindle-cells, each of which is furnished with a process at each end by which it is connected with its fellows on either side. They have no continuity with the stellate cells. The corneal substance, he states, is broken up by clefts circumscribed by a double outline which extend from the

investing epithelium to a varying depth into the fibrillary tissue.
These he believes to indicate a splitting up of the corneal substance
into compartments equivalent to the secondary and tertiary bundles
of tendon. In inflammatory conditions of the cornea, the clefts become widened out and their finer ramifications become visible; moreover, their serous contents show more abundantly the dark granular
substance which results from the reduction of the chloride of gold.
The opening out of the clefts favours the action of this reagent, and
thus preparations are obtained in which the two kinds of flat cells
disposed after the manner of an epithelium can be recognised. These
cells are easily identified as those which can be isolated from the
healthy cornea by a warm solution of caustic potash. He states that
in the cauterised cornea of the frog as well as in the inflamed tongue
of the same animal isolated portions of primary bundles can be seen
lying loose. These generally have a constant length, a uniform
breadth, sharply defined borders, are sometimes puckered transversely, or show a faint longitudinal fibrillation, and are occasionally
cut transversely by straight hyaline lines. The constituent elements of the primary bundles undergo a chemical change in inflammation, as proved by the different depths of tint produced by gold
staining. The only effect which the inflammation produces upon the
elongated flat cells and the quadrangular cells previous to their destruction, is a segmentation of the nucleus into one or more parts. With
regard to the stellate cells, the author contends that there is no
identity between the cell and its processes with the visible protoplasm;
these cells may be compared to the pigment cells of Lister; and the
phenomenon described by the Germans as "*Zusammenballen*" of the
cell-processes, he believes to be due to the aggregation of the protoplasmic particles in the centre of the cell similar to what takes place
in the concentration of the pigment particles in Lister's cells. Osmic
acid, he states, is the only reagent by means of which he has been
able to get satisfactory preparations showing the stellate cells. Red
aniline further differentiates the protoplasm from the cell-processes,
and hæmatoxylin brings out the nuclei. Inflammation produces disintegration of these cells, but occasionally the processes may be seen
to be represented by fine darkly-stained lines on which is a series of
small globular swellings. The nuclei and processes are very stable,
the former rarely showing signs of segmentation. Between the layers
of the superficial epithelium of the cornea is a network of stellate cells,
and these increase greatly in size in inflammatory conditions of the
cornea. The author next describes the changes which the spindle-
cells undergo in inflammation. The cell-protoplasm becomes increased
in amount, and the processes can be easily traced. This swelling of
the protoplasm next extends along the processes from one cell to
another, so that a row of such cells comes to resemble a long column
of protoplasm with slight constrictions at intervals. The last stage
consists in the division of the nucleus which gives rise to nuclear
bodies which may be retained within the cell, or partly expelled from
it. These are identical with red blood-corpuscles. From this, he
concludes that in inflammation these corpuscles become free bodies

equivalent to red blood-corpuscles. Further, he believes that the appearances described by many as white corpuscles in "Spindelform," are nothing else than spindle-cells rendered prominent by the inflammatory process. White blood-corpuscles are to be seen in the wider spaces, in the nerve-channels, and in the tracts between the larger bundles. They are round, or sometimes clavate, and a few may consist of two globular bodies joined by a smooth isthmus. In the latter case a nucleus is observed in each mass, but not in the isthmus. This is a corpuscle undergoing division. Further, from his observations, the author is of opinion that red corpuscles are produced by the escape of nuclei from the white blood-cells. He lastly explains that new capillary vessels are developed in inflamed tissue in the same manner in which they are developed in fœtal tissue. The spindle-cells become enlarged, and contain several nuclei which are identical with the red corpuscles of the blood. The blood-plasma passes into the inter-fibrillary space in which these cells lie, and the nuclei escapes from the cells. Lastly, by a process of diapedesis the formed elements of the blood in the nearest vessels pass into this space, and the circulation is established. The fibrine of the plasma, he believes, solidifies over the outer surface of the current, and to this adhere white corpuscles which in time constitute an epithelium. ——S. Martyn (*British Med. Journ.* June, 1875) contributes an article upon CONJOINED EPITHELIUM. He first narrates the views advanced on this subject by Max Schultze and others, and then gives the results of his own investigations. He endeavours to prove that the appearance of the prickles, spines, or cell-processes, is altogether artificially produced. What really exist are slender bands which, passing from all parts of the cell, connect it directly with the adjacent cells. These communicating bands when broken across constitute the prickles, and further on the surface of the cell-wall they give rise to the appearance of the parallel bars or ribs. He adds a diagram which illustrates well the points he wishes to prove. He next theorizes upon the probable origin of this cell-structure, and states that in his opinion it arises from the fissiparous multiplication or budding of cells—the progeny never completely separating from the parent cells, but remaining joined to them by connecting bands which, when broken across, constitute the prickles.——Julius Arnold communicates observations (*Virchow's Archiv*, LXIV. 203) on the anatomical relations of the CEMENT SUBSTANCE OF EPITHELIUM, and R. Thoma, p. 394, on its physiological relations. The conclusion drawn is, that in the epithelial covering of the mucous membranes and skin, as well as in the glands, a clear fluid or viscous substance lies not only between the cells, but surrounds their deeper parts, and which not only cements the cells together, but serves to distribute the material for their nourishment between them; for injection can be passed from the blood-vessels, through the "juice-canal-system," so as to run between the epithelial cells. The figures in illustration of this opinion closely resemble Fig. 1, Plate v. of T. A. Carter's paper on diaplasmatic vessels in Vol. IV. of this *Journal.*

VASCULAR SYSTEM.—J. Curnow describes (*Trans. Path. Soc.* XXVI.) a specimen of a DOUBLE AORTIC ARCH enclosing the Trachea and Œsophagus, from a female aged 87. The vascular circle was formed in front by a left brachio-cephalic trunk, the first part of the left sub-clavian, and a communicating branch to the posterior aortic arch, joining it about three inches from its summit; and behind by a posterior and larger arch, from which the right carotid and then the right subclavian took origin. The ductus arteriosus joined the right extremity of the communicating vessel, and was impervious. During fœtal life the blood-current must have passed from the arterial duct through the communicating vessel into the descending aorta, whilst after the obliteration of the duct a volume of blood equal to that which was carried into the distal portion of the left subclavian still passed from the first part of the subclavian through the same communi-cation into the aorta. There was no transposition of viscera. The author looked on the specimen as exemplifying a persistently per-vious condition of both 4th vascular arches: the right being posterior to the trachea and œsophagus and forming the main aortic arch, the left being anterior and forming the left brachio-cephalic trunk and the first part of the left subclavian, whilst the left aortic root also persisted as the patent communicating vessel. The analogy of this arrangement to the ordinary reptilian type was pointed out.—— Fanny Berlinerblau describes and figures (*Reichert u. du Bois Reymond's Archiv*, 1875, 177) the direct PASSAGE OF ARTERIES INTO VEINS in the rabbit's ear.

SPLEEN.—E. Klein (*Journ. of Mic. Science*, Oct., 1875) contri-butes OBSERVATIONS ON THE STRUCTURE OF THE SPLEEN. He begins by referring to the various views advanced by different writers on this subject, and points out that the whole doctrine of the function of the spleen rests more or less completely upon anatomical observation. He then enters upon his own investigations into this subject, and states that in the dog the capsule of the spleen consists in its deepest part of two layers of unstriped muscular fibres—the internal layer longitudi-nally, and the external circularly disposed around the organ. The latter is not so continuous as the former, and is imbedded in a matrix of connective tissue and elastic fibres. Nearest the surface the elastic fibres form regular elastic laminæ. The trabeculæ are almost en-tirely formed by muscle. In the human spleen the capsule has few muscular fibres, and only at the points where it is joined by the trabeculæ. The trabeculæ contain muscular fibres. In well pre-pared specimens of the pulp a honey-comb of membranes is observed, which, when seen in profile, have the appearance of fibres. In the dog these membranes contain two kinds of nuclei: large, pale, round flattened nuclei, and smaller oval, spherical or irregularly-shaped nuclei, which take a deeper tint with haematoxylin than the former. In some places the membrane has the appearance of being divided into oblong, irregular, or polygonal territories, each of which contains one of the large pale nuclei. In other words, the membrane at these points seems to be composed of flattened cells very like endothelial cells. The spaces of the honeycomb are of different sizes, and constitute a labyrinth of anastomosing spaces. Many of these contain blood-corpuscles and hæmoglobin lumps. Projecting from the walls into these cavities are numerous buds of granular matter, each containing one of the small darkly-tinted nuclei. These are probably washed off by the blood-current to constitute lymphoid cells. The multinuclear cells observed by Kölliker in the spleen of young animals, are branched flattened structures, containing numerous budding nuclei, and in connection with the membranes of the stroma. The flattened cells of the matrix are in some cases arranged continuously around a blood-sinus, and then they are analogous to its endothelial lining. In profile these cells frequently exhibit a peculiar notching which the author considers to be due to one side of the cell possessing linear thickenings. The venous radicles represent merely a labyrinth of spaces in the splenic parenchyma. The author states that he believes the adenoid tissue of the Malpighian corpuscles to be continuous with that of the arterial sheath—the one passing gradually into the other. In the smaller spaces, in the matrix itself or even in the cells, yellow pigment-lumps are found, and from this fact the author concludes that the destruction of the red blood-corpuscles is effected by the matrix.

THYROID BODY.—M. Poincaré (*Journ. de l'Anat. et de la Phys.*, Sep. et Oct., 1875) gives an account of the INNERVATION OF THE THYROID GLAND. He states that even taking into consideration the great number of vessels in the thyroid body, and the vaso-motor nerves necessary for their supply, the nervous filaments and ganglia are much more plentiful in this organ than we should expect. In addition to the vaso-motors there are present sensitive filaments capable of exerting sympathetic manifestations, and perhaps also secretory nerves corresponding in number, if not to the number of the vesicles, at least to that of the lobules. Of the nervous fibres which run in the intimate structure of the thyroid, there are many which take birth in the gland itself, and which are not the terminations of the nervous cords which the organ receives. This fact is at once apparent by comparing the poverty which the external capsule exhibits in nerves and ganglia with the abundance of the same nervous agents in the connective-tissue partitions of the organ. In the thyroid therefore there is a network of nervous cords circumscribing extremely limited islands of its proper substance, and having a personality quite distinct from that of the afferent nerves. The author compares this arrangement to a sub-marine cable connecting the thyroidean colony with the metropolitan cerebro-spinal system, whilst in the colony itself is an independent telegraphic network of nervous filaments possessing at different intervals its own stations or ganglia. The filaments which pass between the various ganglia arise and end within the organ. He next describes the methods he has adopted in preparing the thyroidean tissue for examination into its nervous anatomy. He points out that the ganglia do not present the same appearance and the same volume. Sometimes they enclose about ten cells; at other times they are composed of one only. In the latter case they most frequently occupy the angle of bifurcation of a little twig, or they may be enclosed in the midst of a large branch. The large ganglia are generally situated at the junction of three or four branches; sometimes however they interrupt the continuity of long branches. He states that he has not been able to determine the connection which exists between the ganglionic cells and the afferent and efferent fibres; nor has he been able to make out the exact termination of the fibres. He has sometimes seen a splitting of the filament into fibrillæ which ended in a vague manner between the epithelial cells of the vesicles. At other times he has lost the fibre in a granular mass in which he was exceptionally able to distinguish the outline of a cell.

EYE-BALL.—J. D. M'Donald describes (*Quart. Journ. Mic. Sc.* July, 1875) the *posterior elastic lamina* of the cornea as distinct from the pectinate ligament of the iris. The peripheral tendon-like processes of the pectinate ligament perforate the lamina, divide dichotomously, intercommunicate, and break up on its anterior surface into a fibrous plexus, the strands of which are concentric rather than radial. On entering the lamina the tendons are enveloped by conical extensions of structureless substance, more like a coat of vitreous enamel than the tubular reflections of a membrane.——L. von Thanhoffer contributes (*Virchow's Archiv*, LXIII. 136) to the HISTOLOGY OF THE CORNEA. He especially describes the relations of the nerves to the corneal tissue and the epithelium. In many respects his descriptions correspond with those made by G. Thin (Reports, VIII. 390. IX. 193), as that observer has pointed out in a letter to Virchow (*Archiv*, LXIV. 136).——F. Poncet (*Arch. de Phys.* Aug. et Sept. 1875) writes upon the TERMINATIONS OF THE NERVES IN THE CONJUNCTIVA. He enters very fully into the history of this subject, and reviews at considerable length the views advanced by various observers. He states that the nerves end in three ways: (1) in large-meshed networks of fine fibres; (2) in the corpuscles of Krause, which exist chiefly in the territory supplied by the lachrymal nerve in the upper and external aspect of the eye; (3) in inter-epithelial swellings.——H. Nicati (*Arch. de Phys.* Aug. et Sept. 1875) describes the distribution of THE NERVE-FIBRES IN THE OPTIC NERVES AND IN THE RETINA. He draws a distinction between the arrangement of these fibres in mammals, and in other vertebrates. In the former he states that the fibres of the optic-nerve before passing through the sclerotic have a rectilinear direction. Anastomoses take place in the following manner: two neighbouring bundles which at first are separated by connective tissue, join into one when about to pass through the sclerotic; and it follows from this, that the bundles are more numerous behind the eye than at the papilla. This diminution in the number of the bundles is accompanied by a corresponding diminution in the quantity of connective tissue. In other vertebrates, such as birds, frogs, &c., the cylindrical optic-nerve flattens out at the eye, and forms a straight elongated papilla. The bundles which constitute the nerve are arranged as they traverse the sclerotic in a unilinear series, and are distributed alternately so as to constitute a true intercrossing. Unlike the mammals, they do not possess a central artery and vein. He next points out the arrangement of the fibres in the retina. From the elongated papilla about ten principal bundles set out on each side. Each of these in spreading out in the retina divides into smaller and smaller bundles in such a manner as to give rise to the appearance of a triangle with the apex at the optic-nerve, and the base at the ora serrata. The united apices form the straight line represented by the papilla. He ends by giving a brief outline of the arrangement of the fibres in the retina of man as described by Michel.

LACHRYMAL APPARATUS.—Robin and Cadiat (*Journ. de l'Anat. et de la Phys.* Sept. et Oct. 1875) describe the STRUCTURE OF THE LACHRYMAL SAC AND ITS DUCTS. They state that the lachrymal ducts are formed of only one tunic which is intimately connected with Horner's muscle except near the sac, where some fat cells interpose. This tunic is composed of elastic fibres, among which run flat fibrous bands and capillary vessels. Towards the inner surface of the ducts the fibres become finer, and their place is taken by a hyaline substance which extends into the wall for one-third of its thickness. This supports the epithelium of the duct, which is squamous, though in its deeper layers the cells are more or less rounded. Small arteries and veins run in all directions on the outer part of the wall, and send capillary networks into the interior. A few small nervous filaments run in a longitudinal direction, and end apparently in the hyaline substance. On these nerves, where they branch, there is generally a nucleated swelling. At the entrance of the lachrymal sac the epithelium abruptly changes to the columnar variety, and the wall of the sac and nasal duct consists merely of mucous membrane, quite unconnected with the periosteum of the bones, and is separated from it by a cellular interval. Over the membranous part of the sac this intervening tissue becomes more strongly developed and fibrous. The mucous membrane throughout the whole length of the canal is smooth. Its chorion is like that of the pituitary membrane—a network of fibro-plastic cells and elastic fibres, among which lie round and oval corpuscles singly or in groups. It is richly supplied with blood-vessels and nerves. There are, however, no glands in the mucous membrane of either the canal or the sac. Close to the lower end of the canal in the submucous cellular tissue, there are some small compound racemose glands whose ducts open on the surface of the mucous membrane.

LARYNX.—J. Disse contributes (*Inaugural Dissertation*, Bonn, 1875) to the anatomy of the *Human Larynx*. He investigates the vestibulum laryngis, the ostium tracheale laryngis, as well as the larynx proper, and publishes a series of figures of transverse sections through the larynx in different planes.

MALE ORGANS OF GENERATION.—G. Gulliver records (*Proc. Zool. Soc.* April 20, 1875) the size of the *Spermatozoa of Petromyzon*,

In *P. marinus* they have a mean length of $\frac{1}{1000}$th inch, and a thickness of $\frac{1}{48000}$. In *P. planeri* they are much larger, and club-shaped, without a distinct head, and with an average length of $\frac{1}{1000}$th inch, and a thickness of $\frac{1}{30000}$.

MAMMARY GLANDS.—C. Gegenbaur (*Morphologisches Jahrbuch*, 1875, 266) investigates the structure of the NIPPLES in *Didelphys* and *Mus*.——W. Gruber records a case of SUPERNUMERARY NIPPLES in a living man (*Virchow's Archiv*, LXIII. 99).

OVARY AND OVA.—De Sinéty (*Arch. de Phys.* Aug. et Sept. 1875) gives the results of his researches upon THE OVARY OF THE FŒTUS AND OF THE NEW-BORN CHILD. He states that in an embryo of three months the ovary is composed of what, later on, will become the cortical substance. The medullary substance formed of vessels and embryonic connective tissue presents on transverse section the appearance of an isolated pedicle of cortical substance with which it only communicates by a very limited space. In this ovary are large cells and small cells. The former are the primordial ovules, and they are placed in small groups irregularly separated from each other by slender bundles of connective tissue, and some embryonic vessels. The most external of the cellular groups communicates with the epithelium of the surface. This — the germinal epithelium — is composed of two kinds of cells, viz. small and cylindrical, and large and round. The latter are the primordial ovules less developed than those situated in the stroma, and the former are probably destined to constitute the elements of the granular membrane. The most developed ovules are, for the most part, situated along the course of the vessels. In that part of the pedicle which is furthest away from the stroma, the author has observed sections of the caniculi of the Wolffian bodies. These were placed in the midst of the embryonic connective tissue, and were lined with cylindrical epithelium. In the embryo of five months the bundles of connective tissue are thicker and more abundant, and begin to form very distinctly the utricles or tubes of Pflüger. The primordial follicles are isolated, and in the ovule surrounded by a row of epithelial cells, and a limiting layer of connective tissue may be observed. In the ovary of an embryo seven months old the isolated primordial follicles are more numerous, and the connective-tissue bundles are thicker. It presents little, if any, difference from the ovary of a mature child. In the ovary of a newly-born mature infant, the germinal epithelium still presents the two kinds of cells, but the round cells are not so numerous. The ovarian tubes anastomosing amongst themselves are, for the most part, separated from the surface-epithelium by a thin layer of connective tissue. In some cases, however, the germinal epithelium is seen to be directly continuous with the contents of the tubes. Some tubes, having rather the form of utricles, contain in their deepest part an ovule. He also points out that at this stage four or five Graafian vesicles may be seen with the naked eye. In these vesicles he observed no fluid, but several layers of cells of the granular membrane surrounding the ovule. The author next describes the ovaries of a mature child three days after birth. He states that they are of unequal size, and the surface is interrupted by several small projecting knobs, some of which are as large as a pea. Some of these on being at once opened allow a little fluid to escape, which, on being examined, is found to contain one or two perfectly preserved ova, some degenerated cells, and some granules and molecules. Beneath the epithelium there are a great number of primordial ovules and utricles of Pflüger separated from the surface by connective-tissue bands. In the deeper layer there are a series of cavities of different sizes, containing in some cases an ovule surrounded by epithelial cells, granules, epithelial débris, and also some bodies like oil-globules, but capable of being coloured by reagents. In a certain number of these large follicles the contained ovule is not in relation, as regards size, to the capacity of the containing cavity; but in those which have attained a medium development, the relations are the same as in the follicles of the same size in the adult female. The large follicles are separated from each other by bands of connective tissue which support the blood-vessels. He states that the Graafian vesicles— even those which project from the surface—are always separated from the surface by a layer of connective tissue which often contains primordial vesicles. Below the large follicles are a certain number of follicles which have begun to atrophy and disappear. At this stage of life, therefore, the author has been able to study both the progressive and retrograde changes in the Graafian vesicles. He concludes that the hypertrophy of the Graafian vesicles should be regarded as a physiological process taking place in the ovary at this epoch; and also that there is an excitation on the part of the internal generative organs in relation to that which takes place in the mamma at this period; and both resemble that which takes place at puberty.——M. Z. Gerbe (*Journ. de l'Anat. et de la Phys.* July et August, 1875) points out THE PLACE WHERE THE CICATRICULE IS FORMED IN OSSEOUS FISHES. He states that it is only after the ovum has been in the water for some time that the separation of its diverse constituents individualizes the cicatricule, concentrates its scattered elements, and forms on the surface of the nutritive globe a little discoid eminence. This phenomenon is independent of fecundation, but in the non-fecundated ovum the process miscarries, and in place of a round prominent cicatricule with distinct borders, there is a cicatricule with vague borders and irregular outline. Moreover the constituent molecules are only slightly coherent, and therefore it easily decomposes. The author further points out that on the external envelope of the ovum there is a little cup-shaped depression in the centre of which is the micropyle. It is through this aperture that the fecundating corpuscles gain access into the ovum, and here is the seat of election for the formation of the cicatricule. He next describes the method by which these phenomena may be observed. The ova are placed in water with the micropyle uppermost. After a time a considerable extent of the hemisphere presents a greater whiteness than it did before the immersion. This change is evidently due to the presence of a thin layer of molecular granules which soon condenses to form a circular area always in relation to the micropyle, and surrounded by a crown of oily granules. Lastly, the cicatricule becomes completely disengaged from the yolk, and forms a small jutting out distinctly circumscribed mamilla. The author, lastly, endeavours to prove that the formation of the cicatricule is not due to the laws of gravity by which the lighter particles would rise to the micropyle or highest point.

UTERUS AND PLACENTA.—John Williams describes (*Obstetrical Journal*, Nov. 1875), the structure of the UTERINE MUCOUS MEMBRANE. He states that it possesses a very highly developed *muscularis mucosæ*. In the roe-deer he describes the muscular wall of the uterus as divided into two layers by a thin layer of connective tissue in which the blood-vessels run; one layer is between this connective tissue and the peritoneum, the other between it and the uterine cavity; on the surface of the latter lie the soft tissue and glands of the mucous membrane: the layer next the mucosa is the muscularis mucosæ, and the connective tissue outside it is the submucous layer. The sheep's uterus has a similar structure. In the human uterus the submucous connective tissue has not been found, though indications of its position have been found in the situation of the larger vascular trunks. The paper concludes with a reply to some objections raised by G. J. Engelmann (*American Journal of Obstetrics*, May, 1875) to the author's previously published views on the removal of the uterine mucosa during menstruation. (Report, IX. 398).——Ad. Pansch gives observations (*Reichert u. du Bois Reymond's Archiv*, 1874, 702), on the POSITION OF THE UTERUS in the pelvis.——G. B. Ercolani publishes (*Mem. dell' Accad. delle Scienze di Bologna*, March, 1875), on MONSTROSITAS PER INCLUSIONE with reference to the question if a PLACENTA is developed. From a microscopic investigation of the structure supposed to be a placenta in a case of this kind, he has come to the conclusion, that it had no analogy with that organ, but consisted of an irregular mass of cartilage, surrounded by muscular fibre, in which are cavities of variable form filled with cellular elements which surrounded vascular networks. On the other hand, in extra-uterine tubal pregnancies, or even in abdominal pregnancies where the intermediate mass between mother and fœtus is developed on the peritoneum, this mass is a true placenta identical in structure with that within a normal uterus.

UMBILICAL CORD.—Lawson Tait notes in *Proc. Roy. Soc. Lond.*, June 17, 1875, the *Anatomy of the Umbilical Cord*. Its external form and mode of growth, covering, substance, vessels, relations to fœtus and placenta, and nutrition, are referred to. He ascribes the spiral twisting chiefly to the vein; states that stomata may be seen on its epithelial covering: divides the alveolated canalicular tissue into three columns. "When the canals are empty they present the appearance of fibrous tissue by the collapse of their walls, and when partially distended they look like stellate cells." The canalicular tissue ends in three cones, one for each column of the cord. If the capillary plexus belonging to the dermal ring be injected, a vascular arrangement will be found in the centre of the cord lying in the firm nucleated tissue which forms the omphalic ring: belonging to it is a sacculated sinus which extends from the omphalic ring for 45 mm. up into the true substance of the cord, giving off thick trunks which rapidly break up into the capillaries; these capillaries do not form loops, but enter directly into the canalicular tissue: ·the sinus seems to originate from the small arteries of the abdominal wall which enter with the vein.

EMBRYOLOGY.—F. M. Balfour compares the EARLY STAGES OF DEVELOPMENT OF VERTEBRATES (*Quart. Journ. Mic. Sc.*, July, 1875). He describes the changes -in the ovum of Amphioxus, the Frog, the Selachian and the Bird. ·He disputes the statement recently made by Kölliker (Report, IX. 402), that the mesoblast is derived from the epiblast, and is convinced that it proceeds from the cells of the lower layer, which he regards as a common feature in vertebrate embryology, and has a wide extension amongst the invertebrates. The explanation of this is, he thinks, that in the ancestors of the vertebrates the body-cavity was primitively a part of the alimentary. He calls attention to the anus of Rusconi, or primitive opening by which the alimentary canal communicates with the exterior. It is well marked in the Amphioxus and Batrachians, less ·marked in Selachians: in Birds no trace can be seen; it does not become the anus of the adult, though the final anus corresponds closely to it in position. He believes, contrary to the opinion expressed by some embryologists, that it never becomes either mouth or anus.——W. His in His and Braune's *Zeitsch. für Anat. u. Entwick.* 1875, publishes his researches on the DEVELOPMENT OF THE BONY FISHES, especially the salmon.——W. Krause figures and describes (*Reichert u. du Bois Reymond's Archiv*, 1875, 215) the ALLANTOIS in the human embryo at the 4th week as a pear-shaped sac.——In the course of an address to the Anatomical and Physiological department of the British Association, Aug. 25, 1875, J. Cleland communicates some observations on the DEVELOPMENT OF THE BRAIN. He states that the 1st cerebral vesicle of the chick in the 2nd day is undifferentiated. At the 36th hour the optic nerves are separated from the rest of the vesicle by distinct elevations of the floor of the brain, reaching inwards to the constriction between the 1st and 2nd vesicles. In a chick of the 3rd or 4th day there can be seen in front of the optic lobes in series from behind forwards a posterior division of the 1st vesicle, an anterior division, the cerebral hemispheres and the olfactory lobes. In the middle of the 3rd day the 3rd cerebral vesicle is divided into a series of five parts separated by slight constrictions; the auditory vesicle lies opposite the constriction between the 4th and 5th parts. During the 3rd day the divisions become more strongly marked, and on the next day they have a more complex appearance; and after that the first compartment alone remains distinct as the cerebellum, whilst the divisions between the others disappear in the thickening of the cerebral walls. These constrictions the author considers warrant belief in the existence of a larger number of segments in the head than 'is usually admitted.

MALFORMATIONS.—Harrison Allen writes a report (Philadelphia, 1875) upon the AUTOPSY ON THE BODIES OF THE SIAMESE TWINS. He states that the connecting band passed from the one body to the other at the junction of the abdominal and thoracic regions immediately below the ziphoid cartilages. It measured nine inches in circumference, and was broader above than below. On the lower surface was a scar where the skin was adherent. This was one inch in length, and was considered to be the umbilicus. In both bodies a mass of fat, one inch in diameter, was found in the position of the normal umbilicus. On reflecting the skin from the band the superficial fascia was exposed, and beneath this on Chang's side was the tendon of the external oblique, and the aponeurotic attachment of the pectoralis major. On Eng's side in addition to this there was a second layer of fascia attached below to the linea alba of Chang. Chang's linea alba was found to be inserted into Eng's ensiform cartilage. The author next describes the structures within the band. Passing into the commissure from each body was a prolongation of hepatic substance. These were each enclosed in a peritoneal pouch, and they took origin from the anterior margin of each liver, and were continuous with each other. Passing through this hepatic prolongation was a branch of Chang's portal vein. Below the hepatic pouches were two other pouches of peritoneum, and each of these communicated with both peritoneal cavities. Towards the middle of the band was a fibrous septum. The ensiform cartilages were found to stretch into the commissure. They were very long (seven and eleven inches respectively), and were united by a symphysis. The vertebral column of each showed a marked lateral curvature, and the ribs of each were only twenty-two in number—seven true and four false.

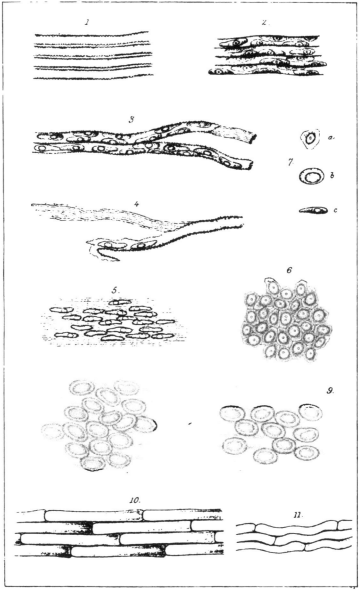

Plate X.

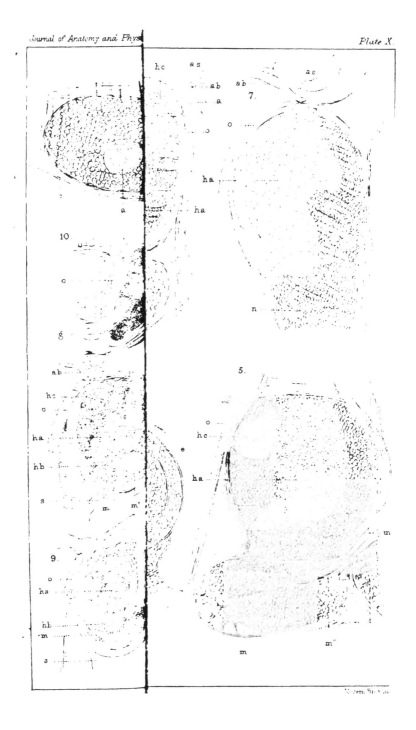

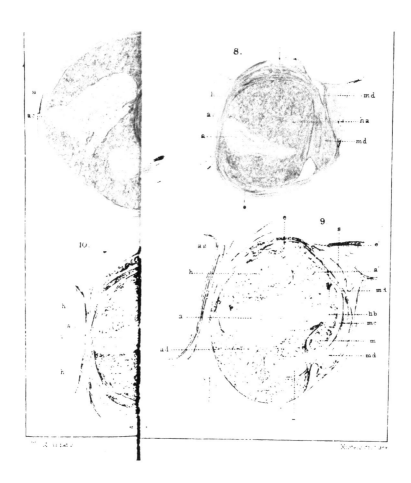

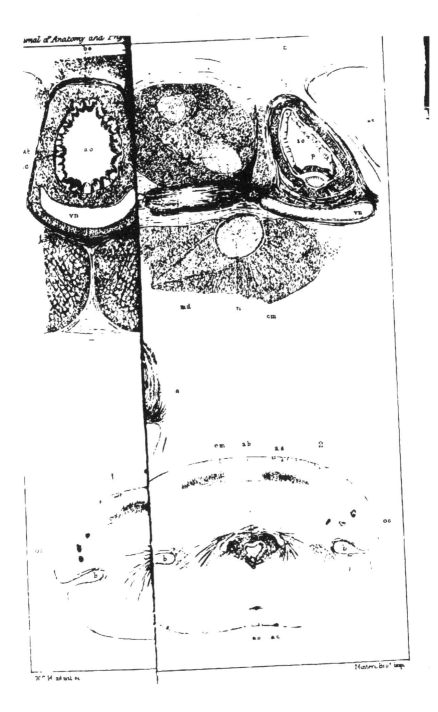

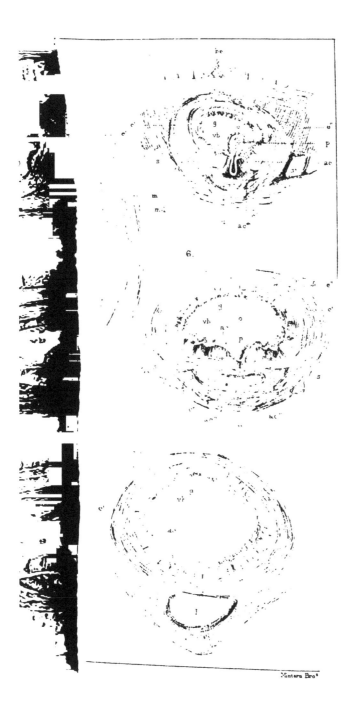

6.

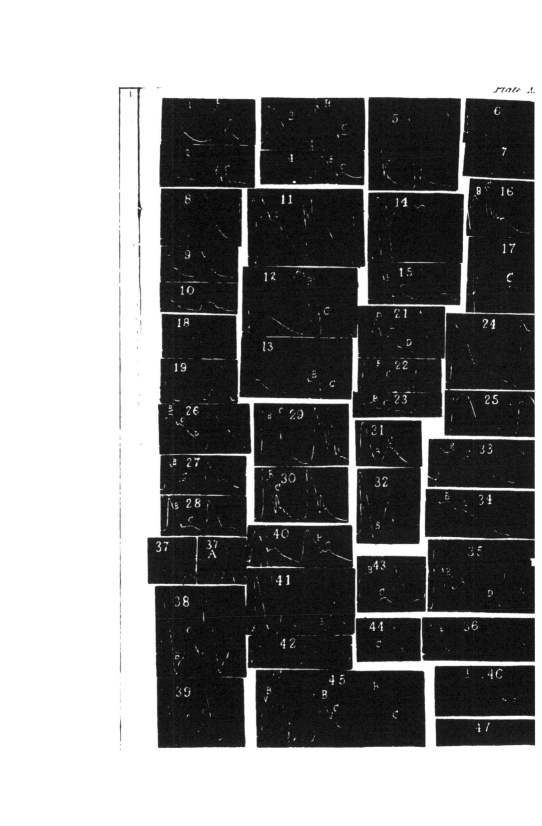

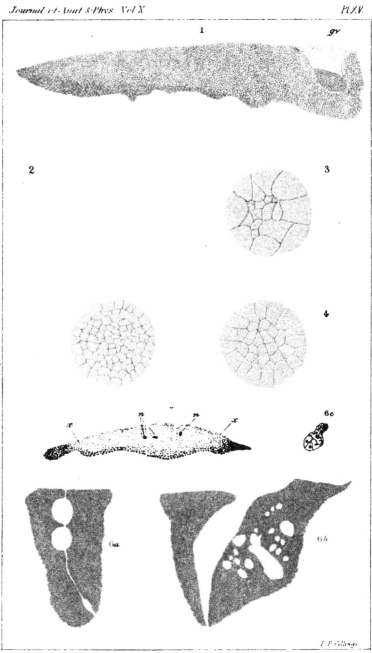

Pl XVI.

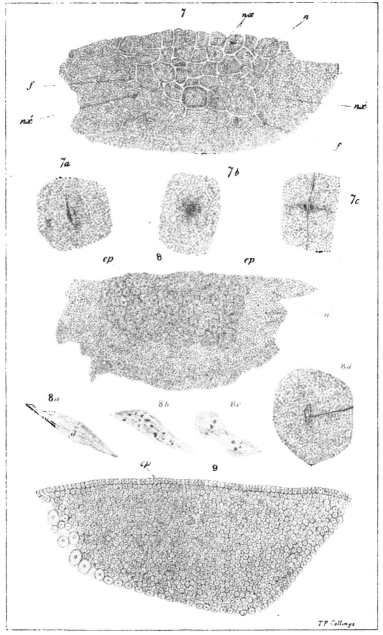

Fig. 1.

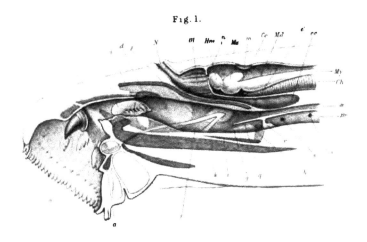

Fig 2.

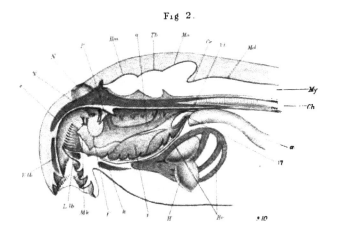

T P Collings.

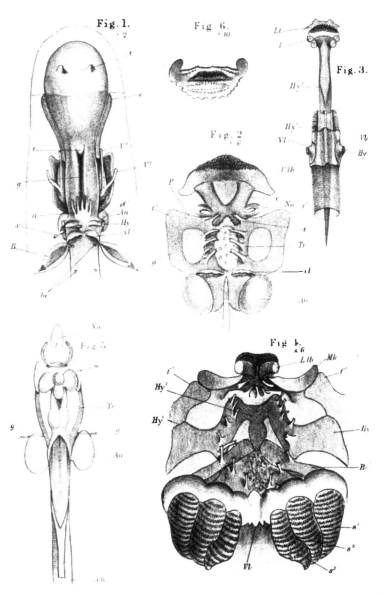

Fig. 1.

Fig. 6.

Fig. 2.

Fig. 3.

Fig. 5.

Fig. 4.

T. P. Collings

Journal of Anatomy and Physiology.

ON THE RELATIVE POWERS OF FRESH AND PRE-VIOUSLY USED PEPSINE IN THE DIGESTION OF ALBUMIN. By ARTHUR RANSOME, M.D., M.A.

IN the year 1868 I attempted to ascertain quantitatively the amounts of peptones produced by the action of different kinds of pepsine upon albumin, by dialysing the digestive fluid, through parchment paper or sheep's gut. It was presumed that the quantities of peptones passing through these septa would give the relative powers of the several samples of pepsine.

In the course of these experiments specimens of pepsine which had been previously used were sometimes compared with similar samples of fresh pepsine, and I was surprised to find in several instances, especially when thin sheep's gut had been used for the septum, that more digested albumin had dialysed from the used than from the fresh pepsine mixture.

The experiments were certainly not free from suspicion on the ground of the uncertain action of the dialysers, or of a possible influence of the pepsine upon the diaphragm. No further attention was therefore paid to the observation at that time.

I have recently been able to obtain the aid of Mr Gibbon, formerly assistant in the physiological laboratory of Owens College, and he has made for me the following experiments for the purpose of more accurately measuring the powers of used and unused pepsine in digestion.

1. Pepsine was prepared from a pig's stomach, by Brücke's method. 2. A standard solution of pepsine was prepared containing 0·02 grammes of the ferment in 100 cc. of water, and 0·2 per cent. of hydrochloric acid. 3. An air-chamber was so arranged as to remain approximately at a temperature of from 35° to 38° cent. 4. 100 cc. of the pepsine solution were now

mixed with 100 cc. of a solution containing 0·2 per cent. of hydrochloric acid, and 2 grammes of albumin, which had been
· dried and dissolved in the acid. The mixture was then put
into the air-chamber for 24 hours, after which it was taken out
and divided into two portions of 100 cc. each. One of these
portions was boiled to destroy the pepsine it contained, 10 cc. of
it were added to 80 cc. of a solution of 1 gramme of albumin
with 0·2 per cent. of hydrochloric acid, and 10 cc. of the standard pepsine solution were added, making the mixture up to
100 cc. in bulk. Of the other unboiled solution 10 cc. were
taken and added to 90 cc. of water containing 0·2 per cent. of
acid and 1 gramme of albumin. These two solutions may be
denoted by the letters A and B respectively, and they differed
from one another in one of them (A) containing pepsine which
had been used, while the other (B) contained fresh pepsine.
Both solutions were placed in the air oven (heated to 37° cent.),
and after the lapse of 6 hours both were tested quantitatively
for albumin.

5. The unchanged albumin was determined by taking
50 cc. of the digestive fluid, cautiously neutralizing with sodium bicarbonate solution; again acidifying slightly with acetic
acid, and then boiling the solution to coagulate the albumin.
The coagulum was washed on a dried and weighed filter
paper—dried thoroughly, and the weight ascertained. After
6 hours of digestion, the amount of unaltered albumin obtained
from solution (A) containing *used* pepsine was 0·87 gramme,
whilst from solution (B) with *fresh* pepsine 0·95 gramme were
recovered. At the end of 40 hours the quantities of uncon-
verted albumin were respectively, in solution A 0·5 gramme, in
solution B 0·75 gramme.

Six separate sets of experiments were now carried out,
having for their object the further investigation of the action of
used and unused pepsine, and the accurate estimation of the
quantities of this substance required for the digestion of albumin. In each set of experiments the digestive fluids to be
compared contained exactly the same materials, in the same
proportions, and only differed in the circumstance of the ferment employed having been used or unused. The period of
time during which each set of fluids was exposed in the air-

oven was varied, and the temperature of the oven varied somewhat, but this variation was presumed not to interfere with the
experiments, as each kind of fluid was subjected to the variation
at the same time.

The successive experiments were conducted as follows:

The digestive fluids, A and B, were again diluted, to form
fluids C and D, so that each 100 cc. contained $\frac{1}{10}$ th the quantity of pepsine that they did at first, namely, 0·001 gramme.
The same process was then followed as in the first experiment,
and at the end of 6 hours, the quantities of unchanged albumin
were, for the used pepsine 0·89 gramme, for the fresh pepsine
0·93 gramme. After the lapse of 20 hours, the numbers obtained were respectively 0·75 and 0·82. On the next dilution
to form fluids E and F, containing 0·0001 gramme of pepsine to
each 100 cc. (i. e. $\frac{1}{100}$ th of the first quantity), after 12 hours of
digestion the numbers obtained were respectively 0·95, 0·985;
and after 24 hours, 0·75 for the used, and 0·85 for the fresh
pepsine, and so on for the whole seven sets of experiments. In
three sets of trials the dilution was carried on until there was
only 0·000001 gramme of pepsine in each 100 cc. of solution,
and with the same kind of result.

Three facts seem to be established by these experiments.
1. The excessively minute quantity of pepsine capable of affecting albumin. 2. Its power of continuous action. 3. The
greater activity of used than of fresh pepsine.

The first of these propositions has been amply proved before:
by the researches of Brücke, von Wittich and others. It is
probably due to the specific attraction of the ferment for one
kind of substance, and for one kind only. So soon as the
albumin with which it is in contact is changed into peptone,
this resulting substance has no longer any affinity for the pepsine, it is at once relinquished and the ferment is free to act
upon fresh portions of material.

2. The power of continuous action is probably to be explained in the same way. In any case the fact has been
fully confirmed by other observers. Thus Vogel remarked, " ce
ferment ne s'épuise pas," and more recently von Wittich shews
that the power of pepsine, when fresh acid is continuously

31—2

added, is truly unlimited ("seine Wirksamkeit in diesem Sinne
allerdings unbegränzt erhalten bleibt[1]."

That the Amylo-lytic ferment of the saliva, a substance
closely allied to pepsine, also possesses this power of continuous
action, has been amply proved by Dr M. Foster, in the first
number of the *Journal of Anatomy and Physiology*[2].

3. As to the remarkable superiority of the used over the
fresh pepsine, it is interesting to note that in this respect again,
ptyaline resembles pepsine. In his experiments on this substance, Dr Foster compared two series of trials, one with used
the other with unused ferment. "In the last members, where
many hours were required for the conversion of the starch, the
result was slightly in favour, not of the latter, but of the former
series." This increased power Dr Foster connects with "the
so-called spontaneous conversion of boiled starch." If this
explanation, however, could be shewn to be true of the Amylo-
lytic ferment, which may reasonably be doubted, it would not
apply to the experiments now under consideration. The solutions of albumin were placed under precisely the same conditions, and any change going on spontaneously in the one ought
equally to have affected the other.

The fact opens up a wide field for speculation, and without
venturing at the present time to offer any hypothesis to account
for it, I would place in connection with it certain observations
that have been made in other directions.

1. The increase of catalytic power gained by many substances when finely divided. Thus even glass and earthenware,
when well pounded and reduced to powder, have been shewn to
possess catalytic powers. The most energetic catalytes are those
in which the active element is not only finely divided, but their
particles are separated by some foreign substance: thus wood
charcoal is inferior in power to animal charcoal, in which the
particles of carbon are separated from one another by earthy
matter[3].

---

[1] *Pflüger's Archiv*, Band v. p. 485.
[2] The denial of this power by Paschaten ("zur Frage über die Wirkung des
Speichels auf Amylum," *Centralblatt*, 1871) is hardly entitled to much weight
in face of the conclusive experiments made by Dr Foster.
[3] See also on this subject a paper by the writer on some conditions of
molecular action. *Phil. Mag.*, May, 1867.

It is perhaps possible that the used pepsine may have become more intimately mixed up with the particles of albumin, and may thus have become more finely divided and able to act with greater vigour.

2. The spontaneous changes that all colloidal substances undergo with lapse of time; as Prof. Graham has shewn, "their condition is a continued metastasis," "the colloidal being a dynamical state of matter, the crystalloidal the statical condition."

It is probable that pepsine and indeed all the "hydrolytic" ferments are colloids. ("Pepsine ist vollkommen indiffusibel," v. Wittich.) It seems possible that when this substance is constantly in presence of the materials upon which its energies are mainly employed in fermentation, then the internal work, by which the drawing together, the *pectisation*, of its own particles is accomplished, is more slowly performed.

The *pectisation* of the unused ferment, on the other hand, in the presence of heat and moisture may be carried on more rapidly, and its fermentative power may thus be gradually destroyed.

The fact that pepsine will still digest albumin when it is partly putrid does not preclude this explanation, since there must remain some portion of the ferment possessing molecular power, and if this power is not exerted upon the fermentescible substance, it may be producing changes in its own constitution.

3. Dr Burdon Sanderson has pointed out to me Haidenhain's recent researches on the allied ferment pancreatin. From these it appears that there exists in the pancreas a substance which he terms "zymogen," from which pancreatin is slowly evolved. That in fact "a glycerine extract of a perfectly fresh pancreas contains either no ferment ready formed, or only the slightest traces thereof, whilst the extract of a gland which has remained a longer time after death acts very powerfully upon fibrin[1]."

It is possible that some such zymogen exists in Brücke's pepsine, and that the transformation into pepsine goes on more rapidly in presence of the substance to be acted upon.

4. The enquiry may be hazarded whether certain peculiari-

[1] *Pflüger's Archiv*, x. 583.

ties in some of the disease-ferments, notably those of typhoid fever and cholera, may not be in some way connected with these facts. It is probable that the poison of typhoid fever becomes much more virulent after a sojourn of several days in contact with decomposing animal matter. Moreover, although Pettenkofer's theory of cholera is in itself more than doubtful, the facts upon which it is founded point to some similar peculiarity in the ferment of this disease.

The observations of Dr Thiersch on the increasing virulence of cholera excreta during a period of four to five days after their emission have been fully confirmed by Dr Burdon Sanderson.

Although the phenomena in question are probably to some extent due to vital action, it seems important that they should be borne in mind in reviewing all the facts bearing upon this difficult subject.

CONTRIBUTIONS TO THE ANATOMY OF THE CUTIS OF THE DOG. By WM. STIRLING, D. Sc., M.D., *Demonstrator of Practical Physiology in the University of Edinburgh.* (Pl. XIX. XX.)

THROUGH the investigations of Emminghaus[1], who collected directly the lymph formed in the skin, it became possible to arrive at more just conceptions respecting the conditions which are concerned in the formation of this juice. The assumptions, by which the observations made by him might be explained, must always remain to some extent uncertain, as long as we are ignorant of the channels which the fluid that has exuded from the vessels has to take, in order to pass from its place of origin into the lymphatics. The completion of this omission could only be obtained by a new anatomical investigation of the skin directed to this point, the anatomical investigation being undertaken with the view of planning new experiments from the results thus obtained.

The anatomical data which are furnished to us by the important works of Rollet, Langer, and Tomsa were not sufficient for this purpose, and specially so, because those properties only of the cutis by which it is rendered fit to be the elastic investment of the trunk and extremities were taken into account by these observers. The researches of the above-named authors were made with special reference to the human skin, but as the physiological experiments were observed on the dog, it must be ascertained how far the structure of the human skin coincides with that of the cutis of the dog.

When however, under the direction of Prof. Ludwig of Leipzig, I began the investigation of the skin of the dog, it soon became apparent that its schema was to be regarded in a different light, from that which was to be expected from the results of Tomsa on the human skin. At the same time, great difficulties presented themselves in the manner in which I

---

[1] Ueber die Abhängigkeit d. Lymphabsonderung vom Blutstrom. H. Emminghaus, *Arbeiten aus d. phys. Anstalt zu Leipzig*, 1874.

wished to pursue my investigations, inasmuch as I did not succeed in preparing the closed lymphatic channels in the cutis of the dog, which are so generally obtained in the human skin, when the necessary methods are employed.

For this reason my investigation is far from having yielded the wished-for results. Still I hope to be able to communicate that which is not altogether valueless[1].

In order to render the true connection of the elements of the skin accessible to the microscope, it was first of all necessary to bring the skin into a sufficient state of softness, to permit sections to be made in all directions, and also to reduce it to such transparency, that comparatively thick sections could be examined with good results.

The method which I found to succeed best, was that of *digesting* the tissue in artificial gastric juice. This method I have already fully described in the *Journal of Anat. and Phys.* x. 185. I found that the skin of the dog was much more easily digested than that of the human subject.

The sections can afterwards be tinged with carmine or logwood without becoming shrivelled up, and can be hardened in bichromate of potash; in short, they can be manipulated in any of the ways known to the microscopist. Skin in which the blood-vessels have been injected can with equal advantage be subjected to the digestive process. I shall describe shortly the method by which I injected the blood-vessels, because by this method I have been able to prepare such complete injections as can be obtained with the ordinary methods only under specially favourable circumstances. The injection fluid I used was a clear watery solution of Berlin blue. The extremities were the parts injected. The pressure under which the fluid was forced into the vessels was constant, between 100 and 200 mm. of mercury. In the limb which was to be injected, a cannula was placed in the large arterial trunk; and after it was properly tied in, the limb was constricted with a strong brass wire immediately behind the cannula. This wire was drawn as tight as possible around the limb by means of strong wire-pincers, such as are used by gas-fitters for screwing tubes.

[1] This research was originally published in the *Berichte d. Math.-phys. Classe der könig. sächs. Gesellschaft d. Wissenschaften*, 1875.

Great care must be taken that this closure is complete, for upon its completeness depends the success of filling the vessels perfectly. After the band has been put on, the coloured fluid is permitted to flow into the cannula which has been placed in the artery, as long as an inward current takes place under the above pressure. This generally lasts many hours though with diminishing rapidity. During this time several hundred cubic centimeters of the fluid can be driven into the limb below the middle of the thigh, the part becoming greatly swollen and turgid. The result is, that the constituents of the fluid are precipitated inside the vessels; so that the latter appear completely stopped up with blue granules, while the water passes through the wall of the vessels into the connective tissue, and produces therein a considerable œdema, and even often flows in fine drops through the epidermis.

I now proceed to the description of what appears to me to be of value in my observations.

1. *The relation of the elastic fibres and the connective tissue corpuscles to the collagenous bundles.* In a section made vertical to the free plane of the skin (Fig. 1), one notices that the space, which remains free between the hair follicles, sebaceous glands, and sweat glands, is filled up by an elastic network, and a large number of scattered cells. The meshes of this network are bounded by arched fibres, which are of unequal thickness. The strongest elastic bands run at great distances from each other; and from them fine branches proceed, subdividing the spaces between the strong fibres. The greatest number of strong bands occurs in the middle of the cutis. On following them into the space between two hair follicles, towards the surface, we find that they proceed, for the most part, from the surface which surrounds the hair follicle, indicating that the latter is covered by a felted arrangement of elastic tissue. From this covering of the hair follicle there stretches in the direction of the so-called erector pili a strong elastic band of fibres, which gradually divides into bundles as it proceeds towards the surface, and the finer bundles are continued directly into the elements of the elastic net-work which pervades all parts of the cutis. As in a section, such as is given in Fig. 1, the meshes

remained distended, the collapse of the limiting fibres must be prevented by a mass lying between them. It therefore follows, that the bundles of connective tissue, which fill up these meshes, are not completely dissolved, but have only become swollen up, and have acquired a very high degree of transparency.

The section given in Fig. 2, which was made parallel to the surface, may serve to complete Fig. 1. It was made from a piece of skin which had not been digested so long as the previous one. The space between the hair follicles is here again seen to be filled by fibres and scattered corpuscles. The elastic fibres are distinguished from the collagenous ones by their well-known characters. Whilst the former are distributed as a net-work over the whole surface, as far as it is not taken up by the hair follicles, the latter are disposed as straight bundles, frequently crossing each other in the spaces between the hair follicles. The plane of these crossings lies, in by far the greater number of cases, parallel with the surface of the cutis, which is shewn by the fact, that although the different horizontal sections were made through the thickness of the skin, no collagenous fibres, cut across vertical to the direction of the bundles, are to be seen. This disposition differs from that in the human cutis; for, while the human skin is made up of collagenous bundles which run obliquely upwards from the subcutaneous connective tissue towards the epidermis, the cutis of the dog, analogous to the cornea, appears to be built up of many layers superimposed upon one another, and held in position by the elastic fibres running between and embracing them.

The original form of the cells scattered in the stroma of the cutis—which from their position are certainly not to be referred to the epidermis, the glands, the sheaths of the nerves or the blood-vessels—is somewhat changed in consequence of the treatment with gastric juice. Still two sorts of corpuscles, differing in their form and appearance, are always to be distinguished. The one of these shews round, the other spindled-shaped nuclei. The cells with round nuclei occur much less frequently than the others, and appear especially at those places which are richly supplied with blood-vessels. They are therefore present in large numbers in the upper layers of the cutis, upon which the epidermis is planted; and here they are seen specially along the course of the capillaries, see Fig. 5. They are still more numerous and much more extensively distributed in the subcutaneous connective tissue than at the above-named level (Fig. 6). As the nuclei appear very similar to the nuclei of the lymph corpuscles, and as in the subcutaneous connective tissue the existence of rich networks of lymphatics can easily be demonstrated, it will scarcely be considered erroneous, if we assume a connection between the cells with round nuclei and the contents of the lymphatics.

With regard to the cells with spindle-shaped nuclei, the cell-plate, like the elastic fibres and keratin, offers a considerable resistance to the dissolving action of the gastric juice; and part of it is retained, together with the nucleus, after very prolonged digestion. Hence, in preparations whose collagenous tissue has completely disappeared, we still find many of these cells in their normal position. When the digestion has reached this stage, the cells may be often seen covering the stronger branches of the elastic network in continuous layers. If the digestion has not proceeded so far, and parts of the collagenous fibres are still visible, the cells then appear between them, the long axis of the spindle-shaped nucleus running in the direction of the bundles of fibre. According to this view, these cells would belong to that species, which Schweigger-Seidel has described on the bundles of fibres of the cornea, and Ranvier on those of tendons.

From the above-described anatomical arrangement we may understand how it is that the skin, although its fibres are in connection with subcutaneous connective tissue, can assume such variable degrees of tension, and can do so independently of the tension of the subcutaneous connective tissue. Forasmuch as the collagenous bundles swell up in watery solutions, much more than the elastic fibres by which they are surrounded, they, according to the composition and quantity of the fluid bathing them, must communicate to the elastic fibres a greater or less, but certainly appreciable, tension. For the elastic fibres oppose the tendency to extension of the collagenous fibres. Besides that imbibed by the tissues, fluid is also contained in the skin in another way. This is shewn by the well-known fact, that fluid can be pressed out of the cutis by a force which is not capable of expelling the water imbibed by the tissues themselves; and this fluid is probably contained in the spaces which must exist, for each of these constituents, by virtue of its elasticity, tends to maintain its peculiar shape, and to reassume it when the external forces which have disturbed it are removed.

For the correct apprehension, however, of the movement of the tissue juices, a more exact investigation of these relations described must be made, and we must consider

2. *The distribution of the blood-vessels.* From a piece of skin which has been injected with Berlin blue and subjected to digestion, a preparation such as is given in Fig. 3 can without great difficulty be obtained. Of course in this case the arteries appear of a blue colour. In order to distinguish the arteries from the corresponding veins, it is best to follow the small branches back to the trunks in the subcutaneous connective tissue, as they lose the rings of non-striped muscular fibre shortly after their entrance into the cutis, instead of retaining them to their termination in the capillaries, as is usual with the small arterial branches in other parts. The course of the vessels is the following :

From the arterial branches in the subcutaneous connective tissue (*A*) fine branches spring, which ascend between each set of four (in the section between each two) hair bundles. From these branches other finer branches proceed, of which one (*a*) goes to the mass of fat and the closed end of the sweat-gland, a second (*b*) to the hair follicle, a third (*c*) runs to the sebaceous glands of the hair—this generally sends fine branches to the erector of the hair. The remainder of the ascending arterial twigs (*d d*) proceed to the superficial aspect of the cutis, *i.e.* to the under surface of the epidermis. Each of these branches passes into a capillary from which a vein springs; and the latter descends, along with the corresponding artery, to its origin. The distribution of the blood-vessels in the skin of the dog corresponds essentially to that which, since the investigations of Tomsa, we are familiar with in the human skin. In the case of the dog also, the capillary vessels are completely absent from the masses of connective tissue which are situated between the fat, the muscles and the glands. In pieces of dog's skin which have been treated with artificial gastric juice the want of the blood-capillaries in the connective tissue is very obvious, because the masses of connective tissue possess the property of swelling up in such a high degree.

I have made some observations on the structure of certain of the constituents of the dog's skin after treatment with artificial gastric juice, which I desire to mention.

3. *The sweat-glands are very numerous.* Between each hair bulb one sweat-gland, at least, is to be found. They begin usually below the masses of fat, which lie under the roots of the hairs. From the coil a twisted tube proceeds, which is continued into a long neck, and ultimately opens with a small funnel-shaped dilatation into the hair bulb. Fig. 1 *b*. The place where the sweat-gland passes into the hair bulb, is, as Chodakowski[1] correctly noted, above the opening of the sebaceous gland. It is, however, a considerable distance from the exit of the hair on the surface of the cutis.

The hair-follicles are arranged like a brush, and, as is known, it is the rule for several to unite before they open on the surface of the skin. A sweat-gland opens into the region of the hair follicle common to several hairs. Although, therefore, the skin of the dog contains much fewer sweat-glands than hairs, it is, as above noticed, by no means poor in sweat-glands. It has long been known that numerous sweat-glands open on the surface of the pad of the dog's foot.

The structure of the sweat-glands of the dog varies in an important point from that of man. According to the observations of Heynold[2] the tunica propria of the human sweat-gland consists of several layers of flat cells, upon the inner side of which the epithelium of the gland-tube is placed. In the dog the wall of the tube, from without inwards, also begins with a layer of flat cells, but this does not lie next the epithelium, but is everywhere separated from it by a uniformly strong layer

---

[1] *Ueber die Hautdrüsen einiger Säugethiere*, Dorpat 1871.
[2] *Virchow's Arch.*, LXI., p. 77.

of homogeneous substance. Fig. 7 *A*. This layer is of so firm
a consistence that it retains its character as a homogeneous
membrane, even when the digestion has proceeded so far that
the cells lying within the tube have been broken down into an
amorphous mass. Fig. 7 *B*.

Inasmuch as the hydrochloric acid increases the size of
many of the constituents of the skin by imbibition, it might
be imagined that this had taken place with the membrana
propria of the sweat-glands. But, even if it had much less
thickness, a homogeneous membrane separating the cavity of
the gland from the surrounding textures would be of importance in an organ whose function essentially consists in the
occasional rapid exudation of watery solutions.

4. A clear notion of the *structure of the erector pili* is
obtained from the digested skin. The cord, which is described
as the erector muscle of the hair, consists, in my preparations,
of a strong bundle of elastic fibres in which a large number of
muscular fibre-cells are imbedded. The elastic fibres take
their origin from the elastic covering of the hair follicle.
From this they proceed, as a compact fibrous mass, in the
ordinary way, towards the free surface of the cutis. Before however it reaches this point, the bundle divides in an irregularly
fan-shaped manner, into a large number of finer threads.
Each of these again divides into fibres which ultimately pass
into the elastic network spread out close under the free surface
of the cutis. (Fig. 9.)

Between these elastic fibres, and especially between those
which form the stem of the elastic bundle, muscular fibre-cells
are more or less richly imbedded (Fig. 8 *a a*). These appear
strongly pronounced after the action of carmine, as the elastic
tissue does not take up this colouring matter.

According to this description the muscle of the hair follicle
does not consist of smooth muscular fibres which spring from
the hair bulb, ascend through the thickness of the cutis as
such, and end in the surface of the latter, but is composed
of an elastic band, the constituents of which can be moved
upon each other by the included muscular fibre-cells.

5. The bundles of connective tissue of the cutis sometimes

undergo a change under the action of the digestive fluid, in consequence of which they appear not unlike transversely striped muscular fibres. (Fig. 8.) One might easily make this mistake were it not for their great numbers, and for certain peculiarities of the individual forms. To the latter belongs the passage of a transversely striped into a completely smooth piece (Fig. 9 *b*), which often, at a twist, shews the excessive thinness, the ribband-like nature, of the structure. Amongst these bands there often occur others, so that the transverse markings depend on a further development of the fibre twisted round them, as was first described by Henle (Fig. 8 *e*). From this arrangement of the fibres it may be assumed that the transverse markings belong to a membrane, or sheath, from the interior of which the collagenous substance has been removed by means of digestion, whilst the markings, being of an elastic material, have resisted the action of the solvent. These sheaths cover not only the large bundles of connective tissue, but extend even to the very fine ones. Often it appears as if the sheaths divided, Fig. 9 *d*. It is however difficult, owing to the fine nature of the structure, to determine whether the division actually takes place, or whether the appearance is due to the superposition of originally separate bundles.

6. In pieces of skin, in which the digestive process has proceeded so far that the superficial layers of the cutis can be easily removed with the razor as a pulpy mass, the blood-vessels present certain peculiarities (Fig. 4). The capillaries appear as exceedingly delicate structures, which are composed of spindle-shaped cells arranged one after the other. In appearance and distribution they so much resemble the network of nerves well known as occurring in other surfaces—*e.g.* cornea —that at first one looks in vain for some character by which to distinguish them. The identity of the network, however, with that of the injected capillary plexus which lies on the most superficial part of the cutis, appears at once when we compare the two. For comparison with Fig. 4 see Fig. 5, which was drawn from an injected preparation, taken from a piece of skin corresponding to Fig. 4. The proof, that in the latter case one has to do with capillary vessels, will be com-

plete when the nerve-fibres themselves are discovered.   In Fig.
4 a the non-injected capillaries present a somewhat different
appearance; their course is straighter, and the nuclei, which
are present in the threads, are placed at long intervals from
each other, and any adjoining two are united by a straight line.

EXPLANATION OF THE PLATES, XIX. AND XX.

Fig. 1.   Vertical section through the skin of the dog—partly schematic.

Fig. 2.   Horizontal section parallel to the surface of the skin.   Hartnack, Obj. 7.   Oc. 3.

Fig. 3.   Blood-vessels injected.   Arteries red, veins blue, corresponding
to *Fig.* 1.

Fig. 4.   Non-injected network of blood-vessels lying in the most superficial layers of the cutis.   Hartnack, Obj. 7.   Oc. 3.

Fig. 5.   The same network of vessels injected.   Hartnack, Obj. 7.   Oc. 3.

Fig. 6.   Horizontal section through a deeper part of the cutis.   Hartnack, Obj. 7.   Oc. 3.

Fig. 7.   *A.* Sweat-gland of the skin of the dog opening into the hair
follicle.   *B.* with the secreting elements inside the membrana propria
destroyed by the digestion.   Hartnack, Obj. 7.   Oc. 3.

Fig. 8.   Termination of the erector pili upwards.   Hartnack, Obj. 8.
Oc. 3.

Fig. 9.   Various forms of connective tissue bundles from the skin after
digestion in artificial gastric juice.   Hartnack, Obj. 7.   Oc. 3.

OBSERVATIONS ON TASTE-GOBLETS IN THE EPIGLOTTIS OF THE DOG AND CAT. By R. H. A. SCHOFIELD, B.A. B.Sc. Pl. XXVII.

IN this paper I propose to describe certain structures occurring in the epiglottis of the Cat and Dog, which in their microscopic characters closely resemble the well-known taste-goblets of the tongue. The observations to be recorded were made under the direction of Dr Klein in the laboratory of the Brown Institution.

Method. The method adopted in preparing the specimens was briefly as follows: The fresh epiglottis was placed in a mixture of two parts of $\frac{1}{4}$ p. c. solution of chromic acid to one of methylated spirit and left there for about a week, after which time it was transferred into absolute alcohol. The organ thus prepared was imbedded, and longitudinal and transverse sections cut, which were stained and mounted in the usual way. The microscopic examination of the sections proves that there exist on the posterior, *i.e.* laryngeal surface, " taste-goblets," the structure of which, as far as it could be ascertained from my specimens, agrees in the main with the description given by Engelmann[1]. I have corroborated the following points, viz.

The goblets occupy cavities in the epithelium, which are somewhat flask-shaped, and whose bases rest on the connective-tissue surface of the mucosa. The taste-goblets which lie in these cavities consist of 15—30 elongated cells which are set like the leaves in a bud; they are closely arranged in several concentric layers around the axis of the goblet.

Of these cells there are two principal kinds. (1) Those forming the external layer (Deckzellen), which are elongated spindle-shaped bodies, enclosing an ellipsoidal, vesicular nucleus placed near the centre. They consist almost entirely of hyaline protoplasm, and appear to possess no cell-wall; they are crescentic in outline with the concavity inwards, and cover in (2) the *axial cells* (Geschmackzellen), which are elongated, thin, almost homogeneous, and highly refractive. Each consists of an ellipsoidal body, which is chiefly occupied by a vesicular nucleus, and which at its upper pole runs out into a

[1] Stricker, *Handb. d. Histolog.* 822—830.

moderately thick, and at its lower pole into a very thin, fine process. The tip of the upper process is truncated, and at its extremity is a little hair-like body. These axial cells are supposed to be connected at their lower extremities with nerve fibres.

Distribution. As to the distribution of the goblets the chief points of interest which I have observed are : (1) They occur only on the posterior (laryngeal) surface of the epiglottis. (2) They are always met with in the lower half of this surface : none were seen near the tip of the epiglottis. (3) They appear to be arranged approximately in rows both horizontally and vertically : for in many of the sections (both horizontal and vertical) they were found at regular intervals. (4) With respect to their number, it may be mentioned that in a small dog's epiglottis (cut longitudinally) which yielded 184 mounted specimens—the usual deduction being made for waste in cutting—goblets were observed in 53 sections (rather less than ⅓ of the specimens), while in another dog's epiglottis (cut horizontally) of which 48 mounted sections were obtained (all those taken near the tip, where goblets are invariably absent, being excluded), goblets were observed in 15—again rather less than ⅓ of the sections. The proportion in the two cases is very nearly the same, whence we may infer that the distribution over that part of the laryngeal surface of the epiglottis is tolerably uniform. (Fig. 1.) 5. The surface of the epithelium just over a goblet is invariably pitted : there is a shallow saucer-like depression with the tip of the goblet abutting against its lowest point. (Fig. 2.) (6) With each goblet is associated the duct of a mucous gland, which opens either at the side of the depression above mentioned, or on the free surface of the epithelium at some little distance. (Fig. 2.) This special relationship between the goblet and duct is of great interest, as Von Ebner[1] has shown that a similar constant relationship obtains between the taste-goblets of the tongue and the ducts of the glands which he terms " serous glands." Verson mentions and figures[2] on the posterior surface of the epiglottis in general certain "bud-like or pyramidal structures," which I have little doubt correspond to our taste-goblets.

[1] *Die Acinösen Drüsen der Zunge*, von V. Ebner, Graz, 1873.
[2] *Stricker.* German Ed. I. 457, fig. 124.

This author mentions that there is a canal in the centre of the goblet, from which a fine duct leads to the free surface of the epithelium.

Finally, as regards the connection of these goblets with nerve-fibrils asserted by some authors[1], I cannot give an opinion, as I have not observed it in any of my sections.

In one case a nerve-fibril was observed close alongside of, but apparently not entering, a goblet.

EXPLANATION OF PLATE.

Fig. 1. From sections of the epiglottis of a dog. (1) *Made in a plane at right angles to the long axis.* Showing five goblets, a, b, c, d, e. N.B. Between b and c in the figure a piece of epithelium equal in length to that between a and b has been (for convenience) omitted. (2) *Made parallel to the long axis of the epiglottis.* Showing four goblets a, b, c, d. Hartnack, Oc. 3, Obj. 4 (about 90 diam.).

Fig. 2. From a section of dog's epiglottis made parallel to its long axis. Showing the pits (g) on the surface of the epithelium (e) into which the ducts (d) of the mucous glands (gl) open; the left-hand duct is seen discharging a mass of mucus on to the free surface. The letter (g) is placed in a pit just over the apex of a goblet. (m m) the mucous membrane. The part of the section figured is from the posterior (laryngeal) aspect of the epiglottis towards its base. Hartnack, Oc. 3, Obj. 5 (about 160 diam.).

Fig. 3. Single goblets more highly magnified to show their structure. (a) Section at right angles to free surface. (b) Oblique. In the former the nuclei of the two sets of cells of which the goblet consists are plainly distinguishable. The nuclei (c) of the axial cells (geschmackzellen) being much more darkly stained than the nuclei (d) of the superficial cells (deckzellen). The latter are semi-transparent, and in many places the nuclei of the deeper cells can be seen through them. The outlines of the cells converging towards the tip of the goblet are clearly seen, and also the small hair-like processes in which they terminate. Hartnack, Oc. 3, Obj. 8, about 400 diam.

[1] Hönigschmied. *Jahresbericht, Virchow und Hirsch*, p. 72.

ON THE PHYSIOLOGICAL ACTION OF THE SALTS OF BERYLLIUM, ALUMINIUM, YTHIUM AND CERIUM. By JAMES BLAKE, M.D., F.R.C.S., *San Francisco, California.*

IN the present paper I shall relate experiments shewing the action of the salts of the above metals when introduced directly into the blood. The molecular properties of some of these substances have not yet been determined, as, with the exception of Aluminium, neither the specific heat of the metals nor the vapour-density of their chlorids has been ascertained; so that physiological investigation here assists us in determining the properties of inorganic substances where chemical methods have failed.

Salts of Beryllium.

Ex. 1. Animal, rabbit, weight 2315 grm.: injection in jugular; manometer connected with carotid. Inject. ·021. Be,b, as sulphate in 4 cc. water, pressure in arteries 110—120; 5″, pressure falling; 10″, 80—90; 15″, 110—120, animal not much affected, oscillation; 1′, 90—110, in pressure greater both at each pulsation; 3′, 90—115, and also in the general range. Inject ·017; 5″, pressure falling; 8″, 80—85; 12″, rising; 15″, 130—140; 45″, 70—85; 1′, 60—70, heart's action irregular, pulse 120, respiration; 2′, 100—115, not affected, animal sensible; 4′, 110—125. Inject. ·021; 4″, pressure falling; 8″, 80—90; 15″, 200—220, respiration stopped; 30″, 180—200; 50″, 130—150, pulse 40; 2′, 125—130, tonic spasm, no respiration; 3′, 90—110, slight spasm, oscillation 10 to 12 cm. at each pulsation; 4′, 40—60, pulse 24; 5′, 20—35, pulse 10, 5′40″, heart stopped. On opening the thorax the auricles were still contracting rhythmically; blood in right cavities dark, in left red but not bright. On examining the blood five hours after death the arterial blood was quite fluid, corpuscles subsided. There was a loose clot on the surface of the venous blood, but fluid underneath[1]. Under the microscope the corpuscles were found mostly angular, some quadrilateral, not crenated; about one-eighth were globular; molecular movements active. In arterial blood two-thirds of the corpuscles were globular.

---

[1] I think this clot on the surface of the blood was connected with its contact with the air, as I noticed some blood that had escaped on the table was coagulated half an hour after death.

Injection of Nitrate of Beryllium into the Arteries.

Ex. 2. A rabbit, weighing 2462 grm. The injection was made into the carotid towards the heart. Manometer connected with the femoral; pressure 115—120 cm. Inject. ·032; 4″, pressure rising; 7″, 180—200; 30″, 140—160, slight convulsive movements; 1′, 125 —130, respiration heavy, slow, 18; 2′, 120—125, animal apparently not much affected. Inject. ·064; 4″, pressure rising; 8″, 200—220, general spasm suspending respiration for 45″; 30″, 180—190; 1′, 150—165; 1′30″, 160—170, respiration slow, 15; 2′30″, 140—155, respiration stopped; 3′, 130—140; 4′, 70—80; 5′, 25—30; 5′50″, 8—10, heart stopped. On opening the thorax auricles beating rhythmically, left cavities full, scarlet blood; veins and right cavities almost empty, blood very dark. Immediately after the last injection the conjunctiva became perfectly white although before it had been quite red. The pupil was dilated to its fullest extent; it contracted slowly after death. Both the venous and arterial blood were coagulated six hours after death.

Salts of Alumina.

Ex. 3. Injection of sulphate of alumina into the jugular; manometer connected with the carotid. A rabbit, weighing 2305 grm.[1] Pressure, 100—110 cm., pulse 120. Inject. ·016 in 4 cc. water; 4″, pressure falling; 15″, 40—50, pulse 65; 30″, rising; 35″, 130—140, pulse 48; 45″, 130—150; 1′, 90—110, pulse 64, respiration slow, 15; 2′, 30—40, pulse 68, respiration stopped; 2′30″, 25—30; 3′, rising; 3′30″, 120—130, no respiration; 4′30″, 105—120; 4′50″, 110—130, pulse 78, respiration again commenced; 6′, 80—90, animal insensible, respiration 18; 7′, 75—85, pulse 75; 8′, 65—75, respiration suspended; 8′30″, 50—65, pulse 78, respiration renewed; 9′, 40—50, pulse 76, respiration 15, mostly costal; 13′, 30—35, pulse 70, respiration 10; 14′, 20—23, respiration stopped; 15′, 10—14, pulse 60, oscillations well marked; 15′30″, heart stopped. On opening the thorax the auricles were found contracting rhythmically. Both cavities of the heart contained blood, right most distended; blood in left red but not bright. Blood from right side coagulated; from left remained fluid for half an hour, then a loose coagulum formed. Six hours after death about one fourth of corpuscles in venous blood deformed, the round ones more globular; in arterial blood more deformed, corpuscles more angular.

Injection of Sulphate of Alumina into Arteries.

Ex. 4. Dog, weighing 7465 grm. Injection tube in left axillary artery pointing towards the heart. Manometer connected with

---

[1] This experiment was one of a series undertaken to compare the physiological action of beryllium and aluminium, in order to determine the molecular relations of the former. The quantity injected represents the weight of the hydrated oxide of alumina dried at 200°C., and dissolved in sulphuric acid.

femoral; pressure 125—130 cm., pulse 116. Inject. ·330 of sulphate
of alumina; 4″, pressure rising; 7″, 175—180, pulse 150; 1′, 150
—160, respiration not affected, no sign of suffering; 5′, 145—150,
pulse 148. Inject. 0·52 grm.; 7″, pressure rising; 20″, 290—305;
pulse 180; 45″, 280—300, respiration arrested; 1′30″, 270—285,
pulse 180, 2 or 3 slight respiratory movements; 2′, 250—260, pulse
150; 4′30″, 215—230; 5′, pressure falling rapidly; 5′30″, 80—90;
6′, 80—85; 7′, 50—53, pulse 64; 8′, 25—27; 8′15″, heart stopped.
On opening the thorax the auricles were contracting. Blood in both
cavities, in left red but not bright[1].

Injection of Sulphate of Alumina into the veins. General symptoms.

Ex. 5. Dog, weighing 3576 grm. A tube was inserted in the
jugular and the animal then set at liberty. Inject 0·25 grm. of the
salt. In 15″ the animal fell on its side and had a tonic spasm; 30″,
cried as if in pain; 1′, respiration arrested; conjunctiva insensible;
respiration suspended for 1′30″, and then again commenced and continued regularly. The animal gradually recovered its sensibility and,
after fifteen minutes, had so far regained its strength as to be able to
raise itself on the fore legs. On injecting 0·25 more of the salt
respiration stopped in one minute, and the animal was dead. The
salt used in the last two experiments was the ordinary sulphate containing about 15 per cent. of the oxide.

Salts of Cerium.

Ex. 6. Rabbit, weighing 2235 grm. Injection into jugular.
Manometer connected with carotid. The salt used was the oxy-
sulphuret, and as it is decomposed by water it was mixed with a
solution of gum arabic by which it was held in suspension. Pressure
in arteries, 120—130 cm. Inject. ·060 in 3 cc. water; 7″, falling;
20″, 50, respiration suspended; 45″, 50—55, respiration 38; 1′10″,
80—85, respiration 30; 1′20″, 150—156; 1′30″, 175—180; 2′, 125
—130, spasm for some seconds, respiration then becoming slow, 10;
3′, 85—90, respiration 8; 4′30″, 50—55, respiration 4, pulse 100;
5′, 25—27, respiration stopped; 5′30″, 10, no pulsation. On opening the thorax the auricles were beating rhythmically. General
vermicular movements in the ventricles. Right cavities distended.
Small quantity of scarlet blood in left cavities. The microscope
showed the corpuscles much changed, crenated; not more than one
in fifty had an unbroken outline. The blood coagulated. The salt
is evidently very poisonous, much more so than the salts of the lower
oxide or the serous salts, of which it requires eight to ten times the
quantity to prove fatal[2]. They differ entirely in their physiological

[1] 173 grm. of blood was collected after death, and analyzed for alumina.
The result gave for the weight of the blood about one-ninth that of the body.
[2] Chemists are not yet decided as to the molecular constitution of the oxides
of cerium; but from the physiological action of the ceric salts, I think the
oxide will be found to have the formula $Ce_2 O_3$, according to the views of
Mendeljerv, with the atomic weight 138.

action from the ceric salts, resembling to a certain extent the substances in the Baryta group in paralyzing the heart, and in causing contraction of the voluntary muscles long after death.

Salts of Ythia.

Ex. 7. Rabbit, weighing 2362 grm. Tubes in jugular and carotid. The salt used was the nitrate of ythia; and the weight indicates the weight of the oxide contained in the injection. Pressure 110—120. Inject. ·05 in 3 cc. water; 7″, falling; 30″, 70—75, respiration suspended; 1′, 70—75; 1′30″, rising; 1′45″, 125—130; 2′, 110—114, respiration renewed, but only three or four respiratory movements; 3′, 50—53; 4′, 23—25; 5′, 10—11. On opening the thorax the left ventricle was found beating rhythmically and continued beating for ten minutes; auricles still. Both cavities were full of blood. Venous blood coagulated. The blood from left side, which was dark, remained fluid for some hours. Corpuscles much altered, irregular amoeboid projections, crenated, oblong.

Injection of Sulphate of Ythia into the Arteries.

Ex. 8. Rabbit, weighing 2715 grm. Tube inserted into carotid. Manometer connected with the femoral. The weight represents the quantity of oxide contained in the solution. Pressure, 110—115, pulse 160. Inject. ·0125; 15″, 80—84, fall probably owing to contraction of pulmonary capillaries; 30″, 110—115; 2′, 110—120. Inject. ·025; 4″, 90; 10″, rising; 15″, 130—140, respiration quickened, cries; 30″, 130—160, great oscillations from heart's action; 1′15″, 140—165; 2′, 120—160; 4′, 135—160, respiration 30, pulse 140. Inject. ·025; 4″, 190—200, cries; 30″, 170—190, pulse 80; 1′, 145 —160, respiration suspended for 30″, then slow and deep: 2′, 140— 160, respiration 6, pulse 50; 3′, 140—150, oscillations caused principally by respiration, those from heart's action hardly perceptible; 4′30″, 110—115; 6′, 110—112, respiration ceased; 7′, 100—104; 8′, 80—83, pulse 42; 9′30″, 50—54, pulse 30; 12′, 25, no oscillations. On opening the thorax the heart was contracting feebly. Both cavities contained blood. The blood in the left side brighter than in right, but not scarlet. It coagulated firmly. The lungs were hepatized, and the bronchial tubes were nearly filled with a serous frothy secretion which must greatly have interfered with the aeration of blood. Coordinated movements of the fore legs continued for more than half an hour after the thorax had been opened. A spectroscopic examination of the blood, soon after death, showed plainly the presence of didymium and, I think, of erbium; so that the above experiments can but imperfectly represent the action of the ythia salts. Unfortunately the chemistry of these compounds is so imperfect that pure preparations of the ythia salts cannot be obtained; they are however evidently very poisonous.

The substances used in the above experiments all agree in the more marked physiological phenomena they give rise to when introduced directly into the blood. Their most striking action is on the systemic and pulmonary capillaries. This is shown as regards the pulmonary capillaries by the sudden fall of the pressure in the arteries that immediately follows their injection into the jugular. When this gives way and the substance passes on into the left side of the heart, and through the arteries to the systemic capillaries, these become contracted, so that, the blood not escaping from the arteries, the pressure rises rapidly, and in a few seconds is increased far above the normal amount. These substances also exert an action on the nervous system, as is shown by the arrest of respiration and its renewal sometimes after an interval of two and a half minutes. The action on the nervous system is more marked with the salts of cerium and ythium, more particularly in changes in the rhythm of the heart's action. This effect on the heart can, I believe, be carried so far as to arrest its pulsations even before the substances are applied to its parietes; at least I have seen the action of the heart arrested within two or three seconds when rather large doses of the cerium and ythium salts had been injected into the arteries. I think the muscular movements that continue so long after death are connected with their action on nerve-tissue. These substances are evidently not heart-poisons, as they are stated to be by Rabuteau, otherwise the heart would not go on contracting under a pressure of 150 cm. to 200 cm., and continue beating long after respiration is arrested. If they exert any influence on the irritability of the heart, it is to increase it, particularly in the muscular fibres of the left auricle, which has been frequently found contracting when every other part of the organ was still. As regards the connection between the atomic weights of these substances and the intensity of their physiological action, a series of experiments with the salts of beryllium and aluminium, conducted expressly to determine this point, shews that, at least as regards these substances, the connection is closer than I had supposed[1].

[1] As my previous experiments had been conducted to determine the general physiological action of inorganic compounds, no great attention was paid to the exact weight of the substances experimented, or to determine the smallest quantity that would prove fatal. Since I have discovered the close connection that exists between the intensity of physiological action and the atomic weight

When injected into the veins in divided doses, the quantity of the oxide of beryllium required to kill 2250 grm. of rabbit is ·06 grm., and of oxide of alumina ·022 grm., and as the atomic weight of beryllium is 9·3, and of aluminium 27·4, the poisoning power of these substances increases in the same ratio as their atomic weights. I think the same ratio will be found between these salts and the ferric salts, which agree with them perfectly in their physiological action (see *this Journal*, III. 24). Owing to the uncertainty of the molecular constitution and the purity of the salts of cerium and ythium, no satisfactory evidence as regards their quantitative relations can be obtained. They are undoubtedly very poisonous, and I think, when pure and definite preparations can be obtained, their physiological activity will be found to be proportional to their atomic weight. There is one fact shown by these experiments which I think deserving the attention of physiologists, and that is the striking manner in which the capillaries are affected by the presence of extremely minute quantities of foreign matter in the blood. The presence of one part of oxide of alumina in twenty thousand of blood was enough to produce strong contraction both in the pulmonary and systemic capillaries.

of the metallic elements in the same isomorphous group, a more accurate determination of the quantities that prove fatal will, I doubt not, shew a closer connection between the intensity of physiological action and atomic weight than my published experiments would indicate. This certainly has been the case in my more recent experiments with beryllium and aluminium.

PHYSIOLOGICAL ACTION OF CONDURANGO. By T. LAUDER BRUNTON, M.D., F.R.S., *Assistant Physician and Lecturer on Materia Medica at St Bartholomew's Hospital.*

THE experiments were made in the autumn of 1871 on frogs and rabbits with a solution of an extract prepared from a pound of the root which yielded one ounce. The solution was injected either under the skin into the peritoneal cavity, or into the jugular vein.

Experiments 1 *to* 7. From one to five grains of extract mixed with a little water were injected under the skin of seven Frogs. In six of these experiments no effect followed the injection, except some temporary restlessness, which, as it occurred immediately after the injection and soon subsided, must have been due to the operation and not to any absorption of the extract. One frog died some time after the injection; but the death can hardly be attributed to the condurango, as the frog was less lively than the others before the injection, remained quite unaffected by it after twenty-four hours, and died five days afterwards.

These experiments show that the extract of condurango injected under the skin of frogs produces no effect, even when so large a dose as five grains is administered.

Experiments 8 *to* 10. These experiments were performed on Rabbits.

Experiment 8. One gramme (15·4 gr.) was dissolved in 20 cub. centim. of water containing ½ per cent. of chloride of sodium, the salt being added to prevent the irritation which water causes when brought in contact with the tissues of the body. This liquid was injected in two equal portions, with an interval of half an hour, into the peritoneal cavity of a rabbit weighing 936 grammes (2 lb. 1 oz.). No effect followed either operation.

In order to test the effect of the chloride of sodium itself, 10 cub. cents. of a solution of the same strength but without any extract were injected into the peritoneal cavity of a rabbit weighing 425 grammes (nearly 1 lb.) and produced no effect.

Experiment 9. One gramme (15·4 grains) was mixed with 10 cub. cent. of water. The solution was not filtered, and contained small undissolved particles. A cannula was placed in the jugular vein of a rabbit weighing 1030 grammes (2 lb. 4 oz.), and about one-third of the fluid was introduced by three injections with intervals of fifteen and eight minutes. No apparent effect was produced by the first injection, but after the second and third the respiration became quicker, the animal.sank down on its fore-paws, turned on its side, drew its head backwards at nearly a right angle with the body, gave one or two convulsive kicks, the eyes became exceedingly prominent, and the cornea became insensible to the touch.

The body was immediately opened. The heart was still pulsating, though feebly. Its right cavities were full, its left cavities were empty.

In this experiment the injection of about one-third of a gramme (5 grs.) of the extract into the jugular vein was followed by death; whereas in Experiment 8 the injection of three times this dose into the peritoneal cavity of a smaller rabbit produced no effect. As drugs are generally absorbed with great rapidity from the peritoneal cavity, the difference between the results of the two experiments could hardly be due to the non-absorption of the condurango in Experiment 8, and it was therefore in all probability due to embolism of the pulmonary vessels caused by suspended particles, and not to the physiological action of the drug.

Experiment 10. One gramme (15·4 grs.) of the extract was dissolved in 20 cub. cent. of warm water, and carefully filtered twice. A cannula was placed in the jugular vein of a rabbit weighing 680 grammes (1½ lb.), and 10 cub. cent. of the filtrate were slowly injected into it.

During the injection the respiration became hurried, and occasionally a convulsive twitch of the limbs occurred. A few minutes afterwards 5 cub. cent. more were injected. The animal was then released. It trembled much and seemed weak, but could move about perfectly well. Its respiration was laboured. Twenty hours afterwards it seemed weak and languid and its respiration somewhat hurried.

In this experiment nearly three-quarters of a gramme were injected into the jugular of a rabbit weighing 680 grammes, with the effect of producing hurried respiration and weakness. In Experiment 9, one-third of a gramme was followed by the death of a rabbit weighing 1030 grammes.

The difference between these experiments consisted in the solution of condurango being filtered in Experiment 10 and not in Experiment 9, and we must therefore conclude that the fatal result in the latter case was due to the presence of large particles in the fluid.

The hurried breathing in Experiment 10 might be due to the small particles which had passed through the filter, but were yet large enough to become impacted in the pulmonary capillaries.

RESULTS OF EXPERIMENTS ON THE GENERAL ACTION OF THE
EXTRACT OF CONDURANGO.

Condurango has no poisonous action. The extract produced
no apparent effect in frogs when given in doses of $\frac{1}{3}$ of a gramme
(5 grains), or on a rabbit in a dose of 1 gramme (15$\frac{1}{2}$ grains), an
amount which is equivalent to 72 grammes (2$\frac{1}{2}$ ounces) of the
extract, or 2$\frac{1}{4}$ lbs. of the root for a man weighing 150 lbs. It
has no paralysing action, and apparently no action of any kind
on voluntary muscles or motor nerves, as the movements of
the animals were unaffected by its administration.

I also made a series of Experiments, XI. to XVII., to ascertain whether condurango had any power of *diminishing reflex
action*, or any effect on *circulation* and *respiration*.

The conclusions at which I arrived are :

I. Condurango has very little effect on reflex action. A
slight diminution of reflex irritability was observed after the
injection of the extract into frogs in doses of 2 to 5 grains, but
the number of experiments was insufficient to show whether or
not this was due to the exhaustion produced by the struggles of
the frog.

II. Condurango has no action on the pressure of the blood
when it is injected into the peritoneum in large doses; the
diminished blood-pressure noticed in some instances might be
due to the animal's strength being diminished by its position on
the board and the effects of the operations. Its effect on the
pulse and respiration is not constant, the injection being sometimes followed by quickening, and sometimes by slowing, and
moreover the results which followed its injection into the
jugular might in some degree be due to the presence of fine
particles in it having a mechanical effect.

For the purpose of ascertaining if condurango has any effect
in contracting or dilating the arterioles and capillaries, the
vagus was irritated so as to stop the heart, and the rapidity
with which the blood-pressure fell during the stoppage of the
heart before the injection was compared with that which was
found after it.

The experiments appeared to indicate that no extensive alterations were produced in the arterioles by the condurango.

The general result of all these experiments is that condurango is physiologically inert, and that Giannuzzi[1], in attributing to it a tetanizing action, probably fell into the same error into which Experiment 9 had almost led me, and ascribed to the physiological action of the drug what were really the effects of embolism of the pulmonary artery.

[1] Ricerche esequite nel Gabinetto di Fisilogia della Universita di Siena, pp. 71—86. Abstracted in the *Centralblatt d. med. Wissenschaften*, 1873, p. 824.

NOTE ON THE ABDOMINAL PORES AND UROGENITAL SINUS OF THE LAMPREY. By J. C. EWART, M.B. (*Edin.*), *University College, London.*

SIR EVERARD HOME in the *Philosophical Transactions* for 1815, in a paper on the Generation of the Lamprey and Myxine, holds that the Lamprey is a hermaphrodite animal, that the ova escape from the abdominal cavity by two small apertures into a tube common to them and the semen, which tube opens just within the verge of the anus. Johannes Müller[1] states that in the Myxinidae, at the end of the abdominal cavity at both sides, lying close to the rectum, is a short canal which, piercing the abdominal walls, opens into a single pore situated behind the anus. This pore along with the anus is enclosed by a fold of skin which forms a kind of cloaca. He adds that the Lamprey only differs from the Myxine in having the pore drawn out to form a papilla.

F. M. Balfour[2] in a paper on the development of the urogenital organs quotes Johannes Müller, and refers to a memoir by Vogt and Pappenheim[3] on the generative organs of the vertebrata, in which two abdominal pores are described in the Lamprey each connected with a short canal. These pores they say are so small that to demonstrate them it is necessary to inject mercury into the abdominal cavity, the mercury after some time appearing as a small drop at both sides of the anal aperture. They were never able to introduce a thread into the openings—threads piercing the tissues and human hair being too flexible. In the same paper Rathke is said never to have been able to find the openings, though convinced of their existence.

The above writers having failed to demonstrate the openings further than by the injection of mercury, it remains to be proved what the real arrangement is, whether or not there are, as Müller describes, short canals leading from the abdominal cavity lying close to each side of the rectum.

On examining the ventral surface of a lamprey a well-marked papilla is seen projecting from a shallow elliptical fossa. At the apex of this papilla a very fine opening is found

---

[1] *Abhandlungen der Königlichen Akademie der Wissenschaften*, zu Berlin, 1848.
[2] *Journal of Anat. and Physiol.*, Vol. x., Part i.
[3] *Annales des Sciences Naturelles*, Vol. xi.

in the male, a wider one in the female. On pressing the abdominal walls with the fingers from before backwards in the fresh Lamprey a quantity of fluid, generally containing blood and sometimes ova, may be seen escaping by this opening at the apex of the papilla, indicating that some of the contents of the abdominal cavity thus reach the exterior.

In front of the papilla, at the bottom of the fossa, in the ventral surface from which the papilla projects, is the anus. When closed its concave anterior margin is in contact with the anterior convex surface of the papilla. Behind the papilla there is no opening; the outer layer of the papilla dips into the fossa and then becomes continuous with the skin. If after making this superficial examination an incision is made through the abdominal wall and water injected into the abdominal cavity, a fine stream rises from the papilla. If the cannula is introduced into the ureter the result is the same. If however it is introduced into the intestine, nothing escapes from the papilla, but a large stream from the anus in front of it. The conclusions arrived at by injecting may be further verified by introducing bristles, carefully guarded at the ends, into the different openings. A bristle introduced into the papilla by the aperture at its apex may enter either the ureter or the abdominal cavity, but never the intestine, unless undue force is used. If after having introduced a bristle from the abdominal cavity into the papilla one side of the abdominal wall is removed—say the right side—the bristle will be found passing through, not a canal but a simple foramen, into a small

Fig. 1.

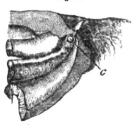

sinus, which it traverses to escape externally by the aperture at the apex of the papilla. The foramen represented in Fig. 1 c is not so small as one would expect from Vogt and Pappenheim's description, being in this case nearly two lines in diameter. It is not a tube formed by the narrowing of the abdominal cavity, for the abdominal cavity extends one line beyond and above the opening, and several lines below; neither is it connected in any way with the rectum, being altogether posterior to it. It is simply a foramen in a membrane not more than half a line in thickness, by which the abdominal cavity communicates with a sinus at the base of the papilla. In the six lampreys which I examined I found this foramen, and without any difficulty introduced bristles through it, both from the abdominal cavity, and *vice versa*, even after the lampreys had been three months in spirit; and in the preparation represented in Fig. 1 the opening was at once apparent on cutting into the abdominal cavity. In the Figure, besides this opening, a part of the intestine and kidney is represented. The intestine of the lamprey is a straight tube which lies free in the abdominal cavity, except near its termination, where several fine bands—the remains of the mesentery—fix it to the under surface of the notochord. Lying at each side of the intestine and between it and the notochord, is the ovary or the testis, and externally lying close to the abdominal walls are the kidneys—one at each side. The right kidney is represented in Fig. 1. The ureter runs along the free margin and ends about one-eighth of an inch from the abdominal pore. On cutting into the ureter it will be found to open into the same sinus as the abdominal cavity.

Figure 2 shows a section near the middle line. The right kidney and ureter along with the right wall of the sinus, seen

Fig. 2.

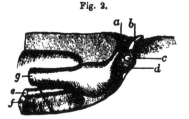

at *c*, Fig. 1, has been removed. The sinus, as seen in this section, is somewhat pear-shaped; at the small end is the external opening of the papilla; at the base are the openings of the ureters, separated from each other by a thin membrane; and in the lateral walls, nearer the base than the apex, are the openings of the abdominal cavity—the abdominal pores. Fig. 1 c shows the pore of the right side, and Fig. 2 c shows the pore of the left. Fig. 2 d indicates the opening of the left ureter. The sinus is thus a small chamber having the right and left ureters opening separately at its base and the abdominal cavity by a pore at each side. Through these the urogenital products enter, and passing through the sinus escape to the exterior by the fifth and last aperture at the apex of the papilla. We may thus look upon this chamber as a urogenital sinus, differing however from the arrangement generally met with in the vertebrata by having the anus in front, the genital apertures behind and inferior.

The anus is seen at a, Fig. 2, partly distended. The same figure shows the intestine dilating about half an inch from its termination into a rectum, which is attached both above and below to the body-wall. The rest of the intestine hangs free in the cavity, except where the narrow bands, already mentioned, bind it to the notochord. There are sometimes not more than four such bands. The mucous membrane of the rectum and last two inches of the intestine is raised into fine parallel longitudinal folds; but the valve found in the upper part is no longer visible. Fig. 1 shows part of the fold of skin which forms the elliptical fossa, out of which the papilla rises. This fossa is what Johannes Müller calls a "kind of cloaca."

That Sir Everard Home was mistaken in supposing the Lamprey to be a hermaphrodite animal has long been well known. Any one by a careful examination of the papilla alone can distinguish the male from the female. In the male the papilla is long, conical, and curved, and the aperture at the apex is very small. In the female it is short, thick, compressed laterally, and almost straight. On account of its shortness, and the fossa in which it lies being much deeper than in the male, the apex seldom projects beyond the general surface of the body; and lastly, there is a large oval external opening. What Sir E. Home believed about the different openings we are unable to say, partly, because the description is very vague, and, partly, on account of the plates representing openings which are not to be found in the Lampreys we have examined, and of which nothing is said in the text.

In reference to what Müller states, it is evident, from what has been already said, that the abdominal cavity does not open by tubes, and even, although the pores were capable of being termed tubes, these would still have no relation to the rectum, for on each side they would be at least one-eighth of an inch nearer the caudal extremity than the rectum. The opening of the papilla Müller considers the abdominal pore—the pore having been drawn out to form a tube. To prevent any misunderstanding, it would perhaps be better to speak of two internal abdominal pores opening into a sinus which communicates with the exterior by a single external aperture or pore at the apex of the papilla.

It is difficult to understand why Rathke, Vogt, and Pappenheim failed to discover the openings otherwise than by injecting mercury. Gulliver[1] found them without much difficulty; and, in Lampreys distended with ova or semen, they are generally at once apparent on cutting into the abdominal cavity. This is easily understood when we consider the amount of stretching the abdominal walls undergo up to the time the ova are shed. The openings enlarge in the same way as a mere puncture in a sheet of india-rubber becomes a large foramen when the india-rubber is expanded. But even in Lampreys with the abdominal walls in a relaxed condition, on careful examination the openings are visible when looked at from the cavity of the sinus; and, when the walls of the sinus are stretched, they become so dilated, that large guarded bristles can be passed through them into the abdominal cavity.

From the description given above it will be evident that in the Lamprey the intestine opens in front of the urogenital sinus, but does not communicate with it, that the ureters and internal abdominal pores open into the sinus situated in the middle line immediately behind the rectum, and that this sinus opens externally by a small aperture at the apex of the papilla.

Since making the above investigations I have had the op-

[1] *Proceedings Zoological Society*, 1870.

portunity of examining the arrangement of the urogenital openings in the Myxine, and find them essentially the same as in the Lamprey. The tubes described by Müller, who specially examined the Myxine, lying along the side of the rectum, I have not been able to find. In the preparations dissected, each side of the abdominal cavity gradually diminished as it neared the urogenital sinus, and then, without contracting to form a canal, simply opened by a somewhat large aperture into the sinus common to it and the ureters. As in the Lampreys, in front of this small sinus, and separated from it by a semilunar fold, is the anus. Enclosing both is the fold of skin which forms the kind of cloaca of Müller. However, it should be borne in mind that, though the fossa formed by this fold may be called a kind of cloaca, there is really no communication between the rectum and the ureters, or between the rectum and the abdominal cavity.

The specimens which I examined were procured by Prof. E. Ray Lankester and dissected at his request for the Zoological Museum of University College.

THE LENGTH OF THE SYSTOLE OF THE HEART, AS ESTIMATED FROM SPHYGMOGRAPHIC TRACINGS. By EDGAR THURSTON, of King's College, London.

VARIOUS observers[1] have in different ways attempted to measure the duration of the Ventricular Systole, both absolutely, and relatively to the whole cardiac period. Perhaps the most important observations are those of Donders[2], who measured the interval between the first and second sounds. The conclusions at which he arrived are briefly as follows:

In different individuals in a condition of repose with different pulse-rates the interval between the first and second sounds is remarkably constant, varying from ·301 to ·327 seconds for ordinary pulses. Relatively therefore to the whole cardiac period it is greater. in the quick than in the slow pulses. In the same individual, when the pulse was quickened, as by change of position, &c., the duration of the interval was absolutely lessened, as from ·327 to ·298 for a rise of from 63·4 to 83·6 per minute, while the proportion of the interval to the whole cardiac period reckoned as 100, rose from 34·8 to 41·5.

The following is an attempt to determine the absolute and relative duration of the ventricular systole by the interpretation of sphygmographic tracings of the radial artery. It is of course necessary first of all to prove the precise nature of the several points observed in such tracings.

In all tracings of healthy pulses, the following main features can be readily distinguished:—a primary uprise nearly vertical, followed by a gradual line of descent, this being interrupted by a well marked secondary uprise, termed the dicrotic wave, from the summit of which the descent is continued onwards to the commencement of the next period. This dicrotic wave is a constant factor, in a greater or less degree of development, of tracings from the radial artery, though it is stated by Marey to be present only in some cases at the wrist. Frequently a swelling is seen on the line of descent before the dicrotic wave, with which it must not be confounded. This is called the tidal or predicrotic wave, and concerning it, I will borrow the words of

[1] Ludwig, Valentin, Volkmann, Marey, Chauveau and Landois.
[2] *Nederlandsch. Arch. v. Genees en Naturkunde*, II. 189 (1865), translated in full in *Dublin Quarterly Journal of Medical Science*, Feb. 1868.

Dr Galabin in his article on 'Secondary Pulse-waves'[1], in which he states that "though waves occur in the tracing which have no separate existence in the pulse, the instrument is more clinically useful than if it followed the artery more closely, for slight differences in form of pulse wave are thus transformed into a form much more manifest to the eye." With this statement I entirely agree, for I have found by experience that the determination of the exact spot at which the dicrotic wave commences is, in cases where that event is not well marked, rendered an easy matter, since it is indicated by the termination of the predicrotic wave, which serves to bring it out with greater distinctness than would otherwise have been possible.

It was first noticed by Marey, and has since been observed by Dr Galabin, that the predicrotic wave may be broken up into two, and this has been stated to be entirely due to the secondary spring which presses on the base of the lever in the old knife-edge instruments. The instrument with which I have always worked possesses no such spring, but I have in several instances noticed the wave split up in the manner described by these two authorities.

With these passing remarks on the predicrotic wave, which do not bear directly on the subject of the present communication, I proceed to the consideration of two important factors of sphygmographic tracings:—(1) The primary uprise of the tracing is caused by the systole of the ventricle. (2) The dicrotic notch is the result of closure of the aortic valves. This latter statement requires further consideration, it being opposed to the views entertained by some physiologists at the present day; but in support of it the following proofs can be brought forward:—

a. From the calculation of cardio-sphygmograph tracings, *i.e.*, tracings taken simultaneously from one of the systemic arteries and the apex of the heart, it has been shewn[2] that the dicrotic wave occurs precisely the same time after the closure of the aortic valves that the primary uprise occurs after the commencement of the systole of the ventricle.

b. It cannot be a secondary result of the primary shock, as there are many reasons for the assumption that the blood-vessels not being exposed tubes do not develope minor undulations, such as are found in experiments with elastic tubes. It is evident that though the arteries are possessed of definite walls, these are from a hydrodynamical point of view formed by all the tissues between their inner coats and the surface of the body, the thus formed compound walls damping from within outwards any undulations transmitted to their tissues by their contained blood. This continual damping reduces the shock, and thus reduces the tendency to the formation of secondary waves.

A careful analysis of the results arrived at by Donders, and a comparison of them with measurements of sphygmographic tracings, tend to throw great light on the true nature of the dicrotic wave.

[1] *Journ. Anat. and Phys.*, Nov. 1873.
[2] *Proc. Roy. Soc.* 1871, p. 318.

According to Donders, the length of interval between the first and second sounds in a state of rest occupies from 0·309 to 0·327 of a second. In one instance with a pulse of average rapidity, *i.e.* 67 per minute, the interval between the primary and dicrotic rises, which I shall in future call the first part of the beat, was found by measurement to be contained 2·8 times in each period, and occupied therefore ·310 of a second, while in a pulse of 73 per minute it occurred 2·6 times in each period, occupying ·316 of a second. The following short Table is copied from the paper by Donders, to which reference has been already made :—

TABLE I.

Pulse rate.	Position.	Length of Systole.	
		In per cent. of the whole period.	In seconds.
68·4	Sitting.	84·8 (40·6 *auct.*)	0·327
83·6	Standing.	41·5	0·298

Subjoined are the results arrived at by myself from the estimation of sphygmographic tracings with similar rapidities of pulse :—

TABLE II.

Pulse rate.	Position.	First part of the period.	
		In per cent. of the whole period.	In seconds.
64	Sitting.	84·5	0·828
84	Sitting.	41·8	0·295

These numbers, derived from two entirely different methods of investigation, agree so closely that they must be considered not as mere coincidences, but as a positive indication that the dicrotic wave corresponds to the moment of closure of the aortic valves.

Professor A. H. Garrod[1] has stated his conclusions derived from the comparison of a large number of sphygmographic tracings in the form of a proposition, viz :—"The length of interval between the commencement of the primary and dicrotic rises is constant for any given pulse rate, and varies as the cube root of the pulse rate, being found from the equation, $xy = 47\sqrt[3]{x}$, where 47 is a constant number, x = the pulse rate, and y = the ratio borne by the above-named part to the whole beat."

[1] *Proc. Roy. Soc.* 1870, p. 851 ; 1875, p. 142.

This proposition is stated to be true not only of the radial, but also of the carotid and posterior tibial arteries, and the length of the interval is therefore constant throughout the larger arteries.

It appears, therefore, that there are two important theories as to the length of the systole of the ventricle, one authority basing his views on the investigation of the cardiac sounds, and the other on the examination of sphygmographic tracings. By both theories it is allowed that the shorter the period is, *i.e.* the more rapid the pulse, the greater is the proportion occupied by the systole to the whole beat. But with regard to the actual time occupied by the systole, whereas Donders believes that it occupies from 0·309 to 0·327 of a second, continuing tolerably equal with different rapidities of pulse, Professor A. H. Garrod maintains on the contrary that it varies as the cube root of the pulse rate.

With a view of proving or disproving the correctness of the statement that the length of systole varies as the cube root of the pulse rate, I have been engaged during the past six months in a series of investigations of sphygmographic tracings.

The form of instrument used in all cases was the rack-work modification, as made by Brequet of Paris. The tracings were taken on highly glazed card, coated with a thin film of soot, by passing it through the flame of a composite candle, and afterwards varnished with ordinary spirit varnish. The frame supporting the card ran its full distance in 7 seconds; consequently by multiplying the number of periods that occurred in a single tracing by 8·57143, the rapidity of the pulse per minute was arrived at. The measurements of the tracings were made with an ordinary springbow, the distance between the points of which can be varied by means of a screw.

If, as is usually the case, the commencement of the primary and dicrotic rises are at a different level on the tracing, the length of the first part of the beat cannot be measured by simply superimposing the springbow on the tracing, but lines must be drawn upwards from the commencement of each period. It is found in practice exceedingly inconvenient to describe lines on the tracing with the lever of the instrument, for, in addition to the fact that it is not an easy matter to arrest the motion of the clockwork at the precise moment when the tip of the lever is opposite the commencement of a period, this process necessitates an extra strain on the machinery. If simply vertical lines are projected upwards, the systole is represented as being shorter than it really is, owing to the fact that the lever moving about an axis describes the arc of a large circle. To obviate this difficulty, a simple piece of apparatus was devised by Professor Garrod, which I have found of great value. This consists of a strip of wood attached to a wooden base, against which the tracing rests. Into the base is fixed a screw in such a position that it is exactly the same height above the trace-supporting piece of wood, that the axis of the lever is above the summit of that portion of the sphygmograph which contains the clockwork. A strong needle is threaded with a piece of cotton of the same length as the lever, which is attached by its opposite end to the screw, the point of the needle passing through a knot in a piece of string of similar length, which is likewise attached around the screw; with this simple contrivance lines can be drawn on the tracing, corresponding accurately to those which would be described by the lever.

In all cases the proportion was taken in each period of the tracing of the relation of the first part to the whole beat, and an average found for the whole tracing, as slight differences in this respect may be found to exist in individual periods, due to trifling imperfection in the clockwork, which should be wound up to its full extent before taking an observation, and also to the effects of the respiratory movements on the circulation. I have frequently noticed that persons not accustomed to the application of the sphygmograph to their arteries involuntarily hold their breath while the tracing is being taken. This should however be, if possible, avoided, inasmuch as the cardiac movements are thereby affected.

In a pulse of 72 per minute the measurements of the first part of the beat were as follows in each period of the tracing: 2·6; 2·6; 2·7; 2·66; 2·75; 2·66; 2·75; 2·66; 2·83; 2·75; 2·66; 2·75; with an average of 2·69.

In determinations of the pulse rate, the most accurate method is to take the mean of the rate as deduced from the actual tracing, and as calculated in the ordinary manner at the wrist immediately before or after the experiment.

If x = the pulse rate, and y = the number of times that the first part is contained in a whole beat, the part of a minute which is occupied by the first part is evidently represented by the expression $\dfrac{1}{xy}$, and it has been already stated, that, according to the view entertained by Professor Garrod, $xy = 47\sqrt[3]{x}$. It should be possible, therefore, if this equation is correct, to determine the time occupied by the first part of the beat by multiplying the cube root of the pulse rate by the constant number 47.

The following Table contains some of my results obtained from measurements of tracings of the radial artery, to which alone my attention has been confined, the rapidity of pulse being indicated, as also the actual length of time occupied by the first part of the beat estimated by (1) measurement with a springbow, (2) calculation from the equation $xy = 47\,\dfrac{3}{x}$.

TABLE III.

Pulse rate.	Length of first part of beat.	
	Determined by actual measurement.	Calculated from equation $xy = 47\sqrt[3]{x}$ (approximately).
47	·8468 of a second	·8378 of a second
60	·317 „	·321 „
70	·310 „	·3102 „
72	·3108 „	·3072 „
84	·300 „	·294 „
85·5	·2898 „	·292 „
90	·286 „	·283 „
128	·2556 „	·2592 „

In the consideration of the results contained in the above Table, it must of course be borne in mind that the actual measurements are subject to a certain amount of experimental error, but they cling so closely to the calculated results as to prove conclusively that the length of the first part of the beat does really vary, and that it varies, not as rapidly as the pulse-rate, but as its cube root.

It must not be supposed, however, that the variation in the length of systole with different pulse rates is capable only of demonstration by mathematical calculation, for it requires no very minute power of perception to recognize with the unaided eye, by comparison of a series of tracings of various rapidities of pulse, that the first part of the beat occupies a considerably greater length of time in slow than in rapid pulses.

With regard to the relative, in contradistinction to the absolute, length of the first part to the whole period, I have arranged the results of my investigations in connection with this point in a tabular form, the rapidity of pulse being indicated, as also the actual relation determined by measurement of tracings, and the relation which should exist if the equation $xy = 47'\sqrt{x}$ is true; and the truth of this equation is, in my opinion, proved beyond all doubt. The results were not obtained from a single individual, but from a number of different persons between the ages of 18 and 35, by which it is shewn that the law is in a state of health a universal one, and not merely an exceptional coincidence of occasional occurrence.

TABLE IV.

Pulse rate.	Ratio of first part to whole period.	
	Found by measurement.	Calculated from equation $xy = 47\sqrt{x}$ (approximately).
48	8·81	8·8
ditto	8·87	ditto
47	8·68	8·68
60	8·15	8·12
66	2·9	2·9
70	2·76	2·76
ditto	2·72	ditto
72	2·718	2·7
72·5	2·698	2·692
ditto	2·707	ditto
78	2·51	2·56
84	2·42	2·4
85·5	2·4	2·88
90	2·335	2·25
ditto	2·24	ditto
98·5	2·29	2·2
187	1·8	1·8

I must now dwell briefly on a point, on which great stress is laid by some of the opponents of the views which I have just held out. They would ask, Is it possible to measure the tracing with such a degree of accuracy as to obtain such figures as 2·693 or 2·76? Further, Can one distinguish. between 3·2 and 3·33 or between 2·7 and 2·75? To such questions I reply, that, in every case in which the commencements of the primary uprise and dicrotic notch are sharply defined, it is quite possible to calculate the number of times that the first part is contained in a whole period to one, if not two, places of decimals, for by continued practice and delicacy of manipulation exceedingly small variations can be detected.

It will be observed, however, that in some instances the calculations have been extended to a third place of decimals, this being derived from the calculation of the average of the different periods, and not from actual measurement. An example will serve best to illustrate this latter statement. In one instance, with a pulse of 137 per minute, the measurements of the individual beats ran as follows: 1·75, 1·8, 1·75; with an average of 1·766.

Finally, the conclusion at which I have been led to arrive is:—

That the law enunciated by Professor Garrod—to the effect that the length of the systole of the heart, as indicated in the radial artery, is constant for any given pulse rate, and varies as the cube root of the rapidity—is the correct expression of the length of the systole as transmitted to the human wrist.

ON THE MODE OF OVIPOSITION OF AMPHIOXUS.

BY A. MILNES MARSHALL, B.A., B. Sc., *Demonstrator of Comparative Anatomy in the University of Cambridge.*

(From the Zoological Laboratory, Cambridge.)

ONLY two observers have, so far as I am aware, actually witnessed the act of oviposition in Amphioxus; and their statements as to the orifice by which the ova leave the body of the female are directly opposed to one another.

Quatrefages[1] states that he has several times witnessed the exit of ova from the abdominal pore. Kowalevsky[2], on the other hand, in his memoir on the development of Amphioxus, states that he saw the ripe ova extruded from the mouth of the female. This account has not been generally accepted by naturalists, and has never, I believe, been confirmed by direct observation ; while its correctness has been called in question by some of the most recent writers on the subject. Thus Stieda, in his Studien über Amphioxus[3], rejects Kowalevsky's account because he could find no communication between the branchial sac and the atrial cavity of sufficient size to allow the passage of the ova. Prof. Wilhelm Müller[4] considers Kowalevsky's statement refuted by "a simple comparison of the diameter of the ova and of the branchial clefts ;" the latter being, according to him, far too small for the ova to pass through. On the other hand, Prof. Lankester[5] considers that "the branchial slits are not too fine for the passage of the ova," and that "the ova probably do escape by the mouth, or else, and more probably, by the two apertures placed one on each side of the mouth," which he has named "hyoidean apertures."

From an examination of a series of sections lately prepared by Mr A. Sedgwick of Trin. Coll., in the course of the class-work of the Zoological Laboratory of the University of Cambridge, I am enabled to confirm Kowalevsky's statement, and to supplement it by some details which I think will prove of interest.

[1] *Annales des Sciences Naturelles.* Serie 3. Tome 4. 1845.
[2] *Memoires de l'Acad. Imp. de St Petersbourg.* Tome 11. 1867.
[3] *Memoires de l'Acad. Imp. de St Petersbourg.* Tome 19. 1873.
[4] *Jenaische Zeitschrift.* Bd. 9. 1875.
[5] *Quart. Journal of Micros. Science.* July, 1875.

The specimens were procured at Naples from the Zoological Station established there by Dr Dohrn, and were taken at a time of year—the commencement of May—when the act of oviposition was being carried on actively.

At the time of full development of the ova the ovaries swell very considerably. The ova are discharged from the ovaries into the atrial cavity, where they lie for the most part in the space between the ovaries of the two sides[1].

The enlarged ovaries and the escaped ova together form a bulky and tightly packed mass, which not only distends the ventral wall of the atrial cavity, so as to smooth out the plaits of the ventral integument and almost obliterate the meta-pleural cavities (in the manner described by Prof. Lankester[2]), but also pushes up the ventral wall of the pharynx, doubling it up against the dorsal wall in such a manner as to reduce the cavity of the pharynx to a comparatively narrow slit, as shown in the annexed figure.

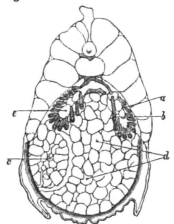

Transverse section through anterior part of pharynx, just behind the point at which the ovaries commence, showing ova passing into pharynx. *a*, dorsal wall of pharynx. *b*, doubled-up ventral wall of pharynx. The branchial bars are seen to be cut more obliquely in the ventral than the dorsal wall. *c*, ovary. *d*, ova lying free in atrial cavity. *e*, ova that have passed into pharynx.

[1] The presence of free ova in the atrial cavity seems to have been first noticed by Retzius in his account of Amphioxus in the *Monatsbericht der Berlin. Academie der Wissenschaften*, November, 1839.

[2] *Loc. cit.* p. 262.

The continued pressure from below causes the ova to pass into the pharynx through the branchial slits of its doubled-up ventral wall. This doubling-up of the ventral wall takes place through a considerable length of the pharynx: posteriorly, it commences at a short distance behind the anterior end of the hepatic cæcum; it then extends forwards through the whole of the anterior half of the pharynx, except at the extreme anterior end, where it is lost, and the ventral wall of the pharynx is almost in contact with the ventral wall of the atrial cavity.

The passage of ova into the pharynx takes place through the branchial slits of this doubled-up ventral wall alone. The majority of the sections through the anterior half of the pharynx show ova in the act of passing through the ventral wall, but none can be seen entering through the dorsal wall, though detached ova often occur between the lateral parts of the dorsal wall and the atrial walls.

It is worthy of notice that while in the dorsal and lateral walls of the pharynx a large number of branchial bars are cut through in transverse sections, owing to the branchial bars sloping from above backwards as well as downwards, yet in the doubled-up ventral wall a much smaller number are cut through—usually only three or four. This is due to the pushing up of the ventral wall of the pharynx by the accumulated ova taking place from behind as well as from below, so that in the ventral wall the branchial bars ultimately lie almost completely tranversely, instead of obliquely. This will manifestly tend to widen the branchial slits and so facilitate the passage of the ova through them.

It must then, I think, be regarded as established that the ova do pass from the atrial cavity through the branchial slits into the pharynx. Ova occur in tolerable abundance in the anterior part of the pharynx in front of the commencement of the ovaries; but whether the actual egress of the ova takes place through the mouth, or through the "hyoidean apertures" at each side of the mouth (as suggested by Lankester), I have been unable to determine.

It is quite possible, however, that the ova may also escape in part through the abdominal pore. In the specimens I have examined I have never detected free ova in the atrial cavity in the immediate neighbourhood of the abdominal pore : but Quatrefages' statement[1], that he has several times witnessed the egress of ova from the abdominal pore, is so positive, that we must, for the present, assume that the ova may escape through the abdominal pore as well as through the mouth.

[1] *Loc. cit.* p. 207.

ON THE STRUCTURE OF THE SNAIL'S HEART. By
FRANCIS DARWIN, M.B. (Cantab.).

IN the important research "On the Behaviour of the Hearts of Mollusks[1]" by Dr Michael Foster, and Mr Dew Smith, the anatomy of the organs studied was necessarily rather briefly dealt with. Dr Foster suggested that I should undertake the investigation of the histology of the snail's heart; the present unfortunately very imperfect note gives the result[2].

The main anatomical fact stated by them is that the snail's heart is destitute of automatic nervous mechanism, and is not connected with the central nervous system in any way[3]. This statement I can, as far as my work goes, confirm; I have searched carefully, but in vain, for ganglion cells, both by staining with chloride of gold, and also with picrocarminate of ammonia, which brings out the nuclei of the ganglia of the central nervous system very brightly. Nor can I trace any nerve passing into the heart. The only reasons which render it at all probable that the heart may possibly have some kind of nervous mechanism, are the following[4]:

(i.) A nerve leaves the suboesophageal splanchnic ganglion and travels along the aorta towards the heart. I was unable to make out how this nerve terminated; the innumerable branching muscle-cells of curiously nerve-like aspect rendering the task difficult. It is possible that it may be merely distributed to the branches of aorta like a nerve described by Albany Hancock[5] in Ommastrephes (Loligo).

(ii.) A nervous supply to the heart is known to exist in some Mollusca: Keferstein[6] describes a nerve supplying the heart in *Tergipes edwardsii* (from Nordmann, 1845). Lacaze-Duthiers[7] describes a branch from the brancho-genital ganglion, supplying the heart in Pleurobranchus. Albany Hancock[8] describes and figures two ganglia at the apex of the heart in *Doris tuberculata*. The same

[1] *Proceedings of Royal Society*, No. 160,
[2] I have the pleasure of thanking Dr Klein for the kind manner in which he has helped me with his valuable advice and opinion.
[3] *Loc. cit.* p. 320.
[4] Mr Romanes has shown (*Proc. Roy. Soc.* No. 165), that in certain Medusæ the removal of the margin of the nectocalyx completely prevents the occurrence of spontaneous rhythm in the mutilated remnant; but that "the smallest atom of this (the margin) when left in situ is frequently sufficient to animate the entire nectocalyx." With this fact before us, we must be cautious how we build on the negative evidence which would declare the snail's heart to be devoid of nervous mechanism.
[5] *Annals and Mag. of Nat. Hist.* x. 1852, p. 8.
[6] *Thierreich*, B. III. Tab. 50.
[7] Quoted in *Thierreich*, B. III. p. 726.
[8] Albany Hancock, *Phil. Trans.* 1852.

author [1] describes a large ganglion on the pericardium in Ommastrephes (Loligo), which supplies the systemic heart and other parts of the vascular system.

My observations have been almost entirely confined to the heart of *Helix pomatia;* from what I have seen of that of the small garden snail I believe the structure to be the same in both. The superficially placed pericardial sac which contains the heart is situated between the kidney and the respiratory membrane. The pericardium is said by Keferstein [2] to be muscular, and is lined by a beautiful endothelial mosaic of cells. The heart consists of two cavities; a globular auricle which receives the pulmonary vein, and a conical ventricle, the apex of which is continuous with the main artery of the body or "aorta". The two cavities communicate by a small orifice guarded by a pair of very efficient semilunar valves, projecting into the cavity of the ventricle. The contractile tissue of the heart consists of unmistakeable striated muscle.

I was not aware until I had completed my research, that Gegenbaur [3] concluded that the tissue of the heart in the Pulmonogasteropoda is analogous with true striped muscle; he does not give any details on the subject, basing his opinion on general grounds. Leydig [4] however has described and figured the striated muscle composing the heart of *Paludina vivipara.* Keferstein [5] merely states that the snail's heart is composed of very granular muscle. Neither Weismann, Boll, Schwalbe, or Margo appear to have studied the heart of the Gasteropoda. The following instances of striated cardiac tissue in the Mollusca may be worth mentioning:—Müller [6] (quoted by Margo): branchial hearts of Cephalopoda (this tissue has been carefully described and figured by Boll [7]).—Margo [8]: heart of Anodon, faint striation.—Ray Lankester [9]: the extraordinary two-celled heart of Appendicularia.

To return to the snail's heart: the muscle is composed of elongated spindle-shaped cells closely superimposed and very intimately attached to each other. Without careful examination the cellular structure is not observable, and one merely sees

---

[1] *Annals and Mag. of Nat. Hist.* x. 1852, p. 10.
[2] *Thierreich*, Bd. III. p. 1206.
[3] Siebold and Kölliker's *Zeitschrift*, Bd. III. p. 389.
[4] Siebold and Kölliker's *Zeitschrift*, Bd. II.
[5] *Thierreich*, Bd. III. p. 1206.
[6] Siebold and Kölliker's *Zeitschrift*, 1853.
[7] Max Schultze's *Archiv*, Bd. v. Supplement.
[8] *Wien. Sitzungsb.* Bd. 89.   [9] *Quart. Jl. of Micros. Sc.* 1874.

bundles of parallel-walled fibres of about ·0022 mm. in diameter. In Paludina, Leydig[1] points out the difference between the ordinary body muscles and those of the heart. In Helix also the fibres of one of the buccal-mass retractors (for instance) are much stronger, less granular, broader, and more easily isolated, than those of the heart.

In some specimens the striation is perfectly clear, distinct light and dark bands alternating with each other with perfect regularity. In others again the muscle cells appear simply granular with Hartnack No. 8 and 9. Between these two extremes intermediate conditions, faintly striated and half-granular half-striated, may be found. These cells exhibit the transition from smooth to striped muscle which was, I believe, first pointed out by Bowman[2], and has been well described by Boll[3]. The muscle elements contain an oval nucleus occupying the broadest portion of the cell. I have not observed the "Axenstrang" or axial chain of granules figured by Boll[3] in the cardiac muscle of Octopus, nor have I made out the branching of the fibres mentioned by Leydig[4].

In the muscles of Anodon, Margo[5] has shown that the striation depends, as in vertebrate muscle, on the alternation of isotropous and anisotropous substances. Through the kindness of Dr Klein I have been enabled to examine the snail's cardiac muscle with polarised light. I did not attempt to repeat Margo's observation, but the examination of unstained sections between crossed Nicolls proved distinctly that the muscle of the snail's heart resembles vertebrate muscle in being uniaxial, with the axis parallel to the direction of the muscle cells.

The arrangement of the larger bundles into which the muscle is massed, offers no points of interest. The ventricle laid open shows them passing in all directions, and interlacing in an elaborate and irregular manner. The valves are attached to the opening by the continuity of their fibres with those of both chambers of the heart.

The connective tissue of the organ presents a few points of interest. The substance which unites cell to cell stains brightly in chloride of gold, and also comes out with nitrate of silver;

[1] Loc. cit.
[2] Loc. cit.   [4] Loc. cit.
[3] Cycl. Anat. and Phys. III. 514, 519.
[5] Loc. cit. (Wiener Sitz. 39).

portions of the valve stained with this reagent exhibit the elongated muscle cells marked out with dark lines.

Sections of the heart stained with hæmatoxylin present a curiously dotted appearance under a low power; this is chiefly due to the connective-tissue corpuscles with which the muscle abounds. These bodies are about ·009 mm. in diameter, and contain a large nucleus embedded in a scanty protoplasmic body which stains in chloride of gold. They are often pyriform in shape, and may be seen in profile projecting from the side of a muscular bundle; or sending fine processes to be intimately distributed among the muscle cells.

The connective tissue is remarkably developed towards the periphery of the heart. If the external surface is stained with nitrate of silver, a mosaic work of polygonal cells becomes apparent. In sections or teased preparations this structure is found to consist of closely packed pyriform connective-tissue corpuscles, whose larger extremities are directed outwards. Dr Foster and Mr Dew Smith mention this structure as an "external tesselated epithelium[1]." The cells composing it are firmly attached to each other, and can be torn off in strips which do in fact exactly resemble a tesselated epithelium. Leydig[2] describes and figures the heart of Paludina as covered by an epithelium of simple rounded cells. In Helix the constituent cells differ from those described by Leydig in being connected together by their processes. In some sections I have observed a kind of honeycomb-tissue, the bodies of the cells having fallen out of the meshes in which they were contained.

If one of the epithelium-like strips torn from the external surface be examined with its internal surface upwards, the curious arrangement of the muscles at the periphery of the heart may be observed. Muscular bundles are seen fraying out into beautiful fans, the fibres losing themselves among the mosaic of cells. I am unable to say what the exact connection between the cells and the muscle fibres is; I have never seen them actually terminate in the processes of the connective-tissue cells in the manner described in the frog's tongue[3]. From the fact of their losing themselves among the cells it is clear

[1] Loc. cit. p. 320.   [2] Loc. cit.
[3] Handbook for the Physiolog. Laboratory, p. 62.

that the muscles do not all form loops at the periphery of the heart, as analogy would have led one to expect.

In what manner the internal surface of the heart is limited I cannot say. The pulmonary vein is lined with a regular endothelium; according to Keferstein, an "epithelium" coats the inner surface of the aorta. Analogy would suggest that the cavity of the heart is limited internally by the same membrane that lines the vessels continuous with it. The valves are certainly covered with a somewhat irregular pavement of cells, but I cannot make out such a structure in other parts of the heart.

Dr Michael Foster and Mr Dew Smith describe a remarkable kind of physiological insulation which exists between the auricle and ventricle, and suggest the explanation that a ring of connective tissue, penetrated by no nerves, separates the two portions of the organ. This statement I can hardly confirm; the muscles at the region of the junction are apparently rather more intimately permeated by connective tissue; but I can find nothing which can be supposed to insulate the two portions of the organ. At the point of junction many fibres from both auricle and ventricle enter the valve, but this arrangement can hardly be efficient in insulating the two chambers from each other. I have convinced myself, by making longitudinal sections from hearts hardened in osmic acid, that there is muscular continuity between the auricle and ventricle; so that I can offer no anatomical explanation of the above mentioned fact. The appearance of a ring of connective tissue might easily be given by a section passing through the peripheral layer of connective tissue covering the narrow neck between the auricle and ventricle, and which at the same time included muscle fibres from both chambers of the heart. These would be cut obliquely, and would be separated from each other or even seem to arise from a region where they were replaced by connective tissue. Dr Foster and Mr Dew Smith[1] describe the curious waves of contraction which can be made to traverse the ventricle in any direction; it is remarkable that this should be the case in muscular tissue which seems to contain a considerable quantity of connective tissue. But the anatomical relations corresponding to physiological conduction and insulation are no doubt at present obscure or unknown.

[1] *Loc. cit.* p. 329.

NOTE ON THE EFFECTS OF DIVISION OF THE SYMPATHETIC NERVE OF THE NECK IN YOUNG ANIMALS. By W. STIRLING, D. Sc., M.D., *Demonstrator of Practical Physiology in the University of Edinburgh.*

THE increased supply of blood to a part, especially if that part be in growing state, is followed by very marked effects.

John Hunter[1] long ago observed that if the spur of a cock be transplanted from the leg to the comb, which is a much more vascular part than that with which it was originally connected, it undergoes an extraordinary augmentation in size. A case of hypertrophy of an entire limb was recorded by John Reid, and several cases of hypertrophy of the fingers were described by Mr Curling. Schiff[2] observed that after section of the nerves of a limb the bones of the latter exhibited hypertrophy of the periosteum. This effect was attributed by him to paralysis of the vessels. In the case of the ear of the rabbit, however, its vaso-motor nerves are contained chiefly in the trunk of the sympathetic in the neck. Division of this nerve in the neck, as Cl. Bernard pointed out, is followed, amongst other effects, by dilatation of the vessels of the ear on that side of the head.

A. Bidder (*Centralblatt für Chirurgie*, No. 7, 1874) excised a piece of the sympathetic nerve in the neck of a young growing rabbit. This was followed by hypertrophy of the ear of the same side. He however made only a *single* experiment, and as yet his observation has remained unconfirmed.

I was led to repeat this experiment. This I did, not only upon young growing rabbits, but also upon a young growing dog. In the dog the sympathetic nerve in the neck is contained in the same fibrous sheath as the vagus, so that in this animal in order to excise a portion of the sympathetic a piece of the vagus must be removed along with it. After the first day or so, the dog seemed to suffer no inconvenience from the excision of a part of its sympathetic and vagus on one side of the neck.

[1] Carpenter's *Physiology*, 7th edition, p. 397.
[2] *Comptes rendus*, p. 1060, 1854.

In all cases, both in the rabbit and dog, excision of a piece of the sympathetic was followed by hypertrophy of the ear on the same side.   Subjoined are the measurements in some of my experiments.

EXPERIMENT I.—*Dec. 26th*, 1874.   Young albino rabbit.   Part of *right* sympathetic excised.   Effects of the operation well marked. Length of ears before the operation, $2\frac{7}{16}$ inches; breadth, $1\frac{9}{16}$ inches.

April 10th, 1875.

	Left.	Right.	Diff.
Length of Ear	$3\frac{1}{2}$	$3\frac{13}{16}$	$\frac{5}{16}$.
Breadth of Ear	2	$2\frac{2}{16}$	$\frac{2}{16}$.

The *hair* also covering the *right* ear was longer and stronger, especially upon the margins of the ear, than that on the left one. The eye-lids remained semi-closed, and the retraction of the eye-ball persisted.

EXPERIMENT II.—*Dec. 24th*, 1874.   Young albino rabbit.   Part of the *left* sympathetic nerve excised in the neck.   Length of ears before operation, $3\frac{1}{4}$ inches; breadth, $1\frac{7}{8}$ inches.

April 10th, 1875.

	Left.	Right.	Diff.
Length of Ear	4	$3\frac{3}{4}$	$\frac{1}{4}$.
Breadth of Ear	$2\frac{5}{16}$	$2\frac{1}{8}$	$\frac{3}{16}$.

The same phenomena were observed in this rabbit as in Experiment I.

EXPERIMENT III.—*Dec. 24th*, 1874.   Young growing dog.   A piece half an inch in length of the *left* vago-sympathetic was excised. Length of ear, $2\frac{1}{2}$ inches; breadth at base, 2 inches.

April 10th, 1875.

	Right.	Left.	Diff.
Length of Ear	$3\frac{1}{2}$	$3\frac{3}{8}$	$\frac{1}{8}$.
Breadth of Ear	$2\frac{1}{2}$	$2\frac{5}{8}$	$\frac{1}{8}$.

The ear on this side was thus distinctly longer and broader, and also somewhat thicker than that on the intact side.   The hair also was longer and stronger on the side operated on, and the ear remained distinctly warmer.

These results afford an excellent instance of the effects of the increased supply of blood to a part, and one which can be produced at will provided young growing animals be selected for the operation.

ON THE STRUCTURE OF THE NON-GRAVID UTE-RINE MUCOUS MEMBRANE IN THE KANGAROO. By PROFESSOR TURNER.

I HAVE elsewhere recorded a series of observations made on the structure of the uterine mucosa in a number of placental mammals[1]. In this communication I intend to describe the characters of the mucous membrane of the uterus in one of the *Marsupialia, Macropus giganteus;* for a specimen of which I am indebted to Prof. A. H. Garrod.

Professor Owen in his *Comparative Anatomy of Vertebrates* states that in all the *Marsupialia* the lining of the uterus "is well organised, not deciduous; it is soft and disposed in many irregular folds, but when these are effaced has a smooth surface. This is a distinct but delicate layer with minute pores, and is connected to the muscular coat by an abundant tissue, consisting of fine lamellæ stretched transversely between the muscular layer and the smooth membrane, the whole being of a pulpy consistence and highly vascular, especially in the unimpregnated state." But he gives no description of its microscopic structure.

In *Macropus giganteus* the mucous membrane of the uterus formed longitudinal folds projecting into the cavity. The thickness of the uterine wall varied from $\frac{1}{8}$th to $\frac{1}{14}$th inch according as it was measured at the summit of a fold or at the bottom of a furrow between two folds. Much the greater part of the thickness was due to the mucous membrane, which was not only relatively but absolutely thicker than in any other uterus I have examined; for the muscular coat formed an extremely thin layer on the outer surface. The free surface of the mucosa was covered by a well-defined layer of columnar epithelium. Here and there an appearance was seen as if the remains of cilia projected from the free ends of the cells. When thin sections parallel to the free surface of the mucosa were examined microscopically, the circular mouths of the glands

---

[1] *Lectures on the Comparative Anatomy of the Placenta.* First series. Edinburgh, 1876.

were seen. In vertical sections through the mucosa multitudes of gland-tubes were observed; as a rule they were transversely or obliquely divided, but occasionally a short portion of a tube longitudinally divided was present. I infer from this that the gland-tubes were arranged so very tortuously in the mucosa that only a small part of the length of a tube came into the plane of section at any given spot. The glands were lined by well-defined nucleated columnar epithelium, the cells being arranged around a central lumen. The glands were characterized by their large size: compared with the long tubular glands in the uterine mucosa of the common bitch they had twice the diameter, and in some cases even somewhat more; they are the largest glands I have yet seen in the uterine mucosa of any mammal. The glands were separated from each other by slender but distinct bands of vascular connective tissue, which formed a framework of anastomosing trabeculæ, in the meshes of which the divided tubes were situated. In the deeper part of the mucosa an important modification in the appearance of the glandular tissue was observed. The cells which occupied the meshes of the trabecular framework, instead of being arranged as a series of columns surrounding a central lumen, were rounded in form, not unlike in appearance *leucocytes*, and filled up the entire space. The form of these cells and their relations to the trabecular tissue bore some resemblance to the arrangement of parts in an organ composed of lymphoid or adenoid tissue. But it is probable that these rounded cells were the contents of the deep, closed end of the tubular glands, which differed in form from the cells nearer the surface, just as the deep cells of the gastric glands differ from the columnar epithelium lining the gland-tube near its mouth. The extreme tortuosity of the glands rendered it impossible to trace the continuity between the deeper and more superficial part of the gland-tubes.

The mucous lining of the vagina was longitudinally folded, but differed in structure in a marked manner from that of the uterus. The epithelial covering was a thick stratified layer. The superficial strata consisted of nucleated squamous cells, but the cells next the fibro-vascular corium were elongated, as is not unfrequently the case on surfaces possessing a stratified squamous epithelium. The corium was elevated into occasional large papillæ, the intervals between which were filled up by the stratified arrangement of epithelial cells.

The great thickness of the uterine mucosa was due to the size and numbers of the tubular glands, and not to the interglandular connective tissue, which was remarkably small in quantity as compared with the uteri of the placental mammals. The large glands have doubtless some special relation to the peculiarity in the utero-gestation in the Kangaroo. There can be little doubt that the glands must be capable of forming a considerable quantity of secretion, sufficient perhaps for the nutritive needs of the embryo during the comparatively short period in which it occupies the uterine cavity. The small proportion of interglandular connective tissue leads one to infer that the production of an interglandular series of crypts, on the free surface of the mucosa, takes place only to a limited extent, if at all, in the gravid kangaroo; and it is possible that the intra-uterine nutrition of the fœtus may be effected solely through the secretion of the tubular glands, and not through that of secreting crypts, newly formed during pregnancy, such as are found in the gravid uteri of the placental mammals[1]. If I am right in supposing that secreting crypts are not formed, we may have the anatomical reason why the intra-uterine nutrition of the embryo can only be carried on for a short period. It would be very desirable to obtain the uterus of a gravid kangaroo, in a suitable condition for examination, to decide these and other as yet imperfectly understood questions connected with the utero-gestation of this animal.

---

[1] See my paper in this *Journal*, October, 1875, for an account of these secreting crypts, and more fully my *Lectures on the Comparative Anatomy of the Placenta*.

ADDITIONAL NOTE ON THE DENTITION OF THE NARWHAL (*Monodon monoceros*). By PROFESSOR TURNER.

IN Vol. VII. of this *Journal* I gave an account of the examination of a very young fœtal Narwhal, 7¼ inches long, in which I found two dental papillæ developed in the gum enveloping the dentary border of each superior maxilla. Each papilla possessed an enamel organ, although, as is well known, the tusk of the Narwhal is destitute of enamel. The early age of the fœtus did not enable me to state which of the papillæ would have developed into the tusk, but I conjectured that "the more anterior would have developed into the maxillary tusk, and the posterior either have disappeared altogether, or formed one of those irregular non-protruding teeth, such as Berthold described some years ago in the skull of a young Narwhal which he examined."

In the month of November, 1875, I received, through the intermediation of my friend Mr C. W. Peach, the gravid uterus of a Narwhal, containing a fœtus 5 ft. 5 in. long, which now enables me to state that my conjecture was accurate. In this specimen each superior maxilla was 8½ inches long, and possessed at the anterior end of its dentary border two well marked alveoli, the one of which opened immediately in front of the other. The anterior alveolus contained a rudimentary cylindriform tusk such as one always finds in the well advanced fœtus of the Narwhal. The posterior alveolus contained an aborted tooth ½ inch long and $\frac{2}{10}$ ths inch wide in its thickest part. It was elongated in form, with short irregular processes projecting from the posterior end, whilst the anterior end was smooth and rounded. The aborted tooth in each superior maxilla was enclosed in a distinct sac of fibrous tissue, which, like the sac of the rudimentary tusk, was firmly united to the fibrous tissue of the gum. The tusk had unquestionably developed from the more anterior of the two dental papillæ, the aborted tooth from the more posterior.

THE DEVELOPMENT OF ELASMOBRANCH FISHES.

By F. M. BALFOUR, B.A., *Fellow of Trinity College,*
Cambridge. (Plates 21—26.)

FORMATION OF THE LAYERS.

(Continued from p. 410.)

IN the last chapter the blastoderm was left as a solid lens-shaped mass of cells, thicker at one end than at the other, its uppermost row of cells forming a distinct layer. There very soon appears in it a cavity, the well known segmentation cavity, or cavity of von Baer, which arises as a small space in the midst of the blastoderm, near its non-embryonic end (Pl. XXI. fig. 1).

This condition of the segmentation cavity, though already[1] described, has nevertheless been met with in one case only. The circumstance of my having so rarely met with this condition is the more striking because I have cut sections of a considerable number of blastoderms in the hope of encountering specimens similar to the one figured, and it can only be explained on one of the two following hypotheses. Either the stage is very transitory, and has therefore escaped my notice except in the one instance ; or else the cavity present in this instance is not the true segmentation cavity, but merely some abnormal structure. That this latter explanation is a possible one, appears from the fact that such cavities do at times occur in other parts of the blastoderm. Dr Schultz[2] does not mention having found any stage of this kind.

The position of the cavity in question, and its general appearance, incline me to the view that it is the segmentation cavity[3]. If this is the true view of its nature the fact should be noted that at first its floor is formed by the lower-layer cells and not by the yolk, and that its roof is constituted by both the

[1] *Qy. Journal of Microsc. Science,* Oct. 1874.
[2] *Centr. f. Med. Wiss.* No. 88, 1875.
[3] Professor Bambeke (*Poissons Osseux, Mém. Acad. Belgique* 1875) describes a cavity in the blastoderm of Leuciscus rutilus, which he regards as the segmentation cavity, but not as homologous with the segmentation cavity of Osseous Fishes, usually so called. Its relations are the same as those of my segmentation cavity at this stage.

lower-layer cells and the epiblast cells. The relations of the floor undergo considerable modifications in the course of development.

The other features of the blastoderm at this stage are very much those of the previous stage.

The embryonic swelling is very conspicuous. The cells of the blastoderm are still disposed in two layers: an upper one of slightly columnar cells one deep, which constitutes the epiblast, and a lower one consisting of the remaining cells of the blastoderm.

An average cell of the lower layer has a diameter of about $\frac{1}{900}$ inch, but the cells at the periphery of the layer are in some cases considerably larger than the more central ones. All the cells of the blastoderm are still completely filled with yolk spherules. In the yolk outside the peculiar nuclei, before spoken of, are present in considerable numbers. They seem to have been mistaken by Dr Schultz[1] for cells: there can however be no question that they are true nuclei.

In the next stage the relations of the segmentation cavity undergo important modifications.

The cells which form its floor disappear almost completely from that position, and the floor becomes formed by the yolk.

The stage, during which the yolk serves as the floor of the segmentation cavity, extends over a considerable period of time, but during it I have been unable to detect any important change in the constitution of the blastoderm. It no doubt gradually extends over the yolk, but even this growth is not nearly so rapid as in the succeeding stage. Although therefore the stage I proceed to describe is of long continuance, a blastoderm at the beginning of it exhibits, both in its external and in its internal features, no important deviations from one at the end of it.

Viewed from the surface (Pl. XXIV. fig. A) the blastoderm at this stage appears slightly oval, but the departure from the circular form is not very considerable. The long axis of the oval corresponds with what eventually becomes the long axis of the embryo. From the yolk the blastoderm is

[1] *Loc. cit.*

still well distinguished by its darker colour; and it is surrounded by a concentric ring of light-coloured yolk, the outer
border of which shades insensibly into the normal yolk.

At the embryonic portion of the blastoderm is a slight
swelling, clearly shewn in Plate XXIV. fig. A, which can easily
be detected in fresh and in hardened embryos. This swelling is
to be looked upon as a local exaggeration of a slightly raised
rim present around the whole circumference of the blastoderm.
The roof of the segmentation cavity (fig. A, s. c.) forms a second
swelling; and in the fresh embryo this region appears of a
darker colour than other parts of the blastoderm.

It is difficult to determine the exact shape of the blastoderm, on account of the traction exercised upon it in opening
the egg; and no reliance can be placed on the forms assumed
by hardened blastoderms. This remark also applies to the
sections of blastoderms of this stage. There can be no doubt
that the minor individual variations exhibited by almost
every specimen are produced in the course of manipulations
while the objects are fresh. These variations may affect even
the relative length of a particular region and certainly the
curvature of it. The roof of the segmentation cavity is especially apt to be raised into a dome-like form.

The main internal feature of this stage is the disappearance
of the layer of cells which, during the first stage, formed the
floor of the segmentation cavity. This disappearance is nevertheless not absolute, and it is doubtful whether there is any
period in which the floor of the cavity is quite without cells.

Dr Schultz supposes[1] that the entire segmentation cavity
is, in the living animal, filled with a number of loose cells.
Though it is not in my power absolutely to deny this, the
point being one which cannot be satisfactorily investigated in
sections, yet no evidence has come under my notice which
would lead to the conclusion that more cells are present in the
segmentation cavity than are represented on Pl. XIII. fig. 1, of
my preliminary paper, an illustration which is repeated on Pl.
XXI. fig. 2.

The number of cells on the floor of the cavity differs
considerably in different cases, but these cases come under the

[1] *Loc. cit.*

category of individual variations, and are not to be looked upon
as indications of different states of development.

In many cases especially large cells are to be seen on the
floor of the cavity (Pl. XXI. fig. 2, $b\,d$). In my preliminary
paper[1] the view was expressed that these are probably
cells formed around the nuclei of the yolk. This view I am
inclined to abandon, and to substitute for it the suggestion
made by Dr Schultz, that they are remnants of the larger
segmentation cells which were to be seen in the previous stages.

Plate XXI. figs. 2, 3, 4 (all sections of this stage) show the
different appearances presented by the floor of the segmentation
cavity. In only one of these sections are there any large number
of cells upon the floor; and in no case have cells been observed
imbedded in the yolk forming this floor, as described by Dr
Schultz[2], but in all cases the cells simply rested upon it.

Passing from the segmentation cavity to the blastoderm
itself, the first feature to be noticed is the more decided differentiation of the epiblast. This now forms a distinct layer
composed of a single row of columnar cells. These are slightly
more columnar in the region of the embryonic swelling than
elsewhere, and become less elongated at the edge of the blastoderm. In my specimens this layer was never more than one
cell deep, but Dr Schultz[3] states that, in the Elasmobranch
embryos investigated by him, the epiblast was composed of
more than a single row of cells.

Each epiblast cell is filled with yolk spherules and contains
a nucleus. Very frequently the nuclei in the layer are arranged in a regular row (vide Pl. XXI. fig. 4). In the later
blastoderms of this stage there is a tendency in the cells to
assume a wedge-like form with their thin ends pointing alternately in opposite directions. This arrangement is, however,
by no means strictly adhered to, and the regularity of it is
exaggerated in Plate XXI. fig. 4.

The nuclei of the epiblast cells have the same characters as
those of the lower-layer cells to be presently described, but
their intimate structure can only be successfully studied in

---

[1] Qy. Journal of Micros. Science, Oct. 1874.
[2] Loc. cit. Probably Dr Schultz, here as in other cases, has mistaken nuclei
for cells.
[3] Loc. cit.

certain exceptionally favourable sections. In most cases the yolk spherules around them render the finer details invisible.

There is at this stage no such obvious continuity as in the succeeding stage between the epiblast and the lower-layer cells; and this statement holds good more especially with the best conserved specimens which have been hardened in osmic acid (Plate XXI. fig. 4). In these it is very easy to see that the epiblast simply thins out at the edge of the blastoderm without exhibiting the slightest tendency to become continuous with the lower-layer cells[1].

The lower-layer cells form a mass rather than a layer, and constitute the whole of the blastoderm not included in the epiblast. The shape of this mass in a longitudinal section may be gathered from an examination of Plate XXI. figs. 3 and 4.

It presents an especially thick portion forming the bulk of the embryonic swelling, and frequently contains one or two cavities, which from their constancy I regard as normal and not as artificial products.

In addition to the mass forming the embryonic swelling there is seen in sections another mass of lower-layer cells at the opposite extremity of the blastoderm, connected with the former by a bridge of cells, which constitutes the roof of the segmentation cavity. The lower-layer cells may thus be divided into three distinct parts:

(1)   The embryo swelling.

(2)   The thick rim of cells round the edge of the remainder of the blastoderm.

(3)   The cells which form the roof of the segmentation cavity.

[1] Prof. Haeckel (Die Gastrula u. die Eifurchung d. Thiere, *Jenaische Zeit-schrift*, Vol. IX.) has unfortunately copied a figure from my preliminary paper (*loc. cit.*) (repeated now), which I had carefully avoided using for the purpose of describing the formation of the layers on account of the epiblast cells in the original having been much altered by the chromic acid, as a result of which the whole section gives a somewhat erroneous impression of the condition of the blastoderm at this stage. I take this opportunity of pointing out that the colouration employed by Professor Haeckel to distinguish the layers in this section is not founded on my statements, but is, on the contrary, in entire opposition to them. From the section as represented by Professor Haeckel it might be gathered that I considered the lower-layer cells to be divided into two parts, one derived from the epiblast, while the other constituted the hypo-blast. Not only is no such division present at this period, but no part of the lower-layer cells, or the mesoblast cells into which they become converted, can in any sense whatever be said to be derived from the epiblast.

These three parts form a continuous whole, but in addition to these there exist the previously mentioned cells, which rest on the floor of the segmentation cavity.

With the exception of these latter, the lower layer is composed of cells having a fairly uniform size, and exhibits no trace of a division into two layers.

The cells are for the most part irregularly polygonal from mutual pressure; and in their shape and arrangement, exhibit a marked contrast to the epiblast cells. A few of the lower-layer cells, highly magnified, are represented in Pl. XXI. fig. 2 a. An average cell measures about $\frac{1}{800}$ to $\frac{1}{900}$ of an inch, but some of the larger ones on the floor attain to the $\frac{1}{415}$ of an inch.

Owing to my having had the good fortune to prepare some especially favourable specimens of this stage, it has been possible for me to make accurate observations both upon the nuclei of the cells of the blastoderm, and upon the nuclei of the yolk.

The nuclei of the blastoderm cells, both of the epiblast and lower layer, have a uniform structure. Those of the lower-layer cells are about $\frac{1}{1800}$ of an inch in diameter. Roughly speaking each consists of a spherical mass of clear protoplasm refracting more highly than the protoplasm of its cell. The nucleus appears in sections to be divided by deeply stained lines into a number of separate areas, and in each of these a deeply stained granule is placed. In some cases two or more of such granules may be seen in a single area. The whole of the nucleus stains with the colouring reagents more deeply than the protoplasm of the cells; but this is especially the case with the granules and lines.

Though usually spherical the nuclei not infrequently have a somewhat lobate form.

Very similar to these nuclei are the nuclei of the yolk.

One of the most important differences between the two is that of size. The majority of the nuclei present in the yolk are as large or larger than an ordinary blastoderm cell; while many of them reach a size very much greater than this. The examples I have measured varied from $\frac{1}{500}$ to $\frac{1}{250}$ of an inch in diameter.

Though they are divided, like the nuclei of the blastoderm, with more or less distinctness into separate areas by a network of lines, their greater size frequently causes them to present an aspect somewhat different from the nuclei of the blastoderm. They are moreover much less regular in outline than these, and very many of them have lobate projections (Pl. XXI. figs. 2a and 2c and 3), which vary from simple knobs to projections of such a size as to cause the nucleus to present an appearance of commencing constriction into halves. When there are several such projections the nucleus acquires a peculiar knobbed figure. With bodies of this form it becomes in many cases a matter of great difficulty to decide whether or no a particular series of knobs, which appear separate in one plane, are united in a lower plane, whether, in fact, there is present a single knobbed nucleus or a number of nuclei in close apposition. A nucleus in this condition is represented in Pl. XXI. fig. 2b.

The existence of a protoplasmic network in the yolk has already been mentioned. This in favourable cases may be observed to be in special connection with the nuclei just described. Its meshes are finer in the vicinity of the nuclei, and its fibres in some cases almost appear to start from them (Pl. XXIII. fig. 12). For reasons which I am unable to explain the nuclei of the yolk and the surrounding meshwork present appearances which differ greatly according to the reagent employed. In most specimens hardened in osmic acid the protoplasm of the nuclei is apparently prolonged in the surrounding meshwork (Pl. XXIII. fig. 12). In other specimens hardened in osmic acid (Pl. XXIII. fig. 11), and in all hardened in chromic acid (Pl. XXI. fig. 2a and 2c), the appearances are far clearer than in the previous case, and the protoplasmic meshwork merely surrounds the nuclei, without showing any signs of becoming continuous with them.

There is also around each nucleus a narrow space in which the spherules of the yolk are either much smaller than elsewhere or completely absent, vide Pl. XXI. fig. 2b.

It has not been possible for me to satisfy myself as to the exact meaning of the lines dividing these nuclei into a number of distinct areas. My observations leave the question open as to whether they are to be looked upon as lines of division, or as

protoplasmic lines such as have been described in nuclei by
Flemming[1], Hertwig[2] and Van Beneden[3]. The latter view appears to me to be the more probable one.

Such are the chief structural features presented by these
nuclei, which are present during the whole of the earlier periods
of development and retain throughout the same appearance.
There can be little doubt that their knobbed condition implies
that they are undergoing a rapid division. The arguments for
this view I have already insisted on, and, in spite of the observations of Dr Kleinenberg showing that similar nuclei of
Nephelis do not undergo division, the case for their doing so in
the Elasmobranch eggs is to my mind a very strong one.

During this stage the distribution of these nuclei in the
yolk becomes somewhat altered from that in the earlier stages.
Although the nuclei are still scattered generally throughout
the finer yolk-matter around the blastoderm, yet they are
especially aggregated at one or two points. In the first place
a special collection of them may be noticed immediately below
the floor of the segmentation cavity. They here form a distinct row or even layer. If the presence of this layer is coupled with the fact that at this period cells are beginning to
appear on the floor of the segmentation cavity, a strong argument is obtained for the supposition that around these nuclei
cells are being produced, which pass into the blastoderm to
form the floor. Of the actual formation of cells at this period
I have not been able to obtain any satisfactory example, so
that it remains a matter of deduction rather than of direct
observation.

Another special aggregation of nuclei is generally present
at the periphery of the blastoderm, and the same amount of
doubt hangs over the fate of these as over that of the previously
mentioned nuclei.

The next stage is the most important in the whole history
of the formation of the layers. Not only does it serve to show,
that the process by which the layers are formed in Elasmobranchs can easily be derived from a simple gastrula type like

[1] Entwicklungsgeschichte der Najaden, *Sitz. d. k. Akad. Wien*, 1875.
[2] *Morphologische Jahrbuch*, Vol. I. Heft 3.
[3] Développement des Mammifères, *Bul. de l'Acad. de Belgique*, XL. No. 12,
1875.

that of Amphioxus, but it also serves as the key by which other meroblastic types of development may be explained. At the very commencement of this stage the embryonic swelling becomes more conspicuously visible than it was. It now projects above the level of the yolk in the form of a rim. At one point, which eventually forms the termination of the axis of the embryo, this projection is at its greatest; while on either side of this it gradually diminishes and finally vanishes. This projection I propose calling, as in my preliminary paper[1], the embryonic rim.

The segmentation cavity can still be seen from the surface, and a marked increase in the size of the blastoderm may be noticed. During the stage last described, the growth was but very slight; hence the rather sudden and rapid growth which now takes place becomes striking, and has, as I hope to show, a not inconsiderable importance.

Longitudinal sections at this stage, as at the earlier stages, are the most instructive. Such a section on the same scale as Pl. XXI. fig. 4, is represented in Pl. XXI. fig. 5. It passes parallel to the axis of the embryo, through the point of greatest development of·the embryonic rim.

The three fresh features of the most striking kind are (1) the complete envelopment of the segmentation cavity within the lower layer cells, (2) the formation of the embryonic rim, (3) the increase in distance between the posterior end of the blastoderm and the segmentation cavity. ·The segmentation cavity has by no means relatively increased in size. The roof has precisely its earlier constitution, being composed of an internal lining of lower layer cells and an external one of epiblast. The thin lining of lower layer cells is, in the course of mounting the sections, very apt to fall off; but I am absolutely satisfied that it is never absent.

The floor of the cavity has undergone an important change, being now formed by a layer of cells instead of by the yolk. A precisely similar change in the constitution of the floor takes place in Osseous Fishes[2].

---

[1] Qy. Journal Microsc. Science, Oct. 1874.

[2] Vide Oellacher, Zeitschrift f. Wiss. Zoologie, Bd. XXIII.; Götte, Der Keim d. Forelleneies, Arch. f. Mikr. Anat. Vol. IX.; Haeckel, Die Gastrula u. die Eifurchung d. Thiere, Jenaische Zeitschrift, Bd. IX.

The mode in which the floor is formed is a question of some importance. The nuclei, which during the last stage formed a row beneath it, probably, as previously pointed out, take some share in its formation. An additional argument to those already brought forward in favour of this view may be derived from the fact that during this stage such a row of nuclei is no longer present.

This argument may be stated as follows :

Before the floor of cells for the segmentation cavity is formed a number of nuclei are present in a suitable situation to supply the cells for the floor ; as soon as the floor of cells makes its appearance these nuclei are no longer to be seen. From this it may be concluded that their disappearance arises from their having become the nuclei of the cells which form the floor.

It appears to me most probable that there is a growth inwards from the whole peripheral wall of the cavity, and that this ingrowth, as well as the cells derived from the yolk, assist in forming the floor of the cavity. In Osseous Fish there appears to be no doubt that the floor is largely formed by an ingrowth of this kind.

A great increase is observable in the distance between the posterior end of the segmentation cavity and the edge of the blastoderm. This is due to the rapid growth of the latter combined with the stationary condition of the former. The growth of the blastoderm at this period is not uniform, but is more rapid in the non-embryonic than in the embryonic parts.

The main features of the epiblast remain the same as during the last stages. It is still composed of a very distinct layer one cell deep. Over the segmentation cavity, and over the whole embryonic end of the blastoderm, the cells are very thin, columnar, and, roughly speaking, wedge-shaped with the thin ends pointing alternately in different directions. For this reason, the nuclei form two rows ; but both the rows are situated near the upper surface of the layer (vide Pl. xxi. fig. 5). Towards the posterior end of the blastoderm the cells are flatter and broader ; and the layer terminates at the non-embryonic end of the blastoderm without exhibiting the slightest tendency to become continuous with the lower layer cells. At the embryonic end of the blastoderm the relations of the epiblast and lower layer cells are very different. At this part, throughout the whole extent of the embryonic rim, the epiblast is reflected and becomes continuous with the lower layer cells.

The lower layer cells form, for the most part, a uniform stratum in which no distinction into mesoblast and hypoblast is to be seen.

. Both the lower layer cells and the epiblast cells are still filled with yolk spherules.

The structures at the embryonic rim, and the changes which are there taking place, unquestionably form the chief features of interest at this stage.

The general relations of these parts are very fairly shewn in Pl. XXI. fig. 5, which represents a section passing through the median line of the embryonic region. They are however more accurately represented in Pl. XXII. fig. 5 a, taken from the same embryo, but in a lateral part of the embryonic rim; or in Pl. XXII. fig. 6, from a slightly older embryo. In all of these figures the epiblast cells are reflected at the edge of the embryonic rim, and become perfectly continuous with the hypoblast cells. A few of the cells, immediately beyond the line of this reflection, precisely resemble in character the typical epiblast cells; but the remainder exhibit a gradual transition into typical lower layer cells. Adjoining these transitional cells, or partly enclosed in the corner formed between them and the epiblast, are a few unaltered lower layer cells (m), which at this stage are not distinctly separated from the transitional cells. The transitional cells form the commencement of the hypoblast (hy); and the cells (m) between them and the epiblast form the commencement of the mesoblast. The gradual conversion of lower layer cells into columnar hypoblast cells, is a very clear and observable phenomenon in the best specimens. Where the embryonic rim projects most, a larger number of cells have assumed a columnar form. Where it projects less clearly, a smaller number have done so. But in all cases there may be observed a series of gradations between the columnar cells and the typical rounded lower layer cells[1].

[1] When writing my earlier paper I did not feel so confident about the mode of formation of the hypoblast as I now do, and even doubted the possibility of determining it from sections. The facts now brought forward are I hope sufficient to remove all scepticism on this point.

In the last described embryo, although the embryonic rim had attained to a considerable development, no trace of the medullary groove had made its appearance. In an embryo in the next stage of which I propose describing sections, this structure has become visible.

A surface view of a blastoderm of this age, with the embryo, is represented on Pl. XXIV. fig. B; and I shall, for the sake of convenience, in future speak of embryos of this age as belonging to period B.

The blastoderm is nearly circular. The embryonic rim is represented by a darker shading at the edge. At one point in this rim may be seen the embryo, consisting of a somewhat raised area with an axial groove (*m g*). The head end of the embryo is that which points towards the centre of the blastoderm, and its free peripheral extremity is at the edge of the blastoderm.

A longitudinal section of an embryo of the same age as the one figured[1] is represented on Pl. XXII. fig. 7. The general growth has been very considerable, though as before explained, it is mainly confined to that part of the blastoderm where the embryonic rim is absent.

A fresh feature of great importance is the complete disappearance of the segmentation cavity, the place which was previously occupied by it being now filled up by an irregular network of cells. There can be little question that the obliteration of the segmentation cavity is in part due to the entrance into the blastoderm of fresh cells formed around the nuclei of the yolk. The formation of these is now taking place with great rapidity and can be very easily followed.

Since the segmentation cavity ceases to play any further part in the history of the blastoderm, it will be well shortly to review the main points in its history.

Its earliest appearance is involved in some obscurity, though it probably arises as a simple cavity in the midst of the lower layer cells (Pl. XXI. fig. 1). In its second phase the floor ceases to be formed of lower layer cells, and the place of these is taken by the yolk, on which however a few scattered cells

[1] Owing to the small size of the plates this section has been drawn on a considerably smaller scale than that represented in fig. 5.

still remain (Pl. XXI. figs. 2, 3, 4). During the third period of
its history, a distinct cellular floor is again formed for it, so
that it comes a second time into the same relations with the
blastoderm as at its earliest appearance. The floor of cells
which it receives is in part due to a growth inwards from the
periphery of the blastoderm, and in part to the formation of
fresh cells from the yolk. Coincidently with the commencing
differentiation of hypoblast and mesoblast the segmentation
cavity grows smaller and vanishes.

One of the most important features of the segmentation
cavity in the Elasmobranchs which I have studied, is the fact
that throughout its whole existence its roof is formed of *lower
layer cells*. There is not the smallest question that the segmentation cavity of these fishes is the homologue of that of
Amphioxus, Batrachians, etc., yet in the case of all of these
animals, the roof of the segmentation cavity is formed of
epiblast only. How comes it then to be formed of lower layer
cells in Elasmobranchii ?

To this question an answer was attempted in my paper,
"Upon the Early Stages of the Development of Vertebrates[1]."
It was there pointed out, that as the food material in the ovum
increases, the bulk of the lower layer cells necessarily also increases; since these, as far as the blastoderm is concerned, are
the chief recipients of food material. This causes the lower layer
cells to encroach upon the segmentation cavity, and to close
it in not only on the sides, but also above; from the same cause
it results that the lower layer cells assume, from the first, a
position around the spot where the future alimentary cavity
will be formed, and that this cavity becomes formed by a
simple split in the midst of the lower layer cells, and not by
an involution.

All the most recent observations[2] on Osseous Fishes tend
to show that in them, the roof of the segmentation cavity is
formed alone of epiblast; but on account of the great difficulty
which is experienced in distinguishing the layers in the blastoderms of these animals, I still hesitate to accept as conclusive
the testimony on this point.

[1] *Quart. Journ. of Microscop. Science*, July, 1875.
[2] Oellacher, *Zeit. f. Wiss. Zoologie*, Bd. XXIII.   Götte, *Archiv f. Mikr.
Anat.* Vol. IX.   Haeckel, *loc. cit.*

In the formation a second time, of a cellular floor for the segmentation cavity in the third stage, the Elasmobranch embryo seems to resemble that of the Osseous Fish[1]. Upon this feature great stress is laid both by Dr Götte[2] and Prof. Haeckel[3]: but I am unable to agree with the interpretation of it offered by them. Both Dr Götte and Prof. Haeckel regard the formation of this floor as part of an involution to which the lower layer cells owe their origin, and consider the involution an equivalent to the alimentary involution of Batrachians, Amphioxus, &c. To this question I hope to return, but it may be pointed out that my observations prove that this view can only be true in a very modified sense; since the invagination by which hypoblast and alimentary canal are formed in Amphioxus is represented in Elasmobranchs by a structure quite separate from the ingrowth of cells to form the floor of the segmentation cavity.

The eventual *obliteration* of the segmentation cavity by cells derived from the yolk is to be regarded as an inherited remnant of the involution by which this obliteration was primitively effected. The passage upwards of cells from the yolk, may possibly be a real survival of the tendency of the hypoblast cells to grow inwards during the process of involution.

The last feature of the segmentation cavity which deserves notice is its excentric position. It is from the first situated in much closer proximity to the non-embryonic than to the embryonic end of the blastoderm. This peculiarity in position is also characteristic of the segmentation cavity of Osseous Fishes, as is shown by the concordant observations of Oellacher[4] and Götte[5]. Its meaning becomes at once intelligible by referring to the diagrams in my paper[6] on the Early Stages in the Development of Vertebrates. It in fact arises from the asymmetrical character of the primitive alimentary involution in all anamniotic vertebrates with the exception of Amphioxus.

Leaving the segmentation cavity I pass on to the other features of my sections.

There is still to be seen a considerable aggregation of cells at the non-embryonic end of the blastoderm. The position of this, and its relations with the portion of the blastoderm which

---

[1] This floor appears in most Osseous Fish to be only partially formed. Vide Götte, *loc. cit.*
[2] *Loc. cit.*   [3] *Loc. cit.*   [4] *Loc. cit.*   [5] *Loc. cit.*   [6] *Loc. cit.*

at an earlier period contained the segmentation cavity, indicate that the growth of the blastoderm is not confined to its edge, but that it proceeds at all points causing the peripheral parts to glide over the yolk.

The main features of the cells of this blastoderm are the same as they were in the one last described. In the non-embryonic region the epiblast has thinned out, and is composed of a single row of cells, which, in the succeeding stages, become much flattened.

The lower layer cells over the greater part of their extent, have not undergone any histological changes of importance. Amongst them may frequently be seen a few exceptionally large cells, which without doubt have been derived directly from the yolk.

The embryonic rim is now a far more considerable structure than it was. Vide Pl. XXII. fig. 7. Its elongation is mainly effected by the continuous conversion of rounded lower layer cells into columnar hypoblast cells at its central or anterior extremity.

This conversion of the lower layer cells into hypoblast cells is still easy to follow, and in every section cells intermediate between the two are to be seen. The nature of the changes which are taking place requires for its elucidation transverse as well as longitudinal sections. Transverse sections of a slightly older embryo than B are represented on Pl. XXII. fig. 8 a, 8 b, and 8 c.

Of these sections a is the most peripheral or posterior, and c the most central or anterior. By a combination of transverse and longitudinal sections, and by an inspection of a surface view, it is rendered clear that, though the embryonic rim is a far more considerable structure in the region of the embryo than else-where (compare fig. 6 and fig. 7 and 7 a), yet that this gain in size is not produced by an outgrowth of the embryo beyond the rest of the germ, but by the conversion of the lower layer cells into hypoblast having been carried far further towards the centre of the germ in the axial line than in the lateral regions of the rim.

The most anterior of the series of transverse sections (Pl. XXII. fig. 8c) I have represented, is especially instructive with reference to this point. Though the embryonic rim is cut through at the sides of the section, yet in these parts the rim consists of hardly more than a continuity between epiblast and lower layer cells, and the lower layer cells show no trace of a division into mesoblast and hypoblast. In the axis of the embryo, however, the columnar hypoblast is quite distinct; and on it a small cap of mesoblast is seen on each side of the medullary groove. Had the embryonic rim resulted from a projecting growth of the blastoderm, such a condition could not have existed. It might have been possible to find the hypoblast formed at the sides of the section and not at the centre; but the reverse, as in these sections, could not have occurred. Indeed it is scarcely necessary to have recourse to sections to prove that the growth of the embryonic rim is towards the centre of the blastoderm. The inspection of a surface view of a blastoderm at this period demonstrates it beyond a doubt (Pl. XXIV. fig. B). The embryo, close to which the embryonic rim is alone largely developed, does not project outwards beyond the edge of the germ, but inwards towards its centre.

The space between the embryonic rim and the yolk (Pl. XXII. fig. 7 al.) is the alimentary cavity. The roof of this is therefore primitively formed of hypoblast and the floor of yolk. The external opening of this space at the edge of the blastoderm is the exact morphological homologue of the anus of Rusconi, or blastopore of Amphioxus, the Amphibians, &c. The importance of the mode of growth in the embryonic rim depends upon the homology of the cavity between it and the yolk, with the alimentary cavity of Amphioxus and Amphibians. Since this homology exists, the direction of the growth of this cavity ought to be, as it in fact is, the same as in Amphioxus, etc., viz. towards the centre of germ and original position of the segmentation cavity. Thus though a true invagination is not present as in the other cases, yet this is represented in Elasmobranchs by the continuous conversion of lower layer cells into hypoblast along a line leading towards the centre of the blastoderm.

In the parts of the rim adjoining the embryo, the lower layer cells, on becoming continuous with the epiblast cells, assume a columnar form. At the sides of the rim this is not strictly the case, and the lower layer-cells retain their rounded form, though quite continuous with the epiblast cells. One curious feature of the layer of epiblast in these lateral parts of the rim is the great thickness it acquires before being reflected and becoming continuous with the hypoblast (Pl. XXII. fig. 8 c). In the vicinity of the point of reflection there is often a rather large formation of cells around the nuclei of the yolk. The cells formed here no doubt pass into the blastoderm, and become converted into columnar hypoblast cells. In some cases the formation of these cells is very rapid, and they produce quite a projection on the under side of the hypoblast. Such a case is represented in Pl. XXII. fig. 8 b, n. al. The cells constituting this mass eventually become converted into the lateral and ventral walls of the alimentary canal.

The formation of the mesoblast has progressed rapidly. While many of the lower layer cells become columnar and form the hypoblast, others, between these and the epiblast, remain spherical. The latter do not at once become separated as a layer distinct from the hypoblast, and, at first, are only to be distinguished from them through their different character, vide Plate XXII. figs. 6 and 7. They nevertheless constitute the commencing mesoblast.

Thus much of the mode of formation of the mesoblast can be easily made out in longitudinal sections, but transverse sections throw still further light upon it.

From these it may at once be seen that the mesoblast is not formed in one continuous sheet, but as two lateral masses, one on each side of the axial line of the embryo[1]. In my preliminary account[2] it was stated that this was a condition of the mesoblast at a very early period, and that it was probably its condition from the beginning. Sections are now in my possession which satisfy me that, from the very first, the mesoblast arises as two distinct lateral masses, one on each side of the axial line.

---

[1] Professor Lieberkühn (*Gesellschaft zu Marburg*, Jan. 1876) finds in Mammalia a bilateral arrangement of the mesoblast, which he compares with that described by me in Elasmobranchs. In Mammalia, however, he finds the two masses of mesoblast connected by a very thin layer of cells, and is apparently of opinion that a similar thin layer exists in Elasmobranchs though overlooked by me. I can definitely state that, whatever may be the condition of the mesoblast in Mammalia, in Elasmobranchs at any rate no such layer exists.

[2] *Loc. cit.*

In the embryo from which the sections Pl. XXII. fig. 8 a, 8 b, 8 c were taken, the mesoblast had, in most parts, not yet become separated from the hypoblast. It still formed with this a continuous layer, though the mesoblast cells were distinguishable by their shape from the hypoblast. In only one section (b) was any part of the mesoblast quite separated from the hypoblast.

In the hindermost part of the embryo the mesoblast is at its maximum, and forms, on each side, a continuous sheet extending from the median line to the periphery (fig. 8 a). The rounder form of the mesoblast cells render the line of junction between the layer constituted by them and the hypoblast fairly distinct; but towards the periphery, where the hypoblast cells have the same rounded form as the mesoblast, the fusion between the two layers is nearly complete.

In an anterior section the mesoblast is only present as a cap on both sides of the medullary groove, and as a mass of cells at the periphery of the section (fig. 8 b); but no continuous layer of it is present. In the foremost of the three sections (fig. 8 c) the mesoblast can scarcely be said to have become in any way separated from the hypoblast except at the summit of the medullary folds (m).

From these and similar sections it may be certainly concluded, that the mesoblast becomes first separated from the hypoblast as a distinct layer in the posterior region of the embryo, and only at a later period in the region of the head.

In an embryo but slightly more developed than B, the formation of the layer is quite completed in the region of the embryo. To this embryo I now pass on.

In the non-embryonic parts of the blastoderm no fresh features of interest have appeared. It still consists of two layers. The epiblast is composed of flattened cells, and the lower layer of a network of more rounded cells, elongated in a lateral direction. The growth of the blastoderm has continued to be very rapid.

In the region of the embryo (Pl. XXII. fig. 9) more important changes have occurred. The epiblast still remains as a single row of columnar cells. The hypoblast is no longer fused with the mesoblast, and forms a distinct dorsal wall for the alimentary cavity. Though along the axis of the embryo the hypoblast is composed of a single row of columnar cells, yet in the lateral part of the embryo its cells are less columnar and are one or two deep.

Owing to the manner in which the mesoblast became split off from the hypoblast, a continuity is maintained between the hypoblast and the lower layer cells of the blastoderm (Pl. XXII. fig. 9), while the two plates of mesoblast are isolated and disconnected from any other masses of cells.

The alimentary cavity is best studied in transverse sections. (*vide* Pl. XXIII. fig. 10 *a*, 10 *b* and 10 *c*, three sections from the same embryo.) It is closed in above and at the sides by the hypoblast, and below by the yolk. In its anterior part a floor is commencing to be formed by a growth of cells from the walls of the two sides. The cells for this growth are formed around the nuclei of the yolk; a feature which recalls the fact that in Amphibians the ventral wall of the alimentary cavity is similarly formed in part from the so-called yolk cells.

We left the mesoblast as two masses not completely separated from the hypoblast. During this stage the separation between the two becomes complete, and there are formed two great lateral plates of mesoblast cells, one on each side of the medullary groove. Each of these corresponds to a united vertebral and lateral plate of the higher Vertebrates. The plates are thickest in the middle and posterior regions (Pl. XXIII. fig. 10 *a* and 10 *b*), but thin out and almost vanish in the region of the head. The longitudinal section of this stage represented in Pl. XXIII. fig. 9, passes through one of the lateral masses of mesoblast cells, and shows very distinctly its complete independence of all the other cells in the blastoderm.

From what has been stated with reference to the development of the mesoblast, it is clear that in Elasmobranchs this layer is derived from the same mass of cells as the hypoblast, and receives none of its elements from the epiblast. In connection with its development, as two independent lateral masses, I may observe, as I have previously done[1], that in this respect it bears a close resemblance to mesoblast in Euaxes, as de-

---

[1] *Quart. Journ. of Microsc. Science*, Oct., 1874.

scribed by Kowalevsky[1]. This resemblance is of some interest, as bearing on a probable Annelid origin of vertebrata. Kowalevsky has also shown[2] that the mesoblast in Ascidians is similarly formed as two independent masses, one on each side of the middle line.

It ought, however, to be pointed out that a similar bilateral origin of the mesoblast had been recently met with in Lymnoeus by Carl Rabl[3]. A fact which somewhat diminishes the genealogical value of this feature in the mesoblast in Elasmobranchs.

During the course of this stage the spherules of food yolk immediately beneath the embryo are used up very rapidly. As a result of this the protoplasmic network, so often spoken of, comes very plainly into view. Considerable areas may sometimes be seen without any yolk spherule whatever.

On Pl. XXII. fig. 7 a, and Pl. XXIII. 11 and 12, I have attempted to reproduce the various appearances presented by this network: and these figures give a better idea of it than any description. My observations tend to show that it extends through the whole yolk, and serves to hold it together. It has not been possible for me to satisfy myself that it had any definite limits, but on the other hand, in many parts all my efforts to demonstrate its presence have failed. When the yolk spherules are very thickly packed, it is difficult to make out for certain whether it is present or absent, and I have not succeeded in removing the yolk spherules from the network in cases of this kind. In medium-sized ovarian eggs this network is very easily seen, and extends through the whole yolk. Part of such an egg is shown in Pl. XXIII. fig. 14. In full-sized ovarian eggs, according to Schultz[4], it forms, as was mentioned in the first chapter, radiating striæ, extending from the centre to the periphery of the egg. When examined with the highest powers, the lines of this network appear to be composed of immeasurably small granules arranged in a linear direction. These granules are more distinct in chromic acid specimens than in

[1] Embryologische Studien an Würmen u. Arthropoden. *Mémoires d. l'Acad. S. Pétersbourg.* Vol. XIV. 1873.

[2] *Archiv für Mikr. Anat.* Vol. VII.

[3] *Jenaische Zeitschrift.* Vol. IX. 1875. A bilateral development of mesoblast, according to Professor Haeckel (*loc. cit.*), occurs in some Osseous Fish.

[4] *Archiv für Mikr. Anat.* Vol. XI.

those hardened in osmic acid, but are to be seen in both. There can be little doubt that these granules are imbedded in a thread or thin layer of protoplasm.

I have already (p. 523) touched upon the relation of this network to the nuclei of the yolk[1].

During the stages which have just been described specially favourable views are frequently to be obtained of the formation of cells in the yolk and their entrance into the blastoderm. Two representations of these are given, in Pl. xxii. fig. 7 a, and Pl. xxiii. fig. 13. In both of these distinctly circumscribed cells are to be seen in the yolk (c), and in all cases are situated near to the typical nuclei of the yolk. The cells in the yolk have such a relation to the surrounding parts, that it is quite certain that their presence is not due to artificial manipulation, and in some cases it is even difficult to decide whether or no a cell area is circumscribed round a nucleus (Pl. xxiii. fig. 13). Though it would be possible for cells in the living state to pass from the blastoderm into the yolk, yet the view that they have done so in the cases under consideration has not much to recommend it, if the following facts be taken into consideration. (1) That the cells in the yolk are frequently larger than those in the blastoderm. (2) That there are present a very large number of nuclei in the yolk which precisely resemble the nuclei of the cells under discussion. (3) That in some cases (Pl. xxiii. fig. 13) cells are seen indistinctly circumscribed as if in the act of being formed.

Between the blastoderm and the yolk may frequently be seen a membrane-like structure, which becomes stained with hæmatoxylin, osmic acid etc. It appears to be a layer of coagulated albumen and not a distinct membrane.

---

[1] A protoplasmic network resembling in its essential features the one just described has been noticed by many observers in other ova. Fol has figured and described a network or sponge-like arrangement of the protoplasm in the eggs of Geryonia. (*Jenaische Zeitschrift*, vol. vii.) Metschnikoff (*Zeitschrift f. Wiss. Zoologie*, 1874) has demonstrated its presence in the ova of many Siphonophoriæ and Medusæ. Flemming (*Entwicklungsgeschichte der Najaden, Sitz. der k. Akad. Wien*, 1875) has found it in the ovarian ova of fresh-water mussels (Anodonta and Unio), but regards it as due to the action of reagents, since he fails to find it in the fresh condition. Amongst vertebrates it has been carefully described by Eimer (*Archiv für Mikr. Anat.*, vol. viii.) in the ovarian ova of Reptiles. Eimer moreover finds that it is continuous with prolongations from cells of the epithelium of the follicle in which the ovum is contained. According to him remnants of this network are to be met with in the ripe ovum, but are no longer present in the ovum when taken from the oviduct.

SUMMARY.

At the close of segmentation, the blastoderm forms a somewhat lens-shaped disc, thicker at one end than at the other ; the thicker end being termed the embryonic end.

It is divided into two layers—an upper one, the epiblast, formed by a single row of columnar cells; and a lower one, consisting of the remaining cells of the blastoderm.

A cavity next appears in the lower layer cells, near the non-embryonic end of the blastoderm, but the cells soon disappear from the floor of this cavity which then comes to be constituted by yolk alone.

The epiblast in the next stage is reflected for a small arc at the embryonic end of the blastoderm, and becomes continuous with the lower layer cells; at the same time some of the lower layer cells of the embryonic end of the blastoderm assume a columnar form, and constitute the commencing hypoblast. The portion of the blastoderm, where epiblast and hypoblast are continuous, forms a projecting structure which I have called the embryonic rim. This rim increases rapidly by growing inwards more and more towards the centre of the blastoderm, through the continuous conversion of lower layer cells into columnar hypoblast.

While the embryonic rim is being formed, the segmentation cavity undergoes important changes. In the first place, it receives a floor of lower layer cells, partly from an ingrowth from the two sides, and partly from the formation of cells around the nuclei of the yolk.

Shortly after the floor of cells has appeared, the whole segmentation cavity becomes obliterated.

When the embryonic rim has attained to some importance, the position of the embryo becomes marked out by the appearance of the medullary groove at its most projecting part. The embryo extends from the edge of the blastoderm inwards towards the centre.

At about the time of the formation of the medullary groove, the mesoblast becomes definitely constituted. It arises as two independent plates, one on each side of the medullary groove, and is entirely derived from lower layer cells.

The two plates of mesoblast are at first unconnected with any other cells of the blastoderm, and, on their formation, the hypoblast remains in connection with all the remaining lower layer cells. Between the embryonic rim and the yolk is a cavity,— the primitive alimentary cavity. Its roof is formed of hypoblast, and its floor of yolk. Its external opening is homologous with the anus of Rusconi of Amphioxus and the Amphibians. The ventral wall of the alimentary cavity is eventually derived from cells formed in the yolk around the nuclei which are there present.

Since the important researches of Gegenbaur[1] upon the meroblastic vertebrate eggs, it has been generally admitted that the ovum of every vertebrate, however complicated may be its apparent constitution, is nevertheless to be regarded as a simple cell. This view is, indeed, opposed by His[2] and to a very modified extent by Waldeyer[3], and has recently been attacked from an entirely new standpoint by Götte[4]; but, to my mind, the objections of these authors do not upset the well founded conclusions of previous observations.

As soon as the fact is recognised that both meroblastic and holoblastic eggs have the same fundamental constitution, the admission follows, naturally, though not necessarily, that the eggs belonging to these two classes differ solely in degree, not only as regards their constitution, but also as regards the manner in which they become respectively converted into the embryo. As might have been anticipated, this view has gained a wide acceptance.

Amongst the observations, which have given a strong objective support to this view, may be mentioned those of Professor Lankester upon the development of Cephalopoda[5], and of Dr Götte[6] upon the development of the Hen's egg. In Loligo Professor Lankester showed that there appeared, in

---

[1] Wirbelthiereier mit partieller Dottertheilung. *Müller's Arch.* 1861.
[2] *Erste Anlage des Wirbelthierleibes.*
[3] *Eierstock u. Ei.*
[4] *Entwicklungsgeschichte der Unke.* The important researches of Götte on the development of the ovum, though meriting the most careful attention, do not admit of discussion in this place.
[5] *Annals and Magaz. of Natural History,* Vol. XI. 1873, p. 81.
[6] *Archiv f. Mikr. Anat.* Vol. x.

the part of the egg usually considered as food-yolk, a number of bodies, which eventually developed a nucleus and became cells, and that these cells entered into the blastoderm. These observations demonstrate that in the eggs of Loligo the so-called food-yolk is merely equivalent to a part of the egg which in other cases undergoes segmentation.

The observations of Dr Götte have a similar bearing. He made out that in the eggs of the Hen no sharp line is to be found separating the germinal disc from the yolk, and that, independently of the normal segmentation, a number of cells are derived from that part of the egg hitherto regarded as exclusively food-yolk. This view of the nature of the food-yolk was also advanced in my preliminary account of the development of Elasmobranchs[1], and it is now my intention to put forward the positive evidence in favour of this view, which is supplied from a knowledge of the phenomena of the development of the Elasmobranch ovum; and then to discuss how far the facts of the growth of the blastoderm in Elasmobranchs accord with the view that their large food-yolk is exactly equivalent to part of the ovum, which in Amphibians undergoes segmentation, rather than some fresh addition, which has no equivalent in the Amphibian or other holoblastic ovum.

Taking for granted that the ripe ovum is a single cell, the question arises whether in the case of meroblastic ova the cell is not constituted of two parts completely separated from one another.

Is the meroblastic ovum, before or after impregnation, composed of a germinal disc in which *all* the protoplasm of the cell is aggregated, and of a food-yolk in which *no* protoplasm is present? or is the protoplasm present *throughout*, being simply *more concentrated* at the germinal pole than elsewhere? If the former alternative is accepted, we must suppose that the mass of food-yolk is a something added which is not present in holoblastic ova. If the latter alternative is accepted, it may then be maintained that holoblastic and meroblastic ova are constituted in the same way and differ only in the proportions of their constituents.

---

[1] *Quart. Journ. of Micr. Science*, Oct. 1874.

My own observations in conjunction with the specially interesting observations of Dr Schultz[1] justify the view which regards the protoplasm as present throughout the whole ovum, and not confined to the germinal disc. Our observations show that a fine protoplasmic network, with ramifications extending throughout the whole yolk, is present both before and after impregnation.

The presence of this network is, in itself, only sufficient to prove that the yolk *may* be equivalent to part of a holoblastic ovum; to demonstrate that it is so requires something more, and this link in the chain of evidence is supplied by the nuclei of the yolk, which have been so often referred to.

These nuclei arise independently in the yolk, and become the nuclei of cells which enter the germ and the bodies of which are derived from the protoplasm of the yolk. Not only so, but the cells formed around these nuclei play the same part in the development of Elasmobranchs as do the largest so-called yolk-cells in the development of Amphibians. Like the homologous cells in Amphibians, they mainly serve to form the ventral wall of the alimentary canal and the blood-corpuscles. The identity in the fate of the so-called yolk-cells of Amphibians with the cells derived from the yolk in Elasmobranchs, must be considered as a proof of the homology of the yolk-cells in the first case with the yolk in the second; the difference between the yolk in the two cases arising from the fact that in the Elasmobranch ovum the yolk-spherules bear a larger proportion to the protoplasm than they do in the Amphibian ovum. As I have suggested elsewhere[2], the segmentation or non-segmentation of a particular part of the ovum depends solely upon the proportion borne by the protoplasm to the yolk particles; so that, when the latter exceed the former in a certain fixed proportion, segmentation is no longer possible; and, as this limit is approached, segmentation becomes slower, and the resulting segments larger and larger.

The question how far the facts of developmental history of the various vertebrate blastoderms accord with the view of the nature of the yolk just propounded, is one of considerable

---

[1] *Archiv f. Mikr. Anat.* Vol. **XXI.**
[2] Comparison, &c., *Quart. Journ. Micr. Science*, July, 1875.

interest. An answer to it has already been attempted from a general point of view in my paper[1] entitled 'the comparison of the early stages of development in vertebrates'; but the subject may be conveniently treated here in a special manner for Elasmobranch embryos.

In the wood-cut, fig. 1 *A, B, C*[2], are represented three diagrammatic longitudinal sections of an Elasmobranch embryo.

Fig. 1.

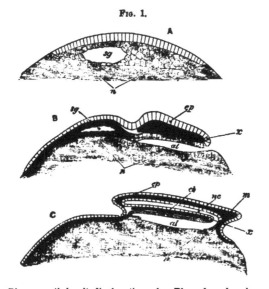

Diagrammatic longitudinal sections of an Elasmobranch embryo.

Epiblast without shading. *Mesoblast* black with clear outlines to the cells. *Lower layer cells* and *hypoblast* with simple shading.

ep. epiblast. *m.* mesoblast. *al.* alimentary cavity. *sg.* segmentation cavity. *nc.* neural canal. *ch.* notochord. *x.* point where epiblast and hypoblast become continuous at the posterior end of the embryo. *n.* nuclei of yolk.

A. Section of young blastoderm, with segmentation cavity in the middle of the lower layer cells.

B. Older blastoderm with embryo in which hypoblast and mesoblast are distinctly formed, and in which the alimentary slit has appeared. The segmentation cavity is still represented as being present, though by this stage it has in reality disappeared.

C. Older blastoderm with embryo in which neural canal has become formed, and is continuous posteriorly with alimentary canal.

---

[1] *Loc. cit.*

[2] This figure, together with fig. 2 and 3, are reproduced from my paper upon the comparison of the early stages of development in vertebrates.

A nearly corresponds with the longitudinal section represented on Pl. XXI. fig. 4, and *B* with Pl. XXII. fig. 7. In Pl. XXII. fig. 7, the segmentation cavity has however completely disappeared, while it is still represented as present in the diagram of the same period. If these diagrams, or better still, the wood-cuts, fig. 2 *A*, *B*, *C* (which only differ from those of the Elasmobranch fish in the smaller amount of food-yolk), be compared with the corresponding ones of Bombinator, fig. 3 *A*, *B*, *C*, they will be found to be in fundamental agreement with them. First let fig. 1 *A*, or fig. 2 *A*, or Pl. XXI. fig. 4, be compared

FIG. 2.

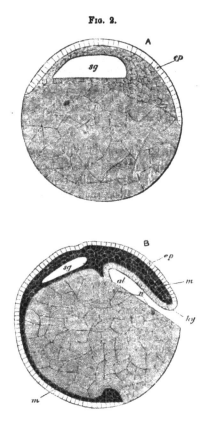

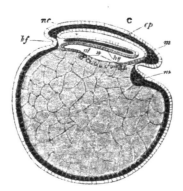

Diagrammatic longitudinal sections of embryo, which develops in the same
manner as the Elasmobranch embryo, but in which the ovum contains far
less food-yolk than is the case with the Elasmobranch ovum.

Epiblast without shading. *Mesoblast* black with clear outlines to the cells.
Lower layer cells and *hypoblast* with simple shading.
    ep. epiblast.   *m.* mesoblast.   *hy.* hypoblast.   *sg.* segmentation cavity.
al. alimentary cavity.   *nc.* neural canal   *hf.* head-fold.   *n.* nuclei of the yolk.
    The stages *A*, *B* and *C* are the same as in figure 1.

with fig. 3 *A*. In all there is present a segmentation cavity
situated not centrally but near the surface of the egg. The
roof of the cavity is thin in all, being composed in the
Amphibian of epiblast alone, and in the Elasmobranch of
epiblast and *lower layer cells*. The floor of the cavity is, in

FIG. 3.

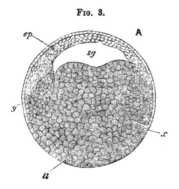

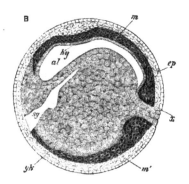

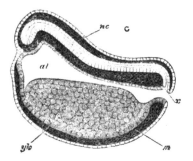

Diagrammatic longitudinal sections of Bombinator igneus. Reproduced with modifications from Götte.

Epiblast without shading. *Mesoblast* black with clear outlines to the cells. *Lower layer cells* and *hypoblast* with single shading.

ep. epiblast. *l.l.* lower layer cells. *y*. smaller lower layer cells at the sides of the segmentation cavity. *m*. mesoblast. *hy*. hypoblast. *al*. alimentary cavity. *sg*. segmentation cavity. *nc*. neural cavity. *yk*. yolk-cells.

A. Is the youngest stage in which the alimentary involution has not yet appeared. *x* is the point from which the involution will start to form the dorsal wall of the alimentary tract. The line on each side of the segmentation cavity, which separates the smaller lower layer cells from the epiblast cells, is not present in Götte's original figure. The two shadings of diagram render it necessary to have some line, but at this stage it is in reality not possible to assert which cells belong to the epiblast and which to the lower layer.

B. In this stage the alimentary cavity has become formed, but the segmentation cavity is not yet obliterated.

x. point where epiblast and hypoblast become continuous.

C. The neural canal is already formed, and communicates posteriorly with the alimentary.

x. point where epiblast and hypoblast become continuous.

all, formed of so-called yolk (Vide Pl. XXI. fig. 4), which in all forms the main mass of the egg. In the Amphibian the yolk is segmented, and, though it is not segmented in the Elasmobranch, it contains in compensation the nuclei so often mentioned. In all the sides of the segmentation cavity are formed by lower layer cells. In the Amphibian the sides are enclosed by smaller cells (in the diagram) which correspond exactly in function and position with the lower layer cells of the Elasmobranch blastoderm.

The relation of the yolk to the blastoderm in the Elasmobranch embryo at this stage of development very well suits the view of its homology with the large cells of the Amphibian ovum. The only essential difference between the two ova arises from the roof of the segmentation cavity being in the Elasmobranch embryo formed of lower layer cells, which are absent in the Amphibian embryo. This difference no doubt depends upon the greater quantity of yolk particles present in the Elasmobranch ovum. These increase the bulk of the lower layer cells, which are thus compelled to creep up the sides of the segmentation cavity till they close it in above.

In the next stage for the Elasmobranch, fig. 1 and 2 *B* and Pl. XXII. fig. 7, and for the Amphibian, fig. 3 *B*, the agreement between the two types is again very close. In both for a small portion (x) of the edge of the blastoderm the epiblast and hypoblast become continuous, while at all other parts the epiblast, accompanied by lower layer cells, grows round the yolk or round the large cells which correspond to it. The yolk-cells of the Amphibian ovum form a comparatively small mass, and are therefore rapidly enveloped; while in the case of the Elasmobranch ovum, owing to the greater mass of the yolk, the same process occupies a long period. In both ova the portion of the blastoderm, where epiblast and hypoblast become continuous, forms the dorsal lip of an opening—the anus of Rusconi—which leads into the alimentary cavity. This cavity has the same relation in both ova. It is lined dorsally by lower layer cells, and ventrally by yolk or what corresponds with yolk; the ventral epithelium of the alimentary canal being in both cases eventually supplied by the yolk-cells.

As in the earlier stage, so in the present one, the anatomical relations of the yolk to the blastoderm in the one case (Elasmobranch) are nearly identical with those of the yolk-cells to the blastoderm in the other (Amphibian). The main features in which the two embryos differ, during the stage under consideration, arise from the same cause as the solitary point of difference during the preceding stage.

In Amphibians, the alimentary cavity is formed by a true ingrowth of cells from the point where epiblast and hypoblast become continuous, and from this ingrowth the dorsal wall of the alimentary cavity is formed. The same ingrowth causes the obliteration of the segmentation cavity.

In the Elasmobranchs, owing to the larger bulk of the lower layer cells caused by the food-yolk, these have been compelled to arrange themselves in their final position during segmentation, and no room is left for a true invagination; but instead of this there is formed a simple split between the blastoderm and the yolk. The homology of this with the primitive invagination is nevertheless proved by the survival of a number of features belonging to the ancestral condition in which a true invagination was present. Amongst the more important of these are the following :—(1) The continuity of epiblast and hypoblast at the dorsal lip of the anus of Rusconi. (2) The continuous conversion of indifferent lower layer cells into hypoblast, which gradually extends backwards towards the segmentation cavity, and exactly represents the course of the invagination whereby in Amphibians the dorsal wall of the alimentary cavity is formed. (3) The obliteration of the segmentation cavity during the period when the pseudo-invagination is occurring.

The asymmetry of the gastrula or pseudo-gastrula in Cyclostomes, Amphibians, Elasmobranchs and, I believe, Osseous Fishes, is to be explained by the form of the vertebrate body. In Amphioxus, where the small amount of food-yolk present is distributed uniformly, there is no reason why the invagination and resulting gastrula should not be symmetrical. In other vertebrates, where more food-yolk is present, the shape and structure of the body render it necessary for the food-yolk to be stored away on the ventral side of the alimentary canal. This, combined with the unsymmetrical position of the anus, which primitively corresponds in position with the blastopore or anus of Rusconi, causes the asymmetry of the gastrula invagination, since it is not possible for the part of ovum which will become the ventral wall of the alimentary canal, and which is loaded with food-yolk, to be invaginated in the same fashion as the dorsal wall. From the asymmetry, so caused, follow a large number of features in vertebrate development, which have been worked out in some detail in my paper already quoted [1].

Prof. Haeckel, in a paper recently published [2], appears to imply that because I do not find absolute invagination in Elasmobranchs, I therefore look upon Elasmobranchs as militating against his Gastræa theory. I cannot help thinking that Prof. Haeckel must have somewhat misunderstood my meaning. The importance of the Gastræa theory has always appeared to me to consist not in the fact that an actual ingrowth of certain cells occurs—an ingrowth which might have many different meanings [3]—but in the fact that the types of early development of all animals can be easily derived from that of the typical gastrula. I am perfectly in accordance with Professor Haeckel in regarding the type of Elasmobranch development to be a simple derivative from that of the gastrula, although believing it to be without any true ingrowth or invagination of cells.

Professor Haeckel [4] in the paper just referred to published his view upon the mutual relationships of the various vertebrate blastoderms. In this paper, which appeared but shortly after my own [5] on the same subject, he has put forward views which differ from mine in several important details. Some of these bear upon the nature of food-yolk; and it appears to me that Professor Haeckel's scheme of development is incompatible with the view that the food-yolk in meroblastic eggs is the homologue of part of the hypoblast of the holoblastic eggs.

The following is Professor Haeckel's own statement of the

---

[1] *Quart. Journ. of Micr. Science*, July, 1875.
[2] Die Gastrula u. Eifurchung d. Thiere, *Jenaische Zeitschrift*, Vol. IX.
[3] For instance, in Crustaceans it does not in some cases appear certain whether an invagination is the typical gastrula invagination, or only an invagination by which, at a period subsequent to the gastrula invagination, the hind gut is frequently formed.
[4] *Loc. cit.*
[5] *Loc. cit.*

scheme or type, which he regards as characteristic of meroblastic eggs, pp. 98 and 99.

Jetzt folgt der höchst wichtige und interessante Vorgang, den ich
als Einstülpung der Blastula auffasse und der zur Bildung der
Gastrula führt (Fig. 63, 64)[1]. Es schlägt sich nämlich der verdickte
Saum der Keimscheibe, der "Randwulst" oder das *Properistom*,
nach innen um und eine dünne Zellenschicht wächst als directe Fort-
setzung desselben, wie ein immer enger werdendes Diaphragma, in
die Keimhöhle hinein. Diese Zellenschicht ist das entstehende Entoderm (Fig. 64 i, 74 i). Die Zellen, welche dieselbe zusammensetzen
und aus dem innern Theile des Randwulstes hervorwachsen, sind viel
grösser aber flacher als die Zellen der Keimhöhlendecke und zeigen
ein dunkleres grobkörniges Protoplasma. Auf dem Boden der Keimhöhle, d. h. also auf der Eiweisskugel des Nahrungsdotters, liegen sie
unmittelbar auf und rücken hier durch centripetale Wanderung
gegen dessen Mitte vor, bis sie dieselbe zuletzt erreichen und nunmehr eine zusammenhängende einschichtige Zellenlage auf dem ganzen
Keimhöhlenboden bilden. Diese ist die erste vollständige Anlage
des Darmblatts, Entoderms oder "Hypoblasts", und von nun an
können wir, im Gegensatz dazu den gesammten übrigen Theil des
Blastoderms, nämlich die mehrschichtige Wand der Keimhöhlendecke
als Hautblatt, Exoderm oder "Epiblast" bezeichnen. Der verdickte Randwulst (Fig. 64 w, 74 w), in welchem beide primäre Keim-
blätter in einander übergehen, besteht in seinem oberen und äusseren
Theile aus Exodermzellen, in seinem unteren und inneren Theile aus
Entodermzellen.

In diesem Stadium entspricht unser Fischkeim einer Amphi-
blastula, welche mitten in der Invagination begriffen ist, und bei
welcher die entstehende Urdarmhöhle eine grosse Dotterkugel auf-
genommen hat. Die Invagination wird nunmehr dadurch vervoll-
ständigt und die Gastrulabildung dadurch abgeschlossen, dass die
Keimhöhle verschwindet. Das wachsende Entoderm, dem die Dotterkugel innig anhängt, wölbt sich in die letztere hinein und nähert
sich so dem Exoderm. Die klare Flüssigkeit in der Keimhöhle wird
resorbirt und schliesslich legt sich die obere convexe Fläche des
Entoderms an die untere concave des Exoderms eng an: die Gastrula
des discoblastischen Eies oder die "Discogastrula" ist fertig (Fig.
65, 76; Meridiandurchschnitt Fig. 66, 75).

Die Discogastrula unsers Knochenfisches in diesem Stadium der
vollen Ausbildung stellt nunmehr eine kreisrunde Kappe dar, welche
wie ein gefüttertes Mützchen fast die ganze obere Hemisphäre der
hyalinen Dotterkugel eng anliegend bedeckt (Fig. 65). Der Ueber-
zug des Mützchens entspricht dem Exoderm (e), sein Futter dem
Entoderm (i). Ersteres besteht aus drei Schichten von kleineren
Zellen, letzteres aus einer einzigen Schicht von grösseren Zellen.
Die Exodermzellen (Fig. 77) messen 0,006—0,009 Mm., und haben
ein klares, sehr feinkörniges Protoplasma. Die Entodermzellen (Fig.

[1] The references in this quotation are to the figures in the original.

78) messen 0,02—0,03 Mm. und ihr Protoplasma ist mehr grobkörnig und trüber. Letztere bilden auch den grössten Theil des Randwulstes, den wir nunmehr als Urmundrand der Gastrula, als "*Properistoma*" oder auch als "RUSCONI'schen After" bezeichnen können. Der letztere umfasst die Dotterkugel, welche die ganze Urdarmhöhle ausfüllt und weit aus der dadurch verstopften Urmund-Oeffnung vorragt.

My objections to the view so lucidly explained in the passage just quoted, fall under two heads.

(1) That the facts of development of the meroblastic eggs of vertebrates, are not in accordance with the views here advanced.

(2) That even if these views be accepted as representing the actual facts of development, the explanation offered of these facts would not be satisfactory.

Professor Haeckel's views are absolutely incompatible with the facts of Elasmobranch development, if my investigations are correct.

The grounds of the incompatibility may be summed up under the following heads :

(1) In Elasmobranchs the hypoblast cells occupy, even before the close of segmentation, the position which, on Professor Haeckel's view, they ought only eventually to take up after being involuted from the whole periphery of the blastoderm.

(2) There is no sign at any period of an invagination of the periphery of the blastoderm, and the only structure (the embryonic rim) which could be mistaken for such an invagination is confined to a very limited arc.

(3) The growth of cells to form the floor of the segmentation cavity, which ought to be part of this general invagination from the periphery, is mainly due to a formation of cells from the yolk.

It is this ingrowth of cells for the floor of the segmentation cavity which, I am inclined to think, Professor Haeckel has mistaken for a general invagination in the Osseous Fish he has investigated.

(4) Professor Haeckel fails to give an account of the asymmetry of the blastoderm; an asymmetry which is unquestion-

ably also present in the blastoderm of most Osseous Fishes, though not noticed by Professor Haeckel in the investigations recorded in his paper.

The facts of development of Osseous Fishes, upon which Professor Haeckel rests his views, are too much disputed, for their discussion in this place to be profitable[1]. The eggs of Osseous Fishes appear to me unsatisfactory objects for the study of this question, partly on account of all the cells of the blastoderm being so much alike, that it is a very difficult matter to distinguish between the various layers, and, partly, because there can be little question that the eggs of existing Osseous Fishes are very much modified, through having lost a great part of the food-yolk possessed by the eggs of their ancestors[2]. This disappearance of the food-yolk must, without doubt, have produced important changes in development, which would be especially marked in a pelagic egg, like that investigated by Professor Haeckel.

The Avian egg has been a still more disputed object than even the egg of the Osseous Fishes. My own investigations on this subject do not in any way accord with those of Dr Götte, or the views of Professor Haeckel[3].

[1] A short statement by Kowalevsky on this subject in a note to his account of the development of Ascidians, would seem to indicate that the type of development of Osseous Fishes is precisely the same as that of Elasmobranchs. Kowalevsky says, *Arch. f. Micr. Anat.* Vol. VII. p. 114, note 5, "According to my observations on Osseous Fishes the germinal wall consists of two layers, an upper and lower, which are continuous with one another at the border. From the upper one develops skin and nervous system, from the lower hypoblast and mesoblast." This statement, which leaves unanswered a number of important questions, is too short to serve as a basis for supporting my views, but so far as it goes its agreement with the facts of Elasmobranch development is undoubtedly striking.

[2] The eggs of the Osseous Fishes have, I believe, undergone changes of the same character, but not to the same extent, as those of Mammalia, which, according to the views expressed both by Professor Haeckel and myself, are degenerated from an ovum with a large food-yolk. The grounds on which I regard the eggs of Osseous Fishes as having undergone an analogous change, are too foreign to the subject to be stated here.

[3] I find myself unable without figures to understand Dr Rauber's (*Central-blatt für Med. Wiss.* 1874, No. 50; 1875, Nos. 4 and 17) views with sufficient precision to accord to them either my assent or dissent. It is quite in accordance with the view propounded in my paper (*loc. cit.*) to regard, with Dr Rauber and Professor Haeckel, the thickened edge of the blastoderm as the homologue of the lip of the blastopore in Amphioxus; though an invagination, in the manner imagined by Professor Haeckel, is no necessary consequence of this view. If Dr Rauber regards the *whole* egg of the bird as the homologue of that of Amphioxus, and the inclosure of the yolk by the blastoderm as the equivalent to the process of invagination in Amphioxus, then his views are practically in accordance with my own.

Apart from disputed points of development, it appears to me that a comparative account of the development of the meroblastic vertebrate ova ought to take into consideration the essential differences which exist between the Avian and Piscian blastoderms, in that the embryo is situated in the centre of the blastoderm in the first case and at the edge in the second.

This difference entails important modifications in development, and must necessarily affect the particular points under discussion. As a result of the different positions of the embryo in the two cases, there is present in Elasmobranchs and Osseous Fishes a true anus of Rusconi, or primitive opening into the alimentary canal, which is absent in Birds. Yet in neither Elasmobranchs[1] nor Osseous Fishes does the anus of Rusconi correspond in position with the point where the final closing in of the yolk takes place, but in them this point corresponds rather with the blastopore of Birds[2].

Owing also to the respective situations of the embryo in the blastoderm, the alimentary and neural canals communicate posteriorly in Elasmobranchs and Osseous Fishes, but *not* in Birds. Of all these points Professor Haeckel makes no mention.

The support of his views which Prof. Haeckel attempts to gain from Götte's researches in Mammalia is completely cut away by the recent discoveries of Van Beneden[3].

It thus appears that Professor Haeckel's views but ill accord with the facts of vertebrate development; but even if they were to do so completely it would not in my opinion be easy to give a rational explanation of them.

Professor Haeckel states that no sharp and fast line can be drawn between the types of 'unequal' and 'discoidal' segmentation[4]. In the cases of unequal segmentation he admits, as is certainly the case, that the larger yolk-cells (hypoblast) are

---

[1] *Vide* Plate xxvi. fig. 1 and 2, and Self, "*Comparison*," &c., *loc. cit.*

[2] The relation of the anus of Rusconi and blastopore in Elasmobranchs was fully explained in the paper above quoted. It was there clearly shown that neither the one nor the other exactly corresponds with the blastopore of Amphioxus, but that the two together do so. Professor Haeckel states that in the Osseous Fish investigated by him the anus of Rusconi and the blastopore coincide. This is not the case in the Salmon.

[3] *Développement Embryonnaire des Mammifères, Bulletin de l'Acad. r. d. Belgique*, 1875.

[4] For an explanation of these terms, *vide* Prof. Haeckel's original paper or the abstract in *Quart. Journ. of Micr. Science* for January, 1876.

simply enclosed by a growth of the epiblast around them; which is to be looked on as a modification of the typical gastrula invagination, necessitated by the large size of the yolk-cells (vide Professor Haeckel's paper, Taf. II. fig. 30). In these instances there is no commencement of an ingrowth in the *manner supposed for meroblastic ova.*

When the food-yolk becomes more bulky, and the hypoblast does not completely segment, it is not easy to understand why an ingrowth, which had no existence in the former case, should occur; nor where it is to come from. Such an ingrowth as is supposed to exist by Professor Haeckel would, in fact, break the continuity of development between meroblastic and holoblastic ova, and thus destroy one of the most important results of the Gastræa theory.

It is quite easy to suppose, as I have done, that in the cases of discoidal segmentation, the hypoblast (including the yolk) becomes enclosed by the epiblast in precisely the same manner as in the cases of unequal segmentation.

But even if Professor Haeckel supposes that in the unsegmented food-yolk a fresh element is added to the ovum, it remains quite unintelligible to me how an ingrowth of cells from a circumferential line, to form a layer which had no previous existence, can be equivalent to, or derived from the invagination of a layer, which exists before the process of invagination begins, and which remains continuous throughout it.

If Professor Haeckel's views should eventually turn out to be in accordance with the facts of vertebrate development, it will, in my opinion, be very difficult to reduce them into conformity with the Gastræa theory.

Although some space has been devoted to an attempt to refute the views of Professor Haeckel on this question, I wish it to be clearly understood that my disagreement from his opinions concerns matters of detail only, and that I quite accept the Gastræa theory in its general bearings.

Observations upon the formation of the layers in Elasmobranchs have hitherto been very few in number. Those published in my preliminary account of these fishes are, I believe, the earliest[1].

Since then there has been published a short notice on the subject by Dr Alex. Schultz[2]. His observations in the main accord with my own. He apparently speaks of the nuclei of the yolk as cells, and also of the epiblast being more than one cell deep. In Torpedo alone, amongst the genera investigated by me, is the layer of epiblast, at about the age of the last described embryo, composed of more than a single row of cells.

[1] I omit all reference to a paper published in Russian by Prof. Kowalevsky. Being unable to translate it, and the illustrations being too meagre to be in themselves of much assistance, it has not been possible for me to make any use of it.

[2] *Centralblatt f. Med. Wiss.* No. 38, 1875.

THE GENERAL FEATURES OF THE ELASMOBRANCH EMBRYO AT SUCCESSIVE STAGES.

No complete series of figures, representing the various stages in development of an Elasmobranch Embryo, has hitherto been published. With the view of supplying this deficiency Plates XXIV. and XXV. have been inserted. The embryos represented in these two Plates form a fairly complete series, but do not all belong to a single species. Those on Pl. XXIV., with the exception of G, are embryos of Pristiurus; G being an embryo of Torpedo. Those on Pl. XXV., excepting K, which is a Pristiurus embryo, are embryos of Scyllium canicula. All the embryos on Pl. XXV. were very accurately drawn from nature by my sister, Miss A. R. Balfour. Unfortunately the exceptional beauty and clearness of the originals is all but lost in the lithographs. To facilitate future description, letters will be employed in the remainder of these pages to signify that an embryo being described is of the same age as the embryo on these Plates to which the letter used refers. Thus an embryo of the same age as L will be spoken of hereafter as belonging to stage L.

A.

This figure represents a hardened blastoderm at a stage when the embryo-swelling (*e. s.*) has become obvious, but before the appearance of the medullary groove. The position of the segmentation cavity is indicated by a slight swelling of the blastoderm (*s. c.*). The shape of the blastoderm, in hardened specimens, is not to be relied upon, owing to the traction which the blastoderm undergoes during the process of removing the yolk from the egg-shell.

B.

B is the view of a fresh blastoderm. The projecting part of this, already mentioned as the 'embryonic rim', is indicated

by the shading. At the middle of the embryonic rim is to be
seen the rudiment of the embryo (*m. g.*). It consists of an
area of the blastoderm, circumscribed on its two sides and at
one end, by a slight fold, and whose other end forms part of
the edge of the blastoderm. The end of the embryo which
points towards the *centre* of the blastoderm is the head end,
and that which forms part of the *edge* of the blastoderm is
the tail end. To retain the nomenclature usually adopted
in treating of the development of the Bird, the fold at the
anterior end of the embryo may be called *the head fold*, and
those at the sides the *side folds*. There is in Elasmobranchs
no tail fold, owing to the position of the embryo at the periphery of the blastoderm, and it is by the meeting of the three
above-mentioned folds only, that the embryo becomes pinched
off from the remainder of the blastoderm. Along the median
line of the embryo is a shallow groove (*m. g.*), the well-known
medullary groove of vertebrate embryology. It flattens out
both anteriorly and posteriorly, and is deepest in the middle
part of its course.

C.

This embryo resembles in most of its features the embryo
last described. It is, however, considerably larger, and the
head-fold and side-folds have become more pronounced structures. The medullary groove is far deeper than in the earlier
stage, and widens out anteriorly. This anterior widening is the
first indication of a distinction between the brain and the
remainder of the central nervous system, a distinction which
arises long before the closure of the medullary canal.

D.

This embryo is far larger than the one last described, but
the increase in length does not cause it to project beyond the
edge of the blastoderm, but has been due to a growth inwards
towards the centre of the blastoderm. The head is now indicated by an anterior enlargement, and the embryo also widens
out posteriorly. The posterior widening (*t. s.*) is formed by a
pair of rounded prominences, one on each side of the middle

line. These are very conspicuous organs during the earlier stages of development, and consist of two large aggregations of mesoblast-cells. In accordance with the nomenclature adopted in my preliminary paper[1], they may be called 'tail-swellings'. Between the cephalic enlargements and the tail-swellings is situated the rudimentary trunk of the embryo. It is more completely pinched off from the blastoderm than in the last described embryo. The medullary groove is of a fairly uniform size throughout the trunk of the embryo, but flattens out and vanishes completely in the region of the head. The blastoderm in Pristiurus and Scyllium grows very rapidly, and has by this stage attained a very considerable size; but in Torpedo its growth is very slow.

E and F.

These two embryos may be considered together, for, although they differ in appearance, yet they are of an almost identical age; and the differences between the two are purely external. E appears to be a little abnormal in not having the cephalic region so distinctly marked off from the trunk as is usual. The head is proportionally larger than in the last stage, and the tail-swellings remain as conspicuous as before. The folding off from the blastoderm has progressed rapidly, and the head and tail are quite separated from it. The medullary groove has become closed posteriorly in both embryos, but the closing has extended further forwards in F than in E. In F the medullary folds have not only united posteriorly, but have very nearly effected a fresh junction in the region of the neck. At this point a second junction of the two medullary folds is in fact actually effected before the posterior closing has extended forwards so far. The later junction in the region of the neck corresponds in position with the point, where in the Bird the medullary folds first unite. No trace of a medullary groove is to be met with in the head, which simply consists of a wide flattened plate. Between the two tail-swellings surface views present the appearance of a groove, but this appearance is deceptive, since in sections no groove, or at most a very slight one, is perceptible.

[1] *Quart. Journ. Micr. Science,* Oct. 1874.

G.

During the preceding stages growth in the embryo is very slow, and considerable intervals of time elapse before any perceptible changes are effected. This state of things now becomes altered, and the future changes succeed each other with far greater rapidity. One of the most important of these, and one which first presents itself during this stage, is the disappearance of the yolk-spherules from the embryonic cells, and the consequently increased transparency of the embryo. As a result of this, a number of organs, which in the earlier stages were only to be investigated by means of sections, now become visible in the living embryo.

The tail-swellings (t. s.) are still conspicuous objects at the posterior extremity of the embryo. The folding off of the embryo from the yolk has progressed to such an extent that it is now quite possible to place the embryo on its side and examine it from that point of view.

The embryo may be said to be attached to the yolk by a distinct stalk or cord, which in the succeeding stages gradually narrows and elongates, and is known as the umbilical cord (so. s.). The medullary canal has now become completely closed, even in the region of the brain, where during the last stage no trace of a medullary groove had appeared. Slight constrictions, not perceptible in views of the embryo as a transparent object, mark off three vesicles in the brain. These vesicles are known as the fore, mid, and hind brain. From the fore-brain there is an outgrowth on each side, the first rudiment of the optic vesicle (op.).

The mesoblast on each side of the body is divided into a series of segments, known as muscle-plates, the first of which lies a little behind the head. The mesoblast of the tail has not as yet undergone this segmentation. There are present in all seventeen segments. These first appeared at a much earlier date, but were not visible owing to the opacity of the embryo.

Another structure which became developed in even a younger embryo than C is now for the first time visible in the living embryo. This is the notochord: it extends from almost the extreme posterior to the anterior end of the embryo.

It lies between the ventral wall of the spinal canal and the dorsal wall of the intestine; and round its posterior end these two walls become continuous with each other (vide fig.). Anteriorly the termination of the notochord cannot be seen, it can only be traced into a mass of mesoblast at the base of the brain, which there separates the epiblast from the hypoblast. The alimentary canal (*al.*) is completely closed anteriorly and posteriorly, though still widely open to the yolk-sac in the middle part of its course. In the region of the head it exhibits on each side a slight bulging outwards, the rudiment of the first visceral cleft. This is represented in the figure by two lines (*1 v. c.*). The visceral clefts at this stage consist of a pair of simple diverticula from the alimentary canal, and there is no communication between the throat and the exterior.

H.

The present embryo is far larger than the last, but it has not been possible to represent this increase in size in the drawings. Accompanying this increase in size, the folding off of the embryo from the yolk has considerably progressed, and the stalk which unites the embryo with the yolk is proportionately narrower and longer than before.

The brain is now very distinctly divided into the three lobes, whose rudiments appeared during the last stage. From the foremost of these, the optic vesicles now present themselves as well-marked lateral outgrowths, towards which there appears a growing in, or involution, from the external skin (*op.*) to form the lens. The opening of this involution is represented by the dark spot in the centre.

A fresh organ of sense, the auditory sac, now for the first time becomes visible as a shallow pit in the external skin on each side of the hind brain (*au. v.*). The epiblast which is involuted to form this pit becomes much thickened, and thereby the opacity, indicated in the figure, is produced.

The muscle-plates have greatly increased in number by the formation of fresh segments in the tail. Thirty-eight of them were present in the embryo figured. The mesoblast at the base of the brain has increased in quantity, and there is still a certain mass of unsegmented mesoblast which forms the tail-swellings. The first rudiment of the heart becomes visible during this stage as a cavity between the mesoblast of the splanchnopleure and the hypoblast (*ht.*).

The fore and hind guts are now longer than they were. A slight pushing in from the exterior to form the mouth has appeared (*m.*), and an indication of the future position of the anus is afforded by a slight diverticulum of the hind gut towards the exterior some little distance from the posterior end of the embryo (*an.*). The portion of the alimentary canal behind this point, though at this stage large, and even dilated into a vesicle at its posterior end (*al. v.*), becomes eventually completely atrophied. In the region of the throat the rudiment of a second visceral cleft has appeared behind the first; neither of them are as yet open to the exterior. The number of visceral clefts present in any given Pristiurus embryo affords a very easy and simple way of determining its age.

I.

A great increase in size is again to be noticed in the embryo, but, as in the case of the last embryo, it has not been possible to represent this in the figure. The stalk connecting the embryo with the yolk has become narrower and more elongated, and the tail region of the embryo proportionately far longer than in the last stage. During this stage the first spontaneous movements of the embryo take place, and consist in somewhat rapid excursions of the embryo from side to side, produced by a serpentine motion of the body.

The cranial flexure, which commenced in stage G, has now become very evident, and the mid-brain[1] begins to project in the same manner as in the embryo fowl on the third day, and will soon form the anterior termination of the long axis of the embryo. The fore-brain has increased in size and distinctness, and the anterior part of it may now be looked on as the impaired rudiment of the cerebral hemispheres.

Further growths have taken place in the organs of sense,

[1] The part of the brain which I have here called mid-brain, and which unquestionably corresponds to the part called mid-brain in the embryos of higher vertebrates, becomes in the adult what Miklucho-Maclay and Gegenbaur called the vesicle of the third ventricle or thalamencephalon. I shall always speak of it as the mid-brain.

especially in the eye, in which the involution for the lens has made considerable progress. The number of the muscle-plates has again increased, but there is still a region of un-segmented mesoblast in the tail. The thickened portions of mesoblast which caused the tail-swellings are still to be seen and would seem to act as the reserve from which is drawn the matter for the rapid growth of the tail, which occurs soon after this. The mass of the mesoblast at the base of the brain has again increased. No fresh features of interest are to be seen in the notochord. The heart is now much more conspicuous than before, and its commencing flexure is very apparent. It now beats actively. The hind gut especially is much longer than in the last specimen; and the point where the anus will appear is very easily detected by the bulging out of the gut towards the external skin at that point (*an.*). The alimentary vesicle, first observable during the last stage, is now a more conspicuous organ (*al. v.*). Three visceral clefts, none of which are as yet open to the exterior, may now be seen.

K.

The figures G, H, I are representations of living and trans-parent embryos, but the remainder of the figures are drawings of opaque embryos which were hardened in chromic acid.

The stalk connecting the embryo with the yolk is now, com-paratively speaking, quite narrow, and is of sufficient length to permit the embryo to execute considerable movements.

The tail has grown immensely, but is still dilated terminally. This terminal dilatation is mainly due to the alimentary vesicle, but the tract of gut connecting this with the gut in front of the anus is now a solid rod of cells and very soon becomes com-pletely atrophied.

The two pairs of limbs have appeared as elongated ridges of epiblast. The anterior pair is situated just at the front end of the umbilical stalk; and the posterior pair, which is the more conspicuous of the two, is situated some little distance behind the stalk.

The cranial flexure has greatly increased, and the angle between the long axis of the front part of the head and of the body is less than a right angle. The conspicuous mid-brain forms the anterior termination of the long axis of the body. The thin roof of the fourth ventricle may in the figure be noticed behind the mid-brain. The auditory sac is nearly closed and its opening is not shown in the figure. In the eye the lens is completely formed.

Owing to the opacity of the embryo, the muscle-plates are only indistinctly indicated, and no other features of the mesoblast are to be seen.

The mouth is now a deep pit, whose borders are almost completely formed by the thickening in front of the first visceral cleft, which may be called the first visceral arch or mandibular arch.

Four visceral clefts are now visible, all of which are open to the exterior, but in a transparent embryo one more, not open to the exterior, would have been visible behind the last of these.

L.

This embryo is considerably older than the one last described, but growth is not quite so rapid as might be gathered from the fact that L is nearly twice as long as K, since the two embryos belong to different genera; and the Scyllium embryos, of which L is an example, are larger than Pristiurus embryos. The umbilical stalk is now quite a narrow elongated structure, whose subsequent external changes are very unimportant, and consist for the most part merely in an increase in its length.

The tail has again grown greatly in length, and its terminal dilatation together with the alimentary vesicle contained in it, have both completely vanished. A dorsal and ventral fin are now clearly visible; they are continuous throughout their whole length. The limbs have grown and are more easily seen than in the previous stage.

Great changes have been effected in the head, resulting in a diminution of the cranial flexure. This diminution is nevertheless apparent rather than real, and is chiefly due to the rapid growth of the rudiment of the cerebral hemispheres. The three main divisions of the brain may still be clearly seen from the surface. Posteriorly is situated the hind-brain, now consisting of the medulla oblongata and cerebellum. At the anterior part of the medulla is to be seen the thin roof of the fourth ventricle, and anteriorly to this again the roof becomes thickened

to form the rudiment of the cerebellum.   In front of the hind-
brain lies the mid-brain, the roof of which is formed by the
optic lobes, which are still situated at the front end of the long
axis of the embryo.

Beyond the mid-brain is placed the fore-brain, whose growth
is rapidly rendering the cranial flexure imperceptible.

The rudiments of the nasal sacs are now clearly visible as a
pair of small pits.   The pits are widely open to the exterior,
and are situated one on each side, near the front end of the
cerebral hemispheres.   Five visceral clefts are open to the
exterior, and in them the external gills have commenced to
appear (L').

The first cleft is no longer similar to the rest, but has commenced to be metamorphosed into the spiracle.

Accompanying the change in position of the first cleft, the
mandibular arch has begun to bend round and enclose the front
as well as the side of the mouth.   By this change in the mandibular arch the mouth becomes narrowed in an antero-posterior
direction.

M.

Of this embryo the head alone has been represented.   Two
views of it are given, one (M) from the side and the other (M')
from the under surface.   The growth of the front part of the
head has considerably diminished the prominence of the cranial
flexure.   The full complement of visceral clefts is now present—
six in all.   But the first has already atrophied considerably, and
may easily be recognised as the spiracle.   In Scyllium, there
are present at no period more than six visceral clefts.   The first
visceral arch on each side has become bent still further round,
to form the front border of the mouth.   The opening of the
mouth has in consequence become still more narrowed in an
antero-posterior direction.   The width of the mouth in this
direction, serves for the present and for some of the subsequent
stages as a very convenient indication of age.

N.

The limbs, or paired fins, have now acquired the general
features and form which they possess in the adult.

The unpaired fins have now also become divided in a

manner not only characteristic of the Elasmobranchs but even of the genus Scyllium.

There is a tail fin, an anal fin and two dorsal fins, both the latter being situated behind the posterior paired fins.

In the head may be noticed a continuation of the rapid growth of the anterior part.

The mouth has become far more narrow and slit-like; and with many other of the organs of the period commences to approach the form of the adult.

The present and the three preceding stages show the gradual changes by which the first visceral arch becomes converted into the rudiments of the upper and of the lower jaw. The fact of the conversion was first made known through the investigations of Messrs Parker and Gegenbaur.

O.

In this stage the embryo is very rapidly approaching the form of the adult.

This is especially noticeable in the fins, which project in a manner quite characteristic of the adult fish. The mouth is slit-like, and the openings of the nasal sacs no longer retain their primitive circular outline. The external gills project from all the gill-slits including the spiracle.

P.

The head is rapidly elongating by the growth of the snout, and the divisions of the brain can no longer be seen with distinctness from the exterior, and, with the exception of the head and of the external gills, the embryo almost completely resembles the adult.

Q.

The snout has grown to such an extent, that the head has nearly acquired its adult shape. In the form of its mouth the embryo now quite resembles the adult fish.

This part of the subject may be conveniently supplemented by a short description of the manner in which the blastoderm encloses the yolk. It has been already mentioned that the growth of the blastoderm is not uniform. The part of it in the immediate neighbourhood of the embryo remains comparatively stationary, while the growth elsewhere is very rapid. From this it results that that part of the edge of the blastoderm where the embryo is attached forms a bay in the otherwise regular outline of the edge of the blastoderm. By the time that one-half of the yolk is enclosed the bay is a very conspicuous feature (Pl. XXVI. fig. 1). In this figure *bl.* points to the blastoderm, and *yk.* to the part of the yolk not yet enclosed by the blastoderm.

Shortly subsequent to this the bay becomes obliterated by its two sides coming together and coalescing, and the embryo ceases to lie at the edge of the yolk.

This stage is represented on Pl. XXVI. fig. 2. In this figure there is only a small patch of yolk not yet enclosed (*yk*), which is situated at some little distance behind the embryo. Throughout all this period the edge of the blastoderm has remained thickened, a feature which persists till the complete investment of the yolk, which takes places shortly after the stage last figured. In this thickened edge a circular vein arises, which brings back the blood from the yolk-sac to the embryo. The opening in the blastoderm (Pl. XXVI. fig. 2 *yk.*), exposing the portion of the yolk not yet enclosed, may be conveniently called the blastopore, according to Professor Lankester's nomenclature.

The interesting feature which characterizes the blastopore in Elasmobranchs is the fact of its not corresponding in position with the opening of the anus of Rusconi. We thus have in Elasmobranchs two structures, each of which corresponds in part with the single structure in Amphioxus which may be called either blastopore or anus of Rusconi, which yet do not in Elasmobranchs coincide in position. It is the blastopore of Elasmobranchs which has undergone a change of position, owing to the unequal growth of the blastoderm; while the anus of Rusconi retains its normal situation. In Osseous Fishes the blastopore undergoes a similar change of position. The possibility of a change in position of this structure is peculiarly interesting, in that it possibly serves to explain how the blastopore of different animals corresponds in different cases with the anus or the mouth, and has not always a fixed situation[1].

---

[1] For a fuller discussion of this question *vide* Self, 'A comparison of the early stages of development in vertebrates.' *Quart. Journ. of Micr. Science*, July, 1875.

EXPLANATION OF PLATE XXI.

ep. epiblast.

l. l. lower layer cell.

m. mesoblast.

hy. hypoblast.

s. c. segmentation cavity.

e. s. embryo swelling.

n. nuclei of yolk.

1. Longitudinal section of a blastoderm at the first appearance of the segmentation cavity.

2. Longitudinal section through a blastoderm after the layer of cells has disappeared from the floor of the segmentation cavity. *bd.* large cell resting on the yolk, probably remaining over from the later periods of segmentation. Magnified 60 diameters. (Hardened in chromic acid.)

The section is intended to illustrate the fact that the nuclei form a layer in the yolk under the floor of the segmentation cavity. The roof of the segmentation cavity is broken.

2 *a.* Portion of same blastoderm highly magnified, to show the characters of the nuclei of the yolk and the nuclei in the cells of the blastoderm.

2 *b.* Large knobbed nucleus from the same blastoderm, very highly magnified.

2 *c.* Nucleus of yolk from the same blastoderm.

3. Longitudinal section of blastoderm of same stage as fig. 2. (Hardened in chromic acid.)

4. Longitudinal section of blastoderm slightly older than fig. 2. Magnified 45 diameters. (Hardened in osmic acid.)

It illustrates (1) the characters of the epiblast; (2) the embryonic swelling; (3) the segmentation cavity.

5. Longitudinal section through a blastoderm at the time of the first appearance of the embryonic rim, and before the formation of the medullary groove. Magnified 45 diameters.

EXPLANATION OF PLATE XXII.

ep. epiblast.

m. mesoblast.

hy. hypoblast.

e. r. embryonic rim.

n. al. cells formed around the nuclei of the yolk which have entered the hypoblast.

c. cell formed around nucleus of yolk.

m. g. medullary groove.

5 *a.* Section through the periphery of the embryonic rim of the blastoderm of which fig. 5 is a section.

6. Section through the embryonic rim of a blastoderm somewhat younger than that represented on Pl. xxiv. fig. B.

7. Section through the most projecting portion of the embryonic rim of a blastoderm of the same age as that represented on Pl. xxiv. fig. B. The section is drawn on a very considerably smaller scale than that on Pl. xxi. fig. 5. It is intended to illustrate the growth of the embryonic rim and the disappearance of the segmentation cavity.

7 *a.* Section through peripheral portion of the embryonic rim of the same blastoderm, highly magnified. It specially illustrates the formation of a cell (*c*) around a nucleus in the yolk. The nuclei of the blastoderm have been inaccurately rendered by the artist.

8 *a.* 8 *b.* 8 *c.* Three sections of the same embryo. Inserted mainly to illustrate the formation of the mesoblast as two independent lateral masses of cells; only half of each section is represented. 8 *a* is the most posterior of the three sections. In it the mesoblast forms a large mass on each side, imperfectly separated from the hypoblast. In 8 *b*, from the anterior part of the embryo, the main mass of mesoblast is far smaller, and only forms a cap to the hypoblast at the highest point of the medullary fold. In 8 *c* a cap of mesoblast is present, similar to that in 8 *b*, though much smaller. The sections of these embryos were somewhat oblique, and it has unfortunately happened that while in 8 *a* one side is represented, in 8 *b* and 8 *c* the other side is figured, had it not been for this the sections 8 *b* and 8 *c* would have been considerably longer than 8 *a*.

9. Longitudinal section of an embryo belonging to a slightly later stage than B.

EXPLANATION OF PLATE XXIII.

ep. epiblast.

m. mesoblast.

hy. hypoblast.

v. p. vertebral plate.

n. a. cells to form ventral wall of alimentary canal which have been derived from the yolk.

l. y. line separating the yolk from the blastoderm.

m. g. medullary groove.

n. nucleus of yolk.

c. cells formed in the yolk around the nuclei of the yolk.

10 *a.* 10 *b.* 10 *c.* Three sections of the same embryo belonging to a stage slightly later than B, Pl. xxiv.

10 *a.* The most posterior of the three sections. It shows the posterior flatness of the medullary groove and the two isolated vertebral plates.

10 *b.* This section is taken from the anterior part of the same embryo and shows the deep medullary groove and the commencing formation of the ventral wall of the alimentary canal from the nuclei of the yolk.

10 *c.* Shows the disappearance of the medullary groove and the thinning out of the vertebral plates in the region of the head.

11. Small portion of the blastoderm and the subjacent yolk of an embryo at the time of the first appearance of the medullary groove × 300. It shows two large nuclei of the yolk (*n*) and the protoplasmic network in the yolk between them; the network is seen to be closer round the nuclei than in the intervening space. There are no areas representing cells around the nuclei.

12. Nucleus of the yolk in connection with the protoplasmic network hardened in osmic acid.

13. Portion of posterior end of a blastoderm of stage B, showing the formation of cells around the nuclei of the yolk.

14. Section through part of a Scyllium egg, about $\frac{1}{12}$th of an inch in diameter.

n. l. protoplasmic network in yolk.

z. p. zona pellucida.

ch. chorion.

f. ep. follicular epithelium.

x. structureless membrane external to this.

EXPLANATION OF PLATE XXIV.

s. c. segmentation cavity.	*so. s.* somatic stalk.
e. s. embryo swelling.	I. *v. c.* 1st visceral cleft.
m. g. medullary groove.	*v. c.* visceral cleft.
t. s. tail-swelling.	*m. p.* muscle-plates.
h. head.	*m.* mouth.
ch. notochord.	*an.* point where anus will appear.
op. eye.	*al. v.* alimentary vesicle.
au. v. auditory vesicle.	*ht.* heart.
al. alimentary cavity.	

A. Surface view of blastoderm of Pristiurus hardened in chromic acid.

B. Surface view of fresh blastoderm of Pristiurus.

C, D, E, and F. Pristiurus embryos hardened in chromic acid.

I, G, H. Pristiurus embryos viewed as transparent objects.

EXPLANATION OF PLATE XXV.

K. Pristiurus embryo hardened in chromic acid.

The remainder of the figures are representations of embryos of Scyllium canicula hardened in chromic acid. In every case, with the exception of the figures marked P and Q, two representations of the same embryo are given; one from the side and one from the under surface.

EXPLANATION OF PLATE XXVI.

yk. yolk.

bl. blastoderm.

a. arteries of yolk sac (red).

v. veins of yolk sac (blue).

x. portion of blastoderm outside the arterial circle in which no blood-vessels are present.

1. Yolk of a Pristiurus egg with blastoderm and embryo. About two-thirds of the yolk has been enveloped by the blastoderm. The embryo is still situated at the edge of the blastoderm, but at the end of a bay in the outline of this. The thickened edge of the blastoderm is indicated by a darker shading. Two arteries have appeared.

2. Yolk of an older Pristiurus egg. The yolk has become all but enveloped by the blastoderm, and the embryo ceases to lie at the edge of the blastoderm, owing to the coalescence of the two sides of the bay which existed in the earlier stage. The circulation is now largely developed. It consists of an external arterial ring, and an internal venous ring, the latter having been developed in the thickened edge of the blastoderm. Outside the arterial ring no vessels are developed.

3. The yolk has now become completely enveloped by the blastoderm. The arterial ring has increased in size. The venous ring has vanished, owing to the complete enclosure of the yolk by the blastoderm. The point where it existed is still indicated (*y*) by the brush-like termination of the main venous trunk in a number of small branches.

4. Diagrammatic projection of the vascular system of the yolk sac of a somewhat older embryo.

The arterial ring has grown much larger and the portion of the yolk where no vessels exist is very small (*x*). The brush-like termination of the venous trunk is still to be noticed.

The two main trunks (arterial and venous) in reality are in close contact as in fig. 5, and enter the somatic stalk close together.

The letter *a* which points to the venous (blue) trunk should be *v* and not *a*.

5. . Circulation of the yolk sac of a still older embryo, in which the arterial circle has ceased to exist. owing to the space outside it having become smaller and smaller and finally vanished.

REMARKS ON THE ANATOMY OF THE ARMS OF THE CRINOIDS. By P. HERBERT CARPENTER, B.A., *Trinity College, Cambridge.*

SINCE the appearance of the now classical memoir[1] by Johannes Müller, " Ueber den Bau des Pentacrinus Caput-Medusæ," comparatively few observers have devoted any attention to the structure of the soft parts of the arms of the Crinoids; and although more recent investigation has shown that in some points his descriptions are not entirely correct, they are still repeated in most of our text-books of Zoology and Comparative Anatomy.

Müller describes the ventral perisome of the arms of Comatula (Antedon) as containing two canals separated by a nerve cord; viz., an inferior one communicating at the edge of the disc with the perivisceral cavity, and a superior one which he called the "tentacular canal" lying beneath the radial furrow, and which, in accordance with Heusinger[2], he describes as extending into the disc almost as far as the mouth, and then opening into the cavities of the spongy substance which occupies the centre of the visceral mass.

Müller believed that the cavities of the tentacles at the sides of the radial furrows of the disc and arms, were in communication with this superior or tentacular canal; but he states in a later memoir[3] that he was never able to assure himself of this by direct observation; further, he points out that the canal is single in Pentacrinus, but that it is divided at many points in the arms of Comatula (Antedon) into two parts by a thin vertical wall. Between these two arm-canals Müller described a cord slightly swollen opposite the attachment of each pinnule, which he recognised as the nerve cord of the arm. Finally, following Heusinger, he described a membranous tube as lying in the axis of the skeletal parts of the arms and pin-

---

[1] *Abhandlungen der Berlin. Akademie*, 1841.
[2] *Zeitschrift für Organ. Physiologie*, III. (1829) 366.
[3] Ueber den Bau der Echinodermen. *Abhandlungen der Berlin. Akademie*, 1853, p. 178.

nules; this he supposed to be a vessel or nutritive canal, as it communicates with a hollow organ lying in the base of the calyx, which he regarded as a heart.

The next paper with which we are concerned is that of Professor Wyville Thomson, on the "Embryogeny of Antedon rosaceus[1]": it contains an account of the development of an oral vascular ring, from which radial vessels extend along the upper (ventral) surface of the arms, while beneath them lie tubular extensions of the perivisceral space. These last were subsequently shown by Dr Carpenter[2] to become very early divided by horizontal partitions; so that the ventral surface of each arm comes to contain three superimposed canals; viz. (1) a *tentacular* canal, giving off lateral diverticula alternately on opposite sides to the crescentic (respiratory) leaves bordering the radial furrows, and the groups of tentacles in connection with them: (2) a *subtentacular* canal, communicating in the young Pentacrinoid with the peristomial portion of the perivisceral space: and (3) an inferior or *cœliac* canal, extended from that portion of the perivisceral space which occupies the hollow of the calyx.

In this first part of his memoir Dr Carpenter also states that the structure described by Müller as a nerve, lying between his two arm-canals, really belongs to the reproductive apparatus; and that the axial cords regarded by Müller as vessels, are solid and probably of a nervous nature.

It will be seen that Dr Carpenter here describes three canals in the ventral perisome of the arms, while Müller only mentions two: this discrepancy is due to the fact that the *real* tentacular canal, i.e. that from which the cavities of the tentacles originate, was overlooked by Müller, his tentacular or superior canal corresponding exactly, as will be seen later on, to the subtentacular canal of Dr Carpenter.

In 1873, M. Edmond Perrier published an elaborate memoir[3] on the structure and regeneration of the arms of Antedon rosaceus; which, while containing much new and valuable information as to their histology, is very misleading as to their

---

[1] *Phil. Trans.*, 1865, Vol. CLV. p. 513.
[2] Researches on the Structure, Physiology, and Development of Antedon rosaceus, part I., *Phil. Trans.*, 1866, CLVI. 723.
[3] *Archives de Zoologie expérimentale*, II. 1873, p. 29.

morphology in consequence of the exclusiveness of the method of study adopted by the author. He does not seem to have made any transverse sections of the arms after decalcification, but to have limited himself to observation of their thinned-out terminations, especially when these were in course of regeneration after fracture ; what he sought being the most favourable opportunity of studying the tentacular apparatus from its ventral aspect.

In this manner he was able clearly to distinguish the longitudinal tentacular canal, and its communications by successive alternating peduncles with the trifid groups of tentacles that fringe the borders of the radial (? ambulacral) furrow. This tentacular canal he obviously supposed to be the tentacular canal of Müller; although he rightly identified it with that which Dr Carpenter[1] had recognised as the representative, in the adult Antedon, of the *true* tentacular canal described by Professor Wyville Thomson as an extension of the oral ring of the Pentacrinoid larva, and which (as stated by Dr Carpenter) lies between the tentacular canal of Müller and the floor of the furrow. M. Perrier recognised the fact[2] that an extension of the perivisceral cavity passes into the arms, as Müller had pointed out, but he affirmed that this is single, and that its existence[3] is limited to a certain period of the growth of the arms and pinnules; refusing to it altogether the character of a canal system. He denied altogether the existence of the third canal specified by Dr Carpenter, and expressed his conviction that no one would ever find it. Yet this third canal, the *subtentacular* one of Dr Carpenter, is really Müller's *tentacular* canal, which can be traced without the smallest difficulty, not to the oral ring (or its representative) in the adult animal, like the tentacular one, nor to the perivisceral space like the cœliac, but to the interior of the central columella of the visceral mass, as described by Müller and Heusinger.

The existence of *three* canals in the arms and pinnules of the ordinary type of Antedon rosaceus, as stated by Dr Carpenter, can no longer be a matter of doubt; as it can be readily demonstrated by any one who will take the trouble to make the thin

---

[1] *Loc. cit.*   [2] *Loc. cit.*, pp. 49 and 57.
            [3] *Loc. cit.*, pp. 48 and 73.

transverse sections which modern methods enable him readily to obtain. I have thus myself satisfied Professor Semper of the correctness of Dr Carpenter's views upon this point. Ludwig has also confirmed them, and Professor Huxley has recently expressed privately his full acceptance of them, as the result of his own use of this method.

The mode of study adopted by M. Perrier, however, led him to recognise a new point in the structure of the arms of Antedon, viz. the existence of a band of fibrils regarded by him as muscular, and extending along the length of the arms and pinnules, beneath the epithelium of the radial furrows, in immediate contact with the ventral wall of the *real* tentacular canal.

Shortly afterwards Professor Semper, who had studied the Crinoids of the Philippine Islands, published a short paper[1] showing from his own independent investigations made some years previously, that, as had been stated by Dr Carpenter, the nervous cord of Müller really belongs to the generative system, of which it is the rachis: while, following Müller, he figured two arm canals (besides the axial cord), an inferior one and a superior one, immediately above which last, he described in his new Philippine species, Actinometra (Comatula) armata (Semper, MSS.), a cord x, which he supposed to be identical with the muscular band of Perrier.

In an addendum[2] to a translation of Professor Semper's paper, Dr Carpenter states that in Antedon rosaceus the tentacular or superior canal of Müller and Professor Semper has no connection with the tentacles, but that the real tentacular canal, from which the tubular tentacles originate, lies above it in the position occupied by the cord x in Actinometra armata. Ludwig[3] has quite recently confirmed Dr Carpenter's statement regarding Müller's tentacular canal in Antedon rosaceus, and says that the radial watervessels correspond to the cord x in Professor Semper's section, adding that Perrier's muscular band is not identical with the cord x, but lies above it, and

---

[1] Kurze anatomische Bemerkungen über Comatula, *Arbeiten aus dem Zool. Zoot. Institut in Würzburg*, Band I. 1874, p. 259.
[2] *Ann. and Mag. of Nat. Hist.*, Sept. 1875.
[3] Zur Anatomie der Crinoidien, *Zeitschrift für Wissenschaftliche Zoologie*, XXVI. 1876, p. 861.

that he believes it to be of a nervous and not of a muscular nature.

Ludwig refers to the real tentacular canals as radial water-vessels; and also states that he has found Antedon rosaceus to possess a true watervascular system of the form typical for all Echinoderms, consisting of a circular oral vessel and radial trunks in connection with it. Accordingly he controverts the opinion expressed by Professor Semper, in his monograph on the Holothurians[1], that our living Crinoids have no *true* water-vascular system. It must be remembered however that, as far as we yet know the development of the oral ring and radial canals in the Crinoids, their origin is very different from that of the watervascular system in the other Echinoderms. Dr Carpenter mentioned long ago[2], and has recently shown at greater length[3], that the vascular ring of the Pentacrinoid larva, from which the oral tentacles arise, is really an extension of the peri-visceral cavity, partly separated from the rest: and Metschnikoff[4], by later independent investigation, has obtained essentially the same results. This last observer, to whom we owe so much of our knowledge of the development of the watervascular system and perivisceral cavity in the other Echinoderms, states expressly that the watervascular system of Comatula Mediterranea (probably = Antedon rosaceus) is developed in a different way from that of the rest of the Echinoderms, and that the larvæ do not possess the so-called lateral discs or their homologues, such as occur in the larvæ of many Echinoderms and in Tornaria.

Further, Professor Wyville Thomson[5] was unable to detect any cilia on the walls of the oral ring of the Pentacrinoid larva, while M. Perrier[6] could not find them in the radial canals of the adult animal; and these are universally present in the watervessels of the other Echinoderms.

Ludwig also asserts the existence of an oral ring in the adult Comatula Mediterranea, with tubuli depending from it

---

[1] *Reisen im Archipel der Philippinen*, II. 1. Holothurien, p. 196.
[2] *Phil. Trans.* 1866, p. 728.
[3] On the Structure, Physiology, and Development of Antedon rosaceus, *Proc. Roy. Soc.*, Jan. 20th, 1876, p. 227.
[4] *Bull. de l'Acad. Imp. de St Pétersbourg*, 1870, pp. 508, 509.
[5] *Phil. Trans.*, 1865, p. 526.
[6] *Loc. cit.* p. 58.

into the plexus of tissue occupying the perivisceral cavity, with which they communicate by their open ends: and he regards them as the homologues of the sand-canals of the other Echinoderms.

Dr Carpenter however has recently stated[1] that the ring canal of the young pentacrinoid of Antedon rosaceus (which is generally regarded as identical with Müller's Comatula Mediterranea) is simply a space between the two folds of the annular lip from which spring the oral tentacles, and that it becomes separated from the perivisceral cavity by a circlet of threads of connective tissue, which passes between the inner and outer folds of the lip.

"The increased development of this connective tissue subsequently constitutes a partition that cuts off the oral ring from the cœlom and finally breaks up the canal itself into an irregular areolation."

Consequently, in the adult Antedon[2], "the thickened annular lip which surrounds the mouth does not contain any distinct ring canal; its original cavity having been entirely filled up by threads and bands of connective tissue, and by a set of cœcal tubuli."

The radial *true* tentacular canals however "still pass towards it, and seem to lose themselves in what may be called its 'interspace system'; this being continuous with that which has come almost entirely to occupy the portion of the perivisceral cavity that originally lay open within the intestinal coil," and which thus forms the central columella of the disc.

Under these circumstances, it would seem that we are scarcely yet justified in regarding the system of tentacular canals in the Crinoids as a *true* watervascular system, homologous with that of the other Echinoderms. We know of no Polian vesicles nor Madreporic canal in connection with it, though Ludwig has recently suggested[3] that the appendages of the ring canal described by him, which appear to correspond to the cœcal tubuli of Dr Carpenter (while the plexus described by him as occupying the perivisceral cavity is the "interspace

---

[1] *Proc. Roy. Soc.*, Jan. 20th, 1876, p. 227.
[2] *Proc. Roy. Soc.*, Jan. 20th, 1876, p. 214.
[3] *Nachrichten von der Königl. Gesellschaft der Wissenschaften und der G. A. Universität zu Göttingen*, No. 5, Feb. 23rd, 1876.

system" of Dr Carpenter), may represent the sand canals (Stein-kanäle) of the other Echinoderms; while he also regards the ciliated funnels in the ventral perisome of the disc which lead into the perivisceral cavity, as collectively comparable to "the apparatus of the Madreporic plate of the Asterids and Echinids": but the justice of these comparisons is yet to be demonstrated. In some points however, but by no means in all, there is a considerable analogy in the functions of the two systems. Dr Carpenter has stated[1] that the "tubular tentacula with which the arms of Antedon are so abundantly furnished, have not in the smallest degree that adhesive power which is possessed by the feet of the Echinida and Asterida," and he is disposed to regard them as "constituting a special respiratory apparatus, and homologous with the tentacular fringe surrounding the mouth of the Holothurida, and with the respiratory tubes of the Asterida and Echinida[2]."

This opinion, common to Professor Semper and Dr Carpenter, seems to be also shared by Professor Huxley; as he has lately stated[3] that no ambulacral vessels have yet been definitely made out in the Crinoidea.

Dr Carpenter has recently described at length[4] the arrangement of the canal systems and of other structures in the arms of Antedon rosaceus; and he shows how his subtentacular canal, although actually double in the arms of the adult animal, is virtually single; the primitively single canal, developed from the upper portion of the tubular extension of the perivisceral space into the arms, beneath the radial tentacular canals of the Pentacrinoid larva, becoming divided into two parts by a vertical perforated partition; while in the disc this canal is actually single, and opens into the axial canal of the central columella, exactly like the tentacular canal of Müller and Heusinger, with which it is in fact identical, the real tentacular canal having been overlooked by these observers.

The last paper which we shall have to notice is a very recent one by Ludwig[5], in which he gives a minute account

[1] Phil. Trans. 1866, p. 699.
[2] Proc. Roy. Soc., Jan. 20th, 1876, p. 224.
[3] Notes on the Invertebrata, Med. Times and Gaz., Aug. 14, 1875, p. 173.
[4] Proc. Roy. Soc., Jan, 20th, 1876, p. 221.
[5] Nachrichten von der Königl. Gesellschaft der Wissenschaften und der G. A. Universität zu Göttingen, No. 5, Feb. 23rd, 1876.

of the histology of the various structures contained in the ventral perisome of the arms and pinnules, basing his description upon their appearance in Antedon Eschrichtii; he has however examined several other species, including Pentacrinus Caput-Medusæ, and finds the differences between them to be merely of subordinate importance. As mentioned above, his description of the canal systems completely confirms Dr Carpenter's views. With regard to Perrier's muscular band, he corrects his previous statement that he believes it to be nervous and not muscular in nature, and points out that in the ventral wall of the tentacular canal, there is a set of longitudinal fibres answering to the description given by Perrier of his muscular band; but that above these, and immediately beneath the epithelium of the radial furrow, is a delicate fibrillar structure, enclosed in a thin sheath of connective tissue which sends in partitions between the fibrils, so as to break them up into small bundles; and it is this structure which he regards as the repre- sentative of the radial nerves of the other Echinoderms. To Ludwig therefore is due the credit of the discovery of this possibly nervous band; although, somewhat later, it was independently observed by myself, and I am glad to be able to testify to the accuracy of his description of it. He points out that, though it is usually continuous across the radial furrow, in Antedon Eschrichtii this is not the case, but that in the centre of the furrow the small fibrillar bundles of the two sides are separated by a short interval in which none occur: he does not tell us, however, how far the band extends on either side. In my specimens I find it usually to end in a rounded and thickened margin, at the point where the thick epithelium of the floor of the groove begins to pass into the thinner layer covering the elevated ridges of the perisome which form its lateral wall; though in rare cases I have noticed that it sends a fine branch upwards under this thinner layer.

Beneath the central part of the epithelium forming the floor of the radial furrow, Ludwig further mentions a small cavity often containing a slight coagulum[1], and occasionally sending off lateral branches in the direction of the groups

---

[1] I have also found a coagulum in the cœliac and subtentacular canals.

of tentacles; and he compares it to the vascular cavities in connection with the radial nerves, described in the Asterids by Greef[1] and other authors. The same view, supposing that the fibrillar subepithelial band is really a nerve, had also occurred to myself. A similar cavity between the radial nerve and the watervessel has been also described by Professor Semper[2] in Cucumaria Japonica.

In the pinnules of Antedon Eschrichtii and of the other species examined by him, Ludwig has found small ciliated diverticula of the cœliac canal, which he has well compared to the ciliated funnels on the mesentery of the Synaptidæ: these however are not limited to the pinnules, for I have found them also in the arms of Antedon Eschrichtii; though they are much less frequent than in the pinnules, being often entirely absent through several successive segments.

Shortly after my arrival in Würzburg towards the close of last year, Professor Semper most liberally placed at my disposal all his specimens of his Actinometra (Comatula) armata, as material for a monograph of the species; and one of the first points to which I directed my attention was to determine the relations of the fibrillar subepithelial band, which I had found in both arms and disc of Antedon Eschrichtii, to the canal system, and to the cord x already described by Professor Semper in this species; and I am now able to show that this cord x is not the representative of the muscular band described by Perrier in Antedon rosaceus, as supposed by Professor Semper, nor on the other hand does it represent the tentacular canal, as suggested by Dr Carpenter and Ludwig: all three, together with the fibrillar subepithelial band (? nerve), coexist in the arms of this species.

The accompanying sections will I hope make this clear. Fig. I. is a transverse section of an arm of Antedon Eschrichtii (in which the relations of the parts are essentially the same as in Antedon rosaceus). On the ventral surface, immediately beneath the epithelium e, of the radial furrow is the delicate fibrillar (? nervous) band, $f. b.$, discovered by Ludwig; and it is

---

[1] Ueber den Bau der Echinodermen, *Marburg. Sitzungsberichte*, 1871. No. 6, 1172.
[2] *Holothurien*, p. 148.

Fig. L.

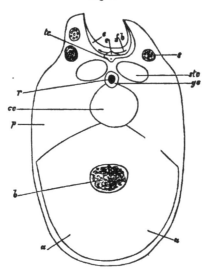

Transverse section of decalcified arm of Antedon Eschrichtii.

aa. Organic basis of calcareous segment.
b. Solid axial cord (nerve?) of Dr Carpenter: 'nutritive canal' of Müller and Heusinger.
e. Epithelium of the radial furrow.
f.b. Fibrillar band regarded as a nerve by Ludwig.
c. Cavity, probably vascular and in connection with this band.
t.c. Tentacular canal of Dr Carpenter and Perrier, with a branch to the tentacular apparatus on the right side: the muscular band of Perrier is in the ventral wall of this canal.
s.t.c. Subtentacular canals of Dr Carpenter: 'tentacular canal' of Müller.
r. Rachis of the generative system: 'nerve' of Müller.
g.c. Genital canal of Dr Carpenter, in which the rachis lies.
c.c. Cœliac canal of Dr Carpenter: 'inferior canal' of Müller.
p. Tissue of ventral perisome.
s. Sacculi ('calcareous glands' of Wyville Thomson) of doubtful (? sensory) nature.

not continuous across the middle line of the furrow, which is occupied by the small, probably vascular cavity, *c*, mentioned above: its shape in this section is triangular, though it is frequently somewhat flattened.   Beneath this again lies the true tentacular canal, *t. c.*, with its branch to the tentacular apparatus on the right side: in the ventral wall of this canal lie the fibres constituting the muscular band of Perrier. Still deeper come the double subtentacular canal, *s. t. c.*, the rachis *r*, of the generative system, lying in its genital canal, *g. c.*, and the cœliac canal, *c. c.*, while in the centre of the skeleton is the axial cord, *b*, regarded by Dr Carpenter as a nerve.

Fig. II. is a similar section of an arm of Actinometra (Comatula) armata, in which the same parts are shown, viz. the fibrillar band, *f. b.*, the tentacular canal, *t. c.*, and its lateral branch, and then beneath this, and projecting into the cavity of the subtentacular canal, which is here actually single as in Pentacrinus, is the body *x*, which is merely a pigmented cellular thickening of the tissue between the tentacular and subtentacular canals; it is not a continuous cord, but occurs at intervals along the whole length of both arms and pinnules: what its function may be, I cannot tell.

In this species, the ovaries, at any rate at the period of sexual maturity, extend themselves, as already stated by Prof. Semper, from the pinnules into the soft portions of the arm, and so their connection with the rachis, the structure of which is very remarkable, becomes visible in transverse sections of the arm, as is seen on the left side in Fig. II. The cœliac canal and axial cord occupy the same positions here as in Fig. I. The muscles, *m, m*, connecting the successive segments of the arms, appear in this section, but not in that of Fig. I., as the former is taken rather nearer the end of a segment than the latter.

It will thus be seen that the arrangement of the canal system in the arms of Actinometra armata, is essentially the same as that described in Antedon rosaceus by Dr Carpenter: the main difference being in the form and simple nature of the subtentacular canal, which Professor Semper, following Müller, described as the tentacular canal; although as he has since informed me, he saw the real one, but did not pay any particular attention to it, as the object of his paper was simply to point out Müller's errors regarding the rachis of the generative system.

The form of the subtentacular canal is, in this species, rather peculiar; for instead of being somewhat compressed vertically, as is the case in the European species, it extends

Fig. II.

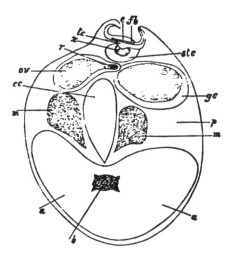

Transverse section of decalcified arm of Actinometra (Comatula) armata
(Semper MSS.).

aa.   Organic basis of calcareous segment.
b.   Solid axial cord (nerve?) of Dr Carpenter: 'nutritive canal' of Müller
and Heusinger, with the commencement of the branches proceeding from it.
e.   Epithelium of the radial furrow.
f.b.   Fibrillar band: 'nerve' of Ludwig.
t.c.   Tentacular canal of Dr Carpenter and Perrier, with a branch to the
tentacular apparatus on the right side: the muscular band of Perrier is in the
ventral wall of this canal.
x.   Pigmented cellular thickening of its floor.
s.t.c.   Single subtentacular canal of Dr Carpenter: 'tentacular canal' of Müller
and Professor Semper.
r.   Rachis of the generative system ('nerve' of Müller) in connection with
the left ovary, *ov.*
g.c.   Genital canal of Dr Carpenter, in which the rachis and ovaries lie.
cc.   Cœliac canal of Dr Carpenter: 'inferior canal' of Müller.
m.   Muscles.
p.   Tissue of ventral perisome.
N.B.   The curious sacculi (Fig. I. *s.*) which occur at the bases of the respiratory leaves in the European Crinoids, do not exist in this species, in which
the extensions of the ovaries into the soft parts of the arms distort the
symmetry of these very considerably.

upwards in the direction of the respiratory leaves; but I
am not yet able to state that there is any connection be-

tween it and the branches of the tentacular canal which they contain.

The small triangular cavity (c, Fig. 1.) between the epithelium of the radial furrow and the tentacular canal in Antedon Eschrichtii, does not appear to exist in Actinometra armata; at any rate I have not yet found it; but it is remarkably distinct and relatively very large in the arms of another new Philippine species, Actinometra nigra (Semper, MSS.), in which its shape is triangular, with the apex directed downwards towards the tentacular canal, and not upwards towards the epithelium as in Antedon Eschrichtii: in this species too, the subtentacular canal is double, as in the European ones.

In both these Philippine species, the structure of the generative rachis is extremely complicated, and by no means corresponds to Ludwig's description of it in the female Antedon Eschrichtii; my sections however enable me to confirm his account as far as the male of this species is concerned.

It is worth notice that the genital canal in which the rachis and genital glands lie, is in the male Antedon Eschrichtii derived from the subtentacular canal of both arms and pinnules; while in the female Actinometra armata a part of the cœliac canal becomes separated off to form the genital canal in which the ovaries lie. Whether this is a specific or merely a sexual difference, is a point for further investigation. It will be seen that my observations have, on the whole, completely confirmed those of Ludwig, but there is one point in which I must differ from him very considerably. He states, in direct opposition to Dr Carpenter's views, that the axial cord in the centre of the skeleton of the arms is not a nerve; but does not give us any reasons for this assertion, nor any further information about this structure. Yet there is more evidence for the nervous nature of this cord than for that of the ventral subepithelial band, which he and I regard as the homologue of the radial nerves of the other Echinoderms; although the possible existence of two nervous cords, a dorsal one, and a ventral one, in the arms of the Crinoids seems morphologically somewhat inexplicable: into that question, however, I will not enter here.

Dr Carpenter's views[1] as to the nervous nature of this cord are based upon both physiological and anatomical considerations: viz. (1) that stimulation of the central quinquelocular organ ('heart' of Müller) contained in the calyx, with which the axial cords of the arms are in connection, is followed by sudden and simultaneous flexion of all the arms, and (2) that it gives off pairs of branches, at the junction of the segments, which ramify upon the muscles.

In both Actinometra armata and Actinometra nigra this cord increases considerably in size in the centre of each ossicle of the arms and pinnules; and in this, the most protected part of each segment, it gives off four main trunks: in the arms, it also gives off here the branches which enter the pinnules; two of these trunks run towards the ventral side and break up into numerous branches, some of which reach the bases (in some cases even the tips) of the respiratory leaves. I have as yet been unable definitely to make out their terminations, though I have some sections which seem to indicate that some of these branches become connected with the ventral sub-epithelial band, the nerve of Ludwig; but this is a point requiring much further investigation, and upon which I cannot by any means speak with certainty. The two inferior or dorsal trunks run towards the surface of the skeleton, and while some of their branches appear to enter into the plexus of tissue forming its organic basis, others seem to become connected with epidermic structures. In some pinnules of Actinometra armata, but not in all, these branches enter into connection with peculiar cellular organs, of very similar structure to the groups of large epidermic cells described in the tactile papillæ of the integument of the Synaptidæ, by Professor Semper[2]. I am inclined to regard these organs also as sensory in function; their position upon the dorsal or most exposed side of the pinnules certainly favours this hypothesis.

I have found similar branches extending from the axial cord in Antedon Eschrichtii, but they are by no means so well developed as they are in the two Philippine species, and

---

[1] *Proc. Roy. Soc.*, Jan. 20th, 1876, pp. 221, 226.
[2] *Holothurien*, pp. 28, 29.

are somewhat difficult to distinguish from connective tissue fibrils.

There are many other points of great interest in the Anatomy of Actinometra armata, into which I cannot enter here, but I hope to publish a detailed account of them in the third volume of the "*Arbeiten aus dem Zool. Zoot. Institut in Würzburg.*"

WÜRZBURG, *March 5th*, 1876.

SOME EFFECTS OF UPAS ANTIAR ON THE FROG'S HEART. By M. FOSTER, M.D., F.R.S.

(From the Physiological Laboratory, Cambridge.)

IN the present paper I wish to call attention to some conclusions of general importance which I think may be drawn from the behaviour of a frog's heart, poisoned by antiar, under the influence of a stimulation of the vagus nerve. With regard to the general action of antiar on the frog's heart, I have little to add to the statements of previous writers. My observations were made in July last on Rana temporaria. The animal was either pithed or curarized, the heart exposed, and the cardiac movements recorded by means of two light levers, placed directly, one on the ventricle and the other on the auricles. One or two experiments were made with endocardial pressure according to Coates' method; but all the main results were gained by help of the simple lever, which, for general purposes, is a most useful instrument deserving to be used much more frequently than it is. The antiar was applied as an aqueous solution of purified antiar extract, kindly given me by Dr Sharpey.

The general effects of the poison on the heart were as follows. After a stage, of some uncertain appearance and duration, in which the rhythm was slightly quickened, with a slight increase of each contraction, the beats became distinctly slow with a tendency to become irregular. The systole of the ventricle was markedly prolonged, its peristaltic nature very pronounced, and the height to which at each beat the lever was raised, after increasing slightly, rapidly diminished. In the diastole succeeding each systole, relaxation of the muscular walls fell short of being complete. Hence the ventricle became more and more permanently contracted as the action of the poison continued; and in proportion as this sort of "tonic" contraction was developed, the extent of each rhythmic contraction diminished; the visible beats becoming smaller and smaller, until at last the ventricle lay apparently perfectly motionless, pale in colour save for some irregular blotches, and firmly contracted into a cone. In this state it passed into rigor mortis.

This condition has been spoken of as a tetanus, and it may fairly be so called, if it be remembered that it is a tetanus brought about by an extraordinary prolongation of the relaxation phase of each contraction, and not by a too rapid sequence of beats. During a certain stage the negative variation of the muscle-current accompanying each contraction is very much increased, and that too at a time when the effect of the contraction in moving the lever has already become markedly less. From a few observations which I made, I am inclined to think that, during this stage, the heat given out at each contraction is also increased; but on this point I am not at present able to speak very decisively.

As the ventricle becomes more and more contracted, the blood, less and less able to enter into it, accumulates in the auricle, which thereby becomes much distended. One effect of this is that the systole of the auricle produces a very large excursion of the lever placed on it. Under no other circumstances have I been able to register with such ease the auricular contractions. Using the lightest levers at my disposal, the curves at the commencement of an experiment generally were: for the ventricular lever a rise at each systole of about 20 to 30 mm., for the auricular lever, slight undulations just distinct enough to be read. Fifteen minutes after the application of a small dose of antiar, the ventricular curve had fallen to about 5 to 10 mm., while the auricular curve had gained the height of from 20 to 30 mm. at each beat. The duration of the systole had by this time become much prolonged, as well in the case of the auricle as of the ventricle.

So long as the beats of the ventricle were visible, each one was preceded in a perfectly regular manner by a beat of the auricle. In the later stages of the poisoning the beats became unfrequent and the times of their recurrence irregular; but so long as the ventricle continued to beat at all, the normal sequence of the ventricular systole upon the auricular systole was maintained.

The ventricular beats becoming feebler and feebler, at last ceased altogether; the auricle, however, continued to beat for some time longer, its beats in turn becoming gradually feebler until they too ceased.

Such are the effects seen, when the dose is of such a strength as to bring the heart to a standstill in from half an hour to an hour, and may, I think, be taken as fairly characteristic of the action of the drug.

The facts all support the view of previous observers, that antiar is essentially a muscular poison, and produces its effects chiefly at least by interfering with the functions of the muscular tissue of the heart. The maintenance of the normal co-ordination between auricle and ventricle directly negatives any large interference with the nervous cardiac mechanisms; in fact, nothing can be more striking than the way in which the ventricle attempts to follow up each auricular systole as long as it can possibly do so.

Schmiedeberg[1] has made the very interesting observation, that in poisoning by antiar, digitalin and the other drugs which produce the same peculiar contracted state of the ventricle, the beats even after they have ceased may be restored by a forcible dilatation of the ventricle, thus indicating that the standstill produced by the drug is caused by the failure of the post-systolic relaxation, and hence presumably is brought about by the poison affecting the elastic qualities of the cardiac tissue.

I have repeated Schmiedeberg's results with success; but found, as apparently he also did, that the restoration was temporary and brief, and, moreover, could only be brought about at a particular phase, i. e. very soon after the ventricle had ceased to beat. When the distension was deferred too long, the result was always a failure. Evidently the lethal action of the drug is of a deeper kind than a simple interference with the elastic qualities of the cardiac muscle.

I was less successful in repeating Neufeld's[2] observation, that the beats can be restored by the application of potassium cyanide (with the view of producing relaxation). Using solutions of the strength applied by him, and also of other strengths, I never succeeded in obtaining a definite relaxation of the contracted ventricle. Frequently at a particular stage,

---

[1] *Beiträge sur Anat. u. Phys. als Festgabe Carl Ludwig gewidmet*, p. ccxxii.
[2] *Studien Phys. Instit. Breslau*, III., p. 97.

a few drops of a strong solution (10 p. c.) poured near the heart produced a few feeble beats; but these seemed to me to be caused by the direct chemical stimulation of the cardiac muscle, and to be the cardiac representatives of the fibrillar contractions observable in striated muscle when the same solution was applied directly to it.

The point to which I wish to draw particular attention is the behaviour of the antiarized heart when the vagus is stimulated.

Exp. 1.   A frog was pithed, the heart laid bare, the lever placed on the ventricle, and a record of the cardiac rhythm taken. The excursions of the point of the lever on the recording surface were about 20mm. in height at each ventricular systole.   The right vagus[1] was laid bare, and a satisfactory inhibition obtained with Du Bois Reymond's machine with the secondary coil at 10 and one Daniell's cell.   A small dose of antiar (two drops of a ·25 p. c. of an aqueous solution of the extract) was given.   In a short time the ventricular systole was so reduced that the recording line was simply marked with hardly visible irregularities.   The vagus was stimulated (1 Daniell, coil at 10) for nine seconds.   During the stimulation the ventricle became perfectly quiescent: it remained so for the next eleven seconds.   Then came a beat, the lever rising 5 mm., followed by a second of 20 mm., and a third of 30 mm., succeeded by a series of about fifty beats, the whole taking up about a minute, and the individual beats raising the lever point about 30 mm., except towards the end of the period, when they gradually became less and at the same time irregular.   The ventricle then became and continued perfectly quiescent.

Exp. 2.   Frog pithed.   Vagus proved to be inhibitory (1 D., coil at 10), and antiar given.   When the movements of the ventricular lever had fallen to 5 mm., the interrupted current (1 D., coil at 10) was sent through the vagus for 10 seconds.   During the first half of this period there were two slight beats each of about 2 mm., during the second half three beats of 14, 18, 22 mm. respectively, followed on the breaking of the current by beats rapidly rising to 70 mm., and becoming very rapid, but subsequently diminishing both in extent and frequency.

Exp. 3.   Frog pithed.   Vagus proved to be inhibitory.   Antiar given.   When all *visible* contractions of the ventricle had ceased, the vagus was stimulated for a few seconds.   During the passage of the current and for some few seconds afterwards no change in the

---

[1] The right was perhaps most generally but not exclusively used, and no marked difference between the right and left was noticed.

ventricle was seen. I looked for, but could not see any signs of relaxation. (The condition of the auricle was not recorded in this case, but in somewhat similar cases, the auricle, still beating spontaneously while the ventricle had become perfectly quiescent, was inhibited by the stimulation of the vagus.) After the lapse of a few seconds, however, the ventricle began to beat, and gave for some little time a series of vigorous beats, gradually but rapidly increasing in strength and rapidity, remaining at a maximum for a while and then gradually fading away.

Thus, under the influence of antiar, the vagus still retains its inhibitory function, though in the range of several observations it was seen that the more profound the action of the drug, the less easily was inhibition obtained. There is nothing special, however, to antiar in this, because inhibition is always more difficult in a slowly beating heart than in one beating with a frequent rhythm. The peculiarity of the antiarized heart is that the inhibitory phase is followed by an augmentative phase, in which the beats become much more rapid, and especially much more forcible, than the beats which are proper to the heart at that stage of the poisoning. They may even become more vigorous than the normal beats of the unpoisoned heart. The augmentative phase may succeed the inhibitory phase while the current is still passing through the vagus, or may not set in till some time after the current has been shut off. As, with the increasing influence of the poison, the inhibitory phase sinks into the background, the augmentative phase becomes more and more pronounced until at least the strange state of things is reached in which, the inhibitory phase being impossible on account of the spontaneous beat having previously ceased, the sole effect of stimulation of the vagus is to produce a prolonged series of vigorous beats.

In all the above observations I took great care to avoid escape of the current on to the heart itself; and from a fact which I shall mention directly, such an escape cannot be considered as in any way explanatory of the results.

I suppose most physiologists, in reading the above, will at once see a ready explanation of the results. "In the vagus of the frog we have accelerator and inhibitory fibres. Antiar slowly and gradually paralyses the inhibitory fibres, but leaves intact, or rather exalts, the accelerator fibres." The question of there being accelerator as well as inhibitory fibres in the vagus generally was I believe first mooted (but decided in the negative) by Rutherford[1]. Their existence has been power-fully urged by Schmiedeberg[2], and illustrated by the action of nicotine. Most subsequent writers have admitted their exist-ence, and quite recently Boehm[3], reviving an old observation of Wundt's[4], finds that in a particular stage of curari poisoning stimulation of the vagus in the mammal brings about distinct acceleration.

I feel, however, great difficulty in accepting this explana-tion, for the following reasons.

1. When atropin is given, previous to the antiar, in suf-ficient dose to prevent the ordinary inhibitory action of the vagus, the curious augmentative effects which I have described *are also absent.* In other words, the inhibitory and augmenta-tive effects are closely correlated. Now atropin is one of the drugs which are said to disclose by their action the existence of the accelerator fibres.

The absence of the effects under atropin also disposes of the explanation that the results are in any way due to escape of the current; they are indubitably caused by some form of activity of the vagus.

2. A somewhat similar augmentation is seen *as an after effect* of the application of the constant current to an antiarized frog's heart.

3. The augmentative effects I have described are merely an exaggerated form of that "reaction" which is always seen in a more or less pronounced degree after stimulation of the vagus. I mean the reaction which may be witnessed in a bloodless heart separated from the body, and which, therefore, cannot be explained by an increased tension of the cavities due to the larger quantity of blood pouring in from the veins during the standstill. If the vagus be stimulated at intervals from the very commencement to the end of the antiar poison-ing, the perfectly gradual way in which the normal "reaction" shifts into the marked "augmentative effects," is very clearly

[1] This *Journal,* Vol. III. p. 408.
[2] *Arbeiten Phys. Anst. Leipzig,* 1870, p. 41.
[3] *Archiv f. Exp. Path. und Pharm.,* IV. 851.
[4] *Verhand. Natur. hist. Vereins, Heidelberg,* 1860, quoted by Rutherford.

seen. Of course I am aware that this very "reaction" is claimed as an effect of "accelerator" fibres; but this I am unwilling to admit, for the reason, that precisely the same reaction may be seen in its most exquisite form, when the snail's heart is inhibited by the direct application of the interrupted current[1]. And one can hardly be expected to add accelerator nerve fibres to the hypothetical structures which have yet to be discovered in that organ. The so-called reaction is really a reaction, is a veritable part of the entire influence of the excited vagus over the heart.

4. Boehm[2] has with great force called attention to the remarkable similarity between the augmentative effects of stimulating the (mammalian) vagus under certain conditions, and the ordinary results of stimulating the special cardiac accelerator nerves. Now it must, I think, have occurred to every one, that the very peculiar features[3] of these accelerator nerves, their enormous latent period, their apparent inexhaustibility, and the impossibility of stimulating them otherwise than by galvanic currents (mechanical and chemical stimulation seeming to have no effect), all point to the conclusion, that the actions which are set up by stimulating those nerves are of a very peculiar kind, and cannot be satisfactorily appealed to as explanatory of similar results occurring elsewhere. The difficulty one falls into by assuming the presence of these accelerator fibres is well shown by a remark of Boehm's[4]. In pointing out that it is only a particular phase of curari poisoning in which the accelerator effects come out, since a too large dose destroys them as well as the inhibitory effects, he goes on to say, that at a stage when the accelerator effects have ceased, "the *irritability* of the accelerator fibres can be instantly *restored* by injecting afresh a small (·005—·01 grm.) dose of curari." Certainly the accelerator fibres are the only nerve fibres whose irritability is at once destroyed and restored by curari.

5. Lastly, I would point out that in my results, in which

---

[1] See Foster and Dew-Smith, *Proc. Roy. Soc.* XXIII. p. 318.
[2] *Loc. cit.* p. 364.
[3] Schmiedeberg, *Arbeit. Phys. Anst. Leipzig*, 1871, p. 84. Bowditch, *ibid.* 1872, p. 259. Boehm, *Archiv f. Exp. Path. und Pharm.* IV. 255.
[4] *Loc. cit.* p. 366.

I have taken care to record not only the frequency but the strength of the beats, the augmentation bears quite as much, at least, on the strength as on the frequency of the beats; hence the augmentation in antiar poisoning is so far distinct from the acceleration gained by stimulation of the mammalian accelerator nerves, in which the strokes while more rapid are less powerful.

These several facts appear to me valid reasons for refusing to call in, in this case, the assistance of any *Deus ex machina*, in the shape of specific accelerator fibres.

On the other hand, there seems to be, if not a solution, at least an opening towards a solution, if it be admitted, on the one hand, that the augmentative effect is an exaggerated reaction after inhibition, and on the other, that the muscular tissue of the heart is affected during inhibition, its quiescence, or slowness of beat, then manifested, being essentially due to changes in itself (in whatever way brought about), and not simply to its ceasing to receive impulses from the automatic ganglia.

Whatever idea we may form of inhibition in spontaneously active tissue, whether simply muscular, or nervous, or neuro-muscular, it seems only natural that the inhibition of the natural movement should be followed by a rebound. The ordinary theory by which a rhythmic effect is supposed to originate from two opposing forces, is based on the supposition of such a rebound occurring, and recurring. Nor is there any insurmountable difficulty in supposing that the rebound may, under certain circumstances, exceed the direct action, so that a case may occur, like that of Exp. 3, in which the inhibition was unable to manifest itself (the absence of visible contractions certainly not excluding the possibility of changes taking place in the direction of inhibition, *i. e.* changes which would be inhibitory if the heart were beating), and yet a well-marked augmentation occurred[1].

The suggestion that the stimulation of the vagus produces changes in the muscular tissue of the heart, may appear to many physiologists a return to an exploded view; but I hope

[1] Does not the so-called "law of exercise" mean that vital reaction is, within limits, *greater* and opposite?

shortly to be able to bring forward some facts which seem to me clearly to prove that the muscular mass of the frog's ventricle is affected during inhibition, in addition to whatever changes may be going on in the ganglia.

If this be admitted, then it becomes intelligible why these peculiar augmentative effects should be so pronounced in the action of a poison so essentially a muscular poison as is antiar; and it follows, that the explanation of the augmentative effects are to be sought for in the changes produced by the drug in the muscular tissue of the heart. What those changes are, I cannot at present with any distinctness say; but it is extremely probable that they are the cause at once both of the characteristic antiar beat, and of the peculiar augmentative phenomena.

As a curious fact I may mention, that (in July) I found toads enjoying a perfect immunity against antiar. In not one of six toads which I tried did I get any satisfactory action on the heart; and to one toad I gave, without any effect, several cubic centimetres of a solution, two drops of which killed a frog in a very short time. Schmiedeberg[1] has already pointed out a difference in their behaviour towards digitalin between Rana temporaria and esculenta.

---

[1] *Loc. cit.* p. ccxxiv.

VARIATIONS IN THE ARRANGEMENT OF THE EXTENSOR MUSCLES OF THE FORE-ARM. *By* JOHN CURNOW, M.D., *Professor of Anatomy in King's College, London.* (Pl. XXVIII.)

THE following observations are taken from my notes and drawings of the arrangement of these muscles in 84 subjects, which have been dissected at King's College during the years 1873—4 and 1874—5. Many of the varieties have been before observed, but some interesting forms have not hitherto been recorded.

1. *Extensor ossis metacarpi pollicis.* A single tendon to this muscle is of very exceptional occurrence. The tendon is usually double, both being inserted side by side into the base of the first metacarpal, or one into this bone, and the other into the trapezium. The attachment to the trapezium is so constant in the anthropoid Apes, that it is most important to ascertain its relative frequency in Man. Flower and Murie say that it is the rule rather than the exception, whilst Macalister estimates its occurrence as only once in every twelve subjects. In twenty-one bodies Wood (*Proc. Roy. Soc.* 1868, p. 511) found the supernumerary tendon attached to the trapezium twice as often as to the metacarpal bone; but in my own observations this only occurred in the proportion of four to five. Occasionally the second tendon had no bony attachment, but gave origin to all or part of the muscular fibres of the short abductor pollicis, or to the lower fibres of the opponens pollicis. This connection with the short muscles of the thumb was however of much more frequent occurrence when the tendon of the extensor ossis metacarpi was still further subdivided. Triple tendons were by no means uncommon. When these existed, two were usually inserted into the metacarpal, and one into the trapezium, or the third tendon joined the short abductor or the opponens pollicis. An insertion into the metacarpal, the trapezium, and the short muscles of the thumb, was less frequently noted. In two cases the third tendon took the place of the extensor primi internodii, and in two others it joined the tendon of this muscle. In one specimen the tendon was quadruple; the outermost division giving partial origin to the abductor, the second and third passing to the metacarpal, and the innermost fusing with the extensor primi internodii at its insertion (Fig. 1). As differentiation of this muscle is an evidence of elevation of type, it is interesting to notice that it had no attachment to the trapezium—the animal arrangement, but formed an additional extensor primi internodii—an exclusively human muscle. I have occasionally seen the muscle double throughout, and in one example it was distinctly triple. One portion was inserted into the trapezium, another into the metacarpal bone, and the third also into the metacarpal beneath the origin of the opponens pollicis, to which it was slightly attached.

2. *Extensor primi internodii pollicis.* The common variations in this muscle are caused either by deficiency, or by more or less fusion with the other long muscles of the thumb. Excess in its development is rare, but it occasionally gives slips to the first metacarpal base, or to the radial side of the ungual phalanx. Very rarely offsets are given to both. In the right fore-arm of a muscular male I found an additional thumb-muscle attached to the proximal phalanx. It arose from the ulna below the origin of the extensor secundi internodii pollicis, and passing through the *same groove* as that muscle, was inserted into the base of the first phalanx, just internal to the ordinary extensor. The extra muscle was quite as large as the normal one, which was also well developed (Fig. 2). I can find no record of a similar variation. As an extensor primi internodii is only found in Man, I carefully noted the other peculiarities which were present with this double extensor. They were (a) a triple tendon to the extensor ossis metacarpi, not attached to the trapezium, but to the metacarpal bone and to the abductor pollicis brevis; (β) cross-slips between the radial extensors; (γ) a double extensor minimi digiti; and (δ) a rudimentary extensor brevis digitorum manûs passing from the radius and posterior carpal ligaments to the ulnar side of the special indicator tendon. The co-existence of this common reptilian muscle, which only survives in a few anomalous mammals of the order Edentata, with an extra extensor primi internodii, seems to me especially noteworthy and very difficult of explanation.

3. *Extensor secundi internodii pollicis.* Duplicity of the muscular belly as well as of the tendon was often noticed. In two cases additional muscles were present. In the first of these, a fleshy fasciculus from, the radial side of the common extensor of the fingers, became tendinous on the back of the radius, and passed through the oblique groove to join the normal extensor secundi on the metacarpal bone[1]. In the second example, in addition to the normal muscle, another fleshy belly of nearly equal size arose from the ulna and the interosseous membrane between the extensor secundi and the extensor indicis. The tendon passed through the *common extensor* groove, and was inserted into the ungual phalanx of the thumb by a separate attachment internal to the usual muscle. This form must be very rare. An abortive example is mentioned in *Guy's Hospital Reports*, 1869, p. 441, and a perfect specimen with similar attachments, but running with the normal extensor secundi, is described by Bradley as having occurred on both sides (*Journal Anat. and Phys.* Vol. VI. p. 421).

4. *Extensor indicis.* I occasionally observed a double muscular belly, and splitting of the tendon was very frequent. In one subject the muscle was absent on the left side only, as has been described by Cheselden, Moser, and Macalister. The common extensor tendon to this finger was single, although those for the middle and ring fingers were split into two. In the right fore-arm of another subject a re-

[1] Compare W. Gruber, Ueber den Musculus extensor, u. s. w., *Reichert's Archiv*, 1875, No. 2, p. 204.

markable subdivision, previously unrecorded, was present. A muscle of the usual size and in the normal position of the extensor indicis divided into three tendons, of which the most external joined the extensor secundi internodii at its insertion, the middle occupied the place of the normal indicator, and the most internal formed a part of the extensor aponeurosis of the middle finger (Fig. 3). The common extensor tendon to the index finger was double. In the Hedgehog "a single muscle takes the place of the extensor secundi internodii pollicis and extensor indicis, and sends a third tendon to the middle digit" (Huxley, *Anat. of Vert. Animals*, p. 446). In the Kangaroo a single muscle also divides into three tendons for the first, second, and third digits, and even a more complete counterpart of the above anomaly is found in Manis, where not only does the representative of the extensor indicis arise from the ulna and give tendons to the terminal phalanges of the first, second, and third digits, but a slip of the extensor pollicis primus passes to the terminal phalanx of that digit, also forming an extensor secundi internodii pollicis (Humphry, *Journ. Anat. and Phys.* Vol. IV. pp. 48, 49). The other slips which were found in connection with the tendon of the extensor indicis were examples of the extensor brevis digitorum manûs.

5. *Extensor medii digiti.* The examples of this muscle were either differentiations of the preceding—in one case with a double tendon—or they were derived from an extensor brevis digitorum manûs.

6. *Extensor quarti digiti vel annularis.* The examples of this muscle were derived from the extensor of the fifth digit or from a short extensor of the fingers, except in one case where a distinct muscle existed, which I have described in this *Journal* (Vol. VII. p. 307[1]).

7. *Extensor minimi digiti.* Splitting of the tendon and occasionally of the muscle was frequently noticed, and in many cases the tendon divided for the little and ring fingers. In one subject there were three tendons to the fifth digit, all derived from this muscle, and in another four tendons were found, of which one only came from the common extensor.

8. *Extensor communis digitorum.* Varying degrees of complexity in the arrangement of this muscle were often met with. The simplest forms were extra tendons to the second, third, or fourth fingers, either singly or in combination. In addition, three rarer forms were observed, and with each of these numerous other peculiarities were present. In the first case, besides the normal tendons to the ring and little fingers, an additional tendon was attached to the vinculum between them, whilst on the tendon to the little finger a small fusiform muscle was developed. It had no attachment whatever to the radius, carpus, or metacarpus. In this fore-arm I found a double extensor minimi digiti, an extra abductor of the same finger, the quadruple extensor ossis metacarpi pollicis, a double insertion to

[1] Cuvier and Laurillard (Pl. 261) figure a long annularis, distinct from the extensor minimi digiti in the Tamandua.

the third lumbricalis, and a remarkable anterior prolongation of the tendon of the extensor carpi ulnaris (q. v.). In the second specimen the common extensor divided into six tendons, of which the first passed to the index finger, the second and third to the middle finger, both the fourth and fifth divided for the middle and ring fingers, whilst the sixth went to the little finger only, and there joined a double tendon from the special extensor, to which it was attached by a broad vinculum. There were thus four distinct tendinous slips on the dorsum of the middle digit. In this fore-arm there were also an extensor carpi radialis accessorius, anomalies of the radial extensors, the supinator longus, and the extensor ossis metacarpi pollicis. In a third case there were eight tendons from the common extensor muscle. Besides the normal tendons, additional ones passed to the middle and ring fingers, and to a vinculum between them but acting on both, as well as to a vinculum between the index and middle fingers, but acting on the latter only. A double extensor minimi digiti, an ulnaris quinti, and a single special indicator, made twelve tendons to the inner four digits, and with the exception of the radial tendon to the index finger they were intimately bound together by vincula. The extensor primi internodii pollicis was single, but the extensor ossis metacarpi and the extensor secundi internodii had double tendons, making altogether seventeen extensor tendons to the digits. This is the most complete differentiation of these tendons that I have met with. Rüdinger is quoted by Macalister as having found eleven tendons to the inner fingers, but the muscles of the thumb are not mentioned. Perrin (*Med. Times and Gazette*, 1873, p. 597) saw eleven tendons to these digits and five to the thumb, whilst in one subject, No. 7, Wood (*Proc. Roy. Soc.*, 1868) describes twelve tendons to the fingers and six to the thumb. The other peculiarities in this arm were a slip from the flexor carpi ulnaris to the fascia, a complex abductor pollicis, and a fusion of the deltoid with the pectoralis major. On the opposite side some slight anomalies were met with, but they did not affect the common extensor.

9. *Extensor brevis digitorum manus.* Several examples of this interesting muscle occurred, but they were all familiar varieties, taking origin from the radius, carpus, or metacarpus, and joining the aponeurosis of the index, middle, or ring fingers.

10. *Extensor carpi radialis longior.* A very large number of variations of this muscle were noted. In many cases the tendon was simply cleft into two, which were inserted side by side into the base of the index-metacarpal. In a very few examples perfect cleavage of the muscle as well as of the tendon was found. In one case the insertion was triple, viz. into both sides of the base of the index-metacarpal, and into the *ulnar* side of the middle metacarpal, and this tendon passed under a normal short radial extensor. In three cases fusion of the bellies of the radial extensors occurred, giving rise in two instances to the ordinary long and short tendons, whilst in the third both tendons were bifurcated at their insertions.

11. *Extensor carpi radialis brevior.* Besides the fusion above mentioned, examples of a double insertion, either side by side into

the middle metacarpal, or into it and the ulnar side of the base of the index-metacarpal, were frequently noticed. Macalister has recorded the absence of this muscle, and quotes Salzmann as having mentioned a similar example, but the latter evidently refers to the *flexor* carpi radialis ("radiæum *internum* omnino defuisse"). I have never seen less than two radial extensor tendons of the carpus, however intimately joined the muscular bellies may have been.

12. *Extensor carpi radialis intermedius* (Wood). The simplest and most common form of this anomaly was by means of a slip— generally tendinous, but occasionally with a few fleshy fibres—from the extensor carpi radialis longior to the brevior or to the radial tendon of a cleft brevior. In other cases cross-slips existed between the tendons, whilst in one instance these slips instead of joining the tendons were inserted into the metacarpal bones, so that the index-metacarpal received the tendon of the longior and a slip from the brevior, and the middle metacarpal the tendon of the brevior and a slip from the longior. In one fore-arm a typical third or intermediate extensor was formed by the fusion of two large muscular bellies derived from the normal extensors, as described by me in this *Journal*, Vol. VII. p. 306. In another case the intermediate extensor arose from the humerus between the supinator longus and the normal long radial extensor, and crossed the latter muscle to be inserted into the middle metacarpal on the radial side of the tendon of the short extensor. The converse arrangement was also seen. A large muscle took origin from the external condyle between the extensor carpi radialis brevior and the extensor communis, and crossed the former to gain the ulnar side of the index-metacarpal. A still more interesting case of complexity occurred in the left arm of a male subject. The radial extensors were fused into one muscle, giving rise to four tendons, of which two corresponded to the normal long and short tendons, the third was attached to the index-metacarpal on its ulnar side, and the fourth, arising highest and most external, crossed beneath the other tendons to join the deep surface of the internal or short tendon in the lower part of the fore-arm.

13. *Extensor carpi radialis accessorius* (Wood). The small tendon passing from the radial side of the extensor carpi radialis longior to the pollex-metacarpal was noticed in three subjects. In one case a small fusiform muscle on the ulnar side of the extensor carpi radialis longior was succeeded by a long tendon which crossed that muscle and joined the extensor ossis metacarpi pollicis about an inch above the lower end of the radius. This is evidently an abortive variety of the same muscle, and in the opposite fore-arm (female) a more complete specimen was found. It arose in common with, but internal to, the extensor carpi radialis longior, and crossed this muscle and the extensor ossis metacarpi superficially to the first metacarpal base, into which it was inserted by a bifid tendon (Fig. 4), through which the extensor primi internodii passed. An ordinary intermediate extensor and a bifurcation of the tendon of the extensor carpi radialis longior co-existed. Another most irregular arrangement of radial extensors of the carpus occurred in the right fore-arm

of a male subject. A large well-developed muscle arising from the external supra-condyloid ridge as well as from the condyle divided into three primary tendons, of which one corresponded to the longior, and another to the brevior, whilst the central or intermediate one was inserted into the ulnar side of the base of the index-metacarpal. From these tendons secondary slips were given off, viz. a cross slip between the longior and brevior, a tendon from the brevior to the index-metacarpal beneath the former, and a third from the longior to the pollex-metacarpal. This complex radial extensor was attached therefore by a single tendon to the first and third metacarpals, and by three tendons to the second metacarpal, and so comprised an extensor carpi radialis accessorius, two extensor carpi radiales intermedii, and the normal radial extensors (Fig. 5).

14. *Extensor carpi ulnaris.* The most numerous examples of variation in this muscle arose either from a bifurcated tendon, or from the presence of the forward prolongation termed the ulnaris quinti. This usually joined the extensor aponeurosis of the little finger, but occasionally it only reached the base of the first phalanx. In two cases prolongations to the *front* surface of the little finger were met with. In one, the tendinous slip was joined by a band from the unciform process and passed in front of the opponens minimi digiti to the base of the first phalanx, where it was split for the passage of the flexor brevis, and was then inserted into the sides of the phalanx (Fig. 6). In the other example, the anterior slip had a single insertion only into the base of the first phalanx. In one instance the ulnaris quinti spread out into the fascia over the fourth dorsal interosseous muscle, and in another a most curious anomaly existed. About $2\frac{1}{2}$ inches above the wrist a tendon separated from the ulnar side of the extensor carpi ulnaris, and on it a small fusiform muscle was developed, which again gave place to a tendon partly inserted into the pisiform bone by the side of the normal extensor, but mainly into the unciform process under the origin of an extra abductor minimi digiti. It very forcibly reminded one of the anomalous extensor ossis metacarpi pollicis, which is occasionally seen. The only double extensor carpi ulnaris (muscle and tendon) which I have seen is recorded in Vol. VII. of this *Journal.*

15. *Extensor pollicis et indicis.* I have not met with any examples of this muscle, which is so common among the carnivora.

EXPLANATION OF PLATE.

Fig. 1. 1. Abductor pollicis brevis detached from origin to bones of carpus. 2. Extensor ossis metacarpi pollicis with quadruple tendon. 3. Extensor primi internodii pollicis.

Fig. 2. 1. Extensor primi internodii pollicis. 2. Extensor secundi internodii pollicis. 3. Additional extensor primi internodii pollicis. 4. Extensor indicis. 5. Extensor brevis digitorum manûs joining tendon of extensor indicis.

Fig. 3. 1. Extensor secundi internodii pollicis.   2. Abnormal extensor indicis.   3. Double tendon to index from extensor communis digitorum.   4. Single tendon to middle digit from extensor communis digitorum.

Fig. 4. 1. Extensor carpi radialis longior.   1 a. Extensor carpi radialis brevior.   1 β. Extensor carpi radialis accessorius.   1 γ. Extensor carpi radialis intermedius.   2. Extensor ossis metacarpi pollicis (triple tendon).   3. Extensor primi internodii pollicis.   4. Extensor secundi internodii pollicis.

Fig. 5. 1, 1 a, 1 β, and 1 γ, as in *Fig.* 4.

Fig. 6. 1. Extensor carpi ulnaris.   2. Ulnaris quinti.   3. Anterior slip from 1.   4. Flexor brevis minimi digiti.   5. Abductor minimi digiti—cut.   6. Flexor carpi ulnaris.

A SIMPLE METHOD OF DEMONSTRATING THE EFFECT OF HEAT AND POISONS UPON THE HEART OF THE FROG.

By T. Lauder Brunton, M.D., Sc. D., F.R.S., *Assistant Physician and Lecturer on Materia Medica at St Bartholomew's Hospital.*

The fact that heat accelerates and cold retards the pulsations of the heart, is one of such fundamental importance, both in regard to a right understanding of the quick pulse, which is one of the most prominent symptoms of fever, and to a correct knowledge of the proper treatment to apply when the heart's action is failing, that for the last year or two I have been accustomed to demonstrate it as a lecture experiment. The apparatus I use is exceedingly simple, but it answers its purpose well, and by its means the pulsations of the frog's heart can be readily shown to several hundred persons at once. I exhibited it at the meeting of the British Medical Association in London more than two years ago, and a description of it appeared in the *British Medical Journal* for August 23, 1873; but as I have reason to believe that few physiologists have seen either the instrument or its description, it may not be amiss to say a few words regarding it here. It consists of a piece of tin plate or glass three or four inches long and two or three wide, at one end of which an ordinary cork cut square is fastened with sealing-wax in such a manner that it projects half an inch or more beyond the edge of the plate. This serves as a support to a little wooden lever about three inches long, a quarter of an inch broad, and one-eighth of an inch thick. A pin is passed through a hole in the centre of this lever, and runs into the cork so that the lever swings freely about upon it as on a pivot. The easiest way of making a hole of the proper size, is simply to heat the pin red hot, and then to burn a hole in the lever with it. To prevent the lever from sliding along the pin, a minute piece of cardboard is put at each side of it, and oiled to prevent friction. A long fine bonnet straw or section of one is then fastened by sealing-wax to one end of the lever, and to the other end of the straw a round piece of white paper cut to the size of a shilling or half-crown, according to convenience, is also fixed by a drop of sealing-wax. The pin, which acts as a pivot, should be just sufficiently beyond the edge of the plate to allow the lever to move freely, and the lever itself should lie flat upon the plate. Its weight too, increased as it is by the straw and paper flag, would now be too great for the heart to lift, and so it must be counterpoised. This is readily done by clasping a pair of bull-dog forceps on the other end. By altering the position of the forceps the weight of the lever can be regulated with great nicety. If the forceps are drawn back as at *c*, Fig. 1, the flag is more than counterbalanced, and does not rest on the heart at all, while the position *a* brings the centre of gravity of the forceps in front of the pivot, and increases the pressure of the lever on the heart. The isolated frog's heart is laid under the lever near the pivot, and as it beats the

lever oscillates upwards and downwards. If the tin plate be now laid on some pounded ice the pulsations will become slower and slower, and if the room be not too warm the heart may stand completely still in diastole. On removing the plate from the ice the pulsations of the heart become quicker. If a spirit-lamp be now held at some distance below it the heart beats quicker and quicker as the heat increases, until at last it stands still in heat tetanus. On again cooling it by the ice its pulsations recommence. At first they are quick, but they gradually become slower and slower. On again applying the spirit-lamp they become quicker, and by raising the temperature sufficiently the heat tetanus is converted into heat rigor. Then no application of cold has the slightest effect in restoring pulsation.

Not only the effects of heat and cold, but the effect of separating the venous sinus or the auricles from the ventricle can readily be shown with this apparatus, as well as the action of various poisons. The best for the purpose of class demonstration is muscaria. A drop of saline solution containing a little of the alkaloid being placed on the heart, it ceases to beat entirely. If a drop of atropia solution be now added the beats recommence. I have seen them do so on one occasion after they had entirely ceased for four hours. When used for demonstrating the action of poisons the wooden lever should be covered with sealing-wax, so as to allow every particle of the poison to be washed off it, and thus prevent any portion from being left behind and interfering with a future experiment. By attaching a small point to the end of the straw in place of the paper flag, tracings may be taken upon smoked paper fixed on a revolving cylinder.

NOTE ON PFLÜGER'S LAW OF CONTRACTION. By GEORGE A. BERRY and Prof. RUTHERFORD.

THE object of the present communication is to point out some of the conditions which appear to influence the phenomena of electrotonus other than the mere strength and direction of the electrotonising current, for we believe that the recognition of these is important, not merely because of the additional light which they throw upon the electrotonic state, but also on account of the assistance which the experimenter may derive from them in explaining the anomalies and failures which not unfrequently attend a demonstration of Pflüger's law.

The following mode of demonstrating this law is a good one, and as it led us to perform other experiments it is here detailed. Two nonpolarisable electrodes are attached to the usual arrangements employed in showing Pflüger's law. The sciatic nerves of two frog's legs are laid across both electrodes in such a manner (Fig. 1), that the current ascends the one while it descends the other, and thus the limbs simultaneously indicate the effect of the ascending and descending current.

Fig. 1.

Double method of showing Pflüger's law of contraction. A, B. Frog's muscles with their nerves (the gastrocnemii only are shown, but in the experiment the whole leg is used). C, D. Clay electrodes.

On transmitting a current from four small Grove's cells the results given by Pflüger as indicating those of a "strong" current are obtained, viz.:

Muscle A.	*Muscle* B.
Close = Rest.	Close = Contraction.
Open = Contraction.	Open = Rest.

A less powerful current gives the effect of Pflüger's "medium" current, viz.:

Muscle A.	*Muscle* B.
Close = Contraction.	Close = Contraction.
Open = Contraction.	Open = Contraction.

A weaker current gives that of Pflüger's "weak" current, viz.:

Muscle A.	*Muscle* B.
Close = Contraction.	Close = Contraction.
Open = Rest.	Open = Rest.

It is known that the probable explanation of these results is to be found: (1) In the fact that at closure the nerve is excited only in the region of the negative pole, while the excitement takes place only at the positive pole when the circuit is opened (Pflüger). (2) In the relation of the poles to the terminal organ affected by the excitement. (3) In the strength of the current. The first of these facts, known as Pfluger's law of electrical stimulation, has been demonstrated by Engelmann's experiment on muscle. (See Hermann's *Physiology*, 1875, p. 330.)

We were curious to see what might be the effect of dividing the nerves where they lay upon the electrodes, and then stimulating them with the currents that had yielded the above results. The position of the nerves after section is shown in Fig. 2.

Fig. 2.

Nerves divided on electrodes. Letters as in Fig. 1.

Notwithstanding this arrangement, Pflüger's results were obtained exactly as before. But the following modification of the experiment brought out a very interesting fact. The portions of the nerves in connection with the muscles were drawn away from each electrode until only about a millimetre of their cut ends rested on the clay close to the detached portions of nerve uniting the electrodes. It was then found that the current which had previously given the effects of Pflüger's "medium" current, now gave those of the "strong" current. On applying a longer piece of nerve to each electrode, the effects were again those of a medium current. Similar results were obtained with metallic electrodes.

The explanation of this effect of the *length* of the piece of nerve traversed by the current was not evident until a considerable number of experiments were performed, into the details of which, however, it is unnecessary to enter, as the following gave the clue to it.

The clay points of Du Bois Reymond's nonpolarisable electrodes were welded together so as to form a continuous roller, and it was found that by simply laying the nerve along this roller it was easy to obtain all the results of Pflüger's law on transmitting a current through the clay. The point where the electricity enters the nerve being the positive pole, while its point of exit is the negative pole.

40—2

The striking fact, however, is that without altering the strength of the current all the phenomena of Pflüger's law could be obtained by transmitting it through a central, middle, or peripheral portion of nerve, at one time in an ascending—at another in a descending direction.   This will be understood by referring to Figs. 3, 4 and 5.

Six small Grove's cells, connected in the usual way to Du Bois Reymond's rheocord.   Plugs of rheocord 1. 1. 2. 5. out.   Mercury bridge at 32.   Sciatic nerve of frog (*n*) laid on a continuous clay electrode (*c*).

Fig. 3.

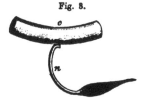

Current Ascending.	Current Descending.
Close = Rest.	Close = Contraction.
Open = Contraction.	Open = Rest.

Fig. 4.

Current Ascending.	Current Descending.
Close = Contraction.	Close = Contraction.
Open = Contraction.	Open = Contraction.

Fig. 5.

Current Ascending.	Current Descending.
Close = Contraction.	Close = Contraction.
Open = Rest.	Open = Rest.

It is evident that the results of stimulating the nerves as arranged in Figs. 3, 4 and 5 are respectively those of Pflüger's "strong," "medium" and "weak" currents.

Seeing that the strength of the current had not been changed, the different effects appear to result from the greater excitability of the central as compared with the peripheral part of the nerve, so that the same current has a "strong" effect on the former, and a "weak" effect on the latter.

From these experiments it is clear that the *degree of excitability* of the nerve traversed by a galvanic stream has a marked influence on the electrotonic state. In addition to this, however, the length of nerve traversed by the electricity is also of importance. One might naturally suppose that by adding a portion of nerve that is highly excitable—and which with a current of given strength exhibits the phenomena of Pflüger's "strong" current—to a portion much less excitable, and giving with the same current the effects of Pflüger's "weak" current, in other words, of laying a piece of nerve stretching from the central to the peripheral end on the roller, the effect would be that of a "medium" current: and further, it would be reasonable to imagine that a portion of nerve having an excitability low enough to give the effect of a "weak" current would, when added to another portion having a still lower excitability, give still weaker effects or no effects at all. That the latter is not the case, however, is seen from the following experiments.

A current from two small Smee's cells was transmitted through 2 mm. in length of the sciatic nerve close to its central end (*a* Fig. 6). The effects were those of a "strong" current. On applying 3 mm. to the roller the effects became those of the "medium," and more definitely so when as much as 15 mm. were applied (*a'*).

The same current was then transmitted through 4 mm. of the nerve about its middle (*b*). The effects were those of a "weak" current, but on applying a few millimetres more of the peripheral part of the nerve the effects of the "medium" current were obtained (*b'*).

Fig. 6.

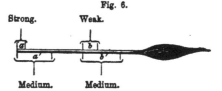

The question now arises, why should the addition of a peripheral portion in the one case (*a – a'*) diminish, while in the other (*b – b'*) it increases the effect? The following appears to be the most reasonable explanation.

When only a small portion of nerve is traversed by a current of moderate intensity the influence of the excitability at that part is more powerful than the specific influence of the current itself, if it may be so called. The effects vary with the degree of excitability of the part stimulated; being "weak" when it is low, "strong" when it is high; but if the electricity be allowed to traverse a longer portion its influence overpowers that of the local excitability, and the effects from such a current as that used by us become invariably medium.

From these experiments it therefore appears that in studying the phenomena that attend the establishment and resolution of the elec-trotonic state it is necessary to take into account not only (1) the strength of the current, and (2) relation of the poles to the terminal organ of the nerve, as has hitherto been done, but also (3) the *ex-citability*, and (4) the *length* of the portion of nerve traversed by the voltaic stream.

NOTE ON THE ACTION OF THE INTERNAL INTER-COSTAL MUSCLES. By PROFESSOR RUTHERFORD.

HAVING at one time, in common with others, fallen into the error of regarding Hamberger's well-known model of parallel bars and elastic bands as illustrative of the action of the intercostal muscles, I believe that this note may be useful to others who have been similarly misled. Everyone who has used the model referred to knows that when the elastic bands are placed in the position of the internal intercostal fibres the wooden ribs and sternum descend; this movement being the only one that permits of a short-ening of the band. Why such a model was ever devised it would be difficult to imagine, for, as Prof. Humphry[1] and others have pointed out, it does not really represent the ribs, these, unlike the bars, being arcs which can rotate around an axis directed from the posterior ends of the ribs to the costal cartilages. It appeared to me that an application of elastic bands to the *ribs themselves* would obviously be the readiest mode of demonstrating their effect when arranged after the manner of the intercostal muscles. Nails were accordingly driven into the ribs and elastic bands placed in the position of fibres of the external intercostal muscles between the four upper ribs. The extent of the resultant costal elevation is shown in Fig. 1.

The bands were then placed between the osseous parts of the same ribs in the position of internal intercostal fibres carefully ascer-tained by reference to a dissection of the chest. The consequent elevation of the second, third, and fourth ribs was evident enough (Fig. 2).

Owing to the circumstance that the second, third, and fourth ribs are not connected one with another, they can, by such bands, be raised individually. When, however, a band is placed between the fourth and fifth ribs there is a simultaneous elevation of the fifth and four succeeding ribs owing to the junction of their cartilages one with another. Fig. 3 shows elastic bands arranged in the position of internal intercostal muscles between the osseous parts of the seven

[1] *Brit. Med. Journ.* 1872. June 29.

Fig. 1.

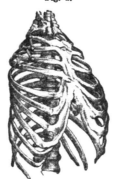

Human chest.   Elastic bands placed in position of external intercostal muscles.
(*From a photograph.*)

Fig. 2.

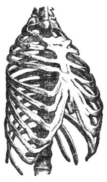

Elevation of ribs by elastic bands placed in the position of internal intercostal
muscles.   (*From a photograph.*)

upper ribs.   The costal elevation and consequent lateral expansion is
very evident when compared with the other side of the chest.

These experiments entirely put an end to the statement founded
on Hamberger's model, that the interosseous parts of the internal
intercostal muscles *must* depress the ribs, because that is the only
costal position which permits of a shortening of these fibres.

Different results are obtained if the elastic bands be placed *only*
between, say the second and third, or between the third and fourth
ribs.   The upper rib is then drawn down while the lower is drawn
up: both when the bands are arranged after the manner of the

Fig. 3.

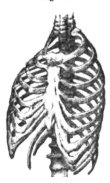

Elévation of ribs and lateral expansion of right side of chest by elastic bands in the position of internal intercostal muscles between osseous parts of ribs. (*From a photograph.*)

internal and of the external intercostals; and if the lower rib of the pair be held immobile the only movement is a drawing down of the upper one. Both intercostals are therefore able to elevate the ribs when the more fixed point is above, and to depress them when the more fixed point is below. The upper rib is more fixed than the others, for it, unlike the others, has no joint between its cartilage and the sternum; it is also less flexible than the lower ribs, and, moreover, it being an arc of a smaller circle than the others is less affected by forces tending to approximate it to the other costal arcs.

Elastic bands were placed in the position of external intercostal muscles between the upper eight ribs on *both* sides of the chest. Great lateral expansion of the chest resulted. The upper end of the sternum was elevated only to the extent of a quarter of an inch. The ensiform cartilage was lifted forwards an inch and three quarters. The elastic bands were then removed, and placed in the position of interosseous parts of internal intercostal fibres between the same ribs. There was neither elevation nor depression of the upper end of the sternum, while the ensiform cartilage was lifted forwards to the extent of three quarters of an inch. How far this may indicate the relative powers of the external and internal intercostal muscles it is impossible to say, for the effect of such bands depends not only on their position but on the degree to which they are put on the stretch, and it does not seem possible in such experiments to arrive at uniformity in this respect.

The object of the foregoing is merely to show the manner in which a practical demonstration of what the internal intercostals *can* do may be readily obtained. Experiment on the living animal has not yet removed the doubt regarding their mode and time of action during life.

THE OXIDATION OF UREA. By JAMES REOCH, M.A., M.B., Professor of Physiology in the University of Durham College of Medicine, Newcastle-on-Tyne.

IN a paper on the Decomposition of Urea in the last number of the *Journal of Anatomy*, the general phenomena attending the decomposition of urea both in the urine and in pure solutions were fully described, but some of the most important points were left somewhat doubtful, because sufficient experiments had not been made at the time of publication. The chief of these questions was whether ozone or nascent oxygen was produced during the fermentation of urea by the fungus, for in some of the experiments referred to in the last paper undoubted evidence of it was found, but in others about equally numerous it was not found. The first thing of course is to select a proper mould, for the fungus being a plant is subject as to its vitality to the general laws of plant-life. In a large pot, into which the drainings of coffee cups had been thrown, and which therefore presumably contained nothing beyond coffee-infusion, milk, and sugar, a large green mould had thickly covered the surface; this was removed, slightly washed, and bottled; after some time it was taken out and thoroughly washed by shaking and decantation, and torn up into shreds of a few milligrammes each. A standard solution of urea was prepared of the strength 2 grms. in 100 cc. of water, and a solution of KI of similar strength, also a solution of dilute H_2SO_4 equal to 2 per cent of soda, or where 100 cc. neutralized a solution containing 2 grms. NaHO. It was then ascertained by repeated experiment that acid of this strength would not liberate iodine from KI so as to strike a blue with starch; no doubt it begins to do so even in the cold in about an hour, and after twenty-four hours it liberates more, but it never gives the blue at once, so that its addition makes no difference in this respect; it is important to state this, because urea turning into carbonate of ammonia, and iodide of starch being very soluble in alkali, it is necessary to add dilute H_2SO_4 as well as starch in testing for free iodine. There need be no mistake, however, as if free iodine be present a deep blue is at once produced, whereas dilute H_2SO_4 would only produce a faint tint after some hours. Now in two experiments some of the KI solution and some starch dissolved in boiling water were added to an ounce or so of the urea solution, and shreds of the fungus placed in each; after two days they both gave blue with ½ cc. dilute H_2SO_4, while the original solutions mixed without any fungus of course gave no reaction whatever; after two days more one gave a very deep blue, but the other gave nothing, and even strong HCl only gave a yellow; the next day the same thing occurred; gradually, however, in two or three days more, the blue colour ceased to be produced in the first as it had ceased in the other, and even strong HCl only gave a yellow indicating free iodine, but to get the deep blue of iodine of starch fresh starch solution required to be added. But starch solution had originally been added to both of them, therefore it must have been decomposed, and as they both gave free iodine at first, which speedily disappeared and did not come out even when fresh starch was added, the inference was plain that the starch was decomposed by the fungus just as the urea was, and that it was transformed into something which destroyed the iodine of starch reaction. To examine the whole question more thoroughly twenty-one test-tubes were divided into seven series of three each and the following additions made :

1st 3. 5 cc. urea sol. + 1 cc. KI sol. + fungus.

2nd 3. 5 cc. urea sol. + 1 cc. KI sol. + fungus + 4 drops starch sol.

3rd 3. 5 cc. urea sol. + 1 cc. KI sol. + fungus + 4 drops starch sol. + 0·1, 0·2, and 0·55 cc. dil. H_2SO_4.

4th 3. 5 cc. urea sol. + drops starch sol. + fungus.

5th 3. 5 cc. urea sol. + fungus.

6th 3. 5 cc. urea sol. + 1 cc. KI sol. + 4 drops starch sol. + 0·1, 0·2, 0·55 cc. dil. H_2SO_4.

7th 3. 5 cc. H_2O + 1 cc. KI sol. + fungus + 4 drops starch sol.

After an hour or two each of the third and sixth series assumed a faint purple tinge, but even after the lapse of two days, though decidedly purple, they were transparent and not properly blue at all. Two days after these solutions were set aside, one of each was treated with ½ cc. dil. H_2SO_4 and a few drops of starch solution, a deep blue was obtained with the one of the first series and a similar though not so deep blue from that of the second series, but none of the others gave any reaction whatever, though of course KI was added to the fourth and fifth as well as starch and dil. H_2SO_4. The next day the same experiment was repeated with the second of each series, and the day after with the third, and in each case with the same positive and gradually increasing result with the first series, and entirely negative with the last five; but as to the second series the second bottle gave no result at first, though very quickly it appeared, while the third bottle gave no result at all even after a considerable time. From these experiments the same result follows as from the first two, there being no result without the urea indicates that no ozone is produced by the fungus without the urea to work upon, while there being no evidence of any unless KI was present the whole time shows that the ozone is not produced all at once, but only gradually, and is therefore absorbed by the KI as it is produced, while again the constant result with urea and KI, but a different result where starch was present all the time (for though the blue was produced in the first bottle it diminished till none was produced in the third) shows that the starch was decomposed, and in this process caused the iodide of starch also to disappear. The want of obtaining the blue where the dil. H_2SO_4 was present shows that the latter hinders the growth of the plant. Another series of experiments of the same nature gave similar results. Twelve test-tubes were taken, and the following additions made :

1st 4. 5 cc. urea sol. + 1 cc. KI sol. + fungus.
2nd 4. 5 cc. H$_2$O + 1 cc. KI sol. + fungus.
3rd 4. 5 cc. urea sol. + 1 cc. KI sol.

The fungus in this case had been allowed to dry in the air, and its vitality was probably diminished thereby, for next day one of each gave no result with starch and ½ cc. dil. H$_2$SO$_4$, and even the day after there was no result, but on the third day, ½ cc. dil. H$_2$SO$_4$ and a few drops of starch being added to each of the twelve test-tubes, there was a reaction in each of the first four, though considerably deeper in two of them than in the other two, indeed, each of the four gave a blue of different depth, showing that from difference of size or otherwise the fungus had not acted with the same vigour in each ; one of the second series of four test-tubes gave a blue at once, and another in a short time, but the other two gave nothing. These are the only cases in which any approach to the decomposition of KI by the fungus alone was made, and if they be admitted, though in many more cases no reaction was obtained, it will only prove that the fungus can produce ozone by the little nutriment it can get from dust floating in the atmosphere, and this would not be surprising, since moulds are found to grow and flourish in the most unlikely situations. None of the third series of four test-tubes gave any result whatever. Various other experiments were made which tended to the same conclusion. But when another fungus was used a similar and yet slightly different result was obtained. From a tall uncovered jar which contained wine about one year old a piece of mould matted together with dirt and other matter was obtained. This was well washed, and shreds of it sown in sixteen test-tubes after the following manner :

1st 4. 5 cc. urea sol.
2nd 4. 5 cc. urea sol. + ½ cc., 1 cc., 1½ cc., 2 cc. KI sol. to each respectively.
3rd 4. 5 cc. urea sol. + 2 drops starch sol. to each.
4th 4. 5 cc. H$_2$O + ½ cc., 1 cc., 1½ cc., 2 cc. KI sol. to each respectively.

No trace of free iodine was obtained from a sample of the most of these a week after, but on the tenth day ½ cc. H$_2$SO$_4$ was added to each without producing blue in any, but on adding a few drops of starch all round the 10th, 11th, and 12th came out very deep blue, and the 6th and perhaps the 9th also showed a slight tint, and the 15th a very faint blue, but none of the others gave anything. Here we have four out of eight tubes containing urea and KI as well as the fungus giving the blue iodide, but none of those containing urea alone, while only one of those containing merely KI and H$_2$O gave any reaction, and that only a trace. It may be asked, why did not all the test-tubes containing urea and KI give a reaction, but on adding a little tincture of litmus it was found that the $\frac{1}{2}$ cc. H_2SO_4 dil. had rendered them strongly acid, they therefore contained little or no carbonate of ammonia; the explanation of want of action therefore lies with the fungus, and there can be little doubt that since last winter was so severe that many of the wines froze in their jars, and that the fungus was lying for a whole year exposed to all the influences which kill or check plant-growth, that its vitality mnst have been very low, and probably much of it was dead. These facts seem to teach that the reaction must not be expected unless the fungus be in a state of perfect vitality; but if this be ascertained the results above described form one of the most beautiful demonstrations which can be shown to a class. Let a series of twelve test-tubes be taken with varying solutions, such as above described, and the fungus being sown in each, then after two or more days let some starch solution and dilute H_2SO_4 be added to each; a deep opaque blue will result at once in those test-tubes in which urea solution and KI solution alone exist. The question what becomes of the starch in the above cases where it is placed in the combined urea and KI solutions and acted on by the fungus, is a very difficult one. Starch can undergo several different fermentations, each with a different product. It does not appear to be changed in this case into glucose, for several experiments where C_4SO_4 and liq. potass. were added to the fluid in which the starch had decomposed gave no trace whatever of suboxide of copper; but whatever may be the product the fact of its decomposition is certain, and it ought therefore never to be added till the last. It being therefore certain that ozone is produced during the decomposition of urea, the question of whether the fungus gains in weight during the process loses much of its importance. The facts described in the last paper all go to negative the idea that urea becomes carbonate of ammonia by assuming H_2O, but the production of nascent oxygen is a crucial point; if that be admitted there can be little doubt that the theory of the continuous oxidation of urea propounded in the last paper is correct. How far this ozone theory may explain other fermentations is another question, and one which in the absence of positive experiments it would be unsafe to speculate on; but doubtless the fungi do not confine their operations to urea. All moulds do not seem to be alike powerful, but the ordinary green mould which so commonly appears on so many domestic articles seems to be the most powerful. In some experiments with the fungus from an ink-bottle a trace of ozone was obtained, but it was perfectly white, and the microscope showed a mass of mycelium matted together, many indeed containing spores, but few of the latter free; yet the spores seem to be the active agent, and probably the ozone is produced by the action of the green colouring matter which they so largely contain.

AN IMPROVED FREEZING MICROTOME. By Richard Hughes, *Resident Medical Officer to the Manchester Hospital for Sick Children.*

The difficulties attendant upon the use of a mixture of ice and salt as a means of freezing tissues for microscopical purposes, have induced me to arrange the instrument about to be described, and of which the accompanying diagram represents a section. It consists of a rectangular brass box closed at the top by a plate of the same material, forming the stage. Near to one end, the box is traversed from top to bottom by a cylinder, provided with a screw-piston and milled head, the arrangement of its working parts being similar to those of Professor Rutherford's microtome. In the end opposite the cylinder is an opening through which ether is introduced when the

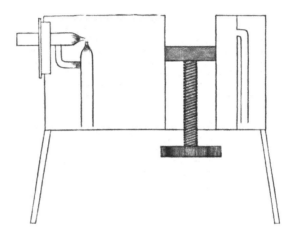

apparatus is used, and which is closed by a perforated cork plug carrying an ordinary glass atomiser. An exit is provided for the ether vapour through a bent tube opening behind the cylinder, and leading into a narrow chamber, in which the razor blade is plunged previous to making a section. The atomiser is worked by a double elastic ball, and the body of the instrument is covered by a non-conducting jacket of flannel and leather: a piece of spongio-piline cut to the shape serves to protect the stage from external warmth during the freezing. In using the instrument, the cork plug is removed, and the box is about half filled with Richardson's anæsthetic ether, care being taken not to incline the box too much, or the ether will run out through the exit tube.   The plug is then replaced, the razor blade is dipped into the narrow secondary chamber, and the elastic ball is worked.   In from two and a half to four minutes a piece of tissue placed in the cylinder will be frozen sufficiently to allow sections to be cut.   The ether used may be "Richardson's anæsthetic ether," made by Robbins of Oxford Street, which costs seven shillings a pint, and which produces a very intense cold, or the cheaper methylated ether may be substituted.   To avoid waste of ether I have arranged a larger instrument, provided with a force-pump and condensing chamber; but the cost and the labour of working are much greater than in the simple apparatus above described, and the increased trouble is not compensated for by the gain of ether.   The apparatus is made by Messrs J. and W. Wood, King Street, Manchester, and is sold for twenty-five shillings.

NOTE ON AN EXAMPLE OF MECKEL'S DIVERTICULUM.

By SIDNEY COUPLAND, M.D, *Lond., Curator of the Middlesex Hospital Museum.*

Diverticula ilei are such common abnormalities, and Meckel's view as to their being fœtal relics—viz. examples of persistency of a portion of the vitelline duct—is so generally adopted, that it may hardly be thought worth while to record any further instances of such structures. But having recently met with an example which presented a curious difference from those I have previously seen, and which is much less frequently met with, I thought a brief account of it might be of some slight value.

When J. F. Meckel[1] first threw light upon the true nature of these structures, he met with considerable opposition, and in replying to his critics he adduced, among other arguments, the fact that, in some cases, the abnormal pouch was adherent to the abdominal wall at the umbilicus, and even had remained open there in cases of congenital non-closure of the abdominal parietes. The case I have to describe is not one of such convincing character as these, but from its position it presents some approach to them.

In making the abdominal incision in the body of a robust male (æt. 41) who died in the Middlesex Hospital of pneumonia, in October, 1875, I was struck by what appeared to be a portion of intestine in the middle line of the body immediately underneath the parietes and below the xiphoid cartilage. On further examination the exact nature of this was ascertained, and also the following disposition of the viscera in the abdomen. The omentum, well laden with fat, concealed the intestinal coils in the left half of the cavity, but in the right half a loop of the ileum passed over the surface of the omentum, and ran up towards the upper part of the abdomen, the two limbs of the loop being lost among the coils of the ileum below. The fold of mesentery seemed lengthened to admit of the loop being formed. From the convexity of the sharp flexure in the centre of the loop, and consequently from the free surface of the intestine forming it, proceeded the structure which had first attracted my attention. This was a diverticulum of the bowel, containing only flatus, 4½ inches in length, freely moveable, and reaching as high as the ensiform cartilage. It thus extended upon the surface of the left lobe of the liver in juxtaposition to the falciform ligament. The liver was slightly grooved by it. Its diameter was about that of the fore-finger, but it could be distended till it nearly equalled in calibre the gut from which it sprang. It was slightly larger and

---

[1] *Beiträge zur vergleichenden Anatomie,* 1808. See, for full discussion of the question, *Manuel d'Anatomie,* par J. F. Meckel, trad. par Jourdain et Brachet, 1825, Vol. III. p. 481.

bulbous at its free extremity, which was further sub-bifid, if I may use the expression : I mean that it was surmounted by a narrow band of fibrous tissue and fat, about 2½ inches in length, on either side of which it projected. At its origin a thin fold of peritoneum passed from it to the surface of the bowel, forming a rudimentary mesentery to the diverticulum. Further examination showed that the appendage arose from the ileum, exactly three feet from the ileo-cœcal valve, a position precisely corresponding with other cases I have seen in the post-mortem room.

The interesting fact in this case lies in the position of the diverticulum upon the upper surface of the liver. It must have long been there, and probably from its close relation to the falciform ligament and the presence of the subperitoneal fat and fibrous tissue at its free extremity, it had once been attached to the umbilical cord or to the abdominal parietes at the umbilicus. In this way only, I think, can the curious looping up of the ileum be explained, for it is almost inconceivable that if the diverticulum had been always as free from attachment as it was when the subject died, it should not have become displaced and drawn deeper into the abdominal cavity by the peristalsis of the intestines.

REPORT ON PHYSIOLOGY. By WILLIAM STIRLING, D.Sc., M.D. (Edinb.), *Demonstrator of Practical Physiology in the University of Edinburgh*[1].

Nervous System, Brain, and Spinal Cord.

ON ELECTRICAL STIMULATION OF THE CEREBRUM.—L. Hermann (*Pflüger's Archiv*, x. 77) upon repeating Hitzig's experiments remarked that, notwithstanding the surface of the brain being exposed for hours together to the air, and also in spite of the surface becoming dry, still the results of the experiment were not affected thereby. The same result was obtained after the destruction (by the repeated application of acetic or nitric acid) of the 'centre,' previously exactly mapped out on the surface. The grey matter of the cerebrum was affected to one-third of its depth by the acid. The experiments were made on middle-sized dogs. The active area or 'centre' was found to be increased by increasing the strength of the current (induced), though the constant current was also employed. In all the other experiments a cork-borer of six or seven millimètres (about ¼ inch) in diameter was pushed into the brain, to a depth of one to one and a half centimètres ('4 to '6 inch); the upper free surface of cylinder of brain-tissue thus produced was formed by the 'cortical centre.' On applying the stimulus (electrical), the results were the same, and even if the electrodes were sunk into the cavity, or placed in its walls, exactly the same results followed. The animals lived four to five weeks afterwards, and were employed for other purposes. The 'area' was always mapped out by a minimum current. In some cases the 'area' was traversed by a small sulcus, in others not. Hermann agrees with the other experimenters that electrical stimulation, by placing the electrodes upon certain convolutions, is followed by distinct movements, but he regards the conclusion, that motor centres lie at these spots, as quite unjustifiable. Further, the author observed in a dog, which lived for several weeks after the operation of extirpation of the so-called 'centre' for the right hind foot, phenomena such as have been described by other experimenters (Duret and Carville), but it was distinguished by having perfect anæsthesia of both extremities of the right side, and incomplete anæsthesia of the left hind leg, and the left side of the trunk. [*Lond. Med. Rec.*, Jan 15, 1876.]

ON CAUTERISATION OF THE CEREBRAL LOBES IN THE GUINEA PIG. —Brown-Séquard (*Progrès Médicale*, 1875, No. 16) observed after deep cauterisation of one side of the brain a peculiar spread-out position of the tips of the toes, but no paralysis of the limbs of the

---

[1] To assist in rendering this report more complete, authors are invited to send copies of their papers to Dr Stirling, Physiological Laboratory, Edinburgh University.

corresponding side. He refers this to a disturbance of the 'muscular sense' of the corresponding side. The injury to the brain, which produced paralytic phenomena of the corresponding (not of the opposite) half of the body, influences, according to the author, remote parts of the brain, upon which the normal functions of the affected extremity depend. The establishment of this hypothesis is wanting.

ON THE MOTOR CENTRES OF THE ENCEPHALON.—Lépine (*Gazette Médicale de Paris*, No. 25, 1875) with Rochefontaine and Tridon, made experiments upon the influence of stimulation of distinct parts of the brain on the blood-pressure, heart, and secretion. Stimulation of the postfrontal convolution of a curarised dog with very weak induction shocks, produced a pronounced increase of the blood-pressure in the crural artery (seven centimètres of mercury). A certain time elapses before the increase of pressure occurs. The same result can be produced by stimulation of the prefrontal convolution and the corresponding part of the sulcus. If the point of the surface of the brain, which in a non-curarised animal discharges movements of the opposite feet, is stimulated with very weak currents, the temperature of this foot rises several tenths of a degree. The temperature of the foot of the opposite side rises, but not so high, whilst that of the rectum remains unchanged. The vessels of the brain and pia dilate and bleed more, but whether in consequence of the vaso-dilator influence of the stimulation on the cerebral vessels, or in consequence of the increased blood-pressure, is uncertain. No such result is observed after stimulation of the surface of the posterior lobes or of the dura mater. Stimulation of the spots which caused increase of the blood-pressure accelerated at the same time the heart-beats. With intact vagi and very strong current, there is a diminution in the number of heart-beats. Stimulation of points, whose locality is not defined, increase the secretion of saliva. [*Lond. Med. Rec.* Feb. 15, 1876.]

CHANGES IN THE CEREBRAL VESSELS UNDER THE INFLUENCE OF THE EXTERNAL APPLICATION OF WATER.—M. Schüller (*Deutsch. Arch. f. Klin. Med.*, XIV. 566, *Centralblatt*, No. 36) trepanned the skull of rabbits, and found that an obstruction to the outflow of venous blood, disturbance of the respiration, or pressure on the abdomen, produced strong injection of the pia mater. After section of the vagi this effect was not produced, ¦on account of the preponderance of the inspiratory movements. Fear, pinching, generally produced narrowing of the vessels, sometimes after previous dilatation. Ice applied to the exposed dura mater produced marked narrowing of the vessels, which was much weaker on the side from which the cervical sympathetic and the ganglion supremum were excised. Cold applied to the abdomen produced an instantaneous dilatation of the vessels of the pia on the uninjured side, and generally no change upon the injured side. A moist warm compress on the abdomen, on the contrary, produced narrowing, which was succeeded by dilatation upon the compress cooling. Complete immersion, as a general rule, acted like a compress. The injection of cold, and generally also of warm water into the rectum dilated the vessels. Packing with the wet sheet, whereby the animals became sleepy, was followed by a very temporary dilatation, which gradually passed into constriction. Similar results were obtained during opium-narcosis, but not by dry packing. Rubbing of the abdomen or back is accompanied with constriction or varying changes in the calibre of the vessels, but in a weaker degree when the sympathetic and the ganglion supremum are extirpated.

The changes above described occurred also in curarised animals, although in this case the filling of the vessels of the brain was somewhat less pronounced. After section of the sympathetic at the second vertebra, there was a pronounced dilatation of all the vessels of the pia mater, and the application of water was without any effect upon it. Cold directly applied to the freely exposed cutaneous sensory nervous trunks which issue upon the back produced constriction; heat, dilatation of the vessels of the pia mater on the same side. Section of individual cutaneous nervous trunks was accompanied by a temporary dilatation of the vessels of the pia on the same side.

The blood-pressure in the carotid, from manometric observations, rose rapidly when cold water was applied to the abdomen, and then fell considerably; with warm water applications, it was just the reverse, with flat variations.

The occurrence of all these phenomena is explained by the author through the changes in the supply of blood to the vessels of the pia in consequence of a constriction or dilatation of the peripheral current-areas in the skin. The movements of the heart and respiration are only indirectly concerned in the result, at one time assisting, at another hindering. The reflex influence of the thermal stimulation of the cutaneous nerves upon the vessels of the pia mater is, according to the author, of subsidiary importance, and acts rather in an inhibitory manner. The second phenomena which occur with long duration of the stimulus, and which are exactly opposite to the initial phenomena, may be explained by the changes in the conditions in the cutaneous vessels and their consequences.

From the results of his experiments the author draws the following conclusions regarding the therapeutical employment of different applications of water to the human organism. It produces (1) a restitution of the normal vascular tonus (specially of the brain); (2) the restoration of normal blood and lymph-currents in the brain; (3) diminution of overfilling of the brain with blood; (4) the restoration of the normal nutrition of the nerve-elements; and (5) of the normal reflex relation between the cutaneous nerves and the brain. In the insane a 'methodical' water-treatment is for the most part not to be trusted, because one cannot say how far the resistance of the cerebral blood-vessels is to be depended upon.

CHANGES IN THE BRAIN IN TRAUMATIC INFLAMMATION.—L. Popoff (*Virchow's Arch.*, XIII. 421, and *Centralblatt*, No. 38, 1875) under v. Recklinghausen's direction examined the brains of twelve individuals who died of abdominal typhus. In all there were changes of an acute active inflammatory character in the vessels, in the neuroglia, and in the ganglionic cells. In the first of these, viz. the vessels, the cells in the walls, or the fat and pigment-cells applied to them, were in a state of proliferation; in the neuroglia division of the nuclei, and in the ganglion-cells, both active proliferation processes and penetration of wandering cells. The former manifested themselves in division and increase in number of the nuclei, then in division of the protoplasm, whereby the individual parts either did or did not possess a nucleus. With regard to the occurrence of wandering cells it is to be remarked that they lay partly around the cells (in the so-called perivascular spaces) and partly also within the nerve-cells, and by the penetration of such cells division of ganglion cells is often brought about. In the preparation these wandering cells fell out of the ganglionic cells, so that these latter appeared as if perforated. Beyond being in and around the ganglion-cells these wandering cells were arranged in rows around the vessels, and here and there along the nerve-fibres, but still preferably on the ganglionic cells.

Essentially the same changes are to be observed in inflammatory processes, and specially in traumatic inflammations, which were produced in a variety of ways upon dogs and rabbits, only here the active changes in the nerve-elements were more pronounced; whilst in typhoid fever the penetration of the wandering cells was in full operation before the proliferation phenomena in the ganglion-cells occurred, and in addition many granule-cells appeared which were quite absent in the case of typhoid fever. Very interesting are the experiments in which the author injected colouring matters, specially China ink, into the brain. This curious result was obtained, that a short time after the injection the chief mass of the pigment lay in the ganglion-cells, which had evidently taken it up by virtue of their own forces, as wandering cells containing pigment which could have accounted for the pigment were absent, and as nothing similar could be produced in dead brains. At this time granule-cells were still absent, but they were present in large quantities, and enclosed the pigment, after the inflammation had lasted longer, whilst the pigment could not, or could only in a very slight degree, be detected in the nerve-cells. The author concludes from this that the granule-cells which generally occur in the brain in acute inflammation are (in part at least) changed nerve-cells.

In another paper in the *Centralblatt* the author records the results of the examination of the brains of three patients who died of exanthematous typhus in the wards of Professor Botkin, of St Petersburg. In this disease also the author finds (1) that there is a similar collection of the wandering cells in the perivascular spaces such as occurs in abdominal typhus; (2) there is also penetration of the wandering cells into the ganglion-cells, and division of nuclei in the latter; (3) infiltration of the neuroglia with young wandering cells; (4) the proliferation phenomena in the walls of the vessels are more pronounced and extensive here than in ileo-typhus. Infiltration of fat and pigment in the vascular walls may also be observed. Capillary extravasations are sometimes to be noted; (5) an interesting, but at the same time very striking result is the formation, in typhus, of small nodules in the substance of the brain. They were found in the cortical substance of the cerebrum, cerebellum, corpus striatum, etc., and were 0·105 – 0·18 millimètre long, and 0·075 – 0·09 millimètre broad; they often had a rounded form. These nodules with a low power presented appearances very similar to miliary tubercle. Like the latter, they were found generally, though not always, next the vessels. With high powers (300 diameters) these nodules were seen to consist chiefly of indifferent newly formed elements which could not be distinguished from lymph-corpuscles or white blood-corpuscles. Sometimes they consisted of such corpuscles alone, and this specially in the peripheral finely granular layers of the cerebrum and cerebellum. Where, however, in fibrous tissue nervous cellular elements were present in considerable proportions, as in other layers of the cerebrum, in the corpus striatum, other elements, nearly as large as the nuclei of the ganglion cells, entered into their composition. The changes already described in the nerve-cells are often very pronounced around these nodules. In the first described form of nodule, consisting of indifferent elements like white blood-corpuscles, there is never a finely granular degeneration of the central part to be observed as is often seen in tubercle. Neither giant-cells nor a special stroma were to be observed. These nodules, from their character and origin, are apparently completely analogous to the nodules described by Wagner as occurring in some parenchymatous organs, such as the liver and kidneys, in abdominal typhus. These nodules were observed in two cases out of the three. The relation of these nodules to the brain-symptoms, owing to the epidemic being at an end, was not made out. In both cases the patient died on the fourteenth day. [*Lond. Med. Rec.*, 1875.]

ON SLEEP PRODUCED BY FATIGUING-STUFFS.—W. Preyer, of Jena, in a preliminary communication to the *Centralblatt*, No. 35, 1875, records the results of some experiments with salts of lactic acid in the production of sleep. He finds, when a concentrated watery solution of lactate of soda is injected subcutaneously, or introduced in large quantities into the empty stomach, that the feeling of fatigue, of drowsiness, and a condition similar to or identical with natural sleep very frequently occurs, provided that strong stimulation of the sensory apparatus is withheld. In many cases also yawning and sleep may occur when no sodic lactate is introduced directly, but when the conditions are present for the abundant formation of this substance in the intestine, as by drinking highly concentrated sugar solution, etc. There are great variations both in the period of occurrence, duration and intensity of the sleep. It differs with the age, size, reflex activity of the animal. [*Lond. Med. Rec.* Dec. 15th, 1875.]

ON THE QUANTITY OF WATER IN THE HUMAN CENTRAL NERVOUS SYSTEM.—M. Bernhardt (*Virchow's Archiv*, LXIV. 297) finds from his analysis that the mean value of the quantity of water in the spinal cord of patients who died from various diseases is, for the cervical part, 73·05 per cent. water; for the lumbar region, 76·04; results which correspond pretty closely with those of Bischoff, but are higher than those of v. Bibra. The cortical portion of the brain gave a mean value of 85·86 per cent., while the white matter only gave 70·08 per cent.; that of the sympathetic, 64·30 per cent.

THE COURSE OF THE MOTOR AND SENSORY PATHS IN THE LUMBAR PORTION OF THE SPINAL CORD OF THE RABBIT.—C. Woroschiloff (*Ludwig's Arbeiten*, IX. 99). The previous observations by Miescher, Nawrocki and Dittmar had shewn that in rabbits the lateral columns of the spinal cord contain the nerve-fibres both afferent and efferent, which are concerned in the production of a reflex contraction of blood-vessels, due to an irritation of the sciatic nerve. To ascertain whether the channels of reflex and voluntary movement of the trunk and limbs also lie in the same region of the cord, Woroschiloff was compelled to modify in several respects the methods adopted by the above-named observers. The necessity of observing the contractions of the voluntary muscles prevented curara being used to render the animal motionless. This was accomplished by means of a double clamp, which was screwed firmly upon the vertebral column above and below the point where the spinal canal was laid open by the removal of the arch of one of the vertebræ. The same clamp also gave support to a small apparatus by means of which delicate blades, cutting in planes parallel to the longitudinal axis of the body, could be adjusted in any desired position, and then thrust with great precision, vertically, horizontally or obliquely, through the substance of the spinal cord. After section the wound was carefully sewed up, and the results of the operation observed. The sections were all made at the level of the last dorsal vertebra.

A systematic study was made of the effect of the mutilation, first, on the production of reflex actions due to irritation by pressure and electricity applied to the feet and ears of the animal; secondly, on the position of the hind-limbs both at rest and in movement; and, thirdly, on the production of movement in the hind-limbs due to irritation by induced currents applied to the cord just below the calamus scriptorius. The animal was usually killed about five hours after the operation, and a careful microscopic examination was made of the part of the cord operated on. The microscopic sections were photographed with a magnifying power of twenty-five diameters. The paper contains a series of beautiful heliotype plates obtained in this way.

After division of the anterior and posterior columns, and nearly the whole of the grey substance, no disturbance of the transmission of motor or sensitive impressions through the cord was detected. The animal sits and moves in a perfectly normal manner.

Section of both lateral columns entirely prevents the transmission of impressions through that region of the cord. Irritation of one hind-leg causes reflex movements of the same leg, or, if the irritation be a strong one, of the opposite leg also. Irritation of a fore-leg causes movement in the anterior but not in the posterior part of the body. Irritation of the cervical cord causes movements in all the muscles of the body except those of the hind-limbs.

Section of the posterior columns alone is absolutely without effect on the condition of the animal. There is nowhere hyperæsthesia nor anæsthesia.

Other modifications of the experiment were made, and the general result shewed that the channels of motor and sensory impressions lie in the lateral and not in the anterior and posterior columns of the cord. This is in opposition to the generally received opinion of physiologists. Many other modifications of the experiment were performed, and the general results may be stated as follows :—

Motor and sensory nerve-fibres are found in all parts of the lateral columns.

Sensory fibres from both hind-limbs are found in each hind-lateral column, but the fibres in either column, which come from the leg on the opposite side, are capable of producing stronger reflex movements in the anterior part of the body than are called forth by excitation of the fibres which come from the leg on the same side (crossed hyperæsthesia).

The centripetal fibres whose excitation produces these strong reflex movements, as well as those whose section on the opposite side gives occasion to them, lie in the middle third of the lateral columns, while the anterior and posterior thirds contain sensory fibres which call forth movements of only moderate intensity in the anterior part of the body.

Motor fibres for both legs are found in each lateral column, but the motor fibres in different parts of these columns are called into activity in different ways. The reflex movements due to irritation of a fore-leg can be excited in a hind-limb only when the anterior half of the lateral column on the same side is preserved.

The coördinated movements of sitting and springing, and those produced by irritation of the cervical cord, are transmitted to each hind-limb through the middle third of the lateral column on the same side.

The fibres which preside over coördinated movements of the hind-limbs, as well as those whose section causes hyperæsthesia on the side of the injury, lie in those parts of the lateral columns which are nearest to the gray substance. The motor fibres of the foot and leg proper seem to lie in the lateral columns externally to those of the thigh. (Fuller report in the *Boston Med. and Surg. Journal*, 1875.)

ON THE SUMMATION OF ELECTRICAL STIMULI APPLIED TO THE SKIN.—Wm. Stirling, *Ludwig's Arbeiten*, IX. 223, and *Journal of Anat. and Phys.* X. 324.

"On Motor Centres in the Cerebral Convolutions." Dalton and others in the *New York Med. Journ.* March, 1875 (Abstract in *Lond. Med. Rec.* No. 125).——"On the Reflex Actions of the Brain, in the Normal and Morbid Conditions of their Manifesta-

tions." Luys, an Abstract in the *Annales Médico-Psychologiques*,
and *Lond. Med. Rec.* No. 121, 1875.——"Physiology of the
Cerebro-Spinal Nervous System," par Dr E. Fournié. Paris,
A. Delahaye.——"On the Vaso-Motor Centres and their Mode of
Action," Masius and Vaulair, *Gaz. Hebdomad.* No. 41, Abstract
in *Lond. Med. Rec.* Nov. 15th, 1875.

"On the difference in the Reflex Actions of the Medulla and
Spinal Cord of the Rabbit." Ph. Owsjannikow, *Ludwig's Arbeiten*,
IX. 308.——"Physiology of the Brain," R. Danilewsky, *Pflüger's
Arch.* XI. 128.——"Experimental Investigations on the simplest
Psychical Processes," S. Exner, *Pflüger's Arch.* XI. 403, and 581.

On the Vaso-Dilator Action of the Glosso-Pharyngeal
Nerve on the Vessels of the Mucous Membrane at the Base
of the Tongue.—A. Vulpian (*Comptes Rendus*, LXXX. 330) shows
that the glosso-pharyngeal is the vaso-dilator nerve for the posterior
parts of the lingual mucous membrane, just as the chorda tympani
is for the anterior part of the tongue. Electrical stimulation of the
glosso-pharyngeal produces intense redness of the part of the tongue
supplied by that nerve. Further, the author has convinced himself
that after destruction of *all* anastomoses, this action of the glosso-
pharyngeal nerve remains the same. [*Lond. Med. Rec.* Nov. 15th,
1875.]

On the Innervation of the Vessels of the Web of the
Frog's Foot.—D. Huizinga (*Pflüger's Archiv*, XI. 207) in his paper,
confirms the results of Goltz, Tarchanoff, Putzeys, &c. As is well
known, the doctrine of the innervation of vessels has undergone
remarkable changes within even a short period. At first it was a
tolerably simple affair, so long as it was imagined that the vaso-motor
centre for the entire body lay in the medulla oblongata and that
the vaso-motor fibres were distributed in all directions through the
sympathetic. Later discoveries have, however, necessitated changes
in this relatively simple theory. First there were the vaso-dilator
nerves; then the vascular reflex action upon sensory stimulation; then
facts that seemed to show that the vaso-motor centres are distributed
throughout the whole spinal cord; and lastly, the view, supported
on more or less good grounds, that the size of the lumen of the vessel
does not depend upon spinal nerve-cells alone, but is also controlled
by nerve-cells situated peripherally.

The author's experiments were made upon the web of freshly
caught, strong examples of Rana temporaria, which is to be preferred
because it contains fewer pigment-cells to obstruct the view of the
vessels. The animals were curarised with a solution of curara,
which took twenty to thirty minutes to operate. Sometimes, though
rarely, a solution of strychnine was employed to increase the reflex
activity; and in an animal previously curarised, a larger dose of
strychnine was required to produce the result. The size of the
arteries was measured either by an eyepiece micrometer, or the

size of the vessels was drawn by means of a camera lucida. The animals must not be too strongly curarised, else the rhythmical movements normally to be observed in the arteries are not to be observed, although the vaso-motor nerves retain their excitability undiminished.

In order to test the independence of the rhythmical contractions in the arteries from the spinal vaso-motor centres, the sciatic plexus and, for greater security, the sciatic nerve on one side, were divided, or even a part of the nerves excised, and the animal left to itself for one or two days to recover from the effects of the operation.

In the web of the foot so operated on, the most beautiful rhythmical contractions are to be observed, whilst the reflex contraction, on pinching the fore-foot, is entirely absent, thus showing that the contraction with the destruction of the spinal centres is completely abolished. On the uninjured side reflex movement occurred. Further, this experiment shows that the rhythmical movements of the vessels do not proceed from the spinal cord, confirming the results of Gunning on the web of the frog's foot, of Roever on the ear of the rabbit, and of Asp on the splanchnics. In addition, the rhythmical contractions do not occur in all the arteries of the web simultaneously. The author is, therefore, of the opinion, like Lister, Goltz, Tarchanoff, Putzeys, that the cause of these movements is to be sought for in local nerve-centres, leaving, however, undecided whether the impulse for these movements proceeds from peripheral nerve-cells placed along the vessels, or arises independently of such organs in the muscles of the vessels.

In addition to the rhythmical movements very pronounced reflex movements of the muscles of the vessels are to be observed. The author finds that when stimuli (mechanical or electrical) are applied to the skin, whether the application is followed by contraction or dilatation of the arteries of the web, depends, on the one hand, on the distance of the stimulated area from the web, and, on the other, on the intensity of the stimulation ; so that with diminishing distance and increasing strength of stimulus dilatation prevails.

In an experiment on a frog whose lumbar portion of the spinal cord was destroyed the day previous, the author observed that a reflex contraction which occurred on stimulating the fore-foot was entirely abolished when the sympathetic nerve accompanying the aorta was divided between the third and fourth ganglia.

In a frog whose spinal cord was destroyed the day previous from below the fourth vertebra, no reflex contraction of the vessels was produced on pinching the posterior extremity, though this result followed pronouncedly when the anterior extremity was pinched. A weak solution of amyl nitrite, suspended in water and applied to the web of the foot, was followed at once by dilatation of the vessels of the web, which reached its maximum in a few seconds, and lasted for ten to twenty minutes. The vessels then returned to their normal calibre. At the maximum of dilatation, pinching the fore-foot was followed by a quite normal reflex contraction of the vessels of the web.

From this experiment the author concludes:

(a)   That amyl nitrite does not paralyse the muscular tonus of the vessels.

(b)   Neither does it paralyse the vaso-motor nerves, nor their end-organs.

(c)   The dilatation of the vessels after the application of amyl nitrite cannot otherwise be explained than that this substance extinguishes the tonus of the peripheral nerve-cells (the so-called local vaso-motor centres) placed along the vessels.   [Lond. Med. Rec. Dec. 15th, 1875.]

ON VASO-DILATOR NERVES.—Goltz, with the co-operation of Drs Freusberg and Gergens (Pflüger's Archiv, Band XI. p. 52), has continued and extended his researches upon this subject (Journ. of Anat. and Phys. IX. 408).   Goltz explains the dilatation which occurs after section of nerves, not by a paralysis of vaso-constrictor nerves, but rather as the result of the stimulation of vaso-dilator fibres.   In accordance with the results of Putzeys and Tarchanoff (Journ. of Anat. and Phys. IX. 408), the author corrects a mistake in his previous observations, and agrees with these authors, that in many cases the effect of electrical and chemical stimulation of the sciatic nerve is at first to cause a contraction of the vessels of the foot, of short duration, before the very pronounced dilatation occurs.   Goltz, however, does not accept the explanation of Putzeys and Tarchanoff.   The method of experimenting is the following.   The spinal cord of a strong young dog is divided in the lumbar region.   After the wound has healed, both sciatic nerves are dissected out and divided, and a period of time is allowed to elapse till the elevated temperature of the hind-limbs has begun to decrease somewhat.   The idea of the author was, that if simple section of the sciatic nerve acts as a stimulus, this stimulus may be perhaps greatly increased, by making methodically a large number of sections of the nerve.   The result showed that the foot, whose nerve was only once divided, was generally ten degrees cooler than the foot whose nerve was cut away or divided in repeated small discs.   Powerful dogs, full-blooded, and with good appetites, are required for these experiments.   According to the author's hypothesis, repeated pinching or section of the nerve produces a stimulation of the vaso-dilator nerves.   The vessels relax at the periphery, similar to the stoppage of the heart's action, when the vagus is stimulated.   When the vessels have relaxed, i.e. when their tonus disappears more and more, there is not necessarily a pronounced dilatation of the lumen of the vessels.   This will only become clearly pronounced when there is a sufficient amount of blood to fill the relaxed vessels corresponding to their diminished elasticity.   The pronounced vascular dilatation and the increase of temperature will accordingly only occur when the general arterial blood-pressure is high and remains high.

What the author means by the term 'stimulation of vaso-dilator nerve-fibres,' is, that simple or repeated section of the nerves produces a peculiar change in the condition of the nerves, in consequence of which a something is propagated in the nerve, which causes a vascular dilatation in that area in which the peripheral expansion of the nerve takes place.

Hammering the nerves by means of Heidenhain's tetanomotor, also produces a similar dilatation of the vessels of the feet, just like methodical pinching of the nerve.

A chemical stimulus, in the form of concentrated sulphuric acid, applied to the sciatic nerve of a guinea-pig prepared as above, gave the same result, though the author could not be certain whether the foot of the stimulated leg became paler at the beginning of the stimulation than before it.

A very curious phenomenon was observed when the sciatic nerve of a kitten was stimulated by induced electricity, viz. the foot began to sweat. The author therefore concludes, that in the sciatic nerve of the cat fibres are present which supply the sweat-glands. Only in two cases, however, was this effect observed in dogs, who, as is well known, possess sweat-glands in the pads of the feet [and also in the skin over the surface of the body generally.—*Rep.*]. In many experiments a cooling, *i. e.* constriction of the vessels of one leg occurred, when in the other leg vascular dilatation was artificially produced. It is very probable, the author thinks, that the vascular constriction is not referable to an increased tonus, but is to be explained on purely physical grounds, by a sudden diminution of the resistance in the vessels of the other leg.

After simple or repeated section of the sciatic nerve in dogs, the author never observed that the dilatation of the vessels was preceded by contraction, although he remarks that if the contraction was sufficiently short it might elude observation. This contraction of the vessels after section of the spinal cord is well established for the frog.

From the rapidity with which the vascular tonus is re-established after section of the nerves, the author concludes, that the tonus must be kept up by arrangements which lie beyond the great nerve-centres, probably in the vascular wall. The tonus was soon re-established in the vessels of the leg of the dog after section of both the sciatic and crural nerves. The author shows in another way, that the tonus of the paralysed limb does not depend on the great nerve-centres.

Section of the spinal cord can produce vascular dilatation, not only in the hinder but also in the anterior part of the body. The whole plexus brachialis in a dog, on one side, was divided, and when the temperature of both fore-limbs had become nearly alike, generally in from seven to fourteen days, the spinal cord was divided between the thoracic and lumbar portions. There then occurred a considerable difference in the temperature of the two fore-feet, that of the paralysed side having become notably diminished. The temperature of the sound side rose or remained the same. From this it appears, that injury to the spinal cord acts similarly both on the fore and hind parts of the body. According to the author, injury to the spinal cord cannot exercise any influence on the tonus of the vessels through nervous channels. The tonus being sustained by the hypothetical end-arrangements in the paralysed part, these being no longer influenced by nervous excitations proceeding from the great centres, still they may not be completely independent of the body. Every change in the blood-pressure in the aorta may alter the diameter of the vessels in the paralysed part, and it seems to the author very probable that the tonus of the vessels of the paralysed side is increased by the sudden diminution of the blood-pressure acting locally on the isolated end-organs. In experiments where double section of the spinal cord was performed at long intervals, e.g. first in the lumbar region, and then in the fore part of the thoracic region, it was found that the hind-feet, immediately after the second section, became greatly cooled, while the fore-feet became warmer. This shows, that in an animal with divided spinal cord, the tonus of the vessels of the hind-feet are governed from foci other than those of the fore-feet. The peculiar reflex actions of the hind part of the body were influenced by the second section. The author thinks that this experiment supports the view that the cause of 'shock' is to be sought in 'inhibitory processes,' which are propagated in nervous channels.

The most pronounced vascular dilatation can be produced by placing the paralysed limb in a freezing mixture. When the foot is taken out of the mixture its temperature is little short of that of the blood itself.

Variations in the temperature of the air of the room, however, produce extreme changes in the diameter of the vessels of the paralysed limbs. The tonus in the paralysed limb varies with a change in certain internal conditions—probably blood-pressure, the quantity of blood, temperature or change in the composition of the blood, may influence it, but these the author has not sufficiently studied.

With regard to the end-arrangements in the sound limb, the author imagines that they are equivalent to the nervous arrangements upon which the activity of the excised heart depends, and, just as Volkmann maintained that the ganglionic cells are the immediate central organs for the cardiac movements, so are the hypothetical ganglionic cells in the walls of the vessels the immediate centres for the vascular tonus. In sound limbs, the activity of these end-arrangements can be powerfully influenced by the great nerve-centres.

The reasons why the author regards the vascular dilatation as a result due to stimulation, which must be explained by an increased activity of the divided nerve, are: (1) the vascular dilatation disappears tolerably rapidly, because the stimulus becomes exhausted; (2) repeated section of the peripheral end of the nerve increases the dilatation; (3) every other form of mechanical stimulation produces the same result; (4) continued chemical or electrical stimulation of the nerve always produces a degree of vascular dilatation, which greatly exceeds that which is present before the stimulation.

Goltz is of opinion, that those cases where occasionally section of the sympathetic in the neck is not followed by its usual results support his hypothesis. He also cites the division of the nervi errigentes, and one on which he lays more weight, viz. Vulpian's experiment of section of the lingual nerve being followed by dilatation of the vessels of the corresponding half of the tongue, and the increase of this dilatation on the peripheral end of the nerve being stimulated ; and he also cites the same author's more recent experiment on section of the glosso-pharyngeal nerve having a similar action on the posterior part of the tongue, as supporting his view (*Journ. of Anat. and Phys.* x. Part III.).

The author concludes that the spinal cord is undoubtedly an independent reflex vaso-motor centre. It has, however, not been proved that this centre continually sends tonic excitations to the vaso-constrictor nerves. The dilatation of the vessels observed after injury to the spinal cord, and which are regarded as a consequence of the interrupted activity of this organ, are much more the result of the stimulation of vaso-dilator nerves. (*Lond. Med. Rec.* Jan. 15th, 1876.)

ON THE REFLEX ACTIONS OF THE TENDONS.—Several papers have appeared upon this somewhat novel subject lately, including those of W. Erb (*Archiv für Psychiatrie*, Band v. p. 792), C. Westphal (*Ibid.* p. 803), and A. Joffroy (*Gazette Médicale de Paris*, Nos. 33 and 35), all of which are reviewed in the *Centralblatt für die Medicinischen Wissenschaften*, No. 54, 1875. Erb noticed these reflex actions of tendons in the healthy subject, but more pronouncedly in many patients suffering from disease of the spinal cord. In the quadriceps, a prompt contraction of the whole muscle, resulting in powerful movement of the leg, is produced by gently touching the region of the ligamentum patellæ, when the knee and thigh are slightly bent. The author calls this a 'patellar tendon reflex,' and shows that it does not proceed from the skin, and can only arrive from the tendon and its direct continuation (no stimulation of the skin, no tapping the surrounding parts, can produce the reflex action), and that it is specially or exclusively the mechanical stimulus which discharges this reflex action. In a patient suffering probably from commencing sclerosis of the lateral columns the author could produce not only reflex action of the patellar tendon, but reflex contractions could also be produced in a similar way, and in many other musculo-tendinous areas of the body, such as from the tendon of the triceps of the upper arm, and most peculiar of all from the tendo Achillis, when suddenly placed upon the stretch by rapidly flexing the foot dorsally. On continuing the pressure, clonic contractions appear, in that the foot being extended by the contraction of the tendo Achillis is brought back into the old position by the continuing passive dorsal flexion, in which the extension, and therewith the stimulus to the contraction of the tendo Achillis, arises. These clonic spasms are not caused to cease by sudden plantar flexion of the great toe (Brown-Séquard), when at the same time there is not a powerful flexion of the whole foot. Then the clonic spasms cease. The physiological and the pathological importance of these tendon-reflexes are only indicated by the author. Westphal, like Erb, also observed the occurrence of the above described phenomena. The phenomena described by Erb as "plantar tendon reflex" is called by Westphal "leg-phenomenon," the clonic contractions of the foot caused according to Erb reflexly by stretching of the tendo Achillis, Westphal calls 'foot-phenomenon.'

From a clinical point of view, Westphal noticed the foot-phenomenon in the hemiplegics (not in quite sound persons) at the earliest during the first week after the attack. In the later stages its occurrence is almost to be regarded as the rule. It also occurs in spinal paraplegics, or in weak persons only in one lower limb from spinal disease, specially in those forms of paraplegia which are accompanied by continued rigidity of the entire muscular structure of the lower extremity. Westphal also regards the assertion of Brown-Séquard, that plantar flexion of the great toe is able to interrupt the clonic spasms, as erroneous, and explains the occurrence of the phenomenon caused by the plantar flexion of the whole foot. The foot-phenomenon is always wanting in the pronounced clinical symptoms of tabes dorsalis. The leg-phenomenon generally occurs where the foot-phenomenon occurs; not unfrequently there are exceptions to this rule. Only this is certain, that it is absent in all undoubted cases of tabes dorsalis. In grey degeneration of the posterior columns, proved by pathological examination, both phenomena are absent, but are to be met with in the most different affections of the brain (embolus, tumour, hæmorrhage), and in the spinal cord in primary myelitis or that caused by disease of the vertebræ, when the grey substance is not too deeply involved, or large parts of the dorsal and lumbar portions of the spinal cord are not completely destroyed. Westphal is inclined to ascribe *the phenomena of the contraction* to a direct mechanical excitation *of the corresponding muscles, by sudden stretching or vibration of the tendon,* for one cannot explain the phenomena as produced reflexly through sensory muscular nerves, as long as the existence of sensory muscular nerves is a subject of controversy, and with regard to reflexes which result from tendons we up to the present know nothing. For the pathological cases the author believes that the continued abnormal condition of contraction of the quadriceps and of the muscles of the calf, which makes them more susceptible of mechanical stimuli, plays a certain part. The author from his anatomico-pathological observations is of opinion that a defect in the conduction of the lateral columns stands in causal connection with the phenomena. This is certain, that, when both phenomena occur, no grey degeneration of the posterior columns stretching to the lower dorsal and lumbar portions of the spinal cord is present. As to the cause of the occurrence, Westphal is inclined to ascribe *the phenomenon of the contraction to a direct mechanical excitation of the corresponding muscles, through tension or vibration of the tendons;* for to assume that they are reflex actions caused through sensory muscular nerves cannot be admitted as long as the existence of sensory muscular nerves is still a controverted question, and up to the present he knew nothing of reflex action resulting from tendons. In pathological cases the author believes that the continued abnormal condition of contraction of the quadriceps and muscles of the calf, which render them more susceptible for mechanical stimuli, plays a certain part.

According to Joffroy, the above-described phenomena are reflex, and they can be produced by actual muscular extension (which the author regards as of more importance for the occurrence of the phenomena than stimulation of the tendons), as well as by stimuli applied to the skin. Drs F. Schultze and P. Fürbringer, stimulated by the above publications, have also experimented upon this subject (*Centralblatt*, No. 54, 1875). The muscles of the thigh and leg of a rabbit were exposed, and tapping of the patellar tendon with a small hammer caused contraction of the right quadriceps and extension of the right foot, and weak contractions in the left quadriceps, and irregular tremor of both extremities. After section of the right crural nerve the contraction of the right quadriceps ceased. Section of the left crural nerve was followed by the same result in the left leg. From these and experiments similar in kind, with section of the cord at the level of the dorsal vertebræ, the authors arrive at the following conclusions: 1. The phenomena in question are not due to mechanical muscular contractions produced directly through the tendons. 2. They are rather to be ascribed to a reflex mechanism discharged by a mechanical stimulation of the tendons, the reflex path for the lower extremities being placed in the lower sections of the spinal cord. 3. Cutaneous reflex actions in the sense of Joffroy do not exist. (*Lond. Med. Rec.*, Jan. 15th, 1876).

General Physiology of Nerves.

STIMULATION OF NERVES BY SOLUTIONS OF INDIFFERENT SUBSTANCES.—H. Buchner (*Zeitschr. f. Biol.* x. 373), in opposition to the older results of Richter, finds that saturated solutions of urea applied to excitable nerves produce contractions, which even may become tetanic with very excitable nerves.

ON THE LAWS OF NERVOUS EXCITATION.—E. Fleischl (*Wiener Sitzungsb*. Dec. 9th, 1875) finds : 1. For chemical stimuli nerves are at all parts of their course alike sensitive. 2. For electrical stimuli they are more sensitive at higher points than at lower, if the electric currents pass downwards; the case is reversed if they pass upwards. 3. The doctrine of an increase (*Anschwellen*) of stimulus in the nerves is untenable.

" Comparative Study of Instantaneous and Continuous Electrical Currents in the case of Unipolar Excitation." M. Chauveau, *Acad. des Sciences*, Dec., 1875. Abst. in *Lond. Med. Rec.*, Dec., 1875.——" Recurrent sensibility of the nerves of the hand." A. Richet (*Lond. Med. Rec.*, Oct. 15th, 1875).——" On the degeneration of nerves separated from their trophic centre." A. Cossy and J. Déjérine (*Arch. de Physiol.*, Aug., 1875). Abstract *Ibid*.

Eye.

" On the importance of specks on the cornea for the origin of squinting." J. Hirschberg (*Centralblatt*, No. 36, 1875).——" On the function of the ciliary muscle." Warlomont and Nuel (*Annales d'Oculistique*, May and June, 1875). Abst. in *Lond. Med. Rec.*, Nov.

15th, 1875.——"On the movements of the iris." Dr Debouzy. Paris;
A. Delahaye.——"On the cause of Keratitis following section of the
trigeminus." Senftleben(*Virch. Arch.* LXV. 69).——"On the semi-decus-
sation of fibres of the optic nerve in man." J. Hirschberg (*Ibid*, 116).
——"On the doctrine of the currents of fluid in the living eye and in
the tissues in general." M. Knies (*Virch. Arch.*, LXV. 401).——"On
the influence of the eye on the metamorphosis in the animal economy."
E. Pflüger, and O. Platen (*Pflüger's Arch.* XI. 263).

Ear.

"Pathological observations on the physiological importance of
high musical tones," and "On the combined occurrence of imperfect
perception of certain consonants together with high musical tones
and their physiological importance." Moos (*Arch. f. Augen. v.
Ohrenheilk.* II. 139 and 165). Abstract in *Centralblatt*, No. 35, 1875.
"On a peculiarity of sounds of lowest intensity." V. Urbantschitsch,
(*Centralblatt*, No. 37, 1875).——"Auditory Vertigo." Brown-Séquard
and Labadie-Lagrave (*Lond. Med. Rec.* No. 123, 1875).——"Which
nerve supplies the tensor tympani?" Voltolini (*Virch. Arch.* LXV. 452).

PHYSICAL EXPERIMENTS ON THE SENSE OF EQUILIBRIUM IN MAN.
—E. Mach (*Wien. Akad. Sitzb.* 3 Abth. LXVIII. 124, LXIX. 44)
describes experiments upon the "sense of equilibrium" which he
made upon a kind of centrifugal apparatus.

A vertical frame of wood, four mètres long and two mètres high,
rotates round a vertical axis. In this is placed a smaller frame, also
rotating round a vertical axis, and which can be placed at any distance from the axis, and in this is placed a stool which rotates round
a horizontal axis. The observer who sits upon this stool can be
placed at any distance from this axis, A, and set in motion in any
direction desired. He may at the same time be enclosed in a paper-
box.

Mach, from his experiments, has arrived at the following results.

If the observer is in the axis he feels the rotation until it has
become quite uniform, then every sensation of rotation ceases. As
soon, however, as the rapidity diminishes, the feeling of rotation in
the opposite direction arises. It is, therefore, not the rapidity, but
the acceleration of the angular rotation which is felt. If the paper-
box is opened when the apparatus is brought to a standstill, one has
the impression as if the visible space turned into an uninterrupted
invisible space. The sense of rotation produced by the angular
rotation has a certain after duration, and can be made to cease by an
opposite angular rotation. The position of the head modifies the
sensation. If during an apparent rotation one observes an opposite
and actual rotation, then he regards the latter as quiescent, and feels
himself in more rapid opposite rotation. If the observer looks towards the axis while he sits vertically at some distance from it, then
he believes himself to lie more upon his back. He feels the direction
of the acceleration and regards this as vertical. Further, he believes

himself to be inclined laterally when he turns one side instead of the face towards the axis. A pendulum suspended in the box, which of course will diverge as the box rotates, the author regards as vertical, himself and the box, however, as oblique.

If an observer is placed in a balance and set in vertical vibration with sufficient rapidity and with closed eyes he feels the vibrations, but always gives the time of return as too early. One therefore does not feel the position or the rapidity, but the acceleration.

For the explanation of these results the author refers to the experiments of Flourens, Goltz, &c., on the semicircular canals. He thinks (like Breuer) that the varying pressure of the endolymph on the ampullary nerves is the cause of the sensation. He acknowledges the difficulty which lies in the great friction owing to the small diameter of the canals. Compare, also, the results of Professor Crum Brown, who, quite independently, from the consideration of the relative position of the semicircular canals and from experiments similar to the above, has arrived at almost similar results and conclusions. [*Journal of Anatomy and Physiology*, VIII. 327.]

Blood and Blood-vascular System.

ON THE RHYTHM OF THE HEART-BEATS.—M. J. Rossbach (*Ludwig's Arbeiten*, IX. 90, *Centralblatt*, No. 28, 1875) investigated first the influence of the systolic cardiac pressure on the formation of groups of beats. Luciani found that a considerable and continued increase of the pressure which the heart had to sustain during the diastole accelerated the occurrence of fatigue, and rendered the pulse curve irregular; as to the influence of the systolic pressure he obtained no results. Rossbach, instead of a mercury, employed a water manometer. The systolic pressure could be diminished or increased by leaving the free end of the manometer open, or by compressing it more or less. In hearts which were fed with clear rabbit's serum the formation of groups occurred both with high and low pressure; but with low systolic pressure the formation of beats was much more irregular than with very great pressure. These irregular groups passed at once into regular ones, when the systolic pressure was diminished. By the term "heart" is meant one whose auricles are ligatured off after the manner of Luciani (*Journ. of Anat. and Phys.* VIII. 191).

On the influence of various filling solutions on grouped succession of the heart-beats. On employing serum as free from corpuscles as possible for filling the heart, the formation of groups occurred in a pronounced manner; the groups disappear when the pure serum is removed and defibrinated blood or blood-serum is employed instead. Also when the heart is filled with a 0·6 per cent. solution of common salt the grouped arrangement of the beats disappears. If the heart is filled with very red serum or defibrinated blood, it never shows a trace of grouped arrangement of its beats. If the defibrinated blood, without renewing it, was allowed to remain in the heart till it had lost its red colour. the formation of groups begins again, and in such a heart the periodic rhythm cannot again be brought to disappear by repeated filling with fresh blood or bloody serum. A heart filled with blood, and beating, and acting regularly, at once forms groups on a weak solution of veratrin being added to its contents. The different conditions in the succession of beats can be produced in one and the same heart. If a heart is filled with solution of salt, its beats follow each other at always regular intervals; if this solution is exchanged for pure serum, the contractions arrange themselves in groups, and if this solution is in turn displaced by serum containing blood-corpuscles, the heart then acts again with a regular succession of beats. The number of beats, however, is not always the same. A heart filled with solution of salt beats six times per minute; with pure serum, in the first group seven, in the second ten times; with blood, on the contrary, twenty-three times. Rossbach cites various explanations for these different actions, and for the phenomena of the periods. It is possible that in pure serum a substance is formed which on the one hand has an influence similar to veratrin, and on the other is destroyed when it comes in contact with bright red blood-corpuscles; or serum acts as a stimulus; under its influence the energy of the conditions under which the grouped arrangement of the beats is active becomes greater, and then diminishes to the stage of the so-called "crisis," during which the heart-beats occur in regular but slower succession. Blood however, is able to stimulate an excitable heart more strongly than serum, and causes the pauses to disappear, whilst solution of salt stimulates less than the serum, so that by its presence the heart, from the first, passes into a state which corresponds to the stage of "crisis" in the serum-heart. The similarity in the action of blood and of salt solution, are therefore only apparent. In the course of his investigations Rossbach often succeeded in restoring apparently dead hearts by increased diastolic pressure and by heating. (*Lond. Med. Rec.* Dec. 15th, 1875.)

"The Characteristic Sign of Cardiac Muscular Movement."— H. Kronecker and W. Stirling, (*Beiträge zür Anat. und Physiol. als Festgabe Carl Ludwig gewidmet von seinen Schülern*, 1874, and *Journal of Anat. and Phys.* IX. 315.)

ON A NEW MEANS OF ARRESTING THE HEART OF A FROG.— J. Tarchanoff (*Archives de Physiologie*, 1875, p. 408) pulled out a loop of intestine with its corresponding piece of mesentery from the left abdomen of a frog, and exposed the loop to the air for a few hours until it became inflamed. The slightest touch of the inflamed loop sufficed at once to produce still-stand of the heart. The time of arrest varies from a few seconds to half a minute. The experiment does not succeed by gently touching the loop of intestine when pulled out at first, and while it is still uninflamed, nor after section of the vagus or poisoning with curara. The arrest takes place through the same mechanism as in the experiments of Goltz and Bernstein.

ON SOME NEW PROPERTIES OF THE WALLS OF THE BLOOD-VESSELS.—A. Mosso (*Ludwig's Arbeiten*, IX. 156), working under Ludwig's direction, has continued and extended the researches upon the artificial circulation of blood in excised organs, begun by Professor

Heger, of Brussels, in the same laboratory (*Journ. of Anat. and Phys.* IX. 417). The complicated apparatus, carefully figured in all its details, consists of several parts. An arrangement constructed on the principle of a Mariotte's flask, so as to give a constant pressure in the principal artery of the organ upon which the experimenter is working, the pressure remaining constant throughout the entire duration of the experiment. A second part of the apparatus is employed for measuring the changes which take place in the volume of the organ experimented on itself, a most essential point, and one which till now has not been taken into consideration. To this part of his apparatus Mosso gives the name of *pléthysograph*. Lastly, there is an exceedingly ingenious arrangement for measuring the velocity with which the blood flows out of the vein. We must refer to the original for the details of the apparatus.

Almost all the experiments were made with the kidneys and liver excised from large dogs, which had been previously bled from the carotids. Of course the blood employed was defibrinated.

Several series of experiments were performed. In the first series Mosso observed the phenomena of the artificial circulation of blood defibrinated in contact with air; in the second he studied the effects of electric currents on the irritability of the vessels; in the third the influence of interruption of the current of blood on the velocity of the blood; then the influence of different gases on the walls of the vessels; fifthly, the action of certain poisons; sixthly, the production of œdema; and lastly, the phenomena of artificial circulation in the liver.

1. After having excised the kidney and completed the necessary arrangements, the artificial circulation is begun. The velocity of the blood from the vein and the increase in size of the organ reach their maximum, and become afterwards successively smaller until their tonus is re-established. Sometimes even with the highest pressure it is impossible to cause defibrinated blood to circulate in the kidneys. This is not due to clots interrupting the circulation, but this arrest is rather to be ascribed to a tetanus of the vessels; in fact, the circulation becomes re-established after a few minutes to be again arrested. There are other mere movements of contraction and dilatation (rapid), which modify the velocity of the current without interrupting it. These phenomena in their rhythm resemble very closely those which Schiff observed in the ear of the living rabbit. These spontaneous movements which modify the rapidity of outflow are not shown by all kidneys. When they exist they are very pronounced a short time after death, and even are not completely absent two days after death.

2. To ascertain whether the cause of the preceding phenomena was to be ascribed to a contraction of the vessels, Mosso stimulated the kidneys with an electric current during the artificial circulation. The kidneys were enclosed in the electrodes of tinfoil. The effect of stimulation of fresh kidneys is a pronounced fall in the rapidity of outflow of the blood. It is very remarkable, however, that the induced current has no influence, whilst the constant current interrupted every second always produces a very obvious contraction.

3. After having interrupted the artificial current for a certain time, on again recommencing the current with the same pressure, it was found that the blood flowed out with a velocity much greater than before, and the muscular tonus was soon re-established. The paralysis of the vessels which was produced during these interruptions is for intervals which are not too large proportional to the length of the interruptions. These phenomena can be observed even on the third day after death.

The rapidity with which the vascular tonus is re-established is in inverse proportion to the length of the interruptions; a short time after death the tonus is re established in a very few minutes, whilst two days after death it requires three-quarters of an hour of artificial circulation to cause the paralysis of the vessels slowly to disappear, and even in this latter case small oscillations are to be perceived in the rapidity of outflow.

4. In this series Mosso studied the influence of the gases of the blood on the circulation in the kidneys. He employed four sorts of defibrinated blood: arterial blood defibrinated without contact with air; venous or asphyxiated blood defibrinated in the same manner under mercury, so as not to lose any of its gases; blood defibrinated in contact with air, which he calls, apnœic; and lastly blood treated with iron filings, which he calls reduced blood, i.e. deprived of its oxygen. In all these researches nitrogen and non-compressed air was employed. These are the results obtained:—Asphyxiated blood circulates with a smaller velocity than the apnœic. The result of this series of researches is that, in artificial circulation, every time that a species of blood is made to follow another which is poorer in carbonic acid, the velocity of outflow increases, in such a way that the vessels, even though independent of the nerve-centres, can regulate the velocity of the blood which passes through them.

5. The researches made in the circulation in the kidneys of poisoned blood, show that many of the phenomena, which up to this time have been ascribed to the vaso-motor centre or the heart, are really due to a special action on the walls of the vessels. Nicotine and atropine, in small doses, produce a contraction which is of short duration; whilst in larger doses there is considerable dilatation, with increase in the velocity of outflow. Hydrate of chloral produces also, at the commencement, a slight contraction of the vessels which is soon followed by a very considerable dilatation. Quite as constant a phenomenon is the tetanus of the vessels, when blood which has been poisoned with chloral is followed by normal blood. In order to demonstrate that all these phenomena were not due to a modification of the blood-corpuscles, the experiments were repeated with poisoned serum, which gave the same results.

6. In passing from these facts to others which have a direct bearing upon pathology, we have another series of experiments on œdema of the kidneys, where the disposition to the production of œdema increases after death. On employing non-poisoned blood for the fresh organs, there is a production of œdema or exudation for a certain time; but several hours after death there are all the characters of a well-pronounced œdema.

7. Lastly the author studied the artificial circulation in the liver. The same phenomena are reproduced in this organ, although the portal system and the hepatic vein are very poor in muscular fibres.

The variations in the velocity, and its increase after each stoppage, show that in the liver, in spite of the absence of muscular fibres in the small vessels, phenomena similar to those in the kidney occur.

The author also demonstrated these results with poisons (nicotine, atropine, chloral, prussic acid, etc.), which yielded the same results with the liver as have already been cited for the kidney. One difference between the vessels of the liver and those of the kidney— a difference of great interest—is that stimulation with a galvanic current, interrupted every second, is followed in the liver by a very pronounced augmentation in the velocity during the irritation; whilst the induced current has no obvious action. We have here the singular case of a stimulus which dilates the vessels without first causing them to contract.

THE CONDITION OF THE WALLS OF THE BLOOD-VESSELS DURING EMIGRATION OF THE COLOURLESS BLOOD-CORPUSCLES.—J. Arnold (*Virchow's Archiv*, Band LXII. p. 487, and *Centralblatt für die Medicin. Wissenschaften*, No. 49), injected a weak solution of silver into the blood-vessels of frogs, in which single parts of the body had been inflamed twenty-four hours previously. The appearances of the so-called endothelial figures varied considerably from the normal. The "cement-lines" appeared as broad, strong, zig-zag lines, or were only indicated by rows of granules.

There are large dark points in them, the stigmata, and in the cement substance as well as in the stigmata are to be found white corpuscles in the act of passing out of the vessels, and this in various stages of their passage. Not unfrequently several colourless corpuscles are to be seen attached to one part of the cement substance, or to one of the stigmata; thus, two corpuscles may be lying within the vessel, and fixed by short processes in a stigma, whilst a third one for the most part has penetrated and only remains in connection with the vessel by means of a short process. Sometimes the accumulation of the white corpuscles on the outer side of the vessel, in the neighbourhood of the stigmata, is so considerable that the sheath of the vessel at the corresponding spot stands out like a little hump from the endothelial layer. From this the author concludes, that the white blood-corpuscles in inflammation pass through the wall at the cement substance, *i.e.* the stigmata; whilst, on the contrary, he never saw that other parts of the wall of the vessel, *e.g.* the endothelium, was permeable to the white blood-corpuscles (see L. Purves, *Journ. of Anat. and Phys.* IX. 423).

From experiments made with gelatine and fine vermilion injections, it seems that, in addition to the pronounced wandering out of colourless blood-corpuscles connected with disturbances in the circulation, other corpuscles also pass through the wall of the vessel, and apparently this also happens at the portion of the stigmata and

cement-substance. When blue-coloured gelatine instead of the
vermilion was employed, the inflamed vessels at numerous spots
seem to be covered with small roundish blue protuberances, or with
more elongated blue processes. From this it results, that the wall
of the vessel permits not only substances in solution to pass through
it at the position of the stigmata and cement-substance, but also a
colloid body (gelatine), and other elements (vermilion.) The masses
which have passed through the vascular walls, penetrate in the direction of the "juice-canals" (Saftcanalsystem) in the tissue, and under
certain circumstances these can be completely filled laterally with
the injection mass. Still the configuration of this system of spaces
when injected is variable, according to the disturbances of the circulation which have occurred in the tissues. Whilst the juice-canals
during venous stasis possess a broad and ampullated form, they
appear smaller and more zig-zag in those disturbances of the circulation which are specially characterised by the exit of white blood-
corpuscles.

ON DIFFUSION BETWEEN BLOOD-CORPUSCLES AND SERUM.—O.
Nasse (Sitzungsb. d. Marburger Ges. Z. B. d. g. N., 1874, No. 4), of
Marburg, investigated how far the presence of carbonic acid and
oxygen influence the diffusion between blood-corpuscles and serum.
The defibrinated blood of a horse was allowed to stand till the corpuscles fell to the bottom, and then mixtures of serum and cruor in
varying quantities were prepared. From one and the same mixture
one portion was treated continually with CO_2, another portion with
O. As the chief result of the CO_2 mixture, increase of the specific
gravity of the serum, diminution of water and NaCl. The extent of
this change increases directly with the quantity of cruor. As the
mean, 1,000 grms. serum weighed 2·5 grms. more, contained 4·45 per
thousand more solid constituents on the whole; on the contrary, 0·57
less of NaCl. Water and NaCl therefore pass into the blood-corpuscles, which thereby increase in volume. The CO_2 also takes part
in the increase of the specific gravity, but still this is not great
enough to account for the entire change. Blood flowing directly
from an artery of a horse was defibrinated at one time with air,
the next time with exclusion of air. The serum of the latter was
0·3 grm. per thousand heavier. Further, the author investigated
how other cells conducted themselves towards CO_2 and serum. He
employed for this purpose minced flesh and freshly-rubbed down
liver. In all cases the specific gravity of the serum increased considerably, nevertheless more pronounced in the presence of CO_2 than
in the presence of air, and the NaCl diminished in the same sense.
The author refers the phenomena to the coagulation of the protoplasm by the acidulation, which must be more pronounced in the
presence of CO_2, as the tissues were all employed fresh.

ON THE INVESTIGATION OF BLOOD-SERUM, WHITE OF EGG, AND
MILK, BY DIALYSIS THROUGH GELATINISED PAPER.—Alex. Schmidt
(Pflüger's Arch., Vol. XI., and Centralblatt für die Medicinischen

Wissenschaften, No. 1, 1876), announces that the so-called English parchment-paper employed by him is not parchment-paper at all, but only a kind of ordinary writing-paper very carefully prepared with alum and gelatine. 100 grammes of the paper yield to boiling water a mean value of 4·11 grammes of gelatine, 0·64 gramme of potash alum, and 0·79 of other soluble salts. The quantities of gelatine and alum which at ordinary temperatures pass into the alkaline solution of albumen or into the diffusate are very small, so that this impurity need not be taken into consideration. The author prefers De La Rue's paper purified by extraction with dilute hydrochloric acid and water, and then covered with a thin coating of gelatine. Steeping for a short time in a one per cent. solution of gelatine is sufficient for this. The author then gives the method he employs for estimating the albumen in blood-serum, etc. The blood-serum is neutralised, coagulated with ten times its volume of alcohol, allowed to stand for twenty-four hours, then boiled, filtered, the coagulum washed with a mixture of ten parts of alcohol and one part of water, then with absolute alcohol, and lastly with ether. The soluble salts contained in the fluid remain in solution, the coagulated albumen only contains the insoluble earthy phosphates. On employing gelatinised paper, the quantity of albumen which passes through the paper is not inconsiderable ; if the dialysis is continued for two or three days, and the water changed very often, then the greater part of the albumen passes through. The author considers 1, blood-serum and egg-albumen ; 2, milk.

1. *Serum and Albumen.*—If diluted serum or solutions of egg-albumen be dialysed, there first occurs a stage at which the solution no longer coagulates on being boiled ; it has an alkaline reaction, and still contains traces of salts. As the dialysis proceeds the reaction becomes neutral, the solution of albumen is free from soluble salts, and on burning leaves only earthy phosphates. If an acidulated solution be subjected to dialysis, it still remains for a time capable of coagulating, even when all salts have been removed from it, and continues so until the last trace of acid has passed out. The quantity of earthy phosphates contained in the albumen continually diminishes with the duration of the dialysis, and not only absolutely, but also relatively to the quantity of albumen—to 0·194 per cent. of albumen. The soluble condition of the albumen therefore depends neither upon the quantity of alkali nor on that of the earthy phosphates. *Albumen is rather a substance soluble by itself in water.* The earthy phosphates pass into the diffusate in combination with a nitrogenous organic body, and remain in solution after the removal of the albumen which has passed into the diffusate.

2. *Milk.*—The former observations of the author upon this subject are here mentioned, that acidity of the milk diffusate also takes place even when paper free from alum is employed, so that it does not depend upon the presence of this substance, but is altogether absent in some cases. A certain part of the diffusate does not undergo the acid reaction ; if, after the removal of all soluble salts and the milk sugar, the diffusate now resulting is separately collected, a fluid is

obtained which, in addition to earthy phosphates, contains only certain organic substances, and shows no tendency to become acid. Like the albumen, a part of the casein also passes through the paper. Milk purified by dialysis conducts itself very peculiarly with regard to rennet. At first the capability of the milk for coagulating is increased by the rennet—*i.e.* the coagulation occurs at a lower temperature, and the cause of this phenomenon is the removal of the alkaline salts, which prevent the coagulation; on continuing the dialysis, however, the milk becomes quite incapable of being coagulated by rennet, therefore by the process of dialysis *some substance must have passed out*, which determined the *coagulation by rennet*. This property of being coagulated by rennet can be again restored through the addi-·tion of the diffusate of milk, nevertheless only the diffusate obtained from milk which has become acid during the dialysis acts in this way, acidity of the diffusate arising spontaneously does not suffice.

In conclusion the author rejects the view of Heynsius, who ascribed the absence of coagulation in the solutions of albumen employed by Schmidt to the presence of alkalis. Schmidt asserts that Heynsius had no solutions quite free from salts, and that completely neutral solutions of albumen when quite free from salts do not coagulate either upon heating or on adding alcohol (see this Report).

ON THE COMPOUNDS OF ALBUMEN, OF BLOOD-SERUM, AND OF WHITE OF EGG.—A. Heynsius (Pflüger's *Archiv*, IX. 514, and *Centralblatt*, No. 30, 1875) in the introduction to his paper reviews the results of Schmidt, Eichwald, Landois, etc. The author justly prefers the method of heating after the addition of acetic acid and a few cubic ·centimètres of a concentrated solution of common salt (or sulphate of magnesia) as the best test for the presence of traces of albumen.

1. The author then attempted to obtain solutions of albumen containing the smallest possible quantity of salts, and for this purpose employed a dialyser with a surface of two square centimètres and of special construction. Rain water was at first employed for the dialysis. White of egg, diluted with water and saturated with a solution of common salt, was dialysed for seven days. A considerable precipitate formed in the dialyser; the fluid filtered from this became turbid at 45°, and the filtrate of this at 48°. If a solution of common salt was added to it, both the temperature at which coagulation took place, and the temperature at which the albumen was completely separated rose. The albumen could not be completely removed by boiling with a smaller quantity of NaCl than 0·4 per cent. The filtrate from this precipitate when further dialysed formed again a precipitate in the dialyser, and the filtrate from this became turbid even at 28°, with a neutral reaction. An addition of NaCl acts just as before. The blood of oxen and horses shows similar phenomena. There is, therefore, present in blood-serum and in white of egg a compound of albumen, which is decomposed at a low temperature. On the author employing distilled water instead of rain-water he could no longer obtain this compound of albumen. From this it appeared that on using rain water the solution of albumen in the dialyser very soon became neutral, but retained its alkaline reaction with distilled water.

The same difference was shown by directly mixing the solution of albumen with water. There must, therefore, be a substance in the rain water which fixes the alkali. The zinc oxide contained in the rain water was probably the substance. The experiments to prevent this action by careful acidulation failed. The author then reviews the results of Schmidt and Aronstein. The solution either free from or at least very poor in salts did not coagulate at 100°, and even when mixed with alcohol showed only a little turbidity, although it had constantly an alkaline reaction, which even in solutions poor in salts exercises a very disturbing influence on the coagulation. The author is of opinion that Schmidt and Aronstein have overlooked this action. As they observed no coagulation in solutions acidulated with acetic acid, this the author ascribes to too strong acidulation. The author records experiments which show the disturbing influence of alkalies and acids in solutions poor in salts.

2. A solution of albumen, which became turbid at 41°, after dialysis was heated to 45°, and brought into solution by the addition of a little alkali in solution (four cubic centimètres of a one-tenth normal solution of potash to 0·568 albumen). A faintly turbid fluid of alkaline reaction, which showed the reactions of paraglobulin, resulted. After precipitation with acetic acid it is only soluble in water and oxygen, as long as upon burning it leaves an ash which has an alkaline reaction. The author made no experiments upon the fibrino-plastic action of this solution. This precipitate in the dialyser gradually passes into insoluble albumen. The alkali-albuminates are varying bodies according to the concentration of the alkali employed for their preparation.

3. Just as the alkalies, so has the concentration of the acids an influence upon the properties of the acid albumen formed. Paraglobulin is scarcely soluble in CO_2; if the solution is allowed to stand, the CO_2 disappears and the albumen is again excreted. If a stream of carbonic acid is passed through a solution of alkali albuminate, the coagulation-temperature is lowered. . If a solution of white of egg is treated with varying quantities of acetic acid, then the temperature at which coagulation occurs rises. These phenomena, however, are only to be well observed when dilute acetic acid and relatively large quantities of albumen are employed. (*Lond. Med. Rec.*, Oct. 15th, 1875.)

"On the preparations and properties of Solutions of Albumen free from Salts." A. Winogradoff (*Pflüger's Arch.* xi. 605).——"On Albumen and its compounds." A. Heynsius, *Ibid.* (xi. 624).

On the Composition and Fate of Fat introduced into the Blood.—A. Röhrig (*Ludwig's Arbeiten*, ix. 1). A. Röhrig remarks that the statement in the ordinary text-books that the blood contains soaps must appear surprising considering the quantity of potash salts present. In fact, even blood-serum treated with soap solution yields a precipitate of chalk-soap which gradually becomes crystallized. Blood-serum therefore does not contain any soaps. The direct test yielded the author only negative results. In order to estimate the amount of fat in the blood-serum, it was shaken with several times its own volume of alcohol, and then completely extracted with alcohol and ether. The residue which remained after evaporating the ether, consisting of the fats, cholesterin and lecithin,'was then weighed, and the cholesterin and lecithin estimated according to the usual methods. The blood was at once treated with its own volume of water, and 2 cc. of a 1 per cent. solution of oxalic acid to prevent coagulation; and then the above process was repeated. Double estimation by this process gave good coinciding results. By this method the author tested the rapidity with which fat when introduced into the blood disappeared from it again. For this purpose an emulsion formed by shaking together oil and water to which a little carbonate of soda had been added, was employed. This emulsion was injected into the peripheral end of an artery. The injection was made by mercurial pressure. The blood was investigated before the injection, immediately after it (from the central end of the artery), and sometime later. In one experiment the blood contained immediately before the injection 0·504 per cent.; immediately after 0·668 per cent.; and thirty minutes later 0·636 per cent. The increase was obvious but small. In the later experiments the time of injection, which in the first injection reached sixty-five minutes, was diminished. The differences became larger, but were not very important. The difficulty of the experiment led the author to adopt another method. After the dogs had fasted, they obtained a large quantity of hog's lard. Four hours after feeding, the thoracic duct, the subclavian vein, and the lymphatic ducts of the chest and neck on both sides, were carefully ligatured. In all cases the quantity of fat rose after feeding, and after ligature of the thoracic duct again fell. The cholesterin shewed a tolerably constant value. In one experiment there was found :

	Fat.	Cholesterin.
Shortly before feeding	0·74	0·11
Immediately after ligature . . .	1·24	0·21
Three hours ,, ,, . . .	0·89	0·19
Eight and a half hours do. . . .	0·52	0·18
Twenty-two hours ,, . . .	0·50	0·19

The cholesterin without doubt arises from the bile which is richly poured out during digestion. After several days fasting, the fat in the blood is 0·5 or 0·7 per cent., and after feeding with fat may rise to 1·25 per cent. The quantity of fat falls after ligature of the ductus thoracicus; at first rapidly, then slowly. The fat probably does not leave the blood without undergoing a change. Supporting this view is the fact of the constancy in the quantity of cholesterin ; and also that the lymphatics of the corresponding extremity do not contain any fat. Röhrig thinks it is probable that the fats are oxidised in the blood with the formation of CO_2. (*Lond. Med. Rec.* 1875.)

ON THE QUANTITY OF OXYGEN ABSORBED BY THE BLOOD AT DIFFERENT BAROMETRIC PRESSURES.—According to P. Bert (*Comptes Rendus*, Tome LXXX. 733, *Centralblatt für die Medicinischen Wissenschaften*, No. 55, 1875), hæmoglobin forms with oxygen a chemical compound, which can be obtained by shaking blood with atmospheric air. Increase of pressure is without influence on this combination. More oxygen is taken up, but only corresponding to the absorption of this gas by the fluid, following Dalton's law. A diminution of the pressure to one-eighth of an atmosphere, at a temperature of 60.8° Fahr. has very little influence on the oxyhæmoglobin; from this limit onwards, however, it rapidly decomposes. The decomposition occurs earlier at a temperature of 104° Fahr. This is a cause of the poverty of oxygen in the blood of animals, which are subjected to an increasing diminution of pressure; a second cause is, that the contact of the blood with the air in the lungs is not sufficiently intimate to influence the degree of saturation with oxygen corresponding to the pressure and temperature. For this reason, the quantity of oxygen in the blood in experiments on animals is always smaller than in experiments outside the body. (*Lond. Med. Rec.* 1876.)

ON HÆMOGLOBIN.—L. Hermann and Th. Steeger (*Pflüger's Arch.* Vol. x.) contribute a paper on this subject. From the observations of Lothar Meyer, Pflüger and Zuntz, we know that by adding an acid to blood only the smallest part of the oxygen of the hæmoglobin can be obtained by a vacuum. The cause of this phenomenon is to be sought for in the decomposition of the hæmoglobin, the oxygen being held fast by a decomposition product. For this purpose arterial blood was passed into a flask with warm water (176° to 194° F.) which was connected with the vacuum. About one-third of its oxygen was obtained by pumping. It was now necessary to see whether oxygen alone showed this property of combining with the decomposition products, or whether other gases which also form compounds with hæmoglobin showed the same property. For deciding this question blood saturated with carbonic oxide, and in another experiment with nitrous oxide, was conducted into the same flask. From the former only 1·7 to 1·8 vols. per cent. of carbonic oxide was obtained; from the latter, 4·9 vols. per cent. nitrous oxide and oxygen. These gases are also fixed by the decomposition products resulting from the rapid decomposition of hæmoglobin.

"Causes and Mechanism of the Coagulation of the Blood." Matthieu and Urbain, *Bulletin Général de Thérapeutique*, Sept. 5th, (Abstract in *Lond. Med. Rec.* Nov. 15th, 1875), published separately by G. Masson, Paris, pp. 285.——"On the Transfusion of Blood." L. Landois. Vogel, Leipsig.——"On certain Physical Phenomena connected with the Circulation, Respiration and Nutrition." Dr G. Johnson, *Brit. Med. Journ.* Jan. 1st, and 8th, 1876.——"On the Accommodation of the Vessels for large quantities of Blood." L. Lesser, *Ludwig's Arbeiten*, IX. 50.——And "Transfusion and Auto-Transfusion." L. Lesser, *Sammlung Klinischer Vorträge*, No. 86, 1875.——"Effect of Lime and Magnesia on the Alkalinity of the Blood." (*Lond. Med. Rec.* July, 1875.) —— "On the Origin of Cells containing Blood-Corpuscles, and the Metamorphosis of Blood in the Lymph Sac of the Frog." O. Lange, *Virchow's Arch.* LXV. 27. ——"On the Relation of the Blood and Lymph-Vessels to the Juice-Canals." P. Foà, *Virchow's Arch.* LXV. 284.——"On the

From
in the
the fu
almost
or in
specially
tinued
Westph
flexion
erroneous
the plant
always war
The leg-ph
occurs; not
is certain, t
In grey deg
examination.
the most diff
rhage), and in
disease of the
involved, or 1
spinal cord are
ascribe *the pho*
tation of the con
of the tendon, f
reflexly through
sensory muscular
to reflexes which
nothing. For th
continued abnorm
the muscles of the
cal stimuli, plays a
logical observation
the lateral columns
This is certain, that
eration of the poste
lumbar portions of t
occurrence, Westph
contraction to a dis
muscles, through tens
that they are reflex
cannot be admitted a
nerves is still a cont
knew nothing of reflex
logical cases the author
dition of contraction of
render them more susc
part.

According to Juffroy
and they can be produce

"On normal digestion in infants." Wegscheider. Hirschwald, ᵗerlin. (Abstract in *Lond. Med. Rec.*, Jan. 15, 1876.)——"On the ιechanism of deglutition," par Dr L. Fiaux, pp. 150, with two plates. aris, P. Asselin.—"The daily Metamorphosis of transfused Albumen," Tschiriew, *Ludwig's Arbeiten*, IX., 292.——"On digestion and ιsorption in the large intestine." F. Falck, *Virch. Arch.*, LXV. 393.

ON THE GASTRIC JUICE.—Rabuteau (*Comptes Rendus*, LXXX., 61) gested the gastric juice of dogs with freshly precipitated and well ιshed quinine, then evaporated, and extracted with absolute alcohol, aporated the alcoholic extract, and then treated the residue with nzol or chloroform, which dissolved the quinine salt formed and on evaporation left it in a crystalline form. This salt proved to hydrochlorate of quinine ; the lactate was not to be found, just as the ordinary method of shaking with ether in the gastric juice. the volumetric method it was found that 1000 parts of gastric ᴄe contained 2·5 parts of HCl, coinciding very closely with the ults of Schmidt—3 per thousand.

ON PEPTONES AND FEEDING WITH THE SAME.—P. Plósz and Györgyai (*Pflüger's Archiv*, Band x., p. 536 ; *Centralblatt für Medicinischen Wissenschaften*, No. 55, 1875) have performed ew experiment on the nutritive value of peptones on an adult which had hungered for several days. They fed it with ιtion of peptones and the necessary nutriment free from nitrogen, ιnated the quantity of nitrogen in the total excreta of the animal, compared it with the known quantity of nitrogen in the food. ·ing the whole experiment there was, nitrogen taken in, 14·451 ᴤ. ; nitrogen excreted, 13·463 grms. ; 0·988 grm. were therefore ιned in the body (under the assumption that the quantity of ᵑgen can be accurately measured under these circumstances !). body-weight rose from 2·531 grms. to 2·790, i.e. ·259 grm. Corᵒnding to the nitrogen deficit in the excreta, one can assume that ng the feeding with peptones an addition of albumen had taken ᵔ, that the peptones can certainly be changed into albumen and ᵒyed for the building up of cells. ⅼnother question of course arises, How far can peptones introᵔⅼ into the body be followed as such ? For the solution of this ᵗιon the authors injected 20 to 30 grms. of a watery solution of ᵓnes into the stomach of dogs which had hungered for forty-eight ᵓᵧ and then killed them after two or four hours. As a test for ᵓnes they employed the reaction of caustic potash and sulphate ᵔᵖper (purple-violet solution), or boiling with Millon's reagent ᵑιtric acid (yellow colouration). The greatest quantity of pep ᵇy this process was found in the blood of the mesenteric veins ᵑ the extract of the mesenterium. The liver contained much less, ιnct traces were found in the blood of the hepatic vein and of the carotid artery. On injecting peptones into the veins, ιppears unchanged in the urine, the larger part remaining in ᵒᵈly, and undergoing further changes. In the blood of the carotid, after three hours, a small quantity of peptone was still traceable. Lastly, the authors conducted blood, to which peptones had been added, in an artificial circulation through the amputated hind quarters and extremities of a large dog just killed (excluding the glands). The circulation lasted four or five hours. Here also the blood lost part of its peptones; in one case 20 grms. of peptones disappeared from the blood. After finishing, the blood was washed out of the vessels with a solution of common salt, and a watery extráct of this as well as of the tissues was then made. Neither peptone, nor any other nitrogenous body which one might expect as a derivative of the albumen, was present.

ON GLYCOGEN IN MARINE ANIMALS.—P. Picard (*Gazette Médicale*, 1874, No. 49) in estimating the amount of glycogen in the liver converted this substance into sugar, whose amount was estimated by titration with Fehling's solution. In different species of osseous fishes the quantity of glycogen in the liver varied from 1·1 to 6·4 per cent.; in cartilaginous fishes from 0·3 to 1·6· The difference is explained by the liver of the latter being more voluminous. In relation to body weight, on the contrary, the quantity of glycogen nearly coincides. In the liver of lobsters there was 0·4 to 0·5 per cent., of crabs, 0·3 per cent. of glycogen. The presence of glycogen was also proved in echinoderms, polyps, and sponges. Sugar was regularly found, still the change of glycogen into sugar proceeds in general slowly in fishes. No sugar was found in crustacea, mollusca, etc.

ON THE PRODUCTION OF GLYCOSURIA BY THE EFFECT OF OXYGENATED BLOOD ON THE LIVER.—F. W. Pavy (*Royal Society*, Nov. 25, 1875) arrives at the conclusion that the amyloid substance found in the liver is a body which tends to accumulate in certain animal structures under the existence of a limited supply of oxygen, and that it is through the liver exceptionally receiving the supply of venous blood it does, that the special condition belonging to it is attributable. It is also shown that the undue transmission of oxygenated blood to that organ at once induces an altered state, which is rendered evident by the production of glycosuria.

ON THE PANCREAS.—R. Heidenhain (*Pflüger's Arch.*, Vol. x., and *Centralblatt für die Medicinischen Wissenschaften*, No. 2, 1876) observed the following appearances were successively presented by the cells of the pancreas at the different stages of digestion.

1. During hunger the granular inner zone occupies the larger, the homogeneous outer zone the smaller part of the cells. 2. In the first period of digestion, during which a plentiful secretion occurs, there is diminution of the entire cells by using up of the granular inner zone, then addition of new materials to the outer zone, so that this becomes enlarged. 3. In the second period of digestion, during which the secretion diminishes and comes to a standstill, there is a new formation of the granular inner zone at the expense of the homogeneous outer zone, most pronounced diminution of the latter, increase of all the cells. 4. With long-continued

hunger there is a gradual increase of the latter to their original dimensions, and therewith slight diminution of the inner zone. During the state of physiological activity there is a continual change in the cells—metamorphosis internally, addition of matters externally. Internally there is conversion of the granules into secretory constituents, externally employment of the nutrient materials for the formation of the homogeneous substance, which again becomes converted into granular masses. The average appearance of the cells depends upon the relative rapidity with which this process occurs. In the first period of change there is a more rapid consumption internally and more rapid addition externally; in the second period the most rapid changes occur at the limit between the outer and inner zones, in that the substance of the former becomes converted into that of the latter. During the condition of hunger the consumption is at a minimum, the addition takes place also slower, but it is still obvious in the apparent increase of the outer zone, which had almost disappeared. Corresponding to these histological changes, the chemical and experimental results of the author shewed that pancreatin ready prepared (albumen-ferment of the pancreas) in the living gland-cells does not exist, but only its peculiar mother-substance : under certain circumstances pancreatin becomes free. This mother-substance, which Heidenhain calls 'zymogen,' is probably a compound of pancreatin with an albuminate. With plentiful secretion from the gland, the quantity of zymogen diminishes, to be again generated during rest of the organ. This process of regeneration does not occur sufficiently in a gland with permanent fistula as soon as the secretion has become continuous.

Further, it is shown that the secretion of water by the gland can be influenced by the medulla oblongata, and from the author's experiments, it is extremely probable that the excretion of the solid constituents of the gland-cells and that of water do not go hand in hand, but each seems to be under the direct influence of the nervous system. The formation of pancreatin is connected with the complicated decompositions in the secreting cells, the development of a free acid playing a part in the process.

ON MECONIUM.—Zweifel (*Archiv für Gynäkologie*, Band VII. 474, and abstract in *Centralblatt für die Medicinischen Wissenschaften*, No. 37, 1875) employed as material the contents of the great intestine of still-born children. The microscopic investigation gave the interesting result, viz. the existence of hæmatoidin crystals with the characteristic Gmelin's reaction. By extracting the meconium with chloroform and evaporating it, bilirubin crystals could be obtained. They are no longer contained in the yellow evacuations, and thus disappear with the other constituents of meconium. As chemical constituents of meconium were found: biliverdin, bilirubin, a bile acid, viz., taurocholic, cholesterin, mucin, traces of formic acid, and the higher volatile fatty acids, and non-volatile fatty acids. Negative results were obtained for grape-sugar, glycogen, paralbumen, leucin, tyrosin, albumen-peptone, lactic acid. The quantity of water in meconium was almost 80 per cent., the ash about one per cent. The fat in fresh meconium was 0·772, and the cholesterin 0.797 per cent. Two analyses of the ash are given. In per cent. the composition is, insoluble substances, 2·1; phosphate of iron, 3·41; sulphuric acid, 23·0; chlorine, 2·53; phosphoric acid, 5·44; lime, 5·7; magnesia, 4·0; potash, 8·6; soda, 41·0. In relation to the excreta of adults the high percentage of sulphuric acid and diminution in phosphoric acid is noticeable.

"Action of Bile on the organism," Feltz and Ritter, in Robin's *Journal de l'Anatomie et de la physiologie*, 1875, also *Acad. de Sciences*, Sept. 1st, 1875. (Abst. in *Lond. Med. Rec.*, Oct. 15th and Dec. 15th, 1875).——"Experiments on the biliary secretion of the dog." Rutherford and Vignal, *Journ. of Anat. and Phys.* x. 253.——"On Cholesterin," V. Krusenstern, *Virch. Arch.* LXV. 410, and Beneke, LXVI. 126.——"On the formation of urea in the liver," I. Munk, *Pflüger's Arch.* XI. 100.——"On the secretion of the liver," N. Socoloff. *Ibid.* XI. 166.——"On Choletelin and hydrobilirubin," L. Liebermann," *Ibid.* XL 181.

THE NERVES OF THE LIVER.—M. Nesterowsky (*Virchow's Archiv*, Band. lxiii. 412) by the gold method and the subsequent addition of ammonium sulphide, succeeded in demonstrating the nerves in fresh sections of frozen injected liver. The nerves of the liver end in the form of networks, which surround the blood capillaries, and the terminations of the nerves have no connection with the liver cells.

Genito-Urinary System.

"Two pathological urinary colouring matters," F. Baumstark, *Pflüg. Arch.* IX. 568. (Abst. in *Centralblatt*, No. 35, 1875).——"Action of Salicylic Acid on the Urine," R. Fleischer, *Berlin Klin. Wochensch.*, Sept. 27th and Oct. 4th, 1875.——"Physiology of the secretion of the Urine." P. Grützner, *Pfüger's Arch.* XI. 370.—— "Estimation of Urea," C. A. Pekelharing, *Pflüger's Arch.* XI. 602.

ON POISONING WITH NITROBENZOL, AND EXCRETION OF SUGAR IN THE URINE.—Dr V. Mering (*Centralblatt für die Medicinischen Wissenschaften*, No. 55, 1876), contradicts the assertion of C. A. Ewald, that the substance excreted in the urine of dogs and rabbits in poisoning with nitro-benzol is sugar, though it reduces alkaline copper solution. It does not undergo fermentation when treated with yest, under the necessary precautions. Further, this body rotates the ray of polarised light to the *left*.

ON THE INFLUENCE OF BOILING DISTILLED WATER ON FEHLING'S SOLUTION.—E. Boisin and E. Loisen (*Comptes Rendus*, Tome LXXIX.) have found that pure distilled water when boiling reduces Fehling's solution employed for testing for sugar. Thus fifty cubic centimètres of water can reduce one cubic centimètre of solution. Ordinary spring water does not do this. A series of substances hinder the reduction, when they are added to the distilled water previously, above all lime salts, ammonia, and soda salts. With a quite pure solution of grape-sugar some chloride of calcium must be previously added to the Fehling's solution. The authors recommend this as a means of testing the purity of distilled water.

ON DIABETUS MELLITUS.—Dr Pawlinoff (*Virch. Arch.* LXIV. 382) says that sugar cannot be oxygenated in the blood, but the muscles can decompose it into substances which can be more easily oxygenated than albumen. In the normal organism, the oxygenation of albumen takes place principally in the arterial blood. By the oxygenation of albumen in the arteries, there is formed urea, while in the veins there carbonic acid is formed by the action of oxygen upon the products of decomposition of sugar. In diabetes, the muscles cease to change sugar into substances which are easily decomposed, in consequence of which the process of oxydation loses its energy, as the albumen is oxydised with greater difficulty.

Milk. "Comparative investigations on human milk, and that of the cow." A. Langgard, *Virch. Arch.* LXV. 1.

Temperature.

"Action of various drugs upon temperature." A. G. Burness, *Brit. Med. Jour.*, Jan. 15th, 1876.——"On the local temperature of paralysed parts." M. Schiff, *Lo Sperimentale*, March and May, Abst. in *London Med. Rec.*, July and August, 1875.——"On the relation of the peripheral to the central temperature in fever." W. Schulein, *Virch. Arch.* LXVI. 109.——"The highest temperature at which life can exist." F. Hoppe-Seyler, *Pflüger's Arch.* XI. 113.

ON THE HIGHEST TEMPERATURE BEARABLE BY THE HAND.—Bloch (*Gazette Hebdomad.* No. 25, 1875) records the following as the highest temperature which the hand could bear in baths of different substances for two minutes at a time:—

Substances: Mercury 48° Cent. Water 49°. Solution of Tannin 49°. Vinegar 51°. Solution of Sodium Carbonate 52°. Alcohol 51°. Milk 52°. Spirits of Turpentine 55°. Glycerine 57°. Oil 60°. Beef suet 65°.

According to the author the reason that Hg. and solution of tannin at low temperatures seems as hot as other things at higher temperatures, is because of the density of the former, and the rapid imbibition by the skin of the latter.

Muscle.

"On the regeneration of striated muscular fibre." Peroncito, *Gazzetta delle Cliniche*, Jan. 26. (Abstract in *Lond. Med. Rec.*, No. 123, 1875).——"On Sommer's Movements" (shortening of the muscles by rigor mortis). H. Storoscheff, *Wien. Acad. Sitzungsb*, 2 Abth. LXX. Abstract in *Centralblatt*, No. 35, 1875.——"Influence of Carbonic oxide on the duration of muscular contractility." M. Rochefontaine, *Gaz. Méd. de Paris*, Dec. 11th, 1875.——"Contractility and double refraction." J. W. Engelmann, *Pflüger's Arch.* XI. 432.——"On the conduction of excitation in the Heart," Engelmann, *Ibid.* XI. 465.——"On electrical tetanisation of Nerves and Muscles." G. Valentin, *Ibid.* XI. 481.

ON THE HEIGHT OF THE MUSCULAR TONE BY ELECTRICAL AND CHEMICAL STIMULATION.—The tetanic contraction of a muscle, whether produced voluntarily or artificially during life, consists in a vibrating motion of small particles in the interior of the muscular fibre. This is shown by the negative variation of the muscular current discovered by Du Bois-Reymond, and by the muscular tone which, according to the investigations of Helmholtz, consists of as many vibrations as the number of stimuli applied. The normal muscular tone, which is heard during voluntary contraction, consists of, according to Helmholtz, only 18—20 vibrations per second, but by means of an electrical tuning-fork much higher tones could be produced. In the human fore-arm he produced a clear tone with 240 vibrations.

J. Bernstein (*Pflüger's Archiv*, xi. 191), investigated the muscular tone produced by a greater number of stimuli per second. A special form of electrical tuning-fork was employed. The shocks were applied to the muscles of the leg of a rabbit, where the muscular tone can be easily heard when a stethoscope is applied over the skin. Three hundred and thirty vibrations gave a tone of the same height, tolerably strong, and of the same *Klangfarbe*; 418, 561 and 748 vibrations per second gave each corresponding tones. With 1,056 vibrations no distinct tone was to be heard in the muscle, although the muscle still contracted; instead of the tone ill-defined noises occurred. When the nerve instead of the muscle directly was stimulated, 1,056 vibrations still gave a distinct tone audible in the muscle, but this tone no longer corresponded with the vibrations of the tuning-fork, but was a fifth, sometimes an octave deeper. It now remained to determine the upper limit to which the muscular tone could be increased by the present apparatus. This was perfectly fixed when the tuning-fork gave the tone $b'' = 933$ vibrations, in which case the muscular tone of similar pitch was softly but distinctly audible.

There is a relation between the muscular tone, contraction, and the negative variation. With 300 vibrations per second the muscular tone was of almost constant moderate pitch. With 300—400 vibrations it began to be weaker. Now as a negative variation of the muscular current corresponds with every vibration of the muscular tone, so it is very probable the weakening of the tone depends upon the duration of the negative variation, which lasts about $\frac{1}{300}$ second. Chemical stimuli (saturated solution of NaCl) applied to the nerve also gave a distinctly audible muscular tone. (*Lond. Med. Rec.* Dec. 15, 1875.)

ON THE ACTION OF EMETICS ON THE TRANSVERSELY STRIPED MUSCULAR FIBRES.—E. Harnack (*Archiv für Experiment. Pathologie* III. 44, and *Centralblatt*, No. 35, 1875) remarks that a large number of the substances which act as emetics and nauseants have also a paralysing action upon the voluntary muscles. To this group belong emetin, apomorphin, tartar-emetic, cyclamin, asclepiadin, and sangui-narin, delphinin, veratrin, and digitalin. The author tested the copper and zinc salts and some drugs obtained from the vegetable kingdom.

We have very few experimental results on the action of the copper salts. To avoid local disturbances by the substance, and specially to avoid coagulation of the albumen, the author chose a soda double salt, viz., tartarate of cupric oxide and soda, for his experiments. In the frog, copper acts eminently in paralysing the voluntary muscles. A quantity of the solution of the copper salt, which contained half to one milligramme of copper oxide injected subcutaneously into a frog, produced, after a few hours, complete paralysis, and with 3 milligrammes CuO the same result after one hour. In rabbits 0·05 grms. CuO injected subcutaneously, or 0·01 to 0·15 into the veins caused death, with the phenomena of paralysis of the heart and respiration, in the former case after a few hours, in the latter after a few minutes. The voluntary muscles, specially those of the posterior extremities, lose their excitability completely, so that the strongest induction shocks are without effect; there is neither dyspnœic breathing nor convulsions. The pupils are dilated. The result is similar in dogs on introducing 0·4 to 0·025 grm. CuO subcutaneously or into the veins. By this method of exhibition vomiting does not occur, which.is in opposition to Orfila's result, but in unison with Daletzky's (St. Petersburg Dissertation, 1857). To produce vomiting 0·6 to 0·7 grm. CuO required to be introduced into the stomach, whereby of course the general action was prevented.

The zinc-salts conduct themselves similarly to the copper-salts. The author employed the phosphate of zinc oxide and soda and the valerianate of zinc. The animals resist the zinc salts better than the copper salts. The minimum dose of ZnO is 0·002 grms. for the pro-duction of complete muscular paralysis in frogs, and with small doses the animals recover. In dogs, 0·4 to 0·5 grm. given subcutaneously or injected into the stomach produce vomiting, 0·1 to 0·12 grm. ZnO injected into the veins, death.

Various other metallic salts, which do not possess a specific emetic action, as the salts of lead, manganese, tin, do not extinguish the muscular irritability when they are not employed in too large lethal doses. Mercury salts, on account of their violent local action, are not well suited for comparison.

In addition to these salts the author investigated asaron and colchicin. The former is derived from *Asarum europœum*. In doses of 0·01 grm. introduced into the stomach it acts as a powerful paralyser of the muscles in frogs. Frogs must receive internally 0·05 grm. of colchicin in order to obtain muscular paralysis. Even then it occurred late, and the heart is also affected late.

Regarding the connection between the emetic and the muscle-paralysing action, only this is certain, that these two phenomena do not stand in the relation of cause and effect.

Bone. "Changes in cartilage before ossification." O. Rosenthal, *Centralblatt* No. 35, 1875.——"On the formation of bone." *Arch. für Mikroscop. Anat.*, 1875.

· *Joints.* "On the hip- and shoulder-joints, and on joints in general." H. Welcker, *Zeitsch. f. Anat. u. Entwicklungs.* ı. 41. Abstract in *Centralblatt,* No. 35, 1875.

Miscellanea.

"Experimental Pathological Studies on the physiological action of toxic and medicinal substances." M. Vulpian, *Le Progrès Médical,* April, 1875 (Abstract in *London Med. Rec.* Nos. 122, 123, 124).——"On Picrotoxine and the antagonism between Picrotoxine and Chloral-hydrate." Crichton Browne, *British Med. Journal,* March 27, April 3, 10, 17 and 24 (Abstract in *Lond. Med. Rec.* No. 122, 1875).——"The mechanism of the action of Quinine on the circulatory system, and its action on muscular fibre in general." V. Chirone, *Lo Sperimentale,* xxvi. 1875.——"On the Graphic Method in the experimental sciences, and on its special application to Medicine." Marey in *Brit. Med. Journ.,* Jan. 1st and 15th, 1876.——"The transference of matters from the Mother to the Fœtus." Benicke, *Allgein. Wiener Med. Zeitung,* No. 40, 1875. (Abstract in *Brit. Med. Journ.,* Dec. 25, 1875).——"Spectrum of Fish-pigment." G. Francis, *Nature,* Dec. 30th, 1875.——"On the action of intra-venous injections of Chloral on the Vaso-motor Nerves." P. Heger and Stiénon, *Annal. de la Societé de Méd. de Gand.* June, 1875.——"A Study of the Movements of the Unimpregnated Uterus." E. van de Warker, pp. 26, New York, 1875.——"On the functions of the Levator Ani, with reference to pathogenesis." J. Budge, *Berlin Klin. Wochensch ,* July 5th. (Abstr. in *Lond. Med. Rec.* Aug. 16th, 1875.)——"Influence of compressed air on Fermentations." P. Bert, *Comptes rendus,* June, 1875.—— "Physiological action of Geselmia." J. Ott.——"On the occurrence of Alcohol in the Organism. A. Rajewsky, *Pflüger's Arch.* xi. 122. ——"On the Amorphous Ferments." O. Nasse, *Ibid.* xi. 138. "Inhibition of the action of Ferments in the living animal." B. Luchsinger, *Ibid.* xi. 502.

On the Action of Chloral Hydrate and Croton Chloral Hydrate.—J. V. Mering (*Archiv für Experiment. Pathologie und Pharmak.* iii. 185, and *Centralblatt,* No. 37, 1875) from his experiments with croton chloral-hydrate, when given in small doses to rabbits before the cessation of the reflex action from the cornea, considerably slows the respiration ; and when the cornea-reflex is extinguished the number of respirations is diminished by a half. The substance (0·6 gramme) was introduced subcutaneously, or divided in scarcely half as large total doses and injected into the veins. Similar results were obtained from parallel experiments with chloral hydrate. Croton chloral-hydrate acts upon the frog's heart similar to chloral hydrate by even small doses (0·025 gramme given subcutaneously) causing cessation of its action. Blood-pressure experiments upon dogs, cats, and rabbits were performed with both substances. Small doses diminish the blood-pressure temporarily, whilst large doses diminish it continuously, till the blood-pressure

. curve reaches the abscissa. The pulse-beats are at first increased by both drugs, and this increase lasts somewhat longer with croton chloral hydrate. At a certain stage of the experiments the blood-pressure remained continually low, in spite of great blood-movements. This seems to show that the croton chloral hydrate extinguishes the arterial tonus, the heart's energy still continuing, this being, as is known, asserted of chloroform and chloral hydrate. From these and former experiments there seems to be a great similarity in the actions of the three above-mentioned drugs.

Lastly, the author combats Liebreich's theory of the action of croton chloral hydrate. According to this theory this substance in alkaline blood splits up into dichlorallylen, hydrochloric acid, and formic acid, and the formdr, analogous to aethyliden chloride, is the active factor.

The author employed only the trichlorcrotonate of soda, which even in the cold in dilute alkaline solutions passes into dichlorallylen. This substance was injected into rabbits, and even when five grammes were injected, it had no effect.

ON THE MODE OF ACTION OF THE MOST FREQUENTLY EMPLOYED EMETICS.—H. Chouppe (*Archiv de physiol. norm. et pathol.*, 1875, 101) on injecting apomorphin into the veins of dogs, produced vomiting, which occurred more rapidly (after two minutes), but not so often, and ceased more quickly than by subcutaneous injection. A watery solution of emetin neutralised by citric acid has a similar action, or if employed as a decoction of the ipecacuanha root, the vomiting occurrs later however (after about a quarter of an hour). The necessary doses are about the same as when employed sub-cutaneously, viz. 0·002 apomorphin as a minimum, 0·02—0·10 emetic, more than 0·10 grm. is fatal; for tartar emetic 0·15 to 0·20 grm. After section of the vagi below the diaphragm, emetin, as D'Ornellus had already found, loses its effect; whilst apomorphin and tartar-emetic act the same as on intact animals. The same dog upon which the largest possible doses of emetin produced no effect vomited with same doses of apomorphin or tartar-emetic. As the animals after section of the vagi often vomited for a short time after the operation, the author always allowed several hours to elapse before the above experiment was performed. Even after complete extir-pation of the stomach the author observed that apomorphin was still active,—as Magendie showed for tartar emetic,—by its pro-ducing violent vomiting movements, which of course in this case were absent when emetin was given.

From these experiments emetin produces its effects by acting on the vomiting centre in the medulla oblongata, by stimulating the fibres of the vagus ending in the gastric mucous membrane. That the action does not occur through the medium of sympathetic fibres is shown by the following experiment. After section of the cord in the neck below the second vertebra, emetin injected into the veins pro-duced strong contractions of the exposed stomach, such as are to be observed during the act of vomiting, and vomiting movements in the pharynx and mouth; an actual evacuation could of course not take place on account of the paralysis of the diaphragm and the abdominal muscles. That the vomiting which follows the introduction of ipecacuana preparations into the stomach occurs more rapidly than when they are injected into the veins agrees very well with the view of the action on the gastric mucous membrane. From analogous grounds it is probable that in the case of apomorphin there is a direct stimulation of the centre in the medulla, still the author admits that there may be an action on the peripheral ends of the vagus. He also assumes the same mode of action for tartar-emetic, only in this case the reflex path (from the gastric, œsophageal or pharyngeal mucous membrane) is the more effective, for as is known, by direct application smaller doses are required than by injection into the veins. With apomorphin the reverse is the case. A direct proof of the action of tartar-emetic on the vomiting centre is not given.

ON THE PHOSPHORESCENCE OF DEAD ORGANISMS.—E. Pflüger, (*Pflüger's Archiv*, XL 222) in a paper partly experimental and largely historical, discusses the cause of the phosphorescence exhibited by dead organisms. On observing by daylight a surface which has been luminous, it is always found to be covered by a whitish dirty slime, which may sometimes be from two to four millimètres in thickness. It is this slime or its constituents which possess the property of phosphorescence. The most important fact connected with this luminosity is, that it depends upon the presence of oxygen. As one experiment made by the author, showing the consumption of oxygen, may be mentioned the following. A solution of crystallised hæmoglobin of such a strength that the two absorption bands were clearly to be observed spectroscopically, was treated with a few cubic centimètres of luminous salt water, and the whole placed in a hatching apparatus and warmed for a few minutes. The oxyhæmoglobin absorption bands totally disappeared, and the single reduction band of Stokes was present. On shaking the mixture a little the two absorption-bands again reappeared. Heat alone, as is well known, only causes the oxygen to disappear from a solution of oxyhæmoglobin after a long time. Here excessively small quantities of O caused an increase in the phosphorescence of luminous water. Experiments were then made with substances which destroyed fermentation and putrefaction by killing the organisms. Alcohol and carbolic acid rendered the luminous water at once dark. Fresh-water fish are not phosphorescent unless they have come in contact with salt-water fish, and in the few cases where they have been observed to be phosphorescent this may be explained by infection from some sea-fish.

The microscopic structure of the slime covering the bodies of phosphorescent fish was then investigated. It was found to consist of lower organisms, the so-called schizomyceten, which are the proper luminous materials. This was shown by a filtration experiment, where the organisms were retained upon the filter (fine thick non-sized printing paper; Swedish filtering paper does not do;) which remained luminous, while the perfectly clear filtrate was absolutely non-luminous, this clearly showing that the small living cells of the schizomyceten are the cause of the. luminosity; and further, the author's experiments furnish strong proofs that the schizomyceten do not arise "spontaneously," but are derived from spores.

In framing any theory of physiological phosphorescence the following chief points must be kept in view.

1. The intensity of the light is generally extraordinarily weak, and therefore the phosphorescence can only be well observed by a retina which has become very exciteable by being long in darkness. The retina is then in such a case of almost endless sensibility.

2. The development of light is caused by oxydation.

3. According to our present knowledge there is no living cellular substance which is so darkly coloured by perosmic acid as the "light cells" (lampyris). Nowhere, therefore, do the atoms which bind the oxygen in the living cell-substance lie so densely as in the light substance. [*Lond. Med. Rec.*, Dec. 15th, 1875.]

THE RELATION OF SULPHINDIGOTATE OF SODA TO THE TISSUES OF THE LIVING BODY.—L. Gerlach, of Erlangen (*Centralblatt für die Medicinische Wissenschaften*, No. 48, 1875), adopted the method of saturating the tissues with this substance for days and even weeks together; the former experimenters, Heidenhain, Kupfer, Von Wittich and Thoma only injected such a quantity of indigo-carmine as remained in the body for a comparatively short time. The author injected indigo-carmine into the lymph-sac of several frogs and killed them at intervals of two days, always renewing the injections. The microscopic examination showed, that the white blood-corpuscles are capable of taking up indigo-carmine.

1. The first traces of this action appear on the third day after the introduction of the colouring matter. After this time both the number of cells which contain the pigment and the quantity of pigment in the individual cells increase.

2. The cells of the connective tissue, *e.g.* of the tendons, take up the colouring matter. This is to be observed from the fourth day.

3. No indigo-pigment is deposited in the bone-cells.

4. The pigment is found in the cartilaginous, *e.g.* the articular cartilage of hip-joint, from the fifth day onwards. None is found in the ground-substance or matrix.

5. The nerve-cells never contain the indigo; only in a few cases was it found in the sympathetic ganglion-cells between the cell-contents and the sheath.

6. The blue colouration of the epithelial cement pointed out by Thoma and Küttner is also true for that of the so-called endothelium. [*Lond. Med. Rec.*, Jan. 15th, 1876.]

ON THE ACTION OF LYCOCTONIA.—This vegetable principle is found in the Aconitum lycoctonium, and an account of some experiments made upon it by Dr Isaac Ott, of Philadelphia, is given in the *Philadelphia Medical Times* of Oct. 16. He says with a dose of ·005 to ·06 gramme subcutaneously in frogs, sensibility remained intact, the spinal cord appeared not directly affected, the motor nerves were paralysed, muscles excitable, pupil contracted, disappearance of respiratory movements, heart beating for a very long time after the absence of every sign of life. In the majority of cases he found the motor nerves completely non-irritable, but in some frogs, without regard to species, and even in the same species, the motor nerves were not completely paralysed. As to its effects upon the circulation, it was found that immediately after its injection into rabbits the pulse waves were considerably greater and steeper. In small doses, like aconitia, its action on the heart causes a delirium cordis, that is, periods when the pulse is greatly slowed whilst the pressure rises and falls in a manner not to be predicted. In larger doses there is reduction of both the pulse and the blood-pressure. The vagi are not paralysed by small doses, while they are so affected by large ones. The following are the conclusions arrived at. 1. Lycoctonia is a weaker toxicant than aconitia. 2. It kills mainly through the respiratory apparatus. 3. It paralyses the motor nerves. 4. It does not affect the sensory nerves, spinal cord, or the striated muscles. 5. It reduces the blood-pressure and pulse without any previous rise of the former as produced by aconitia. 6. The decreased pulse-rate and pressure are due to any action on the intercardiac nervous apparatus. 7. The pneumogastrics are paralysed only by large doses. 8. The delirium cordis produced by small doses is due to a change in the mechanism of the nervous apparatus of the heart.

PHYSIOLOGICAL ACTION OF NITRITE OF AMYL.—Berger (*Allgein Med. Centralzeitung*, 1874) found that, in non-curarised dogs, whose vagi were derived, the division of the spinal cord below the atlas was followed by a lowering of the blood-pressure when the vapour of amyl nitrite was introduced into the lungs by artificial respiration. The lowering of the blood-pressure by nitrite of amyl, is therefore not dependent on the vaso-motor centre in the medulla. The frequency of the pulse in warm blooded animals was always increased; the respiration retarded by fatal doses.——Also Samelsohn, *Berlin Klin. Wochenschr.*, Jan. 14, 1875. (Abstract in *Lond. Med. Rec.*, Aug., 1875.

Text Books.

Physiologie Expérimentale. M. Marey. G. Masson, Paris, 1876.

Handbuch d. physiologisch- und pathologisch-Chemischen Analyse. Dr F. Hoppe-Seyler, 4th edition, 1875, Hirschwald, Berlin.

Nouveaux Éléments de Physiologie, H. Beaunis, 1876, T. B. Baillíere et Fils, Paris.

Guide to the Practical Examination of Urine. J. Tyson, M.D. Trübner and Co., London, 1875.

Carpenter's Physiology, 8th edition, 1876. J. Churchill, London.

Dalton's Physiology, 6th Edition, 1876, J. Churchill, London.

Physiologische Methodik. Dr Richd. Gscheidlen, Part I., 1876. Vieweg and Sohn, Braunschweig.

Methodik der physiologischen Experimente und Vivisectionen, with Atlas of 54 Plates. Prof. E. Cyon, Part I., 1876. J. Ricker-Giessen.

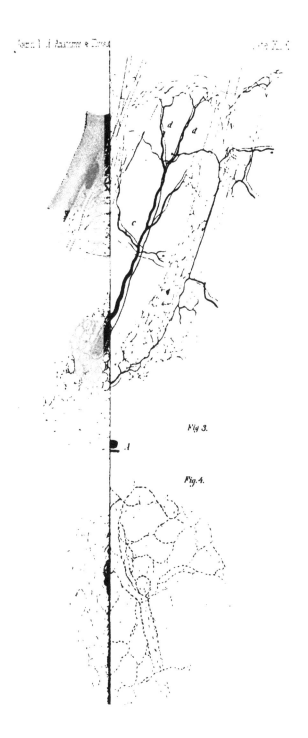

Fig 3.

Fig. 4.

Plate XX

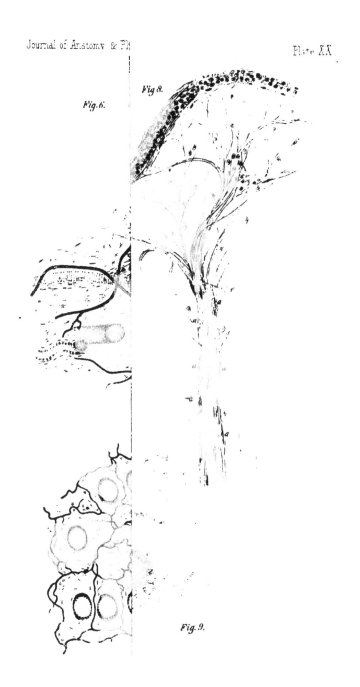

Fig.6.

Fig. 8.

Fig. 9.

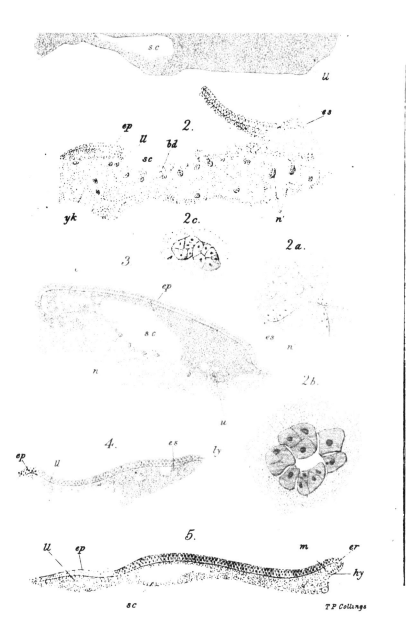

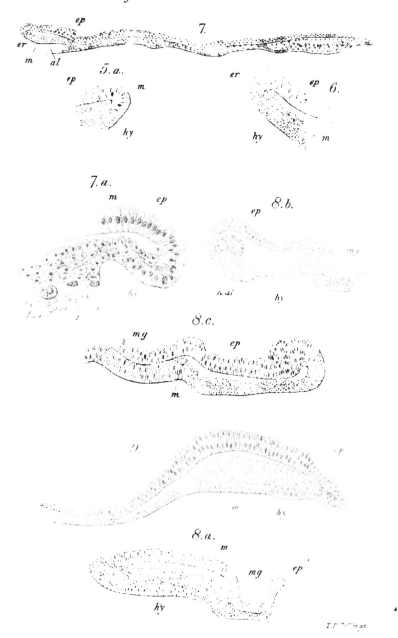

7.

5.a.

6.

7.a.

8.b.

8.c.

8.a.

Pl. XXIII.

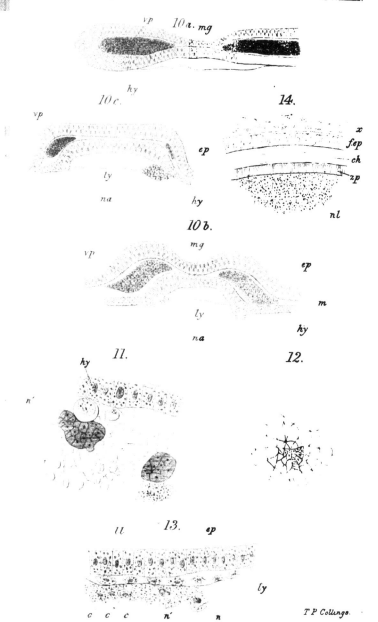

10 a. mg

10 c.

14.

10 b.

11.

12.

13.

T P Collings.

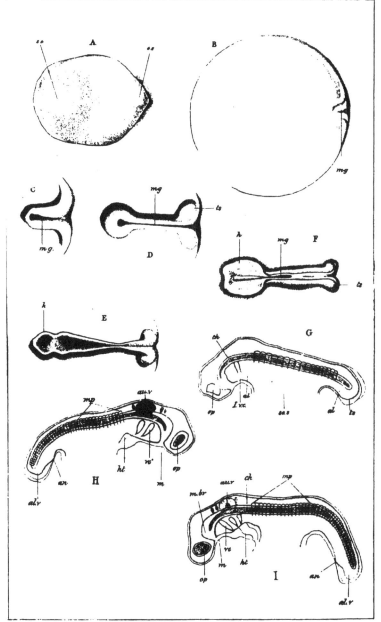

Pl. XII

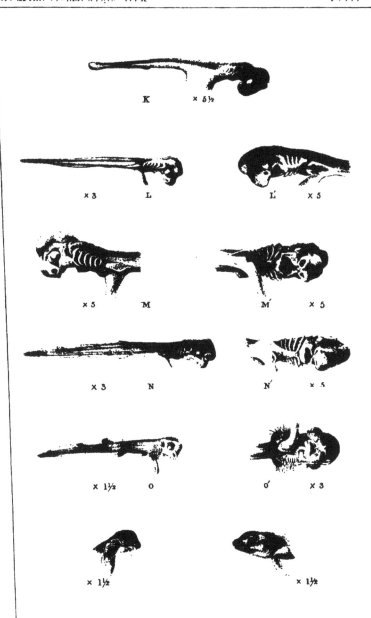

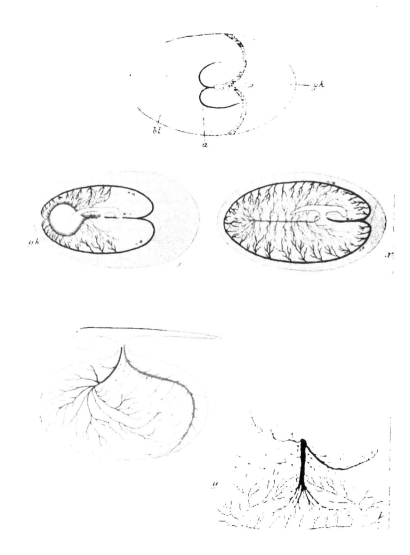

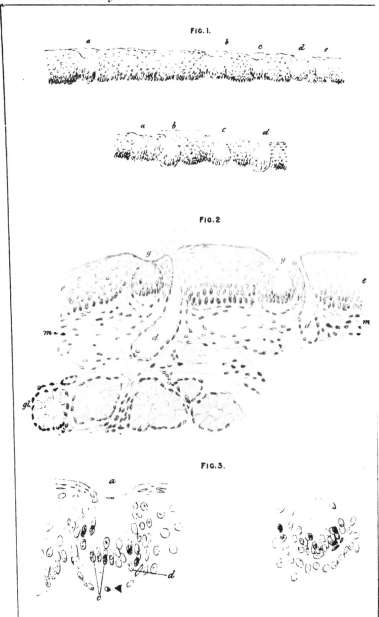

FIG. I.

FIG. 2

FIG. 3.

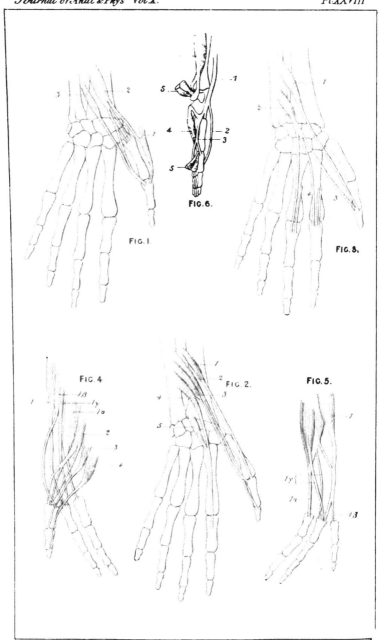

FIG.6.

FIG.1.

FIG.3.

FIG.4.

FIG.2.

FIG.5.

Pl XXVIII

FIG 1

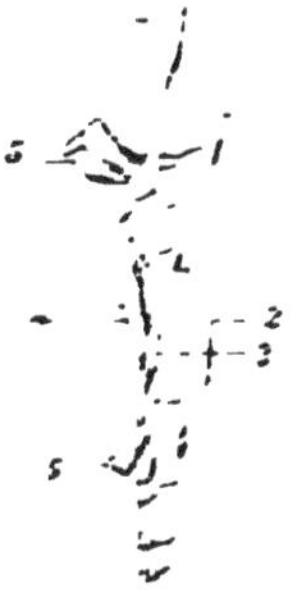

FIG 6

FIG 3

FIG 4

FIG 2

FIG 5

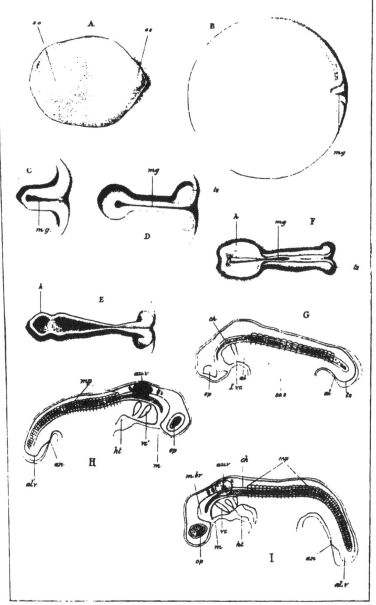

Plate XX.

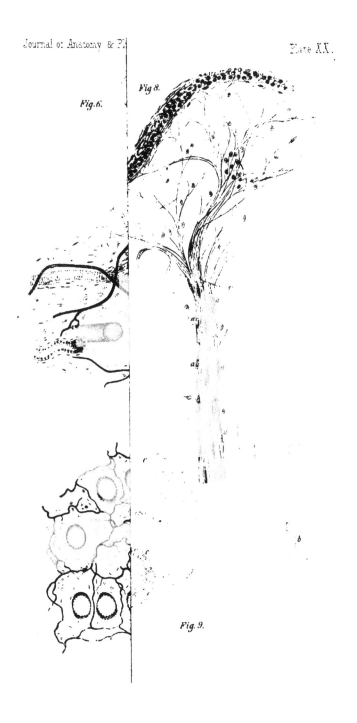

Fig. 6.

Fig. 8.

Fig. 9.

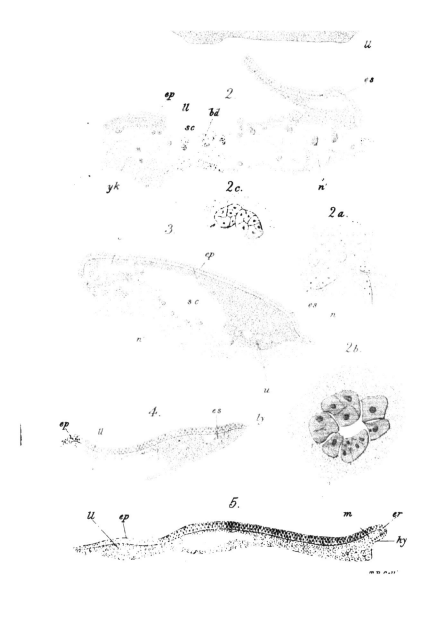

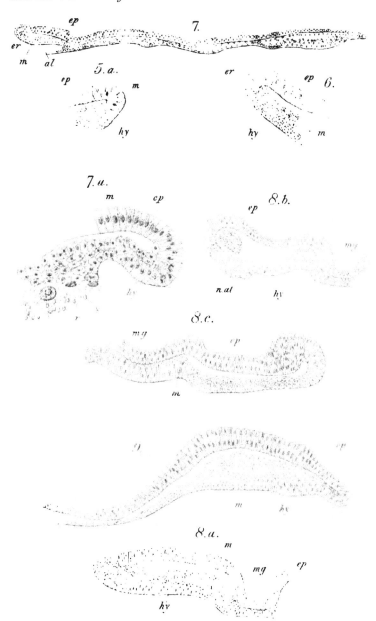

Pl. XXIII.

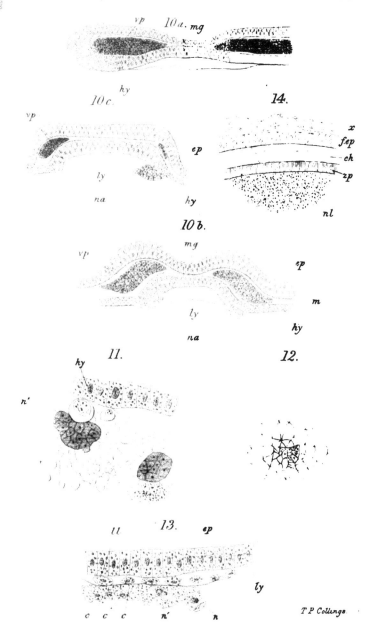

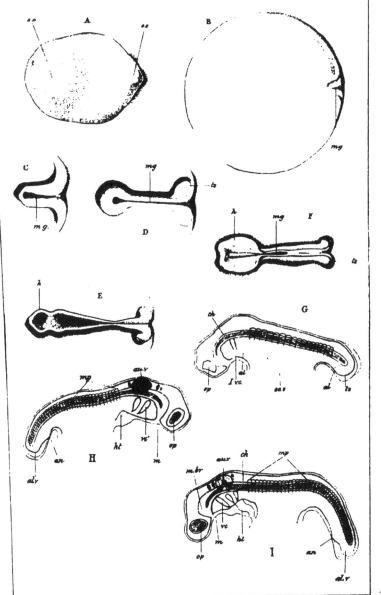

Pl. IX

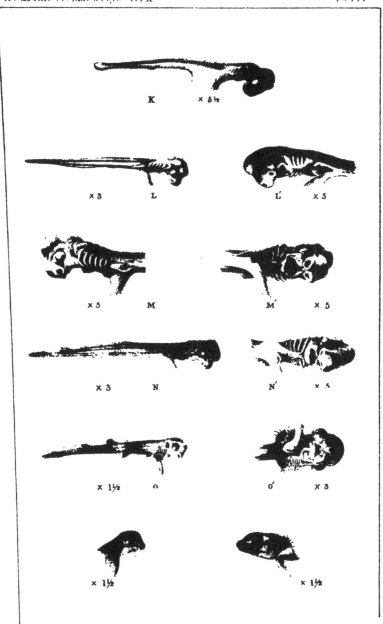

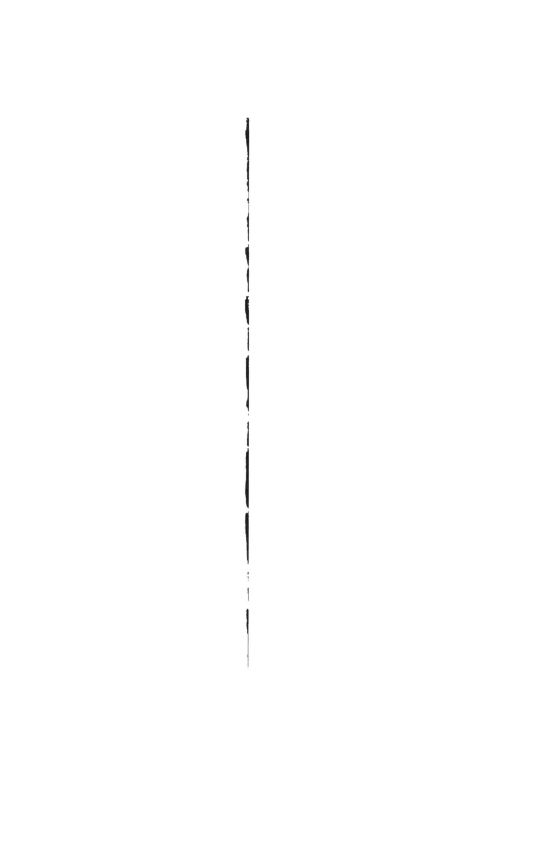

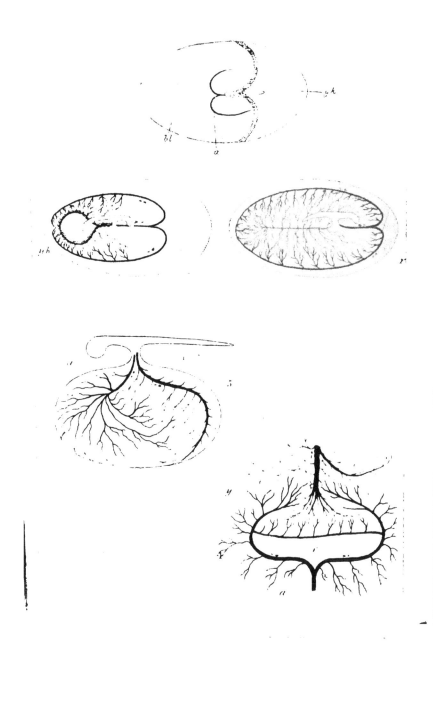

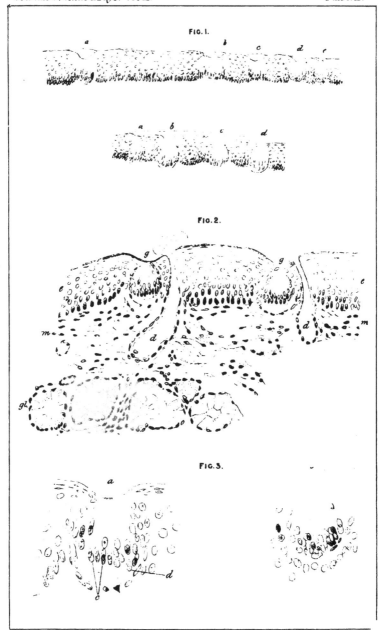

FIG. I.

FIG. 2.

FIG. 3.

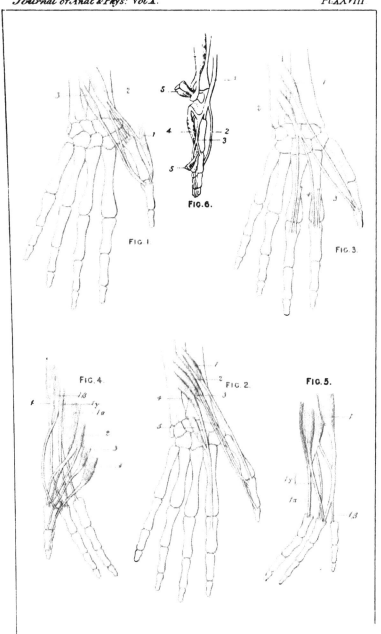

FIG.6.

FIG 1.

FIG. 3.

FIG.4.

FIG. 2.

FIG.5.

ON THE COMPARISON OF THE FORE AND HIND LIMBS IN VERTEBRATES. By Professor Humphry.

The homological comparison of the limbs having been made the subject of an inaugural dissertation in the University of Kiel by Dr Albrecht[1], I think it well to make a few remarks upon it which are, in great measure, a repetition of observations already published in this Journal and elsewhere.

Dr Albrecht first gives a refutation of the torsion theory of the humerus, which, propounded by M. Martins, of Montpellier[2], as an ideal theory affording a clue to the homological comparison of the fore and hind limbs, has been enunciated as the expression of an actual process—an actual twisting of the bone in its length during development—by Gegenbaur[3]. According to this view the lower end of the humerus, and with it the forearm and hand, undergoes a rotation upon the upper end of the bone to the extent of nearly a half circle. The outer condyle and the radius are supposed to have been originally internal, so corresponding in position with the internal condyle of the femur and the tibia; and the internal condyle of the humerus and the ulna are supposed to have been external, so corresponding with the external condyle of the femur and the fibula; but in the course of development, by a rotation or twist taking place in the shaft of the humerus, the radius and its condyle are supposed to have passed gradually backwards, postaxially, behind the ulna and its condyle till they assumed their position on the outer side of the limb. The arguments in favour of this view are, first, the direction of the lines on the shaft of the humerus and of the groove for the radial nerve, which gives the appearance of such a twist having taken place; and, secondly, the observation of Gegenbaur, from measurement, that the angle formed by the transverse axis of the lower end of the humerus with the axis of the upper end, that is a line drawn

[1] *Beitrag zur Torsionstheorie des Humerus und zur morphologischen Stellung der Patella in der Reihe der Wirbelthiere.* Von Paul Albrecht aus Hamburg.
[2] *Mém. de l'Acad. de Montpellier,* III. 1857.
[3] *Jenaische Zeitschrift,* IV.

through the greatest diameter of the upper articular surface (from the supra-spinatus projection to the under prolonged lip), varies during development, and varies in a direction corresponding with the supposed torsion.

With regard to the first argument, Albrecht remarks that the oblique lines are no evidence of a twisting of the bone, any more than similar oblique lines on other bones, for instance the fibula and the femur, are to be regarded as indications of a twisting having taken place in them. With regard to the second argument, he shows that the measurements given by Gegenbaur of the angles formed by the axes of the two ends of the humerus, as a means of estimating the degree of torsion which the bone undergoes in its shaft during the several periods of growth, are far too variable and uncertain to admit of any such theory as this torsion theory of the arm being based upon them. He further shows that although, by resolving or un-twisting this supposed torsion and bringing the parts to their imagi-nary primitive condition the course of the radial nerve would be straightened and simplified, yet this would be a deviation from its course in the simpler limb of the Amphibian, and the several other structures of the arm would be strangely contorted. Not merely the brachial artery and the median and ulnar nerves, but the biceps muscle, would be compelled to curl round the outer side of the limb to its posterior aspect, and the triceps extensor muscle would have to curl round the inner side of the limb to its destination in the olecranon, which would be thrown to the front. Moreover a want of coinci-dence of the axes of the articular ends is no uncommon thing in other bones, and is not regarded as an indication of any such develop-ment or torsion having taken place in them. He shows also that in the vertebrate series no gradually increasing torsion in the humerus is to be distinguished, and points out that the degree of the angle formed between the axes of the upper and lower ends of the bone depends mainly upon the relation of the humerus to the scapula at the shoulder joint, and concludes therefore that, phylogenetically as well as ontogenetically, the evidence of such torsion is wanting.

In this conclusion I fully concur. In my *Observations on the Limbs of Vertebrate Animals*, 1860, p. 22, I make the following objections to M. Martins' theory, which are similar to some of those now adduced by Dr Albrecht.

"Although the relations of the articular surfaces of the humerus to one another, and the direction of the lines upon its shaft, are oblique, and so might be thought to offer some countenance to such a view, yet the same kind of obliquity may be observed in the case of all, or nearly all, the other long bones ; and the lines upon the humerus and other long bones do not appear with any degree of distinctness till the general form of the bones has been acquired and the position of the articular surfaces fixed. Secondly, no such torsion as that contended for has been observed to take place at any period of development. M. Martins is aware of this; and, therefore, describes it as a *virtual*, rather than as an *actual*, torsion. Thirdly, this supposed *virtual* torsion is in an opposite direction to the *real* torsion of the limb, which takes place during development, and which has been referred to in the preceding pages. It does not, therefore, accord with the disposition of the various soft parts. Fourthly, it is improbable that an important principle would have been carried out so completely in the humerus, without there being something of a similar nature observable in the case of the other long bones, more especially in the femur, which M. Martins does not admit to be the case."

Dr Albrecht has therefore, in my opinion, done good service in the first part of his dissertation by giving additional arguments in disproof of the torsion theory of the humerus. Here, however, my agreement with him ceases; for the theory which he endeavours to substitute for it appears to be as untenable, from a morphological, developmental, or any other, point of view, as the torsion theory itself. He conceives that the clue to a correct comparison of the fore and the hind limbs is afforded by the supposition that the radius is primarily situated, internally, upon the inner condyle of the humerus, and gradually passes in front of the ulna, preaxially, to gain its final position upon the outer condyle, that thus in its primitive position upon the inner condyle it occupied a place corresponding with that of the tibia upon the inner condyle of the femur, but subsequently, in process of development, as also in the ascending animal series, it departs from that position and becomes collocated with the outer condyle. In like manner the ulna passes, post-axially, from the outer to the inner condyle of the humerus. He therefore assumes, like M. Martins, a torsion in the fore limb to the extent of a half circle, or nearly so, but differs from him in the part of the limb in which that torsion takes place. The one assumes it to be in the shaft of the humerus, to be a post-axial movement of the radius and outer condyle, and a pre-axial movement of the ulna and inner condyle, and to affect all the limb below the middle of the humerus. The other assumes it to take place at the elbow, to consist of a pre-axial movement of the radius, and a post-axial movement of the ulna, and to affect all the limb below the humerus. The one theory assumes a twist between the two ends of a bone taking place in its shaft; the other assumes the interchange of position of two long bones by their sliding upon one another on the articular end of a third long bone. In both theories the flexor aspect of the arm is considered, primarily, to be continued into the extensor aspect of the forearm and hand, and to correspond with the extensor aspect of the thigh, the *biceps flexor antibrachii* being regarded as serially homologous with the *quadriceps extensor cruris.*

Dr Albrecht supports his view phylogenetically—by arguments drawn from the animal series—,ontogenetically—by arguments drawn from individual development—,and by the observation of the distribution of the nerves in the two limbs.

The argument from the animal series is briefly as follows: Beginning with Ichthyosaurus he finds, in this animal, that the radius lies at the internal condyle of the humerus, and the ulna at the external condyle; that in Plesiosaurus the radius is somewhat more anterior, also in tailed and tailless Amphibians and in Reptiles; and as we ascend to the higher vertebrates, although, in some Mammals, it is anterior, it gradually assumes the external position which it holds in Monkeys and in Man. It is not necessary to go into the details of this argument, as it is evident that the foundation stone of it is the position of the radius and ulna, respectively, at the internal and external condyles of the humerus in the Ichthyosaurus and Plesiosaurus; and unless that is proved, or unless it can be shown that such is the position of these two bones in some other animal, the whole phylogenetical pillar of the structure falls to the ground. I have examined a good many specimens of Ichthyosaurus and Plesiosaurus, and in every one of them the bone designated as radius is in contact with the *anterior* shelving facet, or condyle, at the distal end of the flattened humerus, and the bone designated as tibia occupies a corresponding position with regard to the femur, while the ulna and the fibula are in contact with the respective posterior condyles of the humerus and femur; and I am quite at a loss to discover upon what ground Dr Albrecht describes the radius and the condyle of the humerus upon which it rests as *internal*. Indeed I am not aware of a single instance in the whole animal series in which the radius occupies the position which Dr Albrecht claims for it in the Enaliosaurians. It is possible that he may have been unfortu-

nate in the specimens of these animals selected, and that the
bones in them may have been displaced, as they are liable to be
more or less, from their normal position. Yet in one of Dr
Albrecht's plates, where the fore limbs of these creatures are
contrasted with those of other animals, the parts are correctly
represented, and the connection of the radius with the anterior
condyle of the humerus of the Ichthyosaurus and of the Plesiosaurus, with the anterior condyle of the Salamander, with the
anterior and external condyle of the Crocodile and Dasypus, with
both condyles of the Ruminant, and with the external condyle
of the Monkey, is well shown.

In a review (p. 27) of the position of the bones of the fore-
arm the Ichthyosaurus is omitted. It is said that in Plesiosaurus the radius is internal, the ulna external (it had been
previously stated that this is less markedly so than in Ichthyosaurus); that in Reptiles and in lower Mammals the radius is
anterior and the ulna posterior, and that in higher Mammals
the radius is external and the ulna internal; that between the
first stage in Plesiosaurus and the second in Reptiles and lower
Mammals all intermediate gradations may be found in Amphibians; and that between the second and the third stages connecting links are met with in the lower Mammals. I would
ask Dr Albrecht again carefully to examine the Plesiosaurus as
well as the Ichthyosaurus, of which so many good specimens may
be seen in the various museums of Europe; and I think he
cannot fail to admit that in them, even more markedly than in
Amphibians, the radius is anterior and the ulna is posterior.
If he will also again review the specimens of the animal series
he will, I think, find that the transition of the radius upon the
articular surface of the humerus, if transition it can be called, is
not, as he supposes, from within outwards; but that on the
contrary the radius is primarily and always upon the anterior
condyle of the humerus, and that as that condyle becomes external the radius sometimes maintains an anterior position by
extending from the external condyle, more or less across the
front of the joint, towards, or to, the internal condyle. In so
doing, however, it does not leave the outer condyle (which was
primarily anterior), but retains a very abiding connection with
it. The movement or rather the extension, therefore, is not

from within outwards, but the reverse. It corresponds indeed
with the extension of the tibia from the internal (this was also
primarily anterior) condyle of the femur which has occurred in
so many animals; and the purpose served in each instance is
that of rendering the middle segment of the limb a better
weight-bearing column than it would have been by a more
equal division of the work between the two bones of the segment; and the movement of the ulna, when the radius does
extend across the front of the joint, is in the opposite direction,
namely from within outwards. This however is slight and
scarcely worth dwelling upon.

The momentum of the supposed pre-axial movement of the
radius and post-axial movement of the ulna is regarded to be
the pulling of the supinator muscles in the one case, and of the
triceps muscle in the other, operating through time in the animal
series. Seeing no reason whatever for the belief in such movements, I should not have alluded to this imaginary cause of them
had I not thought that it involved a fallacy of view as to the development forces. We have no evidence of the operation of muscular action in producing such displacements or alterations of
position. On the contrary the tendency to a change of this kind
under the increasing action or increasing force of a muscle is, in
the normal condition, met by increasing resistance to its pull.
Otherwise the radius ought long ago to have been dragged up
the arm, and the humerus should have been drawn from the
glenoid cavity upon the thorax.

With regard to the developmental evidence in favour of
the pre-axial movement of the radius from the internal to the
external condyle of the humerus, Dr Albrecht's investigations
appear to have been confined to the fœtal calf, in which, at the
eighth and tenth week, he finds the radius internal and the
ulna lying behind it and more external than in the grown
animal. The account is briefly and not very clearly given.
Forasmuch as in the Ruminant the radius, being the chief
weight-bearing bone of the forearm, extends quite across the
elbow, articulating with both condyles (as the tibia commonly extends across the knee), the calf is scarcely a good
instance to select for the purpose of demonstrating a developmental change of position in that bone from one condyle to the

other. Till much more conclusive evidence is furnished from the developmental ground, we must regard the ontological pillar of the argument as resting upon no stronger foundation than the phylogenetical.

To the distribution of the nerves I will shortly return, merely now observing that the argument from it in favour of the view we are discussing, will not be found to be any stronger than those drawn from development and the animal series.

I feel, therefore, that neither the theory of the torsion of the humerus, nor that of the movement of the radius and ulna at the elbow, will aid us in instituting a comparison of the two limbs; and I fall back upon the view which I propounded in 1858[1], which subsequent investigation has convinced me is correct, and which certainly has the merit of affording a simple and easily intelligible solution of the difficulties which have been experienced.

That view is as follows. In the primitive and simple condition, as in Ichthyosaurus and Plesiosaurus, and in Cetaceans, the limbs are flat, paddle-like, and projected nearly horizontally from their girdles and from the trunk, at right angles with the median plane. The radial and tibial margins—including the pollex and the radial aspect of the humerus in the one limb, and the hallux and the tibial aspect of the femur in the other—are directed anteriorly, the ulnar and fibular margins are directed posteriorly, the extensor surfaces are dorsal or superior, and the flexor surfaces are ventral or inferior. The coracoid elements of the girdle in the one limb, and the ischiatic and pubic elements of the girdle in the other, are directed inferiorly or ventrally; and the scapular and iliac elements are directed superiorly or dorsally. So that the scapula and the extensor surfaces of the arm, of the fore-arm, and of the hand are in one plane, and correspond, severally, with the ilium and the extensor surfaces of the thigh, of the leg, and of the foot, which are in the same plane: while the coracoid and the flexor surfaces of the arm, of the fore-arm, and of the hand, in like manner, correspond with the ischium and pubes, and with the flexor surfaces of the thigh, of the leg, and of the foot.

[1] In my *Treatise on the Human Skeleton*, also in my *Observations on the Limbs of Vertebrate Animals*, 1860.

We have only to replace the upper and lower, *alias* fore and hind, limbs of Man or any other animal, into this first straight horizontal position, in which they run out parallel with each other and transversely from the trunk, and the several parts in the two limbs are seen in their natural simple homological relations. The *triceps extensor antibrachii* corresponds with the *quadriceps extensor cruris*; the patella, commonly present in the tendon of the latter muscle, corresponds with the supra-anconeal sesamoid body or bone, sometimes (Batrachians, some Birds, and Pteropus) present in the former[1]; the coraco-radial, or biceps, muscle and the humero-ulnar, or brachialis anticus, correspond with the ischio-tibial and femorofibular, or hamstring, muscles; the coraco-brachial and the pectoral-adductors of the arm correspond with the ischiofemoral adductors of the thigh; the deltoid corresponds with the glutæus; the extensor muscles of the carpus of the pollex and of the manual digits correspond with the extensors of the tarsus of the hallux and of the pedal digits; the flexors in the two limbs corresponding in like manner[2].

As the limbs depart from the primitive straight horizontal position and acquire the ordinary Mammalian or land animal form, flexures, and direction, they are first bent upwards at the elbows and knees both alike, and by a second flexure downwards at the wrists and ankles the plantar aspects of both are brought into contact with the ground; but now the difference between the two limbs commences, in that, associated with these flexures, is an inclination in opposite directions of the elbows and knees and of the proximal segments of the two limbs, corresponding with the traction function of the one limb, and the propelling function of the other; while the distal segments in the two limbs still continue to correspond with one another in position and direction for the purpose of

---

[1] The homological relations of the 'femoral patella,' and the supra-anconeal sesamoid or 'brachial patella,' are unmistakably shown in the stump-tailed Lizard of New Holland (*Trachydosaurus rugosus*), of the skeleton of which there is a specimen in the Museum of the Royal College of Surgeons of England.

[2] For fuller details of these homological relationships see my papers on the "Disposition and Homologies of the extensor and flexor Muscles of the leg and fore-arm," Vol. III. of this *Journal*, p. 320, and "On the disposition of the Muscles in Vertebrated Animals," this *Journal*, Vol. VI. p. 851; also in my *Observations in Myology*, p. 169, and elsewhere.

fulfilling their common work of pressing upon the ground
and propelling the body. Thus the knee and thigh are
slanted forwards, and the extensor surface of the leg and foot
is, as a consequence, directed forwards, the plantar surface of
the foot remaining upon the ground. The elbow and arm
are, however, slanted backwards; and at the same time, to
maintain the requisite position of the fore-foot upon the ground,
pronation takes place in the fore-arm. Thus, by a quarter
turn in opposite directions in each of the two limbs, the elbow
and the knee, and the dorsal or extensor surfaces of the two limbs
are brought opposite to one another, that of the fore-limb looking
backwards, and that of the hind-limb looking forwards; while
by the movement of pronation in the fore-arm the dorsal or
extensor surface of the distal part of the fore-limb is, like that
of the hind-limb, still directed forwards. In the fore-limb
two twists occur, a quarter turn of the proximal segment backwards, and a half turn, in pronation, of the distal segment
forwards. The gradations of these changes can be well followed in the limbs of Amphibians and Reptiles. It will be
seen to take place easily without any dislocation or torsion
of the individual parts; and it may be followed on from these
lower animals to Birds and Mammals.

Even a cursory glance at the animal series, beginning with
Ichthyosaurus and Plesiosaurus and passing through Amphibians and Reptiles to Birds and Mammals, will show that as
the fore-limb loses the primitive horizontal position of the
Enaliosaurian and acquires the flexures of the Amphibian,
with the upper segment and the elbow slanted backward,
the anterior condyle, with its ever faithful attendant the
upper end of the radius, and the anterior surface of the
humerus, become directed outwards, away from the median
plane, while the upper or dorsal or extensor surface of the
arm, with the projecting angle of the elbow and the olecranon,
becomes directed backwards: the fore-arm also falls into a
prone position, the palmar surface remaining upon the ground,
and as this takes place the toes become directed forwards.
At the same time the anterior condyle and anterior surface
of the femur, with the tibia, become directed inwards towards
the median plane, while the upper or dorsal or extensor surface

of the thigh, with the projecting angle of the knee and the patella, becomes directed forwards, the toes, which at first pointed outwards, becoming, like those of the fore-limb, pointed forwards.

Moreover, in the lower animals and in many of the upper, and not least in Man, where supination—that is, the power of restoring the fore-arm and hand to their primitive position with regard to the arm—is more complete than in others, it is easy to rotate the limbs back into their primitive position, to extend them horizontally and at right angles from the trunk, with the pollex and radius, and the hallux and tibia forwards, when the homological relations of their several parts are, in most instances, at once recognised, and the dorsal or extensor, and the ventral or flexor surfaces, in the two limbs will be seen to correspond respectively throughout.

The Cheiroptera, as I have shown[1], afford an interesting confirmation of this view by the exceptional rotation which their hind-limbs undergo. In these animals the rotation of the thigh takes place backwards, instead of forwards, and the knee is bent backwards. Hence the hind-limb is turned in the same way as the fore-limb, and a convincing opportunity is afforded for an examination of the serial homologies of the several parts in the two limbs.

Into the interesting and rather more difficult question of the degree in which the shoulder and pelvic girdles participate in these rotations of the limbs, I have already entered[2], and have shown that they are affected, to a greater or less extent, in a manner corresponding with the respective limbs. In the simpler forms of Enaliosaurians and Batrachians, their component parts are directed upwards and downwards in a nearly vertical plane—the scapula and the ilium upwards, the coracoids and the pubes and ischium downwards. In the higher forms, each undergoes a slight rotation upon a transverse axis, traversing the shoulder-joints, in the one instance, and the hip-joints in the other, by which the scapula is thrown a little backwards and the coracoids forwards, while the ilium is thrown a little

---

[1] "On the Myology of the Limbs of Pteropus." This *Journal*, III. 294.

[2] In a paper on the "Comparison of the Shoulder bones and muscles with the Pelvic bones and muscles." This *Journal*, v. 67.

forwards and the pubes and ischium backwards, Poupart's ligament being the representative of the clavicle.

With regard to blood-vessels and nerves, the passage of the femoral artery in front of the ventral elements of the girdle and the disposition of the nerves in three groups, one—the anterior crural, which is the homologue of part of the radial—passing in front of the girdle, a second—the obturator, which is the homologue of part of the musculo-cutaneous—passing through the girdle, and a third—the sciatic, which is the homologue of the remaining nerves of the fore-limb—passing behind the girdle, while the brachial artery and the nerves of the fore-limb pass together behind the coracoid or ventral element of the shoulder girdle, indicate that no very exact deductions can be drawn from them with reference to the comparison of the upper parts of the two limbs. Moreover, the femoral artery and the nerves are separated in their course, while the brachial artery and the nerves are near together; and the ulnar nerve has no distinct counterpart in the hind-limb, its correspondents being incorporated with the sciatic and posterior tibial nerves. There are, however, some points of similarity in the nerves and vessels in the proximal parts of the two limbs which may not, at first, suggest themselves. The great vessel, in each, passes from the extensor, or dorsal part of the limb (the region of the *triceps* in the one and of the *quadriceps* in the other), across the adductor part (over the *coraco-brachialis* in the one, and through the *adductores* in the other), to the flexor part at the elbow in the one and at the knee in the other. The radial and the peroneal nerves take similar spiral courses (one round the ulnar, the other round the fibular aspect of the limb) to their distribution in .the extensor or dorsal regions of the distal segments. In the fore-limb the radial nerve is incorporated with the nerves to the great extensor of the elbow. Hence its spiral turn round the limb is at a higher level than is that of the peroneal nerve, which remains in conjunction with the sciatic till it has passed half way down the thigh; the great extensor of the knee being, in consequence of the different rotation of the proximal segment of the limb, supplied by a nerve (anterior crural) which passes in front of the girdle. In the Cryptobranch the radial, or great extensor, nerve of the fore-limb is conjoined with other extensor branches which are apparently the homologues of the circumflex and dorsalis scapulæ[1]. I can quite understand that the advocates of any theory respecting the comparison of the upper segments of the two limbs may find arguments in its favour from the disposition of the nerves. I do not therefore attach much importance to them. At the same time it may be mentioned that the supply of the *triceps extensor antibrachii* by the radial or musculo-spiral nerve, which supplies the other extensor muscles of the limb, is in accordance with the view which ranges that muscle in the extensor series.

My view, therefore, is that in the primitive position the extensor surfaces of the two limbs are directed upwards, or dorsally, and the radial and tibial surfaces are directed forwards; and in this position the homological relations of their several parts are to be determined. Subsequently by a quarter turn, to some extent participated in by the pelvis, the extensor surface of the hind limb is directed forwards and the tibial surface inwards ; and by a quarter turn in an opposite direction, to some extent participated in by the shoulder girdle, the extensor surface of the upper limb is directed backwards and the radial surface outwards. At the same time pronation occurs in the fore-arm and hand whereby the palm is kept in contact with the ground. These deviations from the primitive position obscure, but do not essentially affect, the homological relationships determined in that position.

I have spoken of the horizontal position of the limbs, at right angles with the median plane and with their extensor surfaces upwards or dorsal, as the primitive position. I have done so to avoid ambiguity, and for simplicity. I have, however, on a former occasion[2], shown or endeavoured to show that their still earlier form is to be sought in the mesial fins of Osseous Fishes, where the representatives of the limbs of the two sides are to be traced in those coalesced extensions of the embryonic

---

[1] On these and other interesting points in the nerves of the Cryptobranch, as also on the distribution of the nerves in the limbs, see my article "on the Muscles and Nerves of the Cryptobranchus Japonicus," in this *Journal*, Vol. vi., also in my *Observations in Myology*, 1872, p. 48.

[2] "On the homological relations to one another of the mesial and lateral fins of Osseous Fishes," this *Journal*, v. 58; also *Observations in Myology*, p. 115, where it is pointed out that the ventral and pectoral fins of Fishes, and their muscles, are formed from the same serial elements as the subcaudal or anal fins and their muscles.

ventral plates in which bones are developed and which constitute the mesial—anal or subcaudal—fin. This fin, though in the mesial plane, and apparently a single organ, is essentially double, or a pair, being a derivative from the ventral laminæ of the two sides, and supplied with muscles and nerves from the two sides. It admits of being split in the middle plane, and each half then forms a series of limbs directed downwards or parallel with the median plane. In the caudal region, where the ventral laminæ of the two sides are in apposition in their whole extent, or nearly so, these fin-extensions of the two sides, with their ray-bones and interspinous bones, are united with one another, and form the one median anal fin. More anteriorly, however, where the ventral laminæ of the two sides are separated by the visceral cavity, their fin-extensions on the two sides are also separated, and grow out at a short distance from the mesial line as lateral fins. Moreover, instead of forming a continuous series as they do in the anal fin, the rays are here clustered and modified into pectoral and ventral fins, and the interspinous bones are developed into shoulder girdle and pelvic bones. The anal interspinous fin-bones may be continuous in plane with the spinous bones which are the caudal representatives of the ribs, or they may alternate with them, being intercalated between them, extending up, that is, into the planes of the ventral laminæ between them. Their pelvic bone representatives, in higher animals, are commonly appended to the distal ends of the truncated extremities of the sacral ribs, and are therefore continuous in plane with them, as is seen in many of the tailed Amphibians where the rib element exists as a separate bone connecting the ilium with the vertebra; whereas the pectoral arch representatives commonly grow up on the external or superficial aspect of the ribs. It may be, however, that, as in the Chelonians, both the pectoral and pelvic arches pass up on the internal or deep aspect of the ribs. These varieties in disposition do not affect their homological relations to one another in the animal series; nor do they in any way affect the rule that the pollex, the radius, and the radial (*alias* external) condyle of the humerus, as well as the hallux, the tibia, and the tibial (*alias* internal) condyle of the femur are, in the primitive position, directed forwards.

THE DEVELOPMENT OF ELASMOBRANCH FISHES.

By F. M. BALFOUR, B.A., *Fellow of Trinity College, Cambridge.* (Plate XXIX.)

From Stages B to G.

(Continued from p. 570.)

THE present chapter deals with the history of the development of the Elasmobranch embryo from the period when the medullary groove first arises till that in which it becomes completely closed, and converted into the medullary canal. The majority of the observations recorded were made on Pristiurus embryos, a few on embryos of Torpedo. Where nothing is said to the contrary the statements made apply to the embryos of Pristiurus only.

The general external features for this period have already been given in sufficient detail[1]; and I proceed at once to describe consecutively the history of the three layers.

General features of the Epiblast.

At the commencement of this period, during the stage intermediate between B and C, the epiblast is composed of a single layer of cells. (Pl. XXIX. fig. 1.)

These are very much elongated in the region of the embryo, but flattened in other parts of the blastoderm. Throughout they contain numerous yolk spherules.

In a Torpedo embryo of this age (as determined by the condition of the notochord) the epiblast presents a very different structure. It is composed of small spindle-shaped cells several rows deep. The nuclei of these are very large in proportion to the cells containing them, and the yolk spherules are far less numerous than in the cells of corresponding Pristiurus embryos.

During stage C the condition of the epiblast does not undergo any important change, with the exception of the layer becoming much thickened, and its cells two or three deep in the anterior parts of the embryo. (Pl. XXIX. fig. 2.)

[1] *Journal of Anatomy and Physiology,* x. 555, et seq.

In the succeeding stages that part of the epiblast, which will form the spinal cord, gradually becomes two or three cells deep. This change is effected by a decrease in the length of the cells as compared with the thickness of the layer. In the earlier stages the cells are wedge-shaped with an alternate arrangement, so that a decrement in their length at once causes the epiblast to be composed of two rows of interlocking cells.

The lateral parts of the epiblast which form the epidermis of the embryo are modified in quite a different manner to the nervous parts of the layer, becoming very much diminished in thickness and composed of a single row of flattened cells, Pl. XXIX. fig. 3.

Till the end of stage F, the epiblast cells and indeed all the cells of the blastoderm retain their yolk spherules, but the epiblast begins to lose them and consequently to become transparent in stage G.

Medullary Groove.

During stage B the medullary groove is shallow posteriorly, deeper in the middle part of the body, and flattened out again at the extreme anterior end of the embryo. Pl. XXIII. fig. 10, a b c.

A similar condition obtains in stage B, but the canal has now in part become deeper. Anteriorly no trace of it is to be seen. In stage C it exhibits the same general features (Plate XXIX. fig. 2 a 2 b 2 c).

By stage D we find important modifications of the canal.

It is still shallow behind and deep in the dorsal region, Plate XXIX. fig. 3 d 3 e 3 f; but the anterior flattened area in the last stage has grown into a round flat plate which may be called the cephalic plate, Plate XXIV. D and Plate XXIX. fig. 3 a 3 b 3 c. This plate becomes converted into the brain. Its size and form give it a peculiar appearance, but the most remarkable feature about it is the ventral curvature of its edges. Its edges do not, as might be expected, bend dorsalwards towards each other, but become sharply bent in a ventral direction. This feature is for the first time apparent at this stage, but becomes more conspicuous during the succeeding ones, and attains its maximum in stage F (Plate XXIX. fig. 5), in which it might almost be supposed that the edges of the cephalic plate were about to grow downwards and meet on the ventral side of the embryo.

In the stages subsequent to D the posterior part of the canal deepens much more rapidly than the rest (*vide* Pl. XXIX. fig. 4, taken from the posterior end of an embryo but slightly younger than F), and the medullary folds unite and convert the posterior end of the medullary groove into a closed canal (Pl. XXIV. fig. F), while the groove is still widely open elsewhere[1]. The medullary canal does not end blindly behind, but simply forms a tube not closed at either extremity. The importance of this fact will appear later.

In a stage but slightly subsequent to F nearly the whole of the medullary canal becomes closed. This occurs in the usual way by the junction and coalescence of the medullary folds. In the course of the closing of the medullary groove the edges of the cephalic plate lose their ventral curvature and become bent up in the normal manner (*vide* Pl. XXIX. fig. 6, a section taken through the posterior part of the cephalic plate), and the enlarged plate merely serves to enclose a dilated cephalic portion of the medullary canal. The closing of the medullary canal takes place earlier in the head and neck than in the back. The anterior end of the canal becomes closed and does not remain open like the posterior end.

Elasmobranch embryos resemble those of the Sturgeon (Acipenser) and the Amphibians in the possession of a spatula-like cephalic expansion: but so far as I am aware a ventral flexure in the medullary plates of the head has not been observed in other groups.

The medullary canal in Elasmobranchs is formed precisely on the type so well recognised for all groups of vertebrates with the exception of the Osseous Fishes. The only feature in any respect peculiar to these fishes is the closing of their medullary canal first commencing behind, and then proceeding to the cervical and cephalic regions. In those vertebrates in which the medullary folds do not unite at approximately the same

---

[1] *Vide* Preliminary Account, etc. *Q. Jl. Micros. Science*, Oct. 1874, Pl. XIV. 8 *a*. This and the other section from the same embryo (stage F) may be referred to. I have not thought it worth while repeating them here.

time throughout their length, they appear usually to do so first in the region of the neck.

Mesoblast.

The separation from the hypoblast of two lateral masses of mesoblast has already been described. Till the close of stage C the mesoblast retains its primitive bilateral condition unaltered. Throughout the whole length of the embryo, with the exception of the extreme front end, there are present two plates of rounded mesoblast cells, one on each side of the medullary groove. These plates are in very close contact with the hypoblast, and also follow with fair accuracy the outline of the epiblast. This relation of the mesoblast plates to the epiblast must not however be supposed to indicate that the medullary groove is due to growth in the mesoblast: a view which is absolutely nega- tived by the manner of formation of the medullary groove in the head. Anteriorly the mesoblast plates thin out and com- pletely vanish.

In stage D, the plates of mesoblast in the trunk undergo important changes. The cells composing them become arranged in two layers (Plate XXIX. fig. 3), a splanchnic layer adjoining the hypoblast, and a somatic layer adjoining the epiblast[1]. Although these two layers are distinctly formed, they do not become separated at this stage in the region of the trunk, and in the trunk no true body-cavity is formed.

By stage D the plates of mesoblast have ceased to be quite isolated, and are connected with the lower layer cells of the general blastoderm.

Moreover the lower layer cells outside the embryo now exhibit distinct traces of a separation into two layers, one con- tinuous with the hypoblast, the other with the mesoblast. Both layers are composed of very flattened cells, and the mesoblast layer is often more than one cell deep, and sometimes exhibits a mesh-like arrangement of its elements.

Coincidentally with the appearance of a differentiation into a somatic and splanchnic layer the mesoblast plates become

---

[1] I under-estimated the distinctness of this formation in my earlier paper, *loc. cit.*, although I recognized the fact that the mesoblast cells became arranged in two distinct layers.

split by a series of transverse lines of division into protovertebræ. Only the proximal regions of the plates become split in this way, while their peripheral parts remain quite intact. As a result of this each plate becomes divided into a proximal portion adjoining the medullary canal, which is divided into *proto-vertebræ*, and may be called the *vertebral plate*, and a peripheral portion not so divided, which may be called the *lateral plate*. These two parts are at this stage quite continuous with each other; and, as will be seen in the sequel, the body-cavity originally extends uninterruptedly to the summit of the vertebral plates.

By stage D at the least ten protovertebræ have appeared.

In Torpedo the mesoblast commences to be divided into two layers much earlier than in Pristiurus; and even before stage C this division is more or less clearly marked.

In the head and tail the condition of the mesoblast is by no means the same as in the body.

In the tail the plates of mesoblast become considerably thickened and give rise to two projections, one on each side, which have already been alluded to as caudal or tail-swellings; vide Pl. XXIV. figs. D, F, and Pl. XXIX. fig. 3 *f* and fig. 4.

These masses of mesoblast are neither divided into protovertebræ, nor do they exhibit any trace of a commencing differentiation into somato-pleure and splanchno-pleure.

In the head, so far as I have yet been able to observe, the mesoblastic plates do *not* at this stage become divided into protovertebræ. The other changes exhibited in the cephalic region are of interest, mainly from the fact that here appears a cavity in the mesoblast directly continuous with the body-cavity (when that cavity becomes formed), but which appears at a very much earlier date than the body-cavity. This cavity can only be looked on in the light of a direct continuation of the body or peritoneal cavity into the head. Theoretical considerations with reference to it I propose reserving till I have described the changes which it undergoes in the subsequent periods.

Figures 3*a*, 3*b* and 3*c* exhibit very well the condition of the mesoblast in the head at this period. In fig. 3*c*, a section taken through the back part of the head, we find the condition of the mesoblast to be nearly the same as in the sections immediately behind. The ventral continuation of the mesoblast formed by the lateral plate has, however, become much thinner, and the dorsal or vertebral portion has acquired a more triangular form than in the sections through the trunk (fig. 3 d and 3 e).

In the section (fig. 3 b) in front of this the ventral portion of the plate is no longer present, and only that part which corresponds with the vertebral division of the primitive plate of mesoblast is present.

In this a distinct cavity, forming part of the body cavity, has appeared.

In a still anterior section, fig. 3 a, no cavity is any longer present in the mesoblast; whilst in sections taken from the foremost part of the head no mesoblast is to be seen (vide fig. 5, taken from the front part of the head of the embryo represented in Pl. XXIV. fig. F).

A continuation of the body-cavity into the head has already been described by Oellacher[1] for the trout: but he believes that the cavity in this part is solely related to the formation of the pericardial space.

The condition of the mesoblast undergoes no important change till the end of the period treated of in this chapter. The masses of mesoblast which form the tail-swellings become more conspicuous (Pl. XXIX. fig. 4); and indeed their convexity is so great that the space between them has the appearance of a median groove, even after the closure of the neural canal in the caudal region. This appearance is nevertheless rather deceptive.

In embryos of stage G, which may be considered to belong to the close of this period, eighteen protovertebræ are present both in Pristiurus and Torpedo embryos.

The Alimentary Canal.

The alimentary canal at the commencement of this period, stage B, is a space between the embryo and the yolk, ending

[1] *Zeitschrift f. wiss. Zoologie*, 1873.

blindly in front, but opening posteriorly by a widish slit-like
aperture, which corresponds to the anus of Rusconi.

The cavity anteriorly has a more or less definite form,
having lateral walls, as well as a roof and floor (Pl. XXIII.
fig. 10b and 10c). Posteriorly it is not nearly so definitely
enclosed (Pl. XXIII. fig. 10a). The ventral wall of the cavity is
formed by yolk. But even in stage B there are beginnings
of a cellular ventral wall derived from an ingrowth of cells from
the two sides.

By stage C considerable progress has been made in the
formation of the alimentary canal. Posteriorly it is as flattened
and indefinite as during stage B (Pl. XXIX. figs. 2b and 2c).
But in the anterior part of the embryo the cavity becomes
much deeper and narrower, and a floor of cells begins to be
formed for it (Pl. XXIX. fig. 2); and, finally, in front, it forms a
definite space completely closed in on all sides by cells (Pl.
XXIX. fig. 2a). Two distinct processes are concerned in effecting these changes in the condition of the alimentary cavity.
One of these is a process of folding off the embryo from the
blastoderm. The other is a simple growth of cells independent
of any folding. To the first of these processes the increased
depth and narrowness of the alimentary cavity is due; the
second is concerned in forming its ventral wall. The combination of the two processes produces the peculiar triangular
section which characterises the anterior closed end of the
alimentary cavity at this stage. The process of the folding off
of the embryo from the blastoderm resembles exactly the
similar process in the embryo bird. The fold by which the
constricting off of the embryo is effected is a perfectly continuous one, but may be conveniently spoken of as composed
of a head-fold and two lateral folds.

Of far greater interest than the nature of these folds is the
formation of the ventral wall of the alimentary canal. This, as
has been said, is effected by a growth of cells from the two
sides to the middle line (Pl. XXIX. fig. 2). The cells for this
are however not derived from pre-existing hypoblast cells, but
are formed spontaneously around nuclei of the yolk. This fact
can be determined in a large number of sections, and is fairly
well shewn in Pl. XXIX. fig. 2. The cells are formed in the

yolk, as has been already mentioned, by a simple aggregation of protoplasm around pre-existing nuclei.

The cells being described are in most cases formed close to the pre-existing hypoblast cells, but often require to undergo a considerable change of position before attaining their final situation in the wall of the alimentary canal.

I have already alluded to this feature in the formation of the ventral wall of the alimentary cavity. Its interest, as bearing on the homology of the yolk, is considerable, owing to the fact that the so-called yolk-cells of Amphibians play a similar part in supplying the ventral epithelium of the alimentary cavity, as do the cells derived from the yolk in Elasmobranchs.

The fact of this feature being common to the yolk-cells of Amphibians and the yolk of Elasmobranchs, supplies a strong argument in favour of the homology of the yolk-cells in the one case with the yolk in the other[1].

---

[1] Nearly simultaneously with my last published paper on the Development of Elasmobranchs, which dealt in a fairly complete manner with the genesis of cells outside the blastoderm, there appeared two important papers dealing with the same subject for Teleostei. One of these, by Professor Bambeke, Embryologie des Poissons Osseux, *Mém. Cour. Acad. Belgique*, 1875, which appeared some little time before my paper, and a second by Dr Klein, *Quart. Jour. of Micr. Sci.* April, 1876. In both of these papers a development of nuclei and of cells is described as occurring outside the blastoderm in a manner which accords fairly well with my own observations.

The conclusions of both these investigators differ however from my own. They regard the finely granular matter, in which the nuclei appear, as pertaining to the blastoderm, and morphologically quite distinct from the yolk. From their observations we can clearly recognize that the material in which the nuclei appear is far more sharply separated off from the yolk in Osseous Fish than in Elasmobranchs, and this sharp separation forms the main argument for the view of these authors. Dr Klein admits, however, that this granular matter (which he calls parablast) graduates in the typical food-yolk, though he explains this by supposing that the parablast takes up part of the yolk for the purpose of growth.

It is clear that the argument from a sharp separation of yolk and parablast cannot have much importance, when it is admitted (1) that in Osseous Fish there is a gradation between the two substances, while (2) in Elasmobranchs the one merges slowly and insensibly into the other.

The only other argument used by these authors is stated by Dr Klein in the following way. "The fact that the parablast has, at the outset, been forming one unit with what represents the archiblast, and, *while increasing has spread* i.e. *grown over the yolk* which underlies the segmentation-cavity, is, I think, the most absolute proof that the yolk is as much different from the parablast as it is from the archiblast." This argument to me merely demonstrates that certain of the nutritive elements of the yolk become in the course of development converted into protoplasm, a phenomenon which must necessarily be supposed to take place on my own as well as on Dr Klein's view of the nature of the yolk. My own views on the subject have already been fully stated. I regard the so-called yolk as composed of a larger or smaller amount of food-material imbedded

The history of the alimentary canal during the remainder of this period may be told briefly.

The folding off and closing of the alimentary canal in the anterior part of the body proceeds rapidly, and by stage D not only is a considerable tract of alimentary canal formed, but a great part of the head is completely folded off from the yolk (fig. 3 a). By stage F a still greater part is folded off. The posterior part of the alimentary canal retains for a long period its primitive condition. It is not until stage F that it begins to be folded off behind. After the folding has once commenced it proceeds with great rapidity, and before stage G, the hinder part of the alimentary canal becomes completely closed in.

The folding in of the gut is produced by two lateral folds, and the gut is not closed posteriorly.

It may be remembered that the neural canal also remained open behind. Thus both the neural and alimentary canals are open behind; and, since both of them extend to the posterior end of the body, they meet there, their walls coalesce, and a direct communication from the neural to the alimentary canal

in protoplasm, and the meroblastic ovum as a body constituted of the same essential parts as a holoblastic ovum, though divided into regions which differ in the proportion of protoplasm they contain. I do not propose to repeat the positive arguments used by me in favour of this view, but content myself with alluding to the protoplasmic network found by Schultz and myself extending through the whole yolk, and to the similar network described by Bambeke as being present in the eggs of Osseous Fish after deposition but before impregnation. The existence of these networks is to me a conclusive proof of the correctness of my views. I admit that in Teleostei the 'parablast' contains more protoplasm than the homologous material in the Elasmobranch ovum, while it is probable that after impregnation the true yolk of Teleostei contains little or no protoplasm; but these facts do not appear to me to militate against my views.

I agree with Prof. Bambeke in regarding the cells derived from the subgerminal matter as homologous with the so-called yolk-cells of the Amphibian embryo.

I have recently, in some of the later stages of development, met with very peculiar nuclei of the yolk immediately beneath the blastoderm at some little distance from the embryo, Pl. xxix. fig. 8. They were situated not in finely subgerminal matter, but amongst large yolk spherules. They were very large, and presented still more peculiar forms than those already described by me, being produced into numerous long filiform processes. The processes from the various nuclei were sometimes united together, forming a regular network of nuclei quite unlike anything that I have previously seen described.

The sub-germinal matter, in which the nuclei are usually formed, becomes during the later stages of development far richer in protoplasm than during the earlier. It continually arises at fresh points, and often attains to considerable dimensions, no doubt by feeding on yolk-spherules. Its development appears to be determined by the necessities of growth in the blastoderm or embryo.

is produced. The process may be described in another way by saying that the medullary folds are continuous round the end of the tail with the lateral walls of the alimentary canal; so that, when the medullary folds unite to form a canal, this canal becomes continuous with the alimentary canal, which is closed in at the same time. In whatever way this arrangement is produced, the result of. it is that it becomes possible to pass in a continuously closed passage along the neural canal round the end of the tail and into the alimentary canal. A longitudinal section shewing this feature is represented on Plate XXIX. fig. 7.

This communication between the neural and alimentary canals, which is coupled, as will be seen in the sequel, with the atrophy of a posterior segment of the alimentary canal, is a feature of great interest which ought to throw considerable light upon the meaning of the neural canal. So far as I know, no suggestion as to the origin of it has yet been made. It is by no means confined to Elasmobranchs, but is present in all the vertebrates whose embryos are situated at the centre and not at the periphery of the blastoderm. It has been described by Goette[1] in Amphibians and by Kowalevsky, Owsjannikow and Wagner in the Sturgeon (Acipenser[2]). The same arrangement is also stated by Kowalevsky[3] to exist in Osseous Fishes and Amphioxus. The same investigator has shewn that the alimentary and neural canals communicate in larval Ascidians, and we may feel almost sure that they do so in the Marsipobranchii.

The Reptilia, Aves, and Mammalia have usually been distinguished from other vertebrates by the possession of a well-developed allantois and amnion. I think that we may further say that the lower vertebrates, Pisces and Amphibia, are to be distinguished from the three above-mentioned groups of higher vertebrates, by the positive embryonic character that their neural and alimentary canals at first communicate pos-

---

[1] *Entwicklungsgeschichte der Unke.*
[2] *Mélanges Biologiques de l'Academie Pétersbourg*, Tome VII.
[3] *Archiv f. mikros. Anat.* Vol. VII. p. 114. In the passage on this point Kowalevsky states that in Elasmobranchs the neural and alimentary canals communicate. This I believe to be the first notice published of this peculiar arrangement.

teriorly. The presence or absence of this arrangement depends
on the different positions of the embryo in the blastoderm.
In Reptiles, Birds and Mammals, the embryo occupies a central
position in the blastoderm, and not, as in Pisces and Amphibia,
a peripheral one at its edge. We can, in fact, only compare
the blastoderm of the Bird and the Elasmobranch, by supposing that in the blastoderm of the Bird there has occurred
an abbreviation of the processes, by which the embryo Elasmobranch is eventually placed in the centre of the blastoderm, as
a result of which the embryo Bird occupies from the first a
central position in the blastoderm[1].

The peculiar relations of the blastoderm and embryo, and
the resulting relations of the neural and alimentary canal,
appear to me to be features of quite as great an importance
for classification as the presence or absence of an amnion
and allantois.

General feature of the hypoblast.

There are but few points to be noticed with reference to the
histology of the hypoblast cells. The cells of the dorsal wall of
the alimentary cavity are columnar and form a single row.
Those derived from the yolk to form the ventral wall are at
first roundish, but subsequently assume a more columnar form.

The Notochord.

One of the most interesting features in the Elasmobranch
development is the formation of the notochord from the hypoblast. All the steps in the process by which this takes place
can be followed with great ease and certainty.

---

[1] Vide Self, *Journal of Anat. and Phys.* Vol. x. page 565, and Pl. xxvi. Fig. 1
and 2, and Comparison, &c., *Qy. Jour. of Micros. Sci.* July, 1875, p. 219. It
seems possible that the primitive streak with the primitive groove (described by
Dursy and myself in Birds and by Hensen in Mammalia), which is situated immediately behind the medullary groove, may be a rudiment at the position
where the edges of the blastoderm coalesced to give to the embryo the central
situation which it has in Birds and Mammals. If the primitive streak has really
this meaning, the fusion of layers which takes place in it to form the axis-cord,
would receive a satisfactory explanation as the remnant of the primitive continuity epiblast and lower-layer cells at the edge of the blastoderm, while the
primitive groove (not the medullary groove) would be the groove naturally left
between the coalescing edges of the blastoderm. In order clearly to understand
the view here expressed, the reader ought to refer to the passages above quoted,
where an account is given of the process by which the embryo Elasmobranch is
constricted off from the edge of the blastoderm.

Up to stage B the hypoblast is in contact with the epiblast immediately below the medullary groove, but exhibits no trace of a thickening or any other formation at that point.

Between stage B and C the notochord first arises.

In the hindermost sections of this stage the hypoblast retains a perfectly normal structure and uniform thickness throughout. In next few sections, Pl. XXIX. fig. 1 c, ch', a slight thickening is to be observed in the hypoblast, immediately below the medullary canal.   The layer, which elsewhere is composed of a single row of cells, here becomes two cells deep, but no sign of a division into two layers exhibited.

In the next few sections the thickening of the hypoblast becomes much more pronounced; we have, in fact, a ridge projecting from the hypoblast towards the epiblast (Pl. XXIX. fig. 1 b, ch').   ·

This ridge is pressed firmly against the epiblast, and causes in it a slight indentation.   The hypoblast in the region of the ridge is formed of two layers of cells, the ridge being entirely due to the uppermost of the two.

In sections in front of this a cylindrical rod, which can at once be recognised as the notochord and is continuous with the ridge just described, begins to be split off from the hypoblast. It is difficult to say at what point the separation of this rod from the hypoblast is completed, since all intermediate gradations between complete separation and complete attachment are to be seen.

Where the separation first appears, a fairly thick bridge of hypoblast is left connecting the two lateral halves of the layer, but anteriorly this bridge becomes excessively delicate and thin (Pl. XXIX. fig. 1 c), and in some cases is barely visible except with high powers.

From the series of sections represented, it is clear that the notochord commences to be separated from the hypoblast anteriorly, and that the separation gradually extends backwards.

The posterior extremity of the notochord remains for a long time attached to the hypoblast; and it is not till the end of the period treated of in this chapter that it becomes completely free.

A sheath is formed around the notochord, very soon after its formation, at a stage intermediate between stages C and D. This sheath is very delicate, though it stains with both osmic acid and hæmatoxylin. I conclude from its subsequent history, that it is to be regarded as a product of the cells of the notochord, but at the same time it should be stated that it precisely resembles membrane-like structures, which I have already described as probably artificial.

Towards the end of this period the cells of the notochord become very much flattened vertically, and cause the well-known stratified appearance which characterises the notochord in longitudinal sections. In transverse sections the outlines of the cells of the notochord appear rounded.

Throughout this period the notochord cells are filled with yolk spherules, and near its close small vacuoles make their appearance in them.

An account of the development of the notochord, substantially similar to that I have just given, appeared in my preliminary paper[1] on the development of the Elasmobranch fishes.

To the remarks which were there made, I have little to add. There are two possible views, which can be held with reference to the development of the notochord from the hypoblast.

We may suppose that this is the primitive mode of development of the notochord, or we may suppose that the separation of the notochord from the hypoblast is due to a secondary process.

If the latter view is accepted, it will be necessary to maintain that the mesoblast becomes separated from the hypoblast as three separate masses, two lateral, and one median, and that the latter become separated much later than the two former.

We have, I think, no right to assume the truth of this view without further proof. The general admission of assumptions of this kind is apt to lead to an injurious form of speculation, in which every fact presenting a difficulty in the way of some general theory is explained away by an arbitrary assumption, while all the facts in favour of it are taken for granted. It is however clear that no theory can ever be fairly tested, so long as logic of this kind is permitted. If, in the present instance, the view is adopted that the notochord has in reality a meso-

[1] *Loc. cit.*

blastic origin, it will be possible to apply the same view to every other organ derived from the hypoblast, and to say that it is really mesoblastic, but has become separated at rather a late period from the hypoblast.

If, however, we provisionally reject this explanation, and accept the other alternative, that the notochord is derived from the hypoblast, we must be prepared to adopt one of two views with reference to the development of the notochord in other vertebrates. We must either suppose that the current statements as to the development of the notochord in other vertebrates are inaccurate, or that the notochord has only become secondarily mesoblastic.

The second of these alternatives is open to the same objections as the view that the notochord has only apparently a hypoblastic source in Elasmobranchs, and, provisionally at least, the first of them ought to be accepted. The reasons for accepting this alternative fall under two heads. In the first place, the existing accounts and figures of the development of the notochord exhibit in almost all cases a deficiency of clearness and precision. The exact stage necessary to complete the series never appears. It cannot, therefore, at present be said that the existing observations on the development of the notochord afford a strong presumption against its hypoblastic origin.

In the second place, the remarkable investigations of Hensen[1], on the development of the notochord in Mammalia, render it very probable that, in this group, the notochord is ' developed from the hypoblast.

Hensen finds that in Mammalia, as in Elasmobranchs, the mesoblast forms two independent lateral masses, one on each side of the medullary canal.

After the commencing formation of the protovertebræ the hypoblast becomes considerably thickened beneath the medullary groove; and, though he has not followed out all the steps of the process, yet his observations go very far towards proving that this thickening is converted into the notochord.

Against the observations of Hensen, there ought, however, to be mentioned those of Lieberkühn[2]. He believes that the

---

[1] *Zeitschrift f. Anat. u. Entwicklungsgeschichte*, Vol. I. p. 366.
[2] *Sitz. der Gesell. zu Marburg*, Jan. 1876.

two lateral masses of mesoblast, described by Hensen (in an earlier paper than the one quoted), are in reality united by a delicate layer of cells, and that the notochord is formed from a thickening of these.

Lieberkühn gives no further statements or figures, and it is clear that, even if there is present the delicate layer of mesoblast, which he fancies he has detected, yet this cannot in any way invalidate such a section as that represented on Pl. x. fig. 40, of Hensen's paper.

In this figure of Hensen's, the hypoblast cells become distinctly more columnar, and the whole layer much thicker immediately below the medullary canal than elsewhere, and this independently of any possible layer of mesoblast.

It appears to me reasonable to conclude that Lieberkühn's statements do not seriously weaken the certainty of Hensen's results.

In addition to these observations of Hensen's on Mammalia, those of Kowalevsky and Kuppfer on Ascidians may fairly be pointed to as favouring the hypoblastic origin of the notochord.

It is not too much to say that at the present moment the balance of evidence is in favour of regarding the notochord as a hypoblastic organ.

This conclusion is, no doubt, rather startling, and difficult to understand. The only feature of the notochord in its favour is the fact of its being unsegmented[1].

Should it eventually turn out that the notochord is developed in most vertebrates from the mesoblast, and only exceptionally from the hypoblast, the further question will have to be settled as to whether it is primitively a hypoblastic or a mesoblastic organ; but, from whatever layer it has its source, an excellent example will be afforded of an organ changing from the layer in which it was originally developed into another distinct layer.

[1] In my earlier paper I suggested that the endostyle of Ascidians afforded an instance of a supporting organ being derived from the hypoblast. This parallel does not hold since the endostyle has been shewn to possess a secretory function. I never intended (as has been imagined by Professor Todaro) to regard the endostyle as the homologue of the notochord.

EXPLANATION OF PLATE.

ep. epiblast.   *hy.* hypoblast.
l p. coalesced lateral and vertebral plate of mesoblast.
m g. medullary groove.   *n c.* neural or medullary canal.
p v. protovertebra.   *so.* somatopleure.
sp. splanchnopleure.   *t s.* mesoblast of tail-swelling.
ch. chorda dorsalis or notochord.
ch'. ridge of hypoblast, which will become separated off as the notochord.
al. alimentary canal.
n. a. cells formed around the nuclei of the yolk to enter into the ventral wall of the alimentary canal.
n. nucleus of yolk.   *yk.* yolk spherules.

Fig. 1 *a.* 1 *b.* 1 *c.*   Three sections from the same embryo belonging to a stage intermediate between B and C, of which fig. 1 *a* is the most anterior. × 96 diameters.

The sections illustrate (1) The different characters of the medullary groove in the different regions of the embryo. (2) The structure of the coalesced lateral and vertebral plates. (3) The mode of formation of the notochord as a thickening of the hypoblast (*ch'*), which eventually becomes separated from the hypoblast as an elliptical rod (1 *a. ch*).

Fig. 2.   Section through the anterior part of an embryo belonging to stage *C.* The section is mainly intended to illustrate the formation of the ventral wall of the alimentary canal from cells formed around the nuclei of the yolk. It also shews the shallowness of the medullary groove in the anterior part of the body.

Fig. 2 *a.* 2 *b.* 2 *c.*   Three sections from the same embryo as fig. 2   Fig. 2 *a* is the most anterior of the three sections and is from a point shortly in front of fig. 2.   The figures illustrate the general features of an embryo of stage *C,* more especially the complete closing of the alimentary canal in front and the triangular section which it there presents.

Fig. 3.   Section through the posterior part of an embryo belonging to stage *D.* × 86 diameters.

It shews the general features of layer during the stage, more especially the differentiation of somatopleure and splanchnopleure.

Fig. 3 *a.* 3 *b.* 3 *c.* 3 *d.* 3 *e.* 3 *f.*   Sections of the same embryo as fig. 3 (× 60 diameters).   Fig. 3 belongs to part of the embryo intermediate between fig. 3 *e* and 3 *f.*

The sections show the features of various parts of the embryo.   Fig. 3 *a.* 3 *b.* and 3 *c.* belong to the head, and special attention should be paid to the presence of a cavity in the mesoblast in 3 *b* and to ventral curvature of the medullary folds.

Fig. 3 *d* belongs to the neck, fig. 3 *e* to the back, and fig. 3 *f* to the tail.

Fig. 4.   Section through the region of the tail at the commencement of stage F.   × 60 diameters.

The section shews the character of the tail-swellings and the commencing closure of the medullary groove.

Fig. 5.   Transverse section through the anterior part of the head of an embryo belonging to stage F. (× 60 diameters).   It shews (1) the ventral curvature of the medullary folds next the head.   (2) The absence of mesoblast in the anterior part of the head.

Fig. 6.   Section through the head of an embryo at a stage intermediate between F and G.   × 86 diameters.

It shews the manner in which the medullary folds of the head unite to form the medullary canal.

Fig. 7.   Longitudinal and vertical section through the tail of an embryo belonging to stage G.

It shews the direct communication which exists between the neural and alimentary canals.

The section is not quite parallel to the long axis of the embryo, so that the protovertebræ are cut through in its anterior part, and the neural canal passes out of the section.

Fig. 8.   Network of nuclei from the yolk of an embryo belonging to stage H.

ON THE SPINAL NERVES OF AMPHIOXUS. — By
F. M. BALFOUR, B.A., *Fellow of Trinity College, Cambridge*.

DURING a short visit to Naples in January last, I was enabled,
through the kindness of Dr Dohrn, to make some observations
on the spinal nerves of Amphioxus. These were commenced
solely with the view of confirming the statements of Stieda on
the anatomy of the spinal nerves, which, if correct, appeared to
me to be of interest in connection with the observations I had
made that, in Elasmobranchs, the anterior and posterior roots
arise alternately and not in the same vertical plane. I have
been led to conclusions on many points entirely opposed to those
of Stieda, but, before recording these, I shall proceed briefly
to state his results, and to examine how far they have been corroborated by subsequent observers.

Stieda[1], from an examination of sections and isolated spinal
cords, has been led to the conclusion that, in Amphioxus, the
nerves of the opposite sides arise alternately, except in the most
anterior part of the body, where they arise opposite each other.
He also states that the nerves of the same side issue alternately from the dorsal and ventral corners of the spinal cord.
He regards two of these roots (dorsal and ventral) on the same
side as together equivalent to a single spinal nerve of higher
vertebrates formed by the coalescence of a dorsal and ventral
root.

Langerhans[2] apparently agrees with Stieda as to the facts
about the alternation of dorsal and ventral roots, but differs
from him as to the conclusions to be drawn from those facts.
He does not, for two reasons, believe that two nerves of Amphioxus can be equivalent to a single nerve in higher vertebrates:
(1) Because he finds no connecting branch between two succeeding nerves, and no trace of an anastomosis. (2) Because
he finds that each nerve in Amphioxus supplies a complete
myotome, and he considers it inadmissible to regard the nerves,

[1] *Mém. Acad. Pétersbourg*, Vol. XIX.
[2] *Archiv f. mikr. Anatomie*, Vol XII.

which in Amphioxus together supply *two myotomes*, as equivalent to those which in higher vertebrates supply *a single myotome only.*

Although the agreement as to facts between Langerhans and Stieda is apparently a complete one, yet a critical examination of the statements of these two authors proves that their results, on one important point at least, are absolutely contradictory. Stieda, Pl. III. fig. 19, represents a longitudinal and horizontal section through the spinal cord which exhibits the nerves arising alternately on the two sides, and represents each myotome supplied by *one nerve*. In his explanation of the figure he expressly states that the nerves of one plane only (*i.e.* only those with dorsal or only those with ventral roots) are represented; so that if all the nerves which issue from the spinal cord had been represented double the number figured must have been present. But since each myotome is supplied by *one* nerve in the figure, if all the nerves present were represented, each myotome would be supplied by two nerves.

Since Langerhans most emphatically states that only *one nerve* is present for *each myotome*, it necessarily follows that he or Stieda has made an important error; and it is not too much to say that this error is more than sufficient to counterbalance the value of Langerhans' evidence as a confirmation of Stieda's statements.

I commenced my investigations by completely isolating the nervous system of Amphioxus by maceration in nitric acid according to the method recommended by Langerhans[1]. On examining specimens so obtained it appeared that, for the greater length of the cord, the nerves arose alternately on the two sides, as was first stated by Owsjannikow, and subsequently by Stieda and Langerhans; but to my surprise not a trace could be seen of a difference of level in the origin of the nerves of the same side.

The more carefully the specimens were examined from all points of view, the more certainly was the conclusion forced upon me, that nerves issuing from the ventral corner of the spinal cord, as described by Stieda, had no existence.

Not satisfied by this examination, I also tested the point by

---

[1] *Loc. cit.*

means of sections. I carefully made transverse sections of a successfully hardened Amphioxus, through the whole length of the body. There was no difficulty in seeing the dorsal roots in every third section or so, but not a trace of a ventral root was to be seen. There can, I think, be no doubt, that, had ventral roots been present, they must, in some cases at least, have been visible in my sections.

In dealing with questions of this kind it is no doubt difficult to prove a negative; but, since the two methods of investigation employed by me both lead to the same result, I am able to state with considerable confidence that my observations lend no support to the view that the alternate spinal nerves of Amphioxus have their roots attached to the ventral corner of the spinal cord.

How a mistake on this point arose it is not easy to say. All who have worked with Amphioxus must be aware how difficult it is to conserve the animal in a satisfactory state for making sections. The spinal cord, especially, is apt to be distorted in shape, and one of its ventral corners is frequently produced into a horn-like projection terminating in close contact with the sheath. In such cases the connective tissue fibres of the sheath frequently present the appearance of a nerve-like prolongation of the cord; and for such they might be mistaken if the sections were examined in a superficial manner. It is not, however, easy to believe that, with well conserved specimens, a mistake could be made on this point by so careful and able an investigator as Stieda, especially considering that the histological structure of the spinal nerves is very different from that of the fibrous prolongations of the sheath of the spinal cord.

It only remains for me to suppose that the specimens which Stieda had at his disposal, were so shrunk as to render the origin of the nerves very difficult to determine.

The arrangement of the nerves of Amphioxus, according to my own observations, is as follows.

The anterior end of the central nervous system presents on its left and dorsal side a small pointed projection, into which is prolonged a diverticulum from the dilated anterior ventricle of the brain. This may perhaps be called the olfactory nerve, though clearly of a different character to the other nerves. It was first accurately described by Langerhans[1].

Vertically below the olfactory nerve there arise two nerves, which issue at the same level from the ventral side of the anterior extremity of the central nervous system. These form the first pair of nerves, and are the only pair which arise from the ventral portion of the cerebro-spinal cord. The two nerves, which form the second pair, arise also opposite each other but from the dorsal side of the cord. The first and second pair of nerves have both been accurately drawn and described by Langerhans: they, together with the olfactory nerve, can easily be seen in nervous systems which have been isolated by maceration.

In the case of the third pair of nerves, the nerve on the right-hand side is situated not quite opposite but slightly behind that on the left. The right nerve of the fourth pair is situated still more behind the left, and, in the case of the fifth pair, the nerve to the right is situated so far behind the left nerve that it occupies a position half-way between the left nerves of the fifth and sixth pairs. In all succeeding nerves the same arrangement holds good, so that they exactly alternate on two sides.

Such is the arrangement carefully determined by me from one specimen. It is possible that it may not be absolutely constant, but the following general statement almost certainly holds good.

All the nerves of Amphioxus, except the first pair, have their roots inserted in the dorsal part of the cord. In the case of the first two pairs the nerves of the two sides arise opposite each other; in the next few pairs, the nerves on the right-hand side gradually shift backwards: the remaining nerves spring alternately from the two sides of the cord.

For each myotome there is a single nerve, which enters, as in the case of other fishes, the intermuscular septum. This point may easily be determined by means of longitudinal sections, or less easily from an examination of macerated specimens. I agree with Langerhans in denying the existence of ganglia on the roots of the nerves.

[1] *Loc. cit.*

ON THE PLACENTATION OF THE CAPE ANT-EATER
(*Orycteropus Capensis*)—By Professor Turner.

THE only observation on the placenta of *Orycteropus* which, so far as I know, has up to this time been recorded, is one made by Professor Huxley in his "Introduction to Classification," p. 104 (London, 1869), where he states that in this genus the placenta is discoidal and deciduate. This observation was made on a specimen preserved in the stores of the Museum of the Royal College of Surgeons, London.

In pursuing the series of investigations on the Comparative Anatomy of the placenta with which I have for some time been engaged, I was desirous of examining the placenta of this animal. With great liberality Professor Flower has placed at my disposal two specimens of its placenta in the stores of the College of Surgeons Museum. As it is desirable that the specimens should be satisfactorily identified as belonging to *Orycteropus*, Professor Flower has kindly given me the following information on the subject. "The viscera have been in the stores of the College from before 1841, the date of the printed catalogue of the store specimens. The jar in which they were contained is marked 85, *Orycteropus*, viscera, two fœtuses: membranes and placenta attached. It contained the tongue, windpipe, thoracic and abdominal viscera of an undoubted female *Orycteropus*." The two fœtuses were well developed Cape Ant-eaters, and of considerable size. One measured 19 inches from the tip of the snout to the end of the tail, and 14 inches from the tip of the snout to the root of the tail: the other was 20 inches long to the end of the tail, and 14 inches to the root. Two placentæ were also present, each contained in an organ, which one had to determine whether it was the whole uterus, or only a part of the uterus; also whether the two fœtuses were from the same or from different animals. The detailed description which I shall give of the specimens will I think shew that each placenta must have been from a different mother, and that I had two uteri therefore before me; for to each was attached a Fallopian tube and ovary, and the specimen, which I have designated A, had, in addition, an undeveloped non-fecundated horn connected with it.

In one of the two specimens, which I shall call A, the two-horned form of the uterus was recognized. The right horn was much distended and contained the placenta and membranes. Its length was $7\frac{1}{2}$ inches, but as it had been cut across above the os uteri, and as a slice had been removed from its fundus, the actual length must have been three or four inches greater. Attached to the left aspect of this cornu were the ovary and its ligament, the Fallopian tube, the broad ligamentous fold of peritoneum, and the round ligament. The left cornu was of small size and the foetal membranes were not prolonged into it. Its anterior part was fusiform, projected from the side of the posterior part of the fecundated horn, and was only $1\frac{8}{10}$ths of an inch in length, whilst its girth at the widest part was $1\frac{4}{10}$ths inch. It was invested, like the fecundated horn, with peritoneum, but the broad and round ligaments, ovary, and Fallopian tube had unfortunately been separated from it. The posterior part of the left cornu was joined for about an inch to the side of the corresponding part of the fecundated horn, so as to be enclosed by a common envelope of peritoneum. Their cavities were separated from each other by a distinct partition, and in all probability, as has been described by Rapp[1] in the non-gravid *Orycteropus*, had opened by independent mouths into the vagina.

The right cornu had been opened into by a longitudinal incision through the posterior wall, the chorion and placenta had been cut through and the foetus removed. The placenta was arranged as a broad zone around the transverse diameter of the cornu. A gap in the continuity of the zone existed across its entire breadth, along a line which corresponded to the free border of the cornu. This gap varied in breadth from $\frac{7}{10}$ths to 1 inch, and in this locality the outer surface of the chorion was non-villous, and in relation with a corresponding smooth surface of the uterine mucosa, and between the two a pouch-like recess was situated. This deficiency in the continuity

---

[1] *Die Edentaten*, p. 102. Tübingen, 1852.

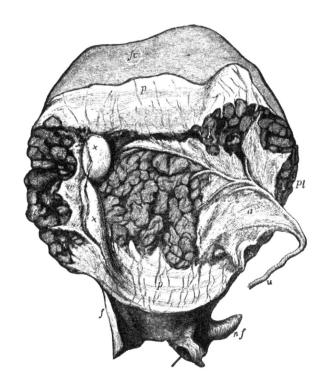

The gravid Uterus and convoluted inner surface of the zonary Placenta of
Orycteropus, specimen *A*. *f*. the fecundated right cornu; (the form of the fundus
of this cornu, *f.c.*, has been drawn from specimen *B*). *n.f.* the non-fecundated
left cornu; a probe has been inserted from below into its cavity. *u*. the umbilical cord, with the amniotic investment *a*. *p.p.* the non-placental poles of
the chorion in which branches of the umbilical vessels ramify. *xx* the gap
in the continuity of the zone, where the villi are absent. *pl*. the placenta.

of the placental zone obviously resembles an arrangement which has been described by Bischoff[1] in the Stone Marten (*Mustela foina*) and Pine Marten (*Mustela martes*). I am unable to say however whether this pouch had contained, as Bischoff found in the Martens, effused blood.

The zonary placenta had an almost uniform breadth of 5 inches. Extending from its lower border for a distance of 2 inches towards the os uteri was one non-placental pole of the chorion, whilst the opposite pole passed from the upper border of the placenta for a similar distance into the fundus of the cornu. The zone of the chorion was intimately united to a corresponding zone shaped surface of the uterine mucosa, though both the upper and lower borders of the placenta were not tied down to the uterine wall, but were covered, for a breadth varying from half an inch to an inch, by a prolongation of the mucous membrane of the uterus reflected for that distance on to them, but this reflected decidua was not continued on to the non-placental part of the chorion. The arrangement of the mucosa at the borders of the placenta bore a strong resemblance to what I have seen at the margin of the zonary placenta of the Grey Seal[2].

The second specimen, which I shall call *B*, consisted only of the left fecundated cornu, which measured 11 inches in length. The fundus, middle and lower portions of the cornu had been preserved, but the vaginal attachment was unfortunately absent. No trace of the right cornu was to be seen. A slender Fallopian tube, enclosed in a peritoneal fold, opened into the left cornu about two inches from the summit of the fundus. The ovary, with its ligament, the round and broad ligaments, were also present. In opening into this cornu for the purpose of removing the foetus the placenta had been cut through, and apparently a portion of its substance cut away. The placenta formed a zone 5¼ inches broad, disposed transversely around the long axis of the cornu. The dilated fundus of the cornu reached for three inches beyond the upper border of the placenta, and into it one smooth non-placental pole of the chorion projected. At the opposite end the cornu rapidly diminished in size to a

[1] *Sitzbericht. Akad. Wissensch.* München, 13 May, 1865, p. 339.
[2] *Trans. Roy. Soc.* Edinburgh. 1875.

tube only half an inch in diameter, which had been divided before the specimen came into my charge, but where the continuity of the cornu with the vagina had probably existed. The other smooth non-placental part of the chorion extended into this part of the horn for two inches beyond the border of the placenta, up to the divided tube just referred to. At the upper and lower borders of the placenta the uterine mucous membrane was reflected on to the placenta, so as to have a similar disposition to the arrangement described in *A;* but whilst the reflected band at the lower border had an average breadth of one inch, that at the upper was only about a quarter of an inch broad. No gap in the continuity of the zone, similar to that present in *A,* was seen in the placenta of this specimen; but it is possible that the portion of the placenta, which had apparently been removed before I obtained it for examination, may have contained this arrangement.

From the examination of these two specimens it is evident that *Orycteropus* possesses a zonary placenta. The proportion of the chorion and uterus occupied by the zone is however considerably greater than in the *Carnivora, Pinnepedia, Hyrax,* and *Elephas.* In the Bitch and Cat the chorion. at the full time is between seven and eight inches in length, but the placental zone does not possess a diameter of more than from 1¼ to 1¾ inch. In a Fox, not at the full time, a chorion five inches long had the placenta only one inch in diameter. In the Grey Seal the chorion was about three feet long, and the placenta varied in diameter from four to nine inches. In Hyrax[1] the chorion was 3½ inches long, and the placenta varied in breadth from a quarter to half an inch. In the Elephant at about the mid-period of gestation, described by Professor Owen, notwithstanding that the chorion was two feet six inches long, the zonary placenta was only from three to five inches in breadth.

In *Orycteropus,* on the other hand, the chorion in *B,* which was 10½ inches in length, had a placenta 5½ inches broad, whilst in *A* the chorion, eight inches in length, had a placenta 5 inches broad. Hence one half, or even more than one

---

[1] *Proc. Royal Soc.* London, Dec. 16, 1875.

half of the longitudinal diameter of the chorion is occupied by the placenta. The placenta in *Orycteropus* may, therefore, as compared with other zono-placentalia, be described as broadly zonular in form.

The inner or free surface of the placenta in both specimens possessed a convoluted appearance, the convolutions being similar in their general disposition to, but smaller in size than, those on the corresponding face of the placenta of the Grey Seal. The largest of the convolutions had a breadth of $\frac{8}{10}$ths inch, whilst the smallest were not more than $\frac{2}{10}$ths inch broad. Adjacent convolutions were continued into each other across the bottom of the intermediate sulci. The umbilical cord, which in *A* was 22½ inches long, though in *B* it was somewhat shorter, bifurcated in *B* seven inches from the placenta, but in *A* only 4½ inches. A blue injection was passed into the umbilical vessels, and, notwithstanding the number of years the specimens had been in spirit, a considerable quantity of injection found its way into the fœtal vessels, the ramifications of which on the inner surface of the placenta could be distinctly seen. A transverse section through the cord displayed two umbilical arteries, two umbilical veins, and a very slender tube situated in the middle of the cord. The central tube was undoubtedly the remains of the allantois, a funnel-shaped prolongation of which was situated in the angle of bifurcation between the two halves of the cord. The cord was invested by the amnion, which was continued along the branches of bifurcation to the inner face of the placenta. Attached to the inner surface of the placenta were several broad folds of translucent membrane which were continuous with the funnel-shaped prolongation of the allantois just referred to. The division of the membranes, which had been made in the removal of the fœtus, prevented me from tracing the precise disposition of the amnion and allantois; but it seemed as if the allantois invested the greater part, if not the whole, of the inner surface of the placenta, and that the amnion was separated from the placenta by the sac of the allantois. Where the amnion and allantois were in contact with each other, the ramifications of slender vessels could be seen between them. I could obtain no satisfactory evidence of the presence of an umbilical vesicle.

The umbilical vessels were not limited in their distribution to the placental part of the chorion, but gave off branches to the non-placental poles, which divided and subdivided in their course. Occasionally the finest of these branches ended in a capillary network, but the injection had not run so minutely as to fill the vessels completely. There can however be little doubt that in *Orycteropus*, as I have elsewhere shewn in the *Carnivora* and *Pinnepedia*[1], a compact capillary plexus is distributed in the non-placental parts of the chorion.

The outer surface of the placenta was adherent to the uterine wall. On taking hold of the placenta with the one hand and the wall of the uterus with the other, and making gentle traction, the placenta could be separated from the uterus by tearing through delicate filamentous tissue, which was continuous with the substance of the placenta. The uterine surface of the placenta was then seen to be fissured and to have a convoluted arrangement somewhat similar to that of the inner surface, only not so distinct. At first I thought that in stripping off the placenta of *Orycteropus*, I had left, as in the Bitch, Fox, and Grey Seal, a large portion of the mucous membrane on the placental zone of the uterus, and only removed, along with the organ, the intra-placental prolongations of the mucosa. I examined microscopically the surface which I had exposed, in the expectation of finding an epithelial investment, such as I had seen in the animals just referred to, but without success; for the tissue exposed consisted of a vascular connective tissue resembling the sub-mucous coat, and not of an epithelial layer, such as would have been possessed by the free surface of a mucous membrane. It was obvious therefore that in stripping off the placenta I had torn through the sub-mucous coat, and had removed, along with the chorion, the mucous lining of the uterus in the placental zone.

I then proceeded to examine the placenta itself, to ascertain if any separation could be made between the chorion and the uterine mucous membrane, but the results which I obtained were by no means uniform. In specimen *B* it was not a difficult matter to draw the chorion away from the mucous mem-

[1] *Lectures on the Comparative Anatomy of the Placenta*, 1st series, pp. 80, 83, 114. Edinburgh, 1876.

brane and to analyse the placenta into its fœtal and maternal
portions, the separation being effected with a little more difficulty than can so easily be done in the Mare, Pig, or Cetacean.
In *A* the chorion and uterine mucosa were more firmly united
to each other, so that a separation could not be artificially
made.

I availed myself of the condition of things displayed by *B*
to examine, independently of each other, the structure of the
chorion and uterine mucous membrane. The placental part of
the chorion presented a velvety villous surface, resembling, in
its naked-eye character, the villous chorion of the Mare and
Cetacean. The villi were in part disposed on sinuous ridges of
the chorion, which sometimes ran parallel to each other, but at
times anastomosed together, and formed a reticulated arrangement. The villous ridges were not crowded together, but were
separated from each other by well-marked intervals. The surface of the chorion between the bases of the ridges gave origin
to crowds of villi, which kept up the continuity of the villous
surface. The villi which arose, both from the ridges and the
intermediate parts of the chorion, divided into numerous minute
stunted villi, some of which had the form of short filaments,
though more usually the free end was truncated. The multitude of these stunted villi contributed materially to the velvety
aspect of the surface. As compared with what I have seen
in the Mare and Cetacea, the villi in *Orycteropus* appeared to
be more uniformly diffused over the placental area of the
chorion, and not to be arranged in clusters, with intermediate
surfaces, either bare of villi or with only short simple villi
springing from them, such as I have described in the former
animals. Well-marked branches of the umbilical vessels ran
along the bases of the ridges, and sent off branches into their
substance, which ended in a capillary network. The capillaries
which entered the minute villi formed simple loops in their
interior. In addition to the intra-villous capillaries a well-
marked extra-villous network was distributed beneath the
general plane of the chorion, and presented an arrangement
similar to what I have seen in the Cetacea and Mare.

The free surface of the uterine mucous membrane presented
a mould of the villous surface of the chorion from which it had

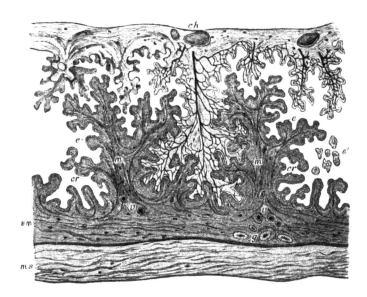

Magnified view of a vertical section through the Placenta and Uterine wall of *Orycteropus;* specimen *B.* *ch.* the chorion, with the villi growing from its outer surface. At the two sides of the figure the villi are seen to be drawn out of the crypts in the mucosa, but in the centre of the figure a villous ridge *r* fits closely between two folds of the mucous membrane, in the crypts of which the villi are lodged. In the villi, to the right and centre, the umbilical vessels are injected; in those to the left they are not. *m.m.* uterine mucosa elevated into folds. *cr. cr.* crypts. *e.e.* epithelial layer belonging to the crypts. *e'.* loose epithelial cells shed from the surface of the mucosa. *g.* glands in transverse section. *v.v.* maternal blood-vessels. *sm.* sub-mucous tissue. *ms.* muscular coat.

been detached. It possessed numerous elongated indentations, into which the villous ridges had fitted, and was punctated by innumerable shallow pits or crypts, which had lodged the stunted villi. The mucous membrane walling in these indentations and crypts sometimes had the form of sinuous trabeculæ, but more frequently, coincident with the multitude of crypts and their communications with each other, these trabeculæ were subdivided into numerous slender papillary elevations, which, when the fœtal and maternal portions of the placenta were in apposition with each other, were lodged between the fœtal villi. This relation of the parts to each other was well seen in vertical sections through the entire thickness of the placenta when its component parts were *in situ*. The villi, which covered the ridges, fitted into crypts situated at the sides and bottom of the indentations in the mucosa, whilst the villi which sprang from the chorion between the bases of the ridges fitted into crypts on the summits of the sinuous trabeculæ or papillary elevations of the mucosa.

Notwithstanding the number of years the placenta had been in spirit I succeeded, in many sections through the placenta, in seeing quantities of cells, which were the epithelial lining of the crypts. In some instances these cells were observed in the crypts themselves; occasionally I saw them attached to the wall of the crypt and presenting the free end to the eye of the observer, though often, where a villus had been drawn out of a crypt, they were detached from the wall and lying free in its cavity. In other instances they were found in large numbers lying loosely in the fluid in which the specimen was examined. They consisted of nucleated masses of granulated protoplasm; sometimes they were columnar in form, at others polygonal, at others rounded, and they closely resembled in shape and appearance the epithelial cells which I have described as lining the uterine crypts in the Mare, Narwhal, and many other mammals. It was by no means uncommon in specimens stained with carmine to find a pink-tinted band on the free surface of the crypts, due, I have little doubt, to staining of the protoplasm of the epithelium cells still adherent to the wall of the crypts. The epithelium rested on a sub-epithelial connective tissue, continuous with the sub-mucous connective tissue coat of the uterus. I made an attempt to inject the uterine vessels so that I might see what their disposition was in the mucous membrane. I failed, however, in the attempt, so that I cannot say whether the vessels of the maternal part of the placenta terminated in a network of ordinary capillaries, as is· the case in the Mare, Pig, Cetacea, and Lemurs, or dilated into colossal capillaries, such as one sees in the Grey Seal, the Fox, and other *Carnivora* and in the Sloth.

In the next place I proceeded to examine the uterine mucous membrane with the view of ascertaining the arrangement of the utricular glands. Portions of the mucosa from the non-placental area of the uterus were first put under the microscope, and tubular glands were seen ramifying in it. As the arrangement of utricular glands can only be satisfactorily ascertained when the epithelium is present, and as the epithelial cells in the glands seemed to be to some extent in a process of disintegration, the arrangement of the glands was not so precisely made out as I should have desired. I ascertained however that the glands occasionally bifurcated, and that they gave origin to short diverticula, in which the secreting cells were as a rule more distinct than in the branches of bifurcation and gland stems. In one preparation I saw two glands close together and parallel to each other, but as a rule they were separated from each other by considerable intervals, occupied by connective tissue, and they were not by any means so numerous as in the Pig, Mare, Cetacea, Lemurs or Manis.

In sections of the wall of the uterus in the placental area sections through the tubular glands were occasionally seen: they were situated in the sub-mucous tissue in close proximity to the true muscular coat: the glands contained an epithelium arranged around a central lumen. The part of the glands, which was observed in this region, was obviously the deeper end, and the sections were transversely through the tubes. I examined a number of sections with the object of tracing the glands through the thickness of the mucosa in the placental region to their openings on the surface of the maternal placenta. I was unsuccessful however in tracing a single gland to its termination, and for undoubtedly the same reason that led to the want of success

in the study of the placenta in the common Cat[1], viz. the
complexity of the surface owing to the multitude of crypts it
contained, and the absence of recognisable areas free from
crypts in which the glands opened, such as I have described
in the Pig, Mare, and most remarkably in the Lemurs. I
examined the crypts with care, but could not determine
that the glands had any direct communication with them.
The number of crypts was so very much greater than that of
the glands, that even if a gland did occasionally open into a
crypt, it is obvious that the crypts could have had no necessary
relation to the glands, and must have been developed quite independently of them. In this respect the examination of the
placenta of *Orycteropus*, imperfect though it to some extent
has been, owing to the length of time the specimen had been
kept, is in accordance with what I have seen in so many other
mammals.

A few words may now be said on the nature of the placentation in this animal, whether it should be referred to the
Deciduate or Non-deciduate group. If specimen *B* alone had
been before me for examination I should have been inclined
to say that *Orycteropus* was a non-deciduate mammal. The
readiness with which the chorion could be separated from the
uterine mucosa, and the ease with which the villi of the former
could be enucleated out of the crypts in the latter, without
drawing away with them maternal tissue, were strongly suggestive of a relation of parts similar to what one is familiar
with in the diffused and non-deciduate placenta of the Mare
and Cetacean. The broadly zonular form of the placenta
in *Orycteropus*, which closely approximates to the zonary
placenta of the deciduate Carnivore, or Seal, is not to be
regarded as an argument of any importance against this view
of its structure, for the non-villous poles of the chorion are
merely an exaggerated condition of the bare spots at the
poles of the chorion in the Mare and Cetacea, and in the Pig,
as was known to Von Baer, and as has been confirmed by
myself and other anatomists, each end of the chorion for
several inches around the pole had a smooth non-villous

---

[1] *Lectures on the Comparative Anatomy of the Placenta*, 1st series, p. 73.
Edinburgh, 1876.

surface. *Orycteropus* differs however from such diffused and non-deciduate mammals, *e.g.* the Mare, Cetacea, and Lemurs, which produce only one fœtus at a birth, in having the chorion confined to the fecundated horn, whereas in the uniparous mammals just referred to, the chorion extends from the end of the fecundated to the end of the non-fecundated cornu. This difference in *Orycteropus* is undoubtedly due to the horns of the uterus opening independently into the vagina, and not communicating with each other through a common corpus uteri.

Specimen *A*, which is I believe the one referred to by Professor Huxley in his Introduction to Classification, did not allow the separation of the chorion from the uterine mucous membrane. These two constituents of the placenta were so closely adapted to each other, that all my attempts to make a clean separation between them failed. Hence the examination of this specimen alone would have justified me in concluding, as had been done by Professor Huxley, that the placenta in *Orycteropus* was deciduate. How the great difference in the separability of the fœtal and maternal elements of the placenta, as exhibited by these two specimens, was occasioned must remain a matter of doubt. It is possible that *A* had been immersed at one time in stronger spirit than *B*, so that the constituent parts of the placenta had been more firmly bound together by the hardening action of the preservative fluids. On the other hand it cannot be said that *B* had been imperfectly preserved, or exhibited any marks of disintegration from putrescence; for the microscopic sections through *B* gave much clearer views of the structure of the tissues, and the fœtal capillaries were more fully injected, a condition of things which inclines me the more to the view that *B* gives one the correct clue to the nature of the placentation. At present, however, a definite statement cannot be made, and the determination of the placentation, as regards its deciduate or non-deciduate nature, must be reserved until some competent anatomist can obtain the perfectly fresh gravid uterus of this animal, or still better, examine its shed placenta.

Orycteropus furnishes an additional illustration to those previously recorded of the modifications in the form of the

placenta exhibited by animals belonging to the order *Edentata.*
The Armadilloes (*Dasypus*), according to Professor Owen, possess a single, thin, oblong, disc-shaped placenta; a specimen,
probably *Dasypus gymnurus*, recently described by Kölliker[1],
had a transversely oval placenta, which occupied the upper
⅔rds of the uterus. In *Manis*, as Dr Sharpey has shewn, the
placenta is diffused over the surfaces of the chorion and uterine
mucosa. In *Myrmecophaga* and *Tamandua*, as MM. Milne
Edwards have pointed out, the placenta is set on the chorion
in a dome-like manner. In the Sloths, as I have elsewhere
described, the placenta is dome-like in its general form, and
consists of a number of aggregated, discoid lobes. In *Orycteropus*, as I have now shewn, the placenta is broadly zonular.

The *Edentata* therefore, so far as their placentation has
been, up to this time, studied, furnish examples of all the
known group-forms of placenta, except the cotyledonary. They
vary also in the intimate relation of their fœtal and maternal
structures, for whilst the Armadilloes and Sloths are deciduate,
Manis, on the other hand, is undoubtedly non-deciduate; whilst
the characters of the Hairy and Cape Ant-eaters, as regards the
decidua, have not yet been precisely determined.

[1] *Entwicklungsgeschichte des Menschen, &c.* 2nd ed., p. 362. Leipzig,
1876.

OBSERVATIONS ON THE GALVANIC EXCITATION OF NERVE AND MUSCLE, WITH SPECIAL REFERENCE TO THE MODIFICATION OF THE EXCITABILITY OF MOTOR NERVES PRODUCED BY INJURY[1]. By G. J. ROMANES, M.A., *Caius College, Cambridge.*

(From the Jodrell Laboratory, University College, London.)

THE object of the investigation of which the following are some of the results, was to determine, as far as opportunity permitted, the electrotonic conditions of muscle permeated by nerve. While conducting this investigation, I happened to observe that injury of a motor nerve is followed by a very marked and very peculiar alteration in its behaviour towards voltaic stimuli. This branch of the enquiry I therefore worked out with some completeness, and as the facts presented by it are of considerable interest in themselves, it seems desirable to treat of them in a separate form. There are likewise one or two other observations which were made in a similarly incidental way; and as these do not stand in any definite relation to my other results, I take this opportunity of publishing them as a sort of appendix to the present paper.

§ 1. Before proceeding to describe the experiments and results which constitute the main subject of this article, it is necessary to state some facts concerning the electrotonic conditions of a frog's gastrocnemius prior to injury of its attached sciatic nerve. If, then, this muscle be placed on non-polarizable electrodes in a horizontal position with its convex surface uppermost, one is generally able to observe that it is somewhat more sensitive to minimal stimulation, supplied by *closure* of the constant current, when the femoral end rests on the kathode than when this end rests on the anode. In other words, the muscle under these conditions is most sensitive to minimal make of the current when the latter is in the ascending direc-

[1] An abstract of this paper was read before the Royal Society on May 4th, 1876.

tion[1]. This appears to be the exact reverse of what is true of nerve when in the "weakest" phase of Pflüger's series; but a little reflection will show that, upon the theory of electrotonus, this is just what we ought in this case to expect. Pflüger's law refers to nerve-trunks outside of muscle, so that, according to the theory of electrotonus, the reason why the descending make constitutes the strongest stimulus, is probably because this is the case in which the kathode, or point of strongest stimulus on making, is nearest the muscle. But if, as in the experiment we are considering, *both* electrodes are placed *in contact with* the muscle, the element of relative proximity of anode and kathode to the muscle is abolished, and the direction of the current which supplies the stronger stimulus will probably be determined by the distribution of the nerve in the substance of the muscle itself. Now the sciatic nerve enters the gastrocnemius near the origin of the latter, and then courses along the muscle, spreading out its peripheral ramifications as it advances. Consequently, in the experiment now under consideration, one electrode is in almost immediate contact with the nerve-trunk where it enters the muscle, while the other electrode supports the part of the muscle that contains only peripheral nervous elements. It is therefore to be expected, upon the theory of electrotonus, that the muscle under these conditions should prove itself most sensitive to the closing shock when the nerve-trunk rests on the kathode—that is, when the current in the muscle is ascending.

It is to be observed, however, that although this expectation is in most cases fulfilled, it is far from being so invariably. Different gastrocnemius muscles, though treated as far as possible in exactly the same way, manifest considerable differences, both in their general sensitiveness to electrical stimulation, and in their relative sensitiveness to closure of the ascending and descending currents respectively. Nay even the same muscle, if rapidly prepared, will generally be found to undergo fluctuations in these respects from minute to minute, and sometimes even from second to second. This fact, I presume, is to be

---

[1] Throughout this paper, when speaking of the gastrocnemius, I shall always use the term "ascending" to denote passage of the current from the tarsal to the femoral end of the muscle, and the term "descending" to denote passage of the current in the opposite direction.

attributed to the unnatural conditions which the experiment imposes upon the processes of nutrition. It may therefore be well to state that I have conducted some observations upon muscles while still in the body of the living frog—taking care not to expose them more than was necessary for the purposes of stimulation. The results yielded by this method, however, were not more uniform than those which I had previously obtained by the less laborious method of rapidly preparing and observing excised muscles.

Notwithstanding, however, these fluctuations of excitability, I think I may confidently assert, as the result of a large number of experiments, that the following should certainly be regarded as the general rule:—*The gastrocnemius muscle of a frog is more sensitive to minimal stimulation supplied by closure of the constant current, when the latter is ascending than when it is descending through the substance of the muscle.*

Similarly, with the same limitations as before, the complementary rule may now be enunciated, viz.:—*The gastrocnemius muscle of a frog is more sensitive to minimal stimulation supplied by opening of the constant current, when the latter is descending than when it is ascending through the substance of the muscle.*

§ 2. It has long been known that when a nerve is cut, or otherwise injured, the parts of the nerve at and near the seat of injury have their excitability increased. Now, if any muscle of a frog be chosen in which the nerve-supply enters at one end, and this muscle be placed on non-polarizable electrodes as described in § 1, I find that section of the nerve-trunk near or within the muscle is attended with very remarkable alterations, not only in the *general* sensitiveness of the muscle, but also and more particularly, in its *relative* sensitiveness to make and to break of the current. But before describing these alterations, it is necessary for me to preface the statement of my results with a brief description of the method by which I obtained them.

The gastrocnemius muscle, being in every respect the best suited for the observations in question, is the muscle which I mainly employed. The sciatic nerve I usually divided at the origin of the femur (never lower down), and then dissected it

47—2

out as far as the tibial end of the femur in the ordinary way, but with more than ordinary care not to injure the nerve in the process. Having previously severed the tibia and divided the tendo Achillis, I next cut through the femur as close to the knee as possible. The nerve-muscle preparation was now placed on non-polarizable electrodes with the flat surface of the muscle downwards, and the nerve supported beyond the electrodes by means of a glass rod. The usual precautions as to temperature, moisture, &c. having been attended to[1], it only remained to ascertain *as quickly as possible*, (1) the minimal descending break, (2) the minimal ascending make, and (3) the minimal descending make. It is very important to take these observations in this order; for, as will immediately appear, the intensity of the current required to cause the earliest response to the descending break is comparatively high; so that, in order to ascertain its exact degree, it is necessary to throw the muscle into a more or less severe state of tetanus every time the current is closed[2]. Now, as it generally requires five or six attempts before the minimal descending break can be determined[3], however short a time the tetanus on each occasion is allowed to last, its repeated occurrence at short intervals cannot but

[1] The Jodrell Laboratory is provided with a moist chamber of a cubical form, and sufficiently large to take in an entire frog. Many of my experiments were therefore made without exposing the sciatic nerve at all. The tendo Achillis being divided and the tibia removed at the knee, the whole frog was introduced into the moist chamber, and there supported on a level with the electrodes. The exposed gastrocnemius muscle was then placed on the latter, and the experiment proceeded with as above.

[2] It may not be superfluous to state that a muscle (whether or not curarized) behaves very differently towards galvanic stimulation when the current is applied directly to itself, and when the current is applied to its attached nerve. For, in the latter case, as is universally known, the muscle only contracts at the moment when the current is being either opened or closed, or when variations are taking place in the intensity of the current while it continues closed. But in the former case the muscle continues in a state of persistent contraction, or tetanus, during the whole of the time that the current is passing. The stronger the current the more severe is such tetanus, and with very feeble currents it is not observable at all, *i. e.* currents which are only just strong enough to cause minimal stimulation on closure, do not cause persistent tetanus while closed.

[3] To those who have not worked at direct voltaic stimulation of muscle, it may seem to require explanation how the minimal break contraction can be observed when the whole muscle is already in a state of tetanus, owing to the passage of the constant current through its own substance. The explanation simply is, that on progressively increasing the strength of the current, although by this means the severity of the tetanus is likewise progressively increased, a point is at last reached where, upon suddenly breaking the current, the muscle, just before relaxing, gives an additional contraction superimposed upon its previous tetanic contraction. This is what I call the minimal break-contraction.

modify, to some extent, the irritability of the muscle. Hence, if the observations on minimal make (which require far less strength of current) were to be made *before* those on minimal break, it is obvious that the comparison of relative sensitiveness to minimal make before and after nerve-section might be seriously vitiated.

It is not of so much consequence which of the two minimal makes is first determined; but I have placed them in the above order for the sake of alluding to an important consideration which must never be lost sight of while conducting any observations on minimal stimulation. The consideration to which I refer involves the fact, that nerve and muscle become rapidly less and less sensitive to electrical stimulation, if this be supplied successively in the same direction. In all experiments, therefore, on minimal voltaic stimulation, it is of the first importance frequently to refresh the tissue by passing the current in a direction the reverse of that whose minimal stimulus we wish to ascertain. In the particular case now before us, the best method to adopt for determining the minimal ascending and the minimal descending make, is to reverse the current at each observation, and so, with the appropriate shifting of the mercury cups at each stimulation, to determine both the minimal makes simultaneously.

It will have been observed that as yet I have said nothing about the minimal ascending break. The reason why I have omitted this phase, is because the strength of current required for its development is usually so great, that the amount of tetanus set up in the muscle in order to procure it is apt to endanger the rest of the experiment by unduly exhausting the muscle[1]. The method, therefore, which I usually adopted in this case was to observe, in a number of muscles, the effects of

---

[1] I may as well state here that the fact of such severe tetanus being required to obtain this minimal, renders it impossible to be quite sure as to the point at which the response to the ascending break first occurs. As the muscle is already in a state of violent contraction, owing to the passage of a strong galvanic current through its substance, it is probably excited by the breaking of the current before the latter is strong enough to cause the contraction consequent on the breaking excitation to assert itself perceptibly over the previously existing tetanic contraction. Of course this is a source of error which it is impossible to avoid, and as all my estimations of this minimal were made, as they only could be made, by noting the point at which the ascending break-contraction first became *perceptible*, it is probable that in the subjoined table this minimal is stated at somewhat below its true value.

nerve-section on the minimal ascending break *alone*, and then
to take an average of all the results. This method I also employed in the case of the minimal descending break, and in
both the minimal makes. The following are the mean results
yielded by a large number of experiments conducted according
to both methods. I think, therefore, that these results may be
regarded as trustworthy in the main, although it would doubtless be difficult, or impossible, to find any one muscle that
would precisely correspond with all the degrees of excitability
here given. Indeed, it is probable that in the case of any given
muscle the correspondence would not even be approximate over
all the phases; for different muscles manifest great variations
in their relative excitability towards the eight cases of stimulation here quoted.

Descending Make.		Ascending Make.		Descending Break.		Ascending Break.	
Before cutting.	After cutting.	Before cutting.	After cutting.	Before cutting.	After cutting.	Before cutting.	After cutting.
24	27	36	46	2	32	1	1½

With regard to this table it is necessary to explain, that
by the word "cutting" is meant section of the gastrocnemius
through its entire substance at its extreme femoral end, and
that the numbers express the relative degrees of sensitiveness
which the muscle exhibits, both before and after cutting, towards the stimuli which are indicated above them[1].

---

[1] These numbers are thus obtained:—Suppose A (see Fig.) to be a battery, B
a set of resistance coils, C a rheochord, D a commutator, and E the muscle. By
removing a plug from B the resistance is increased, and therefore the current
through E is diminished. But the effect of removing a plug from C, although
likewise that of increasing the resistance through the whole circuit, is to *augment* the current passing through E. For, previous to removal of a plug from
C, the current branched at x, and the resistance in E being high as compared
with that in C, the principal part of the current takes the course x, y, C, A.
But if a plug be removed from C, the resistance in C is increased, and a proportional amount of the current takes the direction x, z, E, A. Hence the effect of
removing a plug from B is that of diminishing the current in E, while the
opposite result is effected by removing a plug from C.

Such being the apparatus, in all my experiments I began by removing a
certain number of plugs from B, and keeping these plugs out during the whole

In order to render more readily appreciable these remarkable effects of cutting a nerve, I append to this paper a diagram which is intended to represent, in a graphic form, the numerical relations set forth in the above table.   With reference to this diagram it will only be necessary to explain, that in all the couplets the left-hand vertical line represents the degree of sensitiveness of the muscle to the excitation indicated *before* cutting, while the right-hand vertical line of each couplet represents the degree of sensitiveness of the muscle to the same stimulus *after* cutting.   As in the table, so in the diagram, all the proportions are referred to the ascending break as to a unit—this being the stimulus to which the muscle is least sensitive, and for which, therefore, the strongest current is required in order to elicit a contraction.

With regard to these results I may offer the following observations.   In the first place, it is evident that the increase

course of any one experiment.   I thus worked during the whole course of any one experiment with a current of constant intensity, so far as the whole circuit

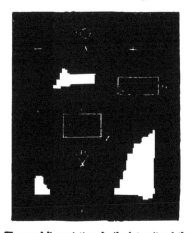

was concerned.   The requisite variations in the intensity of the stimuli were of course effected by the rheochord *C*.   Now the numbers in the above table are obtained by a very simple calculation.   Suppose, for instance, that the minimal ascending break-contraction requires 18 ohms resistance to be thrown into the rheochord, while the minimal ascending make only requires ·5 to be thrown in ; then the relative sensitiveness of the muscle to the ascending break and make would be approximately represented by the numbers 1 : 36.

of excitability shown by the muscle after being cut is affected to an extraordinary extent by the *direction* of the current; and further, that the manner in which it is so affected is very instructive when considered in relation to the known facts of electrotonus. For, just as before cutting, the *normal* sensitiveness of the muscle is greatest to the making excitation when its femoral end (or nerve-trunk) rests on the kathode, and most sensitive to the breaking stimulus when this end rests on the anode; so, after the general sensitiveness of the nerve has been exalted by section, the *exaltation* shows itself in a far higher degree to the making stimulus when the femoral end of the muscle (or severed nerve-trunk) rests on the kathode,

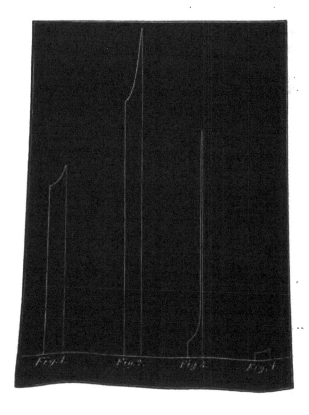

and to the breaking stimulus when this end rests on the anode.
Thus it is that the curves in figs. 2 and 3 are so much
steeper than those in figs. 1 and 4. The only fact, then, that
does not seem to admit of any very satisfactory explanation, is
the altogether *disproportionate* increase of excitability which the
muscle after cutting exhibits to the descending break (fig. 3) as
compared with the ascending make (fig. 2); for it might be
expected, *à priori*, that the increase of excitability due to nerve-
section ought at any rate to be proportional in the case of these
two stimuli—if not that the increase of sensitiveness to the
kathodic make should be greater than that to the anodic break,
seeing that prior to section the normal sensitiveness of the
muscle was much greater in relation to the former than in
relation to the latter stimulus. As the facts, however, so conspicuously negative any such anticipation, I shall devote the
next section to a further consideration of them.

§ 3. Dr Burdon-Sanderson suggested, that if we suppose
the breaking stimulus to be of a more *instantaneous* nature
than the making one, the facts in question might admit of a
probable explanation; because in this case the breaking stimulus
would bear more resemblance to the induction shock than
would the making stimulus, and as it is well known how sensitive nerve is to the induction shock, we might reasonably
conclude that when the sensitiveness of a nerve is increased by
section, it would probably become more than proportionably
increased to the more sudden stimulus. In order to test the
correctness of this hypothesis, Dr Sanderson further suggested
that the period of the muscle's latent stimulation should be
taken both before and after section, and also that the following
experiment should be tried. By means of an appropriate apparatus, the uncut muscle was to have supplied to it a voltaic
stimulus of measured duration, and this duration was to be
graduated down to the point at which the break of the current
succeeded the make with a rapidity just sufficiently great to
prevent the muscle from responding to either stimulus. The
strength of the current remaining unaltered, the nerve was
then to be cut through at the usual place; and, lastly, it was
to be observed whether or not the muscle was thus rendered
more sensitive to stimuli of short duration. The results of

these suggestions promise to be highly interesting; but as yet they have not been fully worked out. So far, however, as this branch of the enquiry has been pursued, these results are as follow.

Section of the nerve—either just above the knee, or immediately after it enters the gastrocnemius—is in all cases attended with a marked increase of excitability towards voltaic stimuli of short duration. In other words, after nerve-section the muscle responds to stimuli of much shorter duration than those which were required to cause responses before nerve-section. As yet no exact quantitative estimations of this effect have been procured; but this will be done as soon as opportunity permits. Meanwhile, then, it must be enough to say, in general terms, that the effect in question is surprisingly great. At first, therefore, it seemed that this experiment was confirmatory of the hypothesis which it was designed to test: it seemed as though the increased sensitiveness after cutting to stimuli of short duration, coincided exactly with the expectation regarding the behaviour of mutilated muscle towards the breaking stimulus. Very soon, however, I observed that the increased sensitiveness to stimuli of short duration was only manifested when the femoral, or cut end, of the muscle rested on the kathode, while it was scarcely, if at all, apparent when this end rested on the anode. This fact, of course, led me to infer that the augmented excitability in question had reference to the closing and not to the opening excitation. Accordingly I fitted up an appropriate arrangement of wires and keys, by which I could at pleasure throw in ordinary closing and opening stimuli, or the rapidly succeeding closing and opening stimuli of short duration[1]. In this way it was easy, by comparing in the two cases

[1] The entire apparatus now consisted of the following parts. The duration of the stimulus was graduated by means of a heavy pendulum, which constituted one pole of the battery, and which while swinging made contact, at the lowest point in its arc, with the other pole. This pendulum could therefore be made to supply a stimulus of any required duration to a muscle interposed in the circuit. For the only conditions required were to keep the points at which the two poles came in contact of uniform size, and to graduate the duration of contact by increasing or diminishing the distance through which the pendulum, or moving pole, was allowed to swing. The pendulum was provided with a smoked plate, on which a lever from the muscle wrote while the plate was swinging past. By putting the key down and letting the pendulum swing, a stimulus of measured duration could be supplied to the muscle, while by holding the moveable pole, or pendulum, in contact with the fixed pole, and then closing or opening the key, ordinary stimuli could be supplied to the muscle.

the nature of the contraction (which in almost every muscle presents some idiosyncric difference on make and break), to obtain optical proof that my inference was correct. When the femoral or cut end of the muscle rested on the kathode, and when, as was usual under these circumstances, the muscle after cutting responded to stimuli of shorter duration than were required to elicit responsive contractions before cutting, I could plainly see that the contractions given in response to stimuli of short duration were always precisely similar to those which the muscle gave when stimulated by *closing* the circuit with the ordinary key. On the other hand, the nature of the contractions upon *opening* the circuit with this key was always markedly different. Therefore, it is impossible to doubt that in the experiment with brief stimulation—*i.e.* with rapidly succeeding make and break—the increased excitability of the muscle after section of its nerve has reference, not to the opening, but to the closing shock.

These facts are, I think, of considerable interest in themselves; but they fail to answer the question originally before us, viz., Why does section of a nerve produce so disproportionate an effect on its sensitiveness to the excitation supplied by the descending break? Nor have I any satisfactory answer to give to this question, unless the following consideration may be deemed so. Before section of the sciatic nerve, the gastrocnemius muscle, as may be seen by a glance at the diagram, is always immensely more sensitive to make than to break of the current. Consequently, when the sensitiveness of the nerve is increased by section, the increase has not so much room (so to speak) for its occurrence in the one case as in the other. For, in the uncut muscle, the minimal break contraction only takes place when the intensity of the current is somewhere from ten to twenty times as great as that which is sufficient to evoke the minimal make contraction. Hence the minimal make contraction before injury occurs at a point so much nearer to zero of the current's intensity than does the minimal break contraction before injury, that when both these minimals are reduced still further by the injury, the latter minimal has a much wider range through which it is free to fall than has the former. Of course this fact need not prevent the lesser fall from being numerically proportional to the greater one, however minute the observed differences may thus require to be. The question, however, is as to how far a strictly *numerical* proportion is in this case a fair one. I think we must certainly hold that the value, as a stimulus of any given increment of current, is determined by the proportion which such increment bears to the intensity of current that is required to produce adequate stimulation. In other words, we must suppose that any given unit of electrical intensity has more influence as an excitant if added to a current of a small number of units—a weak current,—than if added to a current of a large number of units—a strong current; supposing in both cases that the original current is strong enough to cause adequate stimulation. But if this is so, it follows that *subtraction* of a unit from a strong current must have less effect than subtraction of the same unit from a weak current. Now, when the general excitability of the muscle is raised by section, the effect is that the muscle is able, both in the case of the ascending make and in that of the descending break, to afford (as it were) to part with some units of the stimulating influence which were previously required to cause adequate stimulation. But, as the sum of such units which are present before cutting are much less numerous in the case of the make than in that of the break, in the case of the make each unit must be of a correspondingly greater value as a stimulant. Consequently, when both the minimals are reduced by cutting, the minimal make is not able to draw upon, and so cannot part with, so many units as can the minimal break. Hence the reduction in question may take place in a strictly proportional manner; only, if the proportion has reference to the *value of the electrical units as stimulants*, it follows, from what has been said, that there will probably be no *numerical* equality between the two ratios.

In favour of this explanation, it is to be remembered that, as already stated, nerve-section produces much more than a proportional effect in the case of the ascending make as compared with that of the descending break, in respect of increasing the excitability of the muscle *to stimuli of short duration*. In other words, by this mode of stimulation all the above relations are inverted. It is as though the comparatively small number.

ef units of *electrical intensity* by which the minimal make is diminished through nerve-section, represents a great actual increase of excitability *when this is estimated by some other method;* or, to turn to the diagram, it seems as though the small distance through which the curve in fig. 2 passes, as compared with the curve in fig. 3, really represents an increase of excitability much more important than the curve expresses: it seems as though it is just because the difficulty of ascending (so to speak) increases in so rapid a ratio as the curves approach the zero level of the current's intensity, that the steep curve of the descending break terminates at, or below, the point at which the much less steep curve of the ascending make begins. This appears to be so, because, on testing the increase of excitability by means of stimuli of short duration, it is found that the relatively flat curve in fig. 2 represents what would doubtless be a relatively steep curve, if it were possible to institute the numerical comparisons in the case of stimuli of minimal duration, as it is possible to do so in the case of stimuli of minimal intensity[1].

Perhaps it may here be objected to this explanation, that neither the excitability to the descending make (fig. 1), nor that to the ascending break (fig. 4)—the representative lines of which are both low before nerve-section, and so have plenty of room to rise. after nerve-section—correspond in any degree, in the respects we are considering, with the excitability to the descending break (fig. 3). It appears to me, however, that this objection admits of being readily overcome by the very interesting and instructive consideration previously stated. This consideration, it will be remembered, is as follows. Out of the four cases represented in the diagram, the two in which aug-

---

[1] Unfortunately it is not possible to do this with exactness, from the fact that in this mode of stimulation make and break must always necessarily follow one another much too rapidly to admit of the making effect being eliminated from the breaking one. Approximately, however, the desired comparison may be instituted thus :—First observing the minimal duration required to produce adequate stimulation (a) with the uncut femoral end of the gastrocnemius on the kathode, and (b) the uncut femoral end on the anode : and next, (a) the cut femoral end on the kathode, and (b) the cut femoral end on the anode. By now comparing a with a, a with b, and b with b, it would be possible to obtain for responses to stimuli of minimal duration representative curves, such as those in the diagram which have reference to responses to stimuli of minimal intensity. Such approximate estimations I hope to make as soon as opportunity permits.

mentation of excitability after nerve-section occurs in an apparently undue ratio, are the two in which, according to the theory of electrotonus, we should expect it to occur, if it occurs at all. For, in the case of the ascending make (fig. 2), the femoral end of the gastrocnemius is suddenly thrown into kathelectrotonus, while in the case of the descending break (fig. 3) this end is as suddenly thrown into anelectrotonus. Consequently, when the nerve-trunk, as it enters the gastrocnemius at this end, is suddenly rendered abnormally excitable by section, it manifests its increased excitability by responding far more readily than it did before to the appearance of kathelectrotonus (fig. 2), and to the disappearance of anelectrotonus (fig. 3); while, conversely, there is scarcely any change to be perceived in its behaviour towards the appearance of anelectrotonus (fig. 1), or towards the disappearance of kathelectrotonus (fig. 4).

Whether or not, therefore, the explanation I have suggested to account for the absence of numerical proportion between the heights of the corresponding lines in figs. 2 and 3 be deemed satisfactory, I think there can be no question concerning the significance I here assign to the curves in figs. 1 and 4 being so much less steep than those in figs. 2 and 3.

For the sake of clearness, I will now very briefly sum up the results so far as they have as yet been described. Although there are considerable deviations from the rule, the gastrocnemius muscle, when its nerve is uninjured, is upon the whole most sensitive to closure of the current when its femoral end, or nerve-trunk, rests on the kathode, and most sensitive to opening of the current when this end rests on the anode. Similarly, after the general sensitiveness has been exalted by nerve-section, the *exaltation* shows itself most towards the closing stimulus when the femoral end, or severed nerve-trunk, rests on the kathode, and towards the opening stimulus when this end rests on the anode. Thus far the facts agree with the theory of electrotonus. The increase of excitability towards the opening anodic stimulus, however, is out of all proportion greater than that towards the closing kathodic stimulus; and I have sought to account for this fact by the following consideration. As the excitability of the muscle before nerve-section is usually much greater towards the kathodic make than it is towards the anodic break, after the general excitability is increased by nerve-section, the increase is probably better able to assert itself towards the latter than towards the former stimulus. For, as before nerve-section the minimal kathodic make occurs so much nearer to zero of the current's intensity than does the minimal anodic break, after nerve-section every unit of electrical intensity through which the minimal make is reduced must have previously been of much more value as a stimulant than any of the units of electrical intensity through which the minimal break is reduced. Thus it is that although the figures 36 : 46 :: 2 : 32 do not express any *numerical* proportion, they may yet express a *real* proportion, if we have regard, not to the numerical value of the corresponding electrical units, but to the value of these units as stimulants.

This explanation is borne out by the fact, that when we graduate the severity of the stimulus, not by varying its *intensity*, but by varying its *duration*, we find, not only that the above numerical disproportion is abolished, but even that its terms are, in some unknown degree, reversed. That is to say, when we employ a mode of excitation towards which the *normal* excitability of the tissue exhibits a greater equality with regard to the kathodic make and anodic break-stimulations, we thus perceive that the *abnormal elevation* of excitability due to nerve-section is manifested in a higher degree towards the kathodic make than it is towards the anodic break.

§ 4. To the above exposition I must add a few words with reference to a somewhat obscure fact, and one which at first led me to doubt how far the above described effects of cutting the gastrocnemius were to be attributed to the stimulating influence of the operation upon the nerve, and how far to a possibly similar influence upon the muscle. The fact in question is, that if a frog be sufficiently curarized to render its gastrocnemius irresponsive to faradaic electricity applied at any part of the sciatic nerve, and if the gastrocnemius be then removed and one or more incisions made in its own substance, it sometimes, though seldom, happens that the abnormal excitability towards break of the current becomes developed. Further, if a sartorius muscle, however deeply curarized, be cut in a transverse direction through one quarter or one half of its width, it may usually be observed that the effect of introducing the cut is to develop the breaking effect in the neighbourhood of the cut. In view of these facts, therefore, the question arose as to whether the results detailed in the last section might not be due to an increase in the excitability (due to cutting) of the muscular, as distinguished from the nervous, element of the uncurarized gastrocnemius. I therefore made the following experiments with the view of settling this question.

(a) Having carefully dissected out the whole length of the sciatic nerve, I supported it with a glass rod, while I placed the attached gastrocnemius on the electrodes as usual. Having next ascertained the minimal makes and descending break, I severed the sciatic nerve just at the point where it enters the gastrocnemius. Although the gastrocnemius itself was thus in no way injured, the characteristic effects of cutting were as well developed as when in the previous experiments the nerve had been severed after its entry into the muscle. As it thus became interesting to ascertain how far away from the seat of injury the characteristic effects of cutting first become developed, I tried, in several nerve-muscle preparations, successively snipping the sciatic nerve from above downwards, alternating each snip of the nerve with a minimal stimulation of the muscle. In this way I found that the earliest indications of increasing excitability of the muscle (as shown by the alterations in the descending break-contraction) usually occurred when the snipping reached to within two-thirds or one-half the length of the femoral portion of the sciatic. From this point downwards the excitability of the muscle increased with every fresh cutting of the nerve.

(b) In a nerve-muscle preparation set up as just described in paragraph (a), stimulating the attached nerve with acids, alkalies, glycerine, etc. has the same kind of effects on the excitability of the muscle as cutting.

(c) Warming the sciatic nerve is attended with similar results.

(d) Passing a strong descending current through the sciatic nerve, so as to throw the part nearest the gastrocnemius into kathelectrotonus, has a very well marked effect of the same

kind; while, conversely, throwing the same part of the sciatic
into anelectrotonus has the opposite effect: that is to say, when
the sciatic is in kathelectrotonus, the descending break contraction in the gastrocnemius can be obtained with a considerably less powerful current than is required to obtain this
response when no polarizing current is passing through the
nerve; while, if the sciatic is in anelectrotonus, a more powerful
current is required to obtain the descending break contraction
in the muscle than when the nerve is in the normal state. The
inhibiting effect of anelectrotonus, however, is not so pronounced as is the converse effect of kathelectrotonus.

(e) Severe voltaic tetanization of the gastrocnemius—
such as that caused by a Grove's element—is frequently followed by an increase of sensitiveness to the descending break
nearly as remarkable as that which follows cutting. As this
effect of a strong voltaic current does not seem to occur in
curarized muscles, I conclude that it is due to an increase in
the excitability of the intra-muscular nervous elements through
injury caused by the strong voltaic current. The effect in
question is independent of the *direction* of the injuring current,
and persists for a considerable time after this current has been
opened. Soon after the current has been opened, however, there
is a rapid decline in the sensitiveness to the descending break,
after which this decline proceeds in a more gradual manner. All
this is strikingly similar to what occurs after nerve-section.

From these experiments, then, I conclude that it was section
of the *nerve-trunk* in the uncurarized gastrocnemius which in
the former experiments elevated the excitability of the muscle.
It remains, however, to explain why in a curarized gastrocnemius it sometimes happens that cutting serves to develop the
opening contraction. The only explanation I have to offer is,
that in these somewhat exceptional cases, the poison, for some
reason or other, has not succeeded in completely destroying the
function of the intra-muscular nervous elements[1]. As regards
the more usual case of the sartorius when cut in the manner

---

[1] Kühne infers, from his observations on the curarized sartorius, that the
poison does not, except in colossal doses or after long-continued action, entirely
destroy the ultimate intra-muscular nervous elements (*Arch. f. Anat. u.
Physiol.*, 1860, 477). If there is any truth in this view, it would help to explain
the above fact.

before alluded to, I think that the explanation is probably very simple, viz. that a transverse incision made partly across this strap-shaped muscle causes the intensity of the current in the remaining parts near the incision to be considerably increased; and hence, although the strength of the current remains un-altered, these parts naturally become more responsive, both to the opening and to the closing excitations, than they were before. That in any case the effect of cutting a thoroughly curarized sartorius is more *local* than that of cutting an un-curarized one, is not only in most cases evident to the eye, but may be also shown by the following experiment. If one end of an uncurarized sartorius be allowed to overlap the anode, so as to constitute an extra-polar portion, it will often be found that on cutting off this portion, the excitability of all the remaining or intra-polar portion of the muscle to the breaking excitation is greatly increased. Yet the part of the muscle removed by the section (the extra-polar part) could not have previously been concerned in the conduction of the current; so that the *density* of the latter in the remaining, or intra-polar part, must have been the same before and after removal of the extra-polar part. Now I have never been able to obtain this result in the case of a well curarized sartorius.

§ 5. Another method which I employed to test the effects of nerve-injury on excitability was one which, in the first instance, I fell upon accidentally. It consisted in joining up the non-polarizable electrodes with a continuous bridge of clay made perfectly flat on its upper surface. Care being taken to keep this surface uniformly moist with saline solution, the sciatic nerve in a nerve-muscle prepartion was laid upon it; so that where the current passed through the clay bridge, a portion of it also passed through the sciatic nerve, thereby stimulating the attached muscle. The advantage of this method consists in the facility with which different parts of the nerve-length may be stimulated to the exclusion of other parts. By a curious coin-cidence, Prof. Rutherford appears to have been working at this subject at about the same time as myself, though quite inde-pendently of me. I only became aware of this fact a few days before sending in an abstract of the present article to the Royal Society, by observing an article in the last number of this

Journal in which Prof. Rutherford describes his methods and results. As nearly all the latter agree in every particular with those which I obtained, I am now relieved from the necessity of detailing them. ·It is desirable, however, to note that, viewed in the light of my other experiments, these results, if we except the "strong" phase of Pflüger's series, amount to this :—When a few mm. of nerve-length, including the extreme nerve-section, rested on the clay, a much less strength of current was required to produce the *breaking* contraction in the muscle than when any other portion of the nerve of equal length, and similar angle towards the current, was allowed to rest upon the clay. Prof. Rutherford states the case thus :—"*The striking fact, however, is that without altering the strength of the current,* all the phenomena of Pflüger's law could be obtained by transmitting it through a central, middle, or peripheral portion of the nerve, at one time in an ascending, at another time in a descending direction." For my own part I did not satisfy myself with regard to the "strong" phase of Pflüger's law, and am therefore very glad that Prof. Rutherford has been able to make a positive statement upon the subject. Now, as in the descending break of the "strong" phase no contraction occurs, the question arises as to why nerve-section outside of muscle should *diminish* the excitability of the nerve to a strong descending break stimulus, while nerve-section near or inside of muscle should, as we have seen, so greatly *increase* the excitability of the nerve towards this same stimulus. We have here an interesting question, and one to which I am not as yet in a position to give a decisive answer. Further experiments can alone supply such an answer; but meanwhile I may observe that I think it will probably be found to lie in one or other of two directions. The first and most probable explanation is as follows. When a nerve is cut at a distance from muscle, and its excitability in the region of the cut is thereby so far exalted that a strength of current which previously gave the phenomena of the "moderate" phase of Pflüger's series now gives the phenomena of the "strong" phase,—when this is so, the anelectrotonic influence of the positive pole at the central part of the nerve must be much more powerful in the case of the descending current than it

was before the excitability of the nerve had been raised by
cutting. Consequently, as the anodic region is always a region
of low conductivity, the stimulus supplied by the descending
break is not so well able to pass on to the muscle when
the anelectrotonus of the central end of the nerve is severe
(*i.e.* after nerve-section), as it was when the anelectrotonus at
that end was not so severe (*i.e.* before nerve-section). Thus far
the explanation is merely an application of the ordinary theory
regarding the "strong" phase of Pflüger's law to the case
where, in Prof. Rutherford's statement, the "strong" phase may
be induced, not by increasing the strength of the current, but
by approximating the previously "moderate" current towards
the central or cut end of the nerve. Now, if this ordinary
explanation is correct, we should not expect it to apply to the
case of stimulation of nerve *within* muscle; for, just as in the
opening section of this paper it was pointed out that the
electrotonic relations involved by the ascending and descending
directions of a current through a nerve *outside* of muscle, are
abolished when the current passes through a nerve *inside* of
muscle; so, in this case, it is evident that when the anode
itself directly supports the cut end of the nerve-trunk within
the muscle, it is a matter of comparatively little consequence
that the anelectrotonic area of the nerve is in a condition of
low conductivity. For, as this area is actually within the substance of the muscle, there is but little call made upon its
conducting function for purposes of stimulation; and therefore
all the effects of the disappearance of anelectrotonus are asserted
directly upon the contractile, or recording, substance of the
muscle.

The other possible explanation to which I have referred
is that which would suppose the end-plates of the nerve within
muscle to be endowed with somewhat different properties with
regard to electrical stimuli from those of a nerve-trunk. In
favour of this explanation it is to be said, that as nerve-section
produces so extraordinary an increase of excitability with reference to the anodic break in the case of direct stimulation of
muscle, it seems unlikely that the obliterating influence of
anelectrotonus on conductivity should be sufficiently great to
prevent the slightest response on the part of the muscle to the

anodic break stimulus when this is applied at the cut extremity of a nerve outside of muscle. But, as I said before, this matter awaits investigation[1].

It may be worth while to state, as shewing the astonishing excitability of the extreme nerve-section, that if the nerve, while hanging in a vertical direction over the flat surface of the clay bridge, be lowered until the section just touches the flat surface of the clay, it may frequently be observed that the attached muscle responds to make and to break of the current. Yet this must be a case of almost complete transverse stimulation; for, thinking that there might possibly be some passage of the current from the clay into the nerve in a semi-lenticular form, I tried a number of times the effect of ligaturing a nerve with a fine human hair, then with a fine pair of scissors making a transverse section as close beneath the ligature as possible, and lastly, lowering the nerve-section on the clay as before. In no one case, however, did I succeed in obtaining any results similar to those which I obtained with unligatured nerves. It may be stated that in all these experiments with the clay bridge, I graduated the amount of nerve-length to be laid on it by means of a horizontal glass rod firmly fixed to the tube of a microscope. The free end of the rod was pointed, and usually passed between the tendon Achilles and the tibia, the latter having been previously divided at the knee. The sciatic nerve was thus allowed to depend in a vertical direction, and could be very accurately adjusted upon the clay bridge by means of the rack-work which moved the tube of the microscope.

§ 6. I shall now proceed to draw attention to a very interesting phenomenon which is occasionally presented by un-curarized muscle when subjected to the influence of the constant current. As already stated, when a muscle—whether curarized or not—is stimulated by the passage of a constant current through its own substance, it responds to the stimulation, provided this be somewhat more than minimal, by remaining in a state of persistent contraction during the whole of the time that the constant current is passing[2]. Now I have found, in perhaps

[1] In any case I think it is evident, from the results presented by the foregoing sections of this paper, that the method of direct stimulation of nerve within muscle has been far too much neglected.

[2] I repeat this fact because, although it is very easy to verify, it is one which

five per cent. of cases, that when an uncurarized muscle is in a state of moderately strong tetanus from the passage of a rather weak voltaic current, some part or parts of the muscle begin to *pulsate* in a strictly rhythmical manner—the parts concerned alternating their periods of contraction with periods of repose, sometimes at about the rate which is observable in a frog's lymphatic heart, and sometimes faster. This pulsation may perhaps only manifest itself through four or five revolutions; but on such occasions it may often be made to re-commence in a more persistent manner, by graduating the strength of the current to whatever degree is found most suitable. In this way I have obtained more than 100 pulsations without a single intermission, and in perfectly regular time throughout.

That this interesting phenomenon is exclusively due to the nervous element in the substance of the muscle is, I think, conclusively proved by the following facts: (1) Among the many hundreds of curarized muscles which, during the course of my inquiry, I have subjected to the influence of the constant current, I have never once seen the slightest tendency evinced by them to behave in the manner just described. (2) On one occasion, when I had an uncurarized gastrocnemius of a frog on non-polarizable electrodes with the attached nerve supported on another pair of such electrodes, I was fortunate enough to perceive, towards the lower end of the muscle, a very beautiful instance of rhythmical pulsation when the constant current was passing through the muscle in a descending direction. Of course it occurred to me to try, by means of the other pair of electrodes, the effect of throwing the attached nerve first into anelectrotonus and then into kathelectrotonus. The results were most decided. With a current of properly graduated intensity passing through the gastrocnemius, it was always quite easy to inhibit the pulsating effect in the muscle by throwing the attached nerve into anelectrotonus. Every time the key in the nerve circuit was put down, so as to throw the nerve into anelec-

is not generally known. The text-books for the most part ignore it, or even state its exact converse, viz. that muscle, like nerve, only responds to *variations* in the intensity of a stimulating current. Why so great a difference should exist between nerve and muscle in this respect it is not easy to suggest. Perhaps, however, the peculiar action of the constant current on muscle may be due to the continuous production of *chemical* changes in the substance of the muscle, which in turn act as continuous stimuli.

trotonus, the rhythmical pulsations in the muscle were sure to cease; while every time the same key was put up, so as to break the current in the nerve, these pulsations were as sure to recommence. This observation was repeated a great number of times in succession. Conversely, if the attached nerve was thrown into kathelectrotonus, the pulsating effect could be produced in the muscle by means of a current of less intensity than was required to produce this effect when the nerve was either in anelectrotonus or in the normal state.

There can thus be no doubt that the rhythmical pulsation in question is a nervous effect, and I suppose there can be equally little doubt that the phenomenon is to be attributed to the influence exerted by an alternate nutrition and exhaustion of the particular nervous element concerned. It may be well to observe, in passing, that we seem to have here, as it were, the *accidental* beginnings of rhythmical ganglionic action; and that this would doubtless be seized upon by natural selection, if the presence of a rhythmically-discharging centre in the part of the muscular system concerned should happen to be of benefit to the species[1].

§ 7. I will now proceed to describe some experiments on well curarized muscles, which illustrate the major influence of the kathode upon closure, and during the passage, of the constant current.

(a) If the curarized muscle be placed on non-polarizable electrodes in a horizontal direction, and if a capillary glass rod be interposed between the electrodes and then progressively raised until the muscle, falling by its own weight on each side of the rod, just touches the two electrodes with its two extreme ends; under these circumstances it may frequently be observed by a quick eye, and sometimes even by a slow one, that the closing contraction is of the nature of a contractile

---

[1] It may also be observed, that as the effect in question has been proved to be of nervous origin, we must either suppose that nerve inside of muscle so far differs from nerve outside of muscle as to be affected in a very different way by voltaic stimuli of constant intensity; or else, adopting the suggestion in the last footnote to explain the difference between the behaviour of nerve and muscle towards the constant current, we must suppose that the products of the chemical changes act as stimuli to the intra-muscular nervous elements. To me it seems that one or other of these hypotheses must be provisionally accepted, for I do not see that any third one is open.

wave passing from kathode to anode, while the opening contraction is also of the nature of a contractile wave, but passing from anode to kathode. These effects are more perceptible with curarized than with uncurarized muscles, and with minimal than with stronger stimulation. In seeking to obtain them, of course it is desirable to work with long-shaped muscles.

(b) If the sartorius muscle is chosen for the above experiment, and a transverse cut is made half across its width in the line midway between the electrodes, it may generally be observed that on closing the circuit the cut gapes towards the kathode, and with a proper strength of current, continues to do so during all the time the circuit remains closed. This experiment may be very prettily varied by repeatedly reversing the current with the commutator while the key remains down. At each reversal of the current the cut reverses the direction of its gape. These experiments with the sartorius may often be rendered still more striking by making a number of transverse cuts like the one just described, but alternating with one another on opposite sides of the muscle.

(c) If the sartorius muscle is placed on non-polarizable electrodes as before, and is then somewhat stretched in a longitudinal direction by means of weights attached to its two ends, it may almost invariably be observed—especially when the contractions become sluggish by exposure of the muscle[1]—that upon closure of the current, and during all the time of its passage, the substance of the muscle *draws* towards the kathode, while on the kathode itself it heaps up and spreads out in a very beautiful and distinctive manner. On now reversing the current, all the phenomena take place in the reverse way. Hence, by placing any minute body anywhere on the muscle between the poles, this body may be seen to travel some distance towards the kathode every time the current is reversed.

(d) If the last experiment be slightly modified, the major influence of the kathode admits of being shown to an audience. For this purpose, instead of allowing the two

---

[1] Soon after exposure to dry air the contractions of the sartorius begin to become sluggish, and continue to do so more and more until, before the muscle dies, the degree of sluggishness is wonderful.

weights to hang free, they should each be connected with an appropriate index. After a few trials to ascertain the best disposition of the weights in relation to one another and to the strength of the current, a large audience may easily perceive that when the current is passing in one direction the index nearest the kathode is raised, and that when the current is passing in the opposite direction, this index—which is now nearest the anode—falls, while the other index rises. By rapidly alternating the current with the commutator, the two indices may be made to present the appearance of being joined by a rigid but invisible bar, which inclines first one way and then the other as often as the current is reversed.

(e) If the sartorius muscle is submitted through a portion of its length to the polarizing influence of a moderately strong voltaic current, and if the extra-polar portion of the muscle is supplied with minimal stimulation by means of single induction-shocks thrown in close to one of the electrodes of the polarizing current, it may often be observed that the extra-polar portion of the muscle responds to a weaker induction-shock when it is in anelectrotonus than when it is in kath-electrotonus. This at first sight appears anomalous; but the explanation is very simple. It has just been stated that when the sartorius is subjected to the influence of the constant current, the muscle is in a state of stronger contraction in the region of the kathode than anywhere else. Consequently, when a minimal induction-shock is thrown in at the kathodic region, the muscle is less able to respond to the stimulus than when this is thrown in at the anodic region: the substance of the muscle being already heaped up on the kathode by reason of the strength of its contraction at that part, the effect of an additional and feeble stimulus is not so apparent when supplied at that part as it is when supplied at any other part.

§ 8. There is only one other matter concerning which I desire to draw attention at present. If the copper-wire terminals of a Daniell's element be taken one in each hand, and the strength of the current be graduated down to the point at which minimal stimulation is obtained by placing on a fresh muscle first the anode and then the kathode; it may invariably be observed that if this order is reversed, by first laying on the

kathode and then the anode, no contraction will be given unless
the strength of the current is somewhat increased. This curious
fact may be observed equally well on curarized and on uncurarized muscles. It is independent of the direction of the
current, and is not affected by insulating the muscle or by laying
it on a gas-pipe. Thinking, however, that, notwithstanding
these facts, the phenomenon in question might be due to
tension in the substance of the muscle, of positive electricity in
the one case and of a negative electricity in the other, I tried
the effect of first placing on the muscle the anode alone or the
kathode alone in an unclosed circuit from a Grove's cell, and
then experimenting with the Daniell's cell as before. The
presence of the Grove's cell, however, made no perceptible
difference in the results. I may add that the latter are most
easily observed by employing the long muscles of the thigh,
either *in situ* or excised.

At the time when I obtained these results, I was not aware
that any difference between the effects of anodic and of kathodic
closure had been previously observed. My attention, however, has
now been directed to the very interesting statements of Hitzig, to
the effect that such a difference is readily perceptible in the case
of voltaic stimulation of the brain. It is highly remarkable,
however, that the difference between the effect of anodic and
of kathodic closure in the case of brain is the exact converse of that which I have found to be true in the case of
muscle; for while kathodic closure is, as I have said, the more
effective stimulus as regards muscle, it is the less effective,
according to Hitzig, as regards brain. That such an extraordinary inversion of relations cannot be without some important
meaning, must, I think, be held at any rate as probable; and,
in view of Hitzig's other facts, I also think that the present
ones may justly be taken as in a high degree confirmatory of
his views concerning the reversed relations that subsist between
central and peripheral voltaic stimulation. For it is to be
observed, that if we presuppose any difference at all to occur
between the effects of anodic and of kathodic closure, we should
be prepared to expect that when the whole muscle is the
anode in an unclosed circuit, it should be more ready to contract upon closure than when it is the kathode; for, as we

know that the kathode is the point of major stimulation of muscular tissue on closure, it seems probable that . if any difference between the two modes of closure is perceptible at all, it should be in favour of that mode which brings the point of major stimulation to bear upon the tissue with the greatest suddenness—that is, in favour of kathodic closure. But if our anticipation is thus realised in the case of muscle, when in the case of brain we meet with an inversion of the facts, I do not see why we are not to accept an opposite conclusion, and to hold with Hitzig that in voltaic stimulation of the brain the anode, and not the kathode, is the point of major influence.

The question of course arises, What is the behaviour of *nerves* in the respects we are considering? The answer to this question is highly perplexing. So far as I have been able to ascertain, motor nerves, at any rate, do not exhibit the slightest difference in their behaviour towards closure by the anode and closure by the kathode. Even before I became acquainted with Hitzig's results, this fact appeared to me highly remarkable; and since I have become acquainted with these results, the fact in question appears to me even more remarkable than it did before. That there should be any difference between muscle and nerve in the respects we are considering—that while muscle is able to distinguish between anodic and kathodic closure, nerve is not able to do so; this fact, although doubtless a strange one, would not be so strange as is the fact with which we are actually presented; viz. that while both muscle and central nervous matter are able, though in opposite ways, to distinguish between anodic and kathodic closure, peripheral nervous matter is quite unable to do so. Of this fact I have no explanation to suggest; but that it is a fact I can have very little doubt, not only from my own observations, but also from those of an independent investigator, of whose work, however, I was unaware until after my own had been completed. The investigator to whom I refer is Engesser, who, struck with the observations of Hitzig on cerebral stimulation just alluded to, instituted a careful series of experiments on motor nerves, the result of which was to show, as my experiments have likewise shown, that in their case not the slightest difference could be detected between the effects of anodic and of kathodic closure[1].

In now concluding this paper, I should like to record my best thanks to Dr Sanderson for the courteous and valuable assistance which in various ways he has rendered me during the progress of this enquiry. Expression of my gratitude is also due to Mr Page for his ever-ready help in the laboratory.

May, 1876.

[1] See Pflüger's *Arch.* x. p. 147.

THE EFFECTS OF THE CONSTANT CURRENT ON THE HEART. By M. FOSTER, M.D., F.R.S., and A. G. DEW-SMITH, M.A., *Trinity College, Cambridge.*

(From the Physiological Laboratory, Cambridge.)

THE Frog and Toad were chiefly employed in the following observations. Several experiments however were made with the hearts of Tortoises and Dog Fish. For a supply of the latter the authors are deeply indebted to the kindness of Mr Henry Lee, of the Brighton Aquarium, and to the liberality of the Directors of that Institution.

The current was always applied by means of non-polarisable electrodes. Tracings of the heart's beat were generally taken by means of a very light simple lever, or levers, placed directly on the ventricle, or on the auricle and ventricle. Occasionally the endocardial method was employed, but all the main results were obtained by means of the lever. Undoubtedly the contact of even the lightest lever must be regarded as a stimulus; but this stimulus is at its maximum at the moment of application, and very rapidly sinks to zero. Practically there is no difficulty whatever in eliminating the effects of the lever from those of the currents.

The results of the observations may be naturally arranged according to the part of the heart subjected to the current, and according to the condition of the heart previous to the application of the current.

1. *The lower two-thirds of the ventricle.*

This portion was chosen as an especial object of study, because at the time our observations were begun it was generally admitted that a ventricle deprived of its basal third, on the one hand contained no ganglionic apparatus, and on the other never exhibited any spontaneous rhythmic beat. Since our observations were made we have read the interesting observations of Merunowicz[1], who has succeeded in obtaining good spontaneous rhythmic pulsations from this moiety of the ventricle. We cannot but regard his results as corroborative of some at least of our own conclusions.

The results which we obtained on subjecting such a portion of the ventricle to a constant current directed longitudinally, that is from base to apex or from apex to base, differed according to the strength of the current employed.

With very weak currents, sometimes no effect at all is produced; sometimes there is seen a beat at the making of the current, or at the breaking, or at both making and breaking, the tissue during the passage of the current remaining perfectly quiescent. The beat thus brought about is in all its features a normal beat.

In very many cases the making beat was distinctly seen to proceed from the kathode and to travel towards the anode, and the breaking beat to proceed from the anode and to travel towards the kathode. This was most clearly seen when the ventricle was bisected longitudinally almost up to the apex, where a bridge of tissue was left, and the limbs of the V-shaped mass extended into almost a straight line and pinned out in that position. Under favourable circumstances the beat was seen to move as a wave from one pole to the other, from the kathode to the anode or from the anode to the kathode, as the case might be. But the experiment did not always succeed, and occasionally, when probably the bridge left was too small, the two portions acted as two independent masses. We may remark in passing that this experiment quite corroborates Engelmann's[2] views on the *physiological* continuity of the whole ventricular tissue, views which we have fully adopted in our paper on the Snail's Heart[3], where we perhaps ought to have called attention to Engelmann's previous remarks on physiological continuity[4] more specially than we did.

Thus far the cardiac tissue seems to differ in no way from ordinary muscular tissue, and the above results, which remain

[1] *Berichte k. Sächs. Gessellschft. d. Wissenschft.* 1875, p. 254.
[2] Pflüger's *Archiv*, xi. p. 465.
[3] *Proc. Roy. Soc.* xxiii. p. 318.
[4] Pflüger's *Archiv*, ii. p. 243.

the same after the heart has been treated with urari or with atropin, are merely illustrations of Pflüger's law.

When however stronger currents are employed, distinct rhythmic pulsations are set up on the making of the current, continue during the passage of the current, and cease with its cessation. The beats thus produced are in all their characters like to normal spontaneous beats; indeed are indistinguishable from them. We were able to obtain these beats from any piece of the ventricle, however small.

The frequency and force of the beats depend on the strength of the current in relation to the irritability of the heart. Thus if a current which is just strong enough to produce simply a making and a breaking beat, be slightly increased, the result is that the making beat is followed at a considerable interval by a second beat while the current is going on, and this perhaps by a third or fourth; and the breaking beat fails to make its appearance. As the current is still further increased in strength, the beats become more frequent and at the same time more forcible, until a point is reached at which the maximum of pulsation is obtained, *i. e.* at which the beats are at the same time strongest and most rapid. Beyond this the increase of frequency still goes on, but at the expense of the force of the individual beats. The beats then tend to overlap and so to convert the rhythmic pulsation into an ordinary tetanus; but for this a very strong current is required; and indeed we were never able to bring about a complete tetanus, such as could be fairly compared with the tetanus of an ordinary striated muscle.

Of the rhythmic pulsation thus brought about many varieties presented themselves, varieties which we can only refer, without completely explaining them, to the varying irritability of the heart.

The most common type is that which, following the phraseology of the Leipzig school, we may speak of as consisting of an ascending and descending staircase. In this the making of the current is followed immediately or after a short interval by a feeble beat, this by a stronger one, and so on in an ascending series until a maximum is reached, which after being maintained for a variable time gives place to a descending series in which the beats diminish in force and lose in frequency. With currents of short duration the descending limb, as might be expected, is frequently absent.

Very frequently the ascending limb is absent, the pulsations starting at a maximum, and declining either immediately or after a variable period; but, sometimes, even with currents lasting the greater part of a minute, no decline at all is seen, the heart continuing to beat uniformly and regularly during the whole time of the passage of the current.

In nearly all cases, and certainly in all cases where the pulsations are maintained to the end of the passage of the current, the distinct breaking beat is conspicuous by its absence. We have however occasionally but very rarely seen the pulsations survive the application of the current, two or three beats making their appearance after the current had been broken.

This power of the constant current thus to provoke a regular rhythmic pulsation in the part of the ventricle normally devoid of any spontaneous beat was observed long ago by Eckhard[1]; and indeed was the subject of a controversy between that physiologist and Heidenhain[2]. The general impression produced by that controversy seems to have been that the rhythmic pulsations caused by the constant current simply formed a particular case of the tetanus observed by Pflüger[3] as the result of the application of the constant current to an ordinary muscle-nerve.

Without denying that the two sets of phenomena may have a common origin in so far as they may be both due (as Pflüger suggests with regard to one) to electrolytic action, we must confess that in their general features the two are most markedly different. The tetanus set up by the application of the constant current to a nerve is a rapid tetanus; by no adjustment of the strength of the current can the rhythm of the contractions be graduated. With certain strengths of current, the tetanus is absent; and as the strength is increased, a tetanus of rapid rhythm suddenly

---

[1] *Beiträge zur Anat. u. Phys.* Bd. I. 147, also Bd. II. 123.
[2] Müller's *Archiv*, 1858, p. 479.
[3] Virchow's *Archiv*, XIII.

makes its appearance when a certain point is reached. Very striking is the contrast offered by the cardiac muscle under the same constant current. Under favourable circumstances, and by applying the current with great care, a whole series may be obtained ranging, according to the strength of current employed, from the initial simple make- and break-beat, through a rhythm of two or three beats a minute, to one of a beat every two or three seconds, and finally to one every second, or even to a still more rapid rhythm. Nothing like this is ever witnessed in any ordinary muscle. The beats moreover have all the features of normal beats. An observer, however experienced in cardiac experiments, would, on seeing the pulsations, suppose the heart to be beating naturally and spontaneously, if he did not know that the current was being applied. Indeed, the simplest and perhaps the truest mode of stating the facts is to say that the constant current provokes the heart to spontaneous beats.

Nor are these rhythmic contractions witnessed with the constant current alone. As one of us pointed out some time ago[1], when a quiescent lower moiety of the frog's ventricle is submitted to the action of an ordinary interrupted current for some little time, the irregular tetanic contractions which generally speaking first make their appearance, give place to a rhythmic beat, the features of which are in every way normal. On the supposition that the lower two-thirds of the frog's ventricle do not contain any ganglionic structures, these facts seem to point most distinctly to the conclusion that the less differentiated cardiac muscular tissue still retains a power of rhythmic pulsation, which power has been almost entirely lost by those muscles which are more completely under the dominion of the will. We say almost entirely, because in the first place theoretical considerations would lead us to suppose that if the property of rhythmic movement were a fundamental attribute of primitive protoplasm, traces of this property would be visible even in the most differentiated forms, and in the second place, such traces may indubitably be found[2]. But to say that the cardiac

[1] Foster. This *Journal*, Vol. III. p. 400.
[2] See the interesting paper of Mr Romanes which precedes the present

muscular tissue still retains the power of rhythmic pulsation, is very nearly the same thing as saying that the rhythmic pulsation of the heart is a fundamental property of its general muscular tissue and not of any special localized mechanism. Thus although the lower moiety of the ventricle does not under ordinary circumstances exhibit a spontaneous pulsation, its behaviour under the constant current warrants the conclusion that the power of rhythmic pulsation though latent is not absent, and only needs favourable circumstances for its complete manifestation. Those favourable circumstances are provided for by Merunowicz's (*loc. cit.*) mode of experimentation, and he accordingly finds that the lower moiety of the frog's heart beats with a regularity and spontaneity as distinct if not as ready as that of the entire ventricle. In fact, the distinction between the entire ventricle and the moiety is one of degree, not of kind. The entire ventricle will beat under many circumstances; in fact nearly, though not quite, as readily as the ventricle with the auricles attached. The moiety will only beat under certain favourable circumstances, *e. g.* when its fibres are extended by the distension of its cavity, and at the same time supplied with nutrient or invigorating material, or when it is subjected to a constant current of a certain intensity.

Now it may without any risk be affirmed that distinct ganglionic cells, like those which are found in the atrioventricular boundary and elsewhere, are absent from the lower two-thirds of the ventricle. Structures of that size could not be missed by the many acute observers who have searched for them. We may therefore without any fear say that the spontaneous beat of the heart cannot be due to the action of *such ganglionic cells as these*.

Future observers may discover in the lower moiety of the ventricle nervous structures of another order; and in that case it may become necessary to transfer to those structures the attributes which are now monopolized by ordinary gan-

article, and which bears very closely both on this and on many other points. We may perhaps be permitted to point out that, while the rhythmic pulsations observed by Mr Romanes were done away with by urari poisoning, we have not seen any fundamental changes introduced into our results by the exhibition of even large doses of that drug.

glionic cells. For ourselves, we must confess that it seems more in accordance with the tendency of physiological inquiry to suppose that the cardiac muscle-cells (they might be called neuro-muscular cells if there be any advantage in the name) have not yet lost that property of rhythmic movement which is seen to be exercised with varying but progressive regularity in all protoplasm, from that of a bacterium or a vegetable cell upwards.

The fundamental functional homogeneity and at the same time accidental heterogeneity of the cardiac tissue is shewn in the following observations.

When a large piece of the ventricle was employed, for instance the whole of the lower two-thirds, the rhythmic pulsation was always markedly much more easily obtained when the kathode was placed at the base than when the kathode was placed at the tip.

When the piece was small, *e.g.* when the mere tip of the ventricle was employed, rhythmic pulsations were more readily obtained when the kathode was placed at the tip and the anode at the cut surface, than when the current was sent in the contrary direction.

In a piece cut out from any part of the lower two-thirds, rhythmic pulsations were always more readily produced when the current was thrown in one direction than in another; that is, when one part rather than another was made kathodic. By carrying incisions in various directions we could almost at will convert any spot into a point, the kathodization of which appeared more favourable to pulsation than that of other points.

It seems impossible to attribute these results to any arrangement of localized mechanisms, save such a distribution of them as would be coextensive with the muscular tissue itself. Without being able fully to explain the facts, we are inclined to interpret them as being connected with the shape of the ventricle and the consequent unequal distribution of the anodic and kathodic areas.

It is exceedingly probable that the beats are kathodic in character, *i.e.* that they proceed from the kathodic region, and

49—2

are connected with the katelectrotonic phase. We have not however been able to prove this satisfactorily. In many cases, as for instance when the piece of ventricle was divided longitudinally and the parts pinned out, the beats could be distinctly seen to proceed from the kathode. In many other cases, on the contrary, the beats seemed to start in the kathodic and anodic region at the same time. In the snail's heart (*loc. cit.* p. 326) the beats which the constant current evoked in an otherwise quiescent ventricle could very clearly be seen to proceed from the kathode and to travel towards the anode; but when the complicated arrangement of the fibres in the twisted tube which forms the frog's ventricle is compared with that of the fibres in the straight tube of the snail's heart, it is easy to understand why the regular progression from the kathode to the anode, which is so obvious in the latter, should be obscured in the former. On the other hand, the beats which are seen during the passage of the current have all the appearances of being repetitions of the initial beat. But this is a kathodic beat, and we may therefore infer that the following beats are kathodic also. If this be so, then we may go a step further, and conclude that the facility with which beats can be evoked by the constant current, will depend on the ease with which the katelectrotonic phase can be made dominant in the tissue under the influence of the current.

Now, in dealing with the effects of the constant current on nerve-fibres and ordinary striated muscular fibres, we have to do with cylinders of uniform diameter (neglecting, as we may do, the extreme ends of the muscular fibres) throughout, and the relative expanse of the katelectrotonic and anelectrotonic areas will depend entirely on the position of the neutral line, and that again on the strength of the current in relation to the irritability of the tissue. As far as we know, the case has never been considered where the tissue forms a physiologically continuous cone, one electrode being applied to the base and the other to the apex. There is no proof that the anelectrotonic and katelectrotonic phases (*and by these terms we mean the physiological and not the merely electrical conditions of the tissue*) are so related to each other that the movement, whatever it may be, in one direction is accompanied by an equivalent movement in the other direction, the sum of the two remaining null; we are not debarred, by anything that we know, from the supposition that when the kathode is applied broadly to the base of such a cone, the larger portion of the tissue is thrown into the katelectrotonic state, and when the anode is placed at the base, the smaller portion. In the behaviour of the snail's heart, which forms such a cone of physiologically continuous tissue, we thought we found positive indications of an influence of the form of the tissue in this direction; and since Engelmann has shewn that the ventricle of the frog's heart is similarly physiologically continuous, we have some ground for applying a similar hypothesis to that organ also. If we are allowed to do so, then we should say that the lower two-thirds of the ventricle, in spite of the injury caused by the section, still form a cone, so that when the current is applied with the anode at the tip, the larger portion of the tissue is thrown into katelectrotonus, and beats in consequence are more easily evoked than when the kathode is placed at the tip. When, on the contrary, a small piece only of the ventricle is being operated on, the injury at the lines of section, and the directions of the incisions, determine the position in which the electrodes should be placed in order that the katelectrotonic phase should be dominant, and in conse-quence beats more easily evoked. We throw out this view with great diffidence as a mere suggestion, put forward in the hope that it may provoke inquiry into what, on any hypothesis, seems a singular fact, that the beats should be produced more easily by applying the current in one direction than in another, and that the direction depends on the form of the piece of ventricle or at least on the directions of the incisions made.

We naturally turned our attention to the point whether the neutral boundary could be shifted, and the relative areas of katelectrotonus and anelectrotonus in consequence varied by varying the strength of the current.

According to the views just expressed it might be supposed that, since the neutral boundary in ordinary nerve electrotonus moves from the anode towards the kathode as the strength of the current is increased, the ease with which beats could be produced would diminish with the increase in the strength of the current. As a matter of fact the beats are more easily produced with the stronger currents; and this fact seems at first sight distinctly to disprove the hypothesis we have put forward. But, in the first place, it must be remembered that the rhythmically pulsating cardiac tissue differs from a nerve-fibre in this, that with the initial beat, and with each subsequent beat, the condition of the tissue is profoundly altered as the very result of the beat. In the second place, we have in the heart's beat to do with two distinct things; the force and extent of the individual beats, and the rapidity with which the beats are repeated. Between these two things there is a relation, and that relation is in large measure an inverse one. (In the ordinary explanations given of rhythmic action, the two are treated of as being absolutely in an inverse relation. When a gas is bubbling through a fluid the large bubbles come slowly: when they follow rapidly they become small. The problems of the heart would be very simple if this inverse relation were an absolute one for it also; it is not so, for we may have within limits rapid strong beats, and infrequent feeble ones; but nevertheless the inverse relationship does exist though obscured by other influences.)

Now, in operating on the ventricle with a constant current, it is seen that an increase in the strength of the current bears much more upon the rapidity of the rhythm than on the force of the individual beats. We have frequently observed that the beats, few in number, called forth by a weak current are individually as strong as those, many in number, which are produced by a stronger current applied for the same time. Nay more, when a certain limit has been passed the beats lose in force while they gain in rapidity, by increase of the current. In the last matter we have to do probably with the effect of one beat upon another in the shape of exhaustion; and indeed the whole of this part of the subject must remain obscure until we know the limits and conditions of the beneficial and of the injurious effects of any given beat on its successor; for these beneficial effects do exist, as shewn by the labours of Bowditch, and the injurious effects are readily shewn in any overtaxed heart.

˙, If we might make a suggestion it would be in the direction that while the force of the beat is most (but not exclusively) dependent on the area of the katelectrotonus, the rapidity of the rhythm is more immediately connected with the intensity of the electrotonic phases. This would at least enable us to understand why, in spite of the diminished katelectrotonic area, the beats increased in rapidity with an increase in the current[1].

The absence of the break-beat when the current has during its passage evoked a series of pulsations, seems to us a point of interest as contrasted with its very regular presence when the current has simply caused a make-beat, and not given rise to any series. When the beats are maintained during the whole time of the passage of the current, the break-beat is almost invariably absent. It is only when the beats have ceased before the shutting off the current, so that a pause of some length is seen between the last beat and the actual breaking of the current, that the break-beat follows upon a series.

Now the break-beat in many cases was most distinctly seen to be, and in all cases probably is, an anodic beat; it proceeds from the anode, and is due to the disappearance of anelectrotonus. May we infer from this that the effect of a beat is to neutralize the (physiological) electrotonic condition, so that after each explosion neither the kathodic region is so katelectrotonic, nor the anodic so anelectrotonic, as immediately before the beat? that during the interval between that and the succeeding beat the heart is occupied in getting up, so to speak, the katelectrotonic and anelectrotonic phases? This would at least enable us to understand why the larger kathodic area is more favourable for the development of beats.

---

[1] It is impossible to avoid the conviction that the processes concerned in the production of a pulsation spontaneous or otherwise, are capable of being analysed, like the phenomena which belong to an ordinary muscular contraction, into those pertaining to the stimulus wave and those pertaining to the contraction wave. If this be granted, it is more than probable that the two sets of processes would be differently affected by the constant current; and that this difference might explain the results recorded in the text. Our efforts however to lay hold of this difference have been hitherto wholly in vain.

2. *The whole ventricle, when quiescent.*

As is well known, the entire ventricle, from which the auricles have been carefully removed, though still retaining the ganglia situated at the base of the ventricle, frequently remains perfectly quiescent, without offering any spontaneous beats at all.

Such a ventricle when submitted to the action of the constant current behaves exactly as does a ventricle from which the base has been removed. According to the strength of the current in relation to the irritability of the tissue, the effect of the current may be a simple make- and break-beat, a series of infrequent pulsations, or a series of rapid pulsations. The beats are more easily evoked when the kathode is placed at the base than when it is placed at the tip. In fact, we met with no one feature (except a greater irritability, that is, a greater readiness to respond to comparatively weak currents, and a greater endurance, that is, a capability of being submitted to the action of currents for a longer time without losing its activity) which we could point to as distinguishing the ventricle possessing the atrioventricular ganglia from one devoid of those structures. This fact shews very clearly, on the one hand, how entirely subordinate is the influence of the ganglia in question, as far as the intrinsic activities of the ventricle itself are concerned ; and, in the second place, how little permanent damage is done to the ventricle, as a whole, by the rough section needed to remove the ganglia.

3. *The whole ventricle beating spontaneously.*

As is well known also, the frog's ventricle will, in the complete absence of auricles and sinus venosus, frequently go on beating spontaneously for a very considerable time.

When such a spontaneously beating ventricle is submitted to the action of the constant current, the effects are by no means so constant as in the two cases we have already discussed.

With weak currents, that is, with one or two Daniell's cells applied through small non-polarisable electrodes of considerable resistance, no effect whatever was visible. Both in respect to the rapidity of the rhythm and the force and duration of the individual beats, the heart behaved at the making, at the breaking, and during the passage of the current, in a perfectly normal manner.

By stronger currents, such as those supplied by three to six Grove's, the heart was visibly affected, both at the make and break, and during the passage of the current.

The make and break effects were fairly constant, with the exception that they varied according to the phase of the cardiac cycle which was being passed through at the moment when the current was made or broken. Thus, if the current were made during a certain part of the systole of any beat, that beat was followed by one which was at once premature and slight. This in turn was succeeded by an abnormally prolonged diastole ushering in a beat larger than the normal; after which, at least in the cases where the current was of such a strength as only to produce make and break effects, and not to affect the heart to any marked extent during the passage of the current, the pulsations were normal until the break occurred, which if it took place at the same phase of the systole, produced a similar effect to the make. Hence, both at make and break, a normal beat was followed, first by a short diastole and a feeble beat, and then by a long diastole and a strong beat; and in many cases, the long and short diastoles made up together the length of two normal diastoles, and the movements of the lever during the feeble and strong beats were together about equal to the movements of the lever during two normal beats. We do not, however, pretend to say that this compensation was always exact and complete; and it was certainly less exact as regards the force of the beats than as regards the length of the intervals.

When the current was made or broken at certain other phases of the cardiac cycle, the effects we have just described were replaced by others. Thus, sometimes the initial beat was enlarged; sometimes no obvious effect at all was produced. We did not make a sufficient number of observations to work the point out thoroughly, but hope to be able to do so at some future time, since the facts seem not without interest as promising to throw light on the varying conditions of the heart as it passes through its several phases, and to afford a means of

more accurately measuring than has hitherto been done the
duration of the latent period of a natural systole[1].

Far less constant than the above effects were those which
made their appearance during the passage of the current.

The most frequently recurring effect was, that during the
passage of the current the beats were most distinctly lessened,
and with the stronger currents almost completely annihilated,
without *any marked change of the rhythm.* This we have seen
again and again, both when the kathode was placed at the
base and when it was placed at the apex.

We sought, by placing two light levers, one near the base
and the other near the apex, to gain some insight into the relative movements of different parts of the ventricle; but we
found that the varying pressures of the two levers, and a
variety of other circumstances, introduced so many sources
of error, that we were unable to arrive at any satisfactory
conclusion.

We thought that in many cases, especially where moderate
currents were used, we had fairly distinct evidence that when
the kathode was at the base there was an increase of movement at the base and a diminution of movement at the apex
during the beats, and that, on the contrary, when the kathode
was at the apex there was an increase of movement at the
apex and a diminution at the base; there being in both cases a
diminution in the movements of a lever placed midway between
base and apex. But our results were not sufficiently constant
to enable us to lay stress on this, which would point to a tolerably satisfactory explanation of the total lessening effect of the
current in whatever way applied. We may add, that the same
lessening was seen when the current was applied transversely
or obliquely instead of longitudinally.

We also thought we noticed that there was a tendency to a
quickening of the rhythm, though never very pronounced, when
the base was kathodic, and inversely a tendency to retardation
when the apex was kathodic; but on this point again we

---

[1] Since the above observations were made M. Marey (*Comptes Rendus*, 1876)
has published a note in which he describes briefly very similar phenomena, and
promises to deal with them in fuller detail than we have been able to do.

cannot, in face of the inconstancy of our results, assert anything distinct or certain.

Nevertheless there remains the striking fact that, taking the ventricle as a whole, its spontaneous pulsations are diminished by the passage of a constant current of sufficient intensity. So that between a quiescent ventricle and one which is beating spontaneously, there is this marked contrast in their behaviour under the constant current, that whereas the current evokes pulsations in the quiescent ventricle, it stops, or goes far to stop, the pulsations of the pulsating one. Thus the same current acting on the same ventricle, with the electrodes exactly in the same position, may at one moment all but stop pulsations, and a short time afterwards, when the ventricle has ceased to beat spontaneously, call forth pulsations which are in every way like to the pulsations it just before had stopped.

Both these effects were seen not only in a natural ventricle, but in one which previous to excision had been treated with urari or with atropin to a sufficient extent to do away with the inhibitory action of the vagus. And since the exciting effect on the quiescent ventricle has been shewn to be independent of the action of ordinary ganglia, we may fairly infer that the restraining influence on the pulsating ventricle has likewise but little to do with ganglia, the effects of both kinds being due to the direct action of the current on the cardiac tissue.

4. *The ventricle and auricles, removed from the body and beating spontaneously, with the cavities empty.*

The heart after excision was placed on a block of paraffin scooped out slightly so that the heart remained in one position. The electrodes were placed one at the apex of the ventricle and the other at the sinus venosus or at the upper border of the auricles by the side of the bulbus. Sometimes the heart was placed with the anterior surface uppermost, sometimes undermost. No essential difference was observed in its behaviour in the two different positions.

As far as the ventricle was concerned (and we paid no particular attention to the auricle, the movements of which it is extremely difficult to record satisfactorily) the effect of

the current was exactly the same as when the spontaneously
pulsating ventricle was alone operated on. There were the
same break and make effects, and the same diminution of the
beats during the passage of tolerably strong currents. The effects
were essentially the same when urari or when atropin was given,
as without those drugs; in fact we failed to distinguish any
effects which could be attributed to the inhibitory mechanisms.
It seemed as if the influence which the current exercised over
the general cardiac tissue overcame altogether any effect
which might be produced on the purely nervous structures,
and hence ventricle and auricle acted as two organs physiologically isolated, the auricle serving only as a simple con-
ductor of the current to the ventricle, modifying it only by
offering resistance to its passage, but otherwise having no
effect; and *vice versa*. In this point the frog's heart corresponded entirely with the snail's heart, in which we had[1]
previously noticed the same physiological independence.

5. *The ventricle and auricles removed from the body and
    beating spontaneously, but with the cavities distended with
    serum.*

We commenced some observations with the current applied
to the heart, fitted up for registering the endocardial pressure
according to the method of Coates or of Bowditch. The heart
was supplied with rabbit's serum or with rabbit's blood diluted
with a ·75 per cent. solution of sodium chloride, and the electrodes
were applied as usual. We found, however, that the application of the current produced at once such a profound and
lasting change in the rhythm, the beats falling into Luciani
groups immediately after the passage of even a comparatively
weak current, and the groups developing themselves with
such vigour, that all further observations were rendered impossible. We have rarely seen an intermittent rhythm so
markedly shewn as it was under these circumstances.

[1] *l. c.* p. 325.

6. *The whole heart remaining in the body and the circulation maintained intact.*

The animal was sometimes pithed, but more frequently placed under urari. The heart was sometimes left in its natural position, but sometimes a ligature was thrown round the connective-tissue band which passes from the posterior surface of the ventricle to the adjacent pericardial wall, and the heart turned over, so that the apex pointed to the head, and fixed in that position by the ligature. Notwithstanding this unusual position the circulation went on very well; the advantage of the manœuvre lay in the fact that the levers could be more satisfactorily placed on the ventricle, and the electrodes applied to any part of the sinus venosus. One electrode was placed against the apex, and the other either at the upper border of the auricles or at the sinus venosus, or, in order to eliminate the auricle, at the auriculo-ventricular groove.

The main result which we obtained by applying the current under these circumstances was one which seemed to us very striking. Though we employed tolerably strong currents, *ex. gr.* six Grove cells, and as many as twenty-five Leclanché cells, we could produce no other distinct effects than a making and breaking one.

At the make and break we witnessed very frequently, as in 3 and 4, a premature feeble beat followed by a long pause and a strong beat, when the current was thrown or shut off at the appropriate time; but during the passage of the current itself the pulsations of the heart were in no obvious manner different from the normal.

There was perhaps a general tendency for the beats to be increased in force when the kathode was placed at the auricles or at the base of the ventricle, and a similar tendency for the beats to be diminished when the kathode was at the apex of the ventricle; but this was by no means present with sufficient distinctness and certainty to enable us to say that it was a definite effect of the current.

We applied the current again and again, and for several seconds at a time, without producing any other effects. It was only after the lapse of several hours, during which the current had been repeatedly applied, that an intermittence in the beat giving rise to irregular groups made its appearance during and for some time after the application of the current.

When one considers how profound are the effects which a constant current of much less strength than that supplied by six Grove cells produces when applied directly to a nerve, it certainly does seem surprising that the heart should be so little influenced by the constant current. The behaviour towards the constant current of the heart supplied under normal conditions with its proper nutritive fluid, when compared with the behaviour of the same heart, either deprived altogether of blood, or fed with serum only, indicates that the apparent indifference of the former is the result of recuperative influences exercised by the blood-supply, and absent in the case of the latter. Some share in the difference between the two might be referred to the isolation of the excised heart placed on the paraffin block, nearly the whole of the current under these circumstances passing into the heart, whereas when the current is applied to the heart in the body, some of it may escape into the surrounding tissues; but this share can only be a very slight one. The real cause of the difference lies in the fact, that the heart which enjoys a rich and continuous blood-supply can accommodate itself rapidly to the new circumstances in which it is placed by the passage of the current, while the nutrition of the heart without a blood-supply is too slow and too feeble to enable it to do so. That the current did produce an effect during the whole time of its passage, (though its action was at a maximum soon after the make), was shewn by the effect at breaking. During the whole of this time katelectrotonic and anelectrotonic phases were established in the ventricle, otherwise the occurrence of the breaking phenomena would be unintelligible. Yet in spite of this the heart continued to beat during the passage of the current at a rate and with a vigour which careful measurements shewed to differ but very slightly indeed, if at all, from the rate and the vigour which obtained previous to the application of the current. Even in the cases where a distinct break effect was absent (and the absence or character, when present, of the break effect seemed to depend chiefly on the exact phase of the cardiac cycle in which the heart was engaged at the moment when the current was broken), it would be unreasonable to suppose that the current was without effect or had ceased to have any effect; for it can hardly be imagined that the well-nourished and therefore more susceptible heart would be less affected by the current than the ill-nourished and therefore less susceptible heart. It is surely far more in consonance with all the facts to believe that, as we suggested above, the conditions which we know as katelectrotonus and anelectrotonus are developed in connection with spontaneous pulsations, so that, whenever a constant current is applied a struggle takes place between so to speak the natural and the artificial electrotonic conditions, resulting in a defeat of the heart when the heart is weak and the current strong, and in an apparent neglect of the current when the heart is sufficiently active. In this sense we could not speak of any permanent electrotonic condition lasting during the whole time of the passage of the current; since the intensity of the katelectrotonic and anelectrotonic changes would vary during the phases of each cardiac cycle, and thus develope or not a breaking effect, according to the moment at which the current was broken.

An idea presented itself, but only to be rejected, that the pulsations which occurred during the passage of the current were not real spontaneous beats, but artificial beats simulating true ones produced by a current which was strong enough at the same time to place *hors de combat* the ordinary automatic nervous mechanisms. This idea was negatived not only by the fact that the rhythm did not vary with the strength of the current, but also, and more distinctly so, by the fact that, as we subsequently found, the nervous (inhibitory) mechanisms were able to produce, when stimulated, their usual effects in spite of the presence of a strong current.

7. *The ventricles and the auricles brought to a standstill by Stannius' experiment (section of the boundary between auricles and sinus venosus).*

When the heart of the frog is brought to a standstill by this operation, any stimulus applied to the ventricle gives rise to a beat in which the ventricular systole occurs before the auricular. A series of beats in which this reverse rhythm is manifested, the auricle in each case contracting regularly after instead of before the ventricle, may follow upon the application of a single stimulus. This remarkable feature of the Stannius' standstill was observed by that acute observer Von Bezold[1] (whose early death physiology has so often to deplore), but apparently has not distinctly attracted the notice of subsequent investigators.

Bernstein[2] states that when a constant current is applied lengthways to the heart in this condition rhythmic pulsations (beginning with the making and ending with the breaking of the current) are produced in the direction of the current; ex. gr. that the ventricle beats before the auricle when the anode is placed at the apex of the ventricle and the kathode at the auricles, while the auricles beat first when the kathode is placed at the apex and the anode at the auricles.

Bernstein worked with somewhat strong currents, and did not sufficiently vary the strength in different experiments. We find that the result is in close dependence on the strength of the current; and in a series of experiments in which the strength of the current was progressively increased we obtained the following effects.

The animal was generally poisoned with urari in order to eliminate the effects of stimulation of the vagus due to the section from the direct effects of the Stannius' operation; the section was made through the junction of the sinus venosus and auricles; the heart was laid on a paraffin block, and when by its perfect quiescence the operation was seen to have been successful, the current was applied by means of non-polarisable electrodes.

With the weakest currents no effect at all was produced. With somewhat stronger currents rhythmic pulsations were set up on making the current, continued for a shorter or longer time during the passage of the current, and as a rule ceased on the breaking of the current. Both when the base was kathodic and when the apex was kathodic, the beat of

---

[1] *Physiol. d. Herzbewegung.* Virchow's *Archiv*, xiv.
[2] *Nerv und Muskel,* Abschnitt. v. s. 205.

the auricle succeeded instead of preceding the beat of the ventricle.

When still stronger currents were applied the difference between the case where the apex was kathodic and the case where it was anodic became evident. When the apex was anodic, rhythmic pulsations proceeding continuously from the ventricle to the auricle were set up, the rhythm being more rapid than with the weaker currents. The beats continued during the whole time of the passage of the current. When the apex was kathodic the beats were at first in the order ventricle-auricle; then came a pause of variable duration, after which rhythmic pulsations reappeared, but with *the auricles beating before the ventricle*.

With still stronger currents the events when the apex was anodic remained as before. When however the apex was kathodic the reversal of the order of rhythm took place very early, so that after one or two beats the order ventricle-auricle was replaced by the order auricle-ventricle.

The following details of a series of experiments will perhaps put the facts in a clearer light.

Rana esculenta; urari given. Stannius' experiment successful; heart perfectly quiescent.

Exp. 1. Current supplied by 1, 3 and 5 Leclanché cells respectively. No effect at all produced by the current applied in either direction for 30 seconds. Heart remains perfectly quiescent during and after the application of the current.

Exp. 2. Current supplied by 7 Leclanché cells. Current applied ·for 30 secs. V means beat of ventricle only. A, beat of auricles only. VA, beat in which the ventricle precedes the auricle. AV, beat in which the auricle precedes the ventricle. The first column gives in each case the time of each beat measured from the making of the current.

Apex kathodic.			Apex anodic.		
After	·5 sec.	V.	After 12 secs. VA.	Beat of ventricle: very large.
„	1 „	A.		
„	4 „	VA.	„	27 „	VA. „
„	7 „	VA.	„	30 „	current broken.
„	30 „	current broken.	(No breaking beat.)	
(No breaking beat.)				

Exp. 3. Current supplied by 7 Leclanché cells; No. 2 repeated.

Apex kathodic.			Apex anodic.		
After	·5 sec.	V.	After	1·5 sec.	VA.
,,	1 ,,	A.	,,	10 ,,	VA.
,,	13 ,,	VA.	,,	24 ,,	VA.
,,	25 ,,	VA.	,,	30 ,,	break.
,,	30 ,,	break.	(No breaking beat.)		
(No breaking beat.)					

After the experiment with apex kathodic had been made, the heart was lightly touched, once only, with a camel's-hair brush soaked in a ·75 solution of sodium chloride. This was done for the purpose of moistening the surface of the heart. A beat in which the ventricle preceded the auricle immediately took place. This was followed by beats, in which the ventricle similarly preceded the auricle, at intervals of 10, 20, 40, 80, 120 seconds respectively after the application of the brush. The heart then became perfectly quiescent, and the other half of the experiment, *i. e.* with the apex anodic, was proceeded with.

Exp. 4. Current supplied by 10 Leclanché cells.

Apex kathodic.			Apex anodic.		
After	·5 sec.	VA.	After	5 secs.	V.
,,	4 ,,	VA.	,,	5·5 ,,	A.
,,	7 ,,	VA.	,,	16 ,,	VA.
,,	30 ,,	VA.	,,	25 ,,	VA.
			,,	30 ,,	VA.

In this case it was difficult to say if the last beat in each was a simple breaking beat or not. In the case where the apex was kathodic, the long interval preceding (from 7 to 30 secs.) would seem to shew that the beat was a breaking beat; but where the apex was anodic this is not so clear.

Exp. 5. Current supplied by 15 Leclanché cells.

Apex kathodic.			Apex anodic.		
After	·5 sec.	VA.	After	1 sec.	VA.
,,	3·5 ,,	VA.	,,	4 ,,	VA.
,,	13 ,,	VA.	,,	9 ,,	VA.
,,	21 ,,	VA.	,,	13 ,,	VA.
,,	29·5 ,,	VA.	,,	18 ,,	VA.
,,	30 ,,	break.	,,	22·5 ,,	VA.
(No breaking beat.)			,,	26 ,,	VA.
			,,	30 ,,	break.
			(No breaking beat.)		

Exp. 6. Current supplied by 20 Leclanché cells.

Apex kathodic.			Apex anodic.		
After ·3 sec.		VA.	After 1 sec.		VA.
„ 3	„	VA.	„ 4·5	„	VA.
„ 5	„	VA.	„ 8	„	VA.
„ 7·5	„	VA.	„ 11	„	VA.
„ 10	„	VA.	„ 15	„	VA.
„ 13	„	VA.	„ 19	„	VA.
			„ 21	„	VA.
„ 26	„	A V.	„ 24·5	„	VA.
„ 32	„	A V.	„ 28	„	VA.
			„ 30	„	VA.

The last beat registered with the apex anodic was not a strictly breaking beat; the contraction began before the current was actually shut off.

In the case where the apex was kathodic the current was kept on two seconds beyond the half minute, being broken immediately after the commencement of the last beat registered.

Exp. 7. Current supplied by 20 Leclanché cells.

Apex kathodic.			Apex anodic.		
After ·2 sec.		VA.	After 2 secs.		VA.
„ 6	„	VA.	„ 3	„	VA.
			„ 6·5	„	VA.
„ 10	„	AV.	„ 10	„	VA.
„ 14	„	AV.	„ 13	„	VA.
„ 16	„	AV.	„ 17	„	VA.
„ 19	„	AV.	„ 19	„	VA.
„ 22	„	AV.	„ 22	„	VA.
„ 25	„	AV.	„ 24·5	„	VA.
„ 28	„	AV.	„ 28·5	„	VA.
„ 30	„	AV.	„ 30	„	VA.

In the case where the apex was anodic, at the beginning and at the end of the series a distinct interval was visible at each beat between the contraction of the ventricle and that of the auricles; in the middle of the series, on the other hand, the beat of the auricles came so rapidly after that of the ventricle that they appeared almost synchronous, and it became very difficult to say that the ventricle did really precede the auricles.

In this and several preceding experiments, the beat of the ventricle when it preceded that of the auricles was seen to be preceded in turn by a beat of the bulbus arteriosus.

It will be seen from the above, which is one of many experiments having exactly the same general features, that our results in large measure agree with those of Bernstein[1], though they differ to such an extent as to prevent our accepting the interpretation given by that inquirer.

That interpretation, if we understand it aright, is as follows. When the apex is made anodic,—when therefore the current may in relation to the heart be said to be ascending,—the nerves which descend from the atrio-ventricular ganglion to the ventricle, are at their origin from the ganglion thrown into katelectrotonus. This weakens the development of the molecular inhibitory processes in these nerves at their origin from the ganglion, and thus favours the development of a beat. The ventricle, thus assisted, in consequence beats before the auricle. When, on the other hand, the apex is made kathodic, and thus the current descending, the same ventricular nerves are thrown into anelectrotonus, which favours the molecular inhibitory processes. The ventricle thus hampered beats after the auricle. Further, while the nerves descending to the ventricle are thus being affected by the respective currents, the nerves ascending from the ganglion to the auricles are being affected in exactly the converse manner; so that while inhibition is being augmented in the ventricle, it is being decreased in the auricle, and vice versâ. Hence the dependence of the sequence of the rhythm on the direction of the current.

Now, in the first place, this view entirely overlooks the important fact that in the remarkable condition brought about by Stannius' operation the sequence of auricle upon the ventricle is the normal order of the rhythm of the beat. One instance of this has been mentioned in the foregoing experiment (3) ; and it will be observed that the stimulus, itself of the slightest character, was followed not by one but by a series of beats, in each of which the contraction of the auricle followed that of the ventricle. Many more instances of the same kind might be given. No great stress could be laid on a single beat with this abnormal sequence making its appearance; but the fact that a whole series having the same character should be regularly carried on, after being started only by the very slightest stimu-

[1] We failed altogether to observe the making and breaking "simultaneous contractions of all parts of the heart" of which Bernstein speaks. This is probably to be explained by the fact that our currents were in general weaker than those used by him.

lus, shews that the heart must, under the circumstances, be in a peculiar condition. No such change of the order of rhythm is witnessed in ordinary pneumogastric inhibition; and there are many reasons, to which we shall presently add a new one, for concluding that the standstill produced by Stannius' operation is fundamentally different in nature from that produced by stimulation of the pneumogastric.

Hence what needs to be explained is not so much why with the apex anodic (or current ascending) the ventricle-auricle order of rhythm is maintained, as why with the apex kathodic (or current descending) this natural order of rhythm is exchanged for that of auricle-ventricle, which in a heart during a Stannius' standstill is an abnormal rhythm.

In the second place, the reversal of the order is only obtained with comparatively strong currents, and in none of the cases we have had under our notice did it occur on the making of the current, being always preceded by one or more beats in which the order was ventricle-auricle, though we are not prepared to say that with still stronger currents than those we used the reversal might not coincide with the beginning of the application of the current. Bernstein[1] seems to have observed this reversal, but in the opposite sense, and he does not appear to have paid much attention to it. His explanation moreover fails to explain why a current of moderate intensity should produce a reversal in the course of or towards the end of its action; see *antea*, Exp. 6.

No solution of the phenomena can be considered satisfactory which is not at the same time a solution, or an approximation towards a solution, of the difficult problem, why in a normal heart-beat the sequence of the constituent contractions is always such as it is, even in a heart whose cavities are empty, and in which therefore the filling or distension of one cavity by the contraction of another can have no share in the matter. That the sequence is not the result of any fixed molecular constitution of the ganglia is shewn by the very fact of the possibility of its reversal. That the sequence may be changed by circumstances indicates that its normal character is due to a concurrence of circumstances, which concurrence is more readily

[1] *l. c.* p. 223.

brought about than any other arrangement. If we suppose the several parts of the heart, ventricle, auricle, sinus venosus, &c., to be mere passive instruments receiving stimuli or impulses to contraction from some common automatic ganglion (situate in the sinus venosus or elsewhere),—the sequence of the impulses being determined by molecular changes in that ganglion and in that alone,—then changes taking place in the ventricle can only affect the extent and character of its own contraction, and not in any way the sequence of the rhythm; and so with other parts. In the experiments above recorded a definite sequence of one kind or of the other was observed in the entire absence of the sinus venosus. On the above view they must be produced by a ganglion or ganglia situate in the auricles. We must further admit that this ganglion is the seat of the normal sequence in rhythm of the entire heart, or suppose that in the absence of the sinus venosus a ganglion, hitherto having no share in the direction of *the sequence*, comes into play. All of which is very complicated and unsatisfactory.

On the other hand, we have the undoubted fact that the ventricle alone (or even part of the ventricle), the auricles alone, the sinus venosus alone, and the bulbus arteriosus alone, can carry on each by itself a rhythmic pulsation of long duration and wholly like that of the entire heart. This means that each of these several parts of the heart has a rhythm of its own dependent on its own circumstances, including under circumstance everything which affects the nutrition both of its muscular and nervous elements. Now when ventricle and auricle are separated from each other, they beat each with an independent rhythm; but when physiologically connected, they beat in harmony and in sequence. It is impossible to conceive of this harmony being accomplished otherwise than by some mutual action of the two upon each other. We cannot suppose that any event connected with the contraction of the one (such as the negative variation of the natural current) by acting as a stimulus determines the contraction of the other. For, in that case, the systole of the ventricle would provoke a systole of the auricles, as well as the systole of the auricles a systole of the ventricle, and a rhythmic pulsation with long pauses between the whole beat (of both auricles and ventricle) would be impossible. Moreover, a weakly ventricle yoked to strong auricles would soon be driven to exhaustion, and *vice versâ*. We are thus driven to the conclusion that the beat of either organ is dependent not only on its own circumstances, but also on the circumstances of its fellow; that the rhythm of the auricles, for instance, is dependent not only on their own condition, but also on that of the ventricle. That just as the beat of the ventricle or of a part of the ventricle is determined by the condition of the whole of the ventricle or the whole of the part (the physiological continuity of the tissue permitting each fibre or bundle of fibres to influence all the other fibres by a sort of muscular sense, so that each fibre or bundle of fibres, instead of pulsating in a rhythm of its own, joins all the other fibres, or bundles, in a rhythm which is that of the whole tissue), in the same way the condition of the whole ventricle (the summation of the condition of the several fibres) is able, by the nervous continuity of auricle and ventricle, to affect, and in turn be affected by, the condition of the auricles, which again is the summation of the condition of the several auricular fibres. Were it permitted to speak of feeling in the absence of consciousness, we might say that just as each fibre of the ventricle (or auricle) feels the condition of and exerts in consequence an action on all the other fibres, whereby a harmony of the whole ventricle (or auricle) is established, so the auricles feel the condition of and exert an action on the ventricle, and *vice versâ*, whereby the harmony of the two is maintained; the nervous structures connecting them being the agents of the intercourse, the nerve-fibres probably serving simply as conductors, while in the nerve-cells processes may go on which stand in about the same relation to processes taking place in the more purely muscular elements that arithmetical operations on logarithms do to operations on the corresponding numbers.

If it be permitted to hold some such provisional view as that which we have just attempted to sketch, we should be obliged to add, that in the normal heart the nutrition of all parts of the heart is, so to speak, tuned for the production of a beat with the normal sequence of sinus venosus, auricle, ventricle, and bulbus. And further, that though the rhythm is at bottom dependent on the condition of the whole heart, yet each cardiac cycle is *set going* by the contraction of the sinus, the pulsation of that part having just the effect necessary to start the already prepared auricle, and this in turn the ventricle. Hence one might readily imagine that after removal of the sinus venosus (putting aside the effects of the section), the heart, having lost so to speak its leader, would be at a loss how to beat. When it did begin to beat we should expect it to beat in the order auricle-ventricle. The facts, however, that in Stannius' experiment the order ventricle-auricle makes its appearance, and that the ventricle separated from the auricles resumes its pulsations, while the auricles remain quiescent, shew that loss of leadership is not the sole cause of the standstill, but that some inhibitory work (not however of the pneumogastric kind) is going on, which inhibition bears more particularly upon the auricles. Both auricle and ventricle are prepared to beat upon a sufficient stimulus, but they are restrained from spontaneous pulsation by the (at present inexplicable[1]) inhibitory influences started by the section (or ligature) of the sinus venosus, the auricle being more restrained than the ventricle. Hence when a slight stimulus is applied to the heart, a beat, in the order ventricle-auricle, is produced; and that beat, as we know from the researches of Bowditch, being beneficial to the heart, and the inhibitory influences still continuing to work more upon the auricle than the ventricle, the initial beat may be followed by many others having the same sequence of ventricle-auricle.

So, with weak constant currents, which may be regarded as slight stimuli, the same kind of pulsation, the same sequence of ventricle-auricle, is produced, whether the current be ascending or descending.

It is well known that the make and break of a constant current is a more powerful stimulus on muscle than an induction-shock, there being in this point a remarkable differ-

---

[1] We say "at present inexplicable" because the Stannius' standstill has not yet, in spite of all that has been written on it, been fully cleared up. The fact that the standstill takes place when the endings of the vagus fibres have been paralyzed by atropin, proves that the inhibition is not due simply to vagus stimulation. On the other hand the curious results of Pagliani (*Moleschott's Untersuch.* XI. p. 358), shewing that *gradual* separation of the sinus will not produce standstill, disprove the view that the removal of an automatic ganglion in the sinus is the whole cause of the quiescence.

ence between ordinary muscle and ordinary nerve. In the course of our experiments we have been gradually impressed with the view that in applying the constant current to the heart, the effects which are produced (and we cannot help thinking that electrolysis has much more to do with these than is generally admitted) by the action of the current on the more distinctly muscular elements override those which are due to the action of the current on the more distinctly nervous elements. We may indeed go almost so far as to say that the former put the latter altogether on one side. So that in studying the action of the current on the auricles and ventricle, we have been led to consider merely its action on the muscular elements of the ventricle and of the auricles respectively, without paying any attention to either the inhibitory or any other nervous mechanisms present or supposed to be present. We have at least never met with any satisfactory evidence of the excitation of these nervous structures playing any part in the phenomena with which we have had to deal.

Hence in applying the constant current to the auricles and ventricle we have considered only the effects on the muscular tissue of the one and of the other. Now the ventricle, just as it is more muscular, is more susceptible to the action of the current than the auricle. It is more especially in the ventricle, we might say exclusively in the ventricle, that any difference is observable between the effects of the ascending and those of the descending current. In studying the snail's heart we were very much struck with the greater susceptibility of the ventricle as compared with the auricle. The latter is far less readily inhibited by the interrupted current than the former, less easily roused from quiescence into pulsations by the constant current, less easily checked by the constant current when beating spontaneously. Whether these facts are to be explained as mass effects or in some other way, the frog's heart, in spite of the presence of all its nerves and ganglia, acts in these respects very similarly to a snail's heart; and we venture to suggest that this greater susceptibility to extrinsic influences of the ventricle as compared with the auricle has to do both with the inverted order of sequence so characteristic of the heart during the

Stannius' standstill, and with the fact that the same order is also visible in pulsations called forth by weak constant currents.

The reversal which takes place during the action of the descending current is not so easy to explain; and it is with the greatest hesitation that we submit the following suggestions.

In the first place, in the course of many repeated observations, we were struck with the fact that the descending current had distinctly a more *exhausting* effect on the heart than the ascending current. During the minutes which followed upon the application of the descending current for 30 seconds, the heart was less irritable than it was after the application of the ascending current for the same time. So that, unless care were taken to allow sufficient intervals of restorative rest, the primary effects of the action of the current were obscured by the secondary effects of exhaustion. This exhaustion was of course more evident with strong than with weak currents. The fact that the reversal takes place with weaker currents towards the end and with stronger currents towards the beginning of the action of the current, points very distinctly to exhaustion as a prominent factor in its causation.

In the second place, if we may assume, in accordance with Pflüger's results on nerves, that with weak currents the neutral point lies near the anode, then so long as the current is not too strong a large part of the ventricle will always be in the condition of katelectrotonus, so that whether the base or the apex be kathodic, the area of katelectrotonic tissue in the ventricle is sufficient to maintain the greater susceptibility of the ventricle as compared with the auricle. As the current becomes stronger, the neutral line is driven nearer and nearer to the kathode. Under these circumstances, when the current is descending the base becomes largely anodic. This condition of the ventricle, with the base anodic and the apex kathodic, as shewn very distinctly by the phenomena of the snail's heart, and more or less forcibly illustrated by the foregoing observations on the frog's heart, is equivalent to a preponderating anodisation of the entire ventricle, more being lost by the anodic condition of the broad base than is gained by the kathodic condition of the narrow apex. And this anodic condition added to the exhaustion (of which it is probably the cause) so depresses the ventricle that the auricle gains the upper hand and precedes it in each beat; the depression of the ventricle however not being so intense as to prevent it from following the auricle at each beat.

When, on the other hand, a strong current is ascending, and the base therefore is kathodic, however much the neutral line is driven near to the base, there is always left an area of kathodic tissue at the broader pollent base sufficient just to maintain that preponderance of the ventricle over the auricle which is the characteristic of Stannius' standstill; and exhaustion not being produced so readily in this case as in the other, the rhythm ventricle-auricle is carried on in spite of the peculiar condition of the former.

We repeat that we put forward this explanation with much hesitation, but we submit that it is founded on at least no greater assumptions than that of Bernstein. At any rate our view has a certain value in reducing the phenomena to known actions of the constant current on irritable tissues.

8. The heart brought to a standstill by stimulation of the vagus.

So struck were we, in the course of our experiments, with the entire absence of any phenomena, which we could satisfactorily attribute to the action of the constant current on the termination of the vagus fibres, or on the various inhibitory and other mechanisms existing or supposed to exist in the vertebrate heart, the stimulation of which we, on starting our investigations, supposed would render the behaviour of the frog's heart under the current entirely different from that of the nerveless snail's heart, that we were led to suspect that the currents employed being so much stronger than those generally made use of in experimenting on nerves, exhausted on their first application all the nervous elements, and left us dealing with (so to speak) the naked contractile elements.

To have proved that this was the case would have been to bring an additional and strong argument in favour of the thesis of which all our experiments may be regarded as illustrations, that the causes of the rhythmic pulsations of the heart are to be sought for in the properties of contractile tissue. But we found, to our great astonishment, evidence that the most important and active nervous mechanisms of the heart were able fully to exert their influence in spite of a powerful constant current being passed through the heart. Seeing that a portion of the ventricle, or the whole ventricle when quiescent from whatever reason, or the whole heart when quiescent from the experiment of Stannius, is roused into rhythmic pulsations by the constant current, we very naturally expected that the same current would produce rhythmic pulsations when applied to a heart in standstill from stimulation of the vagus.

We found, however, that this was not the case.

Having brought the heart to a standstill by stimulating the pneumogastric with the interrupted current, we threw into the heart constant currents of various strengths, both ascending and descending. But not even with six Grove cells did we succeed in calling forth any rhythmic pulsations. The only effect which we could trace was a very marked reaction, when both the constant current and the vagus stimulation were removed, the heart soon beginning to beat with remarkable vigour and rapidity.

Having applied the constant current, both in the descending and ascending direction, we stimulated the vagus while the current was still passing through the heart. Inhibition was nevertheless produced, and, as far as we could see, took place very much as if no current were passing through the heart. We are not prepared to say that the current made absolutely no difference or that it would not be possible to call forth rhythmic pulsations in pneumogastric inhibition by applying still stronger currents; but we do say that currents distinctly stronger than those which readily rouse into rhythmic pulsations the naturally quiescent ventricle or part of the ventricle, or the whole heart brought to a standstill by Stannius' experiment, failed with us to bring forth pulsations in a heart brought to a standstill by stimulation of the vagus.

From this we draw the following conclusion. We have argued that the pulsations which the constant current calls forth in a piece of the ventricle must, until some hitherto unnoticed nervous elements have been discovered, be considered as pulsations caused by the direct action of the current on the cardiac muscular tissue. Since these pulsations are also seen in the whole ventricle (otherwise quiescent), and indeed in the whole heart (under Stannius' experiment), when the constant current is applied, it is clear that previous division of the ventricle is not necessary to their production, that the continuity of the ganglionless apex of the heart with those portions of the heart which do contain ganglia, is no bar to pulsations arising in the former as the result of the application of the constant current. Now, in the generally accepted theory of inhibition, the action of the pneumogastric is supposed to stop at certain ganglionic centres. In these centres the impulses descending the vagus so exalt, either directly or by the mediation of various mechanisms, the molecular inhibitory forces that the accustomed rhythmic stimuli are no longer set free, and the muscular fibres lie idle till the struggle in the ganglia is over. According to this view then, whether the muscular fibres are removed from the influence of the ganglia by section or by profound urari or other poisoning, or by the ganglia being preoccupied in an inhibitory struggle, the constant current acting directly on the fibre ought in all three cases to produce the same effect, viz. a rhythmic pulsation. As a matter of fact, it does so in the two former cases but not in the third. From this it follows that stimulation of the vagus, in addition to whatever effect it may have on the ganglia, has also an effect of such a kind that the irritability of the cardiac muscular tissue itself is impaired, and the production of rhythmic pulsations hampered in their muscular origin. The depression of irritability thus caused is not so great but that a mechanical stimulus will produce a contraction, and hence the heart, in standstill from stimulation of the vagus, will beat when pricked. Such a method of estimating irritability is however a very rough one. That a mechanical stimulus calls forth a beat, proves that the irritability is not extinguished, but is no evidence that it is not impaired: the more delicate test of the constant current manifests the muscular weakness which stimulation of the vagus has caused.

We are thus led to the conclusion that the pneumogastric, like any other motor nerve, acts when stimulated directly on the muscular tissue with which it is connected; from which conclusion there follows, as a corollary, the view that the peculiar inhibitory effects of stimulation of the pneumogastric are due, not to a specific energy of the nerve itself or of any mechanism in which it terminates, but to the fact that while ordinary nerves are connected with muscles ordinarily at rest, the pneumogastric is connected with a muscle in a state of continued rhythmic pulsation. To which we may add, that the other marked inhibitory nerves, the vaso-dilator nerves, are also in connection with muscles normally in a state of activity (tonic action) which is more closely allied to rhythmic pulsation than to any other form of muscular activity; indeed, in many cases, as in the rabbit's ear, the two merge into each other, and it seems difficult to regard the tonic contraction of blood-vessels in any other light than that of an obscure rhythmic pulsation.

If this view be accepted, the phenomenon of inhibition of the snail's heart by direct application of an interrupted current ceases to be extraordinary; for both that form of inhibition, and the ordinary pneumogastric inhibition, fall into the same category, being both at bottom due to the fact that in them stimulation is brought to bear on a spontaneously active tissue. Against the identity of the two, there may be urged two strong objections, one that the interrupted current applied directly to the vertebrate heart never produces a distinct inhibition similar to that seen in the snail's heart, but a tumultuous irregular sort of tetanus (never however reaching the distinct form of tetanus, and eventually giving rise to a standstill), and the other that stimulation of the pneumogastric never causes (however strong the current) a tetanic contraction of the heart, such as is seen when a too strong interrupted current is applied directly to the snail's heart. Without prolonging the discussion any further, we may be permitted to say that these objections do not seem to us insuperable, and that the study of the vaso-dilator and vaso-constrictor nerves appears likely to afford a solution of the difficulties.

The conclusions then to which our observations point, we do not pretend to say satisfactorily establish, are as follows.

The vertebrate heart, such as that of the frog, behaves towards the constant current in a manner very closely resembling that in which the snail's heart behaves.

The well known, easily recognised, ganglia of the heart play a subordinate part in the production of the heart's spontaneous rhythmic pulsations. The real origin of these is to be sought for in the phenomena of muscular tissue, unless some new form of nervous tissue which has hitherto escaped detection be discovered.

The constant current may according to circumstances call forth or put an end to rhythmic pulsations: calling them forth when they are absent and diminishing or destroying them when they are spontaneously present. Hence, here, as in the case of the snail's heart, stimulation and inhibition are shewn to differ from each other in degree, or according to circumstance, rather than in kind.

Stimulation of the vagus produces an effect on the muscular tissue of the heart; its inhibitory action is not confined to the ganglia; and hence vagus inhibition does not differ so essentially from the inhibition of the snail's heart by direct stimulation as might at first appear.

POSTSCRIPT.

WHILE the above paper was in the printer's hands we received the *Centralblatt f. med. Wissenschaft*, of May 27th (No. 22, 1876), containing a brief communication from Prof. Bernstein *Ueber den Sitz der automatischen Erregung im Froschherzen*.

In it the author relates an experiment in which the ventricle of a frog's heart is violently compressed for a few seconds across its middle with a fine pair of forceps. The line of tissue thus injured breaks the physiological continuity between the upper and lower half of the ventricle; though, there being no actual physical solution of continuity, the apex is still as before supplied with abundance of fresh blood. Since under these circumstances the apex, though irritable towards stimuli, remains perfectly quiescent, never exhibiting any spontaneous pulsations, Bernstein argues that the pulsations witnessed in Merunovicz's experiment are not really automatic, but the result of the rabbit's blood or serum acting as a stimulus "upon certain motor mechanisms in the cardiac muscles, and thus causing in them an intermittent discharge of energy."

Without waiting for the results of the counter experiments which naturally suggest themselves, we should like at once to remark that, if a stimulus so constant in its nature and action as serum or blood must be is capable of producing rhythmic movements so varied in their rate of development and character as those which made their appearance in Merunovicz's experiments, the hypothesis of an automatic centre confined to the sinus venosus needs a fresh definition.

The old view, and the one against which we have in foregoing pages argued, taught that the impulses which caused the heart's beat proceeded *in a rhythmic manner* from the ganglia in the sinus, that the rate and character of the rhythm was determined there, and that the muscular apparatus of the heart had no other task than to respond to those rhythmic impulses according to the measure of its irritability. If however, in accordance with Bernstein's new view, the muscular tissue of the heart with its "motor mechanisms" is capable, when affected by a constant stimulus (whether chemical, as blood and sodium chloride, or electrical, as the constant current), of developing rhythmic pulsations *which in no way, except as far as relates to their causation, differ from normal spontaneous beats,* the need of any intermittence in the action of the automatic ganglia in the sinus is done away with; its presence would shew a wasteful want of economy.

Looking at the matter from an evolution point of view, and seeing that muscular or neuro-muscular tissue is anterior in evolution to strictly differentiated nervous tissue, it is in the highest degree improbable that, supposing the power of generating automatic rhythmic impulses were at some time transferred from undifferentiated protoplasmic or neuro-muscular tissue to purely nervous mechanisms, the muscular remnant, from which spontaneity had been removed, would retain a power of intermittent action which it could never, in actual life, have an opportunity of manifesting, since its intermittence would ever afterwards be determined by its nervous master.

It becomes necessary, therefore, to modify the hypothesis of an automatic centre in the sinus in the sense that the action of that centre is not an intermittent but a continuous one. From this modified view to the one which we ourselves have urged in the foregoing pages, the step is very slight.

We may add that the 20th number of the same *Central-blatt* contains an original communication from Herr Fischer which, while confirming the presence of a considerable number of fine nervous plexuses between the fibres in the ventricle of the dog's heart, throws much doubt on the nervous character of the elements described by Gerlach, as abounding in the striated muscle and in the tissue of the frog's ventricle, elements which might be regarded by some as the long sought for motor mechanisms.

A CONTRIBUTION TO THE HISTORY OF DEVELOP-MENT OF THE GUINEA-PIG. By E. A. SCHÄFER, *Assistant Professor of Physiology in University College, London.*

THE mode of development of the guinea-pig must, unless the anomalies are in some way satisfactorily explained, remain one of the most puzzling problems in the whole range of Morpho-logy. Here is an animal, not apparently differing in any material point of structure from other mammals, in fact so closely allied to some of the most common as to be referred by the zoologist to the Order to which the greater number of existing individuals of the Class belong, yet presenting in its mode of development a total reversal of those relations of the blastodermic membranes which obtain, without so far as is known any exception, not only in the same Class, but throughout the whole Animal Kingdom. It is true that the fact that this difference existed has been announced already a considerable number of years, and has rested on the authority of embryologists no less eminent than Bischoff and Reichert; but their statements on the subject have been suffered to re-main almost unnoticed, or have perhaps been altogether dis-credited. But by aid of the methods which are now extensively employed of making sections of embryos, however small and deli-cate, any observer may readily convince himself of the fact that in the guinea-pig, from quite an early stage of development, the epiblast—or that layer of the blastoderm from which the epider-mis, the central nervous system, the sense organs and the inner layer of the amnion are developed—is the innermost of the blas-todermic layers, and the hypoblast—or the layer which forms the epithelium of the alimentary canal—is the outermost, so that the ordinary positions of these two layers are exactly re-versed. In view of this utterly anomalous aspect of affairs, the question naturally suggests itself, Does this reversal of the ordi-nary conditions obtain *ab initio* and at the time of the arrangement of the cells into distinct layers after the completion of segmentation, or is it brought about, subsequently to the arrangement of the layers in the ordinary position, by the occurrence in some way or other of a complete turning inside out of the blastodermic vesicle? Of the latter no proof has yet been forthcoming, although it has been conjectured by Hensen (*Archiv für Anatomie und Embryologie*, Vol. I. p. 411) that an involution may occur soon after the formation of a segmentation-cavity and before the formation of definite layers. But this conjecture rests upon very insufficient foundation, and the question must still be considered as remaining unsolved.

The developing ova, which are the subject of the present paper, had arrived at a stage of development considerably in advance of that requisite to decide this particular question, but their examination has nevertheless yielded several facts of interest bearing both upon the unusual conditions of development met with in this particular animal, and on the larger question of the mode of origin of some of the earlier formed structures in Mammalia, and perhaps in Vertebrates generally. The ova were two in number, from the same uterus, and, so far as could be judged, were in exactly the same stage of development. The description of the manner in which they were obtained and prepared, I have thought it worth while to give in detail, since the methods adopted are in some respects new, and have throughout yielded good results. The animal from which the embryos were obtained was killed for a totally different purpose, and no note having been taken of its sexual condition, I am not able to state their exact age. But it was obvious even before the organ was opened that they must be at a very early stage of development, and that the greatest possible care must be exercised in obtaining them uninjured. The uterus was accordingly placed in salt-solution, and the muscular coat having first been slit open along the free border over the position of the ova, the mucous membrane was then carefully torn away at the same place bit by bit with forceps, so as freely to expose the small vesicular bodies. They were found to be firmly connected to the mucous membrane at the attached border of the uterus, so that to remove them intact without disturbing this connexion, it was necessary to take away a portion of the uterine wall with them. On examining them thus separated with a low power of the microscope, they were seen to be perfectly spherical, hollow vesicles, about $\frac{1}{15}$ ths of an inch in diameter, closely attached, as before mentioned, at one pole to the mucous membrane of the uterus, and presenting at the other an oval thickening, the rudiment of the embryo (Plate XXX. fig. 1). From one end of this a solid bud-like process could be seen projecting for a short distance into the interior of the vesicle. This process, which on a cursory glance would probably have been taken for the head of the embryo, is, in reality, the commencing allantois. The head-rudiment is seen with a somewhat higher power at the opposite end of the oval in the shape of a curved fold (fig. 2, *h. f.*), near the middle of which a groove—the medullary groove—is seen to commence (*m. g.*), and to be continued backwards along the greater part of the length of the embryo, becoming lost at the posterior part, which is still occupied by the remains of the primitive streak (*p. s.*). Yet another structure could be made out, by careful focussing, in the shape of an excessively delicate, bulging membrane covering that surface of the embryo which is turned towards the interior of the vesicle: this membrane is none other than the amnion, which is thus formed from the first as a completely closed sac.

A careful sketch having been made of these appearances, the ova were then prepared in the following way. One was placed entire in a one-eighth per cent. solution of chromic acid. From the other the hemisphere containing the embryonic thickening was carefully separated and placed in 1 per cent. osmic acid solution, while the remaining uterine portion was placed with the other ovum in chromic acid. These were allowed to remain in the chromic solution for three days; they were then transferred to weak spirit (half water) for twenty-four hours, and then placed in absolute alcohol: in this they were left until time was found to complete the investigation. The other embryo, after having been in osmic acid for three hours, was transferred to water for an hour and then placed in glycerine, in which, like those in alcohol, it was allowed to remain for a considerable time (many months), apparently without the least deterioration. Fig. 2 is a drawing of this embryo viewed with a binocular microscope under a power of about thirty diameters, from the surface which originally looked inwards. After having been studied as carefully as possible intact, both the embryos were cut into sections: the one hardened in chromic acid being used for the production of longitudinal sections, the other for transverse. To obtain thin and complete sections of these exceedingly small and delicate objects the following method was adopted. The embryo was first placed for twenty-four hours or longer in staining fluid. In the case of the embryo hardened in chromic acid an alcoholic solution of magenta was employed, and the embryo when stained sufficiently was transferred directly to oil of cloves. The one that had been in osmic acid and glycerine was placed for twenty-four hours in Kleinenberg's logwood (Foster and Balfour's *Embryology*, p. 248), and transferred through absolute alcohol to oil of cloves. From the oil of cloves the embryos were transferred to cocoa-butter, which was kept just melted, and they were allowed to lie in this for an hour, so that they should become completely permeated by it. They were then placed, with a little of the melted cocoa-butter, in position for cutting sections in the direction desired, in a cavity scooped out at the end of a previously prepared cake of the same substance. When the cocoa-butter which permeates and surrounds the embryo has completely set and hardened, sections may, without any great difficulty, be made with a razor as thin as may be desired: as so obtained they are placed in a drop of oil of cloves on a slide. This slowly dissolves the cocoa-butter and leaves the section free to be mounted in dammar varnish. Some of the sections of the embryo which had been treated with osmic acid were, for the sake of greater distinctness, mounted in glycerine. For this it was necessary, after dissolving out the cocoa-butter with oil of cloves in the manner above mentioned, to transfer the section at first to spirit and then to water before placing it on a slide in a drop of glycerine.

Sections were also made to show the mode of connection of the ovum to the uterus. The description of these it will be convenient to defer until the condition and relations of the blastodermic membranes in the free part of the ovum have been treated of. I may however here mention that I missed altogether the epithelial prolongation, which is described by Reichert (and also by Hensen) as growing from the uterine epithelium around the ovum and forming a complete enclosing capsule, and, in earlier stages at least, a hollow stalk connecting the ovum to the wall of the uterus. Whether it has by this time disappeared, or whether I accidentally removed it in opening the uterus for the purpose of exposing and preserving the ova, I am unable to say.

To describe first the longitudinal sections of the embryo. These show very clearly the relations of the blastodermic layers and the mode in which the amnion and allantois are formed. It is only those which have been taken in or near the longitudinal axis that are of any particular value. A section made close to this axis, but not actually corresponding with it (*i. e.* running just along the side of the medullary groove. but not along its middle), is represented in fig. 3. Here the epiblast (*e.*) is seen to form a completely closed, flattened ring, much thickened at the exterior part, where it is composed of the elongated columnar cells which are characteristic of the developing central nervous system, and thinning off gradually at either end, where it passes round internally to form the inner layer of the amnion. The hypoblast (*h.*) covers and encloses the blastoderm, of which it is seen to form the outermost layer. It is formed of a single layer of columnar or cubical cells, many of which exhibit large vacuoles, and it is prolonged around the vesicular ovum as far as the base of attachment to the uterus; its relations here will be subsequently considered. The mesoblast (*m.*), which is now completely formed, is seen to lie between the hypoblast and epiblast. At the edge of the embryonic area it splits into two parts. One of these (*m'*) passes inwards to form an outer layer to the amnion, the other (*m''*) accompanies the hypoblast round the ovum as far as the attachment to the uterus. The amniotic mesoblast is simply a layer of flattened cells, fusiform in section; the parietal mesoblast consists of a complete layer of similar epithelioid cells internally, and of a number of branched cells, connected together, and undergoing development into blood-vessels and contained blood-corpuscles, lying between this epithelioid layer and the hypoblast. Between the amniotic and parietal mesoblast is a large cavity, which in fact occupies the greater part of the interior of the ovum.

It will be seen that the external wall of this cavity is formed by a layer of hypoblastic cells, and by what I have termed the parietal mesoblast. It is shown in section, at a part where no blood-vessels are as yet developed in it in fig. 4, where *h.* is the hypoblast, *me.* the epithelioid layer of the mesoblast, and *mv.* some of the intermediate mesoblastic cells, which are probably afterwards to undergo transformation into blood-vessels. In fig. 5, on the other hand, a portion is represented as seen on the flat with the hypoblastic layer removed. At this place the development of blood-vessels and blood-corpuscles is much more advanced. The formation of these appears, from a comparison of parts in which different stages of development are exhibited, to be effected in the following manner:—The intermediate mesoblastic cells unite by their processes into a large-meshed network, the nuclei at the same time multiplying and becoming distributed over the cords of the network. At the nodes (corresponding to the bodies of the original cells) the multiplication and accumulation of the nuclei goes on with much greater rapidity than elsewhere, so that a little thickened clump is formed at these places, extending also sometimes a short way into the cords. These nuclei appear to become rounded and transformed into the embryonic blood-corpuscles, the nucleoli forming the so-called nuclei of those bodies. A cavity subsequently forms in which they come to lie, and which extends into the cords of the network. The mode of formation appears therefore to be in most respects similar to that described by Balfour in the chick (*Quart. Journal of Micr. Science,* 1873).

At the posterior end of the embryo a bud-like thickening of the mesoblast (*all.*) projects down into the large cavity above described. This is the commencing allantois. Blood-vessels are already developed in this, but it receives as yet (and even considerably later, Hensen, *l. c.* p. 407) no prolongation of the hypoblast: in fact it consists entirely of mesoblastic tissue. The section represented in fig. 3 passes, as before mentioned, on one side of the middle line of the embryo. The representation of a section passing exactly along the middle of the medullary groove would have had an entirely different appearance except at the ends. For a certain distance at either end the three layers are seen as before, but soon the mesoblast entirely ceases, and the epiblast and hypoblast come into contact and are entirely fused together. From this fused portion, as is still better shown in the transverse sections, the notochord becomes developed.

(*To be continued.*)

NOTE ON THE ACTION OF DILUTED ALCOHOL ON THE BLOOD-CORPUSCLES. By WILLIAM STIRLING, D. Sc., M.D., C.M., *Demonstrator of Practical Physiology in the University of Edinburgh.*

IN recent numbers of the *Archives de Physiologie*[1] Ranvier has called the attention of histologists to the value of a diluted solution of alcohol (1 part spirit of 36° Cartier to two of water) as a reagent for microscopic purposes. On mixing a drop of the blood of the common Rana temporaria with two or three drops of this solution on a glass slide, he has been able to show that the coloured blood-corpuscles of this animal, and also of the axolotl, are provided with nucleoli.

I have had occasion lately to use this mixture pretty extensively. It is easy to confirm the observation of Ranvier. I have found that the coloured blood-corpuscle of the common newt is also provided with a nucleolus, generally only one, placed somewhat excentrically, though sometimes two are to be seen.

The action of this reagent on the coloured corpuscles offers a marked contrast to that of acetic acid, which renders the stroma of the corpuscle very transparent, making the nucleus very obvious, but at the same time causing it to become very irregular in its outline, and often, in addition, to become tinged with the haemoglobin of the corpuscles, so that it is impossible to detect a nucleolus within it. No better example could be cited of the value of a method in histology.

In the colourless corpuscles of the newt, the nuclei, as Ranvier pointed out in the case of the frog's blood, are clearly brought into view by the action of diluted alcohol. The blood of the common newt I find to be better suited for this purpose than that of the frog. I have seen as many as five nuclei clearly brought into view in the finely granular colourless corpuscles of the newt's blood. In addition, each nucleus may show a distinct nucleolus. On carefully watching such a

[1] 1874 and 1875.

colourless corpuscle, however, it shows another peculiarity which is very characteristic. It throws one or it may be more blebs apparently perfectly structureless from its surface. Ranvier also observed these in the case of the frog. They contain none of the granules of the protoplasm, and are quite transparent. They resemble the little blebs of mucin that are sometimes seen to be extruded from the columnar epithelial cells covering the villi of the small intestine. They occur most commonly, indeed constantly, on the large finely granular ordinary colourless corpuscles. I have also observed them on the coarsely granular corpuscles, but their existence here is much more difficult to detect on account of the large granular corpuscles not occurring in such numbers in the blood as the finely granular ones. I have not observed these bullae or blebs to be produced in the case of the smaller finely granular colourless corpuscles which are so abundant in the blood of the newt. On adding the diluted alcohol to human blood I have failed to detect the existence of these blebs in the colourless corpuscles.

ON "MALFORMATION OF THE CARDIAC SEPTA: A TREATISE ON THEIR PATHOLOGICAL ANATOMY. By PROF. ROKITANSKY [1]." Communicated by Prof. STRICKER, *of Vienna.* (Translated into English and prepared for publication by JOHN PRIESTLEY, *Assistant Lecturer on Physiology, The Owens College, Manchester.*)

AMONG the anomalies of internal structure presented by the heart, malformation of the septa is the most important and most frequently met with. Sensible of the great theoretical interest attaching to the phenomena of malformation and their embryonal antecedents, Professor Rokitansky was induced to investigate the rich supply of cases which came under his notice at the Institute of Pathological Anatomy in Vienna; and his labours have led to a more correct and exact description of the irregularity.

In addition Professor Rokitansky has investigated, from the embryological side, the case of transposition of the large arterial trunks, on account of the close connection between that phenomenon and malformation of the cardiac septa.

As regards the statements of fact here given concerning the development of the heart, it may be remarked that they differ essentially from those generally current: indeed the only publication containing similar assertions is a Dissertation by Lindes (Dorpat, 1865), hitherto but slightly known.

The *septum auricularum* does not, as has been said, grow upwards in the form of a sickle so as to leave a *foramen ovale;* but it begins at the roof of the originally simple auricle, and grows thence forwards along the anterior and posterior walls. In the course of this growth, however, a stage occurs in which the septum may be best represented as a curtain stretched between the upper ends of two vertical ridges. Hence the *septum auricularum,* at all events at first, has a foramen in its lower half through which the right and left auricles communicate. The foramen does not continue, for the curtain grows forwards until the septum completely divides the cavity into two.

But before the septum, in its development, has quite reached the roof of the ventricles and become fused therewith by its lower border, certain small apertures may be noticed in it. In the chick this occurs about the fourth day of incubation. They gradually increase in number, so that about the fifth or sixth day of incubation the *septum auricularum* looks like a beautiful lattice through the spaces of which the blood streams from right to left, causing the septum to bulge also to the left. The parietal portion only of the septum remains imperforate, forming a muscular frame which is specially well de-

[1] "Die Defecte der Scheidewände des Herzens, *Pathologisch-anatomische Abhandlung.*"

veloped anteriorly. The lattice, or net, which, as was described above, projects, pouch-like, into the cavity of the left auricle, gradually loses its perforations until but one aperture, large and fissure-like, remains in its upper anterior border. This aperture is the *foramen ovale.*

In a human embryo 25 mm. long the auricular septum was seen still to contain numerous perforations, some of which were of larger size. In a fœtus of three months, and in another of four months, the septum appeared as a cribriform membrane supported on a muscular frame. In a five-months embryo, on the contrary, the membrane— the central portion of the septum—was discovered to have rather the appearance of knitted work, *i. e.* to be extremely thin at numerous points, as if perforated. In fœtuses of six, seven, and eight months, and even in new-born children, the uniformly thickened membrane still exhibited a knitted appearance, and bulged towards the left.

Further information concerning the complete development of the permanent septum, from this, provisional, structure must be gathered from the treatise itself (p. 77).

It is interesting to the comparative anatomist to know that the *septum auricularum* in the fœtus of a horse is stated to have resembled a pouch supported on a muscular frame, projecting into the left auricle, and perforated in several places near its anterior border.

Lastly, it must be remarked that the *septum auricularum* institutes a division of the *vena cava,* a projection above and behind lying from the very beginning in the *vena cava,* in such a manner that a small portion of the vessel to the left is divided off as a *cul-de-sac* opening towards the left auricle, and originating the pulmonary veins.

The ventricular septum begins in the embryo as a crescentic muscular ridge with convexity pointing upwards, inserted into the anterior, inferior and posterior walls of the simple ventricle. While, therefore, the auricular septum presents originally a foramen at its lower border, the ventricular septum exhibits one at its upper, the two foramina being contiguous. Hence there is a stage in development in which the two septa of the heart together form a partition perforated by a figure-eight-shaped foramen. The upper half of the figure eight belongs to the *septum auricularum,* and becomes filled up, as already stated; the lower half, on the other hand, is situated in the *septum ventriculorum,* and *is never filled up, but persists during life, and is, in fact, the permanent ostium aorticum.* More detailed evidence of this very noteworthy relationship must be sought in the original communication (p. 71 *et seq.*). Suffice it here to state, that the *truncus* or *bulbus arteriosus* originally opens entirely into the right ventricle, that it becomes divided by an independent *septum trunci (bulbi)* into an anterior *pulmonalis* and a posterior *aorta*[1], at which stage both the large arterial vessels belong to the right ventricle. The left ventricle would be quite destitute of way of exit, did not the *foramen septi ventriculorum* (afterwards denominated

---

[1] This is the normal arrangement.

ostium aorticum) remain permanently open. At a certain period of development therefore the left ventricle pours its blood into the right, whence mixed blood is driven into both pulmonary artery and aorta. Afterwards the circumference of the *foramen septi ventriculorum* gradually grows on the side facing the right ventricle, as previously indicated; hence it may quite truly be said that the aorta grows *through* the right ventricle.

The author has given special attention to the development, as well as to the normal, perfect condition of the ventricular septum; and the more exact knowledge which he has thus gained has led him to refer certain anomalies of structure to an inhibition of growth. But, although a comprehension of his statements is indispensable in order rightly to understand these failures of growth, the reader must be referred to the detailed account of them, since the complicated relations are not easily grasped without the aid of the necessary drawings. It must, therefore, suffice to state that the *septum ventriculorum* is divided into a posterior (muscular) septum, a *pars membranacea*, and an anterior (muscular) septum, the last mentioned being again separated into a posterior and an anterior portion. The phenomena of development, as well as of malformation, show the importance of this division. For example, it is easily seen, on embryological grounds, that a deficiency of the posterior part of the anterior septum leads to a communication of the two ventricles, while incompleteness in the anterior half is the cause of communication between the two large arterial trunks.

For the sake of those who may be unable to consult the original treatise, we may at this point call attention to a *sequitur* which, though not expressly noticed by Professor Rokitansky, becomes apparent on close inspection. The embryonal crescentic septum does *not* contain the posterior part of the anterior septum, but in its place the *ostium aorticum*. The *ostium* is at this stage itself a communication between the ventricles. As it gradually becomes arched over on the side of the right ventricle by a further development, the portion of the septum so formed is at once the hinder part of the anterior division of the ventricular septum, and the right wall of the aorta.

From the more intimate knowledge gained of the normal development of the septa, the author explains in the first place all the anomalous positions assumed by the great arteries.

Starting from the normal arrangement of the septum of the arterial bulb he passes in review all the variations, and indicates, from the position of the *septum bulbi* and its relations to the anterior division of the interventricular wall, the cause of the anomalies which are actually found.

On account of the very different view of the development of the *septum auricularum* on which they are founded, all previous attempts to describe the malformations of that structure and to explain them on embryological grounds must be regarded as insufficient and leading to incorrect results. Peacock, for example, refers to cases of new-born children in whom the membranous septum had persisted in the form of the embryonal lattice, as examples of a premature closure of the *foramen ovale*.

The author classifies malformations of the auricular septum into two large essentially distinct groups. The first includes all those determined by the course of development of the primary or provisional septum. The septum is entirely wanting, or only so in part, viz. below, having failed to extend its lower border far enough downwards to reach the *septum ventriculorum*, thus leaving a passage over the top of the latter. The second group includes all the defects which arise in the transformation of the provisional septum into the secondary and permanent septum. Several of these have their seat in the membranous septum stretching across the muscular frame described as left in the course of development. The membranous portion may be completely destroyed, owing to the spaces of the lattice-work in the embryonal septum not becoming filled up ; or it may remain within an incompletely developed frame in fragments, due, in part, to a process of destruction and, in part, to regular growth.

A number of instances may be found in each group exhibiting, in addition to the above-mentioned malformations, an anomalous relationship between the large arterial trunks. In these cases the pulmonary artery is very wide, while the aorta is, relatively and absolutely, narrow. Moreover the heart is large, its cavities and the mouths of the veins, especially those of the right side, are dilated, the *sinus venosus* and the cone of the right ventricle particularly being of great size : and the organ often exhibits very typically enlargement of the right side. Occasionally there is seen a narrow pulmonary artery with stenosis and atresia of the mouth.

Although it is beyond the scope of an account like the present one to give the numerous suggestions of the author concerning the causal dependence of the facts discussed, we cannot forbear here mentioning the admirable explanation of the connection between narrowness of the aorta and malformation of the septum which Rokitansky has based on his recent embryological discoveries.

A consideration of what has previously been stated (p. 782) will show that a narrowed aorta must occasion the blood of the left ventricle to accumulate in excessive amount in the right ventricle, since both aorta and pulmonary artery open originally into that cavity. This repletion of the right ventricle must cause a corresponding repletion of the right auricle, in addition to distention of the right ventricle and auricle, and enlargement of the passage of communication between the two auricles. This enlargement would be sufficient to deter the development of the temporary septum or to mar its transformation into the permanent one. If however development proceeded as far as closure of the passage through the *septum ventriculorum* and limitation of the aorta on the side of the right ventricle (p. 782), the condition of repletion would be confined to the cavities of the left heart, and would occasion enlargement in them also, accompanied, as a collateral effect, by distention of the right heart, and especially of the pulmonary arteries, brought about by means of the circulation through the lungs.

Hence it becomes intelligible how, in all cases, enlargement seizes both halves of the heart, and why the right heart, and especially the *sinus venosus* and the cone of the ventricle, exhibit it in so marked a degree.

Deficiency of the *septum ventriculorum*, when complete or almost complete, is frequently accompanied by entire or partial absence of the *septum auricularum*, in which instances we have to do with true bilocular hearts. These cases, also, represent permanent embryonic conditions. If, on the other hand, the auricular septum is developed, we get a trilocular organ with two auricles.

Complete absence of the ventricular septum is, without exception, accompanied by one or more anomalies in position of the large arterial trunks. In addition, malformation of the septum may concern its posterior portion which exists between the auriculo-ventricular openings; or its anterior portion in either of the subdivisions previously noted; but a special description of these cases, as will be evident to the reader, would lead us beyond the limits proper to an account like this.

We may nevertheless regard more closely one very instructive section—that, namely, in which stenoses of the pulmonary artery are considered.

The author divides them into the following groups :

1. Stenoses of the highest kind and atresia of the pulmonary artery.

2. Narrowness of the pulmonary arterial trunk with excessive stenosis, or with atresia of the mouth of the vessel below the valves.

3. Narrowness of the pulmonary arterial trunk with stenosis of the valves.

Professor Rokitansky subjects to special criticism the view hitherto held that these stenoses are caused by inflammations which occur in the fœtal heart. He points out that :

a. Even in cases of very incomplete development of the pulmonary artery there are frequently no signs of inflammatory action to be found.

b. Frequently such signs of inflammation are present while the co-existent atresia bears no relation to them as far as its locality is concerned.

c. Frequently where indications of disease of very intense nature occur, stenosis of the pulmonary artery is of but slight importance.

When on the other hand it is noticed that a dwarfed pulmonary artery is frequently accompanied by other irregularities, *e. g.* of the valves; and that, as a rule, there exist, side by side with it, anomalies both in the structure of the heart and in the arrangement of the arterial vessels, the cause of which must be sought in the earliest periods of embryonic life; we are led to admit that both stenosis and atresia of the pulmonary artery are malformations of developmental origin, and are merely intensified or modified by subsequent inflammations which make their appearance sooner or later.

After the author has discussed the various points which occurred to him, he endeavours to find an explanation iu a theory of abnormal division of the *bulbus arteriosus.* It is in no way improbable, in his opinion, that, just as certain anomalies in division lead to an abnormal arrangement of the vascular trunks, so anomalies may occur of such a nature as to cause an abnormal relationship of size between the two resulting vessels; whence we get irregularity either of the arrangement of the vessels, or of their size, or of both.

But, beside stenoses of all degrees, atresia of the pulmonary artery is also ascribed to the same cause, and therefore considered to be due to irregular development. Atresia occurs whenever the deviation of the septum of the bulb from the normal arrangement is so considerable that the septum (whose convexity is directed towards the pulmonary artery) becomes actually applied to the wall of that vessel and fuses with it as far down as its mouth. In addition many stenoses might result in complete imperforation, or atresia.

As regards the cause of this incomplete division the author agrees with Peacock, whose view is that the abnormal narrowness of the pulmonary arterial trunk is due to imperfect development of the fifth branchial arch.

ATROPHY OF RIGHT HEMISPHERE OF CEREBRUM, LEFT SIDE OF CEREBELLUM, AND LEFT HALF OF BODY. By E. FORSTER BROCKMAN, *Professor of Physiology and Diseases of Eye, Medical College, Madras.*

THE following notes were made from the brain taken from the body of a native female, which was brought to the dissecting-room of the Madras Medical College in October, 1870. They may be of interest in comparison with a similar case recorded by Dr Howden in this *Journal*, IX. p. 288. No history could be obtained of the antecedents of the person. The peculiarities noticed in the brain were the small size of the right cerebral hemisphere, which was about one-half the size of the left hemisphere, due to the atrophy of its component convolutions; so that both its antero-posterior and transverse diameters were materially diminished. The right crus cerebri, corpus albicans and optic nerve were also much reduced in size. The corresponding parts on the left side appeared to be natural. The left cerebellar hemisphere was found to be much smaller than the right. No further dissection was made of the brain, which is now preserved in the Museum of the Medical College.

The following are the dimensions of the skull :—

Externally :—Occipito-frontal diameter ... 6¼ inches.
Occipito-mental   „   . 8   „
Transverse-parietal „   . 4¾   „
Vertex to Basisphenoid ... 5   „

Internally :—Widest transverse diameter between squamous portions of temporal bone   ...   ... 4¼ inches.
Widest antero-posterior diameter 5¹⁄₁₆   „
Oblique from right frontal to left occipital   5   „
Oblique from left ditto to right ditto   ...   ... 5¹¹⁄₁₆   „

Peculiarities are also noticeable in the bony skeleton.

Maxillæ :—Both in the upper and lower jaws as far as the median line the teeth were absent. To the left of the median line in the upper maxilla are to be seen alveolar spaces for nine teeth, corresponding to a similar number of markings in the lower jaw on the same side. No distinct indications of the lodgment of teeth at any prior period were present. The bones were much smaller and atrophied on the left side; and the same was the case with the tongue.

Calvaria :—The bones were more bulky; both internal and external layers and the diploë being much developed at those parts where the contained nervous structures were atrophied.

Thorax :—The right chest was nearly twice as capacious as the left, the left lung being considerably smaller than the right. The ribs on the right side were more bulky and more developed than those on the left side. The vertebral column was bent antero-posteriorly, as well as laterally, on the left side.

Abdomen :—The liver was more bulky antero-posteriorly than laterally. The other organs were normal in appearance.

Pelvis :—The left Os Innominatum was smaller, and more fragile than the right, being somewhat distorted, perhaps by the superincumbent weight.

Extremities :—The bones throughout the upper and lower extremities of the left side were more delicate and less developed than those of the right. The length of the bones on both sides was the same, the difference being chiefly noticeable in their bulk.

It will be observed that throughout the body generally, with the exception of the bones of the face, the deficiency in nutrition was apparent on the left side ; the atrophic condition of the ' brain proper' having been on the right side.

REVIEWS AND NOTICES OF BOOKS.

Entwicklungsgeschichte des Menschen und der Höheren Thiere. Von
ALBERT KÖLLIKER. Zweite Auflage. Erste Hälfte, Leipzig,
1876.

SINCE the appearance of the first edition of Prof. Kölliker's work the
science of Embryology has been progressing with unparalleled rapidity.
The increase in our knowledge of it has been so great that the
necessary alterations introduced into the present edition are sufficiently numerous and important to give to it the character of an
entirely new treatise. In accordance with the plan of the earlier
edition the development of Birds and Mammals alone is dealt
with, but the work is now divided into two parts, of which the first
only has as yet appeared. This contains an account of the formation
of the layers, a general summary of the development of the body, and
a full description of the embryonic membranes. The development of
the individual organs will be dealt with in the second part.

Regarding the work as a whole we have no hesitation in saying
that it contains the most complete and clearest description which has
yet been given of the early development of the Bird and Mammal.
It is copiously illustrated by over two hundred beautifully executed
woodcuts, the majority of them original, which have been admirably
selected to illustrate the more important points described in the
text.

The introduction contains an account of the general scope of
embryology, and a brief but fairly appreciative history of the science.
It concludes with a list of two hundred and sixty-four monographs
and papers referred to in the text. This list comprises, in addition
to numerous papers on invertebrates, a tolerably complete enumera-
tion of the more important publications on vertebrate embryology.

The descriptive part of the work commences with an account of
the ovarian ovum in different groups of animals. The types upon
which the greatest stress is laid are, in accordance with the general
scope, those of the Mammal and Bird, both of which are fairly
described. The remainder of the section is however very meagre,
and we cannot regard the author's classification of ova into *simple*
and *compound* as in any way a satisfactory or even accurate representation of our present knowledge on this subject. Following the
description of the ovarian ovum is a section dealing with segmentation. This section though very insufficient in so far as it treats of
the general aspects of the subject, nevertheless contains the fullest
account which has yet appeared of the segmentation of the Avian
ovum fully illustrated by numerous original drawings.

Prof. Kölliker, in the course of his description, succeeds in showing
that the earliest formed segments of the Avian blastoderm do not
mark out the limits of the germinal disc, but that fresh segments are

continually appearing in the previously unsegmented part of the ovum which adjoins the already formed segments. He confirms Götte's statements as to the manner in which, during the later stages of segmentation, fresh segments are formed by a peculiar process of budding from the surrounding yolk, though he believes that they are derived, not as Götte has stated, from the white yolk, but from the germinal disc (Bildungsdotter). Prof. Kölliker admits however that there is no sharp line to be drawn between the white yolk and germinal disc, and expressly states that what is spoken of as the germinal disc does not exist as such in the unimpregnated ovum, but becomes formed in the course of development. The observations recorded by Prof. Kölliker on the segmentation of the Bird's ovum appear to us distinctly to tend towards the removal of the hard and fast line which has been usually supposed to separate holoblastic and mesoblastic ova.

The description of the segmentation is followed by an account of the formation of the layers and the general evolution of the organs in the Fowl.

This well-worn, but in parts very difficult subject, has been conscientiously and carefully worked over again by the Author, and he has been able not only to give a very fresh and clear description of the whole series of events, but to clear up a number of disputed points and to bring to light several entirely new facts.

At the close of segmentation he finds that the Avian blastoderm is composed of two continuous layers, the epiblast and the hypoblast, but that there is present no trace of a third layer. At the edge of the area pellucida the hypoblast becomes very much thickened, and forms the so-called "germinal wall." At about the sixth hour of incubation there appears in the area pellucida a slight opacity usually known as the germinal streak, but which is rightly interpreted by Prof. Kölliker as the commencing mesoblast. First appearing as a linear mass, the mesoblast rapidly grows out in all directions, and interposes itself as a thick layer between the epiblast and hypoblast.

Up to this point Professor Kölliker's observations on the formation of the mesoblast appear to us quite conclusive: we further think that there can be no question whatever that in the primitive streak a fusion takes place between the epiblast and the mesoblast. Professor Kölliker goes further than this, and asserts that the appearance of fusion between the two layers is due to a proliferation of mesoblast cells from the epiblast, and that the mesoblast is accordingly entirely derived from the epiblast. We cannot accept this conclusion as proved. In the earliest representation (fig. 32) which Prof. Kölliker gives of the developing mesoblast, it already forms a mass of a very considerable size and quite free at its two sides. There is nothing in the figure to prove that the mesoblast may not have been formed of lower layer cells, and afterwards have become fused with the epiblast, and this possibility must not be rejected so long as earlier stages than the one figured remain unknown. We grant that Prof. Kölliker's interpretation of the fusion in the primitive streak is the most natural so long as no better explanation is given of it; but we believe that a more satisfactory explanation of it can be offered, which we shall mention after speaking of his observations on the primitive groove, and its relation to the medullary groove.

To his observations on these points we now pass.

The appearance of the primitive groove and its relation to the medullary groove is worked out in great detail: but we have not space to do more than summarize the results arrived at. The medullary groove arises independently and in front of the primitive groove, and it is stated that in most though not all cases, as Götte has already insisted, the primitive groove is unsymmetrically situated with reference to the medullary groove. In treating of the exact relation of the primitive groove to the formation of the embryo Prof. Kölliker gives it as his view, that though the head of the embryo is formed independently of the primitive groove, and only secondarily unites with this, yet that the remainder of the body is without doubt derived from the primitive groove. With this conclusion we cannot agree, and the very descriptions of Professor Kölliker appear to us to demonstrate the untenable nature of his results. We believe that the front end of the primitive groove at first occupies the position eventually filled by about the third pair of protovertebræ, but that as the protovertebræ are successively formed, and the body of the embryo grows in length, the primitive groove is carried further and further back so as always to be situated immediately behind the embryo. As Prof. Kölliker himself has shown, it may still be seen in this position even later than the fortieth hour of incubation.

Throughout the whole period of its existence it retains a character which at once distinguishes it in sections from the medullary groove.

Beneath it the epiblast and mesoblast are *always fused*, though they are always separate elsewhere: this fact, which was originally shewn by ourselves, has been very clearly brought out by Prof. Kölliker's observations.

The features of the primitive groove which throw special light on its meaning are the following:

(1) It does not enter directly into the formation of the embryo.
(2) The epiblast and mesoblast always become fused beneath it.
(3) It is situated immediately behind the embryo.

Professor Kölliker does not enter into any speculations as to the meaning of the primitive groove, but the above-mentioned facts appear to us clearly to prove that the primitive groove is a rudimentary structure the origin of which can only be completely elucidated by a knowledge of the development of the Avian ancestors.

In comparing the blastoderm of a bird with that of any anamniotic vertebrate we are met at the threshold of our investigations by a remarkable difference between the two. Whereas in all the lower vertebrates the embryo is situated at the *edge* of the blastoderm, it is in Birds and Mammals situated in the centre. This difference of position at once suggests the view that the primitive groove may be in some way connected with the change of position in the blastoderm which the ancestors of Birds must have undergone. If we carry our investigations amongst the lower vertebrates a little further, we find that the Elasmobranch embryo occupies at first the normal position at the edge of the blastoderm, but that in the course of development the blastoderm grows round the yolk far more slowly in the region of the embryo than elsewhere. Owing to this the embryo becomes left in a bay, the two sides of which eventually meet, and coalesce in a linear fashion immediately behind the embryo; thus removing the embryo from the edge of the blastoderm, and forming behind it a linear streak not unlike the primitive streak. We would suggest the hypothesis that the primitive groove is a rudiment which gives the last indication of a change made by the Avian ancestors in their position in the blastoderm, like that made by Elasmobranch embryos when removed from the edge of the blastoderm and placed in a central situation similar to that of the embryo Bird. On this hypothesis the situation of the primitive groove immediately behind the embryo, as well as the fact of its not becoming converted into any embryonic organ, would be explained. The central groove might probably also be viewed as the groove naturally left between the coalescing edges of the blastoderm. Would the fusion of epiblast and mesoblast also receive its explanation on this hypothesis? We are of opinion that it would. At the edge of the blastoderm, which represents the blastopore mouth of Amphioxus, all the layers become fused together in the anamniotic vertebrates. So that, if the primitive groove is in reality a rudiment of the coalesced edges of the blastoderm, we might naturally expect the layers to be fused there; and the difficulty presented by the present condition of the primitive groove would rather be that the hypoblast is not fused with the other layers than that the mesoblast is indissolubly united with the epiblast. The fact that the hypoblast is not fused with the other layers does not appear to us to be fatal to our hypothesis, and in Mammalia where the primitive and medullary grooves present precisely the same relations as in Birds, all three layers are, according to Hensen's account, fused together. This however is denied by Kölliker, who states that in Mammals, as in Birds, only the epiblast and mesoblast fuse together. Our hypothesis as to the origin of the primitive groove appears to explain in a fairly satisfactory manner all the peculiarities of this very enigmatical organ; it also relieves us from the necessity of accepting Professor Kölliker's explanation of the development of the mesoblast, though it does not of course render that explanation in any way untenable.

The remainder of the section of Professor Kölliker's work devoted to Avian development, deals with the general features and first formation of the organs which appear during the first three days of incubation.

The original drawings, both of the embryo as a whole in its various stages and of separate sections of it which fill this part of the work, cannot be praised too much. The description itself, like that in all other parts of the work, is entirely original, and though we can scarcely expect a great number of new facts in a subject which has been so much worked at already, yet Professor

Kölliker has succeeded in adding not a little to our knowledge of the development during this period.

An important section on the fate of the peripheral thickening of the hypoblast deserves special notice.   The subject is one on which Prof. Kölliker and his pupil Dr Virchow have recently made special researches.   According to their statements the thickened edge of the hypoblast (germinal wall) visible at the commencement of incubation, becomes converted into the well-known but peculiar structure which underlies the vascular area, which they consider to be formed of hypoblast cells, but which has usually been regarded as part of the white yolk.   They state that this eventually becomes converted into a columnar epithelium investing the yolk-sac.

Their views have much to recommend them on à priori grounds; but recent observations of Professor His throw considerable doubt upon them.

Professor His attempts to show that the original peripheral thickening of the hypoblast does not become converted into the substance underlying the vascular area, and that the older views which regard this latter as part of the white yolk are correct.   He speaks of it as made up of a protoplasmic network with imbedded spheres of white yolk, and states that there appear in it a number of cells, some of which enter the vascular system, while the remainder become converted into the hypoblastic epithelium investing the yolk-sac.

The manner in which cells are described as being formed in the subgerminal matter is closely paralleled in eggs of Osseous Fish and Elasmobranchii, and a strong support is thereby given to Professor His' statements.

Further observations are much needed to enable us to arrive at a definite conclusion on the nature of the subgerminal substance.

From amongst the observations of Professor Kölliker upon separate organs, we select those upon the heart and allantois as being of especial importance.

The heart is shown to arise as two cavities, one on each side, between the splanchnic layer of the mesoblast and the hypoblast, which are at first more independent than was imagined even by Von Baer.

The observations recorded on the allantois confirm and enlarge the accounts of Dobrynin, Gasser, and others, which prove that the allantois is primitively the terminal dilatation of the alimentary tract.

The most interesting new fact which Prof. Kölliker has brought to light is the existence in the chick of segmental involutions to form the Wolffian bodies similar to those which exist in Elasmobranchs and Amphibians.   Of these he gives a figure, without, however, entering into any great detail with reference to them.   A fuller account of them will, we hope, appear in the second part of his book.   He identifies the segmental tubes with the organ already, though erroneously, described by Romiti as giving rise to the Wolffian duct.

The early Mammalian development, more especially that of the Rabbit, is dealt with in this volume in the same fashion as that of the Bird.

No original account is given of the very first stages of development, but the description commences at a period when the ovum forms a thin-walled vesicle: the wall of one hemisphere of it being formed of two layers of cells, and that of the remainder of a single layer. At the pole of the hemisphere with two layers is situated the germinal area, formed entirely by a thickening of the outer layer or epiblast. Professor Kölliker points out that the germinal area is not to be confounded with the mass of yolk-cells, from which the inner or hypoblastic layer of the vesicle is derived; which occupies, though at an earlier period, the same position as the germinal area.

The descriptions of Professor Kölliker show that Mammals agree closely with Birds in the development of the mesoblast, the formation of the primitive streak, the primitive groove, and the medullary groove; and the observations recorded differ but slightly from those recently published by Professor Hensen. The remarks we have already made with reference to the formation of the mesoblast and the primitive groove apply as completely to Mammals as to Birds.

Of the organs dealt with in the latter part of the Mammalian section, the notochord and the heart appear to us to deserve special notice.

With reference to the development of the notochord the observations recorded by the Author are of great interest.

Professor Hensen has recently stated that in Mammalia the notochord is derived from the hypoblast. This conclusion Professor Kölliker opposes, though without attempting to deny the correctness of the observations upon which Professor Hensen's statements are founded. On the other hand, Professor Kölliker shows, in a fairly satisfactory manner, that the notochord is continuous with the mass of cells which forms the primitive streak; and, since this is mainly mesoblastic, he concludes that the notochord must be regarded as mesoblastic also. This conclusion does not, however, follow, if the view of the primitive streak which we have propounded above be accepted. If the primitive streak really represents the original edge of the blastoderm, where all the layers fuse together, the fact that the notochord there becomes continuous with the mesoblast presents no insuperable difficulty to the view which regards it as split off from the hypoblast throughout the region of the body.

The observations of Kölliker and Hensen, though apparently leading to contradictory results, can, we think, be reconciled on our view of the nature of the primitive streak; and we are inclined to believe that the notochord has in Mammalia, as in Elasmobranchs, an hypoblastic origin; though posteriorly it becomes continuous with the primitive streak, which is no doubt mainly formed of mesoblastic cells.

With reference to the development of the heart Professor Kölliker's observations are very complete. He shows that the heart is formed by the coalescence of two independent tubes, which arise far apart, as splits between the hypoblast and the mesoblast of the splanchnopleure, and in which there is present a delicate endothelial lining. The primitive wide separation and complete independence of the two halves of the heart is certainly surprising, but we are inclined, provisionally at least, to regard it as a secondary condition due to the late period at which closing of the throat takes place in Mammals.

The description of the development of the rabbit is followed by a complete account of what is known of the early development of the human ovum: and the succeeding section contains a description of the development of the embryonic membranes and fœtal appendages in Mammalia. This section is less well illustrated than the remainder, and appears to us also to display a falling-off in clearness as compared with the earlier part of the work.

The volume concludes with some general remarks on modern development theories. The drift of these is in opposition to the theory of descent. On this question we find ourselves completely opposed to Professor Kölliker, every page of whose book appears to us to teem with arguments in favour of the theory which he rejects: at the same time, some of the criticisms contained in the concluding section of the work are perfectly just and fair.

The work as a whole will be found indispensable to all students of embryology ; and from its admirable clearness and numerous illustrations, is well suited to form a text-book for learners.

F. M. BALFOUR,
Trinity College, Cambridge.

METHODIK DER PHYSIOLOGISCHEN EXPERIMENTE UND VIVISECTIONEN. VON E. CYON. *mit Atlas. Giessen, J. Ricker. St Petersburg, Carl Ricker*, 1876.

Every practical physiologist must feel deeply indebted to Prof. Cyon, of St Petersburg, for this most excellent and by far the best treatise on physiological methods which has yet appeared. The text is accompanied by a magnificent Atlas, containing fifty-four plates, of which more than the half are quite new. Special attention has been given to the anatomy of the dog, rabbit and frog, as far as is required for the purposes of the physiologist. The methods for chemico-physiological investigations have been entirely omitted, as they are sufficiently treated of in books specially devoted to that subject.

This book is a great contrast to anything that has yet appeared upon this subject, not only in point of comprehensiveness and in minuteness of detail, but on account of its accuracy and the evident care which has been taken that the description of any experiment shall not have been merely transcribed, as is too often the case, but the author seems to have employed most of the methods himself, and describes them with such minuteness as can only be acquired from a thoroughly practical acquaintance with the subject. Every detail in the process of experimenting is carefully considered, how difficulties are to be overcome is indicated, what parts are to be avoided and what divided, are all pointed out in a masterly manner.

The first chapter is occupied with some general rules concerning experimentation. The object of physiological experiments is indicated, a short sketch of the graphic method is given, and the question of vivisection is touched upon, and the author is most particular in inculcating the use of anæsthetics where such a procedure is permissible.

A special chapter is devoted to the methods for keeping up artificial respiration.

The second chapter treats of Haemodynamics. Special experiments on the passage of fluids through tubes for class demonstrations are given. Thus Ludwig's arrangement for showing the difference of pressure at different points of a tube of uniform diameter deserves special notice. By slight modifications of this apparatus, all of which are figured, the chief factors concerned in the movements of fluids can be demonstrated. These factors are the pressure exerted upon the moving fluid, the rapidity with which it flows out, the resistance which it has to overcome, and the force setting the fluid in motion. The changes in form and place of the heart and the effect of temperature on it are then discussed. A special chapter is devoted to the means of measuring the blood-pressure, the various forms of manometers, and the method of using them as well as of testing their accuracy, are treated very fully. Then follow chapters on Manometers for the frog's heart: the methods of measuring the rapidity of the motion of the blood; of observing the pulse: experiments on the cardiac ganglia, to which is added a very useful table on the poisons which act on the heart, their dose, seat and duration of action. The classic experiments on the cardiac nerves, viz. the Vagus, Accelerans and Depressor, are described very fully. The details for the exposure of each of these are described step by step, there being also special dissections given. Experiments on the vasomotor nerves and on the reflex action of sensory nerves on them, observations on the action of vaso-dilator nerves, and methods of observing the influence of the respiration on the circulation, close this chapter. The third chapter is devoted to the consideration of Respiration. In the fourth, experiments on secretion and excretion are given, including those on the secretion of saliva, urine and lymph; and the formation of gastric, intestinal, pancreatic and biliary fistulæ. Two chapters are devoted to the general physiology of nerve and muscle: the apparatus in most common use being all fully described and figured, including the elements and arrangement of batteries, commutators, rheochord, rheostat, multiplicator, bussole; the measurement of electromotor forces by means of the compensator ; various forms of induction apparatus; and how these last are to be graduated. Next follows the arrangement of experiments for observing the electromotor force of nerve and muscle, electrotonous and the negative variation : for the stimulation of nerve and muscle and testing their irritability : for measuring the rapidity of transmission of excitement along a nerve,

a specially good figure of Fick's pendulum myograph being given.
We miss, however, in the figure the semicircular arc to which the
apparatus for giving a greater or less swing to the pendulum is
attached. The general physiology of nerve and muscle is specially
well done, and the difficult methods of du Bois, Helmholtz and Weber
are for the first time collected and subjected to a critical examination.
The concluding chapter is devoted to *special* nervous physiology, including the experiments on the brain, cerebral nerves, spinal cord,
reflex movement, disturbances of motion and operations on the sympathetic.

Specially useful is the Bibliography given at the end of each
chapter.

The woodcuts of the atlas embrace figures of all the instruments employed in physiological operations, Ludwig's method of
injecting under continuous pressure being figured. A splendid woodcut, showing all the apparatus as used in the physiological laboratory of Leipzig, for obtaining a continuous Kymographic curve,
is given in Plate XIII. Even the latest experimental methods employed by Ludwig's scholars are all figured, including those of Kronecker, Ceradini, Luciani, Woroschilloff, Bowditch, Stirling and Mosso.
The second part of this book, embracing the methods of the physiology of the sensory organs and psychophysic, is promised in the
course of this year. The term Vivisection employed in the title of
this book seems to be quite unnecessary, appearing as it does to us
that "Methods of Physiological Experiments" would have been
amply sufficient, and would not have sounded so repulsive to over-
sensitive English ears. This book is specially intended for the
physiologist, to whom the historical and critical expositions of each
subject will be of the highest value.

INDEX.

Lightning Source UK Ltd.
Milton Keynes UK
UKHW010623110219
337000UK00006B/174/P